Handbook of Culture Media for Food Microbiology

BD

IUMS ICFMH

Vol. 14 (1978) edited by M.J. Bull (1st reprint 1983)
Vol. 15 (1979) edited by M.J. Bull
Vol. 16 (1982) edited by M.J. Bull
Vol. 17 (1983) edited by M.E. Bushell
Vol. 18 (1983) Microbial Polysaccharides, edited by M.E. Bushell
Vol. 19 (1984) Modern Applications of Traditional Biotechnologies, edited by M.E Bushell
Vol. 20 (1984) Innovations in Biotechnology, edited by E.H. Houwink and R.R. van der Meer
Vol. 21 (1989) Statistical Aspects of the Microbiological Analysis of Foods, by B. Jarvis
Vol, 22 (1986) Moulds and Filamentous Fungi in Technical Microbiology, by O. Fassatiová
Vol. 23 (1986) Micro-organisms in the Production of Food, edited by M.R. Adams
Vol. 24 (1986) Biotechnology of Animo Acid Production, edited by K. Aida, I. Chibata, K. Nakayama, K. Takinama and H. Yamada
Vol. 25 (1988) Computers in Fermentation Technology, edited by M.E. Bushell
Vol. 26 (1989) Rapid Methods in Food Microbiology, edited by M.R. Adams and C.F.A. Hope
Vol. 27 (1989) Bioactive Metabolites from Microorganisms, edited by M.E. Bushell and U. Gräfe
Vol. 28 (1993) Micromycetes in Foodstuffs and Feedstuffs, edited by Z. Jesenská
Vol. 29 (1994) *Aspergillus:* 50 years on, edited by S.D. Martinelli and J.R. Kinghorn
Vol. 30 (1994) Bioactive Secondary Metabolites of Microorganisms, edited by V. Betina
Vol. 31 (1995) Techniques in Applied Microbiology, edited by B. Sikyta
Vol. 32 (1995) Biotransformations: Microbial Degradation of Health Risk Compounds, edited by V.P. Singh
Vol. 33 (1995) Microbial Pentose Utilization. Current Applications in Biotechnology, by A. Singh and P. Mishra
Vol. 34 (1995) Culture Media for Food Microbiology, edited by J.E.L. Corry, G.D.W. Curtis and R.M. Baird (second impression 1999)
Vol. 35 (1999) Marine Bioprocess Engineering, edited by R. Osinga, J. Tramper, J.G. Burgess and R.H. Wijffels
Vol. 36 (2002) Biotransformations: Bioremediation Technology for Health and Environmental Protection
edited by V. P. Singh and R.D. Stapleton Jr.

Handbook of Culture Media for Food Microbiology

This is a completely revised edition from 'Culture Media for Food Microbiology' by J.E.L. Corry et al, Progress in Industrial Microbiology volume 34, Elsevier Science Amsterdam, Second impression 1999, isbn 0-444-81498-1

Edited by

Janet E.L. Corry
Department of Clinical Veterinary Science
University of Bristol
Bristol, UK

G.D.W. Curtis
Purbeck
Horton-cum-Studley
Oxford, UK

Rosamund M. Baird
Lyes House, Hummer
Sherborne
Dorset, UK

progress in industrial microbiology

volume 37

2003

ELSEVIER
Amsterdam - Boston - London - New York - Oxford - Paris - San Diego
San Francisco - Singapore - Sydney - Tokyo

ELSEVIER SCIENCE B.V.
Sara Burgerhartstraat 25
P.O. Box 211, 1000 AE Amsterdam, The Netherlands

Completely revised edition from 'Culture Media for Food Microbiology' by J.E.L. Corry et al, Progress in Industrial Microbiology volume 34, Elsevier Science Amsterdam, Second impression 1999, isbn 0-444-81498-1

Library of Congress Cataloging in Publication Data and British Library Cataloguing in Publication Data: A catalog record from both institutes has been applied for.

ISBN: 0-444-51084-2

∞ The paper used in this publication meets the requirements of ANSI/NISO Z39.48-1992 (Permanence of Paper). Printed in The Netherlands.

Acknowledgements

This book is the result of the work of many microbiologists who have taken part in the meetings of the Working Party on Culture Media since its inception in 1978. The International Committee on Food Microbiology and Hygiene of the International Union of Microbiological Societies thanks all those who participated in the meetings and in particular those who prepared monographs for discussion: R.M. Baird, V. Bartl, L.R. Beuchat, R.M. Blood, E. de Boer, F.J. Bolton, R.E. Brackett, G.D.W. Curtis, A.R. Datta, L. Dominguez Rodriguez, E. Elliot, J.M. Farber, G.A. Gardner, W.H. Holzapfel, R. Holbrook, G. Klein, R.V. Lachica, J.V. Lee, W.H. Lee, S.J. Lewis, B.M. Mackey, G.C. Mead, S.C. Morgan Jones, D.A.A. Mossel, P. van Netten, J.D. Oliver, I. Perales, T. Petersen, D.J. Pusch, B. Ralovich, G. Reuter, M. van Schothorst, D.A. L. Seiler, N. P. Skovgaard, B. Swaminathan, G. Wauters, G. Weenk and P. Zangerl. H. Søgaard and M. Jakobsen, as successive Editors-in-Chief of the International Journal of Food Microbiology, have been unfailingly helpful and patient in the publication of proceedings in that journal.

This work has been supported by the following companies whose generosity is gratefully acknowledged: BDH, Becton Dickinson, Difco, Elsevier, Gibco, Lab M (IDG), Merck and Oxoid.

Introduction to the First Edition

The roots of this book are an idea of the past president of the International Committee for Food Microbiology and Hygiene (ICFMH), Professor David Mossel, who formed the Working Party on Quality Assurance of Culture Media in 1978. This group convened its first meeting in Mallorca, Spain in 1979. In those days little attention was paid by many food microbiologists to the possibility that the media they used might not always function optimally. This applied especially to selective media used to isolate pathogens. The Mallorca meeting, funded by the Merck Society for Arts and Science, brought together 35 microbiologists from 13 countries. Topics covered included quality assurance tests used for raw materials for media, microbiological methods of monitoring complete media and particular problems encountered with media developed for specific groups of micro-organisms. Areas identified as requiring particular investigation were: standard methods and choice of reference strains for use in media monitoring; guidance concerning the effect of substrate (e.g. type of food examined); the effects of sublethal damage and competitive flora; information concerning the shelf-life of dehydrated media, rehydrated media and poured plates.

Inhibitors such as bile salts and brilliant green were identified as unsatisfactory because they were poorly defined and methods were needed for monitoring them. The proceedings were published in a book (Corry, 1982).

The sequel to the Mallorca meeting was held in Dallas, USA, in 1981, funded by Oxoid Ltd. Most contributions were concerned with selective media developed for specific groups of food-related pathogens but it was at this time that the possibility of producing a 'pharmacopoeia' of culture media was first discussed in depth following a paper presented by Dr Vladimir Bartl from Prague. The proceedings were published in a special issue of the Archiv für Lebensmittelhygiene (Corry and Baird, 1982).

The third meeting of the Working Party, held in London in January 1984 and funded by Difco Inc., was the first at which all three editors of this pharmacopoeia were present. The proceedings appeared in a special issue of the new ICFMH journal, the International Journal of Food Microbiology (Baird et al., 1988) which set the scene for (i) the format of the information to be included in the monographs for each medium (supplied, appropriately, by Rosamund Baird, whose first degree is in pharmacy): (ii) the protocol for microbiological testing of culture media, devised by a committee of the Working Party chaired by Gordon Curtis and (iii) the media for which monographs were to be written.

There followed two years of hard work editing the pharmacopoeia and organising the fourth meeting of the Working Party. This was held in association with the IUMS 14th International Congress of Microbiology in Manchester, England in 1986 and funded by a consortium of manufacturers and Elsevier Science B.V. It was a highly productive meeting lasting two very full days and considered draft monographs for 42 different media. As a result of this meeting a complete issue of the International Journal of Food Microbiology was published containing all the monographs as well as standard methods of testing the selectivity and productivity of solid and liquid media and a list of standard strains to be used for testing media performance (Baird et al., 1987).

A subsequent meeting in Budapest in 1988 added more monographs, many of which were for *Listeria monocytogenes* media (Baird et al., 1989). This bacterium was by then attracting widespread attention as a 'new' food-borne pathogen.

There followed a period of consolidation while food microbiologists from all parts of the world were requested to monitor their media using the test strains and methods prescribed. In 1992 a four day meeting was held in Heidelberg, funded by Becton Dickinson, at which all the monographs were reviewed. some were added and some deleted. The results of media monitoring using standard strains as well as 'in house' strains of test organisms were presented and analysed by approximately 30 participants and as a result, the numbers of test organisms recommended for monitoring each medium were reduced and the total number of strains recommended for use was rationalised. The new monographs as well as a series of reviews of media for different groups of food micro-organisms were published in the International Journal of Food Microbiology. All current monographs together with these reviews are collected in this new volume.

We hope that microbiologists specialising in food and related areas, particularly those who are members of or who aspire to join a laboratory accreditation scheme, will find this book useful.

We have tried to include all the media most commonly used in food microbiology. Inclusion of a medium, however, implies no endorsement of its superiority over other media, and likewise, there will be good media that are absent from our book. Topics that still need to be addressed include the standardisation of undefined ingredients such as blood, plasma, bile and brilliant green, procedures for resuscitation of sublethally-damaged organisms and the effect of the type of food on the optimal method of examination.

We rely upon readers to use the methods and test organisms suggested and to inform the editors of any errors or ambiguities found. Please send us any comments or suggestions you may have concerning improvements, deletions or additions that can be made in future editions.

Janet E.L. Corry, G.D.W. Curtis and Rosamund M. Baird

Introduction to the Second Edition

The reception given to the first edition of this book, including citation in a draft standard (CEN, 2000) and in the manuals of medium manufacturers, led us to believe that a revision would be of value.

Since the publication of the book in 1995 there have been three meetings of the ICFMH Working Party on Culture Media. A number of papers delivered at the Workshop in Budapest in 1996 was published (Corry, 1998). These included consideration of development of standard microbiological methods, validation systems for new methods and the use of reference materials. The second Workshop was held during the Food Micro '99 meeting in Veldhoven, The Netherlands and the most recent, generously funded by Becton Dickinson, at Temse, Belgium in 2000. Topics considered included updates on media and methods for the various microbe groups covered in the first edition of this book; consideration of the effect of sublethal damage on recovery of microbes from foods using traditional methods, and the development of new medium monographs. This new edition, renamed *Handbook of Culture Media for Food Microbiology* includes some revised, some completely rewritten and some new chapters, mostly derived from these two meetings. There are completely new chapters on stressed organisms, *Shigella* spp., *Alicyclobacillus acidoterrestris,* flavobacteria and bifidobacteria, while the chapters on Enterobacteriaceae, non-sporulating Gram positive spoilage bacteria and the *Aeromonas/Pseudomonas/Plesiomonas shigelloides* groups have been divided to reflect the growing specialised interest in some of these organisms. In particular, the pathogenic Enterobacteriaceae (pathogenic *Escherichia coli, Shigella* spp. and salmonellas) are dealt with separately from the indicator organisms – 'coliforms' and Enterobacteriaceae. New medium monographs include two for *Aeronomas* spp., two for *E. coli* O157:H7, two for *Streptococcus thermophilus,* as well as iron sulphite agar for thermophilic sporeforming spoilage bacteria. A few monographs on media for *Listeria* spp., that are rarely used, have been omitted, and 'half Fraser broth' has been added.

Some monographs exist in draft form, awaiting evaluation before publication. After discussion by members of the Working Party these will be categorised as 'Proposed' or 'Approved' and published in the International Journal of Food Microbiology from time to time. A list of monographs in draft or Proposed form can be obtained by application to gdwcurtis@hcsoxford.fsnet.co.uk. The status of each monograph in this book is specified immediately below the title. ***Approximately one third (31) of the monographs are 'proposed' and await evaluation. We rely upon readers to assist us in evaluating monographs, to use the methods and test organisms suggested in***

this volume, and to inform the editors of any errors or ambiguities found in published or draft monographs or methods. Please send us any comments or suggestions you may have concerning improvements, deletions or additions that can be made in future editions.

Janet E.L. Corry, G.D.W. Curtis and Rosamund M. Baird

References

Baird, R.M., Barnes, E.M., Corry, J.E.L., Curtis, G.D.W. and Mackey, B.M. (editors, 1985) Proceedings of the third international symposium on quality assurance and quality control of culture media. Int. J. Food Microbiol. 2, 1–138.

Baird, R.M., Corry, J.E.L. and Curtis, G.D.W. (editors, 1987) Pharmacopoeia of culture media for food microbiology. Int. J. Food Microbiol. 5, 187–300.

Baird, R.M., Corry, J.E.L., Curtis, G.D.W., Mossel, D.A.A. and Skovgaard, N.P. (editors, 1988) Pharmacopoeia of culture media for food microbiology – additional monographs. Int. J. Food Microbiol. 9, 85–144.

CEN (European Committee for Standardization; 2000) Microbiology of food and animal feeding stuffs – guidelines on preparation and production of culture media – Part 1: General guidelines on quality assurance for the preparation of culture media in the laboratory (ISO/TR 11133-1: 2000).

Corry, J.E.L. (editor, 1982) Quality assurance and quality control of culture media. G.I.T. Verlag, Darmstadt.

Corry, J.E.L. (editor, 1998) Special issue: Laboratory quality assurance and validation of methods in food microbiology. Int. J. Food Microbiol. 45, 1–84.

Corry, J.E.L and Baird, R.M. (editors, 1982) Proceedings of the second international symposium on quality assurance of microbiological culture media. Arch. Lebensmittelhyg. 33, 137–175.

Contents

Part 1

Reviews of Media

Chapter 1

Microbiological assessment of culture media: comparison and statistical evaluation of methods

G.H. Weenk

Nutricia Netherlands, Quality Assurance Dept, PO Box 1, 2700 MA, Zoetermeer, The Netherlands

In this review methods for the quality control of media are compared, taking the following questions as a guideline: i. Which methods are easy to use and give reliable results? ii. Which experimental design should be used in order to obtain reliable data with a minimal input of resources (staff and materials)? These questions can be answered satisfactorily using statistical methods. This review shows that solid media can be assessed with acceptable accuracy using well-established methods like the spread plate technique. In order to assure a minimum of statistical error, at least two plates with an average count of 100 colonies per plate seems to be the best design. This also applies to the ecometric streaking technique, a good alternative to the more quantitative methods. For an accurate assessment of liquid media, large numbers of tubes need to be tested. This is very expensive in terms of laboratory resources and therefore unlikely to be used routinely. Therefore it is proposed to use the serial dilution technique, in which the broths are tested in triplicate (Richard 1982).

The recommendations in this review can be used together with the methods recommended by the International Committee for Food Microbiology and Hygiene, Working Party on Culture Media (ICFMH, WPCM: Appendix I, this volume) to assist laboratories setting up QC tests for culture media.

Introduction

Most quality systems applied in food microbiology laboratories are certified or even accredited according to internationally recognised standards (Bolton, 1998). This implies that these labs use properly validated methods, fit for use. Method validation is generally beyond the scope of most QC labs and is left to the research institutes, universities or internationally recognised bodies like AOAC International, IDF (International Dairy Federation), ISO (International Standards Organisation) and CEN (European Standards Organisation) (Bolderdijk and Milas, 1996; Sartory et al., 1998; Payne et al., 1999; Geissler et al., 2000). Labs are assured that when methods have been issued as an internationally recognised standards, they have been validated in inter- and intra- laboratory trials using pure strains and a range of naturally contaminated food samples (Berg et al., 1994; De Smedt, 1998; Lahellec, 1998; Anon, 1999). The QC lab only has to carry out a small trial in which it shows that the method also works in its own hands.

Researchers will continue to develop faster and better ways to detect and enumerate micro-organisms in food. They make use of new growth substrates, equipment and immunological and genetic techniques (Bennet et al., 1998; Deere et al., 1998; Gutiérrez et al., 1998; Lambert et al., 1998; Thiriat et al., 1998; Veyrat et al., 1999; George et al., 2000; Tanaka et al., 2000). Within Europe this trend has been recognised and the Microval initiative has been taken in order to standardise and quickly certify these emerging and so-called "rapid methods" (Anon, 1998a). Despite these numerous efforts, culture media are still the corner-stone of current food microbiology. Therefore it is very important that QC labs have a system in place and methods available to check the quality of culture media. The Working Party for Culture Media of the International Committee for Food Microbiology and Hygiene (WPCM, ICFMH) was set up to assist QC labs in their efforts to assure medium performance by the issue of standardised protocols (Corry et al., 1995). A panel of experts has issued a series of monographs in which mainstream media are presented together with recommended test strains and quality criteria (Corry et al., 1995). In addition the monographs can be used to: i assess and compare the quality of commercially available dehydrated formula or ready to use plates or tubes or their ingredients (Mossel et al., 1974; Mossel et al., 1979; Curtis and Beuchat 1998); ii check the quality of purchased batches of commercially available media before use (Mossel et al., 1980); iii check on medium preparation procedures (Mossel, 1970; Corry et al., 1986).

In this paper the methods available for checking the quality of microbiological culture media are reviewed. The methods are compared using statistics and a scope and field of application is presented. Statistics are also used to determine the most cost-effective experimental design of culture medium QC in order to obtain results as accurate as possible and with a minimal input of labour and materials.

Standardisation of the inoculum

The media which are to be checked for their performance are inoculated with overnight cultures of pure, well defined and appropriate test strains, preferably obtained from type culture collections. The use of mixed cultures and stressed strains has also been proposed for this purpose and a number of studies on medium performance have been carried out with naturally contaminated samples (Chain and Fung, 1991; Hammack et al., 1997; Curtis and Beuchat, 1998; Entis, 1998). However, such inocula are difficult to standardise and are usually specific to the laboratory or the foods and media which are to be investigated. In addition, the physiological state and stress level can be very diverse within the microbial population present in a food and therefore may require resuscitation or awakening out of a dormant stage like a spore (Ray, 1989; Ahmed and Conner, 1995; Leuschner and Lillford, 1999; Mafart et al., 1999; Bridson and Gould, 2000). The most extreme example of physiological diversity may be the "viable but not culturable" micro-organisms, whose detection seems to require the application of the latest technologies (Nebe-von Caron et al., 1998; Defives et al., 1999; Millet and Lonvaud-Funel, 2000). Variation in physiological status of the microbial cell is undesirable when testing medium quality routinely. Therefore it is important to standardise the inoculum, which may be obtained from an overnight culture originating from an agar slant or freeze-dried stock. The Food Inspection Services in the Netherlands successfully use an overnight culture of *Staphylococcus aureus* to prepare a QC stock by diluting the culture in whole milk and store it at –20C (Mulder and Strikwerda, 1996). Others prepare lenticules, droplets filled with a known number of microorganisms, which can be kept in the fridge (Codd et al., 1998). Standardised inocula are also even commercially available as mixed cultures (Curtis and Beuchat, 1998). In addition, the initiative taken by the European Community Bureau of Reference has resulted in a range of reference and certified reference materials (In 't Veld, 1998). These gelatin capsules contain a known number of colony forming units in dried milk.

QC methods for solid media

The majority of techniques used in QC procedures for solid media rely on colony counting. The most widely used techniques are: the spread plate technique (Hammack et al., 1997); the modified Miles-Misra technique (Miles et al., 1938; Corry, 1982) and the spiral plate technique (Gilchrist et al., 1973).

The productivity of a medium is quantified by using colony counts to calculate the Productivity Ratio (this volume Appendix I; van Netten et al., 1991). Recently a separate expression was proposed for the selectivity of culture media: the Selectivity Factor (SF, Anon, 1998b). As plates are inoculated with overnight cultures when whole spread plates are used more than one plate is needed in order to obtain a reliable count of 30–300 colonies per plate. With the Miles-Misra technique several dilutions are inoculated onto one plate and with the spiral plater a wide range of counts can be assessed by using one plate, so both these methods use relatively few plates. They also are less labour intensive than the traditional spread plate technique. From the four frequently used techniques, the spiral plater has the lowest labour costs (Jarvis et al., 1977; Kramer et al., 1979). However, the high capital cost of the spiral plater may outweigh savings in labour. Colony counting can be done automatically and thus speed up the collection of data and decrease the workload of the laboratory (Chain and Fung, 1991).

The colony morphology and the diagnostic system, summarized by the term electivity (this volume Appendix I), should also be checked in order to get a complete picture of the performance of a medium. When computers are used to facilitate the processing of the data from medium QC, it would be convenient if elective properties could be quantified. The development of an "Indicative Index" was suggested by Curtis and Beuchat (1998) to fill this gap. However, it will take some time before the Indicative Index has a solid basis.

Table 1 summarizes the weaknesses and strengths of the four most frequently used colony count techniques. This qualitative comparison shows a preference for using the spiral plate or Miles-Misra technique. The pour plate technique is not favoured, because it does not allow a proper analysis of the elective properties of a solid medium and colony outgrowth may be restricted. An exception to this rule may be the QC of

Fig. 1.The ecometric streaking template.

Table 1
Strength-weakness analysis of colony counting methods used for the quality control of solid media (adapted from van Netten et al. (1991)).

Technique	Number of plates needed[1]	Visibility of		Cost of		Overall result
		Colonial morphology[2]	Diagnostic system[2]	Labour[1]	Equipment[1]	
Pour plate	–	–	–	–	+	3–
Spread plate	–	+	+	–	+	1+
Miles-Misra	+	+	+	–	+	3+
Spiral plate	+	+	+	+	–	3+

[1] + = low, – = high
[2] + = good, – = poor

media for anaerobes, where Payne and coworkers (1999) showed that the pour plate technique produced higher colony numbers compared to the spread plate technique. This can very likely be attributed to the faster creation of an anaerobic environment with the pour plate technique.

The ecometric streaking method was developed by Mossel and coworkers as a semi-quantitative alternative to the techniques described above (Mossel et al., 1983). It is the only technique which was specially designed to check the quality of solid media. Overnight cultures of test strains are inoculated onto the agar in a defined standardized way, as is shown in Figure 1. One loopful of inoculum only is sequen-

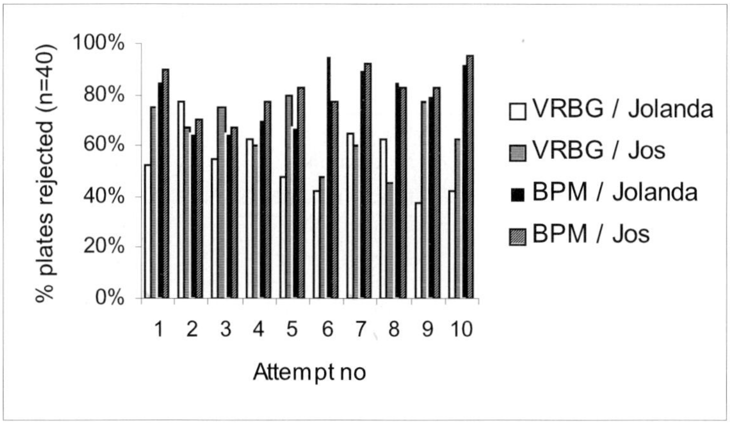

Fig. 2. The percentage of rejected plates (n=40) inoculated by the ecometric streaking technique by two inexperienced technicians. Baird-Parker agar (BPM) was inoculated with *Staphylococcus aureus* ATCC 25923, Violet red bile glucose agar (VRBG) with *Escherichia coli* ATCC 25922.

tially diluted from streak to streak without resterilizing or recharging the loop. Growth on the plates is not recorded as a colony count, but as a score. Five streaks of growth in each quadrant score as one, growth on up to three streaks score 0.5. A maximum score of 5 is obtained when all streaks in the four quadrants show growth and the final streak in the centre of the plate is also colonized. Several reports indicate that the ecometric technique has been applied successfully to validate the quality and properties of solid media or to trace errors in medium preparation (Mossel et al., 1979; Mossel et al., 1983; Corry et al., 1986; van Netten et al., 1991). The main advantages of the technique are that it is not labour intensive and the material costs are low, since the overnight culture is diluted on one plate. However, the ecometric technique requires training in order to obtain consistent results. As diminution in colony numbers should be obtained from streak to streak, slight changes in the angle of the loop or the pressure applied to the loop during inoculation will distort the desired dilution pattern and makes the score unreliable. Figure 2 shows the results obtained by two technicians in our laboratory without any previous experience in ecometric streaking. It shows that when ecometry is applied infrequently, a relatively large number of plates have to be discarded because of streaking errors. The performance also depends on the skill of the technician and the type of microorganism. Better results were obtained with *Escherichia coli* in comparison to *Staphylococcus aureus*. This might be due to the tendency for *Staphylococcus aureus* cells to form clumps. Effective dilution may not always be achieved by streaking when quick quality checks are being carried out (e.g. checking the quality of certified medium batches). However, when the performance of medium suppliers is audited, reliable results should be obtained and thus the criterion should be strictly adhered to. In the latter case well trained or specially skilled microbiologists should carry out the streaking in order to avoid undesirable errors (Curtis, 1985). It is obvious that the scoring of the ecometric plates cannot be automated.

In order to complete the overview of suggested methods for medium QC the stab

Table 2
Sources of error in colony count procedures (from Jarvis (1989)).

Source of error	Includes errors due to
Sampling error	Weighing Maceration
Dilution error	Pipette volumes Diluent volumes
Plating error	Pipetting error Culture medium faults Incubation faults
Distribution error	Non-randomness of propagules Counting errors "Recording" errors
Calculation error	

inoculation technique for testing the quality of mycological media should be mentioned. After inoculation the productivity of the medium is determined by measuring the diameter of the mycelium formation (Seiler, 1985; this volume Appendix I). This method is preferably used when rapidly spreading test strains are used. No information is available from the literature on the performance of the stab inoculation technique (Curtis, 1985). As this method has reliable alternatives (Mossel et al., 1983; Thomson, 1984), it will not be discussed. In several studies agar plating is used to validate other methods. This may imply that alternative methods may be used to validate agar performance, which is not the case. Studies have shown that there may not be a 1:1 relationship between the agar plating method and the alternative method and therefore undesirable variables are introduced when carrying out medium QC (Kirst et al., 1998; Lambert and Ouderaa, 1999). So this approach is not recommended for routine medium testing.

The statistics of methods used for the QC of solid media

In the QC of solid media the performance of the test medium is compared to that of a very nutritious reference medium (e.g. tryptone soy agar + 0.3 % yeast extract). When both media are inoculated, errors can be made and thus should be allowed for when medium quality is assessed. The errors associated with colony counting methods are extensively reviewed and further elaborated by Jarvis (1989) and are listed in Table 2. The contribution of each individual error source to the magnitude of the overall error may be method dependent. Therefore the methods used for the QC of solid media are divided into two groups: the quantitative methods (inoculation by spread plate, plate loop, spiral plate and Miles-Misra technique) and the the semi-quantitative ecometric streaking technique. As they are based on different principles they will be treated separately.

Several authors have shown that there is no statistically significant difference in the

Table 3
Approximate 95% confidence limits for numbers of colonies assuming agreement with Poisson distribution (from Jarvis 1989).

No of colonies counted	Limiting precision (to nearest %)	Approximate 95% Confidence limits of the colony count
500	± 9	455–545
400	± 10	360–440
320	± 11	284–356
200	± 14	172–228
100	± 20	80–120
80	± 22	62–98
50	± 28	36–64
30	± 37	19–41
20	± 47	11–19
6	± 50	8–24
0	± 60	4–16
6	± 83	1–11

performance of the spread plate, spiral plate and Miles-Misra technique (Gilchrist et al., 1977; Jarvis et al., 1977; Kramer and Gilbert, 1978; Donegan et al, 1991; Codd et al., 1998). Therefore they will be regarded in the rest of the paper as one method and generally referred to as surface plating techniques.

The total error associated with colony counts obtained with surface plating techniques can be calculated from Table 2 according to the following formula given by Jarvis (1989):

$$\% \text{ Total error} = \pm \sqrt{(A^2 + B^2 + C^2)}$$

where A = % sampling error; B = % distribution error; C = % dilution error.

The errors introduced by plating and calculation are regarded as less important (Jennison and Wadsworth, 1940). In addition, as the test and reference medium are inoculated from the same culture, the error introduced by sampling can be disregarded. From the errors that remain, the dilution error is assumed to be about 5.5 % (Jarvis, 1989). The error may be lower when the spiral plater is used for inoculation (Reusse, 1982). The distribution error is dependent on the number of colonies which have been counted, and is relatively small at higher counts. The precision of colony counts, assuming random distribution of organisms, was calculated by Jarvis (1989) and is listed in Table 3.

If it is assumed that the counts comply with the Poisson distribution (Jarvis, 1989), the 95% confidence limits can be calculated using the formula given for the overall % of error and the precision for each colony count, as is given in Table 3. For counts higher than 500, the limiting precision was obtained by extrapolation. The results of these calculations are collected in Table 4. They clearly show that the precision of the same count increases with the number of plates that have been used.

The calculated confidence limits are used to determine the optimal combination of numbers of plates and the average number of colonies which have to be used and

Table 4
Approximate confidence limits for colony counts using different numbers of plates.

Total count (Number of plates)	Mean count	Overall error % [a]	95% Confidence limits for the mean count
(n = 2)			
60	30	25.6	14.6–45.4
100	50	20.7	29.2–70.8
200	100	15.0	70.0–130.0
300	150	13.2	110.4–189.6
400	200	11.4	154.4–245.6
600	300	10.5	237.0–363.0
(n = 3)			
90	30	21.7	17.0–43.0
150	50	17.9	32.1–67.9
300	100	13.2	73.6–126.4
450	150	10.5	118.5–181.5
600	200	10.5	158.0–242.0
900	300	8.9	246.6–353.4
(n = 5)			
150	30	17.9	19.3–40.7
250	50	14.1	35.9–64.1
500	100	10.5	79.0–121.0
750	150	9.7	120.9–179.1
1000	200	8.9	164.4–235.6
1500	300	8.1	251.4–348.6
(n = 10)			
300	30	13.2	22.1–37.9
500	50	10.5	39.5–60.5
1000	100	8.9	82.2–117.8
1500	150	8.1	125.7–174.3
2000	200	6.8	172.8–227.2
3000	300	5.8	265.2–334.8

[a] Assumes a dilution error of 5.5 %

counted to minimize the influence of statistical error on the medium QC. In medium QC the performance of the test medium is compared to that of the reference medium, trying to minimize the statistical error of the difference in colony counts obtained on the test and the reference medium. In other words, we try to minimize the difference between the average counts on both media:

$$AVG_{\text{reference medium}} - AVG_{\text{test medium}} \quad \rightarrow 0 \text{ or}$$

$$\log (AVG_{\text{ref.}}) - \log (AVG_{\text{test}}) \quad \rightarrow 0$$

The difference between the logs of average counts on the test and the reference medium are distributed according to the Student t-distribution (Anon, 1989). Their 95% confidence interval can be calculated according to the following equation:

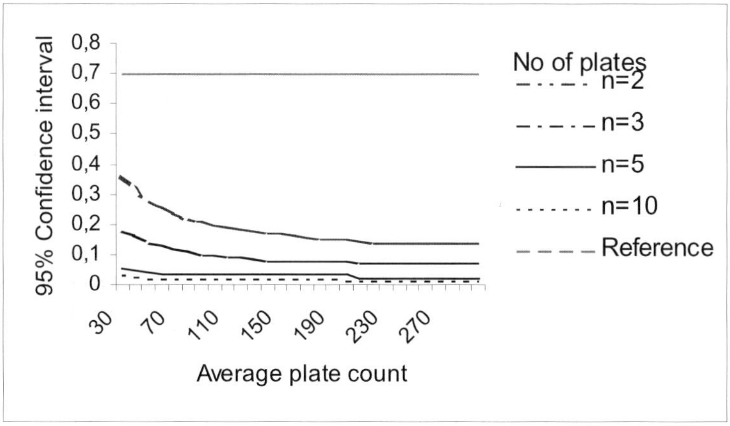

Fig. 3. The 95% confidence interval (in log units) for $\log(AVG_{ref.}) - \log(AVG_{test})$ in relation to the average plate count and the number of plates counted (n).

$$\log(AVG_{ref.}) - \log(AVG_{test}) \pm t \text{ x } s_{comb.} \text{ x } \sqrt{(1/n_{ref.} + 1/n_{test})}$$

Where t = t distribution constant with $n_{ref.} + n_{test}$ – two degrees of freedom; $s_{comb.}$ = the combined standard deviation for the difference in logs of the average counts on the test and reference medium; n = number of plates used with the test and reference medium.

For a given number of plates and average colony count, the values for the different components of the equation can be filled in, except for the combined standard deviation. For s_{comb} following equation is valid:

$$s_{comb.} = \sqrt{(df_{ref.} x s^2_{ref.} + df_{test} x s^2_{test})/(df_{ref.} + df_{test})}$$

When it is assumed that $s_{ref.}$ and s_{test} should in principle be equal, and for the sake of convenience the same numbers of plates are used for the test and the reference media, it can be derived that:

$$s_{comb.} = s_{ref.} = s_{test} = s$$

The standard deviation for the means can be derived from the 95% confidence intervals (CI) for the mean colony counts as they were calculated according to Jarvis (1989) and are collected in Table 4:

$$AVG \pm (\text{overall error \% x 2 x } AVG) = (CI) \quad \text{or}$$

$$(AVG+(CI)) - (AVG-(CI)) = 4 \text{ x } s \quad \text{or}$$

$$\log(AVG+(CI)) - \log(AVG-(CI)) = 4 \text{ x } s \text{ logunits}$$

So from the confidence intervals given in Table 4, the combined standard deviation can be calculated for each set of plates and average number of counted colonies, leading to the 95% confidence intervals for $\log(AVG_{ref.}) - \log(AVG_{test})$. These are shown in Figure 3. As a reference, Figure 3 also includes a set criterion for the productivity of a test medium: a medium is accepted when the (average) count on the test medium does not differ more than 0.7 logunits from the growth on the reference medium (Baird et al., 1987).

Figure 3 clearly shows that the statistical error is minimized when the average count on the plate and the number of counted plates are increased. However, a mean count of more than 100 per plate does not significantly reduce statistical error when more than two plates are used. The use of more than three plates is not worthwhile as the statistical error is only reduced marginally. Therefore it is concluded that the use of three plates with an average count of about 100 colonies per plate is the most optimal combination in respect to minimizing the statistical counting error in the QC of the media. For practical reasons it may however be decided to use two plates per medium. When an average count of about 100 CFU per plate is used, this seems to be acceptable. The maximum difference between test and reference medium introduced by statistical error, given by the sum of the 95% confidence intervals, still remains below the reference value of 0.7 log units.

The 95% confidence interval for the mean count defines the range of means which can be expected on average in 19 tests out of 20 when a mean is calculated from a set of colony counts. The interval is derived from the mean count calculated from the colony counts obtained from a number of samples. In other words the 95% confidence interval gives an estimate of the population within which the true mean count will lie. In one test out of 20 it is possible that the mean lies outside this interval and might therefore result in the conclusion that the mean does not belong to this population. This is the so-called fault of the first order, and defines the chance that a mean may be unjustifiably discarded from the population. In addition, a second order fault defines the chance that a result is regarded as belonging to the population, although in fact it belongs to another one which partly overlaps. In order to calculate this chance, the following situation is taken as an example. The test medium and reference medium are checked for their productivity, and are considered to perform equally well. So according to the criteria proposed by the ICFMH working party (Baird et al., 1987) their colony counts will lie within 0.7 log units from each other. In statistical terms the following chance should be determined:

$$P\{\log(AVG_{ref.}) - \log(AVG_{test}) < 0.7\} \text{ under the condition that}$$

$$\mu_{ref.} - \mu_{test} \neq 0$$

Where $\mu_{ref.}$ and μ_{test} are the population means.

This chance can be rewritten in the following equation:

$$P\{t < (0.7 - (\log(\mu_{ref.}) - \log(\mu_{test})) / (s_{comb.} \sqrt{(1/n_{ref.} + 1/n_{test})})\}$$

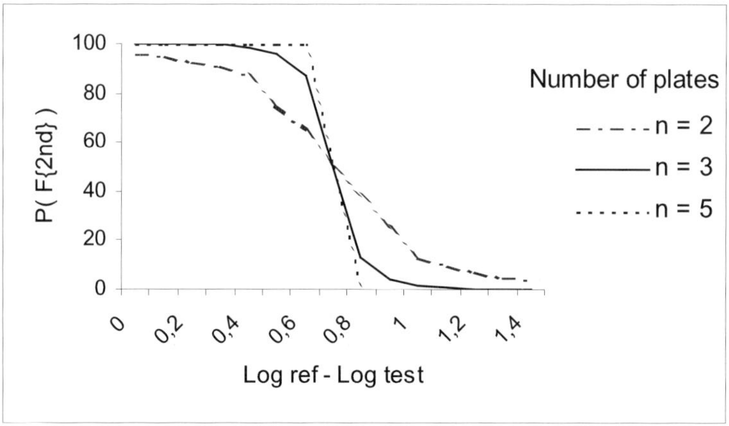

Fig. 4. Probability of a second order fault (P{F 2nd}) for different values of $\log(\mu_{ref.})-\log(\mu_{test})$ and numbers of plates (n) for an average colony count of 100.

For different values of $(\mu_{ref.} - \mu_{test})$ the chance of a fault of the second order can be calculated in relation to the number of plates used and colonies counted. Figure 4 gives an example of a result of such calculations for a given colony count and number of plates. In reference to the fault of the first order, we of course also tried to minimize the chance of a second order fault. Figure 4 shows that for a combination of numbers of plates and colony counts that give the steepest slope around the 50% probability was achieved with the highest number of plates (n=5). This gives the highest chance that when the difference between the population means is indeed more that 0.7 log units, this difference will be recognized and recorded as such. Again, with the combination of three plates and an average colony count of 100, reasonable reliability is obtained. Increasing the number of plates or average number of colonies counted does not improve the chance characteristic significantly (results not shown). So also in respect of minimizing the chance of a second order fault, a combination of three plates per medium on which on average 100 colonies should be counted, is the best choice. When medium assessment is carried out in duplicate, it has to be accepted that the chance of mistakenly approving a batch of medium is significant.

The ecometric streaking technique is semi-quantitative and is based on an arbitrary scoring system (Mossel et al., 1980). The precision of the ecometric data is considerably lower than that of the surface plating techniques discussed above (Mossel et al., 1980). Figure 5 shows the relation between the ecometric score and the colony count as determined for *Staphylococcus aureus* and *Escherichia coli* grown on Baird-Parker agar and VRBG agar respectively. They confirm earlier observations that within the score of five a difference in colony counts of 1 log unit will not be observed (Mossel et al., 1980; van Netten et al., 1991). Mossel and coworkers (1980) calculated the confidence interval from the ecograms they obtained in several experiments and arrived at 0.6 scoring units for a score below 4.6. Compared to the confidence intervals calculated for the conventional surface inoculation techniques (Tables 3 and 4) this is relatively poor. However, the purpose of the medium QC determines whether this is

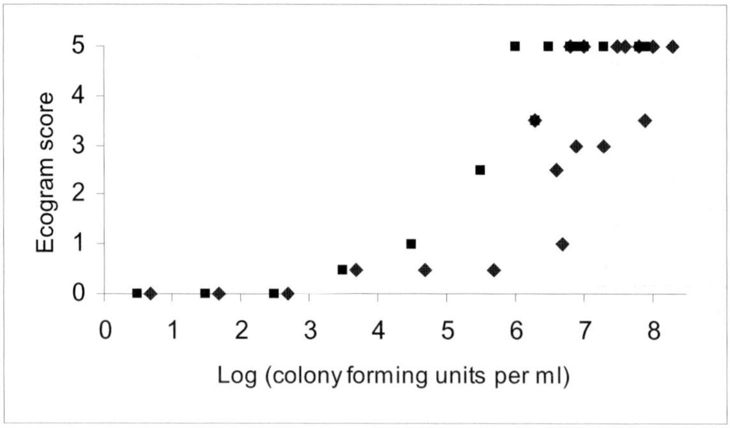

Fig. 5. The relation between the colony count and the score on the ecogram. Each point represents the average score of five accepted plates. *Escherichia coli* on violet red bile agar (VRBG), *Staphylococcus aureus* on Baird-Parker agar (BPA).

important or not. The technique may have sufficient accuracy in medium development trials, where, in general, large differences in medium performance are considered to be relevant.

Several authors have proposed criteria and units of measure for medium assessment, using the ecometric streaking technique (Mossel et al., 1983; Baird et al., 1987; van Netten et al., 1991). The statistics of the ecometric streaking technique, as shown in Figure 5, allow these attempts to be harmonized. Figure 5 clearly shows that consistent scores are obtained at counts higher than 10^7 and below 10^2. At intermediate counts, the ecometric score jumps in steps of 0.5 units from 0 to 5. Assuming that overnight cultures contain 10^8 to 10^9 microorganisms per ml and taking the criteria proposed for surface plating techniques (Baird et al.,1987), as a guide, criteria and units of measure can be derived. They are listed in Table 5 and agree very well with those proposed by van Netten and coworkers (1991). They use the Absolute Growth Index (AGI) as unit of measure.

It is not possible to calculate a statistically acceptable experimental design. However, in line with the statistics of the surface plating techniques, it is assumed that ecometric medium assessment should also be carried out in duplicate or triplicate in order to obtain reliable results.

Comparison of methods used for the QC of solid media

From the qualitative and statistical comparison of the methods which are frequently used for the quality control of media, a scope and field of application for each of the methods can be drawn up. In general, conventional surface plating techniques can be applied in every area of medium QC as they allow the accumulation of relatively precise quantitative (and qualitative) data. The ecometric technique has its application in those areas where semi quantitative data are sufficient, e.g. medium development

Table 5
Criteria for the assessment of solid media for productivity and selectivity, using surface plating (SP) or the ecometric streaking technique (EC).

Medium	Test strain	Criterion SP[a] $(\text{Log}(\text{AVG}_{ref.})\text{–Log}(\text{AVG}_{test}))$	Criterion EC[a] $\text{AGI}_{ref.}\text{–AGI}_{test}$
Non selective	Wanted	≤ 0.7	≤ 1
Selective	Wanted	≤ 1.0	≤ 1
Selective	Not wanted	≥ 5.0	4–5

[a] The test should be carried out in duplicate or triplicate and (for SP) with a count of at least 100 colonies per plate.

Table 6
Scope and field of application for methods for the quality control of solid microbiological media.

Field of application	Scope	SP [a]	EC [b]
Good laboratory practices	Quality check of purchased batches of ready to use media (or their ingredients) before use	+[c]	+
	Check of medium preparation procedures	+	+
	Audit or compare the performance of medium suppliers	+	–
Medium development	Relatively large differences are investigated	+	+

[a] SP = Surface plate (Miles-Misra + Spiral plate + Spread plate) techniques
[b] EC = Ecometric technique
[c] + = method recommended for this purpose
– = method is not recommended for this purpose

where relatively large differences in medium performance are being investigated. However, the ecometric technique can also be the method of choice in the control of a number of good laboratory practices. It should however be stressed that for this purpose conventional surface plating techniques provide more accurate data. In a number of cases more than one method applies. Personal preference and cost will determine in these cases which method a laboratory adopts. The scope and fields of application are summarized in Table 6.

QC methods for liquid media

Liquid media are used to promote (or inhibit) microbial growth, induce the production of visible metabolites or provoke specific reactions. Several methods for the QC of broths have been proposed and can be divided into two groups (Baird et al., 1987; van Netten et al. 1991): methods which determine kinetic parameters and endpoint determinations.

The kinetic parameters relate to the growth curve: the lag phase and/or the growth rate are used as criteria. It has been shown that they can be determined with great accuracy (Papadopoulou and Ionannidis, 1990). However, they have not been used frequently in medium QC and proper validated criteria are lacking (Curtis, 1985; van Netten et al., 1991). In addition they are relatively labour intensive and therefore seem to be more suitable for research purposes than for routine quality checks.

Endpoint determinations focus on the final result of growth: the increase in biomass. This can be determined indirectly by: measuring turbidity or changes in electrical properties, e.g. conductance of the broth; estimating viable numbers within the broth; looking at the production of visible metabolites (e.g. gas) or reactive compounds (e.g. indole).

Adding 15 g agar per litre to broth has also been suggested to determine its quality using techniques developed for solid media (van Netten et al., 1991). As such an approach would limit the range of QC data (e.g. gas production would not be observed), the method is not recommended.

The most widely used method for the QC of liquid media is the serial dilution technique, as proposed by Richard (1982). This method is also used for assessment of recently developed semisolid media for *Salmonella* detection (De Smedt, 1998). A dilution series of an overnight culture is prepared in the test and reference broth, using a 10 fold dilution up to 10^{-12}. The number of positive tubes or the highest titre at which microbial activity is still apparent is scored. The serial dilution technique is not as labour intensive as the kinetic methods and easy to perform on a routine basis. Growth can either be determined by eye, by measurement of the optical density, by changes in conductance or by streaking onto solid media (June et al., 1996). In our hands determination of growth by eye gave, in general, as reliable results as using instrumentation (results not shown). Only opaque media, like Rappaport-Vassiliadis broth, gave problems and confirmation of growth by inoculation onto agar, as described by Stokes and Ridgeway (1979), became necessary. Mossel and coworkers (1974) determined the biomass at the end of growth by colony counting. The same approach was taken by Carvalhal et al. (1991), who applied MPN statistics to microorganisms which were grown in liquid medium and subsequently transferred to agar. The method was shown to give as accurate results as the spread plate and drop plate methods. The use of solid media to determine the microbial yield at the end of growth also allows QC of broths in which mixed strains or naturally contaminated food samples are being used. It permits accurate quantification of growth yields after the test strains have been allowed to express their physiological capacities in the broth. Table 7 gives an overview of the strengths and weaknesses of the methods.

The statistics of methods used for the QC of liquid media

Microbial growth in liquid media is generally considered to comply with the Poisson distribution (Cochran, 1950; Cowell and Morisetti, 1969; Jarvis, 1989). Before the Poisson distribution can be applied the following conditions have to be met: the microorganism should be randomly distributed in the broth: the microorganism is equally likely to be found in any part of the broth and there is no tendency to form pairs, groups or clusters of organisms or to repel one another; one or more microbes

Table 7
Strength-weakness analysis of methods used for the quality control of liquid media.

Technique	Accuracy[a]	Visibility of diagnostic system[a]	Cost of Labour[b]	Cost of Equipment[b]	Overall result
Length of lag phase	–	+	–	–	2–
Growth rate	+	+	–	–	0
Agar addition[c]	+	–	+	+	2+
MPN	+	+	+	+	4+
Agar inoculation[d]	+	+	–	+	2+

[a] + = good, – = poor
[b] + = low, – = high
[c] Agar is added to the liquid medium
[d] Inoculation of agar after growth in the liquid medium.

will result in good (turbid) growth in the medium. In the QC of microbiological media these requirements are generally met, as pure overnight cultures are used. However, naturally contaminated samples have also been proposed for checking the performance of a medium (Mossel, 1985). In such a case the distribution of microorganisms within the food will very likely disagree with the Poisson distribution and therefore other distributions should be applied (Galesloot, 1986; Haas and Heller, 1988; Jarvis, 1989).

The accuracy of growth predictions based on a single inoculated tube is relatively poor (Halvorson and Ziegler, 1933). Therefore the "Most Probable Number" (MPN) concept was developed, using dilution series and multiple tubes per dilution to predict the number of microorganisms (Cochran, 1950; Jarvis, 1989). The MPN became an important tool within food microbiology after ready to use tables became available (de Man, 1977, 1983). As it is not as accurate as colony counts (Thomas, 1955; McCarthy et al., 1958), attempts have been made to improve it (Aspinall and Kilsby, 1979; Reichart, 1991). Townsend and Naqui (1998) used regression analysis to compare MPN results obtained using two different methods and obtained good results. With the aid of computer programs an optimal combination of dilutions, tubes per dilution and repeats can be defined in order to get an accurate MPN result with a minimal input of media, materials and labour (Hurley and Roscoe, 1983; Strijbosch, 1989). Figure 6 illustrates the relation between the standard error of the log(MPN) and the number of tubes and the dilution factor used. It clearly shows that the accuracy of the estimate for the density of the microorganisms in the broth can be improved significantly when the dilution factor is decreased or the number of tubes per dilution are increased. Increasing the number of tubes to more than five tubes per dilution seems not to be worth while.

The experimental design, defined by the dilution factor and the number of replicates, determines the accuracy of the quantitative data obtained from the QC of liquid media. Therefore the criteria for judgement of the quantitative performance of the test medium are also dependent on the experimental design. The following calculations will illustrate this, using the most probable number as estimate for the true density in

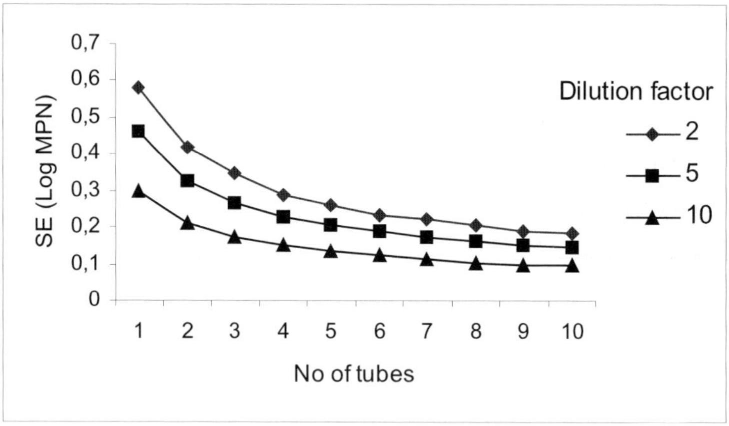

Fig. 6. The relation between the standard error of the $\log_{10}$(MPN) (SE log MPN) and the dilution factor and the number of replicate tubes used in the test (From Jarvis (1989)).

the liquid medium.

The starting point for the statistical analysis is again the Null hypothesis:

H_0 : The productivity of the test and reference medium are the same.

H_1 : The productivity of the test and reference medium are not the same.

The statistical significance of differences between the two MPN estimations can be calculated using the formula given by Jarvis (1989):

$$t = (\log MPN_{ref.} - \log MPN_{test}) / 0.58 \sqrt{}((\log(a_{ref.}/n_{ref.}) + (\log(a_{test}/n_{test}))$$

with t = Student t-test variable with ($n_{ref.} + n_{test} - 2$) degrees of freedom; MPN = most probable number obtained with the test (MPN_{test}) and reference medium ($MPN_{ref.}$); a = dilution factor used in the tube series with test (a_{test}) and reference medium ($a_{ref.}$); n = number of replicate tubes used in the tube series with test (n_{test}) and reference medium ($n_{ref.}$).

When a dilution factor of lower than 10 is used, the formula should use 0.55 instead of 0.58. For each combination of dilution factors, tubes and the resulting t-value, the critical ratio of $MPN_{ref.}$ and MPN_{test} can be calculated. At higher values of the critical ratio the Null hypothesis is rejected at a defined significance level (i.e. p=0.05).

Table 8 illustrates the results of such calculations. For the sake of convenience the same dilution factor and number of tubes were used in the tube series with test and reference medium. Depending on the discriminating requirements of the test, an experimental design can be chosen. For instance, to be able to detect a difference between test and reference medium of 1 titre unit, five replicates per dilution and a dilution factor of 10 would be the best approach.

When the statistics of presence/absence tests are compared with the criteria pro-

Table 8
The critical values for the ratio between $MPN_{ref.}$ and MPN_{test} (i.e. the ratio above which a significant difference between both MPN's is confirmed) at a significance level of p=0.05 (according to the student t-test).
It has been assumed that for the reference and test medium the same dilution factor and number of replicates per dilution have been taken

Number of replicates per dilution	Dilution factor	Degrees of freedom	t-value	$MPN_{ref.}/MPN_{test}$
3	2	4	2.78	4.8
3	5	4	2.78	11.1
3	10	4	2.78	20.6
5	2	8	2.31	2.8
5	5	8	2.31	4.7
5	10	8	2.31	7.0
10	2	18	2.10	1.9
10	5	18	2.10	2.9
10	10	18	2.10	3.5

Table 9
Criteria for the assessment of liquid media for productivity and selectivity, using the serial dilution technique (Richard, 1982).

Medium	Test strain	Criterion[a] Average number of positive tubes (reference – test medium)
Non selective	Wanted	≤ 0.7
Selective	Wanted	≤ 1.0
Selective	Not wanted	≥ 5.0

[a] The test is carried out in triplicate.

posed for medium assessment (Baird et al., 1987) is obvious that large numbers of tubes have to be analyzed in order to achieve an acceptable accuracy. This will impose a significant pressure on the resources of the lab and thus will discourage medium assessment studies. Therefore it seems to be more realistic to follow the serial dilution technique and to carry out the tests in triplicate (Richard, 1982). It has to be accepted that statistical error will significantly interfere with the results. Table 9 summarizes the criteria for the assessment of liquid media.

When the endpoint yield is determined by taking the number of positive tubes (p) out of n inoculated tubes (n ≠ p), growth in the reference and test medium can be assessed and compared according to the Chi squared test (Anon, 1989). The criteria for rejection or acceptance of medium batches again, depend on the number of tubes used. Where growth yields are expressed as CFU's, after subsequent inoculation onto agar (Mossel et al., 1974), the same statistics apply as shown for the QC of solid media.

Table 10
Scope and field of application for methods for the quality control of liquid microbiological media.

Field of application	Scope	KP [a]	ED [b]
Good laboratory practices	Quality check of purchased batches of ready to use media (or their ingredients) before use	– [c]	+
	Check of medium preparation procedures	–	+
	Audit or compare the performance of medium suppliers	–	+
Medium development	Relatively large differences are investigated	+	+

[a] KP = medium assessment using kinetic parameters (lag phase or growth rate)
[b] ED = endpoint determinations
[c] + = method recommended for this purpose
– = method is not recommended for this purpose

Comparison of methods used for the QC of liquid media

With the introduction of growth determination equipment kinetic methods have become available for QC laboratories. Before they can be used properly the experimental design, and the criteria for rejection and release need to be defined. Currently, the methods of choice are therefore endpoint determinations. For this purpose the serial dilution technique is preferred. Good alternatives are growth yield determinations based on the total number of positive tubes or CFU's on agar inoculated from the test tubes. For all these methods, the experimental design and the resulting criteria are available. The proposed scope and fields of application are summarized in Table 10. Endpoint determinations can be (routinely) used for all fields of application. Criteria for rejection and release of liquid media are not yet available for methods using the lag phase or growth rate as unit of measure. Therefore methods using kinetic parameters have not been recommended for use to check good laboratory practices.

Conclusions

In this overview methods for the quality control of media are compared, from the view point of a manager of a quality control laboratory. He or she will ask the following questions before going into extensive testing of media: Which methods are easy to use and give reliable results? Which experimental design should I use in order to obtain reliable data with a minimal input of resources (staff and materials)?

Using statistics as a yardstick, these questions can be answered satisfactorily. This overview shows that solid media can be assessed with acceptable accuracy using well established methods like the spread plate technique. In order to assure minimum statistical error at least two plates with an average count of 100 CFU per plate seems to be the best design. Although less accurate, the ecometric technique provides a good alternative to the more quantitative counting methods. In line with the experimental design proposed for the latter, at least two plates should be tested by ecometry to obtain a reliable result. For an accurate assessment of liquid media, large numbers of tubes need to be tested. This is expensive in laboratory resources. Therefore it is proposed to use the serial dilution technique, in which the broths are tested in triplicate, according to Richard (1982).

It is proposed to complete the criteria, as they have been proposed by the ICFMH Working Party on the Quality Control of Microbiological Media, by adding the required experimental design (Tables 5 and 9). This will help QC laboratories to set up medium assessment as a quality control tool within the framework of Good Laboratory Practices and promote further standardization.

Acknowledgements

The author wishes to thank Prof. B. Jarvis, Dr F.P.G.M. la Fors, Dr P. van Netten and Dr. L.W.G. Strijbosch for helpful discussions on the subject. The technical assistance of Jos Meeuwisse and Jolanda van den Brink is greatly appreciated.

References

Ahmed, N.M. and Conner, D.E. (1995) Evaluation of various media for recovery of thermally-injured *Escherichia coli* O157:H7. J. Food Protect. **58**, 357–360.

Anon. (1989) Introduction and Statistics for Microbiologists, Course Manual. Statistics for Industry, Knaresborough, UK.

Anon (1998a) Microval: a European approach to the certification of new microbiological methods. Int. J. Food Microbiol. **45**, 17–24.

Anon (1998b) A new expression for selectivity of liquid and solid media – statement from the IUMS-ICFMH Working Party on Culture Media. Int. J. Food Microbiol. **45**, 65.

Anon (1999) AOAC International qualitative and quantitative microbiology guidelines for methods validation. J.Assoc. Off. Anal. Chem. **82**, 402–413.

Aspinall, L.J. and Kilsby, D.C. (1979) A microbiological quality control procedure based on tube counts. J. Appl. Bacteriol. **46**, 325–330.

Baird, R.M., Corry, J.E.L. and Curtis G.D.W. (eds.). (1987) Pharmacopoeia of culture media for food microbiology. Proceedings of the 4th International Symposium on Quality Assurance and Quality Control of Microbiological Culture Media. Manchester, 4–5 September. Int. J. Food Microbiol. **5**, 187–299.

Bennet, A.R., Greenwood, D., Tennant, C., Banks, J.P. and Betts, R.P. (1998) Rapid and definitive detection of *Salmonella* in foods by PCR. Lett. Appl. Microbiol. **26**, 437–441.

Berg, C., Dahms, S., Hildebrandt, G., Klaschka, S. and Weiss, H. (1994) Microbiological collaborative studies for quality control in food laboratories: reference matererial and evaluation of analyst's errors. . Int. J. Food Microbiol. **24**, 41–52.

Bolderdijk, R. and Milas, J. (1996) *Salmonella* detection in dried milk products by motility enrichment on Modified Semi-solid Rappaport-Vassiliadis medium: collaborative study. J. Assoc. Off. Anal. Chem. **79**, 441–450.

Bolton, F.J. (1998) Quality assurance in food microbiology – a novel approach. . Int. J. Food Microbiol. **45**, 7–11.

Bridson, E.Y. and Gould, G.W. (2000) Quantal microbiology. Lett. Appl. Microbiol. **30**, 95–98.

Carvalhal, M.L.C., Oliveira, M.S. and Alterthum, F. (1991) An economical and time saving alternative to the most-probable-number method for the enumeration of microorganisms. J. Microbiol. Methods. **14**, 165–170.

Chain, V.S. and Fung, D.Y.C. (1991) Comparison of redigel, petrifilm, spiral plate system, isogrid and aerobic plate count for determining the numbers of aerobic bacteria in selected foods. J. Food Protect. **54**, 208–211.

Cochran, W.G. (1950) Estimation of bacterial densities by means of the "most probable number". Biometrics. **6**, 1105–116.

Codd, A.A., Richardson, I.R. and Andrews, N. (1998) Lenticules for the control of quantitative methods in food microbiology. J. Appl. Microbiol. **85**, 913–917.

Corry, J.E.L. (1982) Quality assessment of culture media by the Miles-Misra method. In: J.E.L. Corry (Ed) Quality Assurance and Quality Control of Microbiological Culture Media. Proceedings of the Symposium held on 6–7 September 1979, Callas de Mallorca, Spain, G.I.T. –Verlag Ernst Geibeler, Darmstadt. pp. 21–38.

Corry, J.E.L., Baird, R. and Terplan, G. (eds). (1982) Proceedings of the second Symposium of Quality Assurance and Quality Control of Microbiological Culture Media. February 27–28 1981, Dallas, USA. Archiv. Lebensmittelhyg. **33**, 137–175.

Corry, J.E.L., Leclerc, M., Mossel, D.A.A., Skovgaard, N., Terplan G. and van Netten, P. (1986) An investigation into the quality of media prepared and poured by an automatic system. Int. J. Food Microbiol. **3**, 109–120.

Corry, J.E.L., Curtis, G.D.W. and Baird, R.M. (editors) (1995) Culture Media for Food Microbiology Progress in Industrial Microbiology vol. 34. Elsevier Amsterdam.

Cowell, N.D. and Morisetti, M.D. (1969) Microbiological techniques – some statistical aspects. J. Sci. Food Agric. **20**, 573–579.

Curtis, G.D.W. (1985) A review of methods for the quality control of culture media. Int. J. Food Microbiol. **2**, 13–20

Curtis, G.D.W. and Beuchat, L.R. (1998) Quality control of culture media – perspectives and problems. Int. J. Food Microbiol. **45**, 3–6.

Deere, D., Vesey, G., Ashbolt, N., Davies, K.A., Williams, K.L. and Veal, D. (1998) Evaluation of fluorochromes for flow cytometric detection of *Cryptosporidium parvum* oocysts labelled by fluorescent *in situ* hybridization. Lett. Appl. Microbiol. **27**, 352–356.

de Man, J.C. (1977) MPN tables for more than one test. Europ. J. Appl. Microbiol. **4**, 307–316.

de Man, J.C. (1983) MPN tables, corrected. Europ. J. Microbiol. Biotechnol. **17**, 301–305.

De Smedt, J.M. (1998) AOAC validation of qualitative and quantitative methods for microbiology in foods. . Int. J. Food Microbiol. **45**, 25–28.

Defives, C., Guyard, S., Oularé, M.M., Mary, P. and Hornex, J.P. (1999) Total counts, culturable and viable, and non-culturable microflora of a French mineral water: a case study. J. Appl. Microbiol. **86**, 1033–1038.

Donegan, K., Matyac, C., Seidler, R. and Portecous, A. (1991) Evaluation of methods for sampling, recovery and enumeration of bacteria applied to the phylloplane. Appl. Environm. Microbiol. **57**, 51–56.

Entis, P. (1998) Direct 24 hour presumptive enumeration of *Escherichia coli* O157:H7 in foods using hydrophobic grid membrane filter followed by serological confirmation: collaborative study. J. Assoc. Off. Anal. Chem. **81**, 403–419.

Galesloot, T.E. (1986) Statistical remarks concerning the limiting dilution test used for the bacteriological testing of milk and milk products. Neth. Milk Dairy J. **40**, 31–40.

Geissler, K., Manafi, M., Amorós, I. and Alonso, J.L. (2000) Quantitative determination of total coliforms and *Escherichia coli* in marine waters with chromogenic and fluorogenic media. J. Appl. Microbiol. **88**, 280–285.

George, I., Petit, M. and Servais, P. (2000) Use of enzymatic methods for rapid enumeration of coliforms in freshwaters. J. Appl. Microbiol. **88**, 404–413.

Gilchrist, J.E., Campbell, J.E., Donelly, C.B., Peeler, J.T. and Delanay, J.M. (1973) Spiral plate

method for bacterial determination. Appl. Microbiol. **25**, 244–252.
Gilchrist, J.E., Donelly, C.B., Peeler, J.T. and Campbell, J.E. (1977) Collaborative study comparing the spiral plate and aerobic plate count methods. J. Assoc. Offic. Anal. Chem. **60**, 807–812.
Gutiérez, R., García, T., González, I., Sanz, B., Hernández, P.E. and Martin, R. (1998) Quantitative detection of meat spoilage bacteria by using the polymerase chain reaction (PCR) and an enzyme linked immunosorbent assay (ELISA). Lett. Appl. Microbiol. **26**, 372–376.
Haas, C.N. and Heller, B. (1988) Statistical approaches to monitoring. In: Mc Feters, A. (Ed). Drinking Water Microbiology. pp 412–427, Springer Verlag. New York.
Halvorson, H.O. and Ziegler, N.R. (1933) Application of statistics to problems in bacteriology. II. A consideration of the accuracy of dilution data obtained by using a single dilution. J. Bacteriol. **26**, 331–339.
Hammack, T.S., Feng, P., Amaguána, M., June, G.A., Sherrod, P.S. and Andrews, W.H. (1997) Comparison of sorbitol McConkey and Hemorrhagic coli agars for recovery of *Escherichia coli* O157:H7 from brie, ice cream and whole milk. J. Assoc. Off. Anal. Chem. **80**, 335–340.
Hurley, M.A. and Roscoe, M.E. (1983) Automated statistical analysis of microbial enumeration by dilution series. J. Appl. Bacteriol. **55**, 159–164.
In 't Veld, P.H. (1998) The use of reference materials in quality assurance programmes in food microbiology laboratories. Int. J. Food Microbiol. **45**, 35–41.
Jarvis, B., Lach, V.H. and Wood, J.M. (1977) Evaluation of the spiral plate maker for the enumeration of micro-organisms in foods. J. Appl. Bacteriol. **43**, 149–157.
Jarvis, B. (1989) Statistical aspects of the microbiological analysis of foods. Progress in Industrial Microbiology, vol. **21**. Elsevier Scientific Publishers B.V. Amsterdam.
Jennison, M.W. and Wadsworth, G.P. (1940) Evaluation of the errors involved in estimating bacterial numbers by the plating method. J. Bacteriol. **43**, 149–157.
June, G.A., Sherrod, P.S., Hammack, T.S., Amaguaña, R.M. and Andrews, W.H. (1996) Relative effectiveness of selenite cystine broth, tetrathionate broth and Rappaport-Vassiliadis medium for recovery of *Salmonella* spp. from raw flesh, highly contaminated foods, and poultry feed: Collaborative study. . J. Assoc. Offic. Anal. Chem. **79**, 1307–1323.
Kirst, von E., Tomforde, M. and Brandt, H. Keimzahl-bestimmung in rohmilch. (1998) DMZ **24**, 1146–1152.
Kramer, J.M. and Gilbert, R.J. (1978) Enumeration of microorganisms in food: a comparative study of five methods. J. Hyg. Camb. **61**, 151–159.
Kramer, J.M., Kendall, M. and Gilbert, R.J. (1979) Evaluation of the spiral plate and the laser colony counting techniques for the enumeration of bacteria in foods. Eur. J. Appl. Microbiol. Biotechnol. **6**, 289–299.
Lahellec, C. (1998) Development of standard methods with special reference to Europe. Int. J. Food Microbiol. **45**, 13–16.
Lambert, R.J., Johnston, M.D. and Simons, E.-A. (1998) Disinfectant testing: use of the Bioscreen Microbiological Growth Analyser for laboratory biocide screening. Lett. Appl. Microbiol. **26**, 288–292.
Lambert, R.J.W. and Oudenraa, van der, M.-L.H. (1999) An investigation into the differences between the Bioscreen and traditional plate count disinfectant test methods. J. Appl. Microbiol. **86**, 689–694.
Leuschner, R.K.G. and Lillford, P.J. (1999) Effects of temperature and heat activation on germination of individual spores of *Bacillus subtilis*. Lett. Appl. Microbiol. **29**, 228–232.
Mafart, P., Mathot, A-G., McMeekin, T.A. and Olley, J (Editors) (1999) Microbial stress and recovery in food. Int. J. Food Microbiol. **55**, 1–298.
McCarthy, J.A., Thomas, H.A. and Delaney, J.E. (1958) Evaluation of the reliability of coliform density tests. Am. J. Public Health. **48**, 1628–1635.
Miles, A.A., Misra, S.S. and Irwin, J.O. (1938) The estimation of the bactericidal power of blood. J. Hyg. **38**, 732–749.

Millet, V. and Lonvaud-Funel, A. (2000) The viable but non-culturable state of wine microorganisms during storage. Lett. Appl. Microbiol. **30**, 136–141.

Mossel, D.A.A., (1970) Microbiological culture media as ecosystems. In: van Bragt, J., Mossel,D.A.A., Pierik, R.L.M. and Veldstra, H. Effects of sterilization on components in nutrient media. H. Veenman & Zonen N.V. Wageningen, pp 15–40.

Mossel, D.A.A., Harrewijn, G.A. and Nesselrooy-van Zadelhoff, C.F.M. (1974) Standardisation of the selective inhibitory effect of surface active compounds used in media for the detection of Enterobacteriaceae in foods and water. Health Lab. Sci. **11**, 260–267.

Mossel, D.A.A., van Rossem, F. and Rantama, A. (1979) Ecometric monitoring of agar immersion plating and contact (AIPC)-slides used in assuring the microbiological quality of perishable foods. Laboratory practice **28**, 470–475.

Mossel, D.A.A., van Rossem, F., Koopmans, M., Hendriks, M., Verouden M. and Eelderink, I. (1980) A comparison of the classical and the so-called ecometric technique for the quality control of selective culture media. J. Appl. Bacteriol. **49**, 439–454.

Mossel, D.A.A., Bonants-van Laarhoven, T.M.G., Lichtenberg-Merkus, A.M.T. and Werdler, M.E.B. (1983) Quality assurance of selective culture media for bacteria, moulds and yeasts: an attempt at standaardisation at the international level. J. Appl. Bacteriol. **54**, 313–327.

Mossel, D.A.A. (1985) Introduction and prospective. Int. J. Food Microbiol. **2**, 1–7.

Mossel, D.A.A., Corry, J.E.L., Struijk, C.B. and Baird, R.M. (1995) Essentials of the microbiology of food. pp 268–274. John Wiley and sons, New York.

Mulder, A.M. and Strikwerda, K. (1996) Microbiological secundary reference material *Staphylococcus aureus* in milk. De Ware(n) Chemicus **26**, 147–150.

Nebe-von Caron, G., Stephens, P. and Badley, R.A. (1998) Assessment of bacterial viability status by flow cytometry and single cell sorting. J. Appl. Microbiol. **84**, 988–998.

Papadopoulou, C. and Ioannidis, K. (1990) Differentation of *S. gallinarum* and *S. pullorum* by means of growth-kinetics analysis. J. Microbiol. Methods. **11**, 247–253.

Payne, J.F., Morris, A.E.J. and Beers, P. (1999) Note: Evaluation of selective media for the enumeration of *Bifidobacterium* sp. in milk. J. Appl. Microbiol. **86**, 353–358.

Ray, B (Editor) (1989) Injured index and pathogenic bacteria: occurrence and detection in foods, water and feeds. CRC Press, Boca Raton, Florida

Reichart, O. (1991) Some remarks on the bias of the MPN method. Int. J. Food Microbiol. **13**, 131–142.

Reusse, U. (1982) The use of the stomacher and spiral plate methods in food microbiology. In: Corry J.E.L. (ed). Quality Assurance and Quality Control of Microbiological Culture Media. Proceedings of the Symposium held on 6–7 September 1979, Callas de Mallorca, Spain, pp 59–61, G.I.T.-Verlag Ernst Giebeler, Darmstadt.

Richard, N. (1982) Monitoring the quality of selective liquid media used in the official serial dilution technique for the bacteriological examination of food. In: Corry J.E.L. (ed). Quality Assurance and Quality Control of Microbiological Culture Media. Proceedings of the Symposium held on 6–7 September 1979, Callas de Mallorca, Spain, pp 51–58, G.I.T.-Verlag Ernst Giebeler, Darmstadt.

Sartory, D.P., Field, M., Curbishley, S.M. and Pritchard, A.M. (1998) Evaluation of two media for the membrane filtration enumeration of *Clostridium perfringens* from water. Lett. Appl. Microbiol. **27**, 323–327.

Seiler, D.A.L. (1985) Monitoring mycological media. Int. J. Food Microbiol. **2**, 123–131.

Stokes, E.J. and Ridgway, G.L. (1979) Media-testing and other techniques. In: Clinical Bacteriology. 5th edition. pp 342–381. Edward Arnold London.

Strijbosch, L.W.G. (1989) Experimental design and statistical evaluation of limiting dilution assays. PhD Thesis. University of Maastricht (The Netherlands).

Tanaka, Y., Yamaguchi, N. and Nasu, M (2000) Viability of *Escherichia coli* O157:H7 in natural river water determined by the use of flow cytometry. J. Appl. Microbiol. **88**, 228–236.

Thiriat, L., Sidaner, F. and Schwartzbrod, J. (1998) Determination of *Giardia* cyst viability in environmental and faecal samples by immunofluorescence, fluorogenic dye staining and differential interference contrast microscopy. Lett. Appl. Microbiol. **26**, 237–242.

Thomas, H.A. (1955) Statistical analysis of coliform data. Sewage Ind. Wastes. **27**, 212–222.

Thomson, G.F. (1984). Enumeration of yeasts and moulds – media trial. Food Microbiol. **1**, 223–227.

Townsend, D.E. and Naqui, A. (1998) Comparison of SimPlate total plate count test with plate count agar method for detection and quantitation of bacteria in food. . J. Assoc. Offic. Anal. Chem. **81**, 563–569.

van Netten, P., Weenk, G. and van der Zee, H. (1991) Quality control of culture media. De Ware(n) Chemicus. **21**, 26–60.

Veyrat, A., Miralles, M.C. and Pérez-Martínez, G. (1999) A fast method for monitoring the colonization rate of lactobacilli in a meat model system. J. Appl. Microbiol. **87**, 49–61.

Chapter 2

Recovery of stressed microorganisms

P.J. Stephens[a] and B.M. Mackey[b]

[a]*Oxoid Ltd., Wade Road, Basingstoke, RG24 8PW, UK*

[b]*Food Microbial Sciences Unit, School of Food Biosciences, The University of Reading, PO Box 226, Whiteknights, Reading, RG6 6AP, UK*

Microorganisms present in processed food, water or the environment may be injured and hence more exacting in their growth requirements. Such organisms may be difficult to detect because they fail to grow on the selective media normally used in their isolation. However, under suitable conditions injured cells can repair cellular damage and recover all their normal properties including virulence. An appreciation of the nature of sublethal injury and its repair is therefore important in detecting and enumerating microbes.

Injured cells may show an extended lag phase, restricted temperature range for growth, and increased sensitivity to selective agents, salt, acidity and oxidative stress. In some cases it has been possible to identify the site of cellular injury leading to the observed phenotype. For example, damage to the Gram-negative outer membrane causes increased sensitivity to bile salts and some antibiotics, whilst disruption of cytoplasmic membrane function is probably related to increased sensitivity to salt and acid. Oxidative stress may arise from reactive oxygen species generated metabolically or present in trace amounts in recovery media. The reasons underlying some effects such as a restricted temperature range for growth are unknown.

To recover injured cells it is necessary to allow resuscitation to take place on non-selective solid or in liquid media. In addition to improving the sensitivity of detection methods, resuscitation improves consistency and reduces variability between laboratories. Recent developments in recovering injured cells have focused on defining the time needed for repair of individual cells within a population and in formulating resuscitation media and/or conditions to avoid exposing cells to oxidative stress. The need to monitor the performance (selectivity and productivity) of selective enrichment media as a quality control exercise is well established. It is now apparent that monitoring the performance of non-selective pre-enrichment or recovery media is equally important. This is particularly true when considering the development and application of new end-point detection systems that employ gene probe or immunological techniques.

Introduction

Within any population of microorganisms surviving exposure to preservation treatments or environmental stresses there exist individual cells that are regarded as sublethally injured. Stress treatments that cause injury include heating, refrigeration, freezing, irradiation, high acid or alkali, high salt levels, preservatives, desiccation, exposure to disinfectants and starvation or nutrient limitation. Individually, cells subjected to these stresses will suffer differing degrees of injury. Some will remain fully intact, exhibiting no signs of damage, some will suffer sufficient damage to lead to a loss of viability and the remainder will be sublethally injured, the latter suffering damage that will lead to a temporary change in physiology. From a methodological point of view injury manifests itself in three important ways: (i) sensitivity to the selective agents used traditionally in microbiological culture media; (ii) sensitivity to low levels of reactive oxygen species that can exist in microbiological culture media and (iii) extended lag times during which repair of damage takes place. Sublethal injury and its repair have been reviewed in numerous articles and book volumes (Busta, 1976; Hurst, 1977; Mossel and Corry, 1977; Beuchat, 1978; Andrew and Russell, 1984; Mackey, 2000).

Despite being in a debilitated physiological state, injured cells can pose a significant spoilage and public health risk; the main concern being that injured cells might undergo resuscitation either in the food or following ingestion. To enable their successful enumeration and/or isolation and detection numerous modifications to methods have been suggested. Following a brief review of cellular changes due to stress and the subsequent manifestations of these changes, some of the most successful and most common modifications to microbiological methods will be discussed. These are divided into those that are applicable to enumeration methods and those that are applicable to presence/absence screening. Lastly, consideration will be given to the optimisation of resuscitation media.

Effects of stress on cell structures and components

Damage to the cytoplasmic membrane

Many workers (Iandolo and Ordal, 1966; Allwood and Russell, 1968; Hurst, 1977) have observed the leakage of small molecular weight compounds, when cells are subjected to stress. This has been associated with damage to the cytoplasmic membrane leading to loss of selective permeability (Mazur, 1966; Gray et al., 1973; Macleod and Calcott, 1976). Damage to the cytoplasmic membrane has also been reported to occur in cold shocked *Clostridium perfringens* (Traci and Duncan, 1974). Lipids and phospholipids in *Staphylococcus aureus* are located entirely in the membrane and their loss during heat stress is indicative of membrane damage (Hurst et al., 1973). By contrast, no lipid loss was detected in *Salmonella typhimurium* heated at 48°C (Tomlins et al., 1972). Visual evidence for membrane damage in *Escherichia coli* has been obtained by using light or electron microscopy. Cells exposed to mild heat released membrane vesicles into the medium, whereas cells exposed to hypotonic stress shed small membrane vesicles into the cytoplasm (Katsui et al., 1982; Schwarz and Koch, 1995). In pressure-treated *Listeria monocytogenes,* visible evidence of membrane perturbations took the form of large clear areas in the cytoplasm adjacent to membranes, suggesting that membrane invaginations had occurred under pressure (Mackey et al., 1994). Electron microscopy has shown fractures in the membranes of heat injured *Bacillus cereus* that would result in the release of cellular material (Silva and Sousa, 1972).

Damage to the Gram-negative outer membrane

In Gram-negative enteric organisms and pseudomonads, resistance to lipophilic dyes, bile salts, and many lipophilic antibiotics is due to their restricted penetration through the outer membrane combined with active efflux through multidrug transporter systems. If the outer membrane is damaged, cells are unable to grow on selective media containing bile salts or dyes (e.g. MacConkey or Brilliant Green agars). This behaviour is typical of cells exposed to heat, cold shock, freezing, drying, aerosolisation and high pressure. The phenotype of such injured cells resembles that of enteric organisms that have defects in the lipopolysaccharide (LPS) component of the outer membrane either arising from mutation or by removal following treatment with EDTA. Loss of LPS has been seen in heat- or freeze-injured *E. coli* and *Pseudomonas* cells and a conformational change in the outer membrane of freeze-injured *E. coli* was indicated by an altered pattern of adsorption of LPS-specific bacteriophages (Hitchener and Egan, 1977; Kempler and Ray, 1978; Ryan et al., 1979).

Damage to Ribosomes and RNA

Ribosome and RNA degradation is known to occur during heat stress (Hurst, 1984). Thermal stress in *S. typhimurium* has been shown to destroy the 16S RNA and the 30S subunit, partially degrade the 23S RNA, and reduce the sedimentation coefficient of the 50S particle (Tomlins and Ordal, 1971). A similar result has been demonstrated following heat treatment of *B. subtilis* and *S. aureus* (Rosenthal and Iandolo, 1970; Miller and Ordal, 1972). Mild heat stress of *Pseudomonas fluorescens* resulted in the release of 260 nm absorbing material which was presumed to be evidence of RNA degradation, and confirmed by colorimetric analysis (Gray et al., 1973). Emswiler et al. (1976) demonstrated that RNA was damaged during heating of *Vibrio parahaemolyticus* and that RNA synthesis was required for recovery. The loss of 260 nm absorbing material also occurs as a result of freeze-injury of Gram negative bacteria (Mossel and Corry, 1977). It has been suggested that freezing may activate ribonuclease activity leading to the degradation of RNA (Morichi, 1969). Allwood and Russell (1968) observed a correlation between loss of RNA and viability during heat stressing of *S. aureus* at temperatures up to 50°C. The extent of degradation was greater at 50°C than 60°C and it was presumed, therefore, that degradation was due to enzyme action, with the enzyme being destroyed at 60°C. The temperatures that cause heat injury in bacteria maybe sufficient to cause some denaturation of ribosomes making them more susceptible to attack by ribonucleases. Lee and Goepfert (1975) showed that heat stressing of *S. typhimurium* resulted in RNA degradation during the initial 10 min of heating, a period during which there was little cell death. Strange and Shon (1964) suggested that ribosome damage following heating was due to loss of the magnesium required for ribosome stability. Indeed, *S. aureus* which lost cellular magnesium during heating in potassium buffer also lost RNA (Tomlins and Ordal, 1976). The presence of Mg^{2+} in the heating menstruum or heating in Tris-magnesium buffer protects RNA from destruction and reduces the extent of cellular injury (Chakraburtty and Burma, 1968; Lee and Goepfert, 1975; Hitchener and Egan, 1977; Stephens and Jones, 1993). Considering the extent of RNA degradation is dependent on the a mount of Mg^{2+} loss, then it is likely that this aspect of cellular injury is governed by the extent of leakage due to membrane damage. This would further emphasise the importance of the membrane in maintenance of viability.

The destruction of the 30S ribosomal subunit that occurs when *S. typhimurium* cells are heated under mild temperature stress conditions and low magnesium concentration is not necessarily lethal and cells can resynthesise the missing component when incubated under favourable conditions (Tomlins and Ordal, 1976). However, studies of cells heated at rising temperatures in the differential scanning calorimeter showed that loss of viability in *E. coli* and *L. monocytogenes* was associated with irreversible destruction of the 30S subunit at higher temperatures (Mackey et al., 1991; Anderson et al., 1991). Loss of the 30S ribosomal subunit can thus be lethal or sublethal depending on heating and recovery conditions. This form of injury is associated with an extended lag time during which RNA and protein are resynthesised.

Damage to DNA

The damage and repair to DNA caused by UV and ionising radiation has been well characterised but it is also apparent that DNA is damaged and subsequently repaired following heat, cold and desiccation stresses (Pierson et al., 1978). *Escherichia coli* can remove up to 850 pyrimidine dimers per genome caused by UV-irradiation and can repair up to 30 single-strand breaks caused by ionising radiation. Although repair of double strand breaks can occur under some circumstances, one double strand break is normally lethal. Many workers have shown that physical stresses can result in DNA strand breakage. Swartz (1971) observed that freezing caused single strand breaks in *E. coli*. Double strand breaks of DNA have been observed in freeze-thawed *E. coli* (Calcott and Thomas, 1981). Such DNA breakage is thought most likely to occur by the induction of deoxyribonucleases following physical stresses and not as a direct effect of the stress (Pierson et al., 1978).

Sato and Takahashi (1970) found that cold shock of *E. coli* resulted in single strand DNA breaks and suggested that loss of Mg^{2+} from cells would lead to a reduction in DNA ligase activity. The cell would then be unable to undergo efficient DNA replication leading to cell death. Experiments utilising DNA repair deficient strains or compounds which prevent DNA synthesis have added to the evidence that DNA is damaged during physical stress and that intact DNA repair mechanisms are needed for cell recovery. Ligase deficient mutants of *E. coli* and *P. aeruginosa* have been shown to be more sensitive to heat and freeze-thaw stress respectively than wild-types that are able to repair DNA damage (Pauling and Beck, 1975; Williams and Calcott, 1982). DNA synthesis during recovery is clearly required, as hydroxyurea and nalidixic acid inhibited repair of heat injured *S. typhimurium* and *V. parahaemolyticus* respectively (Gomez et al., 1973; Emswiler et al., 1976).

MANIFESTATIONS OF INJURY

Sensitivity to selective agents

Many different types of selective media are used in food microbiology; all containing compounds that preferentially or selectively allow for the growth of the target group from a sample containing a mixed population of bacteria. Healthy bacteria of the target group should be resistant to the selective compounds at the concentrations used. Selective agents of choice include surface-active compounds, such as sodium deoxycholate or sodium lauryl sulphate; toxic compounds, such as selenite or tellurite; salts; antibiotics and acids. When cells are injured they frequently become sensitive to such selective compounds. Heat injured *S. aureus* lost the ability to multiply and form colonies on media containing 7.5% salt (Busta and Jezeski, 1963). When the salt concentration was lowered, the cells were able to grow. *Staphylococcus aureus* injured by chilling also lost its salt tolerance (Jackson, 1974). Sensitivity to selective agents was reported in *E. coli* that had been exposed to a disinfectant (Scheusner et al., 1971). Growth was inhibited on lauryl sulphate agar, violet red bile glucose agar and most strongly on deoxycholate agar. Different dyes and salts were also found to inhibit growth. Smith and Dell (1990) demonstrated that media containing selective agents did not permit recovery of heat injured *Shigella flexneri*.

Mackey (1983) showed that *E. coli* injured by heating, freezing, drying or gamma irradiation became more sensitive to hydrophobic antibiotics (e.g. novobiocin, nalidixic acid, bacitracin). Freeze-dried *E. coli*, according to Sinskey and Silverman (1970) became sensitive to chloramphenicol, streptomycin and actinomycin D. Similar results were also found with *Campylobacter jejuni*, which after freezing and heating injury became more sensitive to antibiotics (Ray and Johnson, 1984; Humphrey and Cruickshank, 1985; Humphrey, 1986).

Iandolo and Ordal (1966) described the use of plate counting on trypticase soy agar with or without 7.5% NaCl as a measure of the uninjured and total *S. aureus* populations respectively. The difference between the two counts gave a rapid method of estimating injured populations. Experiments using different types of selective media have led to the wide use of such differential plating techniques to determine the percentage injury within a population (Mossel and Corry, 1977; Ray, 1979). Such an approach was employed by Ray and Speck (1973) whereby they found that 90% of *E. coli* cells surviving freeze treatment were injured, as determined by the ability of cells to form colonies on trypticase soy agar but not on violet red bile agar or deoxycholate-lactose agar. The injured cells were also sensitive to incubation in brilliant green bile broth and lauryl sulphate broth, which prevented subsequent colony formation on non-selective agar. Freeze treatment of *S. anatum* has also been shown to result in 90% injury within viable survivors, in this case by their ability to form colonies on the non-selective xylose-lysine peptone agar but not on agar containing 0.2% sodium deoxycholate (Ray et al., 1972a). Clark and Ordal (1969), using levine eosin methylene blue agar containing 2% NaCl, demonstrated injury within the viable population of *S. typhimurium* after sublethal heating. The heated cells also displayed sensitivity to brilliant green agar, Salmonella-Shigella agar and deoxycholate agar.

Sensitivity to secondary stresses

Injured *C. jejuni* have been shown to be less able to tolerate elevated temperatures, resulting in a significant fall in viable numbers after inoculation into broth. The reduction in numbers was greater at 43°C than at 37°C and was exacerbated by the presence of selective media (Humphrey, 1989). Such inability to tolerate the secondary stress of elevated temperature is characteristic of injured bacteria. Exposure to hot agar during plate pouring reduced the recovery of freeze injured *E. coli* on selective plates by 80% although the count on non-selective agar was less affected (Ray and Speck, 1973).

Fluctuations in pH are another secondary stress to which injured bacteria become more sensitive. Work by Sadovski (1977), during a study on the isolation of *Salmonella* from frozen vegetables, showed that prior to freezing *S. typhimurium* was capable of growth at pH 4.4. A freeze stressed culture, however, which had suffered 48% injury only exhibited growth at pH's above 5.5. The effect of medium pH on the recovery of injured cells was further demonstrated by Blankenship (1981), as *S. bareilly*, injured by acetic acid, took 1.9 h to reach 50% recovery at pH 7.0, but 6.5 h at pH 6.2, and at pH 4.5 there was only 19.5% recovery in 6 h. Increase in sensitivity to alkali has also been observed following chilling of *V. parahaemolyticus* in oyster homogenates, as there was a reduction in viable numbers during incubation in the enrichment broth which had a relatively high pH of 9.0 (Ma-Lin and Beuchat, 1980).

Increased lag period

Injured cells exhibit an extended lag period utilised to regain their integrity (Sinskey and Silverman, 1970), including the ability to multiply in media containing inhibitors (Jackson and Woodbine, 1963; Postgate and Hunter, 1963). With the exception of the repair mechanisms for DNA damaged by various types of radiation (Town et al., 1973; Moseley 1984), little is known of the mechanisms of repair. There is evidence that, under suitable conditions of recovery, heat injured cells synthesise phospholipid (repairing membrane lesions), re-synthesise ribosomal RNA and ribosomes accompanied by protein synthesis. Repair of DNA also occurs (Pierson et al., 1971; Heinis et al., 1978). Similar processes take place in cells damaged by other mechanisms (Ray et al., 1972b). Repair of the cell membrane is probably followed by re-establishment of internal low molecular weight solutes such as K^+, Mg^{2+} and Ca^{2+} (Beuchat, 1978; Pierson et al., 1978).

Taking samples from the recovery conditions and plating onto selective and non-selective media (Mossel and Corry, 1977) has been used to monitor speed of repair. During incubation the count on the non-selective agar remains unchanged but the count on the selective agar increases indicating that repair of the injured cells is occurring. When repair is complete, the two counts are equal and this is determined as the recovery time. This method has been used to measure recovery times in a number of studies. The time for complete repair varies markedly. The resuscitation of injured *Salmonella* has been reported to be complete in 5–8 h (Clark and Ordal, 1969; Ray et al., 1972b; Kafel and Bryan, 1977). Heat stressed *Ps. fluorescens* have been shown to have a lag time of 9 h before regaining resistance to selective agents (McCoy and

Ordal, 1979). It is reported that the duration of the lag phase for injured cells can vary according to the severity of damage (Payne, 1978; McCoy and Ordal, 1979) and lag times of 48–72 h have been recorded (Dabbah et al., 1969). Mackey and Derrick (1982) calculated lag times of *Salmonella* injured by different types of stress. Equivalent treatments showed that heat or freeze injured cells required longer to recover than dried or gamma irradiated cells. They went on to show that in a population of cells with a lag time of 9 h, some cells required up to 14 h to repair thus demonstrating the heterogeneity of the extent of injury within a population. More recently, using a semi-automated method capable of monitoring bacterial growth from individual cells, Stephens et al., (1997) reported broad distributions of lag times with many values in excess of 20 h for heat stressed *S. typhimurium* (Figure 1).

Sensitivity to oxidative stress

It has been known for a long time that injured cells are sensitive to oxidative stress, but the general significance and importance of this phenomenon in recovery of stressed cells has been recognised only relatively recently. For example, the recovery of heat injured cells of *L. monocytogenes, E. coli, S. enteritidis* and *S. aureus*, in rich non-selective media is enhanced by many orders of magnitude by the adoption of strictly anaerobic incubation conditions (Knabel et al.,1990; George et al., 1998; Ugboroghu and Ingham, 1994).

Reactive oxygen species (ROS). The consecutive univalent reduction of molecular oxygen to water produces three active intermediates; superoxide, hydrogen peroxide and the hydroxyl radical. In aqueous solution, superoxide is only moderately reactive but still possesses well-known destructive capability (Fridovich, 1983). The severest toxicity from superoxide arises from its metal catalysed reaction with hydrogen peroxide leading to the production of the hydroxyl radical, the most reactive of all the reduced forms of oxygen. The hydroxyl free radical can instantly oxidise almost any cellular component including proteins, nucleic acids and lipids. Hydrogen peroxide is more reactive than superoxide and readily diffuses across cell membranes, but it is generally held that its toxicity derives mainly from the metal catalysed production of the more reactive hydroxyl radical. Potential sites for hydrogen peroxide induced damage includes cellular membranes, enzymes, nucleic acids and membrane transport processes. Reactive oxygen species as a collective term also includes the non-radical derivatives of oxygen, such as singlet oxygen, hypochlorous acid, lipid peroxide and ozone (Valentine et al., 1995).

Sources of reactive oxygen species. Stephens et al. (2000) proposed that ROS arise from two main sources: intracellular due to endogenous electron transport chain activity and extracellular due to oxidation of culture medium components. The effect of oxidation on culture media was first observed many years ago. Reducing sugars, in the presence of phosphates, were implicated in the auto-oxidation of culture media during the autoclaving process with the subsequent generation of ROS (Lewis, 1930; Baumgartner, 1938). Wright (1933) cited work of earlier researchers on the inhibition

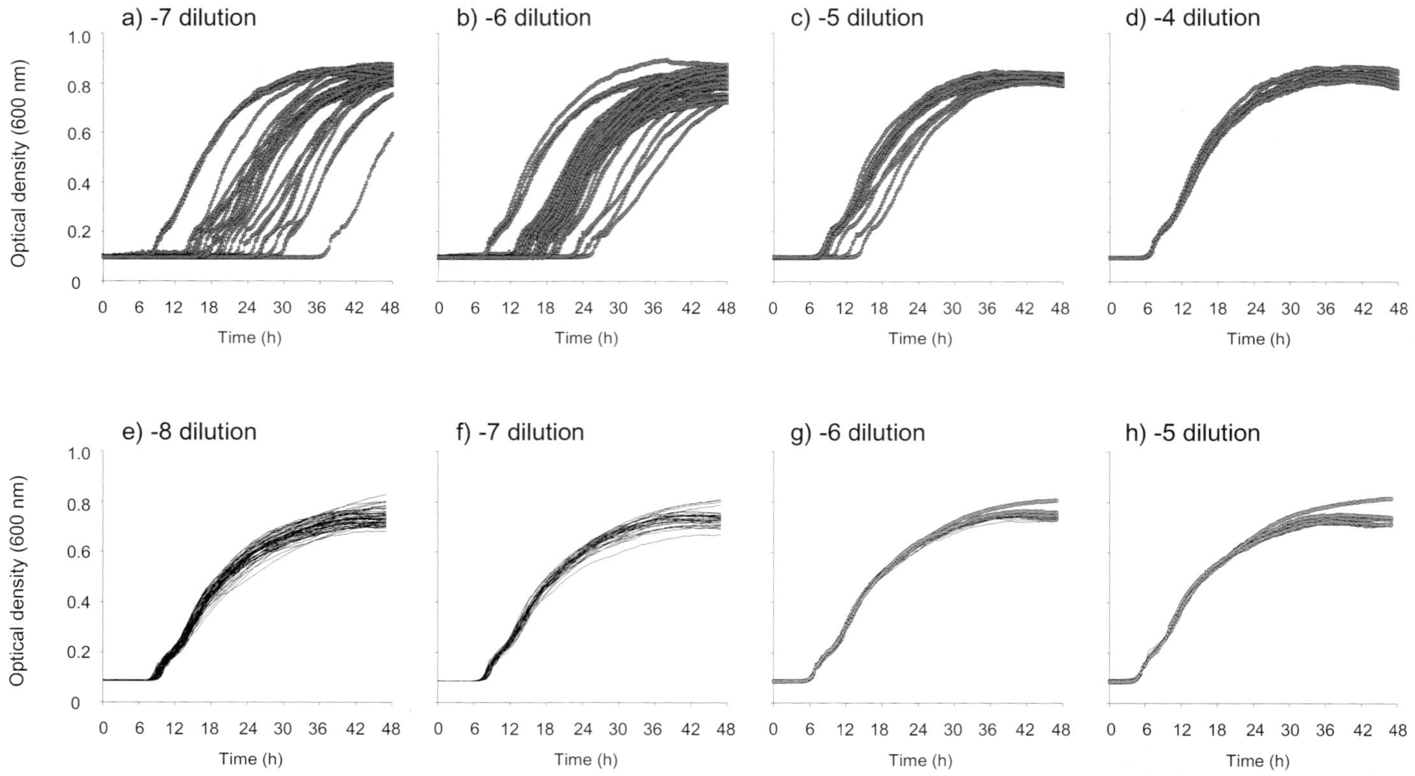

Fig. 1. Recovery and growth of heat stressed (15 min at 53.5°C) (a-d) and unstressed (e-h*) Salmonella typhimurium* (CMCC 3073) at different dilution levels in buffered peptone water incubated at 37°C. The starting inoculum for the 10^{-7} dilution of the heat stressed culture was estimated to be approximately one cell (30 out of 100 inoculated wells positive for growth). Measured using the Bioscreen analyser. (Stephens et al., 1997)

of bacterial growth by culture medium constituents, including his own work from 1929, which showed that the inhibiting factor was an oxidised constituent of peptone. Waterworth (1969) reported the failure of growth of staphylococci on nutrient agar plates to be due to photo-oxidation. Exposure to sunlight for as little as 15 min was found to be sufficient to induce toxicity. Different degrees of inhibition were shown by different medium formulations. Wang (1976), working with tissue culture media found that a major cause of medium deterioration was exposure to fluorescent light in the laboratory. Nixon and Wang (1977) claimed that one of the photoproducts was indeed hydrogen peroxide although Wang and Nixon (1978) eventually proved that only part of the killing effect was due to hydrogen peroxide, accounting for only 40% of the inhibitory activity. Earlier, Webb and Lorenz (1972) demonstrated killing of repair-deficient *E. coli* mutants added to media that had been previously exposed to near-UV light. A component of casamino acids was found to play a role in this toxicity, presumably riboflavin with its ability to act as a photo-sensitiser. McCormick et al. (1976) identified hydrogen peroxide to be a cell lethal product of photo-oxidation and Ananthaswamy and Eisenstark (1976) artificially produced photoproducts that they claimed to be either hydrogen peroxide or organic peroxides, which they used to kill mutants of *E. coli* and a range of bacteriophage. They attributed the killing effect to be due to single stranded DNA breaks. Photoproducts formed from UV-irradiation of tryptophan have been shown to inhibit DNA replication in *E. coli* (Yoakum et al., 1974). Although hydrogen peroxide is produced from UV-irradiation of tryptophan, this did not entirely explain the toxic effects (Yoakum and Eisenstark, 1972). Heat injured DNA repair mutants of *E. coli* were more sensitive than wild-type strains to plating on tryptone soya agar, but this sensitivity was reversed by adding catalase to the medium. Low levels of peroxide (15μM) were found to be present in the medium and similar levels of authentic hydrogen peroxide caused death of heat injured cells (Mackey and Seymour, 1987).

More recently, Edwards et al. (1994) demonstrated an active role for singlet oxygen and hydroxyl radicals in the inhibitory activity of UV irradiated tissue culture media. Because of their transient nature it was thought that other photoproducts must be involved, especially since the toxicity of the medium could last for many days following irradiation. Simpson et al. (1992) added to the evidence that hydrogen peroxide and its closely related ROS were not the only toxic products generated from oxidation reactions. They identified two relatively long-lived ROS including a protein hydroperoxide that consumed key cellular antioxidants. Smith (1975) described how auto-oxidation could also occur at ambient temperatures and that once oxidation had occurred any attempt to reduce the medium did not eliminate oxidised products. Stephens et al. (2000) found that much of the toxicity was attributable to ROS other than hydrogen peroxide and superoxide. Hoffman et al. (1979) measured superoxide levels in Brucella media used to grow *C. jejuni*. Using superoxide dismutase (SOD) as a control they found that it only reduced the toxicity by a maximum of 50% even in the presence of excess enzyme, and concluded that some of toxicity was due to agents other than superoxide. The oxidation events in media caused by exposure to light, oxygen and high temperatures are obviously complex and dynamic. The confusion in the literature over whether nutrient rich or minimal media are best for recovery can be explained by the many different factors that can influence a medium's performance. These include the different conditions of autoclaving, different exposures to oxygen and light, storage conditions and different manufacturers of the same product using different peptone ingredients.

Associated with the normal metabolic reduction of oxygen is the potential for the endogenous generation of ROS. It is considered that between 1 and 5% of all oxygen used in metabolism escapes as ROS (Punchard and Kelly, 1996). In some of the reactions only one pair of hydrogen ions is transferred to molecular oxygen, so giving rise to hydrogen peroxide (Rose, 1968). In addition, thermodynamically there is nothing to prevent earlier components of the electron transport chain from reducing molecular oxygen directly, albeit at a relatively low frequency (Fridovich, 1983). *In vitro* measurements (Imlay and Fridovich, 1992) suggest that one molecule of superoxide is formed for every 1000 electrons delivered to oxygen. Gort and Imlay (1998) showed that unstressed *E. coli* could tolerate only small increases in superoxide production before growth was inhibited. Thus suggesting that *E. coli* constitutively synthesises just enough SOD to defend intracellular components from endogenously generated superoxide.

In injured cells, it may be that membrane damage decouples the electron transport chain, releasing compartmentalised reactions leading to the generation of further ROS. Privalle and Fridovich (1987) heated *E. coli* to 48°C for 1 h and showed an increase in SOD activity. This temperature was more likely to induce a stress response than be lethal to the cell, but the authors proposed that the trigger for the stress response was an increased endogenous superoxide generation which had it been left to continue would quickly have reached lethal levels. They proposed that the increase in superoxide and hydrogen peroxide, was due to a higher rate of flavin related auto-oxidation reactions because of the higher temperature and/or physical disruption of the electron transport assemblies of the plasma membrane. Such disruption could lead to an increased flow of electrons to molecular oxygen to form superoxide. Many of the stress proteins induced in response to stresses such as heat, acid, osmotic stress etc. are antioxidant in nature (Flahaut et al., 1998) suggesting that increasing endogenous ROS maybe an expected occurrence as a result of these stresses.

Physiological basis of enhanced sensitivity to ROS in injured cells. Enhanced sensitivity of stressed cells to ROS may possibly be due to direct inactivation of enzymes protecting against oxidative stress and/or by inactivation of DNA-repair associated enzymes. In addition, the primary physical effect of nearly all stresses is some form of membrane perturbation and it is tempting to assume that this is responsible in some way for increased peroxide sensitivity. Stress-exposed membranes may be more permeable to peroxide which therefore diffuses more rapidly to sensitive sites within the cell. Surface located catalase may be lost during stress exposure. Co-factors required for DNA repair enzymes may be lost from the cell (Sato and Takahashi, 1970). The disrupted membrane itself may be more sensitive to peroxide attack with the resulting lipid peroxides having greater potency towards nucleic acids than the original peroxide. Reduced thiols that protect against oxidative stress may be lost through the more permeable membrane, or a change of the DNA/membrane attachment site(s) may sensitise the DNA to peroxide attack. Injured cells may be unable to synthesise new antioxidant enzymes rapidly (Smith and Archer, 1988).

Methods for detecting and recovering stressed cells

Numerous strategies have been used to avoid underestimating microbial numbers in samples likely to contain stressed cells. These include (i) avoiding the use of selective media altogether (ii) formulating the medium to minimise any inhibitory effects on injured target organisms and (iii) performing a resuscitation treatment to allow repair of injury before inoculation on/into a selective medium.

An example of the first approach is the use of size selection with membrane filters. This technique depends on the ability of small cells to pass through the 0.65 μm pores of a membrane filter placed on non-selective agar. (In practice this is restricted to members of the Campylobacteraceae.) When applied to stool samples of animals and humans this method allowed isolation of *C. upsaliensis* and other species that were initially overlooked because they were inhibited by the usual selective media (see Corry et al., this volume; Kiehlbauch et al., 1996; Linton, 1996).

Another possibility is to use a selective medium that does not inhibit injured target organisms. Unfortunately, there are few examples in current use. Baird-Parker agar (Baird-Parker and Davenport, 1965) supports good recovery of *S. aureus* cells damaged by drying, freezing or heating but there no other examples of such media, presumably because resistance to many selective agents depends on having intact cell membranes, and these are almost always a site of cellular injury.

Where inhibitory selective media cannot be avoided, resuscitation on solid or in liquid media is needed. In addition to improving the sensitivity of detection methods, including a resuscitation step tends to improve consistency and reduce variability between laboratories (van Schothorst and Duke, 1984).

Resuscitation methods for use when estimating microbial numbers

Most probable number (MPN) methods. Incubation of injured cells in a non-selective broth medium allows repair of injury but, once cells have repaired, multiplication begins, which can lead to an overestimation of the microbial load in a sample. To avoid this it is necessary to adopt the MPN approach in which the sample is diluted before resuscitation. One example of this method is the multiple tube MPN method for the enumeration of *E. coli* in water (Anon., 1968). This uses a non-selective resuscitation in minerals modified glutamate medium prior to incubation in bile salts and brilliant green containing media. The method depends only on the proportion of tubes showing growth, and is not thus affected by the extent of multiplication after repair. A similar approach was used to estimate numbers of injured salmonellae in milk powder (Van Schothorst and Van Leusden, 1975). A disadvantage of the MPN method is that it is labour intensive and the precision is poor unless the number of replicate tubes per dilution is very large.

Solid medium repair. Solid medium repair is the method of choice where quantitative estimates of viable numbers are needed. This can be achieved by spreading cells on the surface of a non-selective agar medium and, after allowing for repair to occur, overlaying with a selective medium. A more satisfactory method is to incubate cells on a membrane filter on a non-selective medium before transferring the membrane filter to selective agar. This approach is in routine use for the enumeration of *E. coli* as described by Holbrook et al. (1980). Cellulose acetate membranes are placed on the surface of non-selective minerals modified glutamate agar. The sample is added to the membrane and allowed to dry. Cells are incubated on the membrane at 37°C for 4 h, after which the membrane is transferred to the surface of a selective agar plate (in this case tryptone bile agar) and incubation continued at 44°C for additional selectivity. More recently, McCarthy et al. (1998) used a similar approach for the enumeration of *E. coli* O157:H7 in foods. They used thinner polycarbonate membranes to facilitate transfer of the pH indicator into the developing colony, thus allowing the fermentation reaction to be easily visualised on the surface of the membrane.

Speck et al. (1975) and Hartman et al. (1975) described agar overlay methods for the enumeration of stressed cells whereby cells are plated onto a non-selective agar medium and allowed to resuscitate for 3–6 h. Cells were then overlayered with selective agar. Both this approach and the membrane transfer method have the advantage that should growth of uninjured cells take place during the resuscitation period this will not lead to a falsely high count as the early growth will only contribute to the development of a single colony. More recently, Kang and Fung (1999; 2000) described the use of an agar "underlay" method for the enumeration of stressed *L. monocytogenes* and *S. typhimurium* respectively. In this method, selective medium is overlayered with non-selective medium immediately prior to the sample containing stressed cells being plated onto the surface of the upper non-selective layer. Initially the upper layer remains non-selective allowing resuscitation to take place. Selective agents then diffuse into the upper layer from the selective medium thus suppressing the growth of competing bacterial species.

Some success has been reported from the addition of "recovery supplements" to selective agars to aid the resuscitation of stressed cells in the presence of selective agents. In't Veld and de Boer (1991) and Wood et al. (1996) found that the addition of egg yolk emulsion to Oxford and Palcam formulations improved the recovery of stressed *L. monocytogenes.*

Not all modifications work successfully with all media. McCarthy et al. (1998) highlighted the deficiencies of some of the alternative resuscitation methods for enumeration of *E. coli* O157:H7 on sorbitol MacConkey (SMAC) agar. Firstly, reducing the level of bile salts to a level that was not inhibitory to stressed cells led to overgrowth by competing bacterial species. Incubation of stressed cells in non-selective buffered peptone water (BPW) prior to plating required up to 7 h for full resuscitation, by which time most of the uninjured cells had begun multiplying. The use of sandwich plates, with cells resuscitated on tryptone soy agar prior to being overlaid with SMAC agar, gave rise to an apparent false sorbitol-positive phenotype for *E. coli* O157:H7. Similarly, the addition of egg yolk emulsion led to all *E. coli* O157:H7 producing an acidic pH reaction in the medium.

Methods for presence/absence screening

In cases where the presence or absence of a target organism rather than the number present is of concern, prolonged enrichment can be used because multiplication will not affect the result, and is in fact desirable to increase sensitivity of detection. In the case of *Salmonella*, non-selective pre-enrichment in BPW or lactose broth has now been in place since the early 1970's. Pre-enrichment periods, according to most recognised standardisation bodies, are in excess of 16 h. D'Aoust and Maishment (1979) reported that pre-enrichment for 6 h in various non-selective media failed to identify about half of low and high moisture foods contaminated with *Salmonella*. Later, D'Aoust et al. (1990) reported that shorter incubation times, 3–8 h, in non-selective enrichment media did not result in effective resuscitation of injured organisms, and gave unacceptably high numbers of false negative results.

For the isolation and detection of *L. monocytogenes* there is often an initial 20–24 h primary enrichment in a medium with reduced selectivity, e.g. half Fraser broth. For the isolation and detection of *Campylobacter* spp. there is often an initial short (2–6 h) non-selective incubation followed by the progressive introduction of antibiotics and increase in incubation temperature. The process of delayed addition of selective agents has been simplified by using timed-release capsules to deliver selective agents into non-selective pre-enrichment media without the need for user-intervention (Baylis et al., 2000). With the recent concern regarding *E. coli* O157:H7 there has been much activity in the area of method development. Many enrichment methods still rely on the use of direct selective enrichment with its associated risks of toxicity to stressed cells (Stephens and Joynson, 1998; Blackburn and McCarthy, 2000).

Preventing death of injured cells from oxidative stress during recovery procedures

The recognition that there are two separate sources of oxidative stress in the routine use of conventional culture media has already led to improvements in the detection of low levels of stressed cells from food. The strategies that have been tried are (a) to add supplements to media to remove exogenous ROS, (b) to formulate medium composition such that the production of ROS is minimised and (c) to employ strictly anaerobic conditions to prevent endogenous generation of ROS.

There have been many studies showing that supplements that react with or remove exogenous ROS improve recoveries of injured cells on agar media. Supplements that have been tested include blood, catalase, peroxidase and SOD and also other quenching agents such as mannitol, cysteine and histidine. Often, to gain any benefit from the use of these recovery supplements, it is important to use them during the medium preparation and initial storage when the primary toxic components that they are active against are being generated (Rayman et al., 1978; Humphrey, 1988; Smith and Dell, 1990, Mackey, 2000). In pure culture studies improved recoveries have been shown with cells injured by exposure to a range of stresses including heating, freezing, drying, cold-shock and acidification. Beneficial effects have been reported in many food borne pathogens including *E. coli, C. jejuni, S. enteritidis, S. flexneri, S. aureus* and *L. monocytogenes*. As a minimum precaution when estimating viable numbers of stressed cells on non-selec-

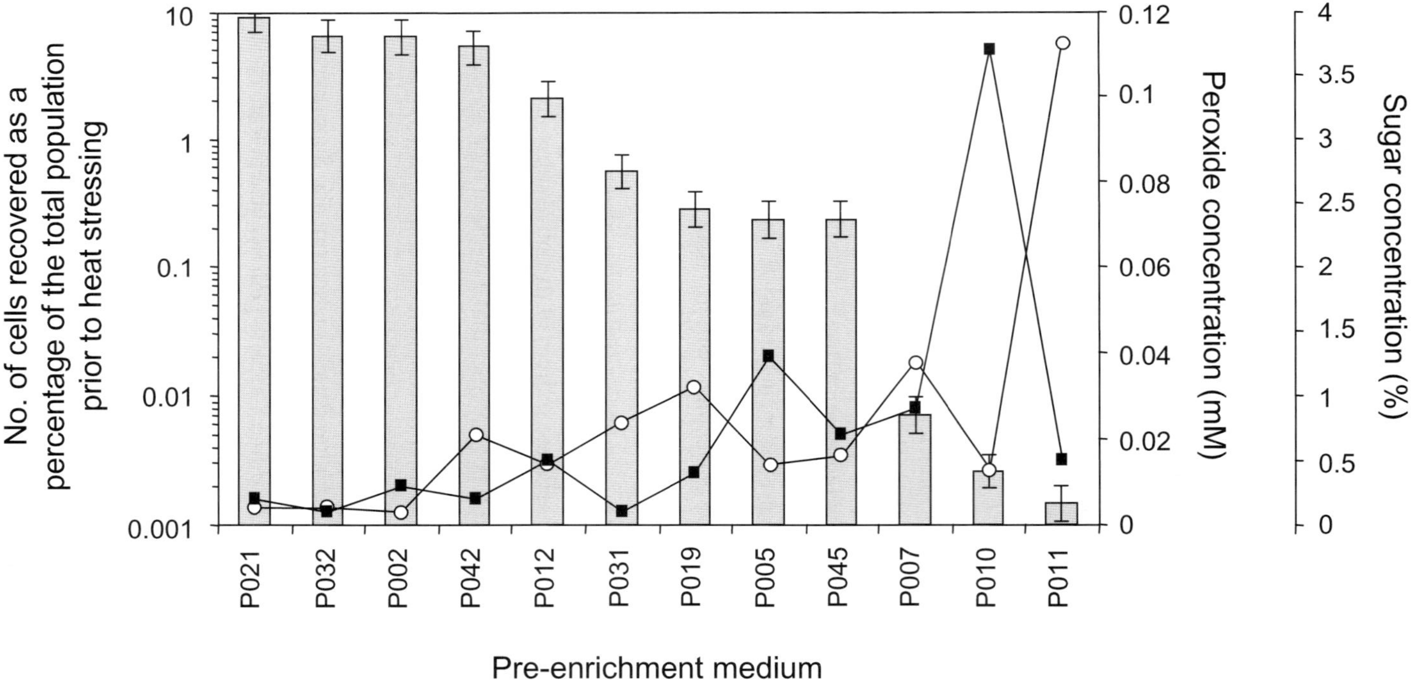

Fig. 2. Recovery of heat stressed (15 min at 53.5°C) *Salmonella typhimurium* (CMCC 3073) at 37°C in buffered peptone water prepared with 12 different peptone types measured against reducing sugar (■) and peroxide (○) concentrations. Peroxide concentrations measured in peptone solutions without buffers and NaCl. Error bars indicate 95% confidence limits on microtitre plate MPN estimates. (Stephens et al., 2000)

tive agar (e.g. in survival studies on pure cultures) media should be supplemented with blood, catalase or pyruvate. The inclusion of sodium pyruvate in Baird-Parker agar contributes substantially to its ability to recover stressed cells, but the addition of pyruvate to other selective media may interfere with performance e.g. where sugar fermentation is used as a diagnostic feature.

The notion that the composition of recovery media can be manipulated to minimise the production of exogenous ROS arose from observations of Stephens et al. (1997) who developed a sensitive technique for studying the resuscitation of low levels of stressed *Salmonella*. Using this technique it was demonstrated that lag times of stressed cells in commercially available non-selective pre-enrichment media were extremely variable and often in excess of 12 h. More significantly, some media recovered up to 3 $\log_{10}$ cycles more stressed cells than others. Later, Stephens et al., (2000) identified components of pre-enrichment media that were influential in the recovery of stressed *Salmonella* (Figure 2). Peptones were proven to vary sufficiently in their constituents to explain the significant differences in recovery reported by Stephens et al. (1997). Thus by appropriate choice of peptone source, ROS production during medium preparation can be reduced and performance improved.

The use of anaerobic conditions to improve recovery of stressed cells has been reported by several workers. In the study of Knabel et al. (1990), heat injured cells could only be recovered in reduced media inside sealed, undisturbed thermal death time tubes or in roll tubes containing pre-reduced media. Ugborogho and Ingham (1994) recovered more heat injured *S. aureus* when the enumeration medium was incubated under strict anaerobic conditions. Xavier and Ingham (1993) showed the same improvement in recovery of heat stressed *S. enteritidis* as did Linton et al. (1992) with *L. monocytogenes*. The failure to demonstrate a benefit from anaerobic conditions achieved using anaerobe jars can be attributed to the need for rapid removal and complete absence of oxygen for recovery of severely injured cells (Fernandez Garayzabal et al., 1986; Dallmier and Martin, 1988; Knabel et al., 1990; Farber et al., 1992). Interestingly, George and Peck (1998) obtained evidence suggesting that a low redox potential was more important for recovering injured cells of *E. coli* O157 than a low oxygen concentration *per se*. Normal resistance to aerobic incubation is recovered during the lag phase when cells are incubated anaerobically (Bromberg et al., 1998).

Strictly anaerobic conditions are usually achieved using the Hungate technique or by working entirely within an anaerobic cabinet. Neither of these is suitable for routine microbiological work. An alternative method of achieving anaerobiosis is to use a crude membrane preparation from *E. coli* that can be added to liquid medium to scavenge oxygen with high affinity (Adler and Crow, 1981). A membrane preparation known as Oxyrase is now available commercially. It has been used primarily for the isolation and manipulation of oxygen sensitive bacteria. Some of these have been relatively aerotolerant organisms such as *C. perfringens* while others have been strict anaerobes such as *C. difficile*. Oxyrase has also been used to promote the recovery of *E. coli* from radiation damage (Adler et al., 1981), and heat damage in *L. monocytogenes* (Yu and Fung, 1991a; Yu and Fung, 1991b; Knabel and Thielen, 1995; Patel et al., 1995), *E. coli* O157:H7 (Thippareddi et al., 1995), *Yersinia enterocolitica* (Thippareddi et al., 1995) and *S. typhimurium* (Doyle et al., 1996). By using a combination of a peptone medium low in ROS and Oxyrase, Baylis et al. (2000) reported significant improvements in the detection of *Salmonella* in spiked ice cream and milk powder. Its possible use in solid media has been less well investigated but supplementation of the SMAC medium for *E. coli* O157 with Oxyrase improved the resuscitation of stressed cells. Its use was not recommended because it led to a loss of the differential property of the medium; all colonies for both fermenting and non-fermenting species appearing positive (McCarthy et al., 1998).

Viable but non-culturable cells

The term viable but non-culturable (VBNC) is commonly used to describe what is believed to be a temporary loss of culturability in bacteria that can normally be cultivated with ease. Interest in the VBNC phenomenon arose from the observations that when *E. coli* and *V. cholerae* were incubated in salt water microcosms they lost their ability to grow on nutrient agar plates but remained metabolically active over long periods (Xu et al., 1982). Metabolic activity of single cells is commonly monitored in terms of their ability to reduce tetrazolium salts or their ability to elongate in broth containing nalidixic acid to inhibit DNA replication. The latter technique is referred to as the Direct Viable Count (DVC) (Kogure, 1979). The putative VBNC state has been reported in many genera, mainly of Gram negative organisms but including the Gram positive *Micrococcus luteus, L. monocytogenes* and *Enterococcus faecalis*. The most studied organisms have been *E. coli, V. cholerae, V. vulnificus, Legionella pneumophila, C. jejuni, S. enteritidis* and *M. luteus*. In most cases the inducing stress is starvation in fresh or sea-water, but VBNC cells have also been reported in response to biocide treatment, osmotic stress, exposure to sunlight, oxidative stress, aerosolisation, and survival in soil (Kell ct al.,1998; Barer and Harwood, 1999; Mackey, 2000)

The existence and nature of the VBNC state excite much controversy. There is also confusion arising from different meanings attached to the term 'viable'. Some proponents of the VBNC hypothesis believe that to demonstrate that a cell is intact and capable of reducing tetrazolium salt or elongating in the Direct Viable Count method is synonymous with it being alive. Others, including the present authors, acknowledge that the persistence of metabolic activity in cells may possibly indicate that they are alive, but believe that this can only be demonstrated unequivocally by showing that the cells can indeed multiply. The VBNC state is commonly interpreted as being a survival mechanism in which cells become dormant in response to adverse conditions. This is seen as being analogous to spore formation and is presumed to be a genetically programmed process. However, the evidence supporting the VBNC state as a survival stratagem is weak. Whilst there is good evidence that cells can become difficult to cultivate even on non-selective media that normally provide ideal growth conditions (see above), there is to date no good evidence that the putative VBNC state represents a unique physiological condition distinct from that in stressed or injured cells.

It should be mentioned that dormancy in bacteria simply represents a state of metabolic quiescence and does not imply non-culturability. However under some conditions dormant cells may become fastidious in their growth requirements. A good example is seen in *M. luteus* in which cells that have been maintained for prolonged periods in stationary phase cultures require a peptide resuscitation factor produced by other cells to grow from small inocula (Mukamalova et al., 1998). There is no doubt however that understanding the basis of the fastidious phenotype of stressed cells represents one of the major challenges facing those interested in the survival of bacteria and those interested in the practical problems of their detection.

Future developments

The need to monitor the performance (selectivity and productivity) of *selective* enrichment media as a quality control exercise is now well established. From the discussions above it is now apparent that monitoring the performance of *non-selective* pre-enrichment or recovery media is equally important. This is particularly true when considering the development and application of new end-point detection systems that employ gene probe or immunological techniques. Because of their specificity, these techniques may sometimes be applied directly to pre-enrichment broths. Differences in performance of non-selective media are usually only apparent with injured cells hence, for testing medium performance, it will be necessary to develop simple and convenient means for producing standardised preparations of injured cells. The productivity of different batches of media with such injured cells could be compared relatively simply by performing most probable number counts using microwell plates as described by Mackey (1985) and Stephens et al. (1997).

The importance of oxidative stress in recovering stressed cells has been emphasised. Substantial progress has been made in simplifying the achievement of strictly anaerobic conditions in liquid media. For the purposes of enumeration it would be helpful also to have a simple means of achieving strictly anaerobic conditions with agar plates. It is likely that the most appropriate recovery conditions will differ depending on the species and type of stress. For example recovery of *L. monocytogenes* was improved by addition of cysteine and sparging with nitrogen, rather than by use of Oxyrase alone (Knabel and Thielen, 1995). Further work to optimise recovery conditions for stressed cells with respect to redox potential and gas atmosphere will be needed.

References

Adler, H.I. and Crow, W.D. (1981) A novel approach to the growth of anaerobic microorganisms. In: S.C. Scott (editor) Biotechnology and Bioengineering Symposium No. 11, New York: John Wiley and Sons Inc. pp. 533–540.

Adler, H.I., Carrasco, A., Crow, W. and Gill, J.S. (1981) Cytoplasmic membrane fraction that promotes septation in an *Escherichia coli* lon mutant. J. Bacteriol. 147, 326–332.

Allwood, M.C. and Russell, A.D. (1968) Thermally induced ribonucleic acid degradation and leakage of substances from the metabolic pool in *Staphylococcus aureus*. J. Bacteriol. 95, 345–349.

Ananthaswamy, H.N. and Eisenstark, A. (1976) Near-UV-induced breaks in phage DNA: sensitization by hydrogen peroxide (a tryptophan photoproduct). Photochem. Photobiol. 24,

439–442.

Anderson, W.A., Hedges, N.D., Jones, M.V. and Cole, M.B. (1991) Thermal inactivation of *Listeria monocytogenes* studied by differential scanning calorimetry. J. Gen. Microbiol. 137, 1419–1424.

Andrew, M.H.E. and Russell A.D. (1984) The Revival of Injured Microbes. Academic press, London.

Anon. (1968) PHLS Standing Committee on Bacteriological Examination of Water Supplies. J. Hyg. Cam. 66, 67–82.

Baird-Parker, A.C. and Davenport, E. (1965) The effect of recovery medium on the isolation of *Staphylococcus aureus* after heat treatment and after storage of frozen or dried cells. J. Appl. Bacteriol. 28, 390–402.

Barer, M.R. and Harwood, C.R. (1999) Bacterial viability and culturability. Adv. Microbiol. Physiol. 41, 93–137.

Baumgartner, J.G. (1938) Heat sterilised reducing sugars and their effects on the thermal resistance of bacteria. J. Bacteriol. 36, 369–382.

Baylis, C.L., MacPhee, S. and Betts, R.P. (2000) Comparison of methods for the recovery of low levels of injured *Salmonella* in ice cream and milk powder. Lett. Appl. Microbiol. 30, 320–324.

Beuchat, L.R. (1978) Injury and repair of Gram-negative bacteria with special consideration of the involvement of the cytoplasmic membrane. Adv. Appl. Microbiol. 23, 219–243.

Blackburn. C. de W. and McCarthy. J.D. (2000) Modifications to methods for the enumeration and detection of injured *Escherichia coli* O157:H7 in foods. Int. J. Food. Microbiol. 55, 285–290.

Blankenship, L.C. (1981) Some characteristics of acid injury and recovery *of Salmonella bareilly* in a model system. J. Food Prot. 44, 73–77.

Bromberg R., George S.M. and Peck, M.W. (1998) Oxygen sensitivity of heated cells of *Escherichia coli* O157:H7. J. Appl. Microbiol. 85, 231–237.

Busta, F.F. (1976) Practical implications of injured microorganisms in food. J. Milk Food Technol. 39, 138–145.

Busta, F.F. and Jezeski, J.J. (1963) Effect of sodium chloride concentration in an agar medium on growth of heat-shocked *Staphylococcus aureus*. Appl. Microbiol. 11, 404–407.

Calcott, P.H. and Thomas, M. (1981) Sensitivity of DNA repair deficient mutants of *Escherichia coli* to freezing and thawing. FEMS Microbiol. Lett. 12, 117–120.

Chakraburtty, K. and Burma, D.P. (1968) The purification and properties of a ribonuclease from *Salmonella typhimurium* extract. J. Biol. Chem. 243, 1133–1139.

Clark, C.W. and Ordal, Z.J. (1969) Thermal injury and recovery of *Salmonella typhimurium* and its effect on enumeration procedures. Appl. Microbiol. 18, 332–336.

Dabbah, R., Moats, W.A. and Mattick, J.F. (1969) Factors affecting resistance to heat and recovery of heat-injured bacteria. J. Dairy Sci. 52, 608–614.

Dallmier, A.W. and Martin, S.E. (1988) Catalase and superoxide dismutase activities after heat injury of *Listeria monocytogenes*. Appl. Environ. Microbiol. 54, 581–582.

D'Aoust, J.-Y. and Maishment, C. (1979) Pre-enrichment conditions for effective recovery of *Salmonella* in foods and feed ingredients. J. Food Prot. 42, 153–157.

D'Aoust, J.-Y., Sewell, A. and Jean, A. (1990) Limited sensitivity of short (6 h) selective enrichment for detection of foodborne *Salmonella*. J. Food Prot. 53, 562–565.

Doyle, M.L., Scantling, M.K., Imel, N.K. and Waldroup, A.L. (1996) A comparison of various media for rapid resuscitation and repair of sub-lethally chlorine sodium hypochlorite-injured *Salmonella typhimurium*. Abstracts of the 96th General Meeting of the American Society of Microbiol., New Orleans, Louisiana, May 19th to 23rd, 1996. ASM Press, Washington.

Edwards, A.M., Silva, E., Jofré, B., Becker, M.I. and De Ioannes, A.E. (1994) Visible light effects on tumoral cells in a culture medium enriched with tryptophan and riboflavin. J. Photochem. Photobiol. B: Biol. 24, 179–186.

Emswiler, B.S., Pierson, M.D. and Shoemaker, S.P. (1976) Sublethal heat stress *of Vibrio*

parahaemolyticus. Appl. Environ. Microbiol. 32, 792–798.
Farber, J.M., Daley, E., Coates, F., Emmons, D.B. and McKellar, R. (1992) Factors influencing survival of *Listeria monocytogenes* in milk in a high-temperature short-time pasteurizer. J. Food Protect. 55, 946–951.
Fernandez Garayzabel, J.F., Dominguez Rodrigez, L. Vazquez Boland, J.A., Blanco Cancelo, J.L. and Suarez Fernandez, G. (1986) *Listeria monocytogenes* dans le lait pasteurisé. Can. J. Microbiol. 32, 149–150.
Flahuat, S., Laplace, J.-M., Frère, J. and Auffray, Y. (1998) The oxidative stress response in *Enterococcus faecalis*: relationship between H_2O_2 tolerance and H_2O_2 stress proteins. Lett. Appl. Microbiol. 26, 259–264.
Fridovich, I. (1983) Superoxide radical: an endogenous toxicant. Ann. Rev. Pharmacol. Toxicol. 23, 239–257.
George, S.M., Richardson, L.C.C., Pol, I.E. and Peck, M.W. (1998) Effect of oxygen concentration and redox potential on recovery of sublethally heat-damaged cells of *Escherichia coli* O157:H7, *Salmonella enteritidis* and *Listeria monocytogenes*. J. Appl. Microbiol. 84, 903–909.
George, S.M. and Peck, M.W. (1998) Redox potential affects the measured heat resistance of *Escherichia coli* O157:H7 independently of oxygen concentration. Lett. Appl. Microbiol. 27, 313–317.
Gomez, R.F., Sinskey, A.J., Davies, R. and Labuza, T.P. (1973) Minimal medium recovery of heated *Salmonella* typhimurium LT2. J. Gen. Microbiol. 74, 267–274.
Gort, A.S. and Imlay, J.A. (1998) Balance between endogenous superoxide stress and antioxidant defenses. J. Bacteriol. 180, 1402–1410.
Gray, R.J.H., Witter, C.D. and Ordal, Z.J. (1973) Characterisation of mild thermal stress in *Pseudomonas fluorescens* and its repair. Appl. Microbiol. 26, 78–85.
Hartman, P.A., Hartman, P.S., Lanz, W.W. (1975) Violet red bile 2 agar for stressed coliforms. Appl. Microbiol. 29, 537–539.
Heinis, J.J., Beuchat, L.R. and Boswell, F.C. (1978) Antimetabolite sensitivity and magnesium uptake by thermally stressed *Vibrio parahaemolyticus*. Appl. Environ. Microbiol. 35, 1035–1040.
Hitchener, B.J. and Egan, A.F. (1977) Outer membrane damage to sublethally heated *Escherichia coli*. Canad. J. Microbiol. 23, 311–318.
Hoffman, P.S., George, H.A., Krieg, N.R. and Smibert, R.M. (1979) Studies of the microaerophilic nature of *Campylobacter fetus* subsp. *jejuni*. II. Role of exogenous superoxide anions and hydrogen peroxide. Can. J. Microbiol. 25, 8–16.
Holbrook, R., Anderson, J.M. and Baird-Parker, A.C. (1980) Modified direct plate method for counting *Escherichia coli* in foods. Food Technol. Austral. 32, 78–83.
Humphrey, T.J. (1986) Injury and recovery in freeze or heat damaged *Campylobacter jejuni*. Lett. Appl. Microbiol. 3, 81–84.
Humphrey, T.J. (1988) Peroxide sensitivity and catalase activity in *Campylobacter jejuni* after injury and during recovery. J. Appl. Bacteriol. 64, 337–343.
Humphrey, T.J. (1989) An appraisal of the efficacy of pre-enrichment for the isolation of *Campylobacter jejuni* from water and food. J. Appl. Bacteriol. 66, 119–126.
Humphrey, T.J. and Cruickshank, J.G. (1985) Antibiotic and deoxycholate resistance in *Campylobacter jejuni* following freezing or heating. J. Appl. Bacteriol. 59, 65–71.
Hurst, A. (1977) Bacterial injury: a review. Canad. J. Microbiol. 23, 936–944.
Hurst, A. (1984) Revival of vegetative bacteria after sublethal heating. In: M.H.E. Andrew and A.D. Russell (editors) The Revival of Injured Microbes London: Academic Press. pp. 77–103.
Hurst, A., Hughes, A., Beare-Rogers, J.L. and Collins-Thompson, D.L. (1973) Physiological studies on the recovery of salt tolerance by *Staphylococcus aureus* after sublethal heating. J. Bacteriol. 116, 901–907.
Iandolo, J.J. and Ordal, Z.J. (1966) Repair of thermal injury of *Staphylococcus aureus*. J. Bacteriol.

91, 134–142.
Imlay, J.A. and Fridovich, I. (1992) Assay of metabolic superoxide production *in Escherichia coli*. J. Biol. Chem. 266, 6957–6965.
In't Veld, P.H. and de Boer, E. (1991) Recovery of *Listeria monocytogenes* on selective agar media in a collaborative study using reference samples. Int. J. Food Microbiol. 13, 295–300.
Jackson, H. (1974) Loss of viability and metabolic injury of *Staphylococcus aureus* resulting from storage at 5°C. J. Bacteriol. 37, 59–64.
Jackson, H. and Woodbine, M. (1963) The effect of sublethal heat treatment on the growth of *Staphylococcus aureus*. J. Appl. Bacteriol. 26, 152–158.
Kafel, S. and Bryan, F.L. (1977) Effects of enrichment media and incubation conditions on isolating salmonellae from ground-meat filtrate. Appl. Environ. Microbiol. 34, 285–291.
Kang, D.-H. and Fung, D.Y.C. (1999) Thin agar layer method for recovery of heat-injured *Listeria monocytogenes*. J. Food. Prot. 62, 1346–1349.
Kang, D.-H. and Fung, D.Y.C. (2000) Application of thin agar layer method for recovery of injured *Salmonella typhimurium*. Int. J. Food. Microbiol. 54, 127–132.
Katsui, N., Tsuchido, T., Hiramatsu, R., Fujikawa, S., Takano, M. and Shibasaki, I. (1982) Heat-induced blebbing and vesiculation of the outer membrane of *Escherichia coli* J. Bacteriol. 151, 1523–1331.
Kell, D.B., Kaprelyants, A.S., Weichart, D.H., Harwood, C.R. and Barer, M.R. (1998) Viability and activity in readily culturable bacteria: a review and discussion of the practical issues. Antonie van Leeuwenhoek 73, 169–187.
Kempler, G. and Ray, B. (1978) Nature of freezing damage on the lipopolysaccharide molecule of *Escherichia coli* B. Cryobiol. 15, 578–584.
Kiehlbauch, J.A., Simon, M.H. and Makowski, J.M. (1996) Use of filtration to isolate *Campylobacter* and related organisms from stools. In: Proceedings of the 8th International Workshop on Campylobacters, Helicobacters and Related Organisms. Winchester, U.K. July 1995 (eds. D.G. Newell, J.M. Ketley and R.A. Feldman) pp 47–49.
Knabel, S.J. and Thielen, S.A. (1995) Enhanced recovery of severely heat-injured thermotolerant *Listeria monocytogenes* from USDA and FDA primary enrichment media using a novel, simple, strictly anaerobic method. J. Food Prot. 58, 29–34.
Knabel, S.J., Walker, H.W., Hartman, P.A. and Mendonca, A.F. (1990) Effects of growth temperature and strictly anaerobic recovery on the survival of *Listeria monocytogenes* during pasteurization. Appl. Environ. Microbiol. 56, 370–376.
Kogure, K., Simidu, U. and Taga, N. (1979) A tentative direct microscopic method for counting living marine bacteria. Canad. J. Microbiol. 25, 415–420.
Lee, A.C. and Goepfert, J.M. (1975) Influence of selected solutes on thermally induced death and injury of *Salmonella typhimurium*. J. Milk Food Technol. 38, 195–200.
Lewis, I.M. (1930) The inhibition of *Phytomonas malvaceara* in culture media containing sugars. J. Bacteriol. 19, 423–433.
Linton, D. (1996) Old and new campylobacters: a review. PHLS Microbiol. Digest. 13, 10–15.
Linton, R.H., Webster, J.B., Pierson, M.D., Bishop, J.R. and Hackney, C.R. (1992) The effect of sublethal heat shock and growth atmosphere on the heat resistance *of Listeria monocytogenes* Scott A. J. Food Protect. 55, 84–87.
Mackey, B.M. (1983) Changes in antibiotic sensitivity and cell surface hydrophobicity in *Escherichia coli* injured by heating, freezing, drying or gamma radiation. FEMS Microbiol. Lett. 20, 395–399.
Mackey, B.M. (1985) Quality control monitoring of liquid selective media used for isolating *Salmonella*. International Journal of Food Microbiology 2, 41–48.
Mackey, B.M. (2000) Injured bacteria. In: B.M. Lund, A.C. Baird-Parker, and G.W. Gould, (editors) Food Microbiology. Aspen Publishers Inc. Maryland. pp. 315–341.

Mackey, B.M. and Derrick, C.M. (1982) The effect of sublethal injury by heating, freezing, drying and gamma-radiation on the duration of the lag phase *of Salmonella typhimurium*. J. Appl. Bacteriol. 53, 243–251.

Mackey, B.M. and Seymour, D.A. (1987) The effect of catalase on the recovery of heat-injured DNA-repair mutants of *Escherichia coli*. J. Gen. Microbiol. 133, 1601–1610.

Mackey, B.M. Miles, C.A., Parsons, S.E. and Seymour, D.A. (1991) Thermal denaturation of whole cells and cell components of *Escherichia coli* examined by differential scanning calorimetry. J. Gen. Microbiol. 137, 2361–2374.

Mackey, B.M. Forestière, K., Isaacs, N.S., Stenning, R and Brooker, B. (1994) The effect of high hydrostatic pressure on *Salmonella thompson* and *Listeria monocytogenes* examined by electron microscopy. Lett. Appl. Microbiol. 19, 429–432.

Macleod, R.A. and Calcott, P.H. (1976) Cold shock and freezing damage in microbes. In: T.R.G. Gray and J.R. Postgate (editors). The Survival of Vegetative Microbes. Twenty-sixth symposium of the Society for General Microbiology, Reading: SGM Press. pp. 81–109.

Ma-Lin, C.F.A. and Beuchat, L.R. (1980) Recovery of chill stressed *Vibrio parahaemolyticus* with enrichment broths supplemented with magnesium and iron salts. Appl. Environ. Microbiol. 39, 179–185.

Mazur, P. (1966) Physical and chemical basis of injury in single celled microorganisms subjected to freezing and thawing. In: H.T. Meryman (editor) Cryobiology. London: Academic Press. pp. 213–315.

McCarthy, J., Holbrook, R. and Stephens, P.J. (1998) An improved direct plate method for the enumeration of stressed *Escherichia coli* O157:H7 from food. J. Food Protect. 61, 1093–1097.

McCormick, J.P., Fischer, J.R., Pachlatko, J.P. and Eisenstark, A. (1976) Characterisation of a cell-lethal product from the photoxidation of tryptophan: hydrogen peroxide. Science 191, 468–469.

McCoy, D.R. and Ordal, Z.J. (1979) Thermal stress of *Pseudomonas fluorescens* in complex media. Appl. Environ. Microbiol. 37, 443–448.

Miller, L.L. and Ordal, Z.J. (1972) Thermal injury and recovery of *Bacillus subtilis*. Appl. Microbiol. 24, 878–884.

Morichi, T. (1969) Metabolic injury in frozen *Escherichia coli*. In: T. Nei, (editor) Freezing and drying of micro-organisms. Baltimore: University Park Press. pp. 53–65.

Moseley, B.E.B. (1984) Radiation damage and its repair in non-sporulating bacteria. In: M.H.E. Andrew and A.D. Russell (editors) The Revival of Injured Microbes. Academic Press, London. pp 147–174.

Mossel, D.A.A. and Corry, J.E.L. (1977) Detection and enumeration of sublethally injured pathogenic and index bacteria in foods and water processed for safety. Alimenta 16, Special issue on microbiology, 19–34.

Mukalamova,G.V., Kaprelyants, A.S., Young, D.I., Young, M and Kell, D.B. (1998) A bacterial cytokine. Proc. Nat. Acad. Sci. USA. 95, 9816–8921.

Nixon, B.T. and Wang. R.J. (1977) Formation of photoproducts lethal for human cells in culture by daylight fluorescent light and bilirubin light. Photochem. Photobiol. 26, 589–593.

Patel, J.R., Hwang, C.-A., Beuchat, L.R., Doyle, M.P. and Brackett, R.E. (1995) Comparison of oxygen scavengers for their ability to enhance resuscitation of heat-*injured Listeria monocytogenes*. J. Food Protect. 58, 244–250.

Pauling, C. and Beck. L.A. (1975) Role of DNA ligase in the repair of single strand breaks induced in DNA by mild heating of *Escherichia coli*. J. Gen. Microbiol. 87, 181–184.

Payne, J. (1978) Damage and recovery in streptococci. In: F.A. Skinner and L.B. Quesnel (editors) Streptococci. SAB Symposium Series No. 7, London: Academic Press. pp. 349–369.

Pierson, M.D., Tomlins, R.I. and Ordal, Z.J. (1971) Biosynthesis during recovery of heat-injured *Salmonella typhimurium*. J. Bacteriol. 105, 1234–1236.

Pierson, M.D., Gomez, R.F. and Martin, S.E. (1978) The involvement of nucleic acids in bacterial injury. Adv. Appl. Microbiol. 23, 263–283.

Postgate, J.R. and Hunter, J.R. (1963) Metabolic injury in frozen bacteria. J. Appl. Bacteriol. 26, 405–414.
Privalle, C.T. and Fridovich, I. (1987) Induction of superoxide dismutase in *Escherichia coli* by heat shock. Proc. Natl. Acad. Sci. USA 84, 2723–2726.
Punchard, N.A. and Kelly, F.J. (1996) Sources of free radicals. In: N.A. Punchard and F.J. Kelly, (editors) Free Radicals, a Practical Approach. IRL Press, Oxford, pp. 2–5.
Ray, B. (1979) Methods to detect stressed microorganisms. J. Food Prot. 42, 346–355.
Ray, B. and Johnson, C. (1984) Sensitivity of cold-stressed *Campylobacter jejuni* to solid and liquid selective environments. Food Microbiol. 1, 173–176.
Ray, B. and Speck, M.L. (1973) Enumeration of *Escherichia coli* in frozen samples after recovery from injury. Appl. Microbiol. 25, 499–503.
Ray, B., Janssen, D.W. and Busta, F.F. (1972a) Characterisation of the repair of injury induced by freezing *Salmonella anatum*. Appl. Microbiol. 23, 803–809.
Ray, B., Jezeski, J.J. and Busta, F.F. (1972b) Isolation of salmonellae from naturally contaminated dried milk products. III. Influence of pre-enrichment conditions. J. Milk Food Technol. 35, 607–614.
Rayman, M.K., Aris. B. and El.Derea. H.B. (1978) The effect of compounds which degrade hydrogen peroxide on the enumeration of heat-stressed cells *of Salmonella senftenberg*. Can. J. Microbiol. 24, 883–885.
Rose, A.H. (1968) Chemical Microbiology. 2nd edition. Butterworths, London, pp. 120–161.
Rosenthal, L. and Iandolo, J. (1970) Thermally induced intracellular alterations of ribosomal ribonucleic acid. J. Bacteriol. 103, 833–835.
Ryan, R.W., Gourlie, M.P. and Tilton, R.C. (1979) Release of rhodanese from *Pseudomonas aeruginosa* and its localisation within the cell. Canad. J. Microbiol. 25, 340–351.
Sadovski, A.J. (1977) Acid sensitivity of freeze injured salmonellae in relation to their isolation from frozen vegetables by pre-enrichment procedure. J. Food Technol. 12, 85–91.
Sato, M. and Takahashi, H. (1970) Cold shock of bacteria: IV. Involvement of DNA ligase reaction in recovery of *Escherichia coli* from cold shock. J. Gen. Appl. Microbiol. 16, 270–290.
Scheusner, D.L., Busta, F.F., Speck, M.L. (1971) Inhibition of injured *Escherichia coli* by several selective agents. Appl. Microbiol. 21, 46–49.
Schwarz, H. and Koch, A.L. (1995) Phase and electron microscopic observations of osmotically induced wrinkling and the role of endocytotic vesicles in the plasmolysis of the Gram negative cell wall. Microbiol. 141, 3161–3170.
Silva, M.T. and Sousa, J.C.F. (1972) Ultrastructural alterations induced by moist heat in *Bacillus cereus*. Appl. Microbiol. 24, 463–476.
Simpson, J.A., Narita, S., Gieseg, S., Gebicki, S., Gebicki, J.M. and Dean, R.T. (1992) Long-lived reactive species on free-radical-damaged proteins. Biochem. J. 282, 621–624.
Sinskey, T.J. and Silverman G.J. (1970) Characterisation of injury incurred *by Escherichia coli* upon freeze drying. J. Bacteriol. 101, 429–437.
Smith, L.D., 1975. The Pathogenic Anaerobic Bacteria, 2nd ed. Charles C. Thomas, Springfield, Illinois, pp. 3–14.
Smith, J.L. and Dell, B.J. (1990) Capability of selective media to detect heat *injured Shigella flexneri*. J. Food Prot. 53, 141–144.
Smith, J.L. and Archer, D.L. (1988) Heat-induced injury in *Listeria monocytogenes*. J. Indust. Microbiol. 3, 105–110.
Speck, M.L., Ray, R., Read, Jr. R.B. (1975) Repair and enumeration of injured coliforms by a plating procedure. Appl. Microbiol. 29, 549–550.
Stephens, P.J. and Jones, M.V. (1993) Reduced ribosomal thermal denaturation *in Listeria monocytogenes* following osmotic and heat shocks. FEMS Microbiol. Lett. 106, 177–182.
Stephens, P.J. and Joynson, J.A. (1998) Direct inoculation into media containing bile salts and antibiotics is unsuitable for the detection of acid/salt stressed *Escherichia coli* O157:H7. Lett.

Appl. Microbiol. 27, 147–151.
Stephens, P.J., Joynson, J.A., Davies, K.W., Holbrook, R., Lappin-Scott, H.M. and Humphrey, T.J. (1997) The use of an automated growth analyser to measure recovery times of single heat-injured *Salmonella* cells. J. Appl. Microbiol. 83, 445–455.
Stephens, P.J., Druggan P. and Nebe-von Caron, G. (2000) Stressed *Salmonella* are exposed to reactive oxygen species from two independent sources during recovery in conventional culture media. Int. J. Food. Microbiol. 60, 269–285.
Strange, R.E. and Shon, M. (1964) Effects of thermal stress on viability and ribonucleic acid of *Aerobacter aerogenes* in aqueous suspension. J. Gen. Microbiol. 34, 99–114.
Swartz, H.M. (1971) Effect of oxygen on freezing damage: II. Physical-chemical effects. Cryobiol. 8, 255–264.
Thippareddi, H., Phebus, R.K., Fung, D.Y.C. and Kastner, C.L. (1995) Use of universal preenrichment medium supplemented with Oxyrase™ for the simultaneous recovery of *Escherichia coli* O157: H7 and *Yersinia enterocolitica*. J. Rap. Meth. Auto. Microbiol. 4, 37–50.
Tomlins, R. I. and Ordal, Z. J. (1971) Precursor ribosomal ribonucleic acid and ribosome accumulation in vivo during the recovery of *Salmonella typhimurium* from thermal injury. J. Bacteriol. 107, 134–142.
Tomlins, R.I. and Ordal, Z.J. (1976) Thermal injury and inactivation in vegetative bacteria. In: F.A. Skinner and W.B. Hugo (editors) Inhibition and Inactivation of Vegetative Microbes. SAB Symposium Series No. 5, London: Academic Press. pp.153–190.
Tomlins, R.I., Vaaler, G.L. and Ordal, Z.J. (1972) Lipid biosynthesis during the recovery of *Salmonella typhimurium* from thermal injury. Canad. J. Microbiol. 18, 1015–1021.
Town, C.D., Smith, K.S. and Kaplan, H.S. (1973) The repair of DNA single-strand breaks in *Escherichia coli* irradiated in the presence or absence of oxygen; the influence of repair on survival. Rad. Res. 55, 334–345.
Traci, P.A. and Duncan, C.L. (1974) Cold shock lethality and injury *in Clostridium perfringens*. Appl. Microbiol. 28, 815–821.
Ugborogho, T.O. and Ingham, S.C. (1994) Increased D-values of *Staphylococcus aureus* resulting from anaerobic heating and enumeration of survivors. Food Microbiol. 11, 275–280.
Valentine, J.S., Foote, C.S., Greenberg, A. and Liebman, J.L. (1995) Active Oxygen in Biochemistry. Blackie Academic and Professional, London.
Van Schothorst, M. and Van Leusden, F.M. (1975) Further studies on the isolation of injured salmonellae in foods. Zentralblatt fur Bakteriologie, Parasitenkunde, Infektionskrankheiten und Hygiene, Abteilung I. Originale A. 230, 186–191.
Van Schothorst, M. and Duke, A.M. (1984) Effect of sample handling on microbial limits laid down in standards. In: M.H.E. Andrew and A.D. Russell (editors) The revival of injured microbes. Academic press, London. pp. 309–328.
Wang, R.J. (1976) Effect of room fluorescent light on the deterioration of tissue culture medium. In Vitro 12, 19–22.
Wang, R.J. and Nixon, B.T. (1978) Identification of hydrogen peroxide as a photoproduct toxic to human cells in tissue-culture medium irradiated with "daylight" fluorescent light. In Vitro 14, 715–722.
Waterworth, P.M. (1969) The action of light on culture media. J. Clin. Path. 22, 273–277.
Webb, R.B. and Lorenz, J.R. (1972) Toxicity of irradiated medium for repair-deficient strains of *Escherichia coli*. J. Bacteriol. 112, 649–652.
Williams, D.L. and Calcott, P.H. (1982) Role of DNA repair genes and an R plasmid in conferring cryoresistance on *Pseudomonas aeruginosa*. J. Gen. Microbiol. 128, 215–218.
Wright, H.D. (1933) The importance of adequate reduction of peptone in the preparation of media for the pneumococcus and other organisms. J. Path. Bacteriol. 37, 257–282.
Wood, M., Holbrook, R. and Stephens, P. (1996) Enhanced recovery of stressed *Listeria monocytogenes* on Oxford and Palcam agars. Proceedings of the Symposium on Food

Associated Pathogens, Uppsala, May 1996 pp. 216–217. Uppsala: SLU.

Xavier, I.J. and Ingham, S. (1993) Increased D-values for *Salmonella enteritidis* resulting from the use of anaerobic enumeration methods. Food Microbiol. 10, 223–228.

Xu, H-S., Roberts, N., Singleton, F.L.Attwell, R.W., Grimes, D.J. and Colwell, R.R. (1982) Survival and viability of nonculturable *Escherichia coli*, and *Vibrio cholerae* in the estuarine and marine environments. Microb. Ecol. 8, 313–323.

Yoakum, G. and Eisenstark, A. 1972. Toxicity of L-tryptophan photoproduct on recombinationless (*rec*) mutants of *Salmonella typhimurium* J. Bacteriol. 112, 653–655.

Yoakum, G., Ferron, W, Eisenstark, A. and Webb, R.B. (1974) Inhibition of replication gap closure in *Escherichia coli* by near-ultraviolet light photoproducts of L-tryptophan. J. Bacteriol. 119, 62–69.

Yu, L.S.L. and Fung, D.Y.C. (1991a) Oxyrase™ enzyme and motility enrichment Fung-Yu tube for rapid detection of *Listeria monocytogenes* and *Listeria* species. J. Food Safety. 11, 149–162.

Yu, L.S.L. and Fung, D.Y.C. (1991b) Effect of Oxyrase™ enzyme on *Listeria monocytogenes* and other facultative anaerobes. J. Food Safety. 11, 163–175.

Chapter 3

Media for the detection and enumeration of clostridia in foods

M.W.J. Bredius and E.M. de Ree

CLF Central Laboratories Friedrichsdorf GmbH, Bahnstrasse 14–30, 61381 Germany

Clostridia are the anaerobic bacteria most frequently associated with foods. They are widely spread in the environment and show an extensive diversity in metabolic activity and nutritional requirements. As they are a rather heterogeneous group, no medium is available which will allow the growth of all clostridia and at the same time exclude the growth of all competitive flora. Most clostridia reduce sulphite to sulphide and hence will produce rather large black haloes of iron sulphide, in iron-containing media under anaerobic conditions. Consequently, most isolation media include sulphite and an appropriate iron salt, so that blackening due to sulphite reduction can serve as a diagnostic test for clostridia.

In addition to their diversity in metabolic activity and nutritional requirements, the enumeration of both vegetative cells and clostridial spores using one method is difficult. Many spores only germinate after heat activation, a process which kills the vegetative cells present in the sample. Thus the suitability of a medium depends on the objective of examination, the competitive flora in the sample and the *Clostridium* species under investigation. The newest media described in the literature are sulphite cycloserine azide medium (SCA) for samples which do not undergo heat treatment before analysis and differential clostridial agar (DCA) for spore counts.

Several media for the detection and enumeration of *C. perfringens* are discussed. Choice will depend on the sample under investigation, the required detection limit and the reason for the test.

Introduction

Of the anaerobic bacteria, the clostridia are the most frequently associated with foods. They have resistant spores and are thus able to survive well under adverse conditions. Many methods have been described which can be used for the isolation of these organisms from food, both when present as spores or as vegetative cells. As clostridia show a wide diversity in metabolic activity and nutritional requirements, it is not easy to develop a (selective) medium or procedure able to recover all *Clostridium* species.

Clostridia are ubiquitous and readily isolated from many natural environments. They are commonly found in soil, sewage, decaying vegetation, animal and plant products, in the intestinal tract of humans and animals, insects and in wounds or soft tissue infections. *Cl. perfringens* in particular, is commonly found in the faeces of many animals. Of over 90 described species, only a few are likely to be encountered in foods. Of particular concern for the public health are *Cl. botulinum*, which forms a deadly toxin in foods, and *Cl. perfringens*, which causes enteritis when present in high numbers. Other species or strains are also known to be toxinogenic or neurotoxinogenic, including *Cl. baratii*, *Cl. butyricum*, *Cl. difficile*, *Cl. sordellii*, *Cl. sphenoides* and *Cl. spiroforme* (Weenk et al., 1991).

Clostridia are also important in relation to spoilage of foods (e.g. vacuum packed chilled products). Although many mesophilic clostridia have been reported to cause food spoilage, spoilage can also be caused by thermophilic (canned foods) and psychrotrophic and psychrophilic species (frozen fish, vacuum packed meat and raw milk) (Attenborough and Scarr, 1957; Bhadsalve et al., 1972; Brocklehurst and Lund, 1982; Dainty et al., 1989). Another kind of food spoilage is caused by *Cl. tyrobutyricum* in 'Gouda-type' cheese, known as 'late-blowing' (Fryer and Halligan, 1976; Senyk et al., 1989).

In some products, low numbers of clostridia can be found (e.g. *Cl. perfringens* in meat and poultry products). Unless the food has been abused, multiplication of clostridia in the food is not common.

The complete group of sulphite reducing clostridia (SRC) may be used as index organisms for *Cl. botulinum* or *Cl. perfringens*, although this index function is not always straightforward. The group can also be used as a marker for General Hygiene Practices, raw material quality or processing deficiencies (Mossel et al., 1995).

This chapter deals with the media described for isolation and enumeration of clostridia and *Cl. perfringens* in food. It shows that the enumeration of spores or vegetative cells of sulphite reducing clostridia in food still constitutes a chain of complex steps, each affecting the numerical level and reliability of the end-result. An overview of the most important media described in this chapter is given in Table 1.

Factors affecting isolation

In foods, both vegetative cells and spores can be present. To count only the clostridial spores, the sample has to be pasteurised to inactivate the vegetative cells. A heat treatment is often also necessary to effect rapid spore germination (Barnes, 1985). Counting both the vegetative clostridial cells and the spores in one method is therefore extremely difficult. Enumeration of the 'total' clostridial load (samples without a heat treatment) estimates lower numbers than a spore count (Gibbs and Freame, 1965: Gibbs, 1971). Vegetative cells of clostridia tend to be oxygen-sensitive and are usually destroyed during processing by heat. This means that in foods where the clostridia are unlikely to germinate and multiply (e.g. dried food ingredients), only a spore count is required. In raw meats, on the other hand, conditions are more favourable for survival of vegetative cells and these tend to outnumber spores. It is clear therefore that for different kind of products, different isolation procedures should be applied. Furthermore, there is no optimal pasteurisation time and temperature for all SRC as strains of clostridia vary with respect to the heating regime needed for optimum germination.

Table 1
Selective and diagnostic systems of media for the enumeration and isolation of SRC and *Cl. perfringens*.

Medium	Selective system	Diagnostic system	Target microorganism	Reference
Iron Sulphite Agar (ISA)	Anaerobic incubation	Sulphite and iron	Thermophilic SRC (canned food)	Attenborough and Scarr, 1957
Differential Clostridium Agar (DCA)	Anaerobic incubation Sample pasteurisation	Sulphite and iron	SRC spores	Weenk et al., 1995
Sulphite Cycloserine Azide Agar (SCA)	Anaerobic incubation Cycloserine Azide	Sulphite and iron	SRC vegetative cells	Eisgruber and Reuter, 1995
Oleandomycin Polymyxin Sulphadiazine Perfringens agar (OPSPA)	Anaerobic incubation Oleandomycin Polymyxin Sulphadiazine	Sulphite and iron	*Cl. perfringens*	Handford, 1974
Rapid perfringens medium (RPM)	Polymyxin Neomycin Incubation temperature (46°C)	Litmus milk	*Cl. perfringens*	Erickson and Deibel, 1978
Tryptose Sulphite Cycloserine Agar (TSC)	Anaerobic incubation Cycloserine	Sulphite and iron	*Cl. perfringens*	Hauschild and Hilsheimer, 1974a, b

Additionally spores germinate at different rates, suggesting the need for a prolonged incubation of the isolation media.

The use of special precautions to guarantee anaerobic conditions during sample preparation is still under discussion. Some are of the opinion that clostridia are generally capable of surviving brief exposure to oxygen. Others however argue that especially in food microbiology, it is important to avoid the loss of stressed and very sensitive cells by over exposure to oxygen. Choosing the correct procedure is difficult, especially as oxygen sensitivity varies considerably amongst the different species (Anderson and Fung, 1983; Barnes, 1985). Other factors influencing the extent of recovery rate of the different clostridia include the nature of the media combined with the varying properties of the different clostridia (Mossel and de Waart, 1968). A lack of nutrients will inhibit the development and metabolic activity of many clostridia. A too nutritious medium, however, will result in copious gas formation and acidification, combined with rapid spread of black haloes, which obscure readings. Liquid media may be preferred, when inclusion of fermentable carbohydrates is deemed necessary, to avoid disruption of the agar (Weenk et al., 1991; Weenk et al., 1995).

Variations in sulphite concentration also influence the extent of recovery and blackening of the various clostridia. In addition to the sulphite concentration, other components can also influence the recovery of specific clostridia species. It was shown that the presence of sodium acetate in the medium inhibited in particular *Cl. beijerinckii*, *Cl. butyricum* and *Cl. perfringens*. (Weenk et al., 1991; Weenk et al., 1995).

Because clostridia are so diverse it is difficult to find selective agents that are equally favourable for all relevant species, whilst still inhibiting other organisms, such as coliforms. Polymyxin seems the only inhibitory compound widely applicable for this purpose, although it is not entirely selective for clostridia (Weenk et al., 1995). For the differentiation between bacilli and clostridia, a combination of resistance to metronidazole and prolific growth on aerobically incubated tryptone soya agar has been shown to be a reliable criterion (Weenk et al., 1991).

Media for sulphite reducing clostridia (SRC)

In 1924, for the first time, Wilson and Blair used the ability of *Cl. perfringens* to reduce sulphite for the development of a medium to isolate this microorganism from water. Since then, several media have been developed to detect, isolate and enumerate *Cl. perfringens* and SRC as a group. About 40 years later, Gibbs and Freame (1965) stated that 'there is still no fully satisfactory medium for selecting the clostridia as a whole' and this remains true today. The media that are presently used include selective and non-selective formulations in both liquid and solid form. The majority of the media use sulphite reduction and the resulting blackening in the presence of iron, as the elective criterion. Sulphite in clostridial media is reduced to sulphide by the enzyme sulphite reductase, produced by the clostridia. The sulphide will then precipitate as a black deposit in the presence of an appropriate iron salt. This is visible as blackening of the liquid media and as black colonies surrounded by a black halo on or in solid media. A double layer technique is often required to achieve a proper blackening of the medium (Mead et al., 1982). On the surface of solid media colonies may appear grey with a black centre.

The medium that was developed by Wilson and Blair was called Iron Sulphite Agar (ISA). It was soon recognised that not only *Cl. perfringens* but also other clostridia were able to reduce sulphite and could cause the typical blackening on media containing iron and sulphite (Prevot, 1948). The medium of Wilson and Blair was from then on used and modified to enumerate clostridia as a whole group (Angelotti et al., 1962).

In addition to clostridia, other microorganisms (e.g. salmonella, proteus) are also able to reduce sulphite. Depending on the selectivity of the medium, other bacteria can therefore grow and blacken on media that are developed for the cultivation of clostridia.

Although clostridia can reduce sulphite, some clostridia are inhibited by it even at low concentrations. Among the species known to include sensitive strains are *Cl. butyricum, Cl. putrificum, Cl. saccharobutyricum, Cl. septicum, Cl. sporogenes* and *Cl. tertium*. Media aimed to detect clostridia as a group, should therefore not contain more than 0.05% sulphite (e.g. Gibbs and Freame, 1965; Mossel and de Waart, 1968;

Weenk et al., 1995).

When the medium becomes acidified, iron sulphide will not longer precipitate. The inclusion of a fermentable carbohydrate in the medium can cause acidification and as a consequence, false negative results. In addition to acidification, fermentation of carbohydrates can lead to the formation of copious gas (which disrupts the agar) and the rapid spread of black haloes which obscure readings. To prevent acidification, gas formation and/or spreading, media should not contain more than 0.1% glucose (Weenk et al., 1991). For some *Clostridium* species however, an optimal glucose concentration is necessary to obtain adequate blackening (Mossel and de Waart, 1968; Eisgruber and Reuter, 1995; Weenk et al., 1995).

Development of media for sulphite reducing clostridia as a group

ISA was the first medium to be used for the enumeration of clostridia and is still used today for this purpose. This medium does not contain selective agents and clostridia are recognised by the black precipitate around the colonies. It is commercially available under the name 'iron sulphite agar' or 'sulphite (iron) agar'. However, not all media called 'iron sulphite agar' have the same formulation and dependent on the formulation, different applications are described. One recipe called ISA is described by Angelotti et al. (1962). This recipe reflects the (improved) medium as originally described by Wilson and Blair. This non selective formula allows other sulphite reducing microorganisms (salmonella, proteus, etc.) in addition to SRC to grow and cause blackening. Another formulation named ISA is recommended by the Nordic Committee on Food Analyses (NMKL) in their method for the determination of SRC in foods (NMKL method No 56, 1994). The ISA recipe given in the NMKL method is actually the basal medium for Shahidi-Ferguson-Perfringens agar (SFP) and Tryptose Sulphite Cycloserine agar TSC (without the addition of antibiotics and egg yolk). SFP and TSC media are described in more detail below. It is currently being discussed whether this medium will be used within the International Standardisation Organisation (ISO) method for the enumeration of SRC. A third ISA formula aims to detect *Desulfotomaculum nigrificans*, formerly *Clostridium nigrificans*. This microorganism can cause sulphide stinker spoilage in non-acid, mainly canned foods and has an optimal growth temperature of 55°C (Donnelly and Graves, 1992). This ISA formulation is nutritionally poorer than those mentioned above. Selectivity is achieved by the high incubation temperature, which inhibits the growth of other sulphite reducing microorganisms.

A medium which could detect the group of SRC as a whole and especially *Cl. butyricum*, *Cl. tyrobutyricum* and *Cl. sporogenes*, was important for the dairy industry. These clostridia are known to cause spoilage of cheese (Hirsch and Grinsted, 1954; Fryer and Halligan, 1976). Hirsch and Grinsted (1954) devised Reinforced Clostridial Medium (RCM) for this purpose. RCM does not require anaerobic conditions during incubation and as this medium is liquid, it is not disrupted by gas formation. Boiling of the tubes just before use and pouring a viscous paraffin or agar seal on the RCM gives good growth of clostridia. The reducing conditions during incubation are maintained by inclusion of cysteine. It was noted by the authors that RCM was a good growth medium, but not selective for clostridia. In 1965 Gibbs and Freame modified RCM by the addition of sodium sulphite and ferric citrate. They called the modified medium differential reinforced clostridial medium (DRCM). Blackening of DRCM due to sulphite reduction shows the presence of sulphite reducing microorganisms. The advantage of DRCM over other SRC media available at that time, was the possibility of anaerobic incubation when filled in screw capped tubes leaving only a small headspace. In that way, the medium was easy to use. Addition of agar to DRCM made the medium suitable for pour plates.

As DRCM was not selective for clostridia, a confirmation step was necessary when a total clostridial count was performed. To do so, DRCM tubes showing blackening after incubation were pasteurised at 75–78°C for 30 min and a new DRCM tube was inoculated with some material from the pasteurised tube. Blackening of the second tube confirmed the presence of clostridia in the sample. Using DRCM for total clostridial counts and spore counts revealed that the spore counts gave higher numbers than the total count. (Gibbs and Freame, 1965; Gibbs, 1971). Two reasons for this difference are suggested. One is that many spores only germinate after a heat shock such as pasteurisation. Another is that for the confirmation procedure, the growth in the first DRCM tube has to be accompanied by good sporulation to survive the pasteurisation step in the confirmation procedure (Barnes et al., 1963; Gibbs, 1971; Barnes, 1985). This research showed that it was very difficult to develop a medium with which vegetative cells and spores could be counted using one method.

Gibbs and Freame (1965) found that when a spore count was made on ISA, some *Bacillus* spp. were also able to grow and blacken the medium. Weenk et al. (1991) reported a substantial increase in the number of black colonies in pasteurised samples when buffered DRCM was incubated more than 2–3 days at 30° C. The colonies that appeared later were generally smaller and called secondary colonies. A large proportion of the secondary colonies was shown to be *Bacillus* spp. To obtain a medium with reduced growth of bacilli, the selectivity of the DRCM was improved. This resulted in a new medium, called differential clostridial agar (DCA). It was shown that the growth and blackening of *B. licheniformis* and *B. pumilus*, was not completely inhibited on DCA. However, the blackening properties of the bacilli were lost upon repeated subculture (Weenk et al., 1991; Weenk at al., 1995).

In the same period, Eisgruber and Reuter (1991; 1995) developed another medium designed to enumerate SRC. This medium was developed to detect and enumerate clostridia in meat and meat products. For such products it is important to detect the vegetative clostridial cells. This requires a medium for which a pasteurisation step is not necessary to obtain sufficient selectivity. The basis for the medium of Eisgruber and Reuter was *Cl. perfringens* selective TSC agar. This medium without its antibiotic cycloserine, is used as a general medium for SRC but allows the growth of other microorganisms as well. By addition of 0.3g/l cycloserine (instead of 0.4g/l, as in TSC) as well as azide, a new selective medium for the enumeration of SRC was developed. This medium is called sulphite cycloserine azide agar (SCA). On this medium, the growth of *Bacillus* spp. is inhibited.

A comparison between DCA and SCA was made by Weenk et al. (1995). The recovery of SRC spores from different food samples was found to be much higher on

Table 2
Comparison of three media for the enumeration of sulphite reducing clostridia. Spore count (in $\log_{10}$ units) in commodity samples (Weenk et al., 1995).

Food sample	Enumeration medium		
	DCA	ISA	SCA
Milk powder	2.6	1.4	1.0
Milk powder	1.9	1.7	<1
Skim milk powder	1.8	<1	<1
Whey powder	1.9	1.7	2.3
Cereal	2.0	1.3	1.8
Curry powder	2.0	2.0	1.6
Chicken meat	2.1	1.6	<1
Starch powder	1.6	1.1	1.2
Caseinate	2.0	1.7	1.8
Caseinate	1.5	0.8	<1
Potato protein	2.5[a]	–	–

[a] Approximately 13% were confirmed as Clostridium species, 87% as *Bacillus licheniformis*.

DCA than on SCA or ISA (Table 2). The presence of selective agents is an important difference between these two media: SCA contains cycloserine and azide, whereas DCA does not contain selective agents. This may indicate that SCA is too selective and inhibits growth of SRC. It was reported that the use of azide in media for clostridia was unsatisfactory as it also inhibits the growth of some *Cl. perfringens* and *Cl. butyricum* strains (Mossel et al., 1956; Gibbs and Freame, 1965; Mossel and de Waart, 1968). However, the azide concentration used in SCA (0.005%) is lower than the azide concentration used by Mossel and de Waart, which was 0.02%.

The presence of selective agents in SCA is important to allow the detection of vegetative SRC cells selectively without the need for pasteurisation of the samples. This is important for meat and meat products, for which SCA was developed. DCA is designed to enumerate SRC spores, which makes it possible to kill the competitive flora by pasteurisation. As a consequence, it possible to leave out selective agents in the DCA medium. At the same time, the pasteurisation step is a heat shock treatment for the SRC spores. As no ideal medium is available to count both vegetative SRC cells and SRC spores, it will depend on the objective of the analysis and the samples under investigation, which medium is more suitable.

Media for *Cl. perfringens*

When it was recognised that other clostridia could grow as well as *Cl. perfringens* on ISA, a medium that was more selective for the detection and enumeration of *Cl. perfringens* was sought. The first selective medium for the detection of *Cl. perfringens* was Sulphite Polymyxin Sulfadiazine agar (SPS), developed by Angelotti et al. in 1962. The medium consists of ISA, to which two antibiotics were added: polymyxin and sulfadiazine. As on ISA, *Cl. perfringens* is visible as black colonies. The medium was widely adopted and became commercially available. It was shown however, that on the commercial SPS, various strains of *Cl. perfringens* did not grow or did not form the characteristic black colonies. On the other hand, other clostridia besides *Cl. perfringens* could grow as black colonies (Shahidi and Ferguson, 1971; Hanford, 1974; Harmon, 1976). In 1965 the first improvement of SPS was published. In the new medium, called Tryptone Sulphite Neomycin medium (TSN), the selective agent sulfadiazine was replaced by neomycin (Marshall et al., 1965). In 1971 Shahidi and Ferguson published an improved selective medium for the enumeration of *Cl. perfringens:* Shahidi-Ferguson Perfringens agar (SFP). This medium uses sulphite reduction together with the lecithinase activity of *Cl. perfringens* on egg yolk as indicators. For a clear lecithinase zone, a spread plate was preferred onto which an overlayer of egg yolk free SFP was poured. The overlayer is needed to obtain clear blackening and egg yolk was left out from the overlayer since it diminished the visibility of the black colonies and did not influence the lecithinase reaction. As some other *Clostridium* species also produce lecithinase, confirmation remains necessary. The recovery of *Cl. perfringens* was higher on SFP compared to SPS and blackening was improved. As reference medium SFP without antibiotics was used.

The SFP agar was further improved by exchanging the polymyxin and kanamycin in SFP for cycloserine. This medium was called Tryptose Sulphite Cycloserine medium (TSC) and was more selective than SFP. In 1974, Handford showed that the recovery on TSN was lower than that on SFP or TSC. Recovery on TSC was a little lower than on SFP, but still satisfactory (Harmon et al, 1971). Some years later the recovery of *Cl. perfringens* on TSC was improved by leaving out the egg yolk (Hauschild and Hilsheimer, 1974a,b; Harmon 1976). In the literature TSC medium is also abbreviated to SC medium or SC agar, which should not be confused with SCA medium (sulphite cycloserine azide agar).

At the same time Handford (1974) developed another selective medium for the enumeration of *Cl. perfringens.* The new medium, Oleandomycin-Polymyxin-Sulfadiazine Perfringens Agar (OPSPA) consists of a base medium that is comparable but poorer in nutrients than TSC and SFP. Oleandomycin, polymyxin and sulfadiazine are used as selective agents. The advantage of this medium compared to SFP and TSC is that the growth of *Cl. bifermentans* is inhibited on OPSPA. Eisgruber and Reuter (1995) however found rather good growth of *Cl. bifermentans* on OPSPA. On the other hand, the growth of enterococci was not inhibited on OPSPA while it was inhibited on TSC. A comparison between the selectivity of the different *Cl. perfringens* selective media and SCA for the whole group of SRC is given in Table 3.

None of the above mentioned media are totally specific for *Cl. perfringens* and confirmation tests remain necessary (Handford, 1974). Marshall et al. (1965) suggested increasing the incubation temperature to 46°C. At this temperature *Cl. perfringens* has very short generation times, whereas the growth of *Cl. bifermentans* is inhibited. Handford (1974) found however that the recovery of *Cl. perfringens* on TSN was decreased at 46°C.

Erickson and Deibel noted in 1978 that all the above mentioned media for *Cl. perfringens* were designed for the isolation of *Cl. perfringens* from highly contaminated foods associated with food poisoning outbreaks, where large numbers of the organ-

Table 3
Comparison of growth of some *Clostridium* species on SCA (selective medium for clostridia as a whole group) and *Cl. perfringens* selective media, in comparison to Brewer medium[a] (Eisgruber and Reuter, 1995).

Microorganism	Number of tested strains	SCA	SPS	TSN	OPSP	SFP	TSC
Cl. perfringens	5	+++	++, +++	+++	+++	+++	+++
	20	+++	n.t.	n.t.	n.t.	n.t.	n.t
Cl. bifermentans	4	+++	++	++	++	++	++
	10	+++	n.t.	n.t.	n.t.	n.t.	n.t
Cl. sporogenes	1	n.t.	++[b]	–	–	–	–
	2	+++	n.t.	n.t.	n.t.	n.t.	n.t
Cl. butyricum	2	++[b], +++[b]	++[b]	+[b], -	+[b]	+[b]	–

[a] Brewer anaerobic agar, pour plating method with overlayer and anaerobic incubation.
[b] Weak or variable sulphite reduction.
+++ full growth ± 0.5 $\log_{10}$; ++ reduced growth about 1 $\log_{10}$; + reduced growth ≥ 2 $\log_{10}$; – no growth; n.t. not tested.

isms are present. They realised that quality control at the manufacturing level required a method that could detect low to moderate numbers of *Cl. perfringens* and designed Rapid Perfringens Medium (RPM), which is based on the stormy fermentation of litmus milk. Selectivity is achieved by polymyxin and neomycin, an elevated incubation temperature and aerobic incubation. RPM, exceptionally is a selective medium which does not use the sulphite reductase system as an elective agent. Using RPM, *Cl. perfringens* was isolated much more frequently from food samples than when TSC was used. These results were confirmed by De Boer and Boot (1983). They also found that heat treatment (30 min, 70°C) reduced the frequency of isolation compared to no heat treatment. During fermentation the pH value in RPM can decrease to 4.0. As confirmation is necessary, a suspect positive sample has to be streaked on an agar plate or transferred to a fresh broth within 18–24 h.

The characteristic stormy fermentation of milk is also used in the Iron Milk Medium (IMM). The selectivity of IMM relies solely on the rapid growth of *Cl. perfringens* at 45°C, which is indicated by the stormy fermentation. In a collaborative study, Abeyta and Wetherington (1994) found a good a recovery of *Cl. perfringens* from shellfish with IMM compared to TSC. Another medium devised for the detection of low numbers of *Cl. perfringens* is lactose sulphite (LS) broth (Beerens et al., 1982). Its selectivity depends on the rapid growth of *Cl. perfringens* at 46°C, its ability to produce gas from lactose and to reduce sulphite. LS is a nutritionally poor medium, to inhibit the growth of competitive flora. Smith (1985) compared the performance of LS broth and RPM and found that RPM was more sensitive than LS. In addition, Schalch et al. (1996) found that not all *Cl. perfringens* strains showed typical growth (blackening and gas formation) in LS medium.

Thus the choice of medium depends on several aspects: the required detection limit, the amount of competitive flora expected, the expected contamination level of the sample and the field of application (routine food samples for quality control vs. outbreak food samples).

Media for *Cl. botulinum*

For food contaminated with *Cl. botulinum,* the concerns are more with the detection of the neurotoxin rather than the detection of the organism itself. Although *Cl. botulinum* spores can be very heat resistant, the toxin is heat labile. To isolate *Cl. botulinum*, an enrichment step may be necessary. Dependent on the *Cl. botulinum* type, cooked meat medium or trypticase-peptone-glucose-yeast broth is used. The possibility of isolating *Cl. botulinum* selectively is improved when spores are present and the competitive flora can be reduced by alcohol treatment or pasteurisation. However, selective media for *Cl. botulinum* are not available and confirmation depends on the detection of the toxin (Kautter et al., 1992)

References

Abeyta, C. and Wetherington, J.H. (1994) Iron milk medium method for recovering *Cl. perfringens* from shellfish: collaborative study. J. Assoc. Off.Anal. Chem. 77, 351–356.

Anderson, K.L. and Fung, D.Y.C. (1983) Anaerobic methods, techniques and principles for food bacteriology: a review, J. Food Protect., 46, 811–822.

Angelotti R., Hall H.E., Foter M.J. and Lewis K.H. (1962) Quantitation of *Clostridium perfringens* in foods. Appl. Microbiol. 10, 193–199.

Attenborough, S.J. and Scarr, M.P. (1957) The use of membrane filter techniques for control of thermophilic spores in the sugar industry. J. Appl. Bacteriol. 20, 460–466.

Barnes, E.M., Despaul, J.E. and Ingram, M. (1963) The behaviour of a food poisoning strain of *Clostridium welchii* in beef. J. Appl. Bacteriol. 26, 415–427.

Barnes, E.M. (1985) Isolation methods for anaerobes in foods. Int. J. Food Microbiol. 2, 81–87.

Beerens, H., Romond, C., Lepage, C. and Criquelion, J. (1982) A liquid medium for the enumeration of *Clostridium perfringens* in food and faeces. *In*: Isolation and identification methods for food poisoning organisms. J.E.L. Corry, D. Roberts and F.A. Skinner (eds), SAB Technical series No 17, Academic Press, London, 137–149.

Bhadsalve, C.H., Shehata, T.E. and Collins, E.B. (1972) Isolation and identification of psychrophilic species of *Clostridium* from milk. Appl. Microbiol. 699–702.

Brocklehurst, T.F. and Lund, B.M. (1982) Isolation and properties of psychrotrophic and psychrophilic, pectolytic strains of *Clostridium*. J. Appl. Bacteriol. 53, 355–361.

de Boer, E. and Boot, M. (1983) Comparison of methods for isolation and confirmation of *Clostridium perfringens* from spices and herbs. J. Food Protect. 46, 533–536.

Dainty, R.H., Edwards, R.A. and Hibbard, C.M. (1989) Spoilage of vacuum-packed beef by a *Clostridium* sp. J.Sci. Food Agric 49, 473–486.

Donnelly, L.S. and Graves, R.R. (1992) Sulfide spoilage sporeformers, p. 317–232. *In*: Vanderzant, C. and Splittstoesser, D.F. (ed.). Compendium of methods for the microbiological examination of foods, 3rd ed. American Public Health Association, Washington D.C.

Eisgruber, H. and Reuter, G. (1991) SCA-ein Selektivnährmedium zum nachweis mesophiler sulfitreduzierender Clostridien in Lebensmittel, speziell für Fleisch und Fleischerzeugnisse, Archiv für lebensmittelhyg. 42, 101–132.

Eisgruber, H. and Reuter, G. (1995) A selective medium for the detection and enumeration of mesophilic sulphite-reducing clostridia in food monitoring programs. Food Res. Int. 28, 219–226.

Erickson, J.E., and Deibel, R.H. (1978) New medium for rapid screening and enumeration of *Clostridium perfringens* in foods. Appl. Environ. Microbiol. 36, 567–571.

Fryer, T.F. and Halligan, A.C. (1976) The detection of *Clostridium tyrobutyricum* in milk. N.Z. J. Dairy Sci. Technol. 11, 132

Gibbs, B.M. and Freame, B. (1965) Methods for the recovery of clostridia from foods. J. Appl. Bacteriol. 28, 95–111.

Gibbs, B.M. (1971) The incidence of Clostridia in poultry carcasses and poultry processing plants. Br. Poult.Sci. 12, 101–110.

Handford, P.M. (1974) A new medium for the detection and enumeration of *Clostridium perfringens* in foods. J. Appl. Bacteriol. 37, 559–570.

Harmon, S.M., Kautter, D.A. and Peeler J.T. (1971) Improved medium for the enumeration of *Clostridium perfringens* Appl. Microbiol. 22, 688–692.

Harmon, S.M. (1976) Collaborative study of an improved method for the enumeration and confirmation of *Clostridium perfringens* in foods. J. Assoc. Off. Anal. Chem. 59, 606–612.

Hauschild, A.H.W. and Hilsheimer, R. (1974a) Evaluation and modifications of media for enumeration of *Clostridium perfringens*. Appl. Microbiol. 27, 78–82.

Hauschild, A.H.W. and Hilsheimer, R. (1974b) Enumeration of food-borne *Clostridium perfringens* in egg yolk free tryptose-sulphite-cycloserine agar. Appl. Microbiol. 27, 521–526.

Hirsch, A. and Grinsted, E. (1954) Methods for the growth and enumeration of anaerobic spore-formers from cheese, with observations on the effect of nisin. J. Dairy Res. 21, 101–110.

Kautter, D.A., Solomon, H.M., Lake, D.E., Bernard, D.T. and Mills, D.C. (1992) *Clostridium botulinum* and its toxins, p. 605–621. *In*: Vanderzant, C. and Splittstoesser, D.F. (ed.). Compendium of methods for the microbiological examination of foods, 3rd ed. American Public Health Association, Washington D.C.

Marshall, R.S, Steenbergen, J.F. and McClung, L.S. (1965) Rapid technique for the enumeartion of *Clostridium perfringens*. Appl. Microbiol. 13, 559–563.

Mead, G., Adams, B.W., Roberts, T.A. and Smart, J.L. (1982) Isolation and enumeration of *Clostridium perfringens*. *In*: Isolation and identification methods for food poisoning organisms. J.E.L. Corry, D. Roberts and F.A. Skinner (eds.), SAB Technical series No 17, Academic Press, London, 99–110.

Mossel D.A.A., Bruin de A.S. van Diepen, H.M.J., Vendrig C.M.A. and Zoutewelle G. (1956) The enumeration of anaerobic bacteria, and of *Clostridium* species in particular, in foods. J. Appl. Bacteriol. 19, 142–154.

Mossel, D.A.A. and de Waart, J. (1968) The enumeration of clostridia in foods and feeds. Ann. Inst. Pasteur (Lille) 19, 13–27.

Mossel, D.A.A., Corry. J.E.L., Struijk C.B. and Baird, R.M. (1995) The microbiological monitoring of foods; the use of marker (index and indicator) organisms and marker biochemical activities. *In*: Essentials of the microbiology of foods, a textbook for advanced studies. John Wiley and Sons, Chichester, 287–309.

NMKL (1994) Sulphite reducing Clostridia determination in foods, NMLK 56: 1994, 3rd ed. Nordic committee on food analyses, Esbo.

Prévot, A.R (1948) Recherches sur la réduction des sulfates et des sulfites minéraux par les bactéries anaérobies. Ann. Inst. Pasteur Lille 1, 517–575.

Schalch, B., Eisgruber, H., Geppert, P. and Stolle, A. (1996) Vergleich von vier Routineverfahren zur Bestätigung von *Clostridium perfringens* aus Lebensmitteln. Archiv. Lebensmittelhyg. 47, 27–30.

Shahidi, S.A. and Ferguson A.R. (1971) New quantitative, qualitative, and confirmatory media for rapid analysis of food for *Clostridium perfringens*. Appl. Microbiol. 21, 500–506.

Senyk, G.F., Scheib, J.A., Brown, J.M. and Ledford, R.A. (1989) Evaluation of methods for determination of spore-formers responsible for the late gas-blowing defect in cheese. J. Dairy Sci., 72, 360–366.

Smith, M. (1985) Comparison of Rapid Perfringens Medium and Lactose Sulphite Medium for detection of *Clostridium perfringens*. J. Assoc. Off. Anal. Chem. 68, 807–808.

Weenk, G., Fitzmaurice, E. and Mossel, D.A.A. (1991) Selective enumeration of spores of *Clostridium* species in dried foods. J. Appl. Bacteriol. 70, 135–143.

Weenk, G.H., van den Brink, J.A., Struijk, C.B. and Mossel, D.A.A. (1995) Modified methods for the enumeration of spores of *Clostridium* species in dried foods. Int. J. Food Microbiol. 27, 185–200.

Wilson, W.J. and Blair, E.M.M.V. (1924) The application of a sulphite-glucose-iron agar medium to the quantitive estimation of *B. welchii* and other reducing bacteria in water supplies. J. Pathol. Bacteriol. 27, 119–121.

Chapter 4

Media for Bacillus spp. and related genera relevant to foods

Dagmar Fritze[a] and Dieter Claus[b]

[a]*DSMZ-Deutsche Sammlung von Mikroorganismen und Zellkulturen GmbH, Mascheroder Weg 1b, 38124 Braunschweig, Germany*
[b]*Chemnitzer Str. 3, 37085 Göttingen, Germany*

Spores are resistant to heat and other means of sterilization, so spore-forming bacteria are of major concern to food microbiologists. Virtually any food can be colonized by these organisms due to their ubiquitous distribution, their diversity in physiological properties and thus in growth requirements. With respect to diagnostics, it is therefore not possible to design a single medium which allows growth of all or most species of this group. Numerous media have been described for the cultivation of individual species of aerobic spore forming organisms, however, most of these are not selective and only some are elective. Thc only largely selective media in the area of aerobic spore forming bacteria acting in the mesophilic/neutrophilic range have been developed for *Bacillus cereus* and related species. The most widely used among those are MEYP (Mannitol Egg Yolk Polymyxin agar) and PEMBA (Polymyxin Egg yolk Mannitol Bromothymol blue Agar)). Another suitable form of selectivity that is easily achieved, is to adjust media (even standard media) to certain pH values or to incubate at certain temperatures. This physiological approach has been successful e.g. with *Alicyclobacillus*. In most cases, however, even under those conditions, groups of species rather than individual species can be expected.

For up-to-date results in food microbiology, it should be understood that the aerobic spore forming organisms are no longer just the traditional genus *Bacillus* but fall into a number of genera. In addition, several of the traditionally well known species such as *B. subtilis*, *B. circulans* or *B. stearothermophilus* have been divided up into newly established species to form more consistent taxa. As it cannot be assumed that the newly described species are of no relevance to food, it is advisable to perform under certain circumstances a more thorough taxonomic identification to complete the diagnostic approach.

Further research into the development of more and better diagnostic media would be worthwhile to ease the detection of members of the large group of aerobic spore forming bacteria in foods. This review aims to describe the physiological peculiarities and relationships within the group in order to support and encourage the development of additional selective or elective media for other aerobic spore forming organisms of relevance to food.

Table 1
The present taxonomic status of aerobic endospore forming bacteria: validly described genera as of June 2000.

Previous Name	Valid Name	No. of Sp./Subsp.
Bacillus	*Alicyclobacillus*	5 / 1
Bacillus	*Aneurinibacillus*	3
Bacillus	*Anoxybacillus*	2
Bacillus	*Bacillus*	75 / 2
Bacillus	*Brevibacillus*	10
Bacillus	*Gracilibacillus*	2
Bacillus	*Geobacillus*	9
Bacillus	*Marinibacillus*	1
Bacillus	*Paenibacillus*	30 / 2
Bacillus	*Salibacillus*	1
Bacillus	*Ureibacillus*	2
Bacillus	*Virgibacillus*	2
Sporolactobacillus	*Sporolactobacillus*	6 / 2
Sporosarcina	*Halobacillus*	3
Sporosarcina/bacillus	*Sporosarcina*	4
Sulfobacillus	*Sulfobacillus*	3
Thermoactinomyces	*Thermoactinomyces*	8
–	*Ammoniphilus*	2
–	*Amphibacillus*	1
–	*Filobacillus*	1
–	*Jeotgalibacillus*	1
–	*Thermaerobacter*	1
–	*Thermobacillus*	1

Systematics of the Group

The term *Bacillus*, though it designates a certain genus, is still today often used synonymously for any 'aerobic endospore-forming bacteria'. Bergey's Manual of Systematic Bacteriology (Sneath et al., 1986) lists four genera and 44 species; however, taxonomic advances since then have distinguished at least 23 genera of aerobic spore forming organisms with more than 170 species (Table 1) and therefore the expression '*Bacillus* and related genera' is more appropriate.

One characteristic used to group all these organisms together is endospore formation. As a consequence, the group is phenotypically heterogenous, versatile and includes a wide variety of physiological specializations. It is therefore not surprising that efforts continue to form more consistent taxa.

Members of the *Bacillus* group have a Gram positive type of cell wall and most species stain Gram positive at least in very young cultures. Formation of spores requires manganese ions so even complex media must be supplemented. With some species or strains it is difficult to observe spore formation in pure culture at all.

The G+C content of their DNA varies from about 34–65%. They can be acidophilic (growth down to pH 1.5), neutrophilic or alkaliphilic (growth up to pH 11), psychrophilic (growth down to minus 5°C) or thermophilic (growth up to 78°C), het-

Table 2
Species of aerobic endospore forming bacteria commonly reported to be of relevance to food.

I. Food spoilage organisms (odour, taste, colour, texture ..)	
Alicyclobacillus acidoterrestris	
Bacillus coagulans	
Geobacillus stearothermophilus	
II. Organisms recorded as causing illness in humans	
Bacillus anthracis	
Bacillus cereus	
Bacillus mycoides	
Bacillus thuringiensis	
Bacillus licheniformis	
Bacillus pumilus	
Bacillus subtilis	
Brevibacillus brevis	
Bacillus circulans	
Bacillus lentus	
Paenibacillus polymyxa	
III. Organisms used for food and feed	
Bacillus subtilis subsp. 'natto'	(fermented food)
Bacillus cereus subsp. 'toyoi'	(probioticum)

erotrophic, chemoautotrophic, N_2-fixing or halophilic, aerobic or facultatively anaerobic. Thus, as far as enrichment, isolation and cultivation of this group of organisms is concerned, it is clear that no medium or set of cultural conditions will suffice for all (or even most) organisms.

Relevance to food

Aerobic endospore forming bacteria (AEFB) can be isolated from almost any natural habitat. However, whether this site represents the real ecological habitat or whether spores have been deposited there by wind or other means cannot usually be determined. The same is true for other habitats like foods, cosmetics and pharmaceuticals. The ubiquitous nature of AEFB results in numerous points of potential entry into a given habitat (e.g. soil, dust, air, water, animals, bedding, foods and feedstuffs, pasture, food handling and processing equipment, transport vessels). Again, these habitats may provide conditions suitable for the growth of AEFB or may only harbour spores, which, due to their remarkable power of resistance and dormancy, may survive in any habitat for longer periods than vegetative organisms.

Obviously, not all species of AEFB are of relevance to food. Only for a relatively small number can food related reports be found. Such reports may cover both positive and negative aspects. The use of AEFB has been described in human as well as animal nutrition. Natto, a well known east Asian food, is based on the metabolic activity of '*Bacillus natto*', a strain of *B. subtilis*, on soy beans. Certain non-toxin producing *B. cereus* strains (e.g. *B. cereus* var. *'toyoi'*) are used as additives for animal feed.

Table 3
Occurence of aerobic spore forming bacteria in food.

Food	Examples of induced effect	Main species reported
Canned products		
low or medium acid food	flat sour, evaporated milk coagulated	*B. coagulans, B. smithii, G. stearothermophilus*
acid food (tomatoes, pears, figs)	flat sour, off odour and flavour	most likely *B. coagulans, P. polymyxa, P. macerans, Alicyclobacillus* spp.
high acid food (fruit juices, apple juice)	off odour and flavour	*Alicyclobacillus acidoterrestris*
Dairy products		
Milk	with or without effect, bitter flavour due to proteolysis, food poisoning	**mesophilic**: *B. cereus, B. circulans, B. firmus, B. megaterium, B. pumilus, B. sporothermodurans, B. subtilis* **thermophilic/thermotolerant**: *B. coagulans, B. kaustophilus, B. licheniformis, G. stearothermophilus* **psychrophilic/psychrotolerant**: *B. cereus, B. weihenstephanensis*
Cream	bitty cream, food poisoning	*B. cereus, B. mycoides*
Other products		
Beer	spoilage	*B. brevis*
beet sugar, cane sugar refinery	transfer of spores to canned products	*B. amyloliquefaciens, B. coagulans, B. licheniformis, B. pumilus, B. smithii, G. stearothermophilus, B. subtilis*
Bakery products, bread	rope	*B. licheniformis, B. subtilis, B. pumilus*
Butter	spoilage	*B. licheniformis*
Cereals, flour, boiled or fried rice	food poisoning	*B. cereus*
Chocolate		*B. subtilis*
Cocoa	transfer of spores to other products	*B. cereus, B. licheniformis, B. megaterium, B. pumilus, G. stearothermophilus, B. subtilis*
fish products	spoilage, food poisoning	*B. cereus, B. coagulans, B. firmus, B. megaterium, B. sphaericus, B. subtilis*
honey		*B. alvei*
meat and meat products	spoilage, food poisoning	*B. cereus, B. coagulans, B. circulans, B. licheniformis, B. macerans, B. polymyxa, B. pumilus, G. stearothermophilus, B. subtilis*
Natto	fermented food	*B. subtilis* = '*B. natto*'
Spices	spices not spoiled when dry, problems in finished food, food poisoning	*B. brevis, B. cereus, B. firmus, B. licheniformis, B. megaterium, B. pumilus, B. subtilis,* and other aerobic spore forming species
starch syrup		*B. cereus, B. coagulans, B. licheniformis, B. megaterium*
vegetable sprouts	food poisoning	*B. cereus*

Interestingly, among the AEFB species *B.cereus* is the predominant causative agent for both food poisoning and food spoilage. On the other hand, a vast range of applications for various industrial uses have been described especially for members of the *B. subtilis* group (Table 2).

According to the nutritional composition of a given substrate, its water content, pH, etc., specific organisms can be expected to grow. Table 3 presents an overview of foods or food additives reported to be spoiled by AEFB or as being sources for later contamination. The types of spoilage and the organisms responsible are also listed. Compared to the high number of validly published species, it is not clear whether food poisoning and spoilage are really caused by only a few of these. Recent taxonomic developments imply that many additional species of aerobic spore forming organisms could be involved in food spoilage. Lack of taxonomic skills amongst researchers and the possibility that enrichment and cultivation procedures may not have been appropriate for those species may well explain such discrepancies.

Detection and enumeration of cells and spores of aerobic spore forming bacteria

The AEFB are of special interest to the food processing industry because their spores can survive periods of drying out or high temperatures, both used for preserving foods.

To detect and enumerate AEFB in a food sample two strategies are generally applied: the detection of vegetative cells and the detection of spores. Vegetative cells dominate in wet samples, and spores often in dry samples. Bacterial spores are biochemically inert and are, therefore, not themselves responsible for food spoilage. Spoilage is caused by vegetative cells developing after germination of spores. Numbers of vegetative cells and spores can be estimated by comparing results from untreated samples and samples treated to select for spores.

Examining food samples for vegetative forms may be useful with canned or other food which had originally only been treated to kill vegetative bacteria. During storage of such food, viable spores may germinate and vegetative cells multiply. It is, however, important to verify the spore forming capability of these isolates because the food sample may also have been contaminated post process by other non-spore forming bacteria. Another important aspect is, that under certain circumstances the conditions in such a spoiled food may become unfavourable to the causative organism, e.g. by a shift in pH, and all vegetative cells may die. In such cases ('auto-sterilization'), no viable cells may be recoverable and searching for viable cells would clearly be an unreliable control method.

Spores of AEFB of different genera are generally detected and enumerated after heating of food suspensions for 10 min at 80°C (pasteurization) and plating serial dilutions on agar media. However, the real number of spores present may be underestimated because spores found in natural environments or in food samples may differ in their heat sensitivity from those studied in pure cultures, and they may have different heat activation requirements for optimal spore germination and outgrowth. To overcome this problem an alternative process using ethanol has been proposed (Koransky et al., 1978). Treatment of samples with about 50% ethanol for one hour is effective for isolating spore forming bacteria. Since commercially available ethanol may itself contain spores, it should be filter sterilized before use. Technical details are briefly described.

Pasteurization of food samples
To 5 g of a homogenized food sample add peptone water (0.1 %) up to 20 g. Mix well and heat the sample in a 300 ml Erlenmeyer flask for 10 min in a water bath at 80°C. Immerse the flask so that the sample is well below surface of the water and heat transmission is optimal. Cool the flask under running water and prepare serial dilutions in peptone water for plating on agar.

Ethanol treatment of food samples
To peptone water (0.2%) add an equal volume of 96% ethanol (final concentration about 50%). Filter the solution through a membrane filter resistant to ethanol.

To 5 g of a homogenized food sample add alcoholic peptone water up to 20 g. Mix well and allow the mixture to stand at room temperature for 1 h. Prepare serial dilutions in peptone water (0.1%) for plating on agar.

Media for detecting and enumerating aerobic spore forming bacteria

Most AEFB will grow on plain nutrient agar (0.8% peptone + 0.3% beef extract) with media containing soya peptone often giving the best results. A few species require glucose or other carbohydrates in nutrient agar or have more specific nutritional requirements (e.g *B. fastidiosus*: uric acid or allantoin; *Sp. pasteurii*: NH_4^+ and alkaline pH; *Halobacillus halophilus*: NaCl, $MgCl_2$). Where spore formation is required, all media should be supplemented with 20–50 mg/l $MnSO_4 \cdot 7H_2O$.

Various media have been devised for different AEFB (see e.g. Atlas, 1995; Atlas and Snyder, 1995; Atlas 1997; DSMZ-Catalogue of Strains, 2001). Most aim to optimise growth of the target organism, or enhance spore formation, while others have been designed to prevent loss of specific properties of the target organism. Generally, these media have no or very low selectivity.

The minimal nutrient requirements of AEFB differ widely. For some species, a mineral base glucose agar with nitrate or ammonium as the nitrogen source is sufficient for growth. For others, the addition of certain vitamins and/or amino acids is required. Although the minimal nutrient requirements of most of the recently described species are not known, the requirements of those thought to cause food spoilage are mostly well documented (Knight and Proom, 1950; Koser, 1968). Species expected to grow in a simple mineral base glucose medium include: *B. amyloliquefaciens*, *B. licheniformis*, *B. megaterium* and *B. subtilis*. *B. pumilus* and *Paenibacillus polymyxa* need biotin; *P. macerans* needs biotin and aneurin. If agar is used in this medium, other species may also grow depending on the purity of the agar.

Selective media for the detection of aerobic spore forming bacteria

There is no general selective medium for all AEFB. However, as described above, some selectivity for spore formers can be achieved by heat or ethanol treatment of samples where vegetative forms of bacteria will be killed and only resistant spores will survive.

Selective media have only been described for *B. cereus*. A thorough review of *Bacillus cereus* media was compiled by van Netten and Kramer (1995), which described their history and their selective and diagnostic systems in detail. Of these, PEMBA (Polymyxin Pyruvate Egg Yolk Mannitol Bromothymol blue Agar) and MEYP (Mannitol Egg Yolk Polymyxin Agar) are most often used in food microbiology.

Both use polymyxin as selective agent, while PEMBA also contains cycloheximide. Both compounds help to suppress most of the unwanted flora. While further selectivity is achieved by the omission of any carbohydrates except mannitol. However, many other organisms besides *B. cereus* can develop under these circumstances. *B. cereus* and related organisms (1) do not produce acid from mannitol and hence do not change the colour of the pH indicator dyes used in the media, and (2) do produce lecithinase whose effect on egg yolk results in a heavy, cloudy precipitate around colonies. Thus, colonies which have not changed the colour of the medium and which are surrounded by a thick cloudy precipitate are presumptive members of the *Bacillus cereus* group. However, lecithinase weak or negative strains are increasingly being reported.

Physiological groups of aerobic spore formers

Although selective diagnostic media for other species of AEFB are not available, it is possible to select for certain physiological groups by choice of culture conditions as shown in Table 4.

AEFB can be divided into smaller groups exhibiting similar physiological requirements. Considering temperature and pH requirements, the following combinations lend themselves for differentiation: for thermophiles 55°C; for mesophiles 30°C, for psychrophiles 5°C. Media should be used at three different pH levels at each temperature: 4.5, 7.0–7.2 and 9.

A complex medium such as Brain Heart Infusion (BHI: Difco) should be chosen to allow growth of as many species of AEFB as possible. BHI even allows the halophilic species to develop because it contains NaCl. A few of the presently known species will not be able to develop on this medium. However, because of their nutritional requirements, these are of no relevance to food.

Table 4 provides an overview of the organisms that may be expected to grow under the conditions of temperature and pH specified above. High pH has not been considered because alkaliphilic AEFB are unlikely to be important in foods. Some species are found in more than one group because of their broader temperature and/or pH ranges. However, species other than those listed may be able to grow under various conditions, because temperature (especially low temperature) and pH tolerance have not been determined for all species. As most research has been done in the mesophilic/

Table 4
Species of aerobic endospore forming bacteria to be expected to grow under certain culture conditions.

a.1. Temperature 55°C/pH 4.5

Alicyclobacillus acidocaldarius	*Alicyclobacillus herbarius*
Alicyclobacillus acidoterrestris	*Bacillus coagulans*
Alicyclobacillus hesperidum	*Sulfobacillus thermosulfidooxidans*

a.2. Temperature 55°C/pH 7.0–7.2

Aneurinibacillus thermoaerophilus	*Geobacillus stearothermophilus*
Bacillus coagulans	*Geobacillus subterreaneus*
Bacillus halodurans	*Geobacillus thermocatenulatus*
Bacillus licheniformis	*Geobacillus thermoglucosidasius*
Bacillus methanolicus	*Geobacillus thermoleovorans*
Bacillus schlegelii	*Geobacillus uzenensis*
Bacillus smithii	*Ureibacillus thermosphaericus*
Bacillus thermoamylovorans	*Brevibacillus thermoruber*
Geobacillus kaustophilus	*Thermobacillus xylanilyticus*
Geobacillus caldoxylosilyticus	

b.1. Temperature 30°C/pH 4.5

Alicyclobacillus acidoterrestris	*Sulfobacillus acidophilus*
Alicyclobacillus cycloheptanicus	*Sulfobacillus disulfidooxidans*
Alicyclobacillus hesperidum	*Sulfobacillus thermosulfidooxidans*
Bacillus coagulans	*Bacillus laevolacticus*
Bacillus naganoensis	

b.2. Temperature 30°C/pH 7.0–7.2

Aneurinibacillus aneurinolyticus	*Bacillus licheniformis*
Aneurinibacillus migulanus	*Bacillus subtilis*
Bacillus cereus	*Bacillus megaterium*
Bacillus circulans	*Bacillus methanolicus*
Bacillus coagulans	*Bacillus mycoides*
Bacillus cohnii	*Bacillus pumilus*
Bacillus clausii	*Bacillus smithii*
Bacillus gibsonii	*Bacillus thermoamylovorans*
Bacillus halmapalus	*Bacillus thuringiensis*
Bacillus halodurans	*Brevibacillus brevis*
Bacillus horikoshii	*Brevibacillus thermoruber*
Bacillus horti	*Paenibacillus polymyxa*
Bacillus niacini	*Thermobacillus xylanilyticus*
Bacillus laevolacticus	*Ureibacillus thermosphaericus*
Bacillus lentus	

Plus around 80 other species of aerobic spore forming bacteria not listed.

c.1. Temperature 5°C/pH 4.5

Organisms able to grow under these culture conditions have not been described to date.

c.2. Temperature 5°C/pH 7.0–7.2

Bacillus insolitus	*Sporosarcina aquimarina*
Bacillus psychrosaccharolyticus	*Sporosarcina globispora*
Bacillus weihenstephanensis	*Sporosarcina psychrophila*
Marinibacillus marinus	*Paenibacillus macquariensis*

neutral range, the group of organisms growing at 30°C/pH 7.0–7.2 is by far the largest, and probably all the AEFB not mentioned here in any other subgroup can be added to the mesophilic/neutral group. However, for the present purpose, only a representative selection of species is presented here.

A special physiological feature of relevance to food is NaCl-tolerance. Aerobic spore forming organisms able to grow without and with up to 10% NaCl are: *B. horti*, *B. pasteurii*, *B. pseudalcaliphilus*, *Virgibacillus pantothenticus* and *V. proomii*. Other organisms tolerate even 12 or 16%: *B. agaradhaerens*, *B. clarkii*, *B. halodurans*, *B. pseudofirmus*, *Gracilibacillus dipsosauri* and *Salibacillus salexigens*. For those organisms which cannot grow without a minimum amount of salt, even higher tolerances are described: 20% *B. haloalkaliphilus*, 30% *B. halophilus*, 25% *B. marismortui*, 20% *G. halotolerans*, 24% *Halobacillus halophilus*, 25% *H. litoralis* or 30% *H. trüperi*. Relevance of these organisms to food spoilage has yet to be established.

Individual species or groups of species in detail

Groups containing most of the food relevant AEFB are described in more detail below.

Mesophiles (and some psychrotolerant derivatives)

The 'cereus group'

Of those AEFB which occur in food, *B. cereus* attracts most attention. *B. cereus* is a common soil saprophyte, is also found in quite high numbers in products of plant origin and may be isolated from several types of foods, such as cereals, spices, but also from milk or milk products, egg products and meat. *B cereus* may be responsible for food spoilage, for food borne infections and intoxication. Food poisoning caused by *B. cereus* is often associated with the consumption of starch containing products like rice (Kramer and Gilbert, 1989).

Two different types of *B. cereus* food poisoning are known. The 'emetic syndrome' type, in which the main symptoms are nausea and vomiting and the 'diarrhoeal syndrome' type causing abdominal pain and diarrhoea. Strains causing the emetic syndrome have been isolated mainly from fried and cooked rice, pasta, pastry and noodles, whereas strains responsible for the diarrhoeal syndrome have been found in dairy products, meat products, soups, sauces, desserts and vegetables. The minimal infective dose for the *B. cereus* diarrhoeal toxin type ranges between 10^5 and 10^7 cfu (total) whereas the emetic syndrome is observed after the ingestion of food containing 10^5 to 10^8 cfug^{-1}. Detailed information on *Bacillus* toxins can be found in the publications of Granum (1994), Granum and Lund (1997) and Beattie and Williams (2000).

A number of other *Bacillus* species are very similar to *B. cereus* phenotypically and genotypically and thus the term '*cereus* group' is often used. Species of this group are *B. anthracis*, *B. mycoides*, *B. pseudomycoides*, *B. thuringiensis* and *B. weihen-*

Table 5
Differentiation of members of the '*cereus*–group'.

	B. anthracis	*B. cereus*	*B. mycoides*	*B. pseudoycoides*	*B. thuringiensis*	*B. weihenstephanensis*
Mannitol[#]	–	–	–	–	–	nd[*]
Lecithinase[#]	w	+	+	+	+	nd[*]
Rhizoid growth	–	–	+	+	–	–
Toxin crystal	–	–	–	–	+	nd[*]
Motility	–	+	–	–	+	nd[*]
Haemolysis	–	+	+	nd	+	nd[*]
Penicillin–resistance	–	+	+	nd	+	nd[*]
Gamma phage	+	–	nd	nd	–	nd[*]

[#] properties commonly used to distinguish the group from other aerobic spore forming bacteria; – negative; + positive; w weak; nd not determined; [*] according to Lechner et al. (1998) *B. weihenstephanensis* is similar to *B. cereus* in cell morphology and substrate utilization.

stephanensis. It has long been debated whether all species of this group should be considered one single species (Helgason et al., 2000), not least because the *B. cereus* enterotoxin HBL has been detected in all these species (Pruss et al., 1999). A case of food poisoning has been reported for *B. thuringiensis* (Jackson et al., 1995) and a number of incidents of intestinal anthrax have been caused by consumption of meat from animals infected with *B. anthracis* (Beattie and Williams, 2000).

For the selective detection and enumeration of *B. cereus* a number of media have been developed which suppress the growth of most other bacteria and allow easy and reliable detection of typical strains of the '*cereus* group'. Differentiation among the members of the group may be made by testing for the characteristics given in Table 5.

A short description of the present members of the '*cereus* group' follows: Cells are > 1 μm in diameter, sporangia are not swollen and spores are ellipsoidal. The organisms are in principle mesophilic and neutrophilic. All of these organisms are placed very close together in 16S rRNA/DNA group 1. Phenotypically they are extremely difficult to distinguish from each other. Non-pathogenic *B. anthracis*, non-rhizoid *B. mycoides* and non-toxin-crystal-carrying *B. thuringiensis* may be judged as *B. cereus* and vice-versa. DNA/DNA hybridizations show clear differentiation between *B. cereus*, *B. mycoides*, *B. pseudomycoides* and *B. thuringiensis* (Nakamura, 1994; Nakamura & Jackson, 1995; Nakamura, 1998). *B. cereus* and *B. mycoides* may comprise psychrotolerant strains. Some of those have been described as *B. weihenstephanensis*. Species with a similar or larger cell diameter (*B. megaterium*, *B. benzoevorans*, *B. fastidiosus*) are easily distinguishable phenotypically. The properties listed in Table 5 have been described as being typical for the '*cereus* group' (no acid production from mannitol, lecithinase positive), and as discriminative between members of the group (toxin crystal formation, rhizoid growth on agar media, motility, haemolysis, penicillin resistance, gamma phage susceptibility). It should be noted that the probability of isolating *B. anthracis* from food is low. Therefore, if a *cereus*-like organism is isolated, for practical reasons, it would be best to chose from the tests those which react positively for non-*anthracis* strains.

Technical details of the differential tests are briefly described.

Acid Formation from Mannitol

No member of the '*cereus* group' is able to form acid from mannitol whereas most other relevant species of AEFB (e.g. *B. megaterium*, the complete *B. subtilis* group, *B. circulans*, *P. polymyxa*, *P. macerans*) will produce acid. On a medium containing bromthymol blue, acid production would be indicated by a yellow colour.

Lecithinase Activity (Egg Yolk Reaction)

The lecithinase reaction is weakly positive for *B. anthracis*, whereas it is usually strongly positive for *B. cereus* and *B. mycoides*. However, it should be noted that occasionally lecithinase-negative or weak strains of *B. cereus* have been isolated. Up to now, no other AEFB has been reported to exhibit lecithinase activity. However, one should be aware that some have active lipases which may liberate fatty acids resulting in a faint, fine crystalline precipitate. The precipitate caused by lecithinase forms a thick milky cloud around positive colonies.

Rhizoid Growth

A pre-dried nutrient agar plate is inoculated centrally and incubated at 30°C for 1–2 d. Rhizoid growth develops as root like structures up to several centimetres from the point of inoculation. Rhizoid growth is characteristic only for *B. mycoides* and *B. pseudomycoides*. However, it should be noted that this property may be lost upon frequent subculturing or other culture conditions. Such strains cannot be discriminated phenotypically from *B. cereus*.

Toxin Crystal Formation

Endotoxin crystals are produced only by *B. thuringiensis*. These may be detected by phase contrast microscopy as bipyramidal, spherical, flat, cubic or pleomorphous bodies. Crystal formation depends on culture conditions. Non crystal forming strains cannot be discriminated phenotypically from *B. cereus.*

Motility

Soft agar with low nutrient concentration (0.1% yeast extract, 0.01% K_2HPO_4, 0.2% agar) is inoculated and observed (24–48 h) for spreading growth developing as a faint confluent veil away from the point of inoculation. *B. anthracis* and the rhizoid strains of *B. mycoides* and *B. pseudomycoides* are non-motile, *B. cereus* and *B. thuringiensis* are motile. As formation of flagella depends on culture conditions and age of culture and as they are easily lost during mounting procedures other methods for detecting motility may be less suitable.

Haemolysis

Tryptic-Soy Sheep-Blood Agar is inoculated by lightly touching the agar surface with a loopful of culture and incubated at 30°C for 24 h. Haemolysis is indicated by a zone of complete clearance around the growth. *B. anthracis* is non-haemolytic or only weakly (visible only beneath the growth), *B. cereus* is strongly haemolytic, *B. mycoides* and *B. thuringiensis* are haemolytic.

Penicillin Resistance

B. anthracis is consistently susceptible to Penicillin G, 10 μg/ml. *B. cereus*, *B. mycoides* and *B. thuringiensis* are resistant.

Gamma Phage Susceptibility

The gamma phage (a derivative of phage W) has been found useful for the detection of non-capsulated strains of *B. anthracis* (Seidel, 1962). *B. cereus* and *B. thuringiensis* are not susceptible.

The '*subtilis* group'

Like members of the *B. cereus* group, *B. subtilis* and related species are widely distributed in the environment and hence may contaminate via dust etc. all types of food and food products. *B. amyloliquefaciens*, *B. atrophaeus*, *B. licheniformis*, *B. mojavensis*, *B. pumilus*, and *B. vallismortis* are phenotypically and genotypically very similar to *B. subtilis* and thus the term '*subtilis* group' is often used for this group of bacteria. *B. subtilis* and *B. licheniformis*, and to a lesser extent *B. pumilus* have all been reported to be implicated in incidences of food-borne illness. Implicated foods include meat and meat products, seafood, pastry products, bread, canned fruit juice and cheese sandwiches. Since no selective isolation medium for members of the '*subtilis* group' is available, the role of this species in food spoilage and food borne infections or intoxications may be underestimated (Te Giffel et al., 1996). A short taxonomic description of the members of the '*subtilis* group' is given in Table 6.

Cells are <1 µm in diameter, sporangia are not swollen and spores are ellipsoidal. The organisms are in general mesophilic with regard to temperature and neutrophilic with respect to pH for growth. All members of the group are placed in 16S rRNA/DNA group 1 and are phenotypically very similar. However, for some of them clearly discriminating features have been determined. *B. pumilus* is starch negative and hippurate positive, *B. licheniformis* is propionate positive, grows up to 55°C and is facultatively anaerobic. For *B. atrophaeus* Nakamura (1989) described pigment formation on tyrosine medium differing from *B. subtilis* from which it is otherwise not distinguishable. For *B. amyloliquefaciens* Nakamura (1987) described a faster acid production from lactose (4–14 d) and slower gluconate utilization (7–14 d) in contrast to *B. subtilis*. *B. subtilis*, *B. mojavensis* and *B. vallismortis* have so far proved to be impossible to distinguish phenotypically and can only be separated by DNA/DNA hybridization. *B. sporothermodurans* may be loosely attached to this group. Strains may grow best on BHI with neutral pH and at 30°C. Strictly, the two species *B. lentus* and *B. firmus* do not belong to this group and are easily distinguishable phenotypically.

Acidophilic and acidotolerants

Flat sour spoilage of acid or medium/low acid food has traditionally been attributed mainly to *B. coagulans*, today also *B. smithii* is identified. High acidic food seemed to be safe from spoilage, but in 1984, Cerny et al. described a hitherto unknown acidophilic and thermophilic *Bacillus* species isolated from spoiled fruit juice. This juice had been subjected to relatively high temperatures during storage. It was later described as *Bacillus acidoterrestris* (Deinhard et al. 1988) and transferred to the new genus *Alicyclobacillus* as *Alicyclobacillus acidoterrestris* (Wisotzkey et al. 1992). Subsequently, this organism was also detected in other juices and spoiled acidic beverages. Other species of the acidophilic, aerobic spore forming bacteria belonging to the genera *Alicyclobacillus* and *Sulfobacillus* have so far not been associated with food spoilage. The optimum pH value for growth of *Alicyclobacillus* is below 5 while the maximum pH values are below 6. All species of *Alicyclobacillus* exhibit a certain

Table 6
Differentiation of members of the '*subtilis* group'.

	B. atrophaeus	*B. amyloliquefaciens*	*B. firmus*	*B. lentus*	*B. licheniformis*	*B. mojavensis*	*B. pumilus*	*B. subtilis*	*B. vallismortis*
Motility		+	v	+	+		+	+	
Anaerobic growth	Identical	–	–	–	+	Identical	–	–	identical
VP reaction		+	–	–	+		+	+	
Max. temp. for growth (°C)	to	50	40–45	35	50–55	to	45–50	45–50	to
Egg yolk reaction		–	–	–	–		–	–	
Growth at pH 5.7	*B. subtilis*	+	–	–	+	*B. subtilis*	+	+	*B. subtilis*
Hydrolysis of starch		+	+	+	+		–	+	
Utilization of citrate		+	–	–	+		+	+	
Utilization of propionate		–	–	–	+		–	–	
Reduction of NO_3 to NO_2		+	+	–	+		–	+	

\+ positive reaction; – negative reaction; v variable reaction.

Table 7
Differentiation of acidophilic and acidotolerant aerobic spore forming bacteria.

	Alicyclobacillus spp.	*B. coagulans*	*B. laevolacticus*	'*B. racemilacticus*'	*Sporolactobacillus* spp.
Catalase	+	+	+	+	–
Growth on NB pH 7.2	–	+	–	–	–
Growth on NB pH 4.5	+	+	–	–	–
Growth on CASO pH 7.2	–	+	–	–	+
Growth on CASO pH 4.5	+	+	+	+	+
Max. temp. for growth (°C)	45–50/60–65	60	40	45	40
DNA hydrolysis	+/–	+	–	–	nd
Voges–Proskauer reaction	–	+	+	+	nd

NB nutrient broth agar (Difco); CASO casein soymeal peptone agar (Merck); + positive reaction; – negative reaction; +/– some species react positive and some negative.

thermotolerance or are thermophilic. The genus *Alicyclobacillus* is covered more thoroughly in chapter 11 of this book.

No acidophilic species is currently found in the genus *Bacillus* and only a few are classified as acid tolerant. *B. coagulans*, *B. laevolacticus* and '*B. racemilacticus*' are able to grow at a pH below 5 as well as at a neutral pH. The genus *Sporolactobacillus* also comprises acid tolerant species (Table 7). *Sporolactobacillus*, a catalase negative group, has been detected in the environment only in low numbers, therefore enrichment techniques and enrichment media are more likely to be successful than direct plating. A suitable enrichment medium is: MRS broth containing alpha-methyl glucoside, potassium sorbate, and bromocresol green indicator adjusted to pH 5.5 with acetic acid, incubated at 37°C, for at least 7 d. Lactobacilli are removed by heat treatment prior to plating onto solid media (Doores and Westhoff, 1983).

Thermophilic and thermotolerants

Thermophilc or thermotolerant species of *Bacillus* able to grow at 55°C may be isolated from a wide range of environments like soil, sewage, compost, water or dust. Most of them are well known as spoilage organisms especially in canned food. With the exception of *B. coagulans* (see above) they prefer a more neutral pH range for growth. *B. smithii* had been separated from *B. coagulans* by its higher temperature maximum for growth and its lower acid tolerance (Nakamura et al., 1988).

B. licheniformis is able to grow anaerobically and can tolerate up to 15 % NaCl. As this organism can reduce both nitrate and nitrite with the production of gas, it may be responsible for 'blowing' in nitrate containing canned products due to the formation of N_2O and N_2.

Geobacillus stearothermophilus may also be responsible for 'flat-sour' spoilage of food. The organism can grow up to 65°C or above and its spores are among the most resistant. The species was long considered to be an extremely heterogenous group, but recently it was recognized that this heterogeneity was due to insufficient taxonomic consistency and in recent years several new species have been described. Their relevance to food is mostly not yet known. As can be seen in Table 4, about 20 thermotolerant or thermophilic species of AEFB have been described to date and allocated to eight genera.

The cell diameter of these organisms is generally <1 μm, spores are ellipsoid (exception *B. schlegelii*; *U. thermosphaericus* may bear both kinds) and sporangia are swollen (exception *B. licheniformis*; *B. coagulans* and *B. smithii* may bear both kinds). Species have been described and separated mainly on the grounds of rRNA sequence similarity or DNA-DNA hybridization studies. Acidophilic/acidotolerant and alkaliphilic/alkalitolerant species are easily distinguishable. A comprehensive study on the phenotypic differentiation within the neutrophilic/mesophilic group is not available. A very good overview of the complete group is given in Sharp et al. (1992).

Other organisms

Due to their widespread distribution and the high resistance of spores to various deleterious agents, most species of the aerobic spore forming bacteria may be found in differing combinations in food samples. Due to the lack of specific diagnostic media, the majority can be detected only by standard methods for the isolation and enumeration of bacteria. Many of the known species can be identified by methods described in Bergey's Manual of Systematic Bacteriology (Sneath et al., 1986). For species described since then, the identification may be more difficult and detailed knowledge of the taxonomic literature is necessary.

Of those organisms relevant to food, two species, *P. polymyxa* and *P. macerans*, may be recognized more easily. Both are clearly distinguishable by their gas production from carbohydrates. Their cell diameter is <1 μm and their ellipsoid spores are formed in bulging sporangia.

References

Atlas, R.M. (1995) Handbook of media for environmental microbiology. CRC Press, Boca Raton. ISBN 0-8493-0603-5

Atlas, R.M. and Snyder, J.W. (1995) Handbook of media for clinical microbiology. CRC Press, Boca Raton. ISBN 0-8493-9497-X

Atlas, R.M. (1997) Handbook of microbiological media. CRC Press, Boca Raton. ISBN 0-8493-2638–9.

Batt, C.A. (2000) *Bacillus cereus*. In: Encyclopedia of Food Microbiology. pp. 119–124. (R.K. Robinson et al., eds.), Academic Press, London.

Beattie, S.H. and Williams, A.G. (2000) Detection of Toxins. In: Encyclopaedia of Food Microbiology. pp. 141–149. (R.K. Robinson et al., eds), Academic Press, San Diego.

Cerny, G., Hennlich, W. and Poralla, K. (1984) Fruchtsaftverderb durch Bacillen: Isolierung und Charakterisierung des Verderbserregers. Z. Lebensm. Unters. Forsch. 179,224–227.

Deinhard, G., Blanz, P., Poralla, K. and Altan, E. (1987) *Bacillus acidoterrestris* sp. nov., a new thermotolerant acidophile isolated from different soils. System. Appl. Microbiol. 10, 47–53.

Doores, S. & Westhoff, D.C. (1983) Selective method for the isolation of *Sporolactobacillus* from food and environmental sources. J. Appl. Bacteriol. 54, 273–280.

DSMZ-Catalogue of Strains (2001) Ed. DSMZ-Deutsche Sammlung von Mikroorganismen und Zellkulturen GmbH. Braunschweig. Germany. (www.dsmz.de)

Granum, P.E. (1994) *Bacillus cereus* and its toxins. J. Appl. Bacteriol. Symp. Suppl. 76, 61–66.

Granum, P.E. and Lund, T. (1997) *Bacillus cereus* and its food poisoning toxins. FEMS Microbiol. Lett. 157, 223–228.

Helgason, E., Ökstad, O.A., Caugant, D.A., Johansen, H.A., Fouet, A., Mock, M., Hegna, I. and Kolstö, A.-B. (2000) *Bacillus anthracis*, *Bacillus cereus*, and *Bacillus thuringiensis* – One Species on the Basis of Genetic Evidence. Appl. Environ. Microbiol. 66, 2627–2630.

Jackson, S.G., Goodbrand, R.B., Ahmed, R. and Kasatiya, S. (1995) *Bacillus cereus* and *Bacillus thuringiensis* isolated in a gastroenteritis outbreak investigation. Lett. In Appl. Microbiol. 21, 103–105.

Knight, B.C.J.G. and Proom, H. (1950) A comparative survey of the nutrition and physiology of mesophilic species in the genus *Bacillus*. J. gen. Microbiol. 4, 508–538.

Koransky, J.R., Allen, S.D. and Dowell, V.R. (1978) Use of ethanol for selective isolation of

sporeforming microorganisms. Appl. Environ. Microbiol. 35, 762–765.
Koser, S.A. (1968) *Bacillus* and *Clostridium*. In: Vitamin Requirements of Bacteria and Yeasts. pp. 379–401. C. Thomas, Springfield, Illinois.
Kramer, J.M. and Gilbert, R.J. (1989) Bacillus cereus and other Bacillus species. In: Foodborne bacterial pathogens, pp. 21–70 (ed. M.P. Doyle). Marcel Dekker Inc. New York.
Lechner, S., Mayr, R., Francis, K.P., Prüß, B.M., Kaplan, T., Wießner-Gunkel, E., Stewart, G.S.A.B. and Scherer, S. (1998) *Bacillus weihenstephanensis* sp. nov. is a new psychrotolerant species of the *Bacillus cereus* group. Int. J. Syst. Bacteriol. 48, 1373–1382.
Nakamura, L.K. (1987) Deoxyribonucleic acid relatedness of lactose-positive *Bacillus subtilis* strains and *Bacillus amyloliquefaciens*. Int. J. Syst. Bacteriol. 37, 444–445.
Nakamura, L.K., Blumenstock, I. and Claus, D. (1988) Taxonomic Study of *Bacillus coagulans* Hammer 1915 with a Proposal for *Bacillus smithii* sp. nov. Int. J. Syst. Bacteriol. 38, 63–73.
Nakamura, L.K. (1989) Taxonomic relationship of black-pigmented *Bacillus subtilis* strains and a proposal for *Bacillus atrophaeus* sp. nov. Int. J. Syst. Bacteriol. 39, 295–300.
Nakamura, L.K. (1994) DNA relatedness among *Bacillus thuringiensis* serovars. Int. J. Syst. Bacteriol. 44, 125–129.
Nakamura, L.K. (1998) *Bacillus pseudomycoides* sp. nov. Int. J. Syst. Bacteriol. 48, 1031–1035.
Nakamura, L.K. and Jackson, M.A. (1995) Clarification of the taxonomy of *Bacillus mycoides*. Int. J. Syst. Bacteriol. 45, 46–49.
Pruss, B.M., Dietrich, R., Nibler, B., Martlbauer, E. and Scherer, S. (1999) The hemolytic enterotoxin HBL is broadly distributed among species of the Bacillus cereus group. Appl. Environ. Microbiol. 65, 5436–5442.
Seidel, G. (1962) Die aeroben Sporenbildner unter besonderer Berücksichtigung des Milzbrandbazillus. Beitr. Hyg. Epidemiol. Heft 17, 1–131.
Sharp, R.J., Riley, P.W. and White, D. (1992) Heterotrophic thermophilic bacilli. In: Thermophilic bacteria (Kristjansson, J.K., Ed.), pp. 19–50. CRC Press, Boca Raton, Florida.
Sneath, P.H.A., Mair, N.S., Sharpe, M.E. and Holt, J.G. (eds.) (1986) Bergey's Manual of Systematic Bacteriology, Vol. 2. Williams and Wilkins, Baltimore.
Te Giffel, M.C., Beumer, R.R., Leijendekkers, S. and Rombout, F.M. (1996) Incidence of *Bacillus cereus* and *Bacillus subtilis* in foods in The Netherlands. Food Microbiology 13, 53–58.
van Netten, P. & Kramer, J.M. (1995) Media for the detection and enumeration of *Bacillus cereus* in foods. In: Progress in industrial microbiology Vol. 34, Culture Media for Food Microbiology, Chapter 3, pp. 35–49. (Eds. J.E.L. Corry, G.D.W. Curtis and R. M. Baird). Elsevier Science B.V., Amsterdam.
Wisotzkey, J. D., Jurtshuk, P. Jr., Fox, G. E., Deinhard, G. and Poralla, K. (1992) Comparative sequences analyses on the 16S rRNA (rDNA) of *Bacillus acidocaldarius*, *Bacillus acidoterrestris*, and *Bacillus cycloheptanicus* and proposal for creation of a new genus, *Alicyclobacillus* gen. nov. Int. J. Syst. Bacteriol. 42, 263–269.

Chapter 5

Culture media and methods for the isolation of *Listeria monocytogenes*

R.R. Beumer[a] and G.D.W. Curtis[b]

[a] *Laboratory of Food Microbiology, Wageningen University, P.O. Box 8129, 6700 EV Wageningen, The Netherlands*

[b] *Purbeck, Horton cum Studley, Oxford, OX33 1BG, United Kingdom*

The recovery of low numbers of *Listeria monocytogenes* from foods and environmental samples requires the use of enrichment cultures followed by selective plating. Media for this task are described, including differential media on which *L. monocytogenes* can be distinguished from other species. Comparisons of media and methods for the detection and enumeration of *L. monocytogenes* are noted and guidance given on the choice of media and methods.

Introduction

Listeria monocytogenes has long been known as a pathogen of man and animals but the role of food in its transmission was not recognised until the 1980s when a number of outbreaks of human listeriosis were traced to various foods (Schlech et al., 1983; Linnan et al., 1988; Bille, 1990). The knowledge that food might play a part in the transmission of this often fatal disease led to many attempts to improve the media and methods available for the isolation of the causative organism from food and food environments. Many media were developed with the aim of excluding the non-listeria flora and providing a diagnostic system for presumptive identification of the target pathogen and some of these were noted in the first edition of this book (Corry et al., 1995; Curtis and Lee, 1995). The more successful of these media have been incorporated into regulatory methods and further improvements have led to new formulations which are discussed in this chapter.

The minimum infective dose of *L. monocytogenes* has not been established and many authorities require that the organism be absent from 25 g of product. This has led to the development of methods, similar to those for salmonellae, following the sequence of pre-enrichment, selective enrichment and diagnostic plating, a procedure that may take four or five days to obtain a negative result. Presumptive positive results may be obtained within two or three days but these must be confirmed which adds further to the time taken to complete the examination.

Recently it has been shown that the incidence of *L. monocytogenes* may be under reported, because presence of this pathogen in enrichment media can be masked by faster growth of other *Listeria* spp. particularly the non-pathogenic *L. innocua* (Petran and Swanson, 1993; McDonald and Sutherland, 1994). In procedures where all species of a genus are considered pathogenic, e.g. *Salmonella*, this will only result in epidemiologically false information and not in failure to identify contaminated samples. In the case of detecting *L. monocytogenes*, the use of isolation media which allow the identification of this pathogen in the presence of high numbers of other *Listeria* spp. is most desirable.

Listeria enrichment broths

Table 1 summarizes the key components of enrichment broths in common use, including those specified by regulatory bodies such as the US Food and Drug Administration (FDA; Hitchins, 1995), the US Department of Agriculture (USDA; Cook, 1998) and the International Organisation for Standardization (ISO; Anonymous, 1996). Common to most of these is the incorporation of acriflavine, nalidixic acid and cycloheximide as selective agents together with a phosphate buffering system. A few broths include lithium chloride as an additional selective agent and sodium pyruvate is also present in some formulations to aid the resuscitation of sublethally injured cells. Primary or pre-enrichment broths usually contain reduced amounts of selective agents (Half Fraser, UVM I) to allow resuscitation of sublethally injured cells. With the combined Listeria Repair Broth (LRB) and the FDA (1995) broth, the basal broth is incubated for four or five hours respectively before the addition of the selective agents, again allowing repair of injured organisms.

Acriflavine inhibits RNA synthesis and mitochondriogenesis (De Vries and Kroon, 1970; Meyer et al., 1972). It is commonly used in enrichment and enumeration protocols in concentrations varying from 10–25 mg/l (Prentice and Neaves, 1988), often in combination with other selective agents. Its use in media for the isolation of *Listeria* was first described by Ralovich et al. (1971) who concluded that *L. monocytogenes* grew well in the presence of acriflavine at 40 mg/l, at which concentration growth of Gram positive cocci was suppressed. Bockemühl et al. (1971) studied twelve acridine dyes as possible selective agents in media for *Listeria.* With three of these (xanthacridin, neutral acriflavine and proflavinehemisulphate) a pronounced inhibition of enterococci was observed. Neutral acriflavine was defined as 2,8-diamino-10-methyl-acridiniumchloride, but is now considered to be a mixture of the hydrochlorides of 3,6-diamino-10-methylacridine and 3,6-diaminoacridine. It is known as acriflavine HCl or acriflavine in the Merck Index (Budvari et al., 1989). According to Beumer et al. (1996), the use of acriflavine in enrichment media for *Listeria* spp. has both direct and indirect effects on the isolation of *L. monocytogenes.* Increasing acriflavine concentrations affect both lag and generation times of *L. monocytogenes*, whereas hardly any effect is observed on *L. innocua*. Acriflavine binds to protein in the samples resulting in a decrease in its activity which, in turn, may allow better growth of *L. monocytogenes.* At pH values < 5.8 more acriflavine is bound but growth promoting effects are limited since growth of this pathogen is restricted at low pH. One may thus expect

Table 1
Listeria enrichment broths (LEBs): selective agents and other additions.

	Acriflavine (mg/l)	Nalidixic acid (mg/l)	NaCl (g/l)	Other	References
Primary LEBs					
Buffered peptone water (PHLS)	0	0	5	0	Greenwood et al., 1991
Half Fraser‡	12.5	10	20	LiCl	Fraser and Sperber, 1988
IDF	10	40	5	cycloheximide	Anonymous, 1995
UVM I‡	12	20	20	0	Donnelly and Baigent, 1986
Universal	0	0	5	$MgSO_4$, pyruvate	Bailey and Cox, 1992
Secondary LEBs					
Fraser‡	25	20	20	LiCl	Fraser and Sperber, 1988
Lovett (PHLS method)	15	40	5	cycloheximide	Lovett et al., 1987
UVM II‡	25	20	20	0	McClain and Lee, 1988
Combined LEB					
FDA (1995)‡	10¶	40¶	5	cycloheximide¶, sodium pyruvate	Hitchins, 1995
Listeria repair broth (LRB)	12*/15†	20*/40†	5	cycloheximide†, glucose, yeast extract, $MgSO_4$, $FeSO_4$, pyruvate	Busch and Donnelly, 1992

All broths are buffered by phosphates. LRB also contains MOPS buffer.

‡ Monographs relating to these media are to be found in Part 2 of this book.
¶ These additions are made after 4 h incubation of the product in the basal broth.
/† These additions are made after 5 h incubation of the non-selective basal repair broth (USDA-FSIS concentrations, † FDA concentrations).

that enrichment protocols employing low acriflavine concentrations with an adequate buffer would favour the isolation of *L. monocytogenes*.

Nalidixic acid inhibits DNA synthesis of cells and was used as a selective agent for *Listeria* by Beerens and Tahon-Castel (1966), and subsequently in many formulations to inhibit Gram negative organisms. In most of these media it is combined with other inhibitory agents (Klinger, 1988; Prentice and Neaves, 1988).

Cycloheximide (actidione) inhibits protein synthesis in eukaryotic cells by binding to 80S rRNA (Obrig, 1971). In enrichment or isolation media it prevents the growth of most yeasts and moulds.

Many studies have shown that *L. innocua* can grow faster than *L. monocytogenes*, notably in non-selective brain heart infusion broth (Due and Schaffner, 1993). Nevertheless, Petran and Swanson (1993) concluded from their experiments with clinical and milk isolates of *L. monocytogenes* and *L. innocua* that growth rates of the two species were comparable. However, in two selective enrichment media, University of Vermont (UVM) broth and Fraser broth, *L. innocua* achieved significantly higher numbers than *L. monocytogenes* strains, approximately 1.0 log unit and 1.9 log unit, respectively. Similar results were obtained in impedance studies in *Listeria* enrichment broth (Oxoid) with strains isolated from raw cows' milk (McDonald and Sutherland, 1994). These authors also reported differences in generation times and suggested that these were probably due to variations in sensitivity to enrichment broth components and incubation procedures because, in non-selective tryptone soya broth (TSB), generation times were similar for both species. In experiments with samples inoculated with both *L. monocytogenes* and *L. innocua*, the latter was recovered from all samples, whereas the recovery rate of *L. monocytogenes* varied from 0–92%, depending on the ratio of the pathogen to *L. innocua* directly after inoculation (from 1:1 to 1200:1). Thus it is clear that the incidence of *L. monocytogenes* may well be underestimated, because enrichment appears to favour the growth of *L. innocua* (Curiale and Lewus, 1994).

Isolation media for *Listeria* spp.

Over the past 50 years there have been many attempts to produce selective agars for the isolation of listeria in general and *L. monocytogenes* in particular. Some of the more important media are summarized in Table 2. Most of the first group of agars differentiate *Listeria* spp. by means of aesculin hydrolysis which, in the presence of iron, forms a black phenolic compound derived from the aglucon. LPM may also be used with the addition of aesculin and iron or, without such additions, may be examined with the illumination system of Henry (1933) which distinguishes *Listeria* spp. from other organisms on the plate. Harlequin agar (Smith et al., 2001) uses a chromogen, CHE-glucoside, with ferrous gluconate producing a black pigment which is retained within *Listeria* colonies.

Until the 1990s it was not possible to differentiate *L. monocytogenes* from other species of *Listeria* by colonial appearance on the selective media specified in the then current standards (LPM, MOX, Oxford and PALCAM). The recommended practice

Table 2
Listeria selective agars.

	LiCl (g/l)	Polymyxin (P) or Colistin (C) (mg/l)	Acriflavine (mg/l)	Other (mg/l)	Indicator system*	References
Agars which do not differentiate *Listeria* spp.						
Harlequin listeria medium	15	C10	2.5	Fosfomycin 5, Cefotetan 1, Cycloheximide 200	CHEg + Fe	Smith et al., 2001
LPM†	5	0	0	Moxalactam 20, Glycine anhydride 10, Phenylethanol 2500	Henry	Lee & McClain, 1986
Oxford†	15	C20	5	Fosfomycin 10, Cefotetan 2, Cycloheximide 400	Ae + Fe	Curtis et al., 1989
Modified Oxford (MOX)†	12	C10	0	Ceftazidime 20	Ae + Fe	Cook, 1998
PALCAM†	10	P10	5	Ceftazidime 30	Ae + Fe, Mann + PR	van Netten et al., 1989
Agars which differentiate species by haemolysis						
EHA	10	P10	5	Ceftazidime 30	MUG, sheep blood	Cox et al., 1991b
LMBA	10	P10	0	Ceftazidime 20	sheep blood	Johansson, 1998
Agars which are specific for pathogenic *Listeria* spp.						
ALOA	10			Cycloheximide, Nalidixic acid	Chrom	Ottaviani et al., 1997
BCM LMPM		*composition not published*			Chrom	Restaino et al., 1999
Rapid' L.mono		*composition not published*			Chrom	Foret and Dorey, 1997

* Ae, aesculin; CHEg, CHE-glucoside; Chrom, chromogenic substrate; Fe, iron salt; Henry, Henry's (1933) method of illumination; Mann, mannitol; MUG, 4-methylumbelliferyl-β-D-glucoside; PR, phenol red.

† Monographs relating to these media are to be found in Part 2 of this book.

of picking five colonies from each plate for further identification does not guarantee the detection of this pathogen, especially when few *L. monocytogenes* colonies are present on the plate. Picking more colonies may reduce the number of false negative results but it is difficult to define how many additional colonies should be picked to ensure no positives are missed.

The development of blood-containing media allowed the separation of the haemolytic species (*L. monocytogenes*, *L. seeligeri* and *L. ivanovii*) from the non-haemolytic and non-pathogenic *L. innocua*, *L. grayi* and *L. welshimeri*. The most important differentiating features for the *Listeria* spp. commonly found in foods and food environments are haemolysis, the CAMP (Christie, Atkins and Munch-Petersen) reaction and xylose fermentation (Seeliger and Jones, 1986). These characteristics were used by Cox et al. (1991a; 1991b) in enhanced haemolysis agar (EHA), which was further improved by Beumer et al. (1997), to distinguish *L. monocytogenes* from other *Listeria* spp. on the basis of haemolysis. The major disadvantage of EHA is the expense of adding sheep blood, sphingomyelinase and 4-methylumbelliferyl-β-D-glucoside to a basal medium with PALCAM supplements. Johansson (1998) described a less expensive medium, *L. monocytogenes* blood agar (LMBA) which performed especially well after enrichment of the sample. It was superior to Oxford and PALCAM agars but a more selective medium would be required for the enumeration of *L. monocytogenes* from samples with a high level of competitive bacteria.

Recently the isolation of *L. monocytogenes* has been greatly improved by the use of chromogenic media such as 'Agar *Listeria* according to Ottaviani and Agosti' (ALOA; Ottaviani et al., 1997), BCM *L. monocytogenes* detection system (Restaino et al., 1999) and Rapid' L. mono^R (Foret and Dorey, 1997). ALOA agar contains the chromogenic compound X-glucoside, a substrate for the detection of β-glucosidase, common to all *Listeria,* which appear as blue coloured colonies. Differentiation of *L. monocytogenes* from other *Listeria* spp. is achieved through the production of a phosphotidylinositol-specific phospholipase C (PI-PLC) by *L. monocytogenes* which hydrolyses the specific purified substrate added to the medium producing an opaque halo around the colonies. Some *L. ivanovii* strains produce a vague halo after 48 h incubation. Selectivity is obtained by the addition of lithium chloride, nalidixic acid and cycloheximide (Vlaemynck et al., 2000). The BCM detection system includes pre-enrichment and selective enrichment broths and an *L. monocytogenes* selective/differential plating medium (LMPM). This agar medium uses a chromogenic substrate to detect PI-PLC which results in the formation of blue colonies by *L. monocytogenes* and *L. ivanovii*. Other *Listeria* spp., lacking the specific enzyme, produce white colonies (Restaino et al., 1999).

Rapid' L. mono agar plates also use the PI-PLC enzyme system but the manufacturers claim that only *L. monocytogenes* produces blue colonies on their product.

Several studies showed these chromogenic media to be superior to Oxford and PALCAM agars (Foret and Dorey, 1997; Restaino et al., 1999; Vlaemynck et al., 2000). We consider that their introduction would enhance the detection and reduce the time and cost of analyses for *L. monocytogenes*.

Methods

The rapid rate of progress in developing new media has led to frequent revisions of the regulatory and other recommended methods for the isolation of *L. monocytogenes*. The protocols of the FDA for dairy products (Lovett et al., 1987) and the USDA for meat (McClain and Lee, 1988) have both been superceded by new recommendations. That of the FDA (Hitchins, 1995) now includes phosphate buffer and a reduced acriflavine content in the LEB. Pyruvate is added to the enrichment broth and the selective agents are added after a preliminary 4 h incubation period to aid the recovery of damaged cells. When plating the enrichment broth, PALCAM may be used along with Oxford agar as an optional replacement for LPM, avoiding the necessity to use Henry's illumination technique when reading the plates. The USDA protocol (Cook, 1998) now requires the streaking of UVM I broth to MOX after 24 h and incubation of Fraser broth (which replaced UVM II) for 48 h if no colour change is observed at 24 h. In 1996 the ISO published a standard (ISO 11290–1) for the detection and enumeration of *L. monocytogenes* in food and animal feeding stuffs. This method was based on a two-stage enrichment procedure with streaking to Oxford and PALCAM agars from the primary enrichment (half Fraser broth) after one day of incubation and from secondary enrichment (Fraser broth) after two days incubation (Anonymous, 1996).

Extensive comparative studies encompassing methods current at the time (Warburton et al., 1991a; 1991b; 1992) together with the study of Beumer et al. (1996) indicated that the two-stage ISO procedure was satisfactory for the detection of *L. monocytogenes*. Owing to the great variety of food and feed products, this method may not be optimal for all types of sample. Moreover, the proposed streaking of the secondary enrichment broth after 48 h incubation appears unnecessary. Future studies with this method should include an isolation step after 24 h incubation of the secondary enrichment broth and also a differential medium for *L. monocytogenes*.

The International Dairy Federation (IDF) standard 143A: 1995 (Anonymous, 1995) uses a one-stage enrichment with 48 h incubation followed by streaking to Oxford and PALCAM agars. In a comparison of this method with the ISO method, Waak et al. (1999) demonstrated the superiority of the latter for the isolation of *L. monocytogenes* from blue veined cheese.

It has been reported that the predominant competitive microflora in enrichment media for *Listeria* spp. consists of lactobacilli, enterococci and Gram negative microorganisms, including pseudomonads (Duffy et al., 1994). Conflicting results have been reported about the effect of these bacteria on *Listeria* growth. Lactobacilli and enterococci may produce compounds such as bacteriocins, some of which are active against listerias, whereas the effect of pseudomonads is both stimulating and inhibitory (Farrag and Marth, 1989). Competition may be the result of the types of organisms, rather than the overall numbers present (Tran et al., 1990; Dallas et al., 1991; Duffy et al., 1994). Yokoyama et al., (1998) demonstrated an inhibitory effect of culture filtrates of *L. innocua* against strains of *L. monocytogenes*. They concluded that most *L. innocua* strains (114 out of 129 in their series) produce a trypsin sensitive bacteriocin-like substance. Curtis and Mitchell (1992) showed that 95 out of 97 *L. monocytogenes* strains were capable of producing monocins active against *L. ivanovii* and that monocins from serovars 1/2, 3 and 7 were active against most serovar 4 strains. However, it is not clear whether these compounds are produced in sufficiently high concentrations in enrichment media to inhibit growth of the susceptible strains.

The frequent changes in culture media formulations and recommended methods, together with the appearance of new media, often render comparative studies redundant before they have been published. It is perhaps significant that few such studies have been published in recent years. Where comparative studies have been reported, no attention was paid to the ratio of *L. monocytogenes* to other *Listeria* spp. in the sample and so information on this cause of under reporting of *L. monocytogenes* is not available.

The success of any protocol for the isolation of *Listeria* depends on four factors, the number and state of organisms in the sample, the selectivity of the media (a balance between inhibition of competitors and inhibition of target organisms), conditions of incubation (time, temperature, presence of oxygen) and the electivity of the isolation medium (ease of distinction between target organism and competitive microflora). The ability to isolate injured *Listeria* from food products or the production environment may also be an important factor (Dykes and Withers, 1999). However, some authors overemphasize the need for resuscitation, especially those who report the survival of heat-stressed cells in artificially contaminated products. The Hazard Analysis Critical Control Points (HACCP) system should be applied to control a pasteurization step, which means that heat-stressed cells should not be present in pasteurized products. The need for recovery of listeriae stressed by other factors such as low pH, presence of preservatives etc. is disputable. In most cases these cells are present due to post process contamination, and therefore may be adapted to the intrinsic parameters of the product. If so, these cells are not stressed, but are able to develop in the product and the shelf life should be short enough to ensure safety i.e. < 100 *L. monocytogenes* cfu/g at point of sale (Lund, 2000).

Several enrichment procedures have been described for the resuscitation of stressed cells using a non-selective broth (Greenwood et al., 1991; Bailey and Cox, 1992), time released addition of selective agents (Busch and Donnelly, 1992) or strict anaerobic incubation (Knabel et al., 1990). Comparisons of these non-selective broths with other enrichment procedures should be made to assess their performance with naturally contaminated products and determine the sensitivity. The use of multiple enrichment media or the investigation of larger samples increases *Listeria* detection (Pritchard and Donnelly, 1999; Duarte et al., 1999), but is more expensive in labour and materials.

The choice of methods for enumeration of *Listeria* depends on the sensitivity required. Where counts of < 100 cfu/g are expected it is necessary to use a Most Probable Number (MPN) technique; for counts above this level surface plating on a selective differential agar may be used. Methods can be found in Cook (1998) and Hitchins (1995). Some authors report that recovery after direct plating is poor compared with the MPN method (Hayes et al., 1991), however the MPN, though more sensitive and capable of adaptation to count sublethally injured cells, is less precise and undoubtedly more expensive. The presence of large numbers of competing microorganisms can sometimes overwhelm the growth of *Listeria* colonies. The addition of extra selective agents may prevent this, but at the price of failure to recover injured cells.

Sub-lethally injured cells may be recovered and enumerated by the procedure of Kang and Fung (1999) in which a thin layer of tryptone soya agar is poured onto a selective agar base immediately before inoculation. Recovery of the injured cells occurs before the diffusion of selective agents to the top layer is complete and listeria showing appearances typical of the selective/indicative agar used may be counted after the usual incubation period for the selective agar employed.

In summary, the current reference methods for the detection of *L. monocytogenes* as proposed by FDA, USDA, ISO and IDF allow the recovery of this pathogen from a variety of foods and food plant environments with relative ease (Asperger et al., 1999). More comparative studies, preferably with naturally contaminated samples, will be needed to select a horizontal method. The ISO procedure forms a good base for such comparisons which, with the improved differential media now available, should result in a simpler and speedier process for the detection of this pathogen.

References

Anonymous (1995) International Dairy Federation Standard 143A: 1995 Milk and Milk products. Detection of *Listeria monocytogenes*. International Dairy Federation, Brussels, Belgium.

Anonymous (1996) Microbiology of food and animal feeding stuffs – Horizontal method for the detection and enumeration of *Listeria monocytogenes*. International Standard ISO 11290-1, International Organization for Standardization, Geneva, Switzerland.

Asperger, H., Heistinger, H., Wagner, M., Lehner, A. and Brandl, E. (1999) A contribution of *Listeria* enrichment methodology – growth of *Listeria monocytogenes* under varying conditions concerning enrichment broth composition, cheese matrices and competing microbial flora. Food Microbiol. 16, 419–431.

Bailey, J.S. and Cox, N.A. (1992) Universal pre-enrichment broth for the simultaneous detection of *Salmonella* and *Listeria* in foods. J. Food Protect. 55, 256–259.

Beerens, H. and Tahon-Castel, M.M. (1966) Milieux à l'acide nalidixique pour l'isolement des streptocoques *D. pneumoniae, Listeria, Erysipelothrix*. Annal. Inst. Pasteur Paris 111, 90–93.

Beumer, R.R., Te Giffel, M.C., Anthony, S.V.R. and Cox, L.J. (1996) The effect of acriflavine and nalidixic acid on the growth of *Listeria* spp. in enrichment media. Food Microbiol. 7, 311–325.

Beumer, R.R., Te Giffel, M.C. and Cox, L.J. (1997) Optimization of haemolysis in enhanced haemolysis agar (EHA) - a selective medium for the isolation of *Listeria monocytogenes*. Lett. Appl. Microbiol. 24, 421–425.

Bille, J. (1990) Epidemiology of human listeriosis in Europe with special reference to the Swiss outbreak. *In:* A.J. Miller, J.L. Smith and G.A. Somtukti, (Eds). Foodborne Listeriosis. Elsevier, Amsterdam, pp. 71–74.

Bockemühl, J., Seeliger, H.P.R. and Kathke, R. (1971) Acridinfarbstoffe in Selektivnahrböden zur Isolierung van *Listeria monocytogenes*. Med. Microbiol. Immun. 157, 84–95.

Budvari, S., O'Neil, M.J., Snite, A. and Heckelman, P.E. (Eds). (1989) Merck Index 11th edn. Rahway, New York, NY.

Busch, S.V. and Donnelly, C.W. (1992) Development of a repair enrichment broth for resuscitation of heat-injured *Listeria monocytogenes* and *Listeria innocua*. Appl. Environ. Microbiol. 58, 14–20.

Cook, L.V. (1998) USDA/FSIS Microbiology Laboratory Guidebook, 3rd edn. Ch. 8 Isolation and identification of *Listeria monocytogenes* from red meat, poultry, egg and environmental samples (Revision 2; 11/08/99) USDA/FSIS, Washington, DC.

Corry, J.E.L., Curtis, G.D.W. and Baird, R.M. (Eds). (1995) Culture Media for Food Microbiology.

Part 2 – Pharmacopoeia of Culture Media. Elsevier, Amsterdam.

Cox, L.J., Siebenga, A., Pedrazzini, C. and Moreton, J. (1991a) Enhanced haemolysis agar (EHA) – an improved selective and differential medium for isolation of *Listeria monocytogenes*. Food Microbiol. 8, 37–49.

Cox, L.J., Siebenga, A. and Pedrazzini, C. (1991b) Performance of enhanced haemolysis agar compared to Oxford medium for the isolation of *Listeria monocytogenes* from food and food enrichment broths. Food Microbiol. 8, 51–62.

Curiale, M.S. and Lewus, C. (1994) Detection of *Listeria monocytogenes* in samples containing *Listeria innocua*. J. Food Protect. 57, 1048–1051.

Curtis, G.D.W. and Lee, W.H. (1995) Culture media and methods for the isolation of *Listeria monocytogenes*. *In:* Culture Media for Food Microbiology, J.E.L. Corry, G.D.W. Curtis and R.M. Baird (Eds). Elsevier, Amsterdam.

Curtis, G.D.W. and Mitchell, R.G. (1992) Bacteriocin (monocin) interactions among *Listeria monocytogenes* strains. Int. J. Food Microbiol. 16, 283–292.

Curtis, G.D.W., Mitchell, R.G., King, A.F. and Griffin, E.J. (1989) A selective differential medium for the isolation of *Listeria monocytogenes*. Lett. Appl. Microbiol. 8, 95–98.

Dallas, H.L., Tran, T.T., Poindexter, C.E. and Hitchins, A.D. (1991) Competition of food bacteria with *Listeria monocytogenes* during enrichment culture. J. Food Safety 11, 293–301.

De Vries, H. and Kroon, A.M. (1970) Euflavin and ethidium bromide; inhibitors of mitochondriogenesis in regenerating rat liver. FEBS Lett. 7, 347–350.

Donnelly, C.W. and Baigent, G.J. (1986) Method for flow cytometric detection of *Listeria monocytogenes* in milk. Appl. Environ. Microbiol. 52, 689–695.

Duarte, G., Vaz-Velho, M., Capell, C. and Gibbs, P. (1999) Efficiency of four secondary enrichment protocols in differentiation and isolation of *Listeria* spp. and *Listeria monocytogenes* from smoked fish processing chains. Int. J. Food Microbiol. 52, 163–168.

Due, Y.H. and Schaffner, D.W. (1993) Modelling the effect of temperature on the growth rate and lag time of *Listeria innocua* and *Listeria monocytogenes*. J. Food Protect. 56, 205–210.

Duffy, G., Sheridan, J.J., Buchanan, R.L., McDowell, D.A. and Blair, I.S. (1994) The effect of aeration, initial inoculum and the meat microflora on the growth kinetics of *Listeria monocytogenes* in selective enrichment broths. Food Microbiol. 11, 429–438.

Dykes, G.A. and Withers, K.M. (1999) Sub-lethal damage of *Listeria monocytogenes* after long-term chilled storage at 4°C. Lett. Appl. Microbiol. 28, 45–48.

Farrag, S.A. amd Marth, E.H. (1989) Behaviour of *Listeria monocytogenes* when incubated together with *Pseudomonas* species in tryptose broth at 7 and 13°C. J. Food Protect. 5, 536–539.

Foret, J. and Dorey, F. (1997) Evaluation d'un nouveau milieu de culture pour la recherche de *Listeria monocytogenes* dans le lait cru. Sci. Aliments 17, 219–225.

Fraser, J.A. and Sperber, W.H. (1988) Rapid detection of *Listeria* spp. in food and environmental samples by esculin hydrolysis. J. Food Protect. 51, 762–765.

Greenwood, M.H., Roberts, D. and Burden, P. (1991) The occurrence of *Listeria* spp. in milk and dairy products; a national survey in England and Wales. Int. J. Food Microbiol. 12, 197–206.

Hayes, P.S., Graves, L.M., Ajello, G.W., Swaminathan, B., Weaver, R.E., Wenger, J.D., Schuchat, A. and Broome, C.V. (1991) Comparison of cold enrichment and US Department of Agriculture methods for isolating *Listeria monocytogenes* from naturally contaminated foods. Appl. Environ. Microbiol. 57, 2109–2113.

Henry, B.S. (1933) Dissociation in the genus *Brucella*. J. Infect. Dis. 52, 374–402.

Hitchins, A.D. (1995) FDA Bacteriological Analytical Manual, 8th edn. Ch. 10 *Listeria monocytogenes*. AOAC International, Gaithersburg, MD.

Johansson, T. (1998) Enhanced detection and enumeration of *Listeria monocytogenes* from foodstuffs and food-processing environments. Int. J. Food Microbiol. 40, 77–85.

Kang, D-H and Fung, D.Y.C. (1999) Thin agar layer method for recovery of heat-injured *Listeria monocytogenes*. J. Food Protect. 62, 1346–1349.

Klinger, J.D. (1988) Isolation of *Listeria*: a review of procedures and future prospects. Infect. 16 (Suppl.2), S98–S105.

Knabel, S.J., Walker, H.W., Hartman, P.A. and Mendonca, A.F. (1990) Effects of growth temperature and strictly anaerobic recovery on the survival of *Listeria monocytogenes* during pasteurization. Appl. Environ. Microbiol. 56, 370–376.

Lee, W.H. and McClain, D. (1986) Improved *Listeria monocytogenes* selective agar. Appl. Environ. Microbiol. 52, 1215–1217.

Linnan, M.J., Mascola, L., Lou, X.D., Goulet, V., May, S., Salminen, C., Hird, D.W., Yonekura, L., Hayes, P., Weaver, R., Auduriier, A., Plikaytis, B.D., Fannin, S.L., Kleks, A. and Broome, C.V. (1988) Epidemic listeriosis associated with Mexican-style cheese. N. Engl. J. Med. 319, 823–828.

Lovett, J., Francis, D.W. and Hunt, J.M. (1987) *Listeria monocytogenes* in raw milk: detection, incidence and pathogenicity. J. Food Protect. 50, 188–192.

Lund, B.M. (2000) *Listeria monocytogenes* in ready-to-eat foods (Letter). Food Safety Express 4, 3–4.

McClain, D. and Lee, W.H. (1988) Development of USDA-FSIS method for isolation of *Listeria monocytogenes* from raw meat and poutry. J. Assoc. Off. Anal. Chem. 71, 660–664.

McDonald, F. and Sutherland, A.D. (1994) Important differences between generation times of *Listeria monocytogenes* and *Listeria innocua*. J. Dairy Res. 61, 433–436.

Meyer, R.R., Probst, G.S. and Keller, S.J. (1972) RNA synthesis by isolated mammalian mitochondria and nuclei: effects of ethidium bromide and acriflavin. Arch. Biochem. Biophys. 148, 425–430.

Obrig, T.G. (1971) The effect of actidione on eukaryotic cells. J. Biol. Chem. 246, 174.

Ottaviani, F., Ottaviani, M. and Agosti, M. (1997) Esperienza su un agar selettivo e differentiale per *Listeria monocytogenes*. Ind. Aliment. 36, 1–3.

Petran, R.L. and Swanson, K.M.J. (1993) Simultaneous growth of *Listeria monocytogenes* and *Listeria innocua*. J. Food Protect. 56, 616–618.

Prentice, G.A. and Neaves, P. (1988) *Listeria monocytogenes* in food: Its significance and methods for its detection. Bull. Int. Dairy Fed. 223, 3–16.

Pritchard, T.J. and Donnelly, C.W. (1999) Combined secondary enrichment of primary enrichment broths increases *Listeria* detection. J. Food Protect. 62, 532–535.

Ralovich, B., Forray, A., Mero, E., Malovics, H. and Szazados, I. (1971) New selective medium for isolation of *Listeria monocytogenes*. Zbl. Bakteriol. I. Abt. Orig. 216, 88–89.

Restaino, L., Frampton, E.W., Irbe, R.M., Schabert, G. and Spitz, H. (1999) Isolation and detection of *Listeria monocytogenes* using fluorogenic and chromogenic substrates for phosphatidylinositol-specific phospholipase C. J. Food Protect. 62, 244–251.

Schlech, W.F., Lavigne, P.M., Bortolussi, R.A.. Allen, A.C., Haldane, E.V., Wort, A.J., Hightower, A.W., Johnson, S.E., King, S.H., Nicholls, E.S. and Broome, C.V. (1983) Epidemic listeriosis: evidence for transmission by food. N. Engl. J. Med. 308, 203–206.

Seeliger, H.P.R. and Jones, D. (1986) The genus *Listeria. In:* Bergey's Manual of Systematic Bacteriology, Vol. 2 (Eds. P.H.A.Sneath, N.S. Mair, M.E.Sharpe and J.A.Holt) pp. 1235–1245. Williams and Wilkins, Baltimore, MD.

Smith, P., Mellors, D., Holroyd, A. and Gray, C. (2001) New chromogenic medium for the isolation of *Listeria* spp. Lett. Appl. Microbiol. 32, 78–82.

Tran, T.T., Stephenson, P. and Hitchins, A.D. (1990) The effect of aerobic mesophilic microfloral levels on the isolation of inoculated *Listeria monocytogenes* strain LM82 from selected foods. J. Food Safety 10, 267–275.

Van Netten, P., Perales, I., Van de Moosdijk, A., Curtis, G.D.W. and Mossel, D.A.A. (1989) Liquid and solid selective differential media for the detection and the enumeration of *Listeria monocytogenes* and other *Listeria* spp. Int. J. Food Microbiol. 8, 299–316.

Vlaemynck, G., Lafarge, V. and Scotter, S. (2000) Improvement of the detection of *Listeria monocytogenes* by the application of ALOA, a diagnostic, chromogenic isolation medium. J. Appl. Microbiol. 88, 430–441.

Waak, E., Tham, W. and Danielsson-Tham, M.L. (1999) Comparison of the ISO and IDF methods for detection of *Listeria monocytogenes* in blue veined cheese. Int. Dairy J. 9, 149–155.

Warburton, D.W., Farber, J.M., Armstrong, A., Caldeira, R., Tiwari, N.P., Babiuk, T., LaCasse, P. and Reed, S. (1991a) A Canadian comparative study of modified versions of the 'FDA' and 'USDA' methods for the detection of *Listeria monocytogenes*. J. Food Protect. 54, 669–676.

Warburton, D.W., Farber, J.M., Armstrong, A., Caldeira, R., Hunt, T., Messier, S., Plante, R., Tiwari, N.P. and Vinet, J. (1991b) A comparative study of the 'FDA' and 'USDA' methods for the detection of *Listeria monocytogenes* in foods. Int. J. Food Microbiol. 13, 105–118.

Warburton, D.W., Farber, J.M., Powell, C., Tiwari, N.P., Read, S., Plante, R., Babiuk, T., Laffey, P., Kauri, T., Mayers, P., Champagne, M-J., Hunt, T., LaCasse, P., Viet, K., Smando, R. and Coates, F. (1992) Comparison of methods for optimum detection of stressed and low levels of *Listeria monocytogenes*. Food Microbiol. 9, 127–145.

Yokoyama, E., Maruyama, S. and Katssube, Y. (1998) Production of bacteriocin-like-substance by *Listeria innocua* against *Listeria monocytogenes*. Int. J. Food Microbiol. 40, 133–137.

Chapter 6

Media used in the detection and enumeration of *Staphylococcus aureus*

Peter Zangerl[a] and Hans Asperger[b]

[a] *Federal Institute of Alpine Dairying BAM, Rotholz, Austria*
[b] *Institute of Milk Hygiene, Milk Technology and Food Science IMML, Veterinary University, Vienna, Austria*

The availability of highly productive and selective media for the detection and enumeration of *Staphylococcus aureus* is of great importance in routine food surveillance programmes. In this chapter recent developments in selective and diagnostic media, especially for stressed *S. aureus* cells and for food samples with high amounts of accompanying flora, are reviewed and an overview of media recommended by international and national standard organisations is given.

Baird-Parker agar is widely accepted as the most satisfactory medium for the enumeration of coagulase-positive staphylococci in foods. Productivity is high especially in the case of stressed cells. However, tellurite reduction and egg yolk reaction are poor diagnostic systems for many foodstuffs and the medium is not completely selective. This may cause unreliable results in cases of samples with high levels of concomitant flora. Attempts have therefore been made to replace the egg yolk by animal plasma in order to differentiate coagulase-positive staphylococci from other competing micro-organisms directly on the plate. Results show that commercially available rabbit plasma fibrinogen agar is a suitable medium for enumerating coagulase-positive staphylococci in foods without the need for cumbersome and time-consuming confirmatory tests. Baird-Parker agar and rabbit plasma fibrinogen medium are recommended by the International Organisation for Standardisation (ISO). Baird-Parker agar is used also in the official AOAC method in the United States. Selective liquid media that use MPN techniques are available for detecting low numbers of coagulase-positive staphylococci (<100 CFU/g). Studies showed that in most samples recoveries of *S. aureus* were similar after enrichment in tryptone soya broth with 10% NaCl and 1% sodium pyruvate (PTSBS), Giolitti-Cantoni broth with Tween 80 (GCBTw) and liquid Baird-Parker (LBP). However, significant differences in selectivity have been observed. Highest numbers of competing flora were observed in PTSBS, thus leading to problems in identifying coagulase-positive staphylococci when streaked onto Baird-Parker agar. Selectivity was superior in LBP than in GCBTw. While PTSBS is used in the official AOAC method, ISO favours the use of GCBTw.

Introduction: The rationale of *S. aureus* methodology

Staphylococcal food poisoning is one of the most common types of foodborne disease around the world. About 20 to 40 % of foodborne illnesses are attributed to the ingestion of stapylococcal enterotoxins (SEs) (Bergdoll, 1983). The enterotoxins are single-chain, low-molecular-weight (25,000–30,000 Da), heat stable proteins produced by some species of staphylococci, primarily *S. aureus*. The pathogenicity of *S. aureus* has been recognised for many years, but other coagulase-positive staphylococci – *S. intermedius* (Hirooka et al., 1988) and coagulase-positive strains of *S. hyicus* (Valle et al., 1990) – may also produce enterotoxins. Furthermore, coagulase-negative staphylococci have also been shown to produce toxins (Genigeorgis, 1989; Valle et al., 1990, Chou and Chen, 1997). Outbreaks due to enterotoxin A producing *S. intermedius* (Bennett, 1996) and *S. epidermidis* (Breckenridge and Bergdoll, 1971) have been reported. Adesiyun et al. (1984) detected enterotoxin production by the coagulase-variable *S. hyicus* (3 strains) and the coagulase-negative *S. chromogenes* (2 strains) when the monkey feeding test was applied, but the toxins did not belong to the then recognised serotypes (A-E).

Ten serologically distinct SEs (SEA, SEB, SEC_1, SEC_2, SEC_3, SED, SEE, SEG, SEH and SEI) have been recognised to date. The enterotoxins G, H and I have been isolated and characterised recently (Su and Wong, 1995; Munson et al., 1998). An SEH producing *S. aureus* strain was found to be the cause of an outbreak of food poisoning resulting from the consumption of cheese (Pereira et al., 1996). Staphylococcal food poisoning is mostly associated with SEA and SED (Halpin-Dohnalek and Marth, 1989; Baird-Parker, 1990). Enterotoxins can be formed in food when enterotoxigenic staphylococci are able to proliferate up to a level of 10^6–10^8 CFU/g food. The amount of enterotoxin required to produce illness varies from 0.1 to 1 μg (Bergdoll, 1990).

Staphylococcal enterotoxicosis is mostly transmitted by foods with neutral pH and reduced water activity, such as hams, cured meats, sweet cream and cream-filled bakery goods (Mossel and van Netten, 1990). The reduced a_w of these products inhibits most Gram-negative rods and other spoilage bacteria, e.g. pseudomonads and lactobacilli, whereas *S. aureus* is able to proliferate under these conditions. Spoilage organisms limit not only the growth of staphylococci by competition but also produce 'warning signals' such as changes in colour, odour, texture or taste, before staphylococci form clinically significant amounts of toxin (Mossel and van Netten, 1990).

As already mentioned, a variety of coagulase-positive and coagulase-negative staphylococci are able to produce enterotoxins, but *S. aureus* still plays the predominant role in staphylococcal food poisoning and therefore its detection and enumeration should provide information about the potential health hazard of a food. Furthermore, the presence of these micro-organisms in heated or pre-cooked ready-to-eat foods – even in low numbers – is an indicator of poor hygiene, since *S. aureus* is inactivated by pasteurisation. Low numbers of *S. aureus* in food are generally not considered hazardous, but conditions for further growth may exist and, consequently low numbers may become dangerous, e.g. presence of *S. aureus* in milk powder for babies in dietetic food. On the other hand low numbers in a product, e.g. cheese, may be the result of a dying off process. In this case not only the content of viable *S. aureus* is of interest, but also the possible content of enterotoxins. The following method is recommended for assessing the health hazard of a food:

- Enumeration of coagulase-positive staphylococci
- Detection of thermonuclease in food and detection of enterotoxins in thermonuclease positive samples

- In the case of high *S. aureus* counts, assessment of the enterotoxigenicity of isolated strains.

If a food is suspected as the cause of a staphylococcal food poisoning outbreak, Bennett (1998) recommended the enumeration of total staphylococci and subsequent detection of the enterotoxigenicity of isolated strains in order to estimate the number of enterotoxigenic staphylococci.

Identification of *S. aureus* is based on the coagulase reaction. However, *S. intermedius* and some strains of *S. hyicus* are also coagulase producers and may also produce enterotoxins. Therefore the ISO (1999 a, b, c) and the International Dairy Federation (IDF; 1997 a, b) have broadened the scope of their standards to include all coagulase-positive staphylococci.

Resuscitation of *S. aureus*

Various intrinsic, processing or extrinsic factors may lead to sublethally injured cells in food. In many cases cell injury leads to underestimation of micro-organisms when using selective media because these cells are inhibited by the selective components of the media which they tolerate in normal conditions. Nevertheless, stressed cells are able to repair injury and subsequently may grow and produce enterotoxins when stress is eliminated or cells are removed from stress-producing environments (Smith et al., 1983). Therefore it is often necessary to apply a repair treatment, known as resuscitation. According to the destruction repair curve presented by Mossel and Corry (1977), a distinction can be made between repair and subsequent multiplication of repaired cells. Both may occur in liquid media and therefore this technique is applicable for presence / absence tests but not for colony counting procedures. For this purpose solid medium repair can be used. In the ISO/TC 34/ SC 9 (ISO, 1981) expert group the need for resuscitation in standard methods for the microbiological examination of foods was stressed and it was concluded that the inclusion of a resuscitation step should be considered for all existing methods and for all methods in development.

Flowers and Ordal (1979) published an overview of current methods to detect stressed staphylococci. Sublethally damaged cells have an increased sensitivity to secondary stress e.g. that caused by selective components of the media. Staphylococci have been shown to be hypersensitive to sodium chloride, tellurite and/or polymyxin used in both solid isolation media and liquid enrichment broths. Up to this time Baird-Parker medium was considered as suitable for the recovery of stressed cells. Possible repair mechanisms for injured cells include the following: (a) an appropriate preincubation period in tryptone soya broth (TSB). For this technique the conditions must be carefully evaluated (time, temperature). (b) preincubation of stressed cells by surface plating on a non-selective medium and then overlaying with an appropriate amount of selective medium. (c) supplementation of the selective media with substances which are able to decompose hydrogen peroxide, i.e. catalase or sodium pyruvate, used in many media to increase productivity.

Several experiments confirmed these proposals. In an IDF/ICMSF methods study the recovery of coagulase-positive staphylococci from dried milk was examined

Table 1
Requirements for media used for the detection and enumeration of *S. aureus* in foods.

•	High selectivity	→ Suppressing accompanying flora as completely as possible
•	High productivity	→ Inhibition of *S. aureus* as little as possible
•	High electivity	→ Easy differentiation between *S. aureus* and accompanying flora
•	Low toxicity	→ No health hazard to workers, no production of hazardous waste
•	Practicability	→ Low on material and time costs

(Chopin et al., 1985). The estimates achieved in Giolitti-Cantoni broth supplemented with Tween 80 (GCBTw) were not different from direct plating on Baird-Parker agar (BPA) but were significantly superior to Giolitti-Cantoni broth without Tween (GCB). Lancette (1986), Lancette et al. (1986) and Lancette and Lanier (1987) verified the higher productivity for heat stressed cells when 1% sodium pyruvate was added to trypticase soy broth (TSB) + 10% NaCl (TSBS). According to van Netten et al. (1990) heat stressed staphylococci could be quantitatively detected on BPA but this was not the case with acid stressed staphylococci. In order to repair the acid stress, six hours incubation at 23°C on a tryptic soya yeast extract egg yolk pyruvate agar was necessary. Sabnis and Kulkarni (1992) proposed a "liquid repair method" for staphylococci after sublethal heating, freezing, freeze drying or osmotic shock process. The incubation of the injured cultures for two hours in brain heart infusion broth (BHI) before streaking on BHI agar or Baird-Parker agar resulted in significantly higher counts compared to MPN estimates in TSBS or colony counts on BPA. Assis et al. (1995) reported that lactic acid concentrations of 0.7 – 0.8% in cheese caused acid injury of *S. aureus*. Between the traditional *S. aureus* counting method on BPA and the method recommended by APHA (liquid media repair system) for the recovery of injured cells, a difference of more than 1.3 log units was detected. Large variations in staphylococcal counts in heat processed food samples such as infant food and dried skim milk were observed by Iyothirmayi et al. (1998). Recovery of staphylococci by direct plating on BPA was low in comparison with the liquid resuscitation using TSB, TSB with 1% sodium pyruvate and BHI.

Selective plating media

For the enumeration of *S. aureus* in food samples a variety of selective agar media has been developed. Table 1 summarises the requirements for a medium for enumerating and/or detecting coagulase-positive staphylococci in foods.

Most of the media used in the past (mannitol salt agar, Vogel Johnson agar, modified Vogel Johnson agar and KRANEP agar) lack productivity and electivity and are only of historical interest today. Baird-Parker agar was found to be one of the most satisfactory media available for enumeration of *S. aureus* in foods (Gilmour and Harvey, 1990) and is the medium now recommended by international standard organisations. However, it is not completely selective and detection of *S. aureus* is limited if the sample is heavily contaminated with competing micro-organisms. Attempts were

Table 2
Selective and diagnostic systems used in agar media for the isolation of *S. aureus*.

Agar medium	Selective system	Diagnostic system	Reference
Baird-Parker agar (BPA)	Potassium tellurite Lithium chloride Glycine	Potassium tellurite Egg yolk	Baird-Parker (1962)
Baird-Parker agar with acriflavine, polymyxins and sulphonamide (DEVA)	Potassium tellurite Lithium chloride Glycine Acriflavine Polymyxin B sulphate Sodium sulphadimidine	Potassium tellurite Egg yolk	Devriese (1981)
Modified Baird-Parker thermonuclease agar (BPTNA)	Potassium tellurite Lithium chloride Glycine	Potassium tellurite DNA-toluidineblue	Lachica (1984)
Baird-Parker agar with pig plasma (BPPA)	Potassium tellurite Lithium chloride Glycine	Potassium tellurite Pig plasma	Devoyod et al. (1976) Stadhouders et al. (1976)
Pig plasma fibrinogen agar (PPFA)	Potassium tellurite Lithium chloride Glycine	Potassium tellurite Pig plasma Bovine fibrinogen	Hauschild et al. (1979)
Rabbit plasma fibrinogen agar (RPFA)	Potassium tellurite Lithium chloride Glycine	Potassium tellurite Rabbit plasma Bovine fibrinogen	Beckers et al. (1984)
Coagulase thermonuclease agar (CTNA)	Potassium tellurite Lithium chloride Glycine	Potassium tellurite Rabbit plasma Bovine fibrinogen DNA- toluidineblue	SLMB (1988)

made to improve selectivity by addition of antimicrobials ("Devriese agar") and electivity was improved by omitting the egg yolk and using thermonuclease and/or coagulase activity instead of the egg yolk reaction as a diagnostic tool. The modifications of Baird-Parker agar are given in Table 2.

Baird-Parker agar (BPA)

The medium contains glycine, lithium and tellurite as selective substances and pyruvate to increase productivity in enumerating stressed cells. The theory that egg yolk also assists in the recovery of damaged cells (Andrews and Martin, 1978; Lachica, 1984) could not be confirmed by Terplan et al. (1982). Since the medium is not completely selective, reduction of tellurite and production of an egg yolk reaction serve as diagnostic systems to differentiate *S. aureus* from other micro-organisms. Typical *S. aureus* colonies are black, shiny and convex, frequently with light-coloured (off-white) margin. The egg yolk reaction gives rise to an opaque zone around the colony, frequently with an outer clear zone. Unfortunately strains from various foods and dairy products also form atypical dark-grey to grey or brown-grey colonies (Stad-

Table 3
Mean log counts/ml of *E. faecalis* and *E. faecium* on three different brands of Baird-Parker agar base (n=2) after cultivation in M17 broth for 24 h at 37°C (Zangerl, 1998).

Strain	Oxoid CM 275	Biokar BK 055	bioMérieux 44003
E. faecalis R 9897	9.3	4.6	5.6
E. faecium ATCC 19434	9.5[a]	9.4[b]	9.4[c]

[a] colony growth after 24 h; colony size after 48 h: 1–1.5 mm.
[b] colony growth only after 48 h; colony size: 0.5 mm.
[c] colony growth only after 48 h; colony size: 1 mm.

houders et al., 1976; Becker et al., 1983; Lancette, 1986; Bennett and Lancette, 1995), frequently without any egg yolk reaction. Only 10 to 50% of *S. aureus* strains isolated from milk and cheese samples yielded a positive egg yolk reaction (Untermann et al., 1973; Stadhouders et al., 1976; Lachica, 1980; Harvey and Gilmour, 1985; Becker et al., 1989; Umeki et al., 1993; Matsunaga et al., 1993; Mueller, 1996). Of 254 strains of coagulase-positive staphylococci isolated from milk and dairy products only 22.4% formed jet black shiny colonies on BPA and 45.7% showed a positive egg yolk reaction (black to grey colonies with a clear halo and/or an opaque zone around the colony on BPA after 48 h at 37°C). It was also shown that the accompanying flora, e.g. coagulase-negative staphylococci, micrococci, enterococci, aerobic sporeformers, *Proteus* spp., can also reduce tellurite (Terplan et al., 1982) and many egg yolk-positive non-*S. aureus* strains have also been isolated (Becker et al., 1983). For that reason reduction of tellurite and the egg yolk reaction are poor diagnostic systems. Consequently all kinds of black to grey colony types, irrespective of egg yolk reaction, must be examined for coagulase production. Owing to this lack of electivity, the estimated *S. aureus* count in food could differ considerably from the real count in which *S. aureus* is only a minor part of the total flora of Gram-positive cocci.

Today a variety of commercially available media are in use. In these media the sterilised Baird-Parker agar bases are supplemented with egg yolk-tellurite emulsion. Although the various brands do not differ in their quantitative composition of basic components, distinct differences in selectivity of the unsupplemented base media have been observed. Table 3 shows the log counts of two strains of enterococci after surface plating on three different brands of agar base without supplements. The colony counts of the *E. faecalis* strain (an isolate from cooked hard cheese) on Biokar and on bioMérieux base were 4.7 and 3.7 log units lower than on Oxoid base. *E. faecium* yielded the same colony count after 48 h incubation (> 9 log CFU/ml) but on Biokar and bioMérieux base delayed growth and smaller colonies were observed compared to Oxoid base. The results indicate that differences in selectivity could be related to the peptones used in the different brands. The quality of the peptones probably influences the inhibitory effect of lithium on competitive flora (Miller, D., personal communication).

Baird-Parker agar supplemented with acriflavine, polymyxins and sulfonamide ('Devriese agar' – DEVA)

Devriese (1981) increased the selectivity of Baird-Parker agar by adding acriflavine, polymyxin B sulphate and sodium sulphamezathine. The high selectivity was confirmed by other authors (Becker et al., 1983; Harvey and Gilmour, 1985; White et al., 1988). Unfortunately recoveries of *S. aureus* from raw milk were significantly lower on DEVA than those obtained on Baird-Parker agar or Baird-Parker agar supplemented with pig plasma (Harvey and Gilmour, 1985). Also Becker et al. (1983) found lower recoveries of thermally stressed cells compared to Baird-Parker agar and pig plasma agar, but with unstressed cells no differences in productivity were established. Ollis et al. (1995) modified Baird-Parker and rabbit plasma fibrinogen agars by the incorporation of acriflavine in a lower concentration (5 instead of 7 μg/ml) for the examination of raw milk samples.

Modified Baird-Parker thermonuclease agar (BPTNA)

In this modified Baird-Parker agar Lachica (1984) replaced the egg yolk by Tween 80 (0.5 g/l) and $MgCl_2.6H_2O$ (1 g/l) to increase productivity. As a diagnostic system, the thermonuclease test is used instead of the egg yolk reaction. After incubation of the inoculated plates *S. aureus* is identified by overlaying the plates with 10 ml of molten toluidine-blue DNA agar and the thermonuclease reaction is read after 3 h at 37°C (Lachica, 1980). Before overlaying, plates with grown colonies are incubated for 2 h at 60°C in order to inactivate heat labile nucleases. Compared to Baird-Parker agar Lachica (1984) found similar recoveries of acid stressed cells when analysing naturally contaminated cheese samples. The procedure does not require cumbersome confirmatory tests and the result is available after 29 h. However, other Gram-positive cocci may also produce thermonuclease causing false positive results (Becker et al., 1984).

Baird-Parker agar supplemented with animal plasma (Baird-Parker pig plasma agar (BPPA); pig plasma fibrinogen agar (PPFA); rabbit plasma fibrinogen agar (RPFA); coagulase thermonuclease agar (CTNA))

Baird-Parker pig plasma agar (BPPA); pig plasma fibrinogen agar (PPFA); rabbit plasma fibrinogen agar (RPFA); coagulase thermonuclease agar (CTNA)

Baird-Parker agar requires time consuming confirmatory tests. Attempts were therefore made to use the coagulase reaction as a diagnostic system by replacing egg yolk in Baird-Parker agar with animal plasma. Such media can differentiate coagulase-positive staphylococci from other micro-organisms directly on the plate due to the formation of haloes of precipitated fibrin around the colonies. Development of plasma media has been lengthy; initially egg yolk was replaced by pig plasma (Devoyod et al., 1976; Stadhouders et al., 1976). Hauschild et al. (1979) improved the medium by including bovine fibrinogen and trypsin inhibitor. Addition of fibrinogen produced more distinct fibrin haloes whilst trypsin inhibitor was necessary to avoid consider-

Table 4
Number of samples positive for coagulase-positive staphylococci[a] (Zangerl, 1999).

Product	n	BPA	RPFA
Milk	20	16 (80%)	20 (100%)
Cheese	36	18 (50%)	22 (61%)
Total	56	34 (61%)	42 (75%)

[a] Theoretical minimum detection level: 5 CFU/ml milk, 50 CFU/g cheese.

able fibrinolysis of precipitation haloes when incubating the plates for more than 24 h. Later Beckers et al. (1984) replaced pig plasma by rabbit plasma. On RPFA less false positive and false negative colonies were obtained than on PPFA. Sawhney (1986) noticed a significantly inferior growth of some test strains of *S. aureus* on RPFA when compared with that on BPA and found the cause to be the high concentration of tellurite. He postulated that the egg yolk in BPA had a protective effect on staphylococci against the inhibitory action of tellurite. Reduction of potassium tellurite concentration from 100 to 25 mg/l in RPFA led to a productivity equal to that of BPA. A modification of RPFA is used as the official method in Switzerland. The so called coagulase thermonuclease agar (SLMB, 1988) with reduced potassium tellurite concentration (73 mg/l) not only detects coagulase but also thermonuclease activity according to a slightly modified procedure of Lachica (1980). Nevertheless an IDF/ICMSF methods study (Chopin et al., 1985) observed that a test for thermonuclease is not necessary when RPFA is used in experienced hands.

Selectivity and productivity experiments and a series of comparative studies of naturally and artificially contaminated foods showed similar counts of *S. aureus* on plasma media compared to Baird-Parker agar (Stadhouders et al., 1976; Hauschild et al., 1979; Terplan et al., 1982; Beckers et al., 1984; Chopin et al., 1985; Harvey and Gilmour, 1985; Sawhney, 1986; Beckers et al., 1987). However, results are not directly comparable since different kinds of plasma and modifications of media were used. The pour plate method with rabbit plasma fibrinogen medium (Beckers et al., 1984) containing reduced tellurite concentration (Sawhney, 1986) is recommended by ISO and IDF for the enumeration of coagulase-positive staphylococci in samples heavily contaminated with competitive flora and when atypical *S. aureus* on BPA are expected, e.g. raw milk cheeses and certain raw meat products.

Plasma media have not been very successful in the past owing to variability in plasma quality giving unreliable results. Recently rabbit plasma fibrinogen media with constant plasma quality have become commercially available. The lyophilised supplement containing rabbit plasma, bovine fibrinogen, trypsin inhibitor and potassium tellurite, is simply added to Baird-Parker agar base after reconstitution. In evaluating the usefulness of this medium for products with high amounts of accompanying flora, 20 raw milk and 36 various ripened and unripened cheese samples made from raw milk were analysed in parallel by surface plating using BPA (Oxoid) and RPFA (Biokar Diagnostics) as selective media (Zangerl, 1999). RPFA allowed easy differentiation of coagulase-positive colonies even on plates with high numbers of accompanying colonies. On BPA a considerably higher amount of competing flora was observed than on RPFA. Consequently, *S. aureus* was detected in more samples on RPFA than with BPA (Table 4). For 31% of the cheese samples tested it could not be accurately determined whether the EU limit for raw milk cheeses of m=1,000 *S. aureus*/g was met due to the intense growth of competing micro-organisms on BPA. In contrast, results were obtained with all cheese samples using RPFA.

In cases where *S. aureus* counts were obtained with both BPA and RPFA (32 out of 56 samples), no significant differences in counts could be detected ($p>0.05$). The relatively high residual standard deviation, $s_{yx} = 0.32$ log units in regression analysis, can be attributed to the uncertainty of the estimation on BPA in the presence of high counts of accompanying flora. Similar results were reported by de Buyser et al. (1998). They also found no statistically significant differences between *S. aureus* counts obtained on two brands of RPFA and BPA ($p > 0.05$) in 57 cheese samples made from raw milk but on RPFA the concomitant bacteria were much more inhibited than on BPA. Since they used the poured plate method with RPFA medium they assumed, in accordance with Beckers et al. (1984), that the reduced oxygen content in poured RPFA plates inhibited aerobic micrococci unlike the surface plating method on BPA. However, this does not explain the distinct differences in selectivity, since in our investigation surface plating was used both with RPFA and BPA in order to achieve equal incubation conditions. Lachica (1984) assumed that the absence of a rich supplement such as egg yolk reduced the outgrowth of competing micro-organisms in egg yolk free media, but our results show that differences in selectivity were primarily attributable to different qualities of Baird-Parker agar bases (Table 3). Experiments with naturally contaminated raw milk cheeses confirmed these findings. Total counts of *S. aureus* and competitive bacteria from six cheese samples were made on Oxoid and Biokar agar bases supplemented with egg yolk-tellurite emulsion (Oxoid) and RPF supplement (Biokar Diagnostics), respectively. Irrespective of the kind of supplement, five out of the six samples yielded counts which were 1–3 log units higher on Oxoid media than on Biokar media (Figures 1 and 2).

Growth intensity also influenced recognition of fibrin haloes. On RPFA with Oxoid base no haloes could be detected after 48 h incubation, whereas distinct haloes were formed with Biokar base (Figure 2). De Buyser et al. (1998) also found differences in selectivity of Biokar Diagnostics RPFA compared to Oxoid RPFA and observed reading difficulties with the latter due to masking of fibrin haloes by fibrinolytic flora. After adjusting the pH of the medium from 6.8 to 7.5, this problem was no longer observed with Oxoid RPFA.

Some manufactures of media recommend 24 h incubation at 37°C. A comparison of log *S. aureus* counts after 24 and 48 h yielded statistically significant differences ($p < 0.0001$). In this experiment 91 raw milk and cheese samples made from raw milk were surface plated on RPFA (Biokar Diagnostics). On average, the counts of 87 samples were only 0.07 log units higher after 48 h than after 24 h. However, at 48 h 12% of the samples showed counts at least twice as high as after 24 h incubation. In four further samples entire fibrin haloes were only visible after 48 h. Therefore counting of colonies with haloes after 24 and 48 h is recommended.

Thus RPFA provides more reliable results and is less labour and time consuming than BPA especially when food samples with high counts of competitive bacteria are

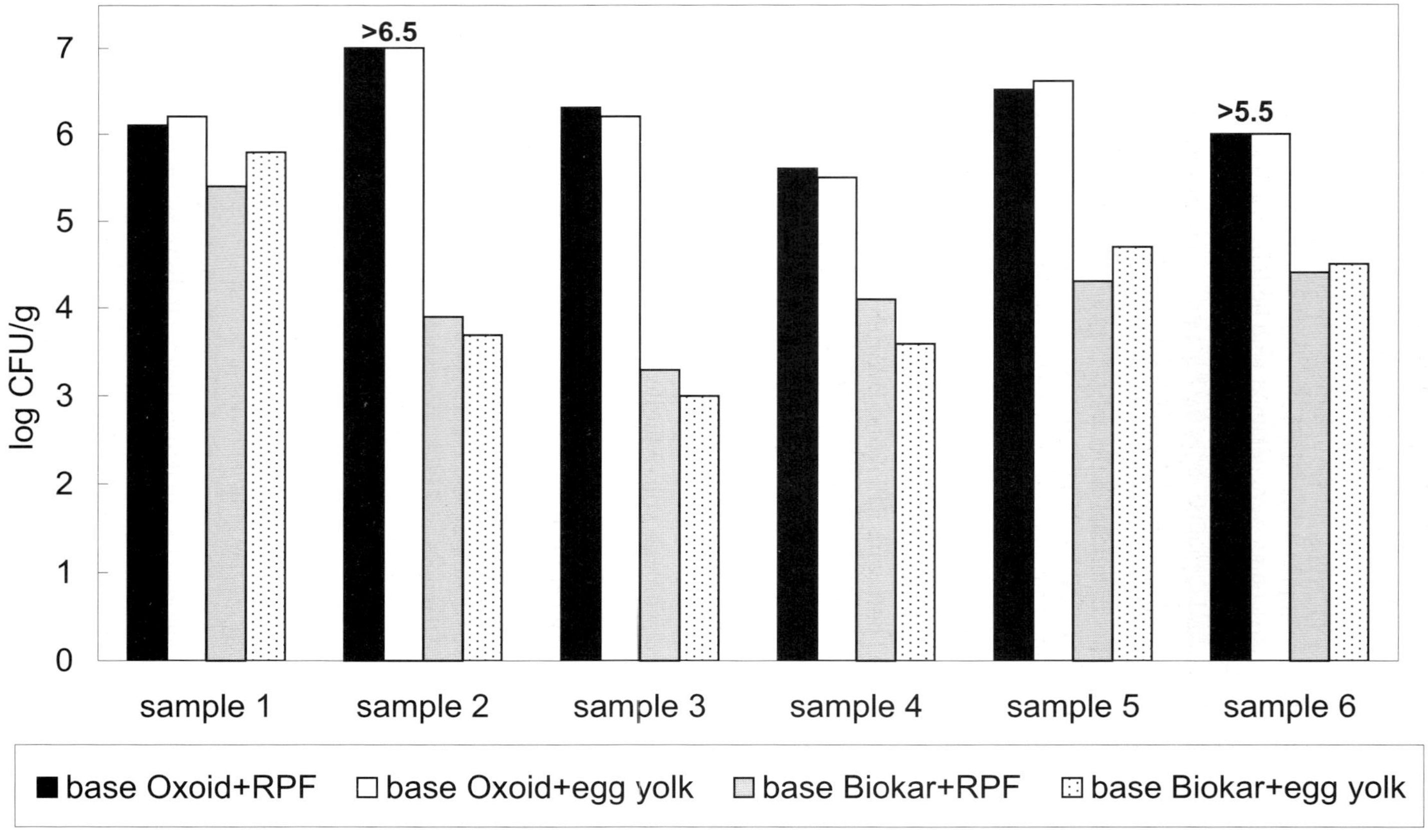

Fig. 1. Influence of agar base brands on total CFU of BPA and RPFA (log CFU/g of six cheeses made from raw milk, 48 h at 37°C) (Zangerl, 1999).

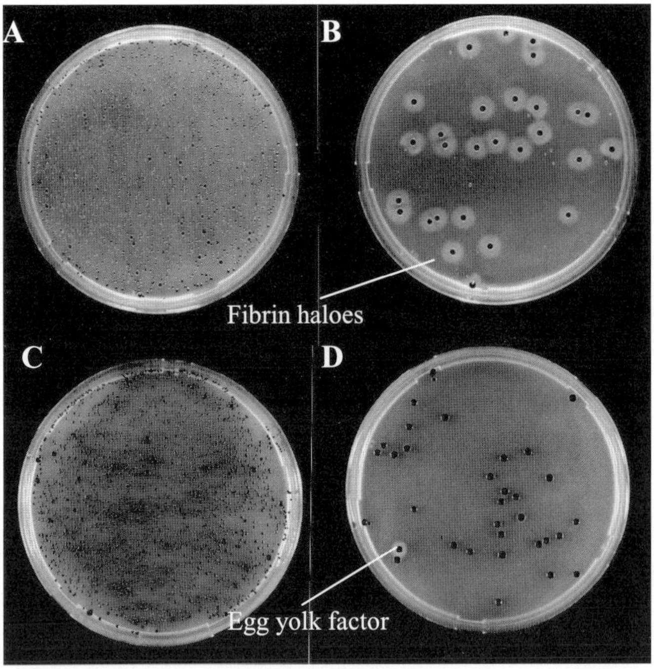

Fig. 2. Colony growth of sample No. 6 on BPA and RPFA using 2 brands of agar base (dilution 10^{-3}; 48 h; 37°C). A: base Oxoid+RPF. B: base Biokar+RPF. C: base Oxoid+egg yolk D: base Biokar+ egg yolk (Zangerl, 1998).

expected. Because of the random selection of colonies for coagulase testing on BPA, results obtained with RPFA are likely to be more accurate and precise than those with BPA. This was confirmed by Beckers et al. (1987) who found better precision parameters for RPFA compared to BPA. The outcome of a collaborative study initiated by CEN expert group TC 275 / WG 6 showed repeatability figures on BPA of r = 0.35 log compared to RPFA of r = 0.25 log and reproducibility figures for BPA of R = 0.59 log compared to RPFA of R = 0.36 log (unpublished results).

Selective liquid media for the detection of low numbers of *S. aureus*

For several foods, e.g. dairy and egg products, ice cream, baby food, information about low numbers of staphylococci (<100/g) is needed. For this reason presence / absence tests with a defined amount of food or most probable number (MPN) techniques are used. In both cases the methodology needs a liquid enrichment procedure as a first step. Moreover, the use of liquid media allows the recovery of stressed cells (known as liquid repair method). Non-selective liquid media were shown to be unsuitable for enrichment procedures, since *S. aureus* may be inhibited by competitive bacteria (Baird and van Doorne, 1982), and therefore selective media were formulated (Table 5).

Table 5
Selective and diagnostic systems used in enrichment media for *S. aureus*.

Enrichment broth	Selective system	Diagnostic system	Reference
Tryptone soya broth (TSB) with 10% sodium chloride (TSBS)	Sodium chloride		Buttiaux and Brogniart (1947) Baer (1966) Gilden et al. (1966)
Tryptone soya broth (TSB) with 10% sodium chloride and 1% sodium pyruvate (PTSBS)	Sodium chloride		Lancette et al. (1986)
Giolitti-Cantoni broth (GCB)	Potassium tellurite Lithium chloride Glycine Anaerobic incubation	Potassium tellurite	Giolitti and Cantoni (1966)
Giolitti-Cantoni broth with 1% Tween 80 (GCBTw)	Potassium tellurite Lithium chloride Glycine Anaerobic incubation	Potassium tellurite	Chopin et al. (1985)
Liquid Baird-Parker (LBP)	Potassium tellurite Lithium chloride Glycine Anaerobic incubation	Potassium tellurite	Baird and van Doorne (1982)

The first of these incorporated 10% sodium chloride as the selective agent in tryptone soya broth (TSBS). However, concentrations of sodium chloride >40 g/l are known to be inhibitory to sublethally injured cells (Iandolo and Ordal, 1966; Hurst et al., 1973; Gray et al., 1974; Smolka et al., 1974; Idziak and Mossel, 1980). In a modified medium (PTSBS) recoveries were markedly improved by supplementing TSBS with 1 % sodium pyruvate (Lancette, 1986; Lancette et al., 1986; Lancette and Lanier, 1987). Potassium tellurite and lithium chloride have also been employed as selective agents for enrichment media. Combined with anaerobic incubation these substances are used in Giolitti-Cantoni broth (GCB). However, poor recoveries of *S. aureus* were reported with GCB compared to other media (Becker et al., 1983; Chopin et al., 1985). An IDF/ICMSF collaborative study (Chopin et al., 1985) showed that productivity was enhanced by addition of Tween 80 (GCBTw). Reduction of tellurite causing blackening of the medium indicates the presence of *S. aureus*, but in this study *S. aureus* was also detected in non-black tubes. Therefore confirmation of all tubes irrespective of colour changes has to be performed. Baird and van Doorne (1982) introduced a liquid Baird-Parker medium (LBP) which did not contain egg yolk and was incubated anaerobically. In comparison with different isolation media they observed highest recoveries of *S. aureus* in LBP followed by streaking on BPA.

ISO working group ISO/TC 34/ SC 9 'food and agricultural products – microbiology' intends to establish a horizontal method for the detection of low numbers of coagulase-positive staphylococci. In the drafting stage comparative studies of three commonly used liquid media (PTSBS, GCBTw, LBP) were organised and the fol-

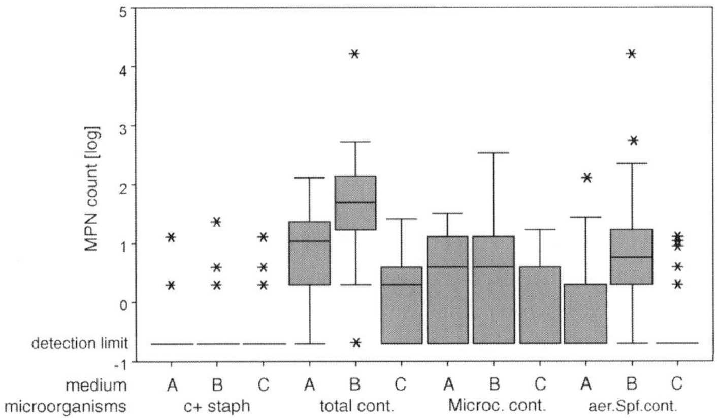

Fig. 3. Comparison of coagulase-postive staphylococci, total contaminants, *Micrococcaceae* contaminants and aerobic sporeformer contaminants in 25 ice cream samples enumerated in GCBTw (medium A), PTSBS (medium B) and LBP (medium C) given in Boxplot-figures (* results of single samples over the detection limit or outliers) (Asperger, unpublished).

lowing results were obtained (ISO, 1999c). At MAFF Central Science Laboratory (UK) 50 naturally contaminated food samples were analysed in parallel with PTSBS and LBP. Results indicated that the two media performed in a similar way. At the IMML Laboratory in Austria *S. aureus* counts of 65 naturally contaminated minced meat and ice cream samples were compared in a 5 tube MPN technique with PTSBS, GCBTw and LBP. After selective enrichment in the three different media (37°C, 48 h, anaerobic conditions in the case of GCBTw and LBP) a loopful of medium from every tube was streaked onto BPA and incubated 24–48 h at 37°C. Colonies were differentiated according to their appearance on BPA plates into micrococcaceae and aerobic sporeformers. The sum of both without *S. aureus* counts was regarded as 'total contaminants'. *S. aureus* was differentiated by the tube coagulase test from 'micrococcaceae' colonies. In ice cream samples the sporadic positive results for *S. aureus* were obtained on all three selective enrichment media, but there was a distinct difference in the background flora present. As shown in Figure 3 'total contaminants' were significantly higher in PTSBS than in GCBTw. The minimal accompanying flora was observed in LBP. The results of minced meat were surprising in some cases. Though in most samples *S. aureus* estimates were similar in all three enrichment media, PTSBS yielded higher *S. aureus* counts (1–2 log units) in 14 samples (35%). We have no explanation for these results. In all cases the background flora was distinctly more apparent in subcultures from PTSBS than in those from GCBTw. Again, a minimum of competing flora was observed when plating from LBP (Figure 4).

Results show that in most cases PTSBS, GCBTw and LBP performed similarly with respect to productivity. In contrast, distinct differences in selectivity were observed. In PTSBS high numbers of micrococcaceae and aerobic sporeformers caused problems in identifying coagulase-positive staphylococci on the following streak on BPA. Compared to PTSBS lower numbers of competitive bacteria were found in GCBTw and minimum background flora was observed in LBP.

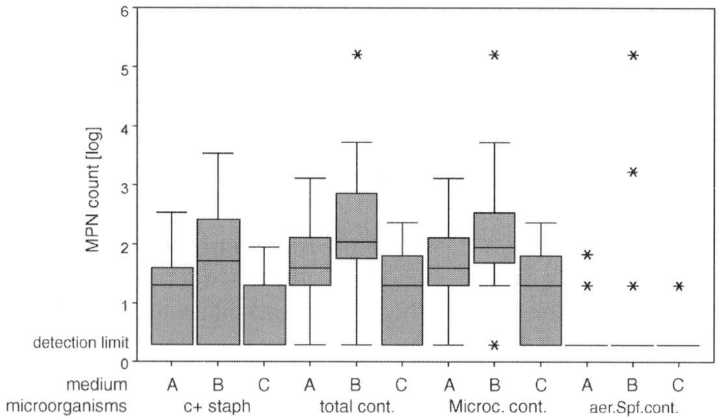

Fig. 4. Comparison of coagulase-positive staphylococci, total contaminants, *Micrococcaceae* contaminants and aerobic sporeformer contaminants in 40 minced meat samples enumerated in GCBTw (medium A), PTSBS (medium B) and LBP (medium C) given in Boxplot-figures (* results of single samples over the detection limit or outliers) (Asperger, unpublished).

Media used in various standard methods for the detection and/or enumeration of *S. aureus*

S. aureus / coagulase-positive staphylococci are very important criteria in routine food surveillance programmes and food poisoning investigations. There is thus a strong need for all relevant standards organisations to have highly productive and selective reference methods available with a precisely defined chapter for each relevant medium.

At present, ISO provides two standard methods for the enumeration of coagulase-positive staphylococci: ISO 6888-1 (1999a) describes the technique using Baird-Parker agar and ISO 6888-2 (1999b) describes the technique using rabbit plasma fibrinogen agar. In the working program a third standard method – ISO 6888-3 Part 3 – is envisaged for a MPN technique for low numbers. It can be expected that Giolitti-Cantoni broth supplemented with Tween 80 will be the enrichment medium of choice. According to the Vienna Agreement CEN (Commité Européen de Normalisation) takes over all horizontal ISO methods for enumeration and detection of coagulase-positive staphylococci in food as European methods.

The 8th edition of the 'Bacteriological Analytical Manual' details the official AOAC (Association of Official Analytical Chemists) methods and was prepared under the direction of the U.S. Food and Drug Administration (FDA). In this manual Baird-Parker agar is used for the direct plate count method and tryptic soy broth containing 10% NaCl and 1% sodium pyruvate (PTSBS) as liquid enrichment medium for MPN procedures (Bennett and Lancette, 1995).

IDF has a lengthy history in standardisation of methods for *S. aureus*. Even in the early seventies a standard method for the detection of *S. aureus* in milk powder was established. The actual versions have a broader scope for all milk and milk based products. Standard 60C (IDF, 1997a) is an MPN method using Giolitti-Cantoni broth with Tween 80 as enrichment medium. Depending on the detection limit a double-strength or single-strength composition is used. Standard 145A (IDF, 1997b) provides a colony counting procedure using either Baird-Parker agar or rabbit plasma fibrinogen medium.

The Nordic Committee on Food Analysis agreed on Method No. 66: 1992 – "*S. aureus* - determination in foods" (NMKL, 1992). This method is performed by surface plating on non-selective blood agar and selective Baird-Parker agar with an appropriate confirmation procedure for colonies from both media.

In Germany several vertical methods for the estimation of coagulase-positive staphylococci in various products are established in the collection of official methods under article 35 of the German Federal Foods Act. Again, Baird-Parker agar is used as the solid medium (DIN, 1988b). For MPN procedures liquid Baird-Parker is applied as enrichment medium in method L 02. 07-2. The procedure is equal to that given in DIN (1988a).

Summarizing it can be seen that a broad consensus exists with regard to choice of agar media. Baird-Parker agar with confirmation of suspect colonies is the medium of choice today. Recently rabbit plasma fibrinogen agar, allowing direct counting of the target micro-organisms, was introduced by ISO and IDF. For detecting low numbers, MPN techniques are used, but selective enrichment media differ in many countries. ISO, CEN, IDF favour Giolitti-Cantoni broth with Tween 80, while AOAC recommends tryptic soya broth containing 10% NaCl and 1% sodium pyruvate. Only in one country has the use of liquid Baird-Parker medium become an official method.

Rapid *S. aureus* identification methods

With the exception of plasma media, all other media lack diagnostic systems for the identification of *S. aureus*. Hence a representative number of colonies has to be taken for confirmatory tests. Identification is done by the tube coagulase test using rabbit plasma with EDTA after overnight incubation in BHI. Since the test takes at least 24 h, alternative methods such as detection of clumping factor, production of thermonuclease and protein A may need consideration. Detection of clumping factor (CF) in suspect colonies is most common and a variety of commercially available tests will detect CF in a few seconds. The German Standard (DIN, 1988b) allows detection of CF as an alternative method to the tube coagulase test, but CF-negative colonies must be submitted to the coagulase test, since CF-negative *S. aureus* strains may occur (Ibrahim and Redford, 1979; Becker et al., 1989). Thermonuclease activity is very sensitive but is not specific for *S. aureus* (Becker et al., 1984), therefore this test cannot be recommended as sole criterion for identification purposes. Compared to coagulase and thermonuclease, protein A was found to be a better marker of *S. aureus* (Chang and Huang, 1995). Recently latex agglutination tests based on CF and protein A detection (Chang and Huang, 1993; 1996) have been developed and become commercially available. They are recommended as rapid identification methods for *S. aureus* by AOAC (Bennett and Lancette, 1995).

For rapid detection and identification of *S. aureus*, enzyme-linked immunosorbent assay (ELISA) procedures (Poutrel and Sarradin, 1992; Chang and Huang, 1994; Bartlett et al., 1996, Hughes et al., 1999) and molecular biological methods such as DNA probes (Kneifel et al., 1992) and polymerase chain reaction (PCR) (Matthews and Oliver; 1994; Matthews et al., 1997; Kahn et al., 1998) may be used.

Conclusions

Baird-Parker agar is the medium of choice for direct plating and enumeration of *S. aureus* today. However, it is not completely selective and reduction of tellurite and egg yolk reaction are poor diagnostic tools. When analysing food samples likely to contain atypical *S. aureus*, i.e. strains without egg yolk factor, or in cases when high amounts of competing micro-organisms are to be expected, rabbit plasma fibrinogen agar provides more reliable results and is less time consuming than Baird-Parker agar. In the analysis of raw milk cheeses or raw meat products usually at least four different suspect colony types appear on BPA, thus approximately 20 to 30 colonies must be submitted to confirmatory tests. Costs of analysis are therefore reduced considerably when using RPFA, although commercially available plasma media are more expensive than Baird-Parker agar. Both media are recommended by ISO and IDF. Since different brands of the media show differences in selectivity, the suitability of commercially prepared media must be checked before use. When liquid media are used, Giolitti-Cantoni broth containing 1% Tween 80 and liquid Baird-Parker seem to be the most promising with regard to productivity and selectivity.

References

Adesiyun, A.A., Tatini, S.R. and Hoover, D.G. (1984) Production of enterotoxin(s) by *Staphylococcus hyicus*. Vet. Microbiol. 9, 487–495.

Andrews, G.P. and Martin, S.E. (1978) Modified Vogel and Johnson Agar for *Staphylococcus aureus*. J. Food Protect. 41, 530–532.

Assis, E.M., Carvalho, E.P. de, Asquiere, E.R., Silva, F.V.da and Robbs, P.G. (1995) [Recovery of *Staphylococcus aureus* after acid injury.] Revista Latinoamericana de Microbiología 37, 127–134.

Baer, E.F. (1966) Proposed method for isolating coagulase-positive staphylococci from food products: report of a collaborative study. J. Assoc. Off. Anal. Chem. 49, 270–273.

Baird, R.M. and van Doorne, H. (1982) Enrichment techniques for *Staphylococcus aureus*. Arch. Lebensmittelhyg. 33, 146–150.

Baird-Parker, A.C. (1962) An improved diagnostic and selective medium for isolating coagulase positive staphylococci. J. Appl. Bacteriol. 25, 12–19.

Baird-Parker, A.C. (1990) The staphylococci: an introduction. J. Appl. Bacteriol., 69, Suppl. 1S-8S.

Bartlett, P.C., Erskine, R.J., Gaston, P., Sears, P.M. and Houdijk, H.W. (1996) Enzyme-linked immunosorbent assay and microbiologic culture for diagnosis of *Staphylococcus aureus* intramammary infection of cows. J. Food Protect. 59, 6–10.

Becker, H., Terplan, G. and Zaadhof, K.-J. (1983) [Suitability of newer selective media for the detection of *Staphylococcus aureus* in foods.] Zbl. Bakt. Hyg, I. Abt. Orig. B 177, 113–126.

Becker, H., El-Bassiony, T.A., Terplan, G. (1984): Zur Abgrenzung der *Staphylococcus aureus*-Thermonuclease von hitzestabilen Nucleasen anderer Bakterien. Arch. Lebensmittelhyg. 35, 114–117.

Becker, H., Gang-Stiller, K. and Terplan, G. (1989) Characterization of *Staphylococcus aureus* strains isolated from raw milk with special reference to the clumping factor. Neth. Milk Dairy J. 43, 355–366.

Beckers, H.J., van Leusden, F.M., Bindschedler, O. and Guerraz, D. (1984) Evaluation of a pour-plate system with rabbit plasma-bovine fibrinogen agar for the enumeration of *Staphylococcus aureus* in food. Can. J. Microbiol. 30, 470–474.

Beckers, H.J., Tips, P.D. and Pateer, P.M. (1987) Collaborative study on the enumeration of *Staphylococcus aureus* in food. Zbl. Bakt. Hyg. B 184, 71–77

Bennett, R.W. (1996) Atypical toxigenic *Staphylococcus* and non-*Staphylococcus aureus* species on the horizon? An update. J. Food Protect. 59, 1123–1126.

Bennett, R.W. (1998) Chapter 13. Staphylococcal enterotoxins. *In:* FDA Bacteriological Analytical Manual, 8th edition, pp. 13.01–13.32.

Bennett, R.W. and Lancette, G.A. (1995) Chapter 12. *Staphylococcus aureus. In*: FDA Bacteriological Analytical Manual, 8th edition, pp. 12.01–12.05.

Bergdoll, M.S. (1983) Enterotoxins. *In*: Staphylococci and staphylococcal infections, Vol 2. C. Easmon and C. Adlam, (Eds) Academic Press, London, pp. 559–598.

Bergdoll, M.S. (1990) Staphylococcal food poisoning. *In*: Foodborne diseases D.O. Cliver, (Ed) Academic Press, San Diego, pp. 85–106.

Breckenridge, J.C. and Bergdoll, M.S. (1971) Outbreak of food-borne gastroenteritis due to a coagulase-negative enterotoxin-producing staphylococcus. N. Engl. J. Med. 284, 541–543.

Buttiaux, R. and Brogniart, R. (1947) Techniques d'isolement des staphylocoques pathogens. Identification des staphylocoques enterotoxiques. Ann. Inst. Pasteur 73, 830–834.

Chang, T.C. and Huang, S.H. (1993) Evaluation of a latex agglutination test for rapid identification of *Staphylococcus aureus* in foods. J. Food Protect. 56, 759–762.

Chang, T.C. and Huang, S.H. (1994) An enzyme-linked immunosorbent assay for the rapid detection of *Staphylococcus aureus* in processed foods. J. Food Protect. 57, 184–189.

Chang, T.C. and Huang, S.H. (1995) Evaluation of coagulase activity and protein A production for the identification of *Staphylococcus aureus*. J. Food Protect. 58, 858–862.

Chang, T.C. and Huang, S.H. (1996) Efficacy of a latex agglutination test for rapid identification of *Staphylococcus aureus*: Collaborative study. J. AOAC Internat. 79, 661–669.

Chopin, A., Malcom, S., Jarvis, G., Asperger, H., Beckers, H.J., Bertona, A.M., Cominazzini, C., Carini, S., Lodi, R., Hahn, G., Heeschen, W., Jans, J.A., Jervis, D.I., Lanier, J.M., O'Connor, F., Rea, M., Rossi, J., Seligmann, R., Tesone, S., Waes, G., Mocquot, G. and Pivnick, H. (1985) ICMSF methods studies. XV. Comparison of four media and methods for enumerating *Staphylococcus aureus* in powdered milk. J. Food Protect. 48, 21–27.

Chou, C.-C. and Chen, L.-F. (1997) Enterotoxin production by *Staphylococcus warneri* CCRC 12929, a coagulase-negative strain. J. Food Protect. 60, 923–927.

De Buyser, M.L., Audinet, N., Delbart, M.O., Maire, M. and Francoise, F. (1998) Comparison of selective culture media to enumerate coagulase-positive staphylococci in cheeses made from raw milk. Food Microbiol. 15, 339–346.

Devoyod, J.J., Millet, L. and Mocquot, G. (1976) Un milieu gélose pour le dénombrement direct de *Staphylococcus aureus*: milieu au plasma du porc pour *S. aureus* (PPSA). Can. J. Microbiol. 22, 1603–1611.

Devriese, L.A. (1981) Baird-Parker medium supplemented with acriflavine, polymyxins and sulphonamide for the selective isolation of *Staphylococcus aureus* from heavily contaminated materials. J. Appl. Bacteriol. 50, 351–357.

DIN. (1988a) Bestimmung Koagulase-positiver Staphylokokken in Trockenmilcherzeugnissen

und Schmelzkäse-Verfahren mit selektiver Anreicherung. DIN 10178, Teil 1. Deutsche Norm, Berlin.
DIN. (1988 b) Bestimmung Koagulase-positiver Staphylokokken in Milch und Milchprodukten – Koloniezählverfahren. DIN 10178, Teil 3. Deutsche Norm, Berlin.
Flowers, R.S. and Ordal, Z.J. (1979) Current methods to detect stressed staphylococci. J. Food Protect. 42, 362–367.
Genigeorgis, C.A. (1989) Present state of knowledge on staphylococcal intoxication. Int. J. Food Microbiol. 9, 327–360.
Gilden, M.M., Baer, E.F. and Franklin, M.K. (1966) Comparative evaluation of a direct plating procedure and an enrichment isolation procedure for detecting coagulase positive staphylococci from foods. J. Assoc. Off. Anal. Chem. 49, 273–275.
Gilmour, A. and Harvey, J. (1990) Staphylococci in milk and milk products. J. Appl. Bacteriol., 69, Suppl. 147S–166S.
Giolitti, G. and Cantoni, C. (1966) A medium for the isolation of staphylococci from foodstuffs. J. Appl. Bacteriol. 29, 395–398.
Gray, R.J.H., Gaske, M.A. and Ordal, Z.J. (1974) Enumeration of thermally stressed *Staphylococcus aureus* MF31. J. Food Sci. 39, 844–846.
Halpin-Dohnalek, M.I. and Marth, E.H. (1989) *Staphylococcus aureus*: Production of extracellular compounds and behavior in foods – a review. J. Food Protect. 52, 267–282.
Harvey, J. and Gilmour, A. (1985) Application of current methods for isolation and identification of staphylococci in raw bovine milk. J. Appl. Bacteriol. 59, 207–221.
Hauschild, A.H.W., Park, C.E. and Hilsheimer R. (1979) A modified pork plasma for the enumeration of *Staphylococcus aureus* in foods. Can. J. Microbiol. 25, 1052–1057.
Hirooka, E.Y., Mueller, E.E., Freitas, J.C., Vicente, E., Yoshimoto, Y. and Bergdoll, M.S. (1988) Enterotoxigenicity of *Staphylococcus intermedius* of canine origin. Int. J. Food Microbiol. 7, 185–191.
Hughes, D., Dailianis, A. and Hill, L. (1999) An immunoassay method for detection of *Staphylococcus aureus* in cosmetics, pharmaceutical products, and raw materials. J. AOAC Internat. 82, 1171–1174.
Hurst, A., Hughes, A., Baere-Rogers, J.L. and Collins-Thompson, D.L. (1973) Physiological studies on the recovery of salt tolerance by *Staphylococcus aureus* after sublethal heating. J. Bacteriol. 116, 901–907.
Iandolo, J.J. and Ordal, Z.J. (1966) Repair of thermal injury of *Staphylococcus aureus*. J. Bacteriol. 91, 134–142.
Ibrahim, G.F. and Radford, D.R. (1979) Confirmation of presumptive coagulase positive staphylococci in dairy products. Australian J. Dairy Technol. 34, 18–20.
IDF. (1997a) Milk and milk-based products – Enumeration of coagulase-positive staphylococci – Most probable number technique. Provisional IDF Standard 60C: 1997. International Dairy Federation, Brussels.
IDF. (1997b) Milk and milk-based products – Enumeration of coagulase-positive staphylococci – Colony count technique at 37°C. Provisional IDF Standard 145A: 1997. International Dairy Federation, Brussels.
Idziak, E.S. and Mossel. D.A.A. (1980) Enumeration of vital and thermally stressed *Staphylococcus aureus* in foods using Baird-Parker pig plasma agar (BPP). J. Appl. Bacteriol. 48, 101–113.
ISO. (1981) The need for resuscitation in standard methods for the microbiological examination of foods. ISO/TC 34/SC 9. Doc. N 118. International Organization for Standardization, Geneva.
ISO. (1999a) Microbiology of food and animal feeding stuffs – Horizontal method for the enumeration of coagulase-positive staphylococci (*Staphylococcus aureus* and other species) – Part 1: Technique using Baird-Parker agar medium. ISO 6888-1:1999. International Organization for Standardization, Geneva.
ISO. (1999b) Microbiology of food and animal feeding stuffs – Horizontal method for the

enumeration of coagulase-positive staphylococci (*Staphylococcus aureus* and other species) by colony-count technique at 35°C/37°C – Part 2: Technique using rabbit plasma fibrinogen agar medium. ISO 6888-2:1999. International Organization for Standardization, Geneva.

ISO. (1999c): Report on the 19th meeting in Paris, April 1999, Annex H: Detection of low numbers of coagulase positive staphylococci by an enrichment technique. ISO/TC 34/SC 9. Doc. N 388. International Organization for Standardization, Geneva.

Iyothirmayi, R., Reddy, V.P., Reddy, I.S. and Sarma, K.S. (1998) A comparative evaluation of the resuscitation methods in recovery of staphylococci from infant foods and dried milks. Microbiologie, Aliments, Nutrition 16, 205–209.

Kahn, M.A., Kim, C.H., Kakoma, I., Morin, E., Hansen, R.D., Hurley, W.L., Tripathy, D.N. and Baek, B.K. (1998) Detection of *Staphylococcus aureus* in milk by use of polymerase chain reaction analysis. Am. J. Vet. Research 59, 807–813.

Kneifel, W., Manafi, M. and Breit, A. (1992) Adaptation of commercially available DNA probes for the detection of *E. coli* and *Staphylococcus aureus* to selected fields of dairy hygiene – an exemplary study. Zbl. Hyg. 192, 544–553.

Lachica, R.V. (1980) Accelerated procedure for the enumeration and identification of food-borne *Staphylococcus aureus*. Appl. Environ. Microbiol. 39, 17–19.

Lachica, R.V. (1984) Egg yolk-free Baird-Parker medium for accelerated enumeration of foodborne *Staphylococcus aureus*. Appl. Environ. Microbiol. 48, 870–871.

Lancette, G.A. (1986) Current resuscitation methods for recovery of stressed *Staphylococcus aureus* cells from foods. J. Food Protect. 49, 477–481.

Lancette, G.A. and Lanier, J. (1987) Most probable number method for isolation and enumeration of *Staphylococcus aureus* in foods: Collaborative study. J. Assoc. Off. Anal. Chem. 70, 35–38.

Lancette, G.A., Peeler, J.T. and Lanier, J. M. (1986) Evaluation of an improved MPN medium for recovery of stressed and nonstressed *Staphylococcus aureus*. J. Assoc. Off. Anal. Chem. 69, 44–46.

Matsunaga, T., Kamata, S.I., Kakiichi, N. and Uchida, K. (1993) Characteristics of *Staphylococcus aureus* isolated from peracute, acute and chronic mastitis. J. Vet. Med. Sci. 55, 297–300.

Matthews, K.R. and Oliver, S.P. (1994) Differentiation of *Staphylococcus* species by polymerase chain reaction-based DNA fingerprinting. J. Food Protect. 57, 486–489.

Matthews, K.R., Roberson, J., Gillespie, B.E., Luther, D.A. and Oliver, S.P. (1997) Identification and differentiation of coagulase-negative *Staphylococcus aureus* by Polymerase Chain Reaction. J. Food Protect. 60, 686–688.

Mossel, D.A.A. and Corry, J.E.L. (1977) Detection and enumeration of sublethally injured pathogenic and index bacteria in foods and water processed for safety. Alimenta, Zurich, 16th Special Issue on Microbiology, 19–34.

Mossel, D.A.A. and van Netten, P. (1990) *Staphylococcus aureus* and related staphylococci in foods: ecology, proliferation, toxinogenesis, control and monitoring. J. Appl. Bacteriol., 69 Suppl. 123S–145S.

Mueller M. (1996) [Detection of *Staphylococcus aureus* and production of staphylococcal enterotoxins in raw milk and soft cheese by immunochemical and genetic methods.] Vet. Med. Thesis Freie Universität Berlin, No. 1947.

Munson, S.H., Tremaine, M.T., Betley, M.J. and Welch, R.A. (1998) Identification and characterization of staphylococcal enterotoxin types G and I from *Staphylococcus aureus*. Infection and Immunity 66, 3337–3348.

NMKL. (1992): *Staphylococcus aureus*. Determination in foods No. 66. Nordic Committee on Food Analysis, Esbo.

Ollis, G.W., Rawluk S.A., Schoonderwoerd, M. and Schipper, C. (1995) Detection of *Staphylococcus aureus* in bulk tank milk using modified Baird-Parker culture media. Can. Vet. J. 36, 619–623.

Pereira, M.L., Carmo, L.S.D., dos Santos, E.J., Pereira, J.L. and Bergdoll, M.S. (1996) Enterotoxin H in staphylococcal food poisoning. J. Food Protect. 59, 559–561.

Poutrel, B. and Sarradin, P. (1992) Diagnosis of *Staphylococcus aureus* intramammary infections by detecting antibodies in milk with an ELISA kit. Lait 72, 321–325.

Sabnis, N. and Kulkarni, P.R. (1992) Liquid repair method for the enumeration of debilitated staphylococci. Lebensmittel-Wissenschaft und Technologie 25, 462–466.

Sawhney, D. (1986) The toxicity of potassium tellurite to *Staphylococcus aureus* in rabbit plasma fibrinogen agar. J. Appl. Bacteriol. 61, 149–155.

SLMB. (1988) Untersuchungsmethode zur Bestimmung von *Staphylococccus aureus* M 7.12. Bundesamt für Gesundheitswesen. Schweizerisches Lebensmittelbuch, Bern.

Smith, J.L., Buchanan, R.L. and Palumbo, S.A. (1983) Effect of food environment on staphylococcal enterotoxin synthesis: A review. J. Food Protect. 46, 545–555.

Smolka, L.R., Nelson, F.E. and Kelly, L.M. (1974) Interaction of pH and NaCl on enumeration of heat stressed *Staphylococcus aureus*. Appl. Microbiol. 27, 443–447.

Stadhouders, J., Hassing, F. and van Aalst-van Maren, N.O. (1976) A pour-plate method for the detection and enumeration of coagulase-positive *Staphylococcus aureus* in the Baird-Parker medium without egg yolk. Neth. Milk Dairy J. 30, 222–229.

Su, Y.-C. and Wong, A.C.L. (1995) Identification and purification of a new staphylococcal enterotoxin H. Appl. Environ. Microbiol. 61, 1438–1443.

Terplan, G., Zaadhof, K.-J. and Becker, H. (1982) Quality assurance of newer media for the enumeration of *Staphylococcus aureus* in food. Arch. Lebensmittelhyg. 33, 142–145.

Umeki, F., Otsuka, G., Yosai, A., Seki, M., Yoshida, M., Takeishi, M., Kozaki, K. and Kozaki, R. (1993) Evaluation of biological character and pathogenicity of *Staphylococcus aureus* isolated from healthy persons, patients, poisoned food and cows. Milchwissenschaft 48, 552–555.

Untermann, F., Kusch, D. and Lupke, H. (1973) [The importance of mastitis staphylococci as a cause of food poisoning.] Milchwissenschaft 28, 686–688.

Valle, J., Gomez-Lucia, E., Piriz, S., Goyache, J., Orden, J.A. and Vadillo, S. (1990) Enterotoxin production by staphylococci isolated from healthy goats. Appl. Environ. Microbiol. 56, 1323–1326.

Van Netten, P., van der Ven, L., Jaisli, F. and Mossel, D.A.A. (1990) The enumeration of sublethally injured populations of *Staphylococcus aureus* in foods. Letters in Appl. Microbiol. 10, 117–121.

White, D.G., Matos, J.S., Harmon, R.J. and Langlois, B.E. (1988) A comparison of six selective media for the enumeration and isolation of staphylococci. J. Food Protect. 51, 685–690.

Zangerl, P. (1999) [Comparison of Baird-Parker agar and rabbit plasma fibrinogen medium for the enumeration of *Staphylococcus aureus* in raw milk and raw milk products.] Arch. Lebensmittelhyg. 50, 4–9.

Zangerl, P. (1998) Baird-Parker agar versus rabbit plasma fibrinogen medium for enumeration of coagulase-positive staphylococci in raw milk and raw milk products. Poster 25d of International Dairy Congres, 21–24 September 1998, Aarhus, Denmark.

Chapter 7

Culture media for enterococci and group D-streptococci

Gerhard Reuter[a] and Günter Klein[b]

[a] *Institute of Meat Hygiene and Technology, Veterinary Faculty, Free University of Berlin, Bruemmerstrasse 10, D-14195 Berlin, Germany*
[b] *BgVV (Federal Institute for Health Protection of Consumers and Veterinary Medicine), Division of Food Hygiene, Diedersdorfer Weg 1, D-12277 Berlin, Germany*

Enterococci, formerly classified as faecal streptococci, belong with few exceptions to the Lancefield group D-streptococci. They are contaminants of various foods, especially those of animal origin. The genus *Enterococcus* currently comprises 26 species. Some of them, and some species of streptococci e.g. *Streptococcus bovis, suis,* and *equinus,* have their habitat in the intestine of man and animals. Serologically based grouping may no longer constitute the best definition for these bacteria. The same applies to the term "faecal streptococci" and it seems reasonable therefore to look for enterococci *sensu stricto.*

Food hygiene monitoring systems use enterococci as indicators for faecal contamination and such use requires reliable methods for selective cultivation and identification of marker strains. More than 100 modifications of selective media have been described in the past for isolating faecal streptococci or enterococci from various specimens. The selection of a medium requires consideration of the specimen type (solid or liquid), the method of cultivation (pre-enrichment/enrichment, plate count or membrane filter technique) and whether the specimen is heavily contaminated with other organisms. The choice of media is made more difficult as commercial versions of the same culture medium may vary in formulation and/or performance from producer to producer. Therefore, reviewing the literature may help in the choice of medium and of confirmatory tests.

Recent research in clinical microbiology suggests glycopeptide-resistant strains of enterococci are a cause of nosocomial infections. Enterococci in the food chain may function as transmitters from animals to man. Hence, the detection and identification of vancomycin-resistant enterococci (VRE) is indicated in food microbiology and special media and procedures are needed.

The selectivity and productivity of some commonly used or cited media are reported here: citrate azide tween carbonate (CATC) agar, kanamycin aesculin azide (KAA) agar, M-Enterococcus (ME) agar, aesculin bile azide (ABA) agar, and thallous acetate tetrazolium glucose (TITG) agar. In clinical literature ABA is cited mostly as Enterococcosel (ECS), and ME as Slanetz and Bartley agar. No medium is completely selective and productive for all strains of enterococci but some media are more selective for some *Enterococcus* species, e.g. *E. faecalis* and *E. faecium*, which serve as indicators of faecal contamination or which are primarily responsible for nosocomial infections.

Confirmatory tests must be performed where sensitivity in the procedure is limited. Selective media for enterococci should be used only after or while simultaneously checking their selectivity (false positive) and productivity (false negative) against appropriate test organisms which are representative of the kind of specimen.

Table 1
Ecology of faecal streptococci in man and animal: Species of the genus *Enterococcus* and *Streptococcus* of intestinal origin (acc. to Reuter, 1992; Leclerc et al., 1996; Manero and Blanch, 1999).

		Human	Cattle	Pigs	Poultry
Enterococcus					
faecalis	(D)	[+]	( )	+	[+]
faecium	(D)	[+]	[+]	+	[+]
durans/hirae	(D)	•	–	•	•
gallinarum	(D)	•	–	–	( )
casseliflavus (=*flavescens*)	(D)	• [a]	–	–	–
cecorum/columbae	–	–	+	+	[+]
Streptococcus					
bovis/equinus	(D)	•	[+]	[+]	–
suis	–	–	–	[+]	–
alactolyticus	(D)	–	–	[+]	[+]
hyointestinalis	–	–	–	[+]	–

occurrence: [+] : usual; + : frequent; ( ) : moderate; • : occasional; – : not mentioned
(D): belonging to Lancefield group D.
[a] possibly of plant origin.

Introduction

Enterococci and group-D streptococci may be considered an essential part of the autochthonous microflora of humans and animals. Some host specificity exists (Table 1). In humans *Enterococcus faecalis* and *E. faecium* are frequent species. In poultry, cattle, and pigs *E. faecium* is a frequent species as well as other species like *E. faecalis* and *E. cecorum. E. cecorum*, originally found in poultry intestines, has also been isolated from pigs, cattle, horses, canaries, and ducks (Devriese et al., 1991a). *E. gallinarum* and *E. durans/hirae* occur occasionally in the intestine of animals and *E. avium* is a very rare species there. Some species like *E. casseliflavus* (identical to *E. flavescens)* and *E. mundtii* are found mostly in plants. Colonies of both are pigmented whilst *E. gallinarum* and *E. casseliflavus* are motile, indicating a great diversity in physiology and ecology existing within the genus.

With this wide distribution it is not surprising that enterococci and other group D-streptococci may occur in many foods, especially those of animal origin. Presence of *E. faecalis* and *E. faecium* is therefore often used to indicate faecal contamination of food. On the other hand, *Streptococcus bovis,* which also occurs in animal faeces in higher amounts, is not generally suitable as a hygienic indicator since it survives poorly in the environment. *S. equinus* has so far not been isolated from humans and is probably of no interest for food hygiene. *S. suis* belongs to the naturally acquired microflora of pigs and also of calves along with some representatives of the Lancefield B-serogroup, *S. agalactiae*, *S. dysgalactiae*, and *S. uberis* (Cruz Colque et al., 1993). These streptococci are not however considered important as indicators of faecal contamination.

Table 2a
Ecology of enterococci in food specimens (acc. to Devriese et al., 1995: isolated with KAA-medium).

	Cheese	Fish and crustaceae	Meat	Cheese-meat combinations
E. faecium	⊞	( )	⊞	+
E. faecalis	•	+	+	•
E. hirae/*E. durans*	( )	–	( )	•
E. mundtii	–	( )	•	–
E. casseliflavus	–	–	•	–
E. gallinarum	–	•	•	•

occurrence: ⊞ : usual; + : frequent; ( ) : moderate; • : occasional; – : not mentioned.

Table 2b
Ecology of enterococci on pork carcasses and in pork sausages (acc. to Knudtson and Hartman, 1993: isolated with fGTC- and KF-media).

	Pork carcasses	Pork sausage fresh	expired	spoiled
E. faecium	( )	–	–	⊞
E. faecalis	⊞	⊞	+	( )
E. malodoratus	•	( )	( )	–
E. hirae/*E. durans*	•	( )	( )	•
E. pseudoavium	•	–	+	–
E. raffinosus/*solitarius*	•	–	( )	–

occurrence: ⊞ : usual; + : frequent; ( ) : moderate; • : occasional; – : not mentioned.
fGTC = fluorescent gentamicin-thallous-carbonate-agar, KF = KF-streptococcus-agar (Reuter, 1985).

Table 2c
Ecology of enterococci in minced meat (acc. to Dazo, 1996; Klein et al.,1998: isolated with CATC- and ABA-media).

	Beef	Pork
E. faecium	( )	( )
E. faecalis	⊞	⊞
E. hirae/*E. durans*	( )	•
E. casseliflavus	–	( )
E. gallinarum	( )	•
E. avium	–	•

occurrence: ⊞ : usual; + : frequent; ( ) : moderate; • : occasional; : not mentioned.

Table 3
The groups and species of the genus *Enterococcus* showing useful reactions for phenotypical preliminary identification at species level. (The most useful traits are in boxes.)

Group and species	Lancefield group D	10°C	6.5% NaCl	Growth properties PA[a]	MDG[b]	yellow pigment	motility	Human clinical relevance
usual reaction:	+	+	+	+	–	–	–	
<u>Faecalis-group</u>								
E. faecalis	[91%]	(0.01% TTC- and 0.04% tellurite-tolerance)						[+]
<u>Faecium-group</u>								
E. faecium	[68%]	(L-(+)-arabinose –, growth at 50°C)						[+]
E. durans	[75%]	(0.01% TTC-tolerance)						
E. hirae	[63%]							( )
E. mundtii						[+]		( )
<u>Avium-group</u>								
E. avium $^{\otimes}$	[32%]	( )	( )					
E. raffinosus	[40%]	[–]						
E. pseudoavium	[–]		( )					
E. malodoratus								
E. asini		( )	[–]					
<u>Gallinarum-group</u>								
E. gallinarum		(0.01% TTC-tolerance)			[+]		[±]	( )
E. casseliflavus (=*E. flavescens*)		( )			[+]	[+]	[+]	( )
<u>Cecorum-group</u>								
E. cecorum	[–]	[–]	[–]	[–]				
E. columbae	[–]	[–]	[–]	[–]				
<u>Other group</u>								
E. solitarius		(microaerophilic, slow growth)						
E. sulfureus	[–]					[+]		
*E. dispar**	[–]	[–]						
*E. saccharolyticus**	[–]	[–]		[–]				

$^{\otimes}$ also Serogroup Q; * no growth at 45°C; + positive; ( ) sometimes positive; ± sometimes negative; – negative.
[a] PA = pyrrolidonyl aminopeptidase-activity [b]MDG = acidification of methyl-α-D-glucopyranoside (Devriese et al., 1996).

The ecology of enterococci in food must be regarded in the light of the new taxonomy of this genus. Some important work has been published (Knudtson and Hartman, 1993; Devriese et al., 1995; Dazo, 1996; Klein et al., 1998; Richter, 1999) and resumées are shown in Tables 2a–c).

Review articles on taxonomic aspects of intestinal enterococci and streptococci were compiled by Devriese et al. (1993), Leclerc et al. (1996), and Franz et al. (1999), as well as for glycopeptide resistant enterococci (GRE) in animals and foods by Butaye et al. (1999). Details of the different taxonomic groups of enterococci and their main features at the present state of knowledge are shown in Table 3.

In food microbiology selective culture media are thus required for representatives of the genus *Enterococcus,* especially for *E. faecalis* and *E. faecium* and closely related species. These species must be cultivated separately from the dominating total count of a food specimen. A wide variety of selective media for the *Enterococcus*- and/or *Streptococcus*-group has been recommended and used in the past. More than 100 modifications have been described, summarized by Barnes (1976). Proposals for specific uses for food of animal origin were made in previous publications (Reuter, 1978; 1985; 1992) and they will be renewed here.

The selective isolation of vancomycin-resistant enterococci (VRE) is a new aspect in clinical microbiology, however a standardized screening method does not yet exist. The method of Willey et al. (1992) was therefore tested by eight laboratories (Swenson et al., 1994) using brain heart infusion (BHI) medium supplemented with 4, 6, and 8 μg/ml vancomycin. More recently work has been carried out with ABA (=ECS) broth and agar, BHI (brain heart infusion) agar, or ME-broth (Van Horn et al., 1996a and b; Ieven et al., 1999). Higher concentrations of vancomycin (32 μg/ml) were used in food microbiology with ECS or CATC medium (Klein et al., 1998; Richter, 1999) or with Slanetz and Bartley (ME) medium (50 μg/ml; Wegener et al., 1997). Other authors did not use agar plates for screening but tested isolates for vancomycin resistance by disk diffusion after non-selective enrichment (Quednau et al., 1998). However, agar screening proved to be more sensitive than disk diffusion in the detection of vancomycin resistance in pure cultures of VRE, especially if low or medium level vancomycin resistance was expected (Endtz et al., 1998).

The detection and enumeration of enterococci as indicators of pollution in drinking water is another special field. It is carried out predominantly by a membrane filtration technique or by enrichment procedures since, in general, only low numbers of enterococci are contained in the specimen. These procedures require a filtration step followed by confirmation either on a selective medium by identification of suspected colonies, e.g. by catalase-test (Leclerc et al., 1996) or with commercially available identification kits, e.g. Chromocult®-broth or Enterolert®-system (Heiber et al., 1998). Although some national standards have been published, ISO-standards (e.g. from ISO TC 147 / 5C4 / WGH 7899/2, 1997) currently exist only as drafts. Collaborative studies (ring-trials) are continuing in that field.

Requirements and composition of selective media for isolating group D-streptococci or enterococci

Selective agents

Various selective agents may be used in media. Commonly used are sodium azide, thallous acetate, and kanamycin or gentamicin and vancomycin in different concentrations to enhance cultivation of glycopeptide-resistant parts of a common enterococcal microflora. Other selective ingredients or factors are crystal violet, bile salts, Tween 80, and carbonate. Special growth conditions may be achieved by low pH (e.g. 6.0 or 6.2) or elevated incubation temperature (42 or 44°C). Increased or decreased concentrations of selective agents will yield different degrees of selectivity or productivity, therefore their correct use is of great importance. For instance, sodium azide is heat sensitive and the solution should be added after the medium has been autoclaved if the culture medium is to be strictly selective for enterococci such as *E. faecalis* and *E. faecium*, as is the case with the CATC-medium. If commercially available media already contain this ingredient, its selective effect must be verified by testing the prepared medium. Misleading results can be caused by neglecting this factor. In cases where vancomycin is used for the determination of glycopeptide-resistant components of the enterococcal microflora, the concentrations of this antimicrobial in a solid or liquid medium depend on the degree of selectivity required by the test procedure. In clinical microbiology where few mixed populations are expected, lower concentrations of vancomycin and the use of a liquid medium are opportune (4, 6, or 8 μg/ml). Confirmatory tests are then performed on solid selective media by plating out from the enrichment medium (Ieven et al., 1999). In the analysis of heavily contaminated specimens, such as food of animal origin, direct plating on solid selective media in combination with higher concentrations of vancomycin, e.g. 32 μg/ml, is a possibility (Klein et al., 1998; Richter, 1999). ABA (ECS) or CATC may be useful in such a context. If stressed microorganisms are expected, a pre-enrichment procedure with 2–4 h of incubation in a non- or low-selective liquid medium (6 μg/ml vancomycin) must precede selective culture.

Indicators

Indicator substances added to the media are useful for the recognition of enterococci and for the rapid identification of single species on the basis of colonial appearance. Triphenyltetrazolium-chloride (TTC)-, or, in special cases, tellurite-reduction may be used, especially for the identification of *E. faecalis* strains. However, some important facts must be considered. In a medium at pH 7.0, most enterococci will reduce tetrazolium to formazan, which becomes visible as a bright red coloured colony; at pH 6.2 or even 6.0, this reaction is evident only for the species *E. faecalis*, while *E. faecium* or other streptococci show only a weak reaction, at most a pale colony with a red centre or a slightly pink colony (Mead, 1985; Reuter, 1985).

Other species may reduce tetrazolium but these grow more slowly. Thus reading of the results should take account of the duration and temperature of incubation. The formation of an aesculin-ferrous complex has also been applied as a criterion. However, aesculin is also split by some other Gram-positive organisms such as lactobacilli, pediococci etc. but incubation at 42°C and reading the results after 18 h of incubation will eliminate such spurious reactions.

Growth factors

Nutritional requirements for group D-streptococci are not very exacting. It is advantageous to use good peptones and beef or yeast extract. Glucose is not necessary in general use. *Enterococcus durans* seems to be somewhat more fastidious. *E. cecorum* requires microaerophilic conditions for good growth on KAA-agar and will not grow at all on ME-medium (Devriese et al., 1991b). The composition and recommendations

Table 4
Composition and handling of culture media for enterococci (group D-streptococci) from food specimens.

Ingredients g or ml/l	TlTG	ME	CATC	ABA (=ECS)	KAA
Target group of organisms	group D-streptococci and related species	formazan producing group D-streptococci	formazan producing enterococci	aesculin splitting group D-streptococci	aesculin splitting enterococci
Tryptone		15.0	15.0	17.0	20.0
Peptone	10.0	5.0		3.0	
Meat extract	10.0				
Yeast extract		5.0	5.0	5.0	5.0
Glucose *	10.0	2.0 / 5.0			
Aesculin				1.0	1.0
Tween 80			1.0		
NaCl				5.0	5.0
K_2HPO_4		4.0			
KH_2PO_4			5.0		
Ferric NH_4-citrate				0.5	0.5
Sodium citrate			15.0	1.0	1.0
Sodium carbonate *			2.0		
Sodium azide *		0.4	0.4	0.25	0.15
Ox bile				10.0	
Kanamycin sulphate					0.02
Thallous acetate *	0.1				
TTC (1% sol.) *	10.0	10.0	10.0		
Agar **	14.0	10.0	15.0	13.5	10.0
Final pH	6.0	7.2	7.0	7.1	7.0
Incubation: (and reading of results)	1d 37°C + 1d b	2d 37°C or 4h 37°C + 1-2d 44°C	1d 37°C + 1d b	1-2d 37°C	18h 42°C or 2d 37°C

Preparation: * added separately according to original descriptions; included in some commercial preparations; ** concentration depends on quality of agar, deleted in case of liquid media; d = day; h = hour; b = on bench (22 ± 2°C); TlTG, thallous acetate tetrazolium glucose agar (Barnes, 1956; Mead, 1985); ME, M-Enterococcus agar (Slanetz and Bartley, 1957); CATC, citrate azide tween carbonate agar (Reuter, 1968); ABA, aesculin bile azide agar (= ECS = Enterococcosel) (Isenberg et al., 1970); KAA, kanamycin aesculin azide agar (Mossel et al., 1978); TTC: triphenyltetrazoliumchloride.

for incubation and reading of five selective agars are shown in Table 4. These media are also described in monographs in Part 2 of this book.

Table 5
Selectivity and productivity of culture media for enterococci and group D-streptococci (Reuter, 1978, 1985; Schotte, 1986; Dazo, 1996; Klein et al., 1998) compared with related and non related species from food microflora.

Group Genus Species	Lancefield Sero-group	TITG	ME	CATC	ABA (=ECS)	KAA
Streptococci						
Enterococcus	D					
faecalis		+ i	+ i	+ i	+ i	+ (i)
faecium / durans / hirae		+ (i)	+ (i)	+ (i)	± i	+ (i)
gallinarum / casseliflavus			+ (i)	+ (i)	+ i	+ (i)
cecorum			–			≈ (i)
Streptococcus						
bovis / equinus	D	+	+	≈	+	+ (i)
agalactiae / uberis	B	+	± (i)	–		± (i)
viridans ssp.			≈	≈ (i)	≈	+
salivarius, ssp. *thermophilus*			± (i)	–	–	≈
Lactococcus spp.	N	≈	≈	–	±	–
Leuconostoc spp.		± (i)	± (i)	–	–	± (i)
Pediococcus spp.		≈ (i)	± (i)	–	±	–
Non-Streptococci						
<u>Gram-positive</u>						
Lactobacillus spp.						
thermophilic		≈	≈	–	–	±
mesophilic		≈	–	–	–	±
sake-curvatus		+	±	–	≈ (i)	≈
Corynebacterium spp.			–	–	+	≈
Listeria spp.					+	–
Erysipelothrix spp.					–	≈
Micrococcus spp.		–	–	–	+	≈
Staphylococcus spp.		≈	–	–	+	≈
Bacillus spp.		±	–	–		–
Yeasts		–	–	–	≈ (i)	–
<u>Gram-negative</u>						
Enterobacteriaceae (*Hafnia*, *Arizona*, *Salmonella*)		≈	–	–	±	±
Pseudomonadaceae (*Ps. aeruginosa*)		≈	–	–	±	–

+ = full growth; ≈ = reduced growth rate or small colonies or delayed growth; ± = further reduction in growth or exceptional growth; – = no growth (at least reduction of more than 2 log cycles); i = indicator reduced (formazan producing or aesculin-splitting) (i = red or black colonies, (i) = pink colonies or brownish halo). See Table 4 for key to media abbreviations.

Results of testing selective media for enterococci from food

Different media have been tested in our laboratories for more than 20 years by comparing the growth of reference strains of enterococci and non-enterococci and of strains from the accompanying food microflora. Additionally, these media have been used for routine investigations of food specimens, including meat and milk and similar products as well as faecal samples from humans and animals.

First results were published (Reuter, 1978) for the following media: ABA, KAA, CATC, ME, TlTG, Streptococcus selective agar (ScS), Crystal violet azide agar (KA). Later (Reuter, 1985) the suitability of ME, KAA agar and CATC medium was proven again for selectivity, productivity, and differentiating features between species. The testing was done qualitatively with the loop-streaking technique and quantitatively with the modified drop-plating technique (Reuter, 1985; Schotte, 1986). CATC and ABA medium, in its commercial form called Enterococcosel (ECS), were later used in investigations on vancomycin-resistant populations of enterococci in minced beef or pork (Dazo, 1996; Klein et al., 1998). CATC was used also for enterococci in poultry meat (Richter, 1999).

CATC, KAA, ME and TlTG media were included in the International Committee on Food Microbiology and Hygiene (ICFMH) pharmacopoeia (Baird et al., 1987). A monograph on ECS (syn. ABA) with and without vancomycin is also included in Part 2 of this book.

It is clear that a highly specific medium for all *Enterococcus* species still does not exist. Some media allow growth of non-enterococci and others partially inhibit single strains of some species. That is more or less the case with ABA, TlTG, ScS and KF media. Single strains of staphylococci, or even yeasts, Enterobacteriaceae (*Hafnia, Salmonella,* and *Arizona*), and Pseudomonadaceae are able to grow, therefore care is necessary in reading results. If a high level of background flora is expected, it is preferable to use media with a high selectivity.

The properties of recommended or commonly used media for the selective cultivation of enterococci in food microbiology with their various advantages and disadvantages can be summarized as follows (see also Table 5).

Aesculin-bile-azide medium (ABA = ECS)

Aesculin-bile-azide medium is a classical selective medium. It was previously used for isolating or confirming pathogenic strains of enterococci from clinical specimens and was widely distributed in the market, especially in the English-speaking area. With increasing interest in GRE in the clinical field, its use was adapted to the identification and confirmation of these strains. It is frequently cited in articles published in clinical microbiology.

In food microbiology this medium was also used in microecological surveys (Klein et al., 1998). These authors confirmed its selectivity for enterococci in specimens from minced beef and pork. Besides enterococci and *Staphylococcus aureus* a few strains of Gram-negatives including *Salmonella* and *Citrobacter* grew but did not split aesculin within 48 h of incubation. The only problem which arose was the growth of a mesophilic *Lactobacillus* strain (*L. curvatus*), which also split aesculin after 24 h of incubation. Reading the results at defined times helps to exclude such false positives. This problem may not arise in clinical microbiology as fewer lactobacilli, especially mesophilic strains, may be present. Confirmatory tests are needed especially if a new habitat has to be investigated with this medium. For confirmatory purposes, e.g. to demonstrate bile salt tolerance, this medium appears to be suitable even after enrichment procedures with vancomycin-containing media in low concentrations. The productivity of this medium for enterococci was as good as that of non-selective media.

Citrate azide tween carbonate agar (CATC)

CATC agar is a highly selective medium for enterococci, used as indicators of faecal contamination, provided plates are read at defined times (24 and about 48 h) (see Table 4). Very good growth and a brilliant formazan reaction can be obtained with *E. faecalis* strains at first reading. Colonies of *E. durans* may also show a distinct red colour but with smaller colony size and with later appearance. Colonies of *E. faecium* as well as *E. gallinarum* and *E. casseliflavus* appear with a weaker formazan reaction, often distinct pink only in the centre of colonies. Single strains sometimes show slightly reduced productivity and smaller colonies. These other species can be separated from *E. faecalis* very easily by colonial appearance. Recovery of *Streptococcus bovis* and *S. equinus* is reduced by more than one $\log_{10}$ cycle. Oral streptococci show weak colour formation and very small colonies. Lactic streptococci and leuconostocs do not grow at all. If plates are read at defined times, i.e. after 24 h of incubation at 37°C, this medium may be considered as highly selective and fully productive for *E. faecalis* and as an adequate medium for *E. faecium* and other *Enterococcus* species. Extending the incubation period for another 24 h at 20–24°C may produce better differentiation of enterococci and may demonstrate a background microflora, which may be of interest and should be verified by confirmatory tests. In order to detect glycopeptide-resistant strains, the addition of vancomycin with 32 or even 50 μg/ml has proved suitable especially for the detection of glycopeptide-resistant *E. faecium* strains (Klein et al., 1998; Richter 1999).

Kanamycin aesculin azide agar (KAA)

KAA medium is widely used in different fields of food microbiology. It has been used for confirmatory tests for enterococci from enrichment procedures and seems essential for cultivating strains of the *E. cecorum* group. However, for some specimens it allows partial growth of mesophilic lactobacilli, some of which also split aesculin; this reaction serves to identify enterococci. Incubation of KAA at 42°C may prevent the growth of aesculin-splitting lactobacilli and some other Gram-positive organisms, such as corynebacteria. The selectivity for growth is not so high (see Table 5) and all colonies on KAA appear to be similar, i.e. small and greyish-white. Confirmatory tests are therefore advisable, especially when investigating new habitats. Increasing the content of sodium azide may eliminate this problem of selectivity but may reduce the recovery rate of enterococci as well. Raising the incubation temperature to 44°C, as proposed for some commercial products, and reading after 24 h may further increase selectivity.

Slanetz and Bartley medium (M-enterococcus; ME)

ME medium lacks some selectivity if used in food microbiology. It does not inhibit the lactococci at all and also fails with some *Lactobacillus* strains. However, other non-streptococci are generally eliminated. The formazan-reaction does not differentiate between *Enterococcus* species as well as with the CATC medium. Whilst it has found favour in clinical microbiology, in food microbiology the lack of selectivity is likely to overestimate the actual number of enterococci. In our experience it should be used in parallel with CATC medium. On the other hand its use in water microbiology in combination with the membrane filter technique is widespread and recommended (Heiber et al., 1998). Sometimes ME agar is used for confirmatory tests of presumptive VRE.

Thallous acetate tetrazolium glucose agar (TlTG)

TlTG medium was designed to be selective for group D-streptococci. Owing to the formazan formation it is suitable for the detection and differentiation of *E. faecalis* and *E. faecium*. The pH of the medium should not exceed 6.2 (Mead, 1985) to allow *E. faecalis* to be read easily. Prolonged incubation of this medium for one or two more days at 20°C (after one day at 30°C) results in the appearance of streptococci, leuconostocs, pediococci and mesophilic lactobacilli when meat products are examined (Reuter, 1968). With such a complex microflora it is difficult to differentiate between the groups of lactic acid bacteria. The incubation regime enables the experienced worker to distinguish between these groups on the basis of colour reactions and different colony appearance, but less experienced workers should check identities using confirmatory tests.

Considerations for the correct use of the selective cultivation procedure

Sub-lethally damaged cells

Resuscitation techniques are required in all cases where enterococci have been subjected to any form of stress such as drying or freezing. Organisms may be incubated on or in a non-selective nutrient medium at 30–37°C (Barnes, 1976; Mossel and Corry, 1977) for 4–6 h, followed by overlaying with a selective medium. Ewald and Eie (1992) obtained good results even after 2 h of incubation. Alternatively, organisms on membrane filters may be incubated first on a non-selective medium and then on the selective medium (Barnes, 1976).

Storage of ready-to-use plates

Following a short drying procedure, plates of selective media may be stored for one week at 0–5°C in plastic bags. Storage at room temperature without prior drying is feasible for up to two weeks if plates are protected from light. If the selective agent is known to be unstable, media should be prepared directly before use.

Confirmatory tests

If new habitats or samples with an unknown microflora composition are being investigated, confirmatory tests are required for colonies grown on selective media. First non-streptococci and then non-enterococci have to be eliminated. Micrococci and corynebacteria can be excluded by a positive catalase or benzidine test on suspected colonies on slides or by flooding the plates with the test solution. Mesophilic lactic streptococci may be excluded by testing liquid subcultures for the ability to grow at 45°C. Further confirmation can be done by checking haemolytic or gelatinase activities or mannitol and arabinose fermentation and growth at 10 ± 0.5 or 50 ± 0.5°C in a water bath, as well as growth in 6.5% NaCl-broth and at pH 9.6. Motility can also be checked. For identification of enterococci at the genus and species level, test kits are available. The computerised test kit Rapid ID32 STREP® (BioMérieux) works quickly and well. Pyrrolidonylarylamidase activity is a common feature for nearly all enterococci (PA-test). Antigens to Lancefield D-serogroup can be verified by the latex slide test. *E. gallinarum* and *E. casseliflavus* can be separated from *E. faecalis* and the *E. faecium*-group by acidification of methyl-α-D-glucopyranoside (MDG) and by a rapid motility test (Devriese et al., 1996; Turenne et al., 1998; Hanson and Cartwright, 1999). Widely used commercial identification systems may fail to identify precisely rare species like *E. avium* and *E. raffinosus*. Here classical biochemical identification methods may be able to verify the correct classification (Wilke et al., 1997). A revised MicroScan method was successfully applied in combination with motility and MDG tests to identify *E. gallinarum* (Iwen et al., 1999). As in clinical microbiology, the aim is to use rapid, accurate tests, therefore several DNA-based methods for the detection of *E. faecalis* and *E. faecium* have been reported. With the determination of 16S rRNA sequences the identification of non-motile *E. gallinarum*, previously identified phenotypically as *E. faecium*, was possible (Patel et al., 1998). Other molecular methods such as pulsed field gel electrophoresis (PFGE) and randomly amplified polymorphic DNA (RAPD)-analysis have been applied (Descheemaeker et al., 1997; Ahrné and Molin, 1999). A PCR-based assay has also been described which is capable of yielding results in only a few hours (Ke et al., 1999). Molecular identification of enterococci was summarised recently by Schleifer (2002) at a symposium in Berlin concluding that genotypic methods are necessary to identify the more recently described new enterococcal species which differ significantly in their physiological and biochemical properties from those of typical enterococci.

References

Ahrné, S. and Molin, G. (1999) Antibiotic resistance of *Enterococcus faecium* strains isolated from chicken meat. Proceedings of the 17th International Conference of ICFMH, Veldhoven, Netherlands, pp. 403–404.

Baird, R.M., Corry, J.E.L. and Curtis, G.D.W. (Eds.) (1987) Pharmacopoeia of Culture Media for Food Microbiology. Int. J. Food Microbiol. 5, 187–300.

Barnes, E.M. (1956) Methods for isolation of faecal streptococci (Lancefield Group D) from bacon factories. J. Appl. Bacteriol. 19, 193–203.

Barnes, E.M. (1976) Methods for the isolation of faecal streptococci. Lab. Pract. 25, 145–147.

Butaye, P., Devriese, L.A. and Haesebrouck, F. (1999) Glycopeptide resistance in *Enterococcus faecium* strains from animal and humans. Rev. Med. Microbiol. 10, 235–243.

Cruz Colque, J.I., Devriese, L.A. and Haesebrouck, F. (1993) Streptococci and enterococci associated with tonsils of cattle. Lett. Appl. Microbiol. 16, 72–74.

Dazo, K.B.R.C. (1996) The antimicrobial susceptibility of *Enterococcus* species isolated from minced meat from an EU approved meat-establishment in Berlin. Master of Science degree-Thesis, Vet. med. Faculty, Free University Berlin, pp. 1–95.

Descheemaeker, P., Lammens, C., Pot, B., Vandamme, P. and Goossens, H. (1997) Evaluation of arbitrarily primed PCR analysis and pulsed-field gel electrophoresis of large genomic DNA fragments for identification of enterococci important in human medicine. Int. J. System. Bacteriol. 47, 555–561.

Devriese, L.A., Ceyssens, K. and Haesebrouck, F. (1991a) Characteristics of *Enterococcus cecorum* strains from the intestines of different animal species. Lett. Appl. Microbiol. 12, 137–139.

Devriese, L.A., Collins, M.D. and Wirth, R. (1991b) The genus *Enterococcus*. In: The Prokaryotes edited by A. Balows, H.Trüper, M. Dworkin, W. Harder and K.-H. Schleifer, 2nd Edn., Vol II. Springer-Verlag, Heidelberg, pp. 1465–1481.

Devriese, L.A., Pot, B. and Collins, M.D. (1993) Phenotypic identification of the genus *Enterococcus* and differentation of phylogenetically distinct enterococcal species and species groups – A review. J. Appl. Bacteriol. 75, 399–408.

Devriese, L.A, Pot, B., Van Damme, L., Kersters, K. and Haesebrouck, F. (1995) Identification of *Enterococcus* species isolated from foods of animal origin. Int. J. Food Microbiol. 26, 187–197.

Devriese, L.A., Pot, B., Kersters, K., Lauwers, S. and Haesebrouck, F. (1996) Acidification of methyl-alpha-D-glucopyranoside: a useful test to differentiate *Enterococcus casseliflavus* and *Enterococcus gallinarum* from *Enterococcus faecium* species group and from *Enterococcus faecalis*. J. Clin. Microbiol. 34, 2607–2608.

Endtz., H.P., Van den Braak, N., Van Belkum, A.., Goessens, W.H ., Kreft, D., Stroebel, A.B. and Verbrugh, H.A. (1998) Comparison of eight methods to detect vancomycin resistance in enterococci. J. Clin. Microbiol. 36, 592–594.

Ewald, S. and Eie, T. (1992) The effect of resuscitation and the incubation-temperature on recovery of uninjured, heat injured and freeze injured enterococci. Int. J. Food Microbiol. 15, 177–184.

Franz, C.M.A.P., Holzapfel, W.H. and Stiles, M.E. (1999) Enterococci at the crossroads of food safety? Int. J. Food Microbiol. 47, 1–24.

Hanson, K.L. and Cartwright, C.P. (1999) Comparison of simple and rapid methods for identifying enterococci intrinsically resistant to vancomycin. J. Clin. Microbiol. 37, 815–817.

Heiber, I., Frahm, E. and Obst, U. (1998) Vergleich von vier Nachweismethoden für Fäkalstreptokokken in Wasser - Comparison of four methods for the detection of fecal streptococci in water. Zbl. Hyg. Umweltmed. 201, 357–369.

Isenberg, H.D., Goldberg, J. and Sampson, J. (1970) Laboratory studies with a selective *Enterococcus* medium. Appl. Microbiol. 20, 433–436.

Ieven, M., Vercauteren, E., Descheemaeker, P., Van Laer, F. and Goossens, H. (1999) Comparison of direct plating and broth enrichment culture for the detection of intestinal colonisation by glycopeptide-resistant enterocci among hospitalized patients. J. Clin. Microbiol. 37, 1436–1440.

ISO. (1997) Water quality – detection and enumeration of fecal enterococci by membrane filtration. Working document, Draft revision: ISO TC 147/SC4/WG4, 7899/2. International Organisation for Standardisation, Geneva.

Iwen, P.C., Rupp, M.E., Schreckenberger, P.C. and Hinrichs, S.H. (1999) Evaluation of the revised MicroScan dried overnight gram-positive identification panel to identify *Enterococcus* species. J. Clin. Microbiol. 37, 3756–3758.

Ke, D., Picard, F.J., Martineau, F., Ménard, C., Roy, P.H., Ouellette, M. and Bergeron M.G. (1999) Development of a PCR assay for rapid detection of enterococci. J. Clin. Microbiol. 37, 3497–3503.

Klein, G., Pack, A. and Reuter, G. (1998) Antibiotic resistance patterns of enterococci and occurence of vancomycin-resistant enterococci in raw minced beef and pork in Germany. Appl. Environ. Microbiol. 64, 1825–1830.

Knudtson, L.M., and Hartman, P.A. (1993) Enterococci in pork processing. J. Food Protect. 56, 6–9.

Leclerc, H., Devriese, L.A. and Mossel, D.A.A. (1996) Taxonomical changes in intestinal (faecal) enterococci and streptococci: consequences on their use as indicators of faecal contamination in drinking water – A review. J. Appl. Bacteriol. 81, 459–466.

Manero, A. and Blanch, A.R. (1999) Identification of *Enterococcus* spp. with a biochemical key. Appl. Environ. Microbiol. 65, 4425–4430.

Mead, G.C. (1985) Isolation media for group D-streptococci: comments. Int. J. Food. Microbiol. 2, 115–117.

Mossel, D.A.A. and Corry, J.E.L. (1977) Detection and enumeration of sublethally injured pathogenic and index bacteria in foods and water processed for safety. Alimenta-Sonderausgabe pp. 19–34.

Mossel, D.A.A., Bijker, P.G.H. and Eelderink, I. (1978) Streptokokken der Lancefield-Gruppe D in Lebensmitteln und Trinkwasser: Ihre Bedeutung, Erfassung und Bekämpfung. Arch. Lebensmittelhyg. 29, 121–127.

Patel, R., Piper, K.E., Rouse, M.S., Steckelberg, J.M., Uhl, J.R, Kohner, P., Hopkins, M.K.,
Cockerill, F.R. and Kline, B.C. (1998) Determination of 16S rRNA sequences of enterococci and application to species identification of nonmotile *Enterococcus gallinarum* isolates. J. Clin. Microbiol. 36, 3399–3407.

Quednau, M., Ahrné, S., Petersson, A.C. and Molin, G. (1998) Antibiotic-resistant strains of *Enterococcus* isolated from Swedish and Danish retailed chicken and pork. J. Appl. Microbiol. 84, 1163–1170.

Reuter, G. (1968) Erfahrungen mit Nährböden für die selektive mikrobiologische Analyse von Fleischerzeugnissen. Arch. Lebensmittelhyg. 19, 53–57 and 84–89.

Reuter, G. (1978) Selektive Kultivierung von Enterokokken aus Lebensmitteln tierischer Herkunft. Arch. Lebensmittelhyg. 29, 128–131.

Reuter, G. (1985) Selective media for group D-streptococci. Int. J. Food Microbiol. 2, 103–114.

Reuter, G. (1992) Culture media for enterococci and group D-streptococci. Int. J. Food Microbiol. 17, 101–111.

Richter, P. (1999) Isolation und Identifikation glykopeptidresistenter Enterokokkenspezies aus Mastgeflügel. Vet. med. Diss. (Thesis), Free University Berlin, pp. 1–92.

Schleifer, K.H. (2002) Molecular classification and identification of enterococci. Symposium "Enterococci in Foods-Functional and Safety Aspects", 30–31 May 2002, BgVV, Berlin, to be published in Int. J. Food Microbiol.

Schotte, M. (1986) Bewertung von 3 semi-quantitativen Prüfverfahren für die Qualitätssicherung von Nährböden in der Lebensmittelmikrobiologie mit Selektivmedien für Milchsäurebakterien. Vet. med. Diss. (Thesis), Free University Berlin, pp. 1–140.

Slanetz, L.W. and Bartley, C.H. (1957) Numbers of enterococci in water, sewage, and feces determined by the membrane filter technique with an improved medium. J. Bacteriol. 74, 591–595.

Swenson, J.M., Clark, N.C., Ferraro, M.J., Sahm, D.F., Doern, G., Pfaller, M.A., Barth Reller, L., Weinstein, M.P., Zabransky, R.J. and Tenover, F.C. (1994) Development of a standardized screening method for detection of vancomycin resistant enterococci. J. Clin. Microbiol. 32, 1700–1704.

Turenne, C.Y., Hoban, D.J., Karlowsky, J.A., Zhanel, G.G. and Kabani, A.M. (1998) Screening of stool samples for identification of vancomycin-resistant *Enterococcus* isolates should include the methyl-alpha-D-glucopyranoside test to differentiate nonmotile *Enterococcus gallinarum* from *E. faecium*. J. Clin. Microbiol. 36, 2333–2335.

Van Horn, K.G., Gedris, C.A. and Rodney, K.M. (1996a) Selective isolation of vancomycin resistant enterococci. J. Clin. Microbiol. 34, 924–927.

Van Horn, K.G., Gedris, C.G., Rodney, K. M. and Mitchell, J.B. (1996b) Evaluation of commercial vancomycin agar screen plates for detection of vancomycin-resistant enterococci. J. Clin. Microbiol. 34, 2042–2044.

Wegener, H.C., Madsen, M., Nielsen, N. and Møller Aarestrup, F. (1997) Isolation of vancomycin resistant *Enterococcus faecium* from food. Int. J. Food Microbiol. 35, 57–66.

Willey, B.M., Kreiswirth, B.N., Simor, A.E., Williams, G., Scriver, S.R., Phillips, A. and Low, D.E. (1992) Detection of vancomycin resistance in *Enterococcus* species. J. Clin. Microbiol. 30, 1621–1624.

Wilke, W.W., Marshall, S.A., Coffman, S.L., Pfaller, M.A., Edmund, M.B., Wenzel, R.P. and Jones, R.N. (1997) Vancomycin-resistant *Enterococcus raffinosus*: molecular epidemiology, species identification error, and frequency of occurrence in a national resistance surveillance programm. Diagn. Microbiol. Infect. Dis. 29, 43–49.

Chapter 8

Culture media for lactic acid bacteria

U. Schillinger and W.H. Holzapfel

Institute of Hygiene and Toxicology, BFE, Haid-und Neu-Str. 9, D-76131 Karlsruhe, Germany

This review deals with culture media for the detection, selective isolation and cultivation of different groups of lactic acid bacteria (LAB). A number of elective and semi-selective media are available and currently used for LAB. Most of them have been developed with the intention to isolate certain groups of LAB from a specific habitat such as meat or dairy products. These media can be rendered more selective by the addition of specific inhibitory agents or by reducing the pH. Members of the genera *Lactobacillus, Leuconostoc, Pediococcus* and *Weissella* (so-called LLPW group) share a number of physiological similarities and generally respond in the same way to conditions or compounds inhibitory to non lactics. Therefore, most culture media developed for the detection of *Lactobacillus* or *Leuconostoc* are not completely selective for the respective genus. Carnobacteria can easily be distinguished from the LLPW group by their non-aciduric nature. However, because of physiological similarities to the genus *Enterococcus* such as ability to grow at pH-values up to 9.5, media developed for the selective isolation of *Carnobacterium* do not suppress growth of enterococci often sharing the same habitat. A number of useful selective media is available for beer pediococci, *Tetragenococcus* and *Oenococcus,* organisms characterised by specific properties associated with their adaptation to special environments. Because of the growing interest in probiotic strains and the inhabitants of the intestine, an increasing number of media have been proposed in recent years for selective isolation of particular species or strains from those habitats typically containing mixed populations of different LAB.

Introduction

The lactic acid bacteria (LAB) are a large and heterogeneous group of Gram-positive bacteria characterised by a strictly fermentative metabolism with lactic acid as the major end product during sugar fermentation. Typical LAB are non-sporing, catalase-negative, aerotolerant and nutritionally fastidious organisms. They are perfectly adapted to environments rich in nutrients and energy sources and their metabolism is aimed at acid production to outcompete other bacteria sharing the same habitat. Representatives of the LAB can be isolated from nearly all types of foods including fresh and processed meat, fish, cereals, vegetables and dairy products. A number of LAB species are normal inhabitants of the mouth and intestine of mammals. Some are specialised to extreme environments such as alcoholic beverages or foods with a high salt content.

The LAB comprise both spoilage and technologically important bacteria such as

Table 1
Recognised genera of lactic acid bacteria (based on Lücke, 1996).

Genus	Relevant to food fermentations	food spoilage
Lactobacillus	++	++
Leuconostoc	+	++
Pediococcus	+	+
Weissella	–	+
Carnobacterium	–	+
Streptococcus	+	–
Enterococcus	(+)	+
Lactococcus	++	–
Tetragenococcus	+	(+)
Oenococcus	+	(+)
Paralactobacillus	?	–
Aerococcus	–	–
Alloiococcus	–	–
Atopobium	–	–
Vagococcus	–	–

starter and probiotic cultures. The role of a particular strain may depend on the specific situation and the product. In one situation it may cause desirable changes of a food product and in another situation the same organism may contribute to spoilage of a product. *Lactobacillus sakei*, for instance, can be used as a starter culture for the production of fermented sausages and on the other hand is one of the most important bacteria responsible for adverse sensory changes of vacuum-packaged processed meat (Mäkelä et al., 1992).

Presently, the LAB comprise more than a dozen different genera (Table 1). The list was recently extended by the genus *Paralactobacillus*, a name introduced by Leisner et al. (2000). The authors isolated this LAB from a Malaysian food ingredient called chili bo and found these strains to be representatives of a hitherto undiscovered taxon distantly related to the *Lactobacillus-Pediococcus* group. Thus, the number of genera and species belonging to the LAB is still increasing and this demonstrates the importance of the development of special culture media for the detection of these organisms that may be part of a very complex microflora.

In this review, culture media for the genera *Lactobacillus, Leuconostoc, Pediococcus, Carnobacterium, Lactococcus, Tetragenococcus* and *Oenococcus* will be considered. The genera *Enterococcus* and *Streptococcus* will be treated elsewhere.

Elective media for lactic acid bacteria

Table 2 presents an overview of the media commonly used for isolation of LAB.

These elective media support growth of the majority of LAB. However, they may be rendered more selective by addition of inhibitory compounds or increasing their concentration, or by reducing the pH.

Table 2
Elective media for lactic acid bacteria (modified from Reuter, 1985; Holzapfel, 1992).

Medium	Groups or species to be cultivated	Application	Reference
APT (pH 6.7)	LAB	cured meats / general	Evans and Niven (1951)
Briggs (pH 6.8)	LAB	milk products, gut, general	Briggs (1953)
MRS (pH 6.2)	LAB	general	De Man et al. (1960)
La (pH 6.8)	LAB	general, meat	Reuter (1970)
TPPY (pH 6.8)	lactobacilli, *S. thermophilus*	yoghurt	Braquart (1981)
M 17 (pH 7.2)	lactococci, *S. thermophilus*	yoghurt	Terzaghi and Sandine (1975)
LSD (pH 6.1)	lactobacilli, lactococci	yoghurt	Eloy and Lacrosse (1976)

All Purpose Tween (APT) medium is an example of a classical medium for LAB. It was originally developed by Evans and Niven (1951) for the isolation of lactobacilli from cured meat products and is still commonly used for the isolation and cultivation of LAB from meat products. Tomato juice was found to stimulate the growth of many LAB and it was included in Briggs agar (Briggs, 1953), a medium frequently used for the cultivation of lactobacilli from milk and dairy products. Modifications of Briggs agar including BL agar which contains liver extract in addition (Mitsuoka et al., 1965) have also been found useful for the detection of presumptive *L.acidophilus* in faecal samples (Mitsuoka, 1969, Von Aulock and Holzapfel, 1987). Early observations by Ochi and Mitsuoka (1958) suggested that typical flat greyish-brown and rough colonies ("ß-colonies") appearing on Briggs or BL agar plates may be *L. acidophilus.*

MRS medium has a similar composition to APT and was developed primarily with the intention of substituting tomato juice by defined growth factors such as Mg^{2+} and Mn^{2+} (De Man et al., 1960). It is probably the most commonly used medium for the cultivation of lactobacilli and other LAB. Reuter (1970) modified MRS medium according to the special nutritional requirements of meat LAB and called it La agar. A number of different elective media were created for the isolation of certain groups of LAB from yoghurt and other fermented dairy products. These include M17 (Terzaghi and Sandine, 1975), L S Differential (LSD; Eloy and Lacrosse, 1976) and Tryptose-Proteose-Peptone-Yeast extract-Eriochrome T (TPPY; Braquart, 1981) agars which have been successfully used for the isolation and differentiation of yoghurt bacteria (Shankar and Davies, 1977).

Selective media

Lactobacillus

The genera *Lactobacillus, Leuconostoc* and *Pediococcus* together with *Weissella* (LLPW group) are physiologically similar organisms often sharing the same habitat

Table 3
Selective media for lactobacilli and related genera (LLPW-group).

Medium	Application	Groups or species to be cultivated	Inhibitor/ Indicator	Reference
Rogosa (pH 5.5)	general	lactobacilli, *Weissella, Leuconostoc, Pediococcus*	acetate, acetic acid	Rogosa et al. (1951)
NAP (pH 5.5)	meat	lactobacilli, lactococci, (*Leuconostoc* ?)	nitrite, actidione, polymyxin B	Davidson and Cronin (1973)
MRS (pH 5.5)	general, meats	lactobacilli, *Weissella*		Baird and Patterson (1980)
MRS-S (pH 5.7)	general, meats	lactobacilli, *Leuconostoc, Weissella*	sorbic acid (0.1 %)	
LaS (pH 5.0)	meats	lactobacilli, (*Leuconostoc, Weissella*)	sorbic acid (0.04 %)	Reuter (1970)
HHD (pH 7.0)	fermented vegetables	homo- and heterofermentative LAB	bromocresol green	McDonald et al. (1987)
M 5 (pH 6.5)	wine	homo- and heterofermentative LAB	bromocresol green	Zúniga et al. (1993)
TJA (pH 5.0)	wine	lactobacilli, pediococci, *Leuconostoc*	cycloheximide, sorbic acid	Yoshizumi (1975)
KPL (pH 5.5)	kefir	*Lactobacillus kefiranofaciens*	white wine	Toba et al. (1986)

and showing relatively similar growth requirements (Holzapfel, 1992). As a consequence, it is difficult to develop a medium which allows the selective isolation of e.g. lactobacilli in the presence of leuconostocs or *Weissella*. On the other hand, marked differences exist among species within each of these genera, with regard to tolerance to reduced pH, acid, aerobiosis/anaerobiosis and inhibitory compounds.

Table 3 summarises the most important selective media used for lactobacilli and the related genera (LLPW). There is no completely selective medium presently available for the genus *Lactobacillus*. The majority of the media designed for the isolation of lactobacilli allow simultaneous growth of leuconostocs and/or pediococci and representatives of *Weissella*. Phase contrast microscopy of typical colonies is necessary to distinguish between these organisms but has only limited value for distinguishing the rod-shaped *Weissella* strains from the lactobacilli or *W. paramesenteroides* (oval morphology) from the leuconostocs.

These selective media typically contain compounds inhibitory to non lactics and to which most LLPW bacteria are resistant. Rogosa agar (Rogosa et al., 1951), for example, includes sodium acetate at a relatively high concentration which is largely undissociated at the reduced pH achieved by the use of acetic acid (Table 3). Furthermore, adjusting the pH of MRS-S and LaS (Reuter, 1970) agar to pH 5.7 and 5.0, respectively, guarantees that a significant proportion of the added sorbic acid is in the undissociated form. Most LLPW strains are also relatively resistant to other inhibitory factors such as sodium nitrite, actidione and polymyxin B used in NAP agar (Davidson and Cronin, 1973). When yeasts are expected to occur in the same habitat, addition of actidione or sorbate to the medium is helpful to suppress their growth. Anaerobic incubation can also be used to inhibit yeasts and moulds. Catalase-positive bacteria such as staphylococci, micrococci and bacilli can be prevented from growing by supplementation of the selective media with sorbate (0.1 – 0.2 %) and the reduction of the pH below 5.8.

Efforts have been made to develop media which allow rapid differentiation between homo- and heterofermentative LAB. In homofermentative-heterofermentative differential medium (HHD; McDonald et al.,1987) and M 5 (Zúniga et al., 1993) glucose was replaced by fructose and bromocresol green was used as a pH indicator. Differences in the colour of the colonies are based on differences in the amounts of acids produced by these bacterial groups. Homofermentative LAB produce more acid from fructose than do heterofermentative LAB which reduce a portion of the fructose to mannitol.

Leuconostoc

A number of media has been proposed for the selective isolation of leuconostocs from foods. Glucose-yeast extract agar (Whittenbury, 1965a), acetate agar (Whittenbury, 1965b), thallous acetate-tetrazolium-sucrose agar (Cavett et al., 1965) and yeast extract-glucose-citrate agar (Garvie, 1967) are examples of media developed for the isolation of leuconostocs from plant material. Most of these media, however, also allow growth of lactobacilli and pediococci. Thus, they are of limited value for the selective isolation of leuconostocs from food commodities with a mixed flora of LAB. In 1993 and 1994 two media were described which were more selective for *Leuconostoc*. The first is the *Leuconostoc* selective medium (LUSM medium; Benkerroum et al., 1993) which was developed for the isolation of leuconostocs from vegetables and dairy products. In addition to tomato juice it contains a number of inhibitory compounds such as sorbic acid (0.05 %), sodium azide (0.25 %), sodium acetate (0.25 %), vancomycin (30 µg/ml) and tetracycline (0.2 µg/ml). The other relatively new selective medium was designated Vancomycin-medium by Mathot et al. (1994) and was developed for the differentiation of dairy starter cultures. It is a modification of MRS or NL agar supplemented with vancomycin (5–200 µg/ml). Lactococci are sensitive to this antibiotic whereas leuconostocs are constitutively resistant.

Pediococcus

Pediococci play an important role in the brewery industry. *Pediococcus damnosus* is the major spoilage organism in beer manufacture. For this reason, several different media have been developed for the detection of pediococci in beer. Some are very complex, e.g. Kirin-Ohkochi-Taguchi (KOT medium; Taguchi et al., 1990) which contains beer, malt extract, liver concentrate, maltose, L-malic acid, cytidine, thymidine, actidione and sodium azide in addition to the components usually used for LAB media. Beer is frequently included in these media as a main component (Nakagawa, 1964; Back, 1978) in order to promote growth of pediococci adapted to this substrate.

Table 4
Selective agents used for the detection of beer spoilage pediococci.

Selective compound	Inhibition of
Cycloheximide, crystal violet, actidione	yeasts
2-phenylethanol, sorbic or acetic acid	yeasts and Gram-negative bacteria
Thallous acetate	most other bacteria
Polymyxin B, sodium azide	Gram-negative bacteria
Vancomycin	Gram-positive bacteria except *Leuconostoc*, *Pediococcus* and some lactobacilli
Hop bitter acids	most non-brewery bacteria

Table 4 lists selective agents commonly used in the development of media for beer pediococci. Inhibition of beer yeasts can be achieved by several agents such as cycloheximide, amphotericin or sorbic acid. Gram-negative bacteria can be suppressed by the addition of compounds such as polymyxin B, acetic acid and thallous acetate.

Carnobacterium

The genus *Carnobacterium* comprises organisms which may be involved in spoilage of various meat products especially unprocessed, vacuum-packaged meat and poultry. Carnobacteria can also be isolated frequently from fish and cow`s raw milk cheeses. They are psychrotrophic organisms which prefer a relatively high pH (8–9) for growth and are characterised by their sensitivity to low pH (< 6.0) and to acetate. Consequently, media with an elevated pH can be used for the semiselective recovery of carnobacteria from food. Such media include modifications of Standard I Nutrient (Merck) and MRS (de Man et al., 1960) agar adjusted to pH 8 or 8.5 and with glucose substituted by sucrose or fructose (Holzapfel and Gerber, 1983). Acetate is omitted from modified MRS-agar (D-MRS) (Schillinger et al.,1993). Tryptone glucose yeast extract agar (TGE, Oxoid) has also been used successfully for the isolation of carnobacteria from refrigerated vacuum-packaged meats (Borch and Molin, 1988), a product in which carnobacteria and/or lactobacilli are frequently dominant.

Media with a greater selectivity may be necessary in food products where carnobacteria are only a minor part of the bacterial population e.g. fermented meats. Thus far, only two media are available for the selective detection and isolation of carnobacteria from such products: cresol red thallous acetate sucrose (CTAS; Holzapfel, 1992) and EBRER agar (Millière and Lefèbvre, 1994).

The selectivity of the CTAS medium is mainly based on its high pH (8.5–9), the presence of thallous acetate and nalidixic acid and a relatively high concentration of sodium citrate. However, it shows some limitations as it is not completely selective, allowing enterococci and some strains of *Leuconostoc* or even *Listeria* to grow. Moreover, not all carnobacteria grow well on this medium. Colonies of *Carnobacterium divergens*, for instance, often develop only to pin-point size (own observations).

EBRER medium was developed for better recovery of carnobacteria from French soft cheeses (Millière and Lefèbvre, 1994). *Carnobacterium piscicola* frequently predominates at the end of the ripening period of Brie cheeses. The most important components of EBRER medium are ribose, aesculin, phenol red and the antibiotics amphotericin and nalidixic acid. The pH of the medium is 7.6. Interestingly, the authors found that the selectivity of EBRER medium is low at 30°C and they therefore recommend incubation at 7°C for 10 days (Millière and Lefèbvre, 1994). Unfortunately the medium also allows growth of enterococci which cannot be differentiated macroscopically from colonies produced by carnobacteria strains.

Lactococcus

Amongst the many media which have been described for the isolation of lactococci from foods, Elliker agar (Elliker et al.,1956) and M 17 (Terzaghi and Sandine, 1975) are most commonly used for the detection of lactococci. They are, however, not completely selective. WACCA-agar (Galesloot et al.,1961) can be used for the collective enumeration of lactococci and leuconostocs in starter and dairy products.

Tetragenococcus

The genus *Tetragenococcus* was created from the former *Pediococcus halophilus*. It is characterised by its very high salt tolerance. Strains of *Tetragenococcus halophilus* are able to grow at a salt concentration > 18 % NaCl and are involved in lactic fermentations of foods containing high concentrations of salt such as soy sauce mashes, pickling brines, salted anchovies and fermented fish sauce ('Shottsuru'). Recently, a second species, *T. muriaticus* isolated from Japanese fermented squid liver sauce was described (Satomi et al., 1997).

Media used for the selective isolation of tetragenococci generally contain high levels of sodium chloride. MRS may be supplemented with 5 % NaCl (Back, 1978) or 10 % NaCl (Gürtler et al., 1998). Other media recommended for selective isolation of tetragenococci include Polypeptone-Acetate-Thioglycolate agar with 15 % NaCl (PAT-15; Uchida, 1982), Nutrient agar supplemented with 0.5 % glucose, 1.2 % NaCl and salted anchovy broth (Villar et al., 1985) and Yeast-Glucose-Peptone (YGP) medium supplemented with 5 % NaCl (Collins et al., 1990) or supplemented with 10 % NaCl, 1 % magnesium sulfate, 0.1 % potassium chloride and 0.5 % calcium carbonate (Satomi et al., 1997).

Oenococcus

Taxonomic studies revealed that *Leuconostoc oenos* has a low phylogenetic relationship to the other leuconostocs. Therefore, this acidophilic branch was separated from *Leuconostoc* and a new genus, *Oenococcus,* was created with the sole species *Oenococcus oeni* (Dicks et al., 1995). These organisms are primarily responsible for the malolactic fermentation in wine and are characterised by their special adaptation to the wine environment. They are acidophilic, prefer an initial growth pH of 4.8, are able to grow in the presence of 10 % ethanol and are relatively resistant to sulphur dioxide. Most strains grow only in the presence of tomato juice, grape juice, apple

Table 5
Media used for isolation of *Oenococcus oeni* from wine.

Medium	pH	Selectivity	Most important components	Reference
Acidic Tomato Broth (ATB)	4.8	semi-selective	tomato juice (25%)	Garvie (1967)
Weiler and Radler medium	5.3–5.4	semi-selective	sodium acetate (0.5%), Tween 80 (0.1%), sorbic acid (0.05%)	Weiler and Radler (1970)
M104 medium	4.5	selective	tomato juice (25%), DL-malic acid (1%), Tween 80 (0.1%)	Barillière (1981)
Tomato juice-Glucose-Fructose-Malate Broth (TGFMB)	5.5	selective	tomato juice (diluted), fructose (0.3 %), malate (0.2 %), Tween 80, cycloheximide (0.005%)	Izuagbe et al. (1985)
Fructose and Tween 80 (FT 80) medium	5.2	semi-selective	fructose (3.5 %), L-malic acid (1%), Tween 80 (0.1%)	Cavin et al. (1988)
Enriched tomato juice broth (ETJB)	4.8	semi-selective	tomato juice broth (Difco) (2%), Tween 80 (0.1%), cysteine (1%)	Britz and Tracey (1990)
UBA-agar		selective	apple juice (20%), cycloheximide (0.005%)	Rodriguez et al. (1990)

juice, pantothenic acid or 4‘0-(ß-glucopyranosyl) D-pantothenic acid (Dellaglio et al., 1995). Media used for the isolation of *O. oeni* from wine include various modifications of MRS agar or other basal media, incorporating tomato juice, apple juice, malic acid, cysteine and a low pH (Table 5). Cycloheximide or sorbic acid are often added to inhibit growth of yeasts.

Enumeration of different LAB in dairy and probiotic products

Yoghurts and other fermented dairy products are examples of foods which usually contain mixtures of different LAB. The selective enumeration of the different strains used as starter or probiotic cultures is essential for the quality control of these products. Novel-type yoghurts and other milk products may contain the classical yoghurt bacteria *L. delbrueckii* ssp. *bulgaricus* and/or *Streptococcus thermophilus* and, in addition, ‘probiotic‘ strains of *L. acidophilus*, *L. johnsonii*, *L. crispatus*, *L. casei* and bifidobacteria. Various media have been suggested to enable differentiation among these types of LAB (Dave and Shah, 1996; Lankaputhra et al., 1996; Nighswonger et al., 1996) and differential plating methodologies developed for that purpose have been reviewed by Charteris et al. (1997). Many of these media do not give satisfactory results because of the lack of recovery of one or more species of the above mentioned bacteria (Lankaputhra and Shah, 1996), lack of selectivity (Lim et al., 1995; Pacher

Table 6
Media proposed for the enumeration of *Lactobacillus acidophilus*, *Bifidobacterium bifidum*, *Lactobacillus delbrueckii* subsp. *bulgaricus* and *Streptococcus thermophilus* in fermented dairy products (Vinderola and Reinheimer,1999).

Organism	Medium[1]	Incubation
L. acidophilus	T-MRS agar	aerobiosis
	Bile-MRS agar	aerobiosis
	G-MRS agar	aerobiosis
B. bifidum	LP-MRS agar	anaerobiosis
	G-MRS agar	anaerobiosis
L. delbrueckii subsp. *bulgaricus*	SM agar	aerobiosis
	LBD agar	aerobiosis
	G-MRS agar	aerobiosis
S. thermophilus	SM agar	aerobiosis
	LBD agar	aerobiosis

[1] T-MRS agar: MRS agar modified by replacing glucose with trehalose;
Bile-MRS agar: MRS agar supplemented with bile (0.15%);
G-MRS agar: MRS agar modified by replacing glucose with galactose;
LP-MRS agar: MRS agar supplemented with lithium chloride (0.2%) and sodium propionate (0.3%);
SM agar: Skim milk agar (Marshall, 1992);
LBD agar: Lactic Bacteria Differential agar (HIMedia Laboratories, Bombay, India).

and Kneifel, 1996) or differentiation between colonies (Kneifel and Pacher, 1993; Ghoddusi and Robinson, 1996; Nighswonger et al.,1996). A differential colony count of *L. acidophilus* and related species, *L. casei*, bifidobacteria, *L. delbrueckii* ssp. *bulgaricus* and *S. thermophilus* on a single medium constitutes a special challenge. Tryptose-Proteose-Peptone-Yeast extract-Eriochrome T agar with added prussian blue (TPPYPB) was shown to allow the visible identification of *L. acidophilus*, a *Bifidobacterium* species, *S. thermophilus* and *L. delbrueckii* ssp. *bulgaricus* (Ghoddusi and Robinson, 1996).

Table 6 gives an overview of the media that may be used for the selective detection of *L. acidophilus* and *Bifidobacterium bifidum* in the presence of yoghurt bacteria. It is based on the results of a comparative study of Vinderola and Reinheimer (1999). Modifications of MRS agar seem to be useful for this purpose. T-MRS, an MRS medium in which glucose is substituted by trehalose and an MRS agar with bile (0.15 %) added, were both recommended by the International Dairy Federation (IDF, 1995) for counting *L. acidophilus* in the presence of yoghurt bacteria and bifidobacteria. Compared to the 'traditional' yoghurt bacteria, only strains of the *L. acidophilus* group are able to ferment trehalose and to grow in the presence of 0.15 % bile salts. Skim Milk (SM) agar was found to be a satisfactory medium for the enumeration of both *L. delbrueckii* ssp. *bulgaricus* and *S. thermophilus* as both bacteria produce easily distinguishable colonies on this medium (Reinheimer and Vinderola, 1997) and the strictly anaerobic bifidobacteria are suppressed under aerobic conditions. The use of Rogosa agar, pH 5.4 was suggested for the selective enumeration of *L. casei* from fermented milk products (Champagne et al., 1997). The authors recommend an anaerobic incubation for 14 days at 15°C. Recently a new medium was described for the selective detection of *L. casei* in yoghurts and fermented milk drinks containing yoghurt cultures and probiotic bacteria (Ravula and Shah, 1998). This medium was called LC agar and is characterised by the presence of ribose as sole fermentable carbohydrate and the relatively low pH of 5.1.

Detection of probiotic LAB in the intestine

Because of the increasing interest in probiotic bacteria and their behaviour in the gastrointestinal tract, there is a strong demand for methods allowing selective detection of these LAB in faeces. Recently, a medium for the selective reisolation of the probiotic *L. casei* Shirota strain from faeces was described (Yuki et al., 1999). It is a modified LBS medium (identical with Rogosa agar) with lactitol substituting glucose as carbon source and added vancomycin which was shown to inhibit intestinal bacteria other than *L. casei* Shirota (Yuki et al., 1999). Hartemink et al. (1997) suggested *Lactobacillus* Anaerobic MRS with Vancomycin and Bromocresol green (LAMVAB) medium for the selective isolation of lactobacilli from faeces. This is a modified MRS medium with an increased selectivity due to the low pH (5.0) and the addition of vancomycin (20 mg/l). In addition, it contains cysteine-HCl to enhance anaerobic conditions and bromocresol green as pH indicator. It allows the detection and isolation of low numbers of lactobacilli in the presence of high numbers of other LAB such as streptococci, enterococci and bifidobacteria which are in general susceptible to vancomycin although vancomycin-resistant enterococci may occur. LAMVAB medium was successfully used for the isolation of LAB from human colonic biopsies (Kontula et al., 2000).

Conclusions

A number of approved culture media are available for LAB and have been successfully used by numerous investigators over many years. Even so, an enhancement of the selectivity of media developed for the detection and isolation of *Lactobacillus, Leuconostoc* and *Carnobacterium* would be desirable. At this stage, it is doubtful whether a suitable medium can be developed for distinguishing members of the genus *Weissella* in presence of strains of lactobacilli and leuconostocs. Furthermore, there is still a need for media allowing easy differentiation of several groups or strains of LAB in complex ecosystems like the intestinal tract. Similarily, the selective detection and enumeration of all LAB strains included in probiotic preparations or products still constitutes a great challenge.

References

Back, W. (1978) Zur Taxonomie der Gattung *Pediococcus*. Phänotypische and genotypische Abgrenzung der bisher bekannten Arten sowie Beschreibung einer neuen bierschädlichen Art: *Pediococcus inopinatus*. Brauwiss. 31, 237–250, 312–320, 336–343.

Baird, K.J. and Patterson, J.T. (1980) An evaluation of media for the cultivation and selective enumeration of lactic acid bacteria from vacuum-packaged beef. Rec. Agric. Res. 28, 55–61.

Barillère, J.M. (1981) Etude de la thermorésistance de souches de levure et de bactéries lactiques isolées de vin, influence de degré alcoolique et de la teneur en SO_2. Thesis, Ecole Nationale Supérieure d'Agronomie, Montpellier.

Benkerroum, N., Misbah, M., Sandine, W.E. and Elaraki, A.T. (1993) Development and use of a selective medium for isolation of *Leuconostoc* ssp. from vegetables and dairy products. Appl. Environ. Microbiol. 59, 607–609.

Borch, E. and Molin, G. (1988) Numerical taxonomy of psychrotrophic lactic acid bacteria from prepacked meat and meat products. Antonie van Leeuwenhoek 54, 301–323.

Braquart, P. (1981) An agar medium for the differential enumeration of *Streptococcus thermophilus* and *Lactobacillus bulgaricus* in yoghurt. J. Appl. Bacteriol. 51, 303–305.

Briggs, M. (1953) An improved medium for lactobacilli. J. Dairy Res. 20, 36–40.

Britz, T.J. and Tracey, R.P. (1990) The combination effect of pH, SO_2, ethanol and temperature on the growth of *Leuconostoc oenos*. J. Appl. Bacteriol. 68, 23–31.

Cavett, J.J., Dring, G.J. and Knight, A.W. (1965) Bacterial spoilage of thawed frozen peas. J. Appl. Bacteriol. 28, 241–251.

Cavin, J.F., Schmitt, P., Arias, A., Lin, J. and Divies, C. (1988) Plasmid profiles in *Leuconostoc* species. Microbiol. Alim. Nutr. 6, 55–62.

Champagne, C.P., Roy, D. and Lafond, A. (1997) Selective enumeration of *Lactobacillus casei* in yoghurt-type fermented milks based on a 15°C incubation temperature. Biotech. Techniques 11, 567–569.

Charteris, W.P., Kelly, P.M., Morelli, L. and Collins, J.K. (1997) Selective detection, enumeration and identification of potentially probiotic *Lactobacillus* and *Bifidobacterium* species in mixed bacterial populations. Int. J. Food Microbiol. 35, 1–27.

Collins, M.D., Williams, A.M. and Wallbanks, S. (1990) The phylogeny of *Aerococcus* and *Pediococcus* as determined by 16S rRNA sequence analysis: description of *Tetragenococcus* gen. nov. FEMS Microbiol. Lett. 70, 255–262.

Dave, R.I and Shah, N.P. (1996) Evaluation of media for selective enumeration of *Streptococcus thermophilus*, *Lactobacillus delbrueckii* ssp. *bulgaricus*, *Lactobacillus acidophilus*, and bifidobacteria. J. Dairy Sci.79, 1529–1536.

Davidson, C.M. and Cronin, F. (1973) Medium for the selective enumeration of lactic acid bacteria from foods. Appl. Microbiol. 26, 439–440.

Dellaglio, F., Dicks, L.M.T. and Torriani, S. (1995) The genus *Leuconostoc*. *In*: Wood, B.J.B. and Holzapfel, W.H. (Eds.). The genera of lactic acid bacteria. Blackie Academic and Professional, London, pp. 235–278.

De Man, J.C., Rogosa, M. and Sharpe, M.E. (1960) A medium for the cultivation of lactobacilli. J. Appl. Bacteriol. 23, 130–135.

Dicks, L.M., Dellaglio, F. and Collins, M.G. (1995) Proposal to reclassify *Leuconostoc oenos* as *Oenococcus oeni* [corrig.]. Gen. Nov. Comb. Nov. Int. J. Syst. Bacteriol. 45, 395–397.

Elliker, P.R., Anderson, W. and Hannesson, G. (1956) An agar medium for lactic acid streptococci and lactobacilli. J. Dairy Sci. 39, 1611–1612.

Eloy, C. and Lacrosse, R. (1997) Composition d'un milieu de culture destine a effectuer le denombrement des micro-organismes thermophiles du yoghurt. Bull. Rech. Agron. Gemblou 11 (1–2), 83–86.

Evans, J.B. and Niven, C.F. (1951) Nutrition of the heterofermentative lactobacilli that cause

greening of cured meat products. J. Appl. Bacteriol. 62, 599–603.
Garvie, E.I. (1967) *Leuconostoc oenos* sp. nov. J. Gen. Microbiol. 48, 431–438.
Galesloot, T.E., Hassing, F. and Stadhouders, J. (1961) Agar media voor het isoleren en tellen van aromabacterien in zuursels. Neth. Milk Dairy J. 15, 127–150.
Ghoddusi, H.B and Robinson, R.K. (1996) Enumeration of starter cultures in fermented milks. J. Dairy Res. 63, 151–158.
Gürtler, M., Gänzle, M.G., Wolf, G. and Hammes, W.P. (1998) Physiological diversity among strains of *Tetragenococcus halophilus*. System. Appl. Microbiol. 21, 107–112.
Hartemink, R., Domenech, V.R. and Rombouts, F.M. (1997) LAMVAB – a new selective medium for the isolation of lactobacilli from faeces. J. Microbiol. Meth. 29, 77–84.
Holzaphel, W.H. (1992) Culture media for non-sporulating Gram-positive food spoilage bacteria. Int. J. Food Microbiol. 17, 113–133.
Holzapfel, W.H. and Gerber, E.S. (1983) *Lactobacillus divergens* sp. nov., a new heterofermentative *Lactobacillus* species producing L(+) lactate. System. Appl. Microbiol. 4, 522 – 534.
IDF (1995) Detection and enumeration of *Lactobacillus acidophilus*. Bulletin No. 306. Int. Dairy Federation, Brussels, Belgium.
Izuagbe, Y.S., Dohman, T.P., Sandine, W.E. and Heatherbell, D.A. (1985) Characterization of *Leuconostoc oenos* isolated from Oregon wines. Appl. Environ. Microbiol. 50, 680–684.
Kneifel, W. and Pacher, B. (1993) An X-glu based agar medium for the selective enumeration of *Lactobacillus acidophilus* in yogurt-related milk products. Int. Dairy J. 3, 277–291.
Kontula, P., Suihko, M.-L., Suortti, T., Tenkanen, M., Mattila-Sandholm, T. and von Wright, A. (2000) The isolation of lactic acid bacteria from human colonic biopsies after enrichment on lactose derivatives and rye arabinoxylo-oligosaccharides. Food Microbiol. 17, 12–22.
Lankaputhra, W.E.V. and Shah, N.P. (1996) A simple method for selective enumeration of *Lactobacillus acidophilus* in yogurt supplemented with *L. acidophilus* and *Bifidobacterium* ssp. Milchwissenschaft 51, 446–451.
Lankaputhra, W.E.V., Shah, N.P. and Britz, M.L. (1996) Evaluation of media for selective enumeration of *Lactobacillus acidophilus* and *Bifidobacterium* species. Food Australia 48, 113–118.
Leisner, J.J., Vancanneyt, M., Goris, J., Christensen, H. and Rusul, G. (2000) Description of *Paralactobacillus selangorensis* gen. nov., sp. nov., a new lactic acid bacterium isolated from chili bo, a Malaysian food ingredient. Int. J. Syst. Evolut. Microbiol. 50, 19–24.
Lim, K.S., Huh, C.S. and Baek, Y.J. (1995) A selective enumeration medium for bifidobacteria in fermented dairy products. J. Dairy Sci. 78, 2108–2112.
Lücke, F.K. (1996) Lactic acid bacteria involved in food fermentations and their present and future uses in food industry. *In*: Bozoglu, T.F. and Ray, B. (Eds) Lactic acid bacteria. Current advances in metabolism, genetics and applications. Springer, Berlin, pp.81–99.
Mäkelä, P., Schillinger, U., Korkeala, H. and Holzapfel, W.H. (1992) Classification of ropy slime-producing lactic acid bacteria based on DNA-DNA homology, and identification of *Lactobacillus sake* and *Leuconostoc amelibiosum* as dominant spoilage organisms in meat products. Int. J. Food Microbiol. 16, 167–172.
Marshall, R.T. (Ed.) (1992) Standard methods for examination of dairy products (16th edition), American Public Health Association, Washington DC, pp. 275
Mathot, A.G., Kihal, M., Prevost, H. and Divies, C. (1994) Selective enumeration of *Leuconostoc* on vancomycin agar media. Int. Dairy J. 4, 459–469.
McDonald, L.C., McFeeters, R.F., Daeschel, M.A. and Fleming, H.P. (1987) A differential medium for the enumeration of homofermentative and heterofermentative lactic acid bacteria. Appl. Environ. Microbiol. 53, 1382–1384.
Millière, J.B. and Lefèbvre, G. (1994) *Carnobacterium piscicola*, a common species of French soft cheeses from cow`s raw milk. Netherl. Milk Dairy J. 48, 19–30.
Mitsuoka, T. (1969) Vergleichende Untersuchungen über die Laktobazillen aus den Faeces von

Menschen, Schweinen und Hühnern. Zentralbl. Bakteriol. Hyg. 1. Abt. Orig. 210, 32–51.

Mitsuoka, T., Sega, T. and Yamamoto, S. (1965) Eine verbesserte Methodik der qualitativen und quantitativen Analyse der Darmflora von Menschen und Tieren. Zentralbl. Bakteriol. Hyg. 1.Abt. Orig. 195, 455.

Nakagawa, A. (1964) A simple method for the detection of beer-sarcinae. Bullet. Brew. Sci. 10, 7–10.

Nighswonger, B.D., Brashears, M.M. and Gilliland, S. E. (1996) Viability of *Lactobacillus acidophilus* and *Lactobacillus casei* in fermented milk products during refrigerated storage. J. Dairy Sci. 79, 212–219.

Ochi, Y. and Mitsuoka, T. (1958) Studies on lactobacilli. Jap. J. Vet. Sci. 20, 71.

Pacher, B. and Kneifel, W. (1996) Development of a culture medium for the detection and enumeration of bifidobacteria in fermented milk products. Int. Dairy J. 6, 43–64.

Ravula, R.R. and Shah, N. P. (1998) Selective enumeration of *Lactobacillus casei* from yogurts and fermented milk drinks. Biotech. Techniques 12, 819–822.

Reinheimer, J.A. and Vinderola, C.G. (1997) Supervivencia de *Bifidobacterium* sp. y *Lactobacillus acidophilus* en yogur. II. Encuentro Bromatológico Latinoamericano, Córdoba, 17 –19 abril.

Reuter, G. (1970) Laktobazillen und eng verwandte Mikroorganismen in Fleisch und Fleischerzeugnissen. 2 Mitteilung: Die Charakterisierung der isolierten Laktobazillenstämme. Fleischwirtschaft 50, 954–962.

Rodriguez, S.B., Amberg, E. and Thornton, R.J. (1990) Malolactic fermentation in Chardonnay: growth and sensory effects of commercial strains of *Leuconostoc oenos*. J. Appl. Bacteriol. 68, 139–144.

Rogosa, M., Mitchell, J.A. and Wisemann, R.F. (1951) A selective medium for thc isolation and enumeration of oral and fecal lactobacilli. J. Bacteriol. 62, 132–133.

Satomi, M., Kimura, B., Mizoi, M., Sato, T. and Fuji, T. (1997) *Tetragenococcus muriaticus* sp. nov., a new moderately halophilic lactic acid bacterium isolated from fermented squid liver sauce. Int. J. Syst. Bacteriol. 47, 832–836.

Schillinger, U., Stiles, M.E. and Holzapfel, W.H. (1993) Bacteriocin production by *Carnobacterium piscicola* LV 61. Int. J. Food Microbiol. 20, 131–147.

Shankar, P.A. and Davies, F.L. (1977) Recent developments in yogurt starters. II. A note on the suppression of *Lactobacillus bulgaricus* in media containing ß-glycerophosphate and application of such media to selective isolation of *Streptococcus thermophilus* from yogurt. J. Soc. Dairy Technol. 30, 28–30.

Taguchi, H., Ohkochi, M., Uehara, H., Kojima, K. and Mawatari, M. (1990) KOT medium, a new medium for the detection of beer spoilage lactic acid bacteria. J. Amer. Soc. Brew. Chem. 48, 72–75.

Terzaghi, B.E. and Sandine, W.E. (1975) Improved medium for lactic streptococci and their bacteriophages. Appl. Microbiol. 29, 807–813.

Toba, T., Abe, S., Arihara, K. and Adachi, S. (1986) A medium for the isolation of capsular bacteria from kefir grains. Agric. Biol. Chem. 50, 2673–2674.

Uchida, K. (1982) Multiplicity in soy pediococci carbohydrate fermentations and its application for analysis of their flora. J. Gen. Appl. Microbiol. 28, 215 – 223.

Villar, M., de Ruiz Holgado, A.P., Sanchez, J.J., Trucco, R.E. and Oliver, G. (1985) Isolation and characterization of *Pediococcus halophilus* from salt anchovies (*Engraulis anchoita*). Appl. Environ. Microbiol. 49, 664–666.

Vinderola, C.G. and Reinheimer, J.A. (1999) Culture media for the enumeration of *Bifidobacterium bifidum* and *Lactobacillus acidophilus* in the presence of yoghurt bacteria. Int. Dairy J. 9, 497–505.

Von Aulock, M.H.M. and Holzapfel, W.H. (1987) Frequency and occurrence of *Lactobacillus acidophilus* in the gut of the pig, as indicated by its presence in faeces. Onderstepoort J. vet. Res. 54, 581–583.

Weiler, H.G. and Radler, F. (1970) Milchsäurebakterien aus Wein und von Rebenblättern. Zentralbl. Bakteriol. Parasitenkd. Infektionskr. Hyg. Abt. 2 Orig. 124, 707–732.

Whittenbury, R. (1965a) The enrichment and isolation of lactic acid bacteria from plant material. Zentralbl. Bakteriol. Parasitenkd. Infektionskr. Hyg. Abt. 1 Suppl. 1, 395–398.

Whittenbury, R. (1965b) A study of some pediococci and their relationship to *Aerococcus viridans* and the enterococci. J. Gen. Microbiol. 40, 97–106.

Yoshizumi, H. (1975) A malo-lactic bacterium and its growth factor. *In*: J.G. Carr, C.V. Cutting and G.C. Whiting (Eds). Lactic Acid Bacteria in Beverages and Food. Academic Press, London pp. 87–102.

Yuki, N., Watanabe, K., Mike, A., Tagami, Y., Tanaka, R., Ohwaki, M. and Morotomi, M. (1999) Survival of a probiotic, *Lactobacillus casei* Shirota, in the gastrointestinal tract: selective isolation from faeces and identification using monoclonal antibodies. Int. J. Food Microbiol. 48, 51–57.

Zúniga, M., Pardo, I. and Ferrer, S. (1993) An improved medium for distinguishing between homofermentative and heterofermentative lactic acid bacteria. Int. J. Food Microbiol. 18, 37–42.

Chapter 9

Culture media for non-sporulating Gram positive, catalase positive food spoilage bacteria

G A Gardner

Beechwood Laboratories, 120 Ballymena Road, Doagh, Ballyclare, BT 39 0TL, UK

Conditions for the isolation of Gram positive, catalase positive food spoilage bacteria are discussed and media for the selective isolation of Micrococcaceae and *Brochothrix thermosphacta* described. No selective media are available for the isolation of *Microbacterium, Kurthia, Brevibaterium* or *Propionibacterium* spp.

An outline scheme for identification is proposed.

Introduction

Organisms associated with food spoilage are generally recognised by their predominance in numbers or as a percentage of the microbiological flora of the food. The total process of isolation which encompasses sample preparation, choice of culture medium and incubation conditions (time, temperature, atmosphere) may not completely reflect the conditions in the food being examined. Thus colonies that develop on a total count medium may not be wholly representative of the organisms responsible for spoilage of the food.

Conventional bacteriological techniques used in the examination for spoilage organisms include performance of a 'total' colony count, isolation of known spoilage organisms using selective media and identification of suspect colonies.

Total colony counts

The principles involved in the detection of micoorganisms in foods by culture have been reviewed by Holbrook (2000). The choice of medium and the associated cultural procedures should reflect the original conditions in the food to minimize, as far as possible, the creation of a result owing more to laboratory artefacts than the true microbiological situation in the food.

Many media have been described for the determination of total colony counts in foods. Most used are Plate Count Agar, Nutrient Agar or Blood Agar Base but the present author's experience and preference is for Total Count Medium (Gardner,

1968) on which all of the organisms described in this chapter grow well. This medium consists of (g/l) peptone (Oxoid L37; 10), Lab Lemco (Difco; 10), yeast extract (2), sodium chloride (5), glucose (1). It should be incubated aerobically at 22°C for 4 days.

Colonies to be identified should be selected in such a way as to represent accurately the flora of the food being tested.

Selective isolation media

Micrococcaceae

Members of this family have long been recognised as spoilage bacteria, particularly in salted foods such as bacon (Kitchell, 1962; Gardner, 1983) and dry-cured hams. Their source may be either pig or human skin (Baird-Parker, 1962) or the brine or salt used in the curing process (Cordero and Zumalacárregui, 2000). Foods with a high (5–10%) salt content select for members of this family which have a high salt tolerance.

For the selective isolation of these bacteria, a general purpose total colony count medium with added salt, Tryptone-Yeast-Glucose agar with 6% NaCl (Cavett, 1962), has been used successfully, as has Mannitol Salt Agar which contains 7.5% NaCl (Eddy and Ingram, 1962; Cordero and Zumalacarreglui, 2000). Aerobic incubation should be carried out at 20–30°C for 3–5 days. It is important that the diluents used in such analyses contain 4–6% NaCl.

Such systems are not totally selective and confirmation of the identity of isolates is required. All Gram positive, catalase positive colonies can be regarded as members of the Micrococcaceae.

Brochothrix spp.

This genus, originally comprising only one species, called '*Microbacterium thermosphactum*', was originally isolated and described by McLean and Sulzbacher (1953) and subsequently allocated to a new genus, *Brochothrix* (Sneath and Jones, 1976). Its importance in meat spoilage has been reviewed by Gardner (1981) and Skovgaard (1985).

Selective isolation may be made on the Streptomycin Thallous Acetate Actidione (STAA) medium of Gardner (1966) which is described in Part 2 of this book. Strains from non-meat sources have been isolated on STAA to which nalidixic acid and oxycillin were added (Talon et al., 1988). In their study a new species, *Brochothrix campestris* was described. Holzapfel (2002) reported a further modification of this medium.

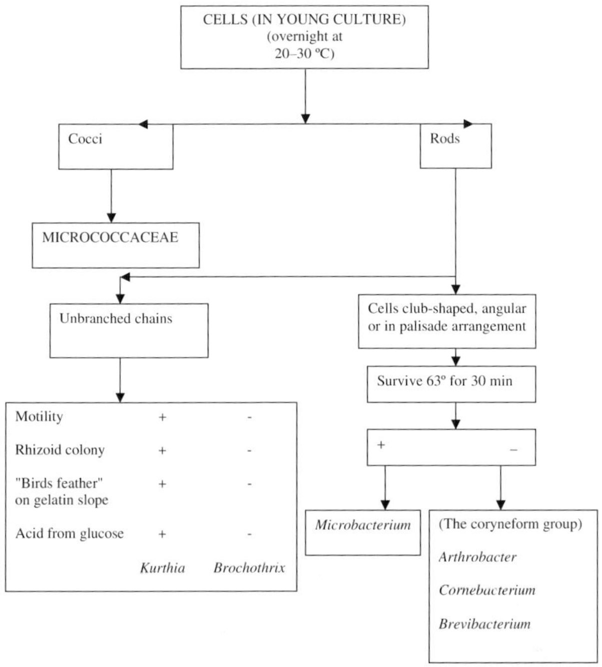

Fig. 1. Simplified scheme for the identification of Gram-positive, catalase-positive, non-sporing bacteria.

Other genera

There are currently no definitive selective media for *Microbacterium*, *Kurthia*, *Brevibacterium* or *Propionibacterium* spp. Many of these can be found in a food spoilage situation and they may be encountered as a small proportion of the flora in fresh food. Most, however, are regarded as beneficial, particularly in the production of cheeses (Teuber, 2000) and, in such situations, members of other genera such as *Arthrobacter* and *Corynebacterium* may also be found.

Identification

As stated above, none of the media described can be regarded as specific for a particular genus or species. A scheme for identification is shown in Figure 1. Further details of methods of identification are given in Collins and Keddie (1986) for *Microbacterium* spp., in Gardner (1968) and Keddie and Jones (1992) for *Kurthia* spp., in Jones and Keddie (1986) and Collins (1992) for *Brevibacterium* spp. and in Malik et al. (1968), Britz and Holzapfel (1973) and Cummins and Johnson (1992) for

Propionibacterium spp. It should be noted that Gram stain interpretation and age of culture at the time of testing is of paramount importance. Both young (< 24 hr) and old (> 3 days) cells should be examined for Gram reaction and morphology on surface colonies on a nonselective agar incubated at 20–30°C.

References

Baird-Parker, A.C. (1962) The occurrence and enumeration, according to a new classification of micrococci and staphylococci in bacon and on human and pig skin. J. Appl. Bacteriol. 25, 352–361.

Britz, T.J. and Holzapfel, W.H. (1973) The suitabillity of different media for the enumeration of propionibacteria. S. Afr. J. Dairy Technol. 5, 213–216.

Cavett, J.J. (1962) The microbiology of vacuum packed sliced bacon. J. Appl. Bacteriol. 25, 282–289.

Collins, M.D. (1992) The genus *Brevibacterium. In:* Balows, A.,. Truper, H.G., Dworkin, M., Harder, W. and Scheiffer, K.H. (Eds). The Prokaryotes, 2nd edn. Vol. 11. Springer Verlag. New York, NY. pp 1351–1354.

Collins, M.D. and Keddie, R.M. (1986) Genus *Microbacterium. In:* P.H.A. Sneath, N.S. Mair, M.E. Sharpe and J.G. Hold (Eds) Bergey's Manual of Determinative Bacteriology, Vol. 2. Williams and Wilkins, Baltimore, MD. pp 1322–1323.

Cordero, M.R. and Zumalacárregui, J.M. (2000) Characterisation of Micrococcaceae isolated from salt used for Spanish dry-cured ham. Lett. Appl. Microbiol. 31, 303–306.

Cummins, C.S. and Johnston, J.L. (1992) The genus *Propionibacterium. In:* Balows, A.,. Truper, H.G., Dworkin, M., Harder, W. and Scheiffer, K.H. (Eds). The Prokaryotes, 2nd edn. Vol. 11. Springer Verlag. New York, NY. pp 834–849.

Eddy, B.P. and Ingram, M. (1962) The occurrence and growth of staphylococci on packed bacon with special reference to *Staphylococcus aureus*. J. Appl. Bacteriol. 25, 237–247.

Gardner, G.A. (1966) A selective medium for the enumeration of *Microbacterium thermosphactum* in meat and meat products. J. Appl. Bacteriol. 29, 455–460.

Gardner, G.A. (1968) Effects of pasteurisation or added sulphite on the microbiology of stored vacuum packed bacon burgers. J. Appl. Bacteriol. 31, 462–478.

Gardner, G.A. (1969) Physiological and morphological characteristics of *Kurthia zopfii* isolated from meat products. J. Appl. Bacteriol. 371–380.

Gardner, G.A. (1981) *Brochothrix thermosphactum (Microbacterium thermosphactum)* in the spoilage of meats: A review. *In:* Roberts, T.A., Hobbs, G., Christian, J.H.B. and Skovgaard, N. (Eds) Psychrotrophic Microorganisms in Spoilage and Pathogenicity. Academic Press, London. pp 139–173.

Gardner, G.A. (1983) Microbial spoilage of cured meats. *In:* Roberts, T.A. and Skinner, F.A. (Eds) Food Microbiology: Advances and Prospects. Academic Press, London. pp 179–202.

Holbrook, R. (2000) Detection of micro-organisms in foods - principles of culture methods. *In* Lund, : B.M., Baird-Parker A.C. and Gould, G.W. (Eds) The Microbiological Safety and Quality of Food . Aspen Publishers, Gaithersburg MD. pp 1761–1790.

Holzapfel, W.H. (1995) Culture media for non-sporulating Gram-positive food spoilage bacteria. In: Corry, J.E.L., Curtis, G.D.W. and Baird, R.M. (eds) Culture Media for Food Microbiology. Progress in Industrial Microbiology, volume 34, Elsevier, Amsterdam, ISBN 0 444 814 98.

Jones, D. and Keddie, R.M. (1986) Genus *Brevibacterium. In:* P.H.A. Sneath, N.S. Mair, M.E. Sharpe and J.G. Hold (Eds) Bergey's Manual of Determinative Bacteriology, Vol. 2. Williams and Wilkins, Baltimore, MD. pp 1301–1313.

Keddie, R.M. and Jones, D. (1992) The genus *Kurthia. In:* A. Balows, H.G. Truper, M. Dworkin, W.

Harder and K.H. Scheiffer (Eds). The Prokaryotes, 2nd edn. Vol. 11. Springer Verlag. New York, NY. pp 1654–1662.

Kitchell, A.G. (1962) Micrococci and coagulase negative staphylococci in cured meats and meat products. J. Appl. Bacteriol. 25, 416–431.

Malik, A.C., Reinhold, G.W. and Vendamuthu, E.R. (1968) An evaluation of the taxonomy of *Propionibacterium*. Can. J. Microbiol. 14, 1185–1191.

McLean, R.A. and Sulzbacher, W.L. (1953) *Microbacterium thermosphactum* spec. nov: a non heat resistant bacterium from fresh pork sausage. J. Bacteriol. 65, 428–433.

Skovgaard, N. (1985) *Brochothrix thermosphactum*: comments on its taxonomy, ecology and isolation. Int. J. Food Microbiol. 2, 71–79.

Sneath, P.H.A. and Jones, D. (1976) *Brochothrix*, a new genus tentatively placed in the family Lactobacillaceae. Int. J. Syst. Bacteriol. 26, 102–104.

Talon, R., Grimont, P.A.D., Grimont, F., Gasser, F. and Boefgras, J.M. (1988) *Brochothrix campestris* sp. nov. Int. J. Syst. Bacteriol. 38, 99–102.

Teuber, M. (2000) Fermented milk products. *In:* B.M. Lund, Baird-Parker, A.C. and Gould, G.W. (Eds) The Microbiological Safety and Quality of Food, Aspen Publishers, Gaithersburg MD. pp 537–589.

Chapter 10

Media for the detection and enumeration of bifidobacteria in food products

Denis Roy

Food Research and Development Centre, Agriculture and Agri-Food Canada, 3600, Casavant Blvd. W. St-Hyacinthe, Quebec, Canada J2S 8E3

Bifidobacteria are commonly used for the production of fermented milks in combination with other lactic acid bacteria. It is important that bifidobacteria claimed to be present in a dairy product should survive in relatively high viable cell numbers (>10^6 per g) until consumption. Hence, the need exists for rapid, reliable methods for enumeration of bifidobacteria, both to routinely determine the initial inoculum and to estimate the time bifidobacteria remain viable. Plate count methods are still preferable for quality control measurements in dairy products. A selective medium is therefore, necessary that promotes the growth of bifidobacteria, but suppresses other bacteria. This chapter reviews media and methods, including summaries of published comparisons between different selective media. Culture media for bifidobacteria may be divided into basal, elective, differential and selective. Non selective media such as RCM (Reinforced Clostridial Medium) and MRS (de Man, Rogosa and Sharpe) agar, are useful for enumeration of bifidobacteria when present as pure cultures in non-fermented milk or in fermented milks made only with bifidobacteria. Many selective or differential isolation media have been described for enumeration of bifidobacteria in the presence of other lactic acid bacteria. There is no standard medium for the detection of bifidobacteria. However, Columbia agar base medium supplemented with lithium chloride and sodium propionate and MRS medium supplemented with neomycin, paromomycin, nalidixic acid and lithium chloride can be recommended for selective enumeration of bifidobacteria in food products.

1. Introduction

The use of bacteria of intestinal origin has resulted in a new generation of 'functional' food products, which exploits the beneficial effects of these bacteria on intestinal metabolism (Mitsuoka, 2000). A 'functional food' contains one or a combination of components which have positive cellular or physiological effects on the body (Gibson and Fuller, 2000). Dairy products are functional foods because they are one of the best sources of calcium. Fermented dairy products are also important functional foods because they include probiotics, prebiotics and synbiotics. Probiotics are defined as "live microbial feed supplements which beneficially affect the host animal by improving its intestinal microbial balance" (Fuller, 1989) whereas a prebiotic is defined as "a nondigestible food ingredient that beneficially affects the host by selectively stimulating the growth and/or activity of one or a limited number of bacteria in the colon" (Gibson and Fuller, 2000). Synbiotics are combinations of prebiotics and probiotics. Compared to probiotics, prebiotics and synbiotics are increasing in importance in the world nutraceutical market.

Strains of the *Lactobacillus acidophilus* and *Lactobacillus casei* complex are well represented in commercial probiotic products, followed by *Bifidobacterium* spp. (*B. animalis, B. bifidum, B. breve, B. infantis* and *B. longum*), some other lactic acid bacteria (lactococci, leuconostocs, enterococci) and non-lactic acid bacteria (propionibacteria, yeasts) (Holzapfel et al., 1998). The consumption of these micro-organisms may affect the composition of indigenous microflora and may have several beneficial effects on the human health such as the maintenance of a balanced flora, alleviation of lactose intolerance symptoms, resistance to enteric pathogens, immune system modulation, antihypertensive effect as well as certain anti-carcinogenic effects. Probiotics can also prevent or ameliorate diarrhoea through their effects on immune system (De Roos and Katan, 2000). However, it is important to note that any postulated benefits from consumption of probiotics should be accepted as facts only after extensive testing in human clinical studies (Rolfe, 2000).

Bifidobacteria are one of the most important groups of probiotic intestinal organisms. *Bifidobacterium bifidum, B. breve, B. longum* and *B. animalis* are commonly used for the production of fermented milks, in combination with other lactic acid bacteria. High viable counts and survival rates during passage through the stomach are necessary to allow live bifidobacteria from the fermented milk products to play a biological role in the human intestine. The technological criteria for selection of strains include ability to survive in relatively high viable cell numbers, retain metabolic activity and provide desirable organoleptic qualities (Holzapfel et al., 1998; Gibson and Fuller, 2000).

In general, the food industry aims to achieve over 10^6 bifidobacteria/g at the time of consumption. This target seems to have been adopted because it is technologically attainable and cost-effective rather than to achieve a specific health effect in humans (Sanders et al., 1996). Rapid, reliable methods for enumeration of bifidobacteria are needed to determine the initial inoculum and to estimate the time post-production that these organisms remain viable. Plate count, molecular genetic or enzymic methods could be used. Plate count methods are still preferred for food products. It is, therefore, necessary to have a medium that selectively promotes the growth of bifidobacteria, and suppresses other bacteria. Many media have been proposed and used over the past sixty years for the isolation, cultivation and enumeration of bifidobacteria from the intestinal tract or faeces, and for the detection and enumeration of bifidobacteria in food products, mainly fermented milks. This chapter reviews media and methods including summaries of published comparisons between different media.

2. Food products containing bifidobacteria

For a food product to contain over 10^6 bifidobacteria/g at the time of consumption it must contain between 1 and 100 million bifidobacteria/g when placed on store shelves. Freeze-drying or lyophilization, is a method of maintaining product stability and usually does not cause significant harm to the organisms. Products that are not freeze-dried, such as liquid probiotic supplements or yoghurt, have a much shorter shelf life. Viability and survival of bifidobacteria is one of the important points to evaluate in using freeze-dried cultures. Dairy products containing bifidobacteria are made with pure cultures, alone or in combination with other lactic acid bacteria such as *Streptococcus thermophilus*, *Lactobacillus delbrueckii* subsp. *bulgaricus*, and *Lb. acidophilus*- or *Lb. casei*-related taxa. Dairy-related bifidobacteria are already used in a wide variety of probiotic dairy products including milk, bifidus milk, cheese, frozen yogurt and ice cream. Yogurt-type products are made using a single genus of bifidobacteria in combination with other lactic acid bacteria. Bifidobacteria can be incorporated into yogurt-type products before or after fermentation. The survival of bifidobacteria in fermented dairy products depends on various factors such as the strain of bacteria used, fermentation conditions, storage temperature, and preservation methods. Post-acidification of yogurt by *Lb. delbrueckii* subsp. *bulgaricus* also contributes to increased reduction of viable counts of bifidobacteria. Hence, freeze-dried cultures and dairy products which claim to contain bifidobacteria, raises the need for appropriate methods to enumerate them.

3. Description of bifidobacteria

It is estimated that over 400 species of bacteria inhabit the human gastrointestinal tract. The *Bifidobacterium* species belong to the dominant anaerobic flora of the colon and are capable of exerting some probiotic effects. Tissier (1990) first described bifidobacteria and noted that they were the dominant bacteria in the stools of breast-fed infants and were more numerous than in the stools of bottle-fed infants. Although breast milk does not contain bacteria, it does contain sugars that promote the growth of bifidobacteria.

Bifidobacteria are Gram-positive, nonsporeforming, strictly anaerobic and pleomorphic fermentative rods, often Y-shaped. The G+C content of DNA varies from 55–67 mol %. The optimum growth temperature for bifidobacteria is 37 to 41 °C (minimum growth temperature: 25–28 °C; maximum growth temperature: 43–45 °C) and the optimum pH is 6.5 to 7.0 (no growth at 4.5–5.0 or 8.0–8.5). Glucose is degraded exclusively and characteristically by the fructose-6-phosphate shunt. Fructose-6-phosphate phosphoketolase is the characteristic key enzyme of the bifid shunt that cleaves fructose-6-phosphate into acetyl phosphate and erythrose-4-phosphate. Acetic and lactic acid are formed primarily in the molar ratio of 3:2 (Scardovi, 1986).

Thirty-three species of bifidobacteria are now included in this genus and twelve of these species have been found in the intestine of man and/or as human clinical isolates. The remaining 21 species have been isolated from fermented milk, the alimentary tracts of various animals and honeybees and from sewage and anaerobic digesters. Bifidobacteria constitute 5 to 10% of the total faecal flora of healthy children and adults. In the days following birth, the intestinal flora is dominated by bifidobacteria. With age and changes in dietary habits, bifidobacteria tend to be suppressed by other micro-organisms, with even lower numbers in the elderly (Mitsuoka, 2000). The main species present in the human colon are *Bifidobacterium adolescentis, B.bifidum, B.infantis, B.breve* and *B.longum*. Although conclusive clinical evidence substantiating the benefits of consuming bifidobacteria is not available, there is accumulating evidence that bifidobacteria are associated with beneficial health effects. It has been claimed that ingestion of specific bifidobacteria could contribute to reestablishment of a bifidobacterial flora in humans after antibiotic therapy; alleviation of constipation; prevention of diarrhoea and other gastrointestinal infections and alleviation of the symptoms of lactose intolerance (O'Sullivan and Kullen, 1998).

4. Culture media

Culture media used for the detection and enumeration of bifidobacteria may be divided into complex, selective, semisynthetic and synthetic, as well as commercial, and can be classified as non-selective media with elective carbohydrates, media with antibiotics, media with sodium propionate and lithium chloride and media with elective substance and/or low pH. Media belonging to more than one group are also used (Hartemink and Rombouts, 1999). No standard medium for the detection of bifidobacteria has emerged from the large number of media published. However, the availability of easy and inexpensive methods for detection, identification, and enumeration of *Bifidobacterium* spp. is important in food microbiology.

The selection of an adequate culture medium for bifidobacteria should be based on the following parameters: supply of nutritive substances to produce optimal growth; low redox potential; maintenance of pH value during growth by an effective buffering capacity; final pH of the prepared medium; optimal growth medium. Anaerobic conditions are also an important factor in detecting and enumerating bifidobacteria. The success of bifidobacteria detection in an optimal growth medium is mainly dependent upon the following factors : a) if the culture medium has no selective effect, non-bifidobacteria may outgrow bifidobacteria; b) the ease of macroscopic identification of bifidobacteria colonies which may be facilitated using indicators; c) the freshness of the ingredients of the medium; d) the composition of the culture medium which should allow the growth of different bio-types present in the material investigated (Rasic and Kurmann, 1983).

4.1. In food products when bifidobacteria are the only fermenting microorganisms present

Non-selective media are useful for routine enumeration of bifidobacteria when present in pure culture in non-fermented milks and in milk fermented with bifidobacteria in order to determine the initial inoculum and to ascertain the length of time these organisms remain viable during storage. Scardovi (1986) observed that, as bifidobacteria vary widely in their physiological requirements for growth, no one selective medium is appropriate for all species. Preference should be given to substrates that permit satisfactory growth of the largest number of bifidobacteria types presently known. The ingredients of choice are trypticase and phytone that are used in a complex, non-selective medium (Trypticase-Phytone-Yeast extract or TPY) which will allow the growth of all *Bifidobacterium* species as well as other lactic acid bacteria (Scardovi, 1986).

Table 1 shows basal and elective culture media recommended for the cultivation of bifidobacteria. Numbers of bacteria detected with synthetic or low-pH media can be lower than in broths with a complex composition and in those with higher pH values (6.8–7.4) (Pacher and Kneifel, 1996). The preparation of liver infusion for Blood-Liver (BL) broth and Blaurock medium is a messy task and cannot, therefore, be recommended for routine use. Modified de Man Rogosa and Sharpe (mMRS) medium (MRS containing cysteine-HCl) can provide optimal overall growth conditions for the bifidobacteria. L-cysteine is added to lower the redox potential and provides better anaerobic conditions for the growth of bifidobacteria. This amino acid is also regarded as an essential nitrogen source for bifidobacteria

The solid non-selective media commonly used for detection of bifidobacteria are: reinforced clostridial agar (RCA) and RCA plus lactose and sheep blood; BL agar, BL agar without blood, Rogosa's modified agar (RM agar), Tryptone phytone yeast extract agar (TPY), modified MRS agar. One of the first useful media for the enumeration of bifidobacteria was Lactobacillus agar (Wijsman et al., 1989; Reuter, 1990). Mitsuoka suggested the use of BL agar for non-selective enumeration of bifidobacteria from dairy products and intestinal materials (Teraguchi et al., 1978).

Studies by Arroyo et al. (1994) found that the commercially available media, mMRS and RCA and laboratory prepared medium, mBL agar, supported excellent growth of bifidobacteria, but mBL agar was not as time or cost effective as mMRS and RCA. Hence, as they are commercially available, RCA and MRS would likely be the media of choice for industrial quality control laboratories (Arroyo et al., 1994).

Culture media can be modified in order to improve the growth conditions for bifidobacteria by using elective substances such as the short-chain fatty acids, propionate and butyrate (Table 1). Beerens (1990) noted that the addition of propionic acid to Columbia agar at pH 5.0 enhanced the growth of bifidobacteria. Although Hartemink et al. (1996) found that butyrate did not increase selectivity whereas the addition of valerate could inhibit bifidobacteria. Certain components such as yeast extract, peptone, cysteine and starch are essential for these microorganisms. Yeast extract is considered an excellent growth promoter and several growth promoting substances derived from human and bovine whey can also be added to culture media, for instance α-lactalbumin and β-lactoglobulin (Ibrahim and Bezkorovainy, 1994).

The exploitation of certain enzymatic properties of bifidobacteria by the inclusion of chromogenic substrates (particularly those detecting α-galactosidase) in media offers a promising tool for their specific and rapid detection (Chevalier et al., 1991; Pacher and Kneifel, 1996).

Oligosaccharides are used as bifidogenic factors to stimulate the growth of bifidobacteria. It is thought that these carbohydrates are selectively fermented by bifidobacteria. Oligosaccharides with bifidogenic activity are: lactulose, trans-galactosyl oligosaccharides (TOS), fructo-oligosaccharides (FOS), isomalto-oligosaccharides, raffinose and soybean oligosaccharides. α-Galactosaccharides such as melibiose

Table 1
Basal and elective (non selective) culture media used for bifidobacteria.

Medium abbreviation	Medium name	Final concentrations of additives	Comments	References
MRS	de Man Rogosa Sharpe	None	Basal	de Man et al., *1960*
TPY	Tryptone Phytone Yeast	None	Basal	Scardovi, 1986
BL	Blood Liver	None	Basal	Mitsuoka et al., 1965
CLB	Columbia	None	Basal	Ellner et al., 1958
LCL	Liver Cysteine Lactose (Blaurock)	None	Basal	Rasic and Sad, 1990
RCM	Reinforced Clostridial Agar	None	Basal	Hirsch and Grinsted, 1954
mMRS	Modified MRS	L-cysteine- HCL, 0.05%	Elective	Sykes and Skinner, 1973
mMRS + blood	Modified MRS	L-cysteine- HCL, 0.05% Sheep blood 10 ml	Elective	Pacher and Kneifel, 1996
X-α-Gal*	MRS	X-α-Gal	Elective	Chevalier et al., 1991
mBL	Modified BL without blood	L-cysteine- HCL, 0.05%	Elective	Arroyo et al., 1994
mRCM	Modified RCM	Lactose 1.0% Human blood 50 ml	Elective Differential	Reuter, 1990
RCPB	RCM with Prussian Blue	Prussian Blue 0.03%	Differential	Onggo and Fleet, 1993

* X-α-Gal = 5-bromo-4-chloro-3-indolyl-α-D-galactopyranoside.

and raffinose are metabolised by bifidobacteria and, because α-galactosidase activity of bifidobacteria participates in the first degradation of these saccharides, TOS can stimulate growth of bifidobacteria in a similar way to raffinose but growth of *B. bifidum* is better with TOS. However, TOS is as yet not available commercially in large quantifies and high purity whereas raffinose is easily available.

4.2. Media for use on food products where bifidobacteria occur with other lactic acid bacteria

Several media have been developed for differential enumeration of bifidobacteria from other lactic acid bacteria. Reuter (1990) proposed RCA (Reinforced Clostridial Agar) plus 1% lactose supplemented with 5% blood or 2% erythrocyte concentrate for total anaerobic counts and for the separation of bifidobacteria in yoghurt-like products. Acidified media such as MRS or Rogosa agar have been used successfully by some research groups for enumeration of bifidobacteria in mixed cultures, while others have reported their limited selectivity (Tamime et al., 1995). Reinforced Clostridial Prussian Blue Agar (RCPB) has been used for the differential enumeration of *Lactobacillus delbrueckii* subsp. *bulgaricus*, *Streptococcus thermophilus* and bifidobacteria. Bifidobacteria appeared as white colonies while *Lb. delbrueckii* subsp. *bulgaricus* and *S. thermophilus* formed pale blue colonies surrounded by wide royal blue or thin light blue zones, respectively. However, in some cases it was extremely difficult to distinguish one species from another when mixed cultures were used (Playne et al., 1999).

4.3. Culture media for enumeration of bifidobacteria

For the purpose of proper and correct enumeration, various attempts have been made to obtain a selective medium for isolating *Bifidobacterium* spp. from other dairy-related microorganisms, especially streptococci and lactobacilli. Methods to detect and enumerate bifidobacteria in products containing a mixed bacterial population are based on the use of more or less traditional media for lactic acid bacteria or anaerobes, supplemented with various selective agents or a single carbon source that inhibit or reduce the growth of other lactic acid bacteria.

These selective media often contain antibiotics which reduce bifidobacteria counts and usually contain lactose or glucose as a carbohydrate substrate, which enhances the growth of all the bacteria used for the production of yogurts. In many cases, the enumeration of low numbers of bifidobacteria is still subject to interference by *S. thermophilus*, which appears as pinpoint colonies. Generally, selective media for bifidobacteria control the growth of lactic acid bacteria at dilutions higher than 10^{-5}. The most important are listed in Table 2.

Scardovi (1986) reviewed and discussed several complex media, and media containing a wide variety of antibiotics for enumeration of *Bifidobacterium* species. Various selective agents are used, including kanamycin, neomycin, paromomycin, sodium propionate, lithium chloride; sorbic acid and sodium azide. Few media are truly selective for bifidobacteria and some of them are time-consuming to prepare.

Table 2
Selective media used for the enumeration of bifidobacteria in fermented dairy products.

Media	Base	Final concentrations of non-selective additives	Selectivity based on,	mg/l	References
NPNL agar	BL	None	NPNL solution*		Teraguchi et al., 1978
MRS-NPNL	MRS	None	NPNL solution*		Dave and Shah, 1996
TOS-NPNL	N/A	None	neomycin	30	Wijsman et al., 1989
			paromomycin	60	
			nalidixic acid	4.5	
			LiCl	900	
TPY + NPNL	TPY	None	neomycin	30	Ghoddussi and Robinson 1996
			paromomycin	30	
			nalidixic acid	4.5	
			LiCl	900	
BL-OG	BL	None	Oxgall,	300	Lim et al., 1995
			Gentamicin,	40	
RMS + PPNL	Rogosa agar	None	Sodium propionate	15000	Samona and Robinson, 1991
			Neomycin	200	
			Paromomycin	50	
			LiCl	3000	
MRS + Dic	MRS	None	Dicloxacillin	2	Sozzi et al., 1990
TPY + Dic	TPY	None	Dicloxacillin	2	Sozzi et al., 1990
mCAB	Modified Columbia agar base	Glucose, 0.5% pH adjusted to 5.0	Propionic acid, 5 ml		Beerens, 1990, 1991
Ecobion 2	CAB	Tween 0.1%	Neomycin	100	Chapon and Kiss, 1991
		Glucose 2%	LiCl	3000	
		L-cysteine, 0.3%;			
DP	CAB	glucose, 0.5%	Dicloxacillin	2	Bonaparte, 1997
		L-cysteine, 0.05%; pH adjusted to 6.8	Propionic acid, 5 ml		
LP	LCL	None	LP mixture*		Lapierre et al., 1992

Table 2. Continued.

Media	Base	Final concentrations of non-selective additives	Selectivity based on,	mg/l	References
RAF 5.1	CAB	L-cysteine, 0.05%; raffinose, 0.5%; pH adjusted to 5.1	LP mixture*		Roy et al. 1997
RB		L-cysteine, 0.05%; raffinose, 0.75% sodium caseinate, 0.5% bromocresolpurple 0.015%	LiCl Sodium propionate	5000 3000	Hartemink et al., 1996
BFM	N/A	Lactulose 0.5% pH adjusted to 5.1	LiCl Methylene blue Propionic acid ,5 ml	2000 16	Nebra and Blanch, 1999
GL agar	N/A	Galactose 1.0%	LiCl	400	Iwana et al., 1993
BIM-25	RCA	None	Nalidixic acid Polymycin B Iodoacetic acid 2,3,5-triphenyltetrazolium chloride	200 8.5 50 25 25	Munoa and Pares, 1988
AMC	RCA	None	Nalidixic acid Polymycin B Iodoacetic acid 2,3,5-triphenyltetrazolium chloride LP mixture*	200 8.5 50 25 25	Arroyo et al., 1994, 1995

* Made from stock solution as follows:
LP mixture = LiCl, 2 g L^{-1}; sodium propionate, 3 g L^{-1}.
NPNL solution = (L^{-1}) Neomycin sulphate, 100 mg; Paromomycin, 200 mg; Nalidixic acid, 15 mg; LiCl, 3g.

Reliable enumeration of bifidobacteria seems to be achieved only if a microbiologist knows which particular *Bifidobacterium* strain was used in the product (Pacher and Kneifel, 1996).

NPNL (neomycin, paromomycin, nalidixic acid, lithium chloride)-based media

Teraguchi et al. (1978) developed an improved selective medium (NPNL) for the enumeration of bifidobacteria in dairy products which contained mixtures of bifidobacteria, lactobacilli and streptococci. This medium contained BL-agar with added neomycin sulphate, paromomycin sulfate, nalidixic acid and lithium chloride. NPNL agar is considered to be the reference medium for the isolation of bifidobacteria from fermented dairy products because it seems that these bacteria grow well in this medium. Neomycin sulphate and nalidixic acid were included as selective agents to inhibit growth of Gram-positive and Gram-negative rods, respectively. Lithium chloride is a substance commonly used for the selective isolation of bifidobacteria, although its mechanism of action is still poorly understood. Rasic and Sad (1990) found that recovery of bifidobacteria on NPNL agar with blood was around 90% and recovery of lactic acid bacteria was less than 2% as compared with BL agar without blood. Although the International Dairy Federation (IDF; Rasic and Sad, 1990) suggested the use of NPNL agar for isolation of bifidobacteria from fermented milks, the medium was not included in the comparative study of Bonaparte (1997). This was because (1) preliminary trials showed that the recovery rates (productivity) of some *Bifidobacterium* species on NPNL were low; (2) it appeared to have only weak inhibitory effect on *S. thermophilus*; (3) NPNL agar is time-consuming to prepare because it contains twenty-four ingredients, some of which must be filter-sterilized. It should be noted that careful measurement of ingredients and treatment of the NPNL medium is essential for consistent and reliable results.

MRS-NPNL agar (Dave and Shah, 1996) is widely used for the recovery and enumeration of bifidobacteria by researchers and quality control laboratories. However, the counts obtained on MRS-NPNL agar may not be representative of the viable cells that are present in the product, which suggests a need to check the efficacy of this medium with pure cultures before the medium is adopted for enumeration purposes.

Blood Liver – Oxgall Gentamicin (BL-OG) agar

Many authors have developed selective media based on the use of commercial medium or with a simpler composition than NPNL agar. Lim et al. (1995) proposed the use of BL agar containing oxgall (0.2mg/ml) and gentamicin (30 μg/ml) (BL-OG agar) which is much simpler to prepare than NPNL agar and gives higher recovery of bifidobacteria in yogurts. Oxgall was used to inhibit other non-intestinal lactic acid bacteria. Gentamicin, in studies on the antibiotic sensitivity of 37 strains of bifidobacteria and 58 strains of lactic acid bacteria, was found to have optimal selectivity for bifidobacteria (Lim et al., 1995). Recovery of bifidobacteria on BL-OG agar was around 90% or higher compared with that on BL agar. All tested strains of *Lb. acidophilus, Lb. delbrueckii* subsp. *bulgaricus* and *Lb. casei* were inhibited in BL-OG and NPNL. Strains of streptococci were inhibited on BL-OG whereas NPNL agar was less inhibitory to *S. thermophilus* strains. BL-OG also showed good inhibition to yoghurt culture and *Lb. acidophilus.* However, as for NPNL, the major disadvantage of this medium is that it is time-consuming to prepare.

Other antibiotic-based media

The antimicrobial activity of 30 antibiotics toward lactic acid bacteria was examined by Sozzi et al. (1990) using MRS and TPY media. Only nafcillin and dicloxacillin were found to inhibit streptococci and lactobacilli. Wijsman et al. (1989) found that MRS-nafcillin agar showed no inhibition of streptococci and lactobacilli whereas Sozzi *et al.* (1990) indicated that MRS and TPY media supplemented with 2 µg/L dicloxacillin behave similarly to NPNL agar medium.

Modified Columbia agar

Beerens (1991) demonstrated that propionic acid added to Columbia agar (mCAB) at pH 5.0, enhanced the growth of bifidobacteria. This agar was described by Beerens (1991) as selective and elective for bifidobacteria. CAB medium with lithium chloride and neomycin sulphate (Ecobion 2) was found by Chapon and Kiss (1991) to be more selective that mCAB. However, mCAB is more appropriate for the enumeration of a wider range of strains of bifidobacteria.

Bonaparte (1997) proposed a new combination called DP medium which was based on the use of Columbia agar with propionic acid (5 ml) and dicloxacillin (2 mg/l). The pH was adjusted to 6.8 with NaOH. This medium allowed the growth of bifidobacteria, showing good productivity, especially with a commercial strain of *B. longum* whereas *Lb. acidophilus, S. thermophilus* and *Lb. delbrueckii* subsp. *bulgaricus* were inhibited.

Lithium chloride-sodium propionate (LP) – based media (Table 2)

A combination of 2 g/l of lithium chloride and 3 g/l of sodium propionate was added by Lapierre *et al.* (1992) to liver-cystine-lactose (LCL) agar to suppress the growth of lactic acid bacteria and to allow the selective enumeration of bifidobacteria from commercial products. Sodium propionate was used as selective agent for bifidobacteria. With NPNL agar as reference medium, bifidobacteria were isolated from all dairy products tested with LP agar in number similar to those obtained on NPNL agar. LP agar was hence comparable with NPNL agar in terms of the numbers of bifidobacteria enumerated from dairy products, but it had the advantage of being simpler to prepare than the latter medium (NPNL) (Lapierre et al., 1992). Similar growth of bifidobacteria was observed on LP agar and on non-selective Tomato Juice agar, demonstrating that the concentrations of lithium chloride and sodium propionate chosen had no inhibitory effect. In contrast, all strains of lactobacilli were completely inhibited but some mesophilic strains of lactococci were resistant to the concentrations of lithium chloride and sodium propionate used. Lapierre et al. (1992) proposed the incubation of plates at 40 °C to prevent the growth of these strains. However, in their comparison of six selective media for the enumeration of three species of bifidobacteria in pure culture, Playne et al. (1999) found incubation at 37 °C preferable due to the poor recoveries observed at the temperature recommended by Lapierre et al. (1992). In addition, there was significant recovery of yoghurt culture organisms. Playne et al. (1999) concluded that LP was a poor medium for selective enumeration of the strains of *B. longum* and *B. adolescentis* used in their study. Sanders et al. (1996) also evaluated LP agar in order to enumerate each species of bifidobacteria selectively from composite mixtures in milk. Bifidobacteria appeared to be somewhat more sensitive to the selective agents; four of the five strains tested showed statistically significant reduction of 24 to 77% in counts on LP agar than on non selective RCA medium. In products formulated with frozen concentrate cultures, freeze injury of the cells could result in even more dramatic differences in counts between selective and nonselective media. Significantly lower numbers of colonies of psychrotrophs were observed on LP agar at 4 and 10 °C than plate count agar. They concluded that the medium was not ideally selective but should be satisfactory when used to enumerate lactic cultures differentially in products within one week of manufacture or in products that had not been stored at 10 °C.

BIM-25

Munoa and Pares (1988) developed a selective medium called BIM-25 for isolation and enumeration of *Bifidobacterium* spp. from water samples based on RCA supplemented with nalidixic acid, polymyxin B sulphate, kanamycin sulfate, iodoacetic acid and 2,3,5-triphenyltetrazolium chloride. Iodoacetate inhibits glyceraldehyde-3-phosphate dehydrogenase and reduces the growth of nonbifidobacterial contaminant colonies. Silvi et al. (1996) observed that BIM-25 was very selective, inhibiting six of the nine non-bifidobacteria strains tested, but it also inhibited the growth of *Bifidobacterium* strains.

Arroya, Martin and Cotton (AMC) agar

AMC agar (Arroya et al. 1995) was developed by adding the selective components in LP agar to mBIM-1 agar which is a modification of BIM-25 agar. The BIM-25 was modified by halving the iodoacetate concentration. *Lb. acidophilus* was not inhibited by mBIM-1 but completely inhibited on LP agar. AMC supported the growth of bifidobacteria while inhibiting the growth of *Lactococcus lactis subsp. lactis, Lb. delbrueckii subsp. bulgaricus and S. thermophilus.* It was, however, found to allow slight growth *of Lb. acidophilus* (Arroyo et al., 1994, 1995; Playne et al., 1999).

On the basis of performance as a selective medium for the enumeration of bifidobacteria, and ease of preparation, Playne et al. (1999) concluded that AMC was the best choice for the routine enumeration of bifidobacteria in mixed culture in dairy products. Finally, it is important to be aware that the plating technique can make a significant difference to the results. It is advisable to compare spread- and pour-plate techniques in order to select the combination of medium and plating technique giving the most accurate representation of the bifidobacterial viable count (Playne et al., 1999). In conclusion, additional research in this area is required because several bifidobacteria showed significant differences between the selective and non-selective media.

References

Arroyo, L., Cotton, L. N. and Martin, J.H. (1994) Evaluation of media for enumeration *of Bifidobacterium adolescentis, B. infantis and B. longum* from pure culture. Cult. Dairy Prod. J. 29, 20–24.

Arroyo, L., Cotton, L.N. and Martin, J.H. (1995) AMC agar- A composite medium for selective enumeration of *Bifidobacterium lonqum.* Cult. Dairy Prod. J. 30, 12–15.

Beerens, H. (1990) An elective and selective isolation medium for *Bifidobacterium spp* Lett. Appl. Microbiol. 11, 155–157.

Beerens, H. (1991) Detection of bifidobacteria by using propionic acid as a selective agent. Appl. Environ. Microbiol. 57, 2418–2419.

Bonaparte, C. (1997) Selective isolation and taxonomic position of bifidobacteria isolated from commercial fermented dairy products in central Europe. Technischen Universität Berlin, pp. 1–199.

Chapon, J.L. and Kiss,K. (1991) Numération des bifidobactéries dans les laits fermentés. Proposition pour une méthode microbiologique. Trav. Chim. Aliment, Hung. 82, 264–277.

Chevalier, P., Roy, D. and Savoie, L. (1991) X-α-gal based medium for simultaneous enumeration of bifidobacteria and lactic acid bacteria in milk. J. Microbiol. Methods 68, 619–624.

Dave, R.I. and Shah, N.P. (1996) Evaluation of media for selective enumeration of *Streptococcus thermophilus, Lactobacillus delbrueck* ssp. *bulgaricus i. delbrueckii, Lactobacillus acidophilus* and bifidobacteria. J. Dairy Sci. 79, 1529–1536.

De Roos, N.M. and Katan, M.B. 2000. Effects of probiotic bacteria on diarrhea, lipid metabolism, and carcinogenosis: a review of papers published between 1988 and 1998. Am. J. Clin. Nutr. 71, 405–411.

Ellner. P.D., Stoessel, C.I., Drakeford, E. and Mack, E.G. (1958) A new culture medium for medical bacteriology. Am. J. Clin. Path. 29, 181–183.

Fuller, R. (1989) Probiotics in man and animals. J. Appl. Bacteriol. 66, 365–378.

Ghoddussi, H.B. and Robinson, R.K. (1996) Enumeration of starter cultures in fermented milks. J. Dairy Res. 63, 151–158.

Gibson, G.R and Fuller, R. (2000) Aspects of in vitro and in vivo research approaches directed toward identifying probiotics and prebiotics for human use. J. Nutr. 130, 391S–395S.

Hartemink, R. and Rombouts, F.M. (1999) Comparison of media for the detection of bifidobacteria, lactobacilli and total anaerobes from faecal samples. J. Microbiol. Methods 36, 181–192.

Hartemink, R., Kok, B.J., Weenk, G.H. and Rombouts, F.M. (1996). Raffinose-Bifidobacterium (RB) agar, a new selective medium for bifidobacteria. J. Microbiol. Methods 27, 33–43.

Hirsch, A. and Grinsted (1954) Methods for the growth and enumeration of anaerobic sporeformers from cheese, with observations on the effect of nisin. J. Dairy Res. 21, 101–110.

Holzapfel, W.H., Haberer, P., Snel, J., Schillinger, U., and Huis in't Veld, J. H. J. (1998) Overview of gut flora and probiotics. Int. J. Food Microbiol. 41, 85–101.

Ibrahim, S. and A. Bezkorovainy (1994) Growth-promoting factors for *Bifidobacterium longum.* J. Food Sci. 59, 189–191.

Iwana, H., Masuda, H., Fujisawa, T., Suzuki, H. and Mitsuoka, T. (1993) Isolation and identification of *Bifidobacterium* spp. in commercial yoghurts sold in Europe. Bifidobacteria Microflora 12, 39–45.

Lapierre, L., Underland, P. and Cox, L.J. (1992) Lithium chloride-sodium propionate agar for the enumeration of bifidobacteria in fermented dairy products. J. Dairy Sci. 75, 1192–1196.

Lim, K.S., Huh, S., Baek, Y.J. and Kim, H.U. (1995) A selective enumeration medium for bifidobacteria in fermented dairy products. J. Dairy Sci. 78, 2108–212.

Mitsuoka, T. (2000) Significance of dietary modulation of intestinal microflora and intestinal environment. Biosci. Microflora 19, 15–25.

Mitsuoka, T., Sega, T. and Yamamoto, S. (1965) Eine verbesserte Methodik der qualitativen und

quantitativen Analyse der Darmflora von Menschen und Tieren. Zbl. Bakt. I. Abt. Orig. 195, 455–469.
Munoa, F.J. and R. Pares (1988) Selective medium for isolation and enumeration of *Bifidobacterium* spp. Appl. Environ. Microbiol. 54, 1715–1718
Nebra, Y. and Blanch, A.R. (1999) A new selective medium for *Bifidobacterium* spp. Appl. Environ. Microbiol. 65, 5173–5176.
Onggo, 1. and Fleet, G.H. (1993) Media for the isolation and enumeration of lactic acid bacteria from yoghurts. Australian J. Dairy Technol. 48, 89–92.
O'Sullivan, D.J. and Kullen, M.J. (1998) Tracking of probiotic bifidobacteria in the intestine. Int. Dairy Journal 8, 513–525.
Pacher, B. and Kneifel, VV. (1996) Development of a culture medium for the detection and enumeration of bifidobacteria in fermented milk products. Int. Dairy Journal 6, 43–64.
Playne, J.F., Morris. A.E.J. and Beers, P. (1999) Note: Evaluation of selective media for the enumeration of *Bifidobacterium* sp. in milk. J. Appl. Microbiol. 86, 353–358.
Rasic, J.L. and Kurmann, J.A. (1983) Bifidobacteria and their role. Birkhauser Verlag, Basel, Switzerland.
Rasic, J.L. and Sad. N. (1990) Culture media for detection and enumeration of bifidobacteria in fermented milk products. Bulletin of the IDF 252, 24–34.
Reuter, G. (1990) Bifidobacteria cultures as components of yoghurt-like products. Bifidobacteria Microflora 9, 107–118.
Rolfe, R.D. 2000. The role of probiotic cultures in thee control of gastrointestinal health. J. Nutr. 130:396S–402S.
Roy, D., Mainville, I. and Mondou, F. (1997) Selective enumeration and survival of bifidobacteria in fresh cheese. Int. Dairy Journal 7, 785–793.
Samona, A. and Robinson, R.K. (1991) Enumeration of bifidobacteria in dairy products. J. Soc. Dairy Technol. 44, 64–66.
Sanders, M.E., Walker, D.C., Walker, K.M., Aoyama, K., and Klaenhammer, T.R. (1996) Performance of commercial cultures in fluid milk applications. J. Dairy Sci. 79, 943–955.
Scardovi, V. (1986) *Bifidobacterium.* In: Bergey's Manual of Systematic Bacteriology. 9th Ed. Volume 2, P.H. Sneath, N.S. Mair, M.E. Sharpe and J.G. Holt (Eds.) p. 1418. Williams and Wilkins Publishers, Baltimore, MD.
Silvi, S. Rumney, C.L. and Towland, I.R. (1996) An assessment of three selective media for bifidobacteria in faeces. J. Appl. Bacteriol. 81, 561–564.
Sozzi, T., Brigidi, P., Mignot, 0. and Matteuzzi, D. (1990) Use of dicloxacillin for the isolation and counting of Bifidobacteria from dairy products. Lait 70, 357–361.
Sykes, G. and Skinner, F.A. (1973) Techniques for the isolation and characterization of *Actinomyces* and *Bifidobacterium* species, report of a panel discussion. Pp. 327–333. In: Soc. Appl. Bacteriol. Symp. Ser. No. 2, Academic Press, New York.
Tamime, A,Y., Marshall, V.M.F., Robinson, R.K. (1995) Microbiological and technological aspects of milk fermented by bifidobacteria. J. Dairy Research 62, 151–187.
Teraguchi, S., Uehara, M., Ogasa, K. and Mitsuoka T,. (1978) Enumeration of bifidobacteria in dairy products. Jpn. J. bacterial. 33, 753–761.
Tissier, H. 1900. Recherches sur la flore intestinale des nourrissons (état normal et pathologique). Paris, Thèse, pp 1–253.
Wijsman, M.R., Hereijgers, J.L.P. and de Groote, J.M.F.H. (1989) Selective enumeration of bifidobacteria in fermented dairy products. Neth. Milk Dairy J. 43, 395–405.

Chapter 11

Media for the detection and enumeration of *Alicyclobacillus acidoterrestris* and *Alicyclobacillus acidocaldarius* in foods

J. Baumgart

Laboratory of Food Microbiology, University of Applied Sciences, Liebigstr. 87, D-32657 Lemgo, Germany

Of the four known species of the genus *Alicyclobacillus* only *Alicyclobacillus acidoterrestris* is of significance for the beverage industry. *Alicyclobacillus acidoterrrestris* is a Gram-positive, obligately aerobic rod-shaped bacterium which is able to multiply even at pH 2.2. The heat resistance of the spores is extremly high ($D_{95°C}$ 1.0–14.5 min, z-value= 6.4–11.3°C). *Alicyclobacillus acidoterrestris* can be detected using various methods, including flow cytometry, PCR, detection of fatty acids, ribotyping and cultural methods with biochemical identification. For routine analysis a cultural method is recommended because it is more reliable and simple.

Introduction

The genus *Alicyclobacillus* belongs to the group of aerobic acidothermophilic sporeforming bacteria. Unlike the other genera of this group (*Brevibacillus, Aneurinibacillus, Sporolactobacillus, Amphibacillus, Bacillus* and *Paenibacillus*), only members of the genus *Alicyclobacillus* multiply under aerobic conditions at pH 3.0 and form ω-alicyclic fatty acids.

The fact that strictly acidophilic organisms can be found in neutral environments is explained by the existence of acidic micro-habitats in the substrate, such as on the surface of sand granules or in areas around plant roots.

Of the four known species of this genus, *A. acidoterrestris, A. acidocaldarius* and *A. hesperidum* contain cyclohexyl fatty acids in their membranes (Albuquerque et al., 2000) while *A. cycloheptanicus* produces cycloheptyl fatty acids and, unlike the other species, is also dependent on growth substances (Table 1).

According to present knowledge, only *A. acidoterrestris* is of significance for the beverage industry. This species has been isolated from different types of soil (Deinhard et al., 1987; Baumgart et al., 1997; Eguchi et al., 1999) which is the natural reservoir of the genus.

High counts in the range of 10^4–10^6 cfu/g have been encountered in soil samples

Table 1
Species of the genus *Alicyclobacillus* (Farrand et al., 1983; Deinhard et al., 1987; Wisotzkey et al., 1992; Deinhard and Poralla,1996; Baumgart et al., 1997; Walls and Chuyate, 1998).

Characteristics	*A. acidoterrestris*	*A. acidocaldarius*	*A. hesperidum*	*A. cycloheptanicus*
ω-Cyclohexyl fatty acids C_{17} and C_{19}	+	+	+	– (ω-Cycloheptyl fatty acids C_{18} and C_{20})
growth factors required	–	–	–	+ Vit B 12, Pantothenate, Isoleucine
Acid from erythritol	+	–	–	not tested
pH range for growth	2. 2–5. 8	2. 0–6. 0	3. 5– <6. 0	3. 0–5. 5
Temperature range for growth °C	23–55	45–70	> 35– < 60	40–53

from an orange grove (Eguchi et al., 1999) and on the surface of the fruit and leaves (counts of < 10 to 10^2 cfu/ kg product).

A. acidoterrestris can be distinguished easily from other acid-tolerant spore formers (Table 2). Cerny et al. (1984) were the first to isolate this bacterium, which was named *Bacillus acidocaldarius*, from apple fruit nectar. Later it was classified as *Bacillus acidoterrestris* (Deinhard et al., 1987), and, in 1992, it was reclassified as the new genus *Alicyclobacillus* (Wisotzkey et al., 1992). *A. acidoterrestris* is a Gram-positive, obligately aerobic rod-shaped bacterium which is able to multiply even at pH 2.2. However, the minimum growth temperature is dependent on the composition of the respective product. For apple tea beverage, the minimum temperature is 23 °C.

A. acidoterrestris is present in many beverage bases and pasteurized fruit juices, but usually at levels below 100/g or ml (Baumgart et al., 1997; Pettipher et al., 1997; Pinhatti et al., 1997; Splittstoesser et al., 1998; Wisse and Parish, 1998; Eguchi et al., 1999). The heat resistance of the spores is extremly high ($D_{95°C}$ 1.0–14.5 min , z-value 6.4–11.3 °C (Splittstoesser et al., 1994; Walls 1997; Baumgart et al., 1997; Previdi et al., 1997; Eiroa et al., 1999; Silva et al., 1999).

The usual pasteurization process for fruit juices, will not kill all spores. Depending on the product, storage temperature and initial count/ml, off-flavour or odour can occur as a result of multiplication of the bacteria and formation of several phenol-type substances, such as guaiacol (Yamazaki et al., 1996a; Pettipher et al., 1997) or 2,6-dibromphenol (Baumgart et al., 1997; Borlinghaus and Engel, 1997).

Table 2
Characteristics of sporeforming, thermoacidophilic rod-shaped bacteria (Baumgart et al., 1997).

Microorganism	Aerobic growth at 46°C in BAM-broth, pH 3. 0 (HCl)	Aerobic growth at 46°C on plate count agar, pH 7. 0	ω -Cyclohexyl fatty acids C 17 and C 19
A. acidoterrestris DSM 2498, 3922-2924	+	–	+
A. acidocaldarius DSM 446, 448	+	–	+
Paenibacillus (P.) macerans DSM 24	–	+	–
P. polymyxa DSM 36	–	+	–
B. coagulans DSM 1	–	+	–
B. licheniformis DSM 13	–	+	–

Table 3
Methods for the detection of *Alicyclobacillus acidoterrestris.*

Method	Detection time	Remarks	References
Flow cytometry	10 hours	method not validated	Borlinghaus and Engel,
Reverse transcription - polymerase chain reaction (RT-PCR)	some hours or 1 day	sensitivity 1–2 cells/ml after enrichment of 15 hours	Yamazaki et al., 1996b
Cultural method with chromatographic detection of fatty acids	3–5 days	specific	Deinhard et al., 1987
Cultural method with biochemical detection of isolated strains	3–5 days	specific	Baumgart and Menje, 2
Cultural method and identification by ribotyping	3–5 days	specific	Walls and Chyate,1998; Nijs et al., 2000

Detection and enumeration of *Alicyclobacillus acidoterrestris*

A.acidoterrestris can be detected using various methods (Table 3). For routine analysis, a cultural method is recommended because it is more reliable and simple (Figure 1).

Media used for the selective isolation of acidothermophilic sporulated bacteria are complex, consisting of heterotrophic sources of energy such as glucose (carbon and energy source), yeast extract (source of vitamins and other growth factors), salts which act as buffers, N, P and K sources, and supplement of metallic ions (Co, Fe, Mo and

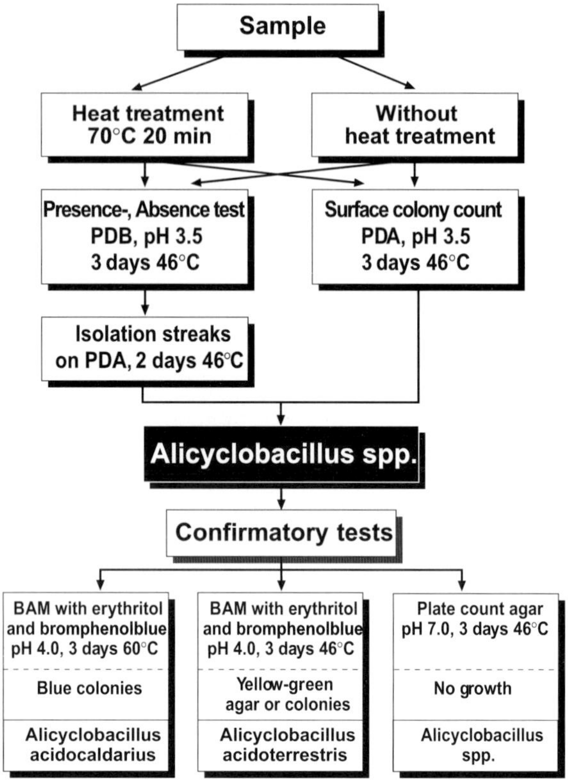

Fig. 1. Schematic diagram for detection, isolation and identification of *Alicyclobacillus acidoterrestris* and *Alicyclobacillus acidocaldarius*.

Mn), among others. *Alicyclobacillus* strains can grow in some acidified media such as Potato Dextrose Agar, Orange Serum Agar, Thermoacidurans Agar, Malt Extract Agar, Wort Agar (Eguchi et al., 1999) or on K-medium (Walls and Chuyate, 1998) or *Bacillus acidocaldarius*-Medium= BAM (Darland and Brock, 1971; Deinhard et al., 1987; Pettipher et al., 1997; Pinhatti et al., 1997; Baumgart et al., 1997).

Enrichment method

Potato Dextrose Broth (PDB, pH 3.5) has been well proven (Baumgart and Menje, 2000) as an enrichment medium. It is based on commercially available Potato Dextrose Broth, adjusted to pH 3.5 using 10% tartaric acid solution.

Incubate 100 ml juice or 10 ml concentrated juice (unheated samples and samples heated to 70 °C for 10 min) in 100 ml PDB (500 ml culture bottle, large surface) at 46 °C for 3 days, if possible in a shaking water bath. Unconcentrated juice (100ml) should be tested in double strength PDB.

Selective plating

Plate the enrichment with a loop on a solid acidic medium. No strictly selective media are available but in general Potato Infusion Agar or Potato Dextrose Agar acidified with 10% tartaric acid to a pH of 3.5 are suitable (PDA, pH 3.5). The agar is based on commercially available Potato Infusion Agar (Difco) or Potato Dextrose Agar (Merck). Incubate the PDA for 2 days at 46 °C. After that time, confirmatory tests should be conducted on colonies found on this medium.

Confirmatory tests

Confirmation of *A. acidoterrestris* is based on the utilization of erythritol on BAM agar (Baumgart et al., 1997; Baumgart and Menje, 2000). While *A. acidoterrestris* will change bromphenol blue to yellow-green, *Alicyclobacillus acidocaldarius* will form blue colonies due to lack of acid formation from erythritol. Neither species will grow on Plate Count agar at pH 7.0. Alternatively confirmation of isolates can be made using gas chromatography for detection of the ω-cycloalkyl fatty acids $C_{17:0}$ and $C_{19:0}$ typical of *A.acidoterrestris* and *A. acidocaldarius* or by ribotyping (Walls and Chyate, 1998; de Nijs et al., 2000).

References

Albuquerque, L., Rainey, F.A., Chung, A.P., Sunna, A., Nobre, M.F., Grote, R., Antranikian, G. and da Costa,
M.S. (2000) *Alicyclobacillus hesperidum* sp. nov. and a related genomic species from solfataric soils of São Miguel in the Azores. Int. J. Syst. Evolutionary Microbiol. 50, 451–457.

Baumgart, J., Husemann, M. and Schmidt, C. (1997) *Alicyclobacillus :* Vorkommen, Bedeutung und Nachweis in Getränken und Getränkegrundstoffen. Flüssiges Obst 64, 178–180.

Baumgart, J. and Menje, S. (2000) The impact of *Alicyclobacillus acidoterrestris* on the quality of juices and soft drinks. Fruit Processing 10, 251–254.

Borlinghaus, A. and Engel, R. (1997) *Alicyclobacillus* incidence in commercial apple juice concentrate (AJC) supplies-method development and validation. Quality Control 7, 262–266.

Cerny, G., Hennlich W. and Poralla, K. (1984) Fruchtsaftverderb durch Bacillen: Isolierung und Charakterisierung des Verderbserregers. Z. Lebens Unter Forsch 179, 224–227.

Darland, G. and Brock, T.D. (1971) *Bacillus acidocaldarius* sp. nov., an acidophilic thermophilic spore-forming bacterium. J. Gen. Microbiol. 67, 9–15.

Deinhard, G., Blanz, P., Poralla, K. and Altan, E. (1987) *Bacillus acidoterrestris* sp. nov., a new thermotolerant acidophile isolated from different soils. System. Appl. Microbiol. 10, 47–53.

Deinhard, G. and Poralla, K. (1996) Vorkommen, Biosynthese und Funktion Omega-alicyclischer Fettsäuren bei Bakterien. Biospektrum 2, 40–46.

De Nijs, M., van der Vossen, J., van Osenbruggen, T. and Hartog, B. (2000) The significance of heat resistant spoilage moulds and yeasts in fruit juices – A review. Fruit Processing10, 255–262.

Eguchi, S.Y., Manfio, G.P., Pinhatti, M.E., Azuma, E. and Variane, S.F. (1999) Acidothermophilic sporeforming bacteria (ATSB) in orange juice: detection methods, ecology, and involvement in the deterioration of fruit juices. Report of the research Project ABE Citrus, Campinas (SP), Brazil, Rua Pedroso Alvarenga, 1284 2° andar, CEP 04531-913- Sao Paulo/SP Brasil.

Eiroa, M.N.U., Junqueira, V. C. A. and Schmidt, F.L. (1999) *Alicyclobacillus* in orange juice: Occurrence and heat resistance of spores. J. Food Protect. 62, 883–886.

Farrand, S.G., Linton, J.D., Stephenson, R.J. and McCarthy, W.V. (1983) The use of response surface analysis to study the growth of *Bacillus acidocaldarius* throughout the growth range of temperature and pH. Arch. Microbiol. 135, 272–275.

Pettipher, G.L., Osmundson, M.E. and Murphy, J.M. (1997) Methods for the detection and enumeration of *Alicyclobacillus acidoterrestris* and investigation of growth and production of taint in fruit juice and fruit juice-containing drinks. Letters Appl. Microbiol. 24, 185–189.

Pinhatti, M.E.M.C., Variane, S., Eguchi, S.Y. and Manfia, G.P. (1997) Detection of acidothermophilic bacilli in industrialized fruit juices. Fruit Processing 7, 350–353.

Previdi, M.P., Quintavalla, S., Lusardi, C. and Vicini, E. (1997) Thermoresistenza di spora di *Alicyclobacillus* in succhi di frutta. Industria Conserve 72, 353–358.

Silva, F.M., Gibbs, P., Vieira, C. and Silva, C. L.M. (1999) Thermal inactivation of *Alicyclobacillus acidoterrestris* spores under different temperature, soluble solids and pH conditions for the design of fruit processes. Int. J. Food Microbiol. 51, 95–103.

Splittstoesser, D.F., Churey, J.J. and Lee, C.Y. (1994) Growth characteristics of aciduric sporeforming bacilli isolated from fruit juices. J. Food Protect. 57,1080–1083.

Splittstoesser, D.F., Lee, C.Y. and Churey, J.J. (1998) Control of *Alicyclobacillus* in the juice industry. Dairy, Food Environ. Sanitation 18, 585–587.

Walls, I. (1997) *Alicyclobacillus* – an overview. Session 36–1 presented at 1997 Institute of Food Technologists Annual Meeting in Orlando, Fl., 14–18 June.

Walls, I. and Chuyate, R. (1998) *Alicyclobacillus* historical perspective and preliminary characterization study. Dairy Food Environ Sanitation 18, 499–503.

Wisse, C.A. and Parish, M.E. (1998) Isolation and enumeration of sporeforming, thermoacidophilic, rod-shaped bacteria from citrus processing environments. Dairy Food Environ. Sanitation 18, 504–509.

Wisotzkey, J.D., Jurtshuk, P. Jr., Fox, G.E., Deinhard, G. and Poralla, K. (1992) Comparative sequence analyses on the 16S rRNA (rDNA) of *Bacillus acidocaldarius*, *Bacillus acidoterrestris*, and *Bacillus cycloheptanicus* and proposal for creation of a new genus, *Alicyclobacilus* gen. nov. Int. J. Syst. Bacteriol. 42, 263–269.

Yamazaki, K., Teduka, H. and Shinano, H. (1996a) Isolation and identification of *Alicyclobacillus acidoterrestris* from acidic beverages. Biosci. Biotech. Biochem. 60, 543–545.

Yamazaki, K., Teduka, H., Inoue, N. and Shinano, H. (1996b) Specific primers for detection of *Alicyclobacillus acidoterrestris* by RT-PCR. Lett. Appl. Microbiol. 23, 350–354.

Chapter 12

Media for detection and enumeration of 'total' Enterobacteriaceae, coliforms and *Escherichia coli* from water and foods

M. Manafi

Hygiene Institute, University of Vienna, Kinderspitalgasse 15, A-1095 Vienna, Austria

Recent developments in enrichment and selective media for the isolation and enumeration of 'total' Enterobacteriaceae, coliforms, faecal coliforms and *E. coli* from water and foods are described and effects of time and temperature of incubation discussed. Factors to be considered in the selection of media for this group of organisms are noted based on the results of comparative studies. Coliforms and *E. coli* are both important indicators of water and food contamination; both therefore need to be detected in the same medium. Several attempts have been made to detect coliforms and *E. coli* simultaneously and novel methods have been introduced, based on the detection of ß-D-galactosidase (ß-GAL) and ß-D-glucuronidase (GUD) using enzymatic methods. Results of some comparative studies are discussed here.

Introduction

Culture media for the isolation and enumeration of 'total' Enterobacteriaceae, coliforms, faecal coliforms and *E. coli* were reviewed by Blood and Curtis (1995) and de Boer (1998). The present review gives a short description of conventional media and recent developments and their application in food and water microbiology.

Enterobacteriaceae

Members of genera belonging to the Enterobacteriaceae family have earned a reputation placing them among the most pathogenic and frequently encountered organisms in microbiology. This family consists of Gram-negative, facultatively anaerobic, non-spore-forming rods which are usually associated with intestinal infections, but can be found in almost all natural habitats. They are oxidase negative and all ferment glucose and reduce nitrate. Those which ferment lactose are coliforms, some of which are associated with pathogenicity (Table 1; Blood and Curtis, 1995). Non-lactose utilizers such as *Salmonella, Shigella* and some strains of *Yersinia* are usually pathogenic. The family is subdivided into a number of genera, based on biochemical characteristics, genetic and antigenic properties as follows:- *Escherichia, Shigella, Salmonella, Cit-*

Table 1
The Enterobacteriaceae and the 'coliform' group (lactose-positive strains) (from ICMSF, 1978).

Genus	Detected by 'coliform test'	Faecal origin	Enteropathogenic for man
Citrobacter	yes[a]	no[b]	no
Edwardsiella	no	yes	some[e]
Enterobacter	yes	no[a]	no
Erwinia	no[c]	no[d]	no
Escherichia	yes	yes	some[e]
Hafnia	no[c]	no[a]	no
Klebsiella	yes	no[a]	some[e]
Proteus	no	no[a]	some[e]
Salmonella	no	yes	yes
Serratia	no	no	no
Shigella	no	yes	yes
Yersinia	no	yes	no

[a] Except slow lactose-positive strains.
[b] Some strains of faecal origin but also grow in other environments.
[c] Except occasional strains.
[d] Except strains adapted to rapid growth near 37°C.
[e] Some serotypes contain enteropathogenic strains.

robacter, Klebsiella, Enterobacter, Erwinia, Serratia, Hafnia, Edwardsiella, Proteus, Providencia, Morganella, Yersinia and others.

Indicator and index concept

Microbial contamination is still considered to be the most critical risk factor in water and food. It is in general impractical to monitor water or food samples for every possible microbial pathogen. Mossel (1982) defined the term 'marker organism' which refers to two different functions, index and indicator. 'Index organisms' are related, directly or indirectly, either to the health hazards or to the presence of pathogens. On the other hand, 'indicator organisms' are related only to the effects of treatment processes or control of water quality. 'Indicator organism' requirements are less restrictive than those for an index organism: their resistance to the environment need be no more than that of pathogens and they require more simple laboratory methods.

Total coliforms, faecal coliforms and E. coli

Total coliform bacteria belong to the family Enterobacteriaceae, and include *E. coli* as well as various members of the genera *Enterobacter, Klebsiella* and *Citrobacter;* all ferment lactose with gas and acid production in 48 h at 35–37 °C. The term 'total coliform' is based only on lactose hydrolysis; however some members of the coliform group originating from different non-enteric environments can only ferment lactose slowly (after 72 h). Faecal coliforms are the subset of total coliforms that are more closely associated with faecal pollution. These bacteria conform to all criteria used to define total coliforms, but in addition they ferment lactose with production of gas and acid at 44.5 ± 0.2°C within the first 48 h of incubation. For this reason, the term 'thermotolerant coliforms' may be more appropriate to this group. Thermotolerant coliforms comprise strains of the genera *Klebsiella* and *Escherichia,* associated with faecal contamination arising from warm-blooded animals. *E. coli* is a most useful microbial indicator of water quality because it is more specific for the presence of faecal contamination than the faecal coliform group (Geldreich, 1997).

Media for the specific group

An ideal medium for use in food microbiology, whether broth or agar, would selectively grow only the target organism or group of organisms. With a group of organisms as closely related as the Enterobacteriaceae it is unlikely that such a medium could be developed for a single genus such as *Escherichia* or even the less well defined coliform group, thus selectivity has inevitably relied upon the incorporation of substrates to provide the means of differentiating target from unwanted strains. Even where selective indicator media have been formulated specifically with Enterobacteriaceae in mind, steps may need to be taken to prevent the growth of unwanted organisms by methods such as overlaying of the inoculated surface with agar. In spite of such precautions, recovery of Enterobacteriaceae will still be affected by the incubation temperature and the nature of the organisms present in the inoculum i.e. psychrophilic, mesophilic or thermophilic (Mossel et al., 1979). The difficulty of recovering sublethally injured organisms on selective media must also be considered when choosing suitable media for these bacteria.

Conventional media

Liquid media

Table 2 shows the main features of liquid media formulated for the growth of Enterobacteriaceae. Selectivity is provided by surface active agents, alone or in combination, in all but three of these broths. Lactose broth contains no selective agents or indicator and is recommended only for use in testing normally sterile material such as eggs and egg products (AOAC, 1990). Minerals modified glutamate (MMG) broth is a chemically defined, elective medium derived from formate lactose glutamate medium (Gray, 1964). Selectivity in boric acid lactose broth is obtained by the use of boric acid. The triphenylmethane dye, brilliant green, is used to inhibit Gram-positive organisms and the combination of sodium deoxycholate and sodium citrate in GN broth inhibits Gram-positive and some Gram-negative organisms. Various differential systems are found in these broths. With the exception of Enterobacteriaceae enrichment (EE) broth which contains glucose, all the media incorporate lactose, with or without a pH indicator to demonstrate acid production. Other carbohydrates readily utilised by the target organisms are also found in some formulations. Production of gas from carbohydrate fermentation may be detected with the aid of an inverted Durham's tube. Lauryl tryptose broth (LTB) or lauryl sulphate tryptose (LST) broth (Mallman and Darby, 1941) is used in the presumptive phase of the standard total coli-

Table 2
Principal features of liquid media for Enterobacteriaceae, *E. coli* and coliforms.

Medium	Target Group[a]	Incubation °C	h	Selective agents	Fermentable carbohydrate(s)/ differential system(s)	Reference
A-1	fc	35 followed by 44.5	3 21	Triton-X 100	lactose, salicin, gas production	Andrews et al., 1981
Boric acid lactose broth	Ec	43	48	boric acid	lactose	Vaughn et al., 1951
Brilliant green bile broth (syn. Brilliant green lactose bile broth; BGBB)	cfm Ec	35 44	48 18	brilliant green, bile	lactose, gas production	Dunham & Schoenlein, 1926
EC broth	cfm Ec	37 45.5	48 48	bile salts	lactose, gas production	Hajna & Perry, 1943
EC-MUG broth	Ec	44.5		bile salts	lactose, MUG hydrolysis	Rippey et al., 1987
Enterobacteriaceae enrichment (EE) broth	Ent[b]	4 32 44	10d 24–48 18	brilliant green, ox bile	glucose	Mossel et al., 1963
GN broth-modified	EEc	37	18–24	sodium citrate sodium desoxycholate	lactose, arabinose	Hajna, 1955
Lactose broth	cfm	35	24–48	(none)	lactose	AOAC., 1990
Lauryl tryptose broth, LTB, syn. lauryl sulphate broth, LST)	cfm Ec	35 44	24–48	lauryl sulphate	lactose, gas & indole production	Mallman & Darby, 1941
Lauryl tryptose - MUG broth	Ec	35	24–48	lauryl sulphate	lactose, gas and indole production, MUG hydrolysis	Feng & Hartman, 1982
MacConkey broth	cfm	30/37	48	ox bile	lactose, gas production	MacConkey, 1905
Minerals modified glutamate (MMG)broth	cfm	30	48	(chemically defined elective medium)	lactose, gas production	Abbiss et al., 1981

[a] cfm = coliforms, Ec = *E. coli*, EEc = enteropathogenic *E. coli*, Ent = Enterobacteriaceae, fc = faecal coliforms.
[b] Incubation conditions will determine which group of Enterobacteriaceae is detected.

form fermentation method in the examination of water. This broth is used widely in standard methods for coliform detection in water and food (Christen et al. 1992).

The examination of shellfish and shellfish-growing waters may require a special technique due to the salinity of the natural environment and the adverse effect this may have on the target organisms. Medium A-1 (Andrews and Presnell, 1972) was designed for this purpose and the A-1 technique has been adopted as an approved method by the AOAC for the estimation of faecal coliforms in shellfish growing waters (Andrews et al., 1981). The medium contains, in addition to lactose, tryptone and sodium chloride, the surfactant Triton X-100 and salicin which is a carbohydrate readily utilised by *E. coli.*

Solid media

The origins of today's media go back to the early part of the 20th century when MacConkey, reporting experiments to determine the distribution of 'Bacillus coli' in nature, published papers on lactose fermenting bacteria in faeces and on the use and advantages of bile salt media in some bacteriological examinations (1905). In his paper of 1908, MacConkey referred to his earlier studies in 1897 using potato juice medium with bile salts, and another medium containing bile salts (0.5%), peptone (2%), lactose (1%), neutral red (0.0025%) and agar (1.5%) in tap water. He found it was essential to use bile salts which were neutral in reaction to neutral red, if acid production was to be easily recognised. It is perhaps worth noting the absence of sodium chloride from these early formulations. MacConkey concluded that the various lactose-fermenting organisms found in faeces could be further distinguished by their action on other carbohydrates, incorporated either singly or in combination in bile salts-neutral red media, and thus the foundations of the selective-indicator media in use today were laid. On MacConkey agar, bacterial colonies that can ferment lactose turn the medium red due to the response of neutral red indicator to the acidic environment created by fermenting lactose. Organisms that do not ferment lactose do not cause a colour change.

The principal features of agars formulated for this group of organisms are shown in Table 3. As with broths, bile salts are used in many of these media to inhibit Gram positive organisms but dyes are also used, alone or in combination with bile salts or sodium sulphite. The intrinsic toxicity of some batches of media to non-stressed Enterobacteriaceae is well recognised (Mossel et al., 1979) but can be overcome by careful selection of crystal violet and, particularly, bile salts. The most common differential system is a fermentable carbohydrate together with a pH indicator. Tryptone bile agar (TBA) is the only medium not containing a fermentable carbohydrate. Lactose is incorporated in all the others, except violet red bile glucose (VRBG) agar which is designed for use where 'total Enterobacteriaceae' are sought using such MacConkey type agars.

Mossel et al. (1962) modified the lactose-containing VRB agar which was based on the orginal formula of MacConkey (1905), by adding glucose to improve the recovery of Enterobacteriaceae. It was later shown (Mossel et al., 1979) that lactose could be omitted. Violet red and bile salt inhibit the growth of Gram-positive bacteria. The fermentation of lactose produces acid which changes the colour of neutral red to red

Table 3
Principal features of solid media for Enterobacteriaceae.

Medium	Incubation °C	h	Selective agents	Fermentable carbohydrate(s)/ differential system(s)	Reference
Deoxycholate agar	37	24	sodium deoxycholate	lactose	APHA, 1984
Deoxycholate hydrogen sulphite lactose agar	37	24–48	sodium deoxycholate	lactose, sucrose, H_2S production	Sakazaki et al., 1960; 1971
Endo agar	37	24	sodium sulphite, fuchsin	lactose, sucrose, fuchsin sheen	Endo, 1904
Eosin methylene blue agar	37	24	eosin, methylene blue	lactose, sucrose	Holt-Harris & Teague, 1916
Faecal coliform agar (FCA)	35 followed by 44.5	2 22	bile salts No.3	lactose, CO_2 production	Chen and Wu, 1992
Gassner's agar	37	24	metachrome yellow	lactose	Gassner, 1918
MacConkey's agar	37	24	bile salts	lactose	MacConkey, 1905
Min. mod. glutamate (MMG) agar[a]	37 followed by	4	(none)	lactose	Holbrook et al., 1980
Tryptone bile agar (TBA)	44	18–24	bile salts No.3	indole production	
Resuscitant agar[b]	30 followed by	0.5	(none)	(none)	Hall, 1984
Yeast extract lactose tryptone bile agar	44	18–20	bile salts No.3	lactose	
Violet red bile glucose agar (VRBG)	4[c] 32 42–44	10d 24–48 18	bile salts No.3, crystal violet	glucose	Mossel et al., 1978
Violet red bile (VRB) agar	4[c] 32 42–44	10d 24–48 18	bile salts No.3, crystal violet	lactose	APHA, 1984
Violet red bile MUG agar	37	18–24	bile salts No.3, crystal violet	lactose, MUG hydrolysis	Feng and Hartman, 1982

[a] inoculum is spread over the surface of an acetate membrane overlaid on MMG agar. After the initial incubation period the membrane is transferred to TBA.
[b] inoculum is mixed with molten resuscitant agar which is allowed to set. After the initial incubation period the plates are overpoured with yeast extract lactose tryptone bile agar.
[c] incubation conditions will determine which group of organisms is detected.

and is responsible for the precipitation of bile salts around a positive colony. Lactose positive bacteria grow as purple colonies 0.5–2 mm in diameter after 24 h and may be surrounded by a purple zone. Lactose negative bacteria grow as pale colonies which can be surrounded by a greenish zone. Overlaying the plates with the same medium assures anaerobic conditions, suppressing non-fermenting Gram negative bacteria and encouraging the fermentation of glucose. Non-Enterobacteriaceae grow as colourless colonies.

Other differential systems are found in some media. Hydrogen sulphide production is detected in deoxycholate hydrogen sulphite lactose agar by the blackening of H_2S-producing colonies due to the formation of iron sulphide. Endo (1904) developed a medium using a fuchsin-sulphite indicator to differentiate lactose fermenting (red colonies) from lactose-nonfermenting organisms (colourless colonies). In Endo agar coliform bacteria produce aldehyde and acid from lactose. The aldehyde liberates fuchsin from the fuchsin-sulphite compound resulting in a red colouration of the colonies due to the fuchsin, a reaction so intense with *E. coli* that the fuchsin crystallizes out and giving the colonies a characteristic green metallic sheen.

Production of gas at 44.5°C by faecal coliform organisms is demonstrated in an unusual way in faecal coliform agar (FCA; Chen and Wu, 1992). Calcium lactate in the medium reacts with CO_2 to form precipitates producing yellow to yellow-green colonies surrounded by pale yellow zones. The detection of indole production by *E. coli* in the system of Holbrook et al. (1980) requires a special technique (see below). In the pour-plate with overlay method of Hall (1984) the inoculum is mixed with a small volume of resuscitant (non-selective) agar containing yeast extract, tryptone, sodium pyruvate, sodium glycerophosphate and magnesium sulphate. This is subsequently overlaid with the selective agar which is formulated so that the lactose content, whilst adequate for acid production, does not interfere with the utilization of tryptophan. Bromothymol blue is used as the pH indicator to avoid interference with the indole test which is done using a rapid micro-tube technique.

Eosin methylene blue agar (EMB) is a differential medium used in identification and isolation of Gram-negative enteric rods. EMB agar also inhibits the growth of Gram-positive organisms. The differential basis of this medium involves two indicator dyes, eosin and methylene blue, that distinguish between lactose fermenting and non-lactose fermenting organisms. Lactose fermenters form colonies with dark centres and transparent, colourless peripheries while the non-lactose fermenters form completely colourless colonies.

Detection and enumeration of E. coli *and coliforms using enzymatic methods*

E. coli

About 94–96% of *E. coli*, including many anaerogenic (non-gas-producing) strains, produce the enzyme ß-D-glucuronidase (GUD) and the activity is measured by using different chromogenic and fluorogenic enzyme substrates such as *p*-nitrophenol-ß-D-glucuronide (PNPG), or 5-bromo-4-chloro-3-indolyl-ß-D-glucuronide (XGLUC) and fluorogenic 4-methylumbelliferyl-ß-D-glucuronide (MUG). A wide range of media is now available for detection and enumeration of *E. coli* using different enzyme substrates (Hartman, 1989; Frampton and Restaino, 1993; Manafi 1996; 2000). PNPG is suitable for liquid media enumeration of *E. coli*, e.g. when applying a MPN technique, but not as suitable for use in solid medium because of the extensive diffusion of yellow colour. When XGLUC is used, colonies with enzymatic activity show a blue colour without diffusion; it is therefore most suitable for incorporation in solid medium. MUG is hydrolyzed by GUD yielding 4-MU, which shows blue fluorescence when irradiated with long-wave UV light (365 nm). MUG has been incorporated into both liquid and solid media including lauryl sulphate broth (Feng and Hartman, 1982; Moberg et al., 1988), lactose broth and m-Endo broth (Alvarez, 1984), EC broth (Koburger and Miller, 1985), violet red bile agar (Alvarez, 1984) and m-FC agar (Mates and Schaffer, 1989). The minimum concentration sufficient for differentiation was 50 µg/ml. However, the concentration necessary for acceptable detection of *E. coli* depends on the other constituents of the media. As the fluorescence of 4-MU is pH dependent, the pH of growth media containing MUG should be slightly alkaline; otherwise alkaline solution needs to be added to reveal fluorescence. MUG can be sterilized together with other medium ingredients without loss of activity; furthermore, no inhibitory effect on *E. coli* growth has hitherto been observed. The disadvantage of incorporating MUG into solid media is that fluorescence diffuses rapidly from the colonies into the surrounding agar. Some foods, such as shellfish, contain natural GUD activity. The problem of endogenous β-D-glucuronidase in molluscan shellfish, which causes a high rate of false positive results, can be overcome by incorporation of the fluorescent substrate into the confirmatory broths rather than in the primary growth medium (Koburger and Miller, 1985; Rippey et al., 1987). Also, all glass tubes should be examined for fluorescence before use. Cerium oxide, which is sometimes added to glass as a quality control measure, will fluoresce under UV light and interfere with the MUG assay (Hartman, 1989).

Coliforms

The most relevant test used for enumeration of the coliform group is the hydrolysis of lactose. The breakdown of this disaccharide is catalyzed by the enzyme ß-D-galactosidase. Both monosaccharides – galactose after being transformed into glucose by further biochemical reactions – are further metabolized through the glycolytic and citrate cycles. The metabolic end products of these cycles are acids and/or CO_2. The determination of ß-galactosidase is accomplished by using several fluorogenic and chromogenic substrates. Buerger (1967) described the use of *p*-nitrophenol-galactopyranoside and 6-bromo-2-naphtyl-β-D-galactopyranoside. Using 4-methylumbelliferyl-β-D-galactopyranoside (MUGAL) and following the MF technique, Berg and Fiksdal (1988) detected as few as one faecal coliform cfu/100 ml water within 6 h. Manafi (1995) described the use of 5-bromo-4-chloro-3-indolyl-ß-D-galactopyranoside (XGAL) in both solid and liquid media. Coliform strains including *E. coli* produce well-defined blue (XGAL) colonies on agar due to the splitting of indolyl substrates and conversion of the liberated aglycone to insoluble indigo dye which does not affect the viability of the colonies. Broths containing XGAL turn blue-green indicating the presence of coliforms. James and Perry (1996) described the synthesis of two new substrates for the detection of β-D-galactosidase, 8-hydroxychinoline-β-

D-galactoside and cyclohexenoesculetin-β-D-galactoside.

Simultaneous detection of *E. coli* and coliforms

Many studies have been undertaken to develop and evaluate media able to detect total coliforms and *E. coli* simultaneously. Approximately 25 media are commercially available using a variety of enzyme substrates for detection of ß-D-galactosidase and ß-D-glucuronidase (Manafi, 2000). The efficiency and rapidity of the detectable reactions make these media a very useful tool in routine water and food microbiology.

Liquid media

Colilert, Colisure (Idexx, Branford, Conn.), LMX broth and Readycult coliforms (Merck, Germany) are commercially available media, which permit rapid simultaneous detection of *E. coli* and coliforms in water. In an evaluation of a number of Presence/Absence tests for coliforms and *E. coli,* Lee et al. (1995), compared four P/A tests with UK Standard methods and found that more coliforms were detected by LMX broth and Colilert than with the membrane filtration technique. Another study by Landre et al. (1998) reported the false positive coliform reaction mediated by aeromonas in the Colilert system. Schets et al. (1993) compared Colilert with Dutch standard enumeration methods for *E. coli* and coliforms in water and found that Colilert gave false-negative results in samples with low numbers of *E. coli* or total coliforms. Manafi and Rosmann (2000) showed that 99% of *E. coli* isolated from drinking water were ß-glucuronidase positive, 98% showed ß-D-galactosidase activity and 99% gave a positive indole reaction. These enzymatic assays may constitute an alternative specific, sensitive and rapid method for enumerating *E. coli* and coliforms in foods (Hahn and Wittrock 1991) and in waters (Betts et al., 1994; Lee et al., 1995; Manafi and Rosmann, 1998).

Solid media

The use of media containing chromogenic and fluorogenic substrates for the enzymes ß-galactosidase (LAC) and ß-glucuronidase (GUD) for simultaneous detection of coliforms and *E. coli* is increasing. Some authors evaluated agar media incorporating XGAL (Ley et al., 1993; Jermini et al., 1994). MI agar, containing indoxyl-ß-D-glucuronide and 4-methylumbelliferyl-ß-D-galactopyranoside for the simultaneous detection of *E. coli* and total coliforms, was described by (Brenner et al., 1993). Grant (1997) described the m-Coliblue system, which detects total coliforms (red colonies) and *E. coli* (blue colonies) simultaneously. COLI ID (bioMérieux) is a new medium for the detection of coliforms and *E. coli*. Coliforms produce blue colonies whilst *E. coli* are rose with a rose zone around the colonies. Other Gram negative bacteria appear bright rose, small and without a surounding zone. Gram-positive bacteria and yeasts are inhibited. On Chromocult coliform agar (CCA) which contains chromogenic Salmon-GAL and X-GLUC, the growth of coliforms, including sublethally damaged cells, is possible due to use of peptone, pyruvate, sorbitol and phosphate buffer. Gram-positive bacteria are inhibited by Tergitol 7. On CC agar, non-*E. coli* faecal coliforms (*Klebsiella, Enterobacter* and *Citrobacter*) were identified by the production of a salmon to red colour from cleavage of the substrate Salmon-GAL, while *E. coli* colonies were detected by the blue/violet colour. Similarly on Coliscan Easygel, Coliscan and CHROMagar ECC, *E. coli* colonies are blue-violet and other coliforms are red.

Detection methods using membranes

A procedure for the estimation of *E. coli* in foods using cellulose membranes was developed by Anderson and Baird-Parker (1975) from a method described by Delaney et al. (1962). This was subsequently modified (Holbrook et al., 1980) to include a resuscitation step involving pre-incubation of the inoculated membrane on a non-inhibitory agar before transfer to TBA, thereby reducing the carbohydrate content in the inoculum by allowing diffusion into the non-inhibitory agar. This modification not only enabled resuscitation of damaged cells but also improved the reliability of the test for indole production after incubation on TBA. The modified method is as follows. A cellulose acetate membrane is laid on the surface of a pre-dried plate of MMG agar and the inoculum is spread over the membrane. After the membrane surface has dried, the plate is incubated at 37°C for 4 h and the membrane is then transferred to a pre-dried TBA plate and incubation continued for a minimum of 18 h at 44.5°C. At the end of this second incubation the membranes are examined. Those with growth are removed from the agar into a Petri dish lid containing 2 ml of indole reagent (Vracko and Sherris, 1963) so that the whole of the lower suface is wetted. The excess reagent is removed after five min and the reaction developed in bright sunlight or under a UV lamp. Indole positive colonies, which appear pink, may then be counted to give the total *E. coli* count. Further confirmation of identity is generally regarded as unnecessary. The cost of the examination can be greatly reduced if cellulose nitrate membranes are used instead of cellulose acetate. Holbrook and Anderson (1982) specify either.

A similar system is the hydrophobic grid membrane filter technique (Entis, 1984). This uses a membrane printed with hydrophobic lines in a grid pattern which acts as a barrier to the spread of colonies. Food dilutions are filtered through the membrane which is then placed on the appropriate agar plate. A resuscitation step can be included by incubating for a short time on non-selective agar prior to overnight incubation on selective agar; testing for indole production can be performed according to Holbrook et al. (1982). The number of squares occupied by colonies is counted and by use of a formula converted to an MPN. By the use of different media and incubation conditions counts may be made of total coliforms, faecal coliforms or *E. coli* (AOAC, 1990; APHA, 1992). Similar systems, the Petrifilm ™ Coliform Count Plate and the Petrifilm ™ *E. coli* Count Plate Methods are described in the AOAC (1995) standard methods.

Incubation conditions

There are many recommendations for incubation temperatures and times. Apart from refrigeration temperatures used to isolate psychrotrophs, the choice is of temperatures between 30 and 45.5°C. When examining dairy products for coliform organ- … ndard practice to incubate at 30°C rather than 35 or 37°C. With most other … 35°C (USA) and 37°C (Europe). A common practice is … d confirmatory tests for coliforms at 35–37°C … 44–45.5°C, depending on the medium … nay be encountered, e.g. *E. coli* O157: … eratures.

… aecal coliforms and *E. coli* using LST at … on in EC at 44.5, 45.0 and 45.5°C. Only … iples gave equal recovery rates of faecal … s production was the criterion of positiv- … without gas production, was the criterion … the meat samples showed no difference in … oncluded that incubation at 44.5°C resulted … the use of 45.5°C gave the greatest specifi- …

… nan (1984) stated that the time may be lim- … riology, or extended to 48 h, as in the classi- … The longer period will recover those *E. coli* … thus the 48 h test generally produces higher counts (Mous… t al., 1978; Bindschedler et al., 1981). Weiss et al. (1983) found that confir… form MPN counts at 24 and 48 h differed by more than 0.6 log cycle in 23% of raw milk and 48% of raw meat samples. Methods specifying a total of 48 h incubation often recommend examination of tubes at 24 h when confirmatory tests on presumptive positive tubes can be initiated. Additional confirmatory tests may need to be set up if extended incubation reveals more presumptive positive tubes.

Temperatures and duration of incubation recommended for plated media are less diverse than those for broths, although similar reasoning has been applied. Pre-incubation for 2–4 h at a lower temperature prior to overnight incubation at 44 or 44.5°C is said to allow recovery of injured cells on FCA and TBA. Incubation of VRBG and VRB agars at 4°C for ten days is recommended for optimal growth of psychrotrophic coliforms whilst the same agars should be incubated at 32°C for 24–48 h or 42–44°C for 18 h for mesophilic or thermotrophic coliforms respectively (Anon., 1987). Mossel et al. (1979) have shown that incubation at 30°C led to higher confirmed colony counts in minced meat than at 37°C but this was not the case when samples of drip from frozen broilers were examined. This difference was explained by the predominance of psychrotrophic types (up to 75% of isolates) in minced meat whereas in chicken drip *E. coli* was the predominant type with psychrotrophs making up only 30% of isolates. A further study (Mossel et al., 1986) showed that psychrotrophs were not markedly inhibited at 37°C although growth was variable at this temperature. They recommended the use of 30°C for incubation of violet red bile agars where psychrotrophs are sought and 42–43°C where their growth is to be suppressed.

Comparative studies

Numerous comparative studies of media have been made but their value is limited by the differences in sample materials, conditions of testing and formulations surveyed. In the examination of frozen food, both Shelton et al. (1962) and Hall (1964) preferred LST broth to lactose broth for detection of coliforms using the MPN technique, gas positive tubes being subcultured to brilliant green bile lactose broth (BGBB) for confirmation. Further comparisons of media for the isolation of coliform organisms from a wide variety of dehydrated and deep frozen foods were reported by Moussa et al. (1973). They found that for the detection of coliforms and Enterobacteriaceae, MMG broth was significantly better than BGBB, LST and EE broths for all types of samples tested except frozen poultry where the difference was not significant. For the detection of faecal coliforms MMG broth was again significantly better than the other media both for dehydrated products and deep frozen foods. For deep frozen poultry the best results were obtained with lactose broth but this was not a good medium for deep frozen products as a whole. In their hands BGBB performed poorly in the determination of coliforms and Enterobacteriaceae as did EE broth for faecal coliforms. In the examination of ground beef Pierson et al. (1978) found LST broth gave higher coliform counts than either MacConkey broth or BBGB whilst EC broth and BGBB did not differ significantly in their ability to detect faecal coliforms. Further studies on raw meats and poultry were made by Rayman et al. (1979) who found the membrane method of Anderson and Baird-Parker (1975) was less variable, more rapid and economical than the AOAC MPN method using LST broth. When examining frozen samples they incorporated a resuscitation step by spreading the inocula on the surface of membranes overlying tryptic soy agar and incubating for 4 h at 35–37°C. The membranes were then aseptically transferred to TBA and incubation continued for 24 h at 44.5°C. The inclusion of this resuscitation step in the membrane method may account for the significantly higher counts which were obtained compared to the AOAC MPN technique.

In a collaborative study involving five laboratories (Sharpe et al., 1983) membrane filter methods gave significantly higher counts of *E. coli* in ground beef, cheese and cut green beans than an MPN technique using LST broth at 35°C with confirmation in EC broth at 44.5°C. Results for the membrane techniques were obtained within 24 h whereas the MPN method, which required confirmation using IMViC tests, took 10–14 days. Membrane filtration was, however, considered unsuitable for bean or alfalfa sprouts, since these often contained high levels of *Klebsiella* spp.

Soft cheeses, cooked meats and pate were examined by Abbiss et al. (1981) using MMG broth, LST, MacConkey broth and BGBB. They found that with an MPN technique MMG broth was superior in sensitivity for coliform counts and compared favourably in specificity to other broths. Other dairy samples were examined by Cooke and Jorgenson (1977) who found BGBB equal in sensitivity to MMG broth and better than lactose broth for the detection of coliforms in dried milk and butter. Work with dairy and cocoa products (Bindschedler et al., 1981) comparing BGBB, LST, lactose broth, EE broth and MMG broth for the determination of coliforms and *E. coli* showed that MMG broth had the highest sensitivity for coliform detection. However it was also the medium which gave rise to the most false positive results and confirmatory tests were therefore required. In general, these studies showed that whenever a significant difference existed in the sensitivity to determine *E. coli*, MMG broth was better than the other media. Frequently EE broth was less satisfactory for the detection of Enterobacteriaceae than the other media. None of these methods included a specific resuscitation step although Bindschedler et al. (1981) carried out their first incubation at 30°C for 48 h and MMG is an elective medium which should allow recovery of damaged cells (Holbrook et al., 1980).

Andrews et al. (1981) evaluated the A-1 and LST broths for faecal coliform detection in a variety of foods and showed that the efficiency of the A-1 procedure was dependent on the type of food analyzed. They recommended it only for use as a screening test in food examinations.

Motes et al. (1984) compared BGBB, LST, direct plating on membranes placed on TBA and a roll tube technique with MacConkey agar for the examination of shellfish. The best recoveries were obtained using LST and BGBB with the lowest recovery on direct plating. The membrane method did not include the resuscitation step now in common use (Holbrook et al., 1980) and the poor success of the method may have been due to the failure to recover stressed organisms exposed directly to the selective TBA. A similar explanation was suggested by West and Coleman (1986) for the poor performance of direct plating methods in their comparison of techniques for the detection of *E. coli* in shellfish. An MPN method using MMG broth gave higher recoveries of confirmed *E. coli* than either a roll tube or pour plate technique with MacConkey agar No.3 even when a 2 h resuscitation at a lower temperature was introduced. They concluded that for both heavily (>1000 *E. coli*/100 g) and lightly (100–500 *E. coli*/100 g) contaminated shellfish, MPN procedures had greater sensitivity and that the MPN is technically superior to plate methods for *E. coli* since it determines gas as well as acid production. Nevertheless they recognised that confirmation was required. By inoculating tryptone water in parallel with BGBB and incubating both at 44°C, their system allowed the completion of the examination within 48 h.

A collaborative study of coliform determination by MPN techniques (Silliker et al., 1979) compared three methods, i) presumptive determination of coliforms using LST broth with subsequent confirmation of gas-positive tubes, ii) determination of coliforms with MacConkey broth and iii) presumptive determination with BGBB followed by confirmation. Differences between the methods were relatively small when compared with the differences between laboratories using the same methods and following a common protocol. There appeared to be no basis for selecting a single procedure as the 'best' method.

In a variety of fresh and processed foods examined without resuscitation, Oblinger et al. (1982) found no significant differences between recoveries on VRB and VRBG agars. The most productive temperature for incubation was 20°C followed by 35, 7, 45 and 1°C. VRBG was not significantly better than VRB for recovery of total Enterobacteriaceae but typical colonies were more easily recognised and enumerated on VRBG than on VRB. Recovery of total Enterobacteriaceae on both agars at 35°C was similar to that at 20°C. Klein and Fung (1976) showed that on VRB agar incubated at 44.5°C, faecal coliforms from sewage or water could be quantified and accurately separated from non-faecal coliforms by differences in colony size. They found no significant differences between recoveries using the elevated temperature VRB method, a membrane filtration method and a five tube MPN system with EC broth. A comparison of pour-plate with overlay and spread plate techniques for counting Enterobacteriaceae in minced meat using VRBG agar (Murthy and Bachhil, 1982) demonstrated significantly higher recoveries with the spread plate technique. Improved performance of the spread plates was thought to be due to the oxygen requirements of the bacteria as suggested by the work of Hechelmann et al. (1973) who reported low Enterobacteriaceae counts under anaerobic conditions. However, Mossel et al. (1978) considered that the overlay procedure, whilst establishing a sufficiently reduced oxygen tension to suppress the growth of strictly aerobic Gram negative organisms, also enhanced anaerobic glucose utilization by the Enterobacteriaceae resulting in colonies of more characteristic appearance.

The problem of resuscitation of stressed organisms was addressed by the VRB-2 pour-plate procedure (Hartman et al., 1975) which used a base layer of plate count agar (PCA) overlayered with double strength VRB (VRB-2) agar. Recoveries using this system were greater than with the conventional system using VRB agar for both layers, increases of 31% with raw milk, 70% with ice cream and 61% with cottage cheese being reported from a multicentre study (Marshall et al., 1978). Reber and Marshall (1982) compared the performance of a pour-plate system using VRB for both layers and PCA/VRB-2 agar for the recovery of stressed coliforms from stored acidified half-and-half (a cream product) and showed that PCA/VRB-2 recovered 20% more coliforms than VRB agar. Comparisons of recoveries of heat- and chlorine-injured *E. coli* cells by an MPN method using lactose tryptone-MUG broth and the VRB-2 method incubated at 35°C for 24–48 h showed the MPN method to be superior (Feng and Hartman, 1982).

A comparison of two plate methods incorporating resuscitation steps with direct counting on VRB agar and MPN estimations in MacConkey broth together with confirmation tests (Hall, 1984) showed that the membrane transfer method of Holbrook et al. (1980) gave recoveries almost as high as the pour-plate with overlay method of Hall. The membrane method has the advantage of direct indole testing on all colonies in one simple operation whereas Hall's method involves removal of the agar from the base of the Petri dish, its inversion into the lid and subsequent selection of colonies for individual testing. The sometimes anomalous results of the indole reaction may have accounted for the slightly lower counts obtained by the membrane method in Hall's study when compared with his medium. The relative simplicity of the membrane method makes it less time consuming. Hall's medium provides information on lactose fermentation, but the exposure of organisms which may have been stressed to molten agar at 45°C before resuscitation is a disadvantage. Both two-stage methods gave higher recoveries of *E. coli* than the other methods, with counts approximately three times higher than on VRB agar and twice the MPN estimation.

Chen and Wu (1992) in their report on FCA (Table 3) found that recovery of freeze-stressed *E. coli* on FCA was about 1 log cycle lower than on nonselective media. Their technique involved preincubation on a selective medium at 35°C for 2 h prior to overnight incubation at 44.5°C, possibly accounting for the reduction in the count on FCA compared to PCA. Compared to the FDA MPN method (FDA, 1984), which is without any resuscitation step, counts on 32 food samples were not significantly different.

Bredie and de Boer (1992) compared commercially available β-D-glucuronidase based methods with the ISO standard MPN (1991a) method and the Anderson and Baird-Parker (1975) procedure for enumeration of *E. coli* in naturally contaminated foods of animal origin. The latter and the commercially produced Petrifilm^R *E. coli* methods were found useful for routine counting of *E. coli* in raw meat, poultry and meat products. The MPN procedure was more sensitive but impractical and considerably more expensive. A collaborative study in 24 laboratories (Entis, 1989) examined a hydrophobic grid membrane filter method incorporating the use of MUG for the enumeration of total coliform and *E. coli* in foods by comparing its performance against the AOAC 3-tube MPN method. The total coliform methods did not differ significantly in the examination of raw milk, raw ground poultry, whole egg powder and cheese powder but the MPN method detected a significantly higher number of organisms in ground black pepper. The hydrophobic grid membrane method detected significantly higher numbers of *E. coli* present in egg powder, cheese powder and ground black pepper samples while not differing significantly from the 3-tube method for the raw milk and raw ground poultry samples.

Huang et al. (1997) compared the IMViC (indole, methyl red, Voges-Proskauer and citrate utilization) tests with the beta-glucuronidase (GUD) assay for the identification of suspect *E. coli* on Levine's eosin-methylene blue (EMB) agar. After testing 258 suspect *E. coli* colonies from raw meat and meat products, 163 and 44 were found to be *E. coli* and non-*E. coli*, respectively, by both methods. The sensitivities for the identification of *E. coli* on EMB were 80.9% (169/209) and 97.1% (203/209), respectively, by the IMViC tests and GUD assay; whereas the specificities were 93.9% (46/49) and 95.9% (47/49), respectively, by the IMViC tests and GUD assay. It is proposed that the GUD assay can be an effective alternative to the conventional IMViC tests for the identification of suspect *E. coli* on EMB.

The Petrifilm *E. coli*/Coliform (EC) Count Plate in foods, has been compared with the AOAC-MPN method (Gangar et al., 1999). Mean log counts for the Petrifilm plate procedure were not significantly different from those for the MPN procedure for cooked fish samples inoculated with low or high inocula levels, for samples of raw turkey inoculated at medium level, and for beef inoculated at low, medium, and high levels. Repeatability and reproducibility variances of the Petrifilm EC Plate method recorded at 24 h were as good as or better than those of the MPN method. The Petrifilm method for enumerating confirmed *E. coli* in poultry, meats, and seafood has been adopted first action by AOAC.

Turner et al. (2000) compared CCA with Petrifilm *E. coli* count plate (PEC) for identifying coliforms and *E. coli* in a variety of meat products. The overall respective confirmation percentages (CFU/g) for the PEC and the CCA methods were 93.1

and 93.7% for coliforms and 99.8 and 98.1% for *E. coli*, although the CCA method yielded significantly ($P < 0.001$) higher mean CFU/g values for both coliforms and *E. coli*. Regression analyses of these data indicated that a strong positive linear relationship existed between the two methods over a wide CFU/g range for both coliforms and *E. coli*. The respective correlation coefficients obtained for coliforms and *E. coli* of 0.89 and 0.86 indicated that the CCA method provided a reliable optional method for these determinations in meat products.

An assessment of media containing MUG and a chromogenic substrate, 5-bromo-4-chloro-3-indolyl-β-D-galactoside (XGAL), demonstrated that cleavage of XGAL is a quicker and more sensitive parameter for total coliforms than gas production and that the combination of fluorescence and indole production is slightly superior to fluorescence and gas production for the identification of *E. coli* (Hahn and Wittrock, 1991). Ley et al. (1993) found that XGAL medium, in addition to providing a rapid test for coliforms, also detected ß-galactosidase-positive aeromonads and nonsheen-forming members of the Enterobacteriaceae on m-Endo agar. They reported the low sensitivity of m-Endo for detecting *Aeromonas* spp. which are considered ubiquitous waterborne organisms and should not be present in drinking water (Moyer, 1987). Chromocult coliforms agar was compared with the Standard Methods membrane filtration faecal coliform (mFC) medium for faecal coliform detection and enumeration (Alonso et al., 1998). Statistically, there were no significant differences between faecal coliform counts obtained with the two media (CC agar and mFC agar) and two incubation procedures (2 h–37°C plus 22 h–44.5°, and 44.5°C) as determined by variance analysis. In this study *E. coli* represented, on average 70.5–92.5% of the faecal coliform population. A high incidence of false negative *Klebsiella, Enterobacter* and *Citrobacter* (19.5%) and *E. coli* (29.6%) colonies was detected at 44.5°C. Two *E. coli* GUD negative phenotype upon reinoculation into CC agar were GUD positive. A total of 31 *Klebsiella, Enterobacter* and *Citrobacter*, LAC-colonies were streaked onto CC agar and incubated at 37°C. Of these, 29 *Klebsiella, Enterobacter* and *Citrobacter* strains that failed to produce ß-galactosidase at 44.5°C were found to produce the enzyme at 37°C. The physiological condition of the faecal coliform isolates could be responsible for the nonexpression of ß-galactosidase and ß-glucuronidase activities at 44.5°C. Geissler et al. (2000) compared the performance of LMX® broth, Chromocult Coliform®-agar (CC) and Chromocult Coliform®-agar plus cefsulodin (10µg/mL) (CC-CFS) with Standard Methods multiple tube fermentation (MTF) for the enumeration of total coliforms and *E coli* from marine recreational waters. Background interference was reduced on CC-CFS providing a more accurate total count.

The MI agar method was compared with the approved method by the use of wastewater-spiked tap water samples (Brenner et al., 1996). The USEPA-approved membrane filter method for *E. coli* requires two media, an MF transfer, and a total incubation time of 28 h. Overall, weighted analysis of variance (significance level, 0.05) showed that recoveries of total coliforms and *E. coli* on the new medium were significantly higher than those on mEndo agar and nutrient agar plus MUG (4-methylumbelliferyl-ß-D-glucuronide), respectively, and the background counts were significantly lower than those on mEndo agar (< 5%).

Recovery of total coliforms and *E. coli* on m-Coliblue was evaluated and conducted according to a US Environmental Protection Agency (USEPA) protocol (Grant, 1997). For comparison, this same protocol was used to measure recovery of total coliforms and *E. coli* with two standard MF media, m-Endo broth and mTEC broth. Comparison of specificity, sensitivity, false positive error, undetected target error, and overall agreement indicated *E. coli* recovery on m-ColiBlue24 was superior to recovery on mTEC for all five parameters. Recovery of total coliforms on this medium was comparable to recovery on m-Endo.

Alonso et al. (1999) compared the performance of CHROMagar ECC (CECC), and CECC supplemented with sodium pyruvate (CECCP) with the membrane filtration lauryl sulphate-based medium (mLSA) for enumeration of *E. coli* and non-*E. coli* thermotolerant coliforms (KEC). In order to establish that the maximum *Klebsiella, Enterobacter, Citrobacter* and *E. coli* population was recoverable, two incubation temperature regimens (41 and 44.5° C) were compared. CECCP agar incubated at 41° C proved most efficient for the simultaneous enumeration of *E. coli* and KEC from river and marine waters.

Japanese and U.S. Food and Drug Administration standard methods, as well as two agar plate methods, were compared with the three commercial kits using enzyme substrates (Venkateswaran et al., 1996). Isolation of *E. coli* on the basis of the ß-glucuronidase enzyme reaction was found to be good. Levine's eosine methylene blue agar, which has been widely used in various laboratories to isolate *E. coli* was compared with 4-methylumbelliferyl-ß-D-glucuronide (MUG)-supplemented agar for isolation of *E. coli*. Only 47% of the *E. coli* was detected when eosine methylene blue agar was used; however, when violet red bile (VRB)-MUG agar was used, the *E. coli* detection rate was twice as high. Of the 200 *E. coli* strains isolated, only two were found to be MUG negative, and the gene responsible for ß-glucuronidase activity (uidA gene) was detected by the PCR method in these two strains. Of the 90 false-positive strains isolated that exhibited various *E. coli* characteristic features, only two non-*E. coli* strains hydrolyzed MUG and produced fluorescent substrate in VRB-MUG agar. However, the PCR did not amplify uidA gene products in these VRB-MUG fluorescence-positive strains.

West and Coleman's (1986) work on the isolation of *E. coli* from shellfish included a comparison of various solid media used as pour plates, spread plates, sandwich plates and roll tubes. With some systems a resuscitation step consisting of 2 h incubation at 30°C prior to overnight incubation at 44°C was included. No significant differences were found between recoveries on the agar media, whether used with or without this resuscitation step, although counts were higher with the methods that used resuscitation. Failure to achieve a significant increase in the counts despite the pre-incubation at a lower temperature was probably due to the presence of bile salts in the medium on which resuscitation was attempted, or possibly to the use of molten agar to overlay plates (Murthy and Bachhil, 1982).

Table 4
Media recommended by ICMSF (1978) for examinations for coliforms and Enterobacteriaceae in foods (details of media are shown in Tables 2, 3).

Organisms or group sought (technique)	Medium (incubation temperature)				
	Coliform isolation	Coliform confirmation	Faecal oliform confirmation	*E. coli* confirmation	Enterobacteriaceae isolation
Coliforms (MPN: N. America)	LST broth (37–37°C)	BGBB (35–37°C)	EC broth (44.5°C)		
Coliforms (MPN: British)	MacConkey broth (35–37°C)		BGBB Peptone water[a] (44°C)	IMViC tests	
Coliforms (MPN: confirmed)	BGBB (35–37°C)	VRB or Endo agar (35–37°C)	BGBB Peptone water[a] (44°C)	IMViC tests	
Coliforms (colony count)	VRB agar[b] (35–37°C)			IMViC tests	
Enterobacteriaceae (presence/absence; preenriched in BPW[c])					EE broth[d] (35–37°C) (35–37°C)
Enterobacteriaceae (colony count)					VRBG agar[b] (35–37°C)

[a] for indole test.
[b] with overlay.
[c] buffered peptone water.
[d] confirm by oxidase test and mode of attack on glucose.

Table 5
Media recommended by ISO for examinations for coliforms, *E. coli* and Enterobacteriaceae in foods and animal feeding stuffs (details of media are shown in Tables 2 and 3). For the present status of ISO standards for Enterobacteriaceae see de Boer (1998).

ISO Reference/ Commodity	Organisms or group sought (technique)	Medium (incubation temperature)			
		Coliform isolation	Coliform confirmation	*E. coli* confirmation	Enterobacteriaceae isolation
1991a, Foods and animal feeding stuffs	Coliforms (MPN)	LST broth (30, 35 or 37°C)	BGBB (30, 35 or 37°C)		
1991b, Foods and animal feeding stuffs	Coliforms (colony count)	VRB agar[a] (30, 35 or 37°C)	none		
1993a, Foods and animal feeding stuffs	*E. coli* (presumptive; MPN)	LST broth (35 or 37°C)		EC broth Tryptone water[b] (45°C)	
1988, Meat and meat products	*E. coli* (Membrane method)			MMG agar (37°C) followed by TBA (44°C)[c]	
1993b, Foods and animal feeding stuffs	Enterobacteriaceae - (colony count) Enterobacteriaceae (MPN)				VRBG agar[a] (35 or 37°C) EE broth[d] (35 or 37°C)
1991c, Foods and animal feeding stuffs	Enterobacteriaceae (presence/ absence; preenriched in BPW[e])				EE broth[d] (35 or 37°C)

[a] with overlay
[b] for indole test
[c] combined isolation and confirmation test
[d] confirm by oxidase test and mode of attack on glucose
[e] buffered peptone water

Table 6
Chromogenic/fluorogenic media recommended by different standards for examinations of *E. coli*, coliforms and Enterobacteriaceae in water and foods.

Standards	medium and title
Water quality	
ISO 1998	EC/MUG: Detection and enumeration of *E. coli* and coliform bacteria in surface and waste water-Part 3 : Miniaturized method (MPN) by inoculation in liquid medium, ISO 9308-3 First edition 1998-11-15
AOAC 1995	Colilert: Total Coliforms and *E. coli* in Water, AOAC Official Method 991.15
APHA 1998	Standard Methods for the Examination of Water and Wastewater 20th Edition 1998: 9221 F, EC/MUG: *E. coli* Procedure (PROPOSED) Confirmatory test after prior enrichment 9222 G, Nutrient agar with MUG medium and EC broth with MUG MF Partition *E. coli* Methods .Verification of presence of *E. coli* from a total-coliform-positive MF on Endo-type media . 9223, ONPG or CPRG for total coliform bacteria and MUG for *E. coli* Enzyme Substrate Coliform Test 9260 F, Fluorocult Laurylsulfate broth for differentiation Pathogenic *E. coli* 0157:H7
Milk and milk products	
ISO 1997	LST/MUG- Enumeration of presumptive *E. coli* . Part 2 : MPN technique. ISO 11866-2 First edition 1997
Food and animal feeding stuffs	
ISO 1999 a	TBX-Agar: Microbiology of food and animal feeding stuffs-Horizontal method for the enumeration of presumptive *E. coli* - Part 1:Colony count technique at 44°C using membranes and 5-bromo-4-chloro-3-indolyl-ß-D-glucoronic acid, ISO/DIS 16649-1 1999
ISO 1999 b	TBX-Agar: Microbiology of food and animal feeding stuffs-Horizontal method for the enumeration of presumptive *E. coli* - Part 2: Colony count technique at 44°C using 5-bromo-4-chloro-3-indolyl-ß-D-glucoronic acid, ISO/DIS 16649-1 1999
AOAC 1995	Petrifilm *E. coli*: Coliform and *E. coli* Counts in Foods, AOAC Official Method 991.14 Colicomplete: Confirmed Total Coliform and *E. coli* in All Foods, AOAC Official Method 992.30 Hydrophobic Grid Membrane Filter /MUG: Total Coliform and *E. coli* counts in Foods, AOAC Official Method 990.11 LST/MUG: *E. coli* in Chilled or Frozen Foods, AOAC Official Method 988.19
DIN 1992	ECD/MUG agar: Microbiological Investigation of Meat and Meat products. Determination of *E. coli* . Fluorescence optical colony count method using membranes . Spatula method (reference method), German DIN Norm 10110, 1992

Concluding remarks

There can be little surprise that in the course of a hundred years of bacteriology, during which time basic techniques have changed very little, numerous formulations of culture media have been recommended for growing the most studied of all organisms, *E. coli*. Perhaps the greatest surprise is that a formula first published by MacConkey in 1905 should still be in use in a recognisable form today. Many authorities, e.g. ICMSF (Table 4) and ISO (Table 5), offer a choice of more than one technique for the enumeration of this group of organisms detailing both MPN and solid media methods. In deciding which type of method to use consideration should be given to the fact that, where low numbers are sought, MPN techniques are more sensitive than plating methods. They are also subject to greater variation, slower to arrive at a confirmed result and more expensive and labour intensive. Plating methods, particularly when they incorporate a rapid confirmation step, as in the technique of Holbrook et al. (1980), produce confirmed results more speedily at the expense of some sensitivity. Some of the current media for *E. coli* and coliforms based on the ß-D-glucuronidase and ß-galactosidase activities are now widely used in food and water microbiology. There is now a wide range of media available for detection and enumeration of indicator organisms in water and foods. Some of these media have been adopted by various Standards organizations (Table 6) and their use has led to improved accuracy and faster detection of target organisms, often reducing the need for isolation of pure cultures and confirmatory testing. Use of a resuscitation step is considered essential if injured organisms are likely to be encountered. This may simply involve pre-incubation of a selective medium at a lower temperature before the definitive incubation at an elevated (44–45.5°C) temperature or, preferably, the use of one of the membrane transfer methods with pre-incubation on a non-selective agar at 35 or 37°C for 2–4 h followed by transfer to selective agar at 44–45.5°C for overnight incubation.

Acknowledgement

This chapter is an updated version of 'Media for total Enterobacteriaceae, coliforms and *E. coli*' by Blood and Curtis published in the first edition of Culture Media for Food Microbiology. The present author thanks the authors of the first edition for giving him the opportunity to update their manuscript.

References

Abbiss, J.S., Wilson, J.M., Blood, R.M. and Jarvis, B. (1981) A comparison of minerals modified glutamate medium with other media for the enumeration of coliforms in delicatessen foods. J. Appl. Bacteriol. 51, 121–127.

Alonso, J.L., Soriano, A., Carbajo, O., Amoros, I., and Garelick, H. (1999) Comparison and recovery of *Escherichia coli* and thermotolerant coliforms in water with a chromogenic medium incubated at 41 and 44.5° C. Appl. Environ. Microbiol. 65, 3746–3749.

Alonso, J.L., Soriano, K., Amoros, I., and Ferrus, M.A. (1998) Quantitative determination of E. coli and faecal coliforms in water using a chromogenic medium. J. Environ. Sci. Health 33, 1229–1248.

Alvarez, R.J. (1984) Use of fluorogenic assays for the enumeration of Escherichia coli from selected seafoods. J. Food Sci. 49,1186–1187.

Anderson, J.M. and Baird-Parker, A.C. (1975) A rapid direct plate method for enumerating *Escherichia coli* biotype I in foods. J. Appl. Bacteriol. 39, 111–117.

Andrews, W.H. and Presnell, M.W. (1972) Rapid recovery of *Escherichia coli* from estuarine waters. Appl. Microbiol. 23, 521–523.

Andrews, W.H., Wilson, C.R., Poelma, P.L., Bullock, L.K., McClure, F.D. and Gentile, D.E. (1981) Interlaboratory evaluation of the AOAC method and the A-1 procedure for the recovery of faecal coliforms from foods. J. Assoc. Off. Anal. Chem. 64, 1116–1121.

Anon. (1987) Pharmacopoeia of culture media for food microbiology. Int. J. Food Microbiol. 5, 187–300.

AOAC. (1990) Official methods of analysis. 14th edition Association of Official Analytical Chemists, Arlington, VA.

AOAC. (1995) Official methods of analysis. 16th edition Association of Official Analytical Chemists, Gaithersburg, MD.

APHA. (1984) Compendium of methods for the microbiological examination of foods. 2nd edition,. American Public Health Association, Washington, DC.

APHA. (1992) Compendium of methods for the microbiological examination of foods. 3rd edition, American Public Health Association, Washington, DC.

APHA. (1998) Standard Methods for the Examination of Water and Wastewater, 20th edition,. American Public Health Association, Washington, DC.

Berg, J.D. and Fiksdal, L. (1988) Rapid detection of total and faecal coliforms in water by enzymatic hydrolysis of 4-methylumbelliferone-ß-D-galactoside. Appl. Environ. Microbiol. 54, 2118–2122.

Betts, R., Murray, K., and MacPhee, S. (1994) Evaluation of Fluorocult LMX broth for the simultaneous detection of total coliforms and *E. coli*. Technical Memorandum No. 705, Campden Food & Drink Research Association, UK.

Bindschedler, O., deMan, J.C. and Curiat, G. (1981) Comparative study of several culture media to determine coliforms and *E. coli* in dairy and cocoa products. Zbl. Bakteriol. II. Abt. 136, 146–151.

Blood, R.M. and Curtis, G.D.W. (1995) Media for "total" Enterobacteriaceae, coliforms and *Escherichia coli*. *In*: Corry, J.E.L., Curtis, G.D.W. and Baird, R. M. (Eds.), Culture Media for Food Microbiology, Elsevier, Amsterdam, pp. 163–185.

Bredie, W.L. and de Boer, E. (1992) Evaluation of the MPN, Anderson - Baird-Parker, Petrifilm *E. coli* and Fluorocult ECD method for enumeration of *Escherichia coli* in foods of animal origin. Int. J. Food Microbiol. 16, 197–208.

Brenner, K., Rankin, C.C., Roybal, Y.R., Stelma J.R., G.R., Scarpino, P.V. and Dufour, A.P. (1993) New medium for simultaneous detection of total coliforms and *E. coli* in water. Appl. Environ. Microbiol. 59, 3534–3544.

Brenner, K.P., Rankin, C.C., Sivaganesan, M. and Scarpino, P.V. (1996) Comparison of the recoveries of *E. coli* and total coliforms from drinking water by the MI agar method and the U.S. environmental protection agency-approved membrane filter method. Appl. Environ. Microbiol. 62, 203–208.

Buerger, H. (1967) Biochemische Leistungen nicht-profilierender Mikroorganismen II. Nachweis von Glycosid-Hydrolasen, Phosphatasen, Esterasen und Lipasen. Zbl. Bakteriol. I Orig. 202, 97–109.

Chen, H.C. and Wu, S.D. (1992) Agar medium for enumeration of faecal coliforms. J. Food Sci. 57, 1454–1457.

Christen, G.L., Davidson, P.M., McAllister, J.S., and Roth, L.A. (1992). Coliform and other indicator bacteria. *In*: R.T. Marshal (Eds.), Standard methods for the examination of dairy products. 16th ed. American Public Health Association, Washington, D.C.

Cooke, B.C. and Jorgenson, M. (1977) An evaluation of modified minerals glutamate medium for use in the presumptive coliform test on dairy products. N.Z. J. Dairy Sci. Technol. 12, 272–273.

de Boer, E. (1998) Update on media for isolation of Enterobacteriaceae from foods. Int. J. Food Microbiol. 45, 43–53.

Delaney, J.E., McCarthy, J.A. and Grasso, R.J. (1962) Measurement of *Escherichia coli* type 1 by the membrane filter. Water and Sewage Works 109, 289–294.

DIN (1992) German DIN Norm 10110, Microbiological Investigation of Meat and Meat products. Determination of *E. coli* . Fluorescence optical colony count method using membranes. Spatula method .

Dunham, H.G. and Schoenlein, H.W. (1926) Brilliant green bile media. Stain Technol. 1, 129–134.

Endo, S. (1904) Ueber ein Verfahren zum Nachweis von Typhusbacillen. Centralbl. Bakt. I. Orig. 35, 109–110.

Entis, P. (1984) Enumeration of total coliforms, faecal coliforms and *Escherichia coli* in foods by hydrophobic grid membrane filter: collaborative study. J. Assoc. Anal. Chem. 67, 812–823.

Entis, P. (1989) Hydrophobic grid membrane filter/MUG method for total coliform and *Escherichia coli* enumeration in foods: collaborative study. J. Assoc. Anal. Chem. 72, 936–950.

FDA. (1984) Bacteriological Analytical Manual, 6th edition Food and Drugs Administration, Washington, DC.

FDA. (1995) Bacteriological Analytical Manual, 8th edition Food and Drugs Administration, Washington, DC.

Feng, P.C.S. and Hartman, P.A. (1982) Fluorogenic assays for immediate confirmation of *Escherichia coli*. Appl. Environ. Microbiol. 43, 1320–1329.

Frampton, E.W. and Restaino, L. (1993) Methods for *Escherichia coli* identification in food, water and clinical samples based on beta-glucuronidase detection. J. Appl. Bacteriol. 74, 223–233.

Gangar, V., Curiale, M.S., Lindberg, K. and Gambrel-Lenarz, S. (1999) Dry rehydratable film method for enumerating confirmed *Escherichia coli* in poultry, meats, and seafood: collaborative study. J. Assoc. Anal. Chem. 82, 73–78.

Gassner, G. (1918) Ein neuer Dreifarbennaehrboden zur Typhus-Ruhr-Diagnose. Centralbl. Bakt. I. Orig. 80, 219–222.

Geissler, K., Manafi, M., Amorós, I. and Alonso J.L. (2000) Quantitative determination of total coliforms and *E. coli* in marine waters with chromogenic and fluorogenic media. J. Appl. Bacteriol. 88, 280–285.

Geldreich, E.E. (1997) Coliforms: A new beginning to an old problem. *In*: Kay, D. and Fricker, C. (Eds.), Coliforms and *E. coli*, Problem or solution?, Athenaeum Press, UK, pp. 3–11.

Grant, M.A. (1997) A new membrane filtration medium for simultaneous detection and enumeration of *E. coli* and total coliforms. Appl. Environ. Microbiol. 63, 3526–3530.

Gray, R.D. (1964) An improved formate lactose glutamate medium for the detection of *Escherichia coli* and other coliform organisms in water. J. Hyg. 62, 495–508.

Hahn, G. and Wittrock, E. (1991) Comparison of chromogenic and fluorogenic substances for differentiation of coliforms and *Escherichia coli* in soft cheese. Acta Microbiol. Hung. 38, 265–271.

Hajna, A.A. (1955) A new enrichment broth medium for gram-negative organisms of the intestinal group. Public Health Lab. 13, 83–89.

Hajna, A.A. and Perry, C.A. (1943) Comparative study of presumptive and confirmative media for bacteria of the coliform group and for faecal streptococci. Am. J. Publ. Hlth. 33, 550–556.

Hall, H.E. (1964) Methods of isolation and enumeration of coliform organisms. *In*: Examination of foods for enteropathogenic and indicator bacteria; Review of methodology and manual of selective procedures. Edited by K.H.Lewis and R.Angelotti. Division of Environmental

Engineering and Food Protection, U.S. Department of Health Education and Welfare, Public Health Service Publication No. 1142. Washington, DC.

Hall, L.P. (1984) A new direct plate method for the enumeration of *Escherichia coli* in frozen foods. J. Appl. Bacteriol. 56, 227–235.

Hartman, P.A. (1989) The MUG glucuronidase test for *E. coli* in food and water. *In*: Turano, A. (ed.), Rapid Methods and Automation in Microbiology and immunology, Brixia Academic Press, Brescia, Italy, pp. 290–308.

Hartman, P.A., Hartman, P.S. and Lanz, W.W. (1975) Violet red bile 2 agar for stressed coliforms. Appl. Microbiol. 29, 537–539.

Hechelmann, H., Rossmanith, E., Periec, M. and Leistner, L. (1973) Untersuchung zur Ermittlung der Enterobacteriaceae zahl bei schlachtgelflugel. Fleischwirtschaft 53, 107–113.

Holbrook, R. and Anderson, J.M. (1982) The rapid enumeration of *E. coli* in foods by direct plating method. In: Isolation and identification methods for food poisoning organisms. SAB Technical Series No. 17, edited by J.E.L. Corry, D. Roberts and F.A. Skinner. Academic Press, London, U.K.

Holbrook, R., Anderson, J.M. and Baird-Parker, A.C. (1980) Modified direct plate method for counting *Escherichia coli* in foods. Food Technol. Austral. 32, 78–83.

Holt-Harris, J.E. and Teague, O.A. (1916) A new culture medium for the isolation of *Bacillus typhosus* from stools. J. Infect. Dis. 18, 596–600.

Huang, S.W., Chang, C.H., Tai, T.F. and Chang, T.C. (1997) Comparison of the beta-glucuronidase assay and the conventional method for identification of *Escherichia coli* on eosin-methylene blue agar. *J. Food Protect.* 60, 6–9.

ICMSF. (1978) Microorganisms in foods I. Their significance and methods of enumeration. 2nd ed. University of Toronto Press.

ISO. (1988) Meat and meat products – Enumeration of *Escherichia coli* – Colony count technique at 44°C using membranes. ISO 6391–1988. International Organisation for Standardisation, Geneva.

ISO. (1991a) Microbiology – General guidance for the enumeration of coliforms – Most probable number technique. ISO 4831–1991. International Organisation for Standardisation, Geneva.

ISO. (1991b) Microbiology – General guidance for the enumeration of coliforms – Colony count technique. ISO 4832–1991. International Organisation for Standardisation, Geneva.

ISO. (1991c) Microbiology – General guidance for the detection of Enterobacteriaceae with pre-enrichment. ISO 8523–1991. International Organisation for Standardisation, Geneva.

ISO. (1993a) Microbiology – General guidance for enumeration of presumptive *Escherichia coli* – Most probable number technique. ISO 7251–1993. International Organisation for Standardisation, Geneva.

ISO. (1993b) Microbiology – General guidance for the enumeration of Enterobacteriaceae without resuscitation – MPN technique and colony count technique. ISO 7402–1993. International Organisation for Standardisation, Geneva.

ISO. (1997) Milk and milk products, LST/MUG- Enumeration of presumptive *E. coli* . Part 2 : MPN technique. ISO 11866-2-1997. International Organisation for Standardisation, Geneva.

ISO. (1998) Water quality, Detection and enumeration of *E. coli* and coliform bacteria in surface and waste water-Part 3: Miniaturized method (MPN) by inoculation in liquid medium ISO 9308-1998. International Organisation for Standardisation, Geneva.

ISO. (1999a) Food and animal feeding stuffs, Microbiology of food and animal feeding stuffs-Horizontal method for the enumeration of presumptive *E. coli* – Part 1:Colony count technique at 44°C using membranes and 5-bromo-4-chloro-3-indolyl-ß-D-glucoronic acid ISO/DIS 16649-1 1999. International Organisation for Standardisation, Geneva.

ISO. (1999b) Microbiology of food and animal feeding stuffs-Horizontal method for the enumeration of presumptive *E .coli* – Part 2: Colony count technique at 44°C using 5-bromo-4-chloro-3-indolyl-ß-D-glucoronic acid ISO/DIS 16649-2 1999. International Organisation for

Standardisation, Geneva.

James, A.L. and Perry J.D. (1996) Evaluation of cyclohexenoesculetin-ß-D-galactoside and 8-hydroxychinoline-ß-D-galactoside as substrates for the detection of ß-D-galactosidase. Appl. Environ. Microbiol. 62, 3868–3870.

Jermini, M., Domeniconi, F. and Jaggli, M. (1994) Evaluation of C-EC-agar, a modified mFC-agar for the simultaneous enumeration of faecal coliforms and *E. coli* in water. Lett. Appl. Microbiol. 19, 332–335.

Klein, H. and Fung, D.Y.C. (1976) Identification and quantification of faecal coliforms using violet red bile agar at elevated temperature. J. Milk Food Technol. 39, 768–770.

Koburger, J.A. and Miller, M.L. (1985) Evaluation of a fluorogenic MPN procedure for determining *Escherichia coli* in oysters. J. Food Protect. 48, 244–245.

Landre, J.R., Anderson, DA., Gavriell, A.A. and Lambl, A.J. (1998) False positive coliform reaction mediated by *Aeromonas* in the colilert defined substrate technology system. Lett. Appl. Microbiol. 26, 352–354.

Lee, J.V., Lightfoot, N.F. and Tillett, H.E. (1995) An evaluation of presence/absence tests for coliform organisms and *E. coli*. Int. Conf. on Coliforms and *E. coli*, Problem or Solution? Leeds, UK.

Ley, A.N., Barr, S., Fredenburgh, D., Taylor, M. and Walker, N. (1993) Use of 5-bromo-4-chloro-3-indolyl-ß-D-galactopyranoside for the isolation of ß-D-galactosidase-positive bacteria from municipal water supplies. Can. J. Microbiol. 39, 821–825.

MacConkey, A. (1905) Lactose fermenting bacteria in faeces. J. Hyg. Camb. 5, 333–379.

MacConkey, A.T. (1908) Bile salt media and their advantages in some bacteriological examinations. J. Hyg. Camb. 8, 322–334.

Mallman, W.L. and Darby, C.W. (1941) Use of a lauryl sulphate tryptose broth for the detection of coliform organisms. Am. J. Publ. Hlth. 31, 127–134.

Manafi, M. (1995) New medium for the simultaneous detection of total coliforms and *E. coli* in water. Abstracts of the 95th Meeting of the American Society for Microbiology. Washington, DC. Abstr.P-43, P. 389.

Manafi, M. (1996) Fluorogenic and chromogenic substrates in culture media and identification tests. Int. J. Food Microbiol. 31, 45–58.

Manafi, M. (2000) New developments in chromogenic and fluorogenic culture media . Int. J. Food Microbiol. 60, 205–218.

Manafi, M. and Rosmann, H. (1998) Evaluation of Readycult presence-absence test for detection of total coliforms and *E. coli* in water. Abstr. Q 263, pp. 464. Abstracts of the 98th Meeting of the American Society for Microbiology, Atlanta, USA.

Manafi, M. and Rosmann, H. (2000) Identification of bacterial strains isolated from drinking water by means of fluorocult LMX® and Readycult® coliforms. Abstr. Q 346, pp. 620. Abstracts of the 100th Meeting of the American Society for Microbiology, Los Angeles, USA.

Marshall, R.T., Hartman, P.A., Cannon, R.Y., Lambeth, L., Richardson, G.H., Spurgeon, K.R., Weddle, D.B., Wingfield, M. and White, C.H. (1978) Group comparative study of VRB-2 agar in the recovery of coliforms from raw milk, ice cream and cottage cheese. J. Food Protect. 41, 544–545.

Mates, A., and Schaffer, M. (1989) Membrane filtration differentiation of *E. coli* from coliforms in the examination of water. J. Appl. Bacteriol. 67, 343–346.

Mehlman, I.J. (1984) Coliforms, faecal coliforms, *Escherichia coli* and enteropathogenic *E. coli*. *In*: Compendium of methods for the microbiological examination of foods. 2nd edition edited by M.L.Speck. APHA Washington, DC.

Moberg, L.J., Wagner, M.K. and Kellen, L.A. (1988) Fluorogenic assay for rapid detection of *Escherichia coli* in chilled and frozen foods: collaborative study. J. Assoc. Off. Anal. Chem. 71, 589–602.

Mossel, D.A.A. (1982) Marker (index and indicator) organisms in food and drinking water.

Semantics, ecology, taxonomy and enumeration. Antonie van Leeuwenhoek 48, 609–611.
Mossel, D.A.A., Eelderink, I., Koopmans, M. and van Rossem, F. (1978) Optimalisation of a MacConkey-type medium for the enumeration of Enterobacteriaceae. Lab. Pract. 27, 1049–1050.
Mossel, D.A.A., Eelderink, I., Koopmans, M. and van Rossem, F. (1979) Influence of carbon source, bile salts and incubation temperature on recovery of Enterobacteriaceae from foods using MacConkey-type agars. J. Food Protect. 42, 470–475.
Mossel, D.A.A., Mengerink, W.H.J and Scholts, H.H. (1962) Use of modified MacConkey agar medium for the selective growth and enumeration of Enterobacteriaceae. J. Bacteriol. 84, 381.
Mossel, D.A.A., van der Zee, H., Hardon, A.P. and van Netten, P. (1986) The enumeration of thermotrophic types amongst the Enterobacteriaceae colonizing perishable foods. J. Appl. Bacteriol. 60, 289–295.
Mossel, D.A.A., Visser, M. and Cornelissen, A.M.R. (1963) The examination of foods for Enterobacteriaceae using a test of the type generally adopted for the detection of salmonellae. J. Appl. Bacteriol. 26, 444–452.
Motes, M.L., McPhearson, R.M. and DePaola, A. (1984) Comparison of three international methods with APHA method for enumeration of *Escherichia coli* in estuarine waters and shellfish. J. Food Protect. 47, 557–561.
Moussa, R.S., Keller, N., Curiat, G. and de Man, J.C. (1973) Comparison of five media for the isolation of coliform organisms from dehydrated and deep frozen foods. J. Appl. Bacteriol. 36, 619–629.
Moyer, N.P. (1987) Clinical significance of *Aeromonas* species isolated from patients with diarrhoea. J. Clin. Microbiol. 25, 2044–2048.
Murthy, T.R.K. and Bachhil, V.N. (1982) Comparison of pour plate with overlay and spread plating for Enterobacteriaceae count in minced meat. J. Food Sci. Technol. India 19, 37–39.
Oblinger, J.L., Kennedy, J.E. and Langston, D.M. (1982) Microflora recovered from foods on violet red bile agar with and without glucose and incubated at different temperatures. J. Food Protect. 45, 948–952.
Pierson, C.J., Emswiler, B.S. and Kotula, A.W. (1978) Comparison of methods for estimation of coliforms, faecal coliforms and enterococci in retail ground beef. J. Food Protect. 41, 263–266.
Rayman, M.K., Jarvis, G.A., Davidson, C.M., Long, S., Allen, J.M., Tong, T., Dodsworth, P., McLaughlin, S., Greenberg, S., Shaw, B.G., Beckers, H.J., Qvist, S., Nottingham, P.M. and Stewart, B.J. (1979) ICMSF methods studies. XIII. An international comparative study of the MPN procedure and the Anderson – Baird-Parker direct plating method for the enumeration of *Escherichia coli* biotype I in raw meats. Can. J. Microbiol. 25, 1321–1327.
Reber, C.L. and Marshall, R.T. (1982) Comparison of VRB and VRB-2 agars for recovery of stressed coliforms from stored acidified half-and-half. J. Food Protect. 45, 584–585.
Rippey, S.R., Chandler, L.A. and Watkins, W.D. (1987) Fluorimetric method for enumeration of *Escherichia coli* in molluscan shellfish. J. Food Protect. 50, 685–690, 710.
Sakazaki, R., Namioka, S., Osada, A. and Yamada, C.A. (1960) A problem on the pathogenic role of Citrobacter of enteric bacteria. Japan. J. Ex. Med. 30, 13–22.
Sakazaki, R., Tamura, K., Prescott, L.M., Benzic, Z., Sanyal, C. and Sinha, R. (1971) Bacteriological examination of diarrheal stools in Calcutta. Indian J. Med. Res. 59, 1025–1034.
Schets, F.M., Medema, G.J. and Havelaar, A.H. (1993) Comparison of Colilert with Dutch standard enumeration methods for *E. coli* and total Coliforms in water. Lett. Appl. Microbiol. 17,17–19.
Sharpe, A.N., Rayman, M.K., Burgener, D.M., Conley, D., Loit, A., Milling, M., Peterkin, P.I., Purvis, U. and Malcolm, S. (1983) Collaborative study of the MPN, Anderson – Baird-Parker direct plating, and hydrophobic grid-membrane filter methods for the enumeration of *Escherichia coli* biotype I in foods. Can. J. Microbiol., 29, 1247–1252.
Shelton, L.R., Leininger, H.V., Surkiewicz, B.F., Boer, E.F., Elliott, R.P., Hyndman, J.B. and Kramer, N. (1962) A bacteriological survey of the frozen pre-cooked food industry. U.S. Department of

Health, Education and Welfare, Food and Drug Administration. Washington, DC.

Silliker, J.H., Gabis, D.A. and May, A. (1979) ICMSF methods studies. XI. Collaborative/comparative studies on determination of coliforms using the most probable number procedure. J. Food Protect. 42, 638–644.

Turner K.M., Restaino L, and Frampton E.W. (2000) Efficacy of chromocult coliform agar for coliform and Escherichia coil detection in foods. *J. Food Protect.* 63, 539–541.

Vaughn, R.H., Levine, M. and Smith, H.A. (1951) A buffered boric acid lactose medium for enrichment and presumptive identification of *Escherichia coli*. Food Res. 16, 10–19.

Venkateswaran, K., Murakoshi, A. and Satake, M. (1996) Comparison of commercially available kits with standard methods for the detection of coliforms and *E. coli* in foods. Appl. Environ. Microbiol. 62, 2236–2243.

Vracko, R. and Sherris, J.C. (1963) Indole spot test in bacteriology. Am. J. Clin. Pathol. 39, 429–432.

Weiss, K.F., Chopra, N., Stotland, P., Riedel, G.W. and Malcolm, S. (1983) Recovery of faecal coliforms and of *Escherichia coli* at 44.5, 45.0 and 45.5°C. J. Food Protect. 46, 172–177.

West, P.A. and Coleman, M.R. (1986) A tentative national reference procedure for isolation and enumeration of *Escherichia coli* from bivalve molluscan shellfish by most probable number method. J. Appl. Bacteriol. 61, 505–516.

Chapter 13

Media for the isolation of *Salmonella*

H. van der Zee

Inspectorate for Health Protection and Veterinary Public Health, De Stoven 2, Postbus 202, 7200 AE, Zutphen, The Netherlands

Conventional isolation of *Salmonella* in food microbiology is accomplished by using cultural methods. Media for pre-enrichment, selective enrichment and isolation have been developed. Several official organisations for standardisation have developed reference methods for the isolation of *Salmonella*. In general these use one pre-enrichment medium, two different selective enrichment media and two or more isolation media. In this paper the main media used for conventional isolation of *Salmonella* are reviewed. Also some recent developments in culture media for Salmonella are described. Modified pre-enrichment media are sometimes required for specific cases, and can be accomplished by addition of supplements as ferrioxamine E or Oxyrase™ to standard media or through the development of new media such as Universal Purpose Broth. For the selective enrichment procedure, motility enrichment in semi-solid media shows equal or better results than the use of the standard liquid selective media. Recently developed isolation media use different selective and diagnostic properties, such as glucuronate fermentation, acid formation from propylene glycol, fermentation of glycerol and addition of Tergitol 4 as selective agent. There seems to be a trend in the official Standards Organisations to follow the more recent developments, but progress is slow.

Introduction

Cultural procedures used for the conventional isolation of *Salmonella* generally have four distinct phases: i) pre-enrichment in a non-selective medium to allow resuscitation of any injured cells and multiplication of the target organism and others present in the sample; ii) selective enrichment to allow the survival or growth of the *Salmonella* spp., while inhibiting accompanying organisms in the selective broth; iii) isolation using selective agar media that restrict growth of bacteria other than salmonellae, in order to produce presumptive isolates; iv) confirmation where isolates are subjected to a variety of biochemical and serological tests to confirm that they are *Salmonella* and to determine their serovar. In general, each individual step needs at least 16 h up to a maximum of 48 h. The whole procedure takes 4–7 days to complete and is therefore laborious and labour intensive.

Since contaminated food samples generally harbour only low levels of *Salmonella* spp., all four phases of the cultural procedure have to be performed. In these samples the salmonellae may also be distributed unevenly, and/or have undergone some sort

Table 1
Non-selective pre-enrichment media used in food microbiology for *Salmonella* detection.

Medium	Commodity	Standard Organisation[a]
Buffered Peptone Water (BPW)	General purpose	ISO, IDF
BPW + Casein	Chocolate	ISO, APHA, AOAC/FDA
Lactose broth (LB)	Eggs, frog legs	APHA, AOAC/FDA
LB + tergitol 7 or Triton X-100	Coconut, meat	APHA, AOAC/FDA
Skim milk + brilliant green	Cacao, chocolate, candy	AOAC/FDA
Tryptone Soya Broth (TSB)	Spices, dried yeast	AOAC/FDA
TSB + 0,5% potassium sulphate	Onion , garlic powder etc.	AOAC/FDA
Water + brilliant green	Milk powder	AOAC/FDA

[a] ISO = International Standard Organisation; IDF= International Dairy Federation: APHA = American Public Health Association; AOAC = American Association of Analytical Chemists; FDA = Food and Drugs Agency.

of injury through exposure to an unsuitable A_w or pH or presence of anti-microbial substances and therefore special attention should be paid to sampling procedures.

Organisations such as the International Standard Organisation (ISO), AOAC International, the American Public Health Association (APHA) and the International Dairy Federation (IDF) have developed reference methods for the isolation of *Salmonella*. Generally these use one pre-enrichment medium, two different selective enrichment media and two or more isolation media, followed by confirmation procedures.

Pre-enrichment

As mentioned earlier, pre-enrichment allows the growth of *Salmonella* and also facilitates their survival against the toxic effects of selective agents in enrichment media. A non-selective liquid medium is incubated for 16–20 h at 37°C to allow growth of *Salmonella* to a level of 10^5 cfu/ml.

In Europe, Buffered Peptone Water (BPW) is the medium of choice for the majority of food products according to ISO-6579 (Anon., 2002). However, alternative media for pre-enrichment are in use for various foods and these are shown in Table 1.

Sometimes for specific cases, modified pre-enrichment procedures are required, e.g. when attempting to shorten the pre-enrichment phase (Fung et al, 1993) or investigating eggs and egg products for *S. enteritidis* (Van der Zee, 1994). This can be accomplished either by addition of supplements, e.g. ferric ammonium citrate, ferrioxamine or Oxyrase™ to existing media or through the development of new media such as Universal Purpose Broth (Bailey and Cox, 1992), as described below.

Addition of supplements to existing media

The most recent ISO-Standard (Anon., 2002) allows also, in contrast to the preceding version (Anon., 1993), for cacao and cacao containing products specific BPW preparations containing casein (50 g/l) or skim milk powder (100 g/l) eventually combined with brilliant green (0.018 g/l) and for acidic and acidifying foodstuffs double buffered peptone water should be used. The addition of ammonium-iron(III)-citrate to BPW, facilitated the isolation of salmonellae from eggs and egg products, and in particular *S. enteritidis* from the yolk of fresh hen-eggs. Omitting the ferric ammonium citrate led to negative results (Dolzinsky and Kruse, 1992). In the presence of the iron-chelating albumen protein ovotransferrin, pre-enrichment of 20 ml pooled egg contents in 180 ml Trypticase Soy Broth (TSB), supplemented with 35 mg/l ferrous sulphate has been recommended to ensure adequate iron availability to support bacterial growth (Gast, 1993; 1995).

For the same reason Reissbrodt and Rabsch (1993) recommended supplementation of BPW with 1 μg/ml ferrioxamine E, a trihydroxamate-type siderophore.

Reducing the pre-enrichment phase by adding Oxyrase™, a commercially available product consisting of sterile oxygen-reducing membrane fragments from *Escherichia coli*, has been reported to stimulate the growth of facultative anaerobic food pathogens in liquid broths. Heat-injured cells are also reported to show an improved recovery in media supplemented with Oxyrase™. This is thought to be due to a reduction or removal of the oxidative stress which hinders the recovery of injured cells (Fung et al., 1993).

Development of new media

Universal Purpose Broth (UPB) was developed as a medium for pre-enriching food products. It allows the simultaneous recovery of *Salmonella, Listeria* and *E. coli* O157 (Jiang et al., 1998). The medium is low in carbohydrates and highly buffered to prevent a rapid drop in the pH of the broth. It also contains ferric ammonium citrate (0.5 g/l) and sodium pyruvate (0.2 g/ l) thus allowing injured bacteria to resuscitate and multiply, even in the presence of high levels of naturally occurring microflora.

The pre-enrichment phase is likely to become an important field of future research, because many of the alternative novel detection systems for *Salmonella* depend on pre-enrichment. New ways to enhance the productivity of specific pathogens in the presence of background microbial contamination in primary enrichment broths need further investigation (Patel, 1997), and include techniques like ion-exchange extraction, aqueous two-phase partitioning and immobilisation with metal hydroxides (Lucore et al. 2000).

Selective enrichment

Standard media

The objective of selective enrichment is to select salmonellae from all the other organisms present in the pre-enriched sample. The ideal selective enrichment should repress competing organisms and allow the salmonellae to multiply without restriction.

Three 'families' of selective enrichments are in common use, i) tetrathionate media, ii) selenite based media and iii) Rappaport. The selective agents in these media are

Table 2
Selective agents (g/l) in the main media for enrichment of *Salmonella*.

Selective agent	Selenite Cystine Broth (SC)	Tetrathionate broth	Rappaport	Rappaport Vassiliadis Broth (SC)	Rappaport Vassiliadis Soya Peptone (RVS)
		USP	MK		
$NaHSeO_3$	4.0	–	–	–	–
Na–thiosulphate	–	30.0	40.7	–	–
$MgCl_2$	–	–	–	28.6 ($6H_2O$)	13.4 (anhyd.)
Malachite green oxalate	–	–	–	0.036	0.036
Brilliant green	–	(0.01)	0.01	–	–
Bile salts	–	1.0	4.75	–	–

summarised in Table 2.

i) Tetrathionate. There are several formulations of selective enrichments based on tetrathionate (TT). These include Mueller-Kauffmann (MKTT) as formulated by Mueller (1923) and modified by Kauffmann (1935), who added brilliant green dye and bile salts as selective agents. Tetrathionate was further modified by adding yeast extract as a growth stimulant (Hajna, 1965), which became known as the United States Pharmacopoeia (USP) formulation. Other modifications, such as addition of novobiocin (Jeffries, 1959), have been reported but their use has not become widely accepted. The selectivity of tetrathionate media depends on the presence of the enzyme tetrathionate reductase in the selected organism. *Salmonella* spp. possess this enzyme, as do some other organisms such as *Proteus* spp., so their growth has to be suppressed by the addition of novobiocin and/or brilliant green. The incubated pre-enrichment sample is added to the tetrathionate enrichment in a ratio of 1:9, and further incubated at 42±1°C for 24 and 48 h.

ii) Several modifications to the original selenite-based medium as described by Leifson (1936) have been proposed. Well known are selenite F, selenite cystine (SC), dulcitol selenite, and selenite brilliant green. The performance of these media has been investigated but with no overall agreement for use with food products (Carlson and Snoeyenbos, 1974; Fagerburg and Avens, 1976). Selenite enrichment is most frequently used for the isolation of salmonellae when direct enrichment of the sample in selective media is recommended e.g. with faecal or clinical samples. A drawback for the use of this type of medium is that the selective agent, sodium acid selenite, is a safety hazard, which has been shown to be toxic to embryos, and produce growth abnormalities and damage to kidneys, liver and spleen (Robertson, 1970).

iii) The original Rappaport medium contained malachite green and magnesium chloride as selective agents (Rappaport et al., 1956), and was modified by Vassiliadis (Vassiliadis et al., 1981). This medium has become known as RV medium and is now widely used in Europe. Differences in performance between commercial RV media have been observed due to differences in the concentration of magnesium chloride used. As was pointed out by Peterz et al. (1989) this was due to the method used by Rappaport and Vassiliadis to prepare the medium, which has led to differences in interpretation and also the use of anhydrous or hexahydrate magnesium chloride can

Table 3
Selective enrichment media used in standard reference methods.

Standard Organisation[a]	Medium	Commodity
ISO	RV + SC broth (100 ml)	All products
AOAC/FDA	TTB (USP) + RV	Raw flesh, high contaminated foods
	TTB (USP) + SC broth (10 ml)	All other products
IDF	RV + SC broth (100ml)	All dairy products

[a] ISO = International Standard Organisation; IDF= International Dairy Federation; AOAC = American Association of Analytical Chemists; FDA = Food and Drugs Agency.

cause misinterpretation. The final concentration in the medium is critical and should be 28.6 g/l $MgCl_2 \cdot 6H_2O$ or 13.4 g/l when anhydrous magnesium chloride is used. A modification, using soya peptone instead of tryptone, was reported to improve recovery rates of *Salmonella* (Van Schothorst and Renaud, 1983; 1987) and is in use as Rappaport-Vassiliadis Soya peptone (RVS) broth.

Reference methods as recommended by ISO, AOAC and IDF all contain at least two selective media belonging to one of the families mentioned above. In use at present are RV, SC and TTB (MK as well as USP formulation), as shown in Tables 2 and 3.

Standards Organisations are, however tending to replace the selenite based media by members of other families, owing to their toxic effects and the poorer performance of these media in food microbiology. Several trials were performed by the AOAC, substituting RV for SC broth (June et al., 1996; Hammack et al., 1999). ISO investigations resulted in replacement of RV by RVS and SC by MKTT + 0.05 g/l novobiocin (MKTTn) in the most recent ISO 6579 Standard (Anon., 2002).

Modifications

Modification of selective enrichment procedures can include the use of 'motility enrichment media' like the semi-solid *Salmonella* media Modified Semisolid Rappaport Vassiliadis Medium (MSRV) and Diagnostic *Salmonella* Medium (DIASALM) and test kits based on this principle (De Smedt et al., 1986; Holbrook et al., 1989).

The use of motility enrichment on MSRV and DIASALM is widely recognised as an effective procedure for identifying products contaminated with salmonellae (Van der Zee and Van Netten, 1992; Davis and Wray, 1994b; Wiberg and Norberg, 1996). This method, based on the motility of salmonellae, indicates *Salmonella* by a swarm zone after inoculation with pre-enrichment cultures and incubation at 41.5–42°C for only 18–24 h in a Petri dish.

It was observed that the use of semi-solid agar as a selective enrichment seems to favour the development of *S. enteritidis* (Svastova et al., 1984; Perales and Erikiaga, 1991; Van Netten et al., 1991; Poppe et al., 1992). The reason for this is still not clear.

Table 4
Selective agents (g/l) used in plating media for *Salmonella*.

Selective agents	BS	BGA	DCA	Hektoen agar	MLCB agar	SS–agar	XLD
Standard							
– Acid fuchsin	–	–	–	0.1	–	–	–
– Bile salts	–	–	–	9.0	–	5.5–8.5	–
– Brilliant green	0.016	0.0125–0.7	–	–	0.0125	0.00033	–
– Bismuth ammonium citrate	1.85	–	–	–	–	–	–
– Cristal violet	–	–	–	–	0.01	–	–
– Deoxycholate	–	–	2.0–5.0	–	–	–	1.0
– Sodium citrate	–	–	5.0–8.5	–	–	10.0	–
– Sodium sulphite	6.15	–	–	–	–	–	–
– Sodium thiosulphate	5.0	–	5.4	5.0	4.0	8.5	6.8
Optional	Novobiocin	Novobiocin Sulfonamides Sulfacetamide Na–mandalate		Novobiocin	Novobiocin		Novobiocin

Table 5
Appearance of *Salmonella* and other organisms on selective agars (Source: Wray and Davies, 1994).

Medium	*Salmonella*	*Proteus* spp.		Coliforms	
	Appearance	Appearance	Growth[a]	Appearance	Growth
BS	Black, metallic sheen	Black	1	Brown-green	3
BGA	Red	Red	3	Yellow-green	2
DCA	Colourless, BC[b]	Colourless, BC	1	Pink	2
HE	Blue-green, BC	Blue-green	1	Pink	3
XLD	Red, BC	Yellow, BC	1	Yellow	2
SS	Colourless, BC	Colourless, BC	2	Pink	2
MLCB	Purple, BC	Colourless, BC	1	Colourless	2

[a] Key: 1 = good, 2 = fair to good, 3 = poor, 4 = absent to poor.
[b] BC = Black centre due to H_2S production.

Selective plating

Standard media

Selective plating agars rely on different selective agents, indicator systems or differential characteristics (like carbohydrate fermentation and hydrogen sulphide production) to distinguish *Salmonella* spp. from many related enteric bacteria (Arroyo and Arroyo, 1995). The main components of selective isolation media include:

- Nutrient substrates: soya or meat peptone, yeast extract and/or carbohydrates (lactose, mannitol, xylose etc).
- Selective agents: bile salts, deoxycholate, bismuth sulphite, brilliant green, crystal violet, malachite green, novobiocin, mandelic acid, sulphadiazine, sulphacetamide and sulphamethazine (see also Table 4).
- Dyes and indicators: neutral red, bromothymol blue and phenol red.
- Detectors: iron salts, which stain colonies black by reacting with hydrogen sulphide produced by the bacteria from sulphur-containing substrates.

Because none of the selective plating agars is ideal when used singly, it has been recommended to use at least two media. Each should contain different selective agents (see also Table 4) and indicator systems or differential characteristics in order to detect a wide range of *Salmonella* serotypes from different situations (Fricker, 1987; Andrews, 1996).

Bismuth Sulphite agar (BS), Brilliant green agar (BGA) (including modifications in which 1 mg/l sodium sulfapyridine (BGS) or 20 mg/l novobiocin (BGN) is incorporated), Deoxycholate citrate agar (DCA), Hektoen enteric agar (HE) and Xylose lysine desoxycholate agar (XLD), both including a modification incorporating novobiocin (HEN and XLDN respectively), and *Salmonella-Shigella* agar (SS) are commonly used for these purposes. Mannitol lysine crystal violet brilliant green agar (MLCB) which incorporates mannitol and lysine together with sodium thiosulphate and iron (III) ammonium sulphate, has been developed for the detection of lactose positive salmonellae. Plating media should be incubated at 35–37°C for 18–24 hours,

Table 6
Minimum inhibitory concentrations (MICs) of brilliant green for *Salmonella* spp. (Source: Curtis and Clarke, 1994).

Salmonella serotype	NCTC number	MIC (g/l)
Dublin	9676	0.0128
Typhimurium	74	0.0064
Typhimurium	12190	0.2048
Enteritidis	5188	0.1024
Gallinarum	9240	0.0512
Virchow	5742	0.1024

unless specified otherwise. Colonial appearances and growth characteristics for *Salmonella* and some non-*Salmonella* on these media (modifications not included) are listed in Table 5.

One reason why some plating media are not ideal for isolating all salmonellae may be the sensitivity of some strains to brilliant green. This is a well known phenomenon for *S. typhi* and *S. paratyphi* strains, but also some *S. typhimurium* and *S. dublin* strains resemble *S. typhi* in their sensibility to this triphenylmethane dye, as shown in Table 6 (Curtis and Clarke, 1994). This supports the requirement for use of a second medium containing different selective agents at all times.

It remains puzzling why bismuth sulphite agar, which contains at least 0.016 g/l brilliant green, will allow growth of *S. typhi* when this serotype is considered not to grow on brilliant green media. Also the other selective agent, the so-called "bismuth sulphite indicator" is not very well defined and because it is difficult to get clear information from the manufacturers, questions remain about the actual effective bismuth content. According to Schwab et al. (1984) the medium contains 1.85 g/l ammonium bismuth citrate and 6.15 g/l sodium sulphite, corresponding to 4mM bismuth and about 50mM sulphite, resulting in formation of bismuth sulphite and additional Na_2SO_3 in the medium, while during preparation also a precipitate is formed, which may contain part of the bismuth (Busse, 1995).

Modifications

Recently, some new selective and differential plating media based on enhanced selectivity and other abilities of *Salmonella* species to produce specific identifying characteristics have been developed. This has been achieved either by the addition of new selective supplements to existing media in order to enhance their selectivity or the development of new media.

An example of the first option is XLT-4, a modification of Xylose-lysine-deoxycholate (XLD) agar. The surfactant Tergitol 4 (7-ethyl-2-methyl-4-undecanol hydrogensulphate sodium salt) is added as a selective inhibitor of *Proteus* species and other non-salmonellae to the xylose-lysine agar base (Miller et al., 1991: 1995). On the XLT-4 medium *Salmonella* spp. appear as black colonies, and can be easily differentiated from colonies of *Citrobacter* spp. which appear as yellow colonies and reduced

Table 7
Appearance of *Salmonella* and some other organisms on XLT-4, Rambach and SM-ID agar.

	XLT-4	RAMBACH	SM-ID
Salmonella spp.	Red, BC	Red/crimson	Pink
S. typhi/paratyphi	–	Colourless	Pink/red
Proteus spp.	–	Colourless	Colourless
Citrobacter spp.	Yellow	Violet	Blue
Coliforms	Yellow	Blue-Violet	Purple/mauve

in size.

In recent years new media have been developed incorporating chromogenic substances to create a better differentiation of target organisms from the accompanying flora on agar media (e.g. red colonies for salmonellae). Rambach agar uses acid formation from propylene glycol (PG) by *Salmonella* spp., which results in red colonies and ß-D-galactosidase production by other Enterobacteriaceae, which after reaction with the chromogenic substance, X-Gal (5-bromo-4-chloro-3-indol-ß-D-galactopyranoside) results in formation of blue-green colonies (Rambach, 1990). However all strains of *Salmonella* subspecies IIIa, IIIb and V produce ß-D galactosidase and therefore also appear as blue-green colonies on this medium (Kühn et al., 1994). These particular subspecies may well be present in samples from sheep, turkeys and cold blooded animals. On the other hand *S. typhi* and *S. paratyphi* A and B failed to produce acid from PG, resulting in colourless colonies on Rambach agar.

Salmonella IDentification (SM-ID) agar is based on the fact that glucuronate is metabolized by *Salmonella* on this medium. Combined with absence of ß-galactosidase activity, this also results in specific pink/red colonies (Dusch and Altwegg, 1993). Since *S. typhi* and *S. paratyphi* B also form acid from glucoronate, colonies belonging to these serotypes can also be detected on SM-ID medium.

Although both media are not very selective and may perform better after selective enrichment (Dusch and Altwegg, 1993), they have been successfully used in differentiating *Salmonella* spp. from other enteric bacteria in clinical isolates (Frydiere and Gille, 1991; Gruenewald *et al.*, 1991; Davies and Wray, 1994a). The appearance of *Salmonella* and other organisms on these two media is described in Table 7.

The most recent media in this group are three chromogenic media specific for *Salmonella*, CHROMagar *Salmonella* Medium (CAS), ABC-Medium and Chromogenic Ester Agar Medium (CSE), and a chromogenic medium called CHROMagar Orientation Medium (CHROM) which was not developed specifically for detection of *Salmonella*, but is suitable for this purpose when used in combination with semi-solid selective enrichment.

The identities of the chromogenic substrates of CAS-Medium have not been disclosed by the manufacturer. They result in mauve colonies for *Salmonella* spp., *S. typhi* and *S. paratyphi* included. Some strains of *Aeromonas hydrophila*, *Candida albicans* and, especially, *Pseudomonas aeruginosa* strains can give similar colonies on this medium. To inhibit growth of *Ps. aeruginosa* the use of cefsoludin (10 mg/l) is advised. Other members of the Enterobacteriaceae appear as blue or uncoloured colonies (Gaillot et al., 1999).

ABC-Medium consists of Modified Desoxycholate Citrate agar base (acc. to Hynes) with two chromogenic substrates. The first is 3,4-cyclohexenoesculetin-ß-D-galactoside (CHE-GAL) which shows production of ß-galactosidase as black colonies in the presence of iron. The second substrate, 5-bromo-4-chloro-3-indonyl-α-D-galactopyranoside (X-α-Gal) is hydrolysed by strains of *Salmonella*, which results in green colonies, including *S.typhi* and *paratyphi* A and B. Lactose positive salmonellas however will appear as black colonies because of their ß-galactosidase production, and may not be detected as salmonellae (Perry et al., 1999).

Besides peptones and nutrient extracts CSE-Medium contains, as a key component, 4-[2-(4-octanoyloxy-3,5-dimetoxyphenyl)-vinyl]-quinolium-1-(propan-3-yl carboxylic acid) bromide, SLPA-octanoate in bromide form, a newly synthesised ester, which is hydrolysed by *Salmonella* spp. to produce burgundy coloured colonies. Non-Salmonella spp. appear as white, yellow or transparent colonies (Cooke et al., 1999).

The CHROM-Medium was originally developed for detecting infections in the urinary tract. It does not contain any specific selective ingredients, but relies on the activity of chromogenic substrates, details of which are not disclosed by the manufacturer. Non-salmonellae such as *Citrobacter* spp. and *Enterobacter* spp. appear as blue, *E.coli* as pink-red colonies and *Proteus* spp, as beige coloured colonies on this medium. *Salmonella* spp., lactose and sucrose positive species included, appear as clear white colonies. As practically no selective agents are present in this medium, it should not be used for isolation from liquid enrichment media. Combined with semi-solid enrichment it will, however, serve its purpose, because isolation from such enrichment media only takes place from a motility zone. Inoculation from the edge of such a zone will produce in most cases a pure culture, which on this medium can give a more reliable indication for presence of *Salmonella* spp. with diverging biochemical profiles (Van Velzen and Verberkt, 1999).

Another modification of conventional plating methods is the Hydrophobic Grid Membrane Filter (HGMF) technique. For *Salmonella* detection there is a test, which uses conventional pre-enrichment, followed by a 6 h selective enrichment. A portion of the selective enrichment is filtered through a HGMF and incubated on EF-18 agar, a medium containing magnesium sulphate, bile salts, sulfapyridine, bromothymolblue and novobiocin as selective agents and uses sucrose and lysine decarboxylase utilisation for its differential reactions (Entis and Boleszczuk, 1991). Negative results are obtained in 42 h. Presumptive positive results (green colonies on EF-18 agar) need an additional 24 h for confirmation. The performance of the test for egg products was evaluated against the US Department of Agriculture *Salmonella* method. Both methods performed identically in frozen, liquid and dried egg products (Entis, 1996). In another study, including poultry products and feed samples, problems with the method resulted from the inability to isolate colonies of *Salmonella* on the HGMF due to small colony size, abnormal colony coloration and overgrowth by competitors (Warburton et al., 1994).

Concluding remarks

For conventional detection of *Salmonella* spp. more media are available than those mentioned in the Standard reference methods that are still considered to be the 'Gold Standard'. Sometimes it can be advisable to use modifications, because the 'pure' Standard Method appears not to be the optimal method for some types of food samples. This can often be the case in the non-selective enrichment phase.

Although progress is slow, there seems to be a trend in the official Standards Organisations to optimise their methods and follow the more recent developments. However these developments concentrate mostly on the selective enrichment and isolation phases. ISO recently replaced RV by RVS and SC by MKTTn and AOAC is planning to replace SC by RV. Modifications such as semi-solid enrichment have not yet been taken into consideration, in spite of all the benefits for the laboratory staff and proven equivalence to liquid enrichment media. For the isolation phase ISO has substituted the mandatory use of BGA by the use of XLD, and the use of a second medium of own choice. Besides the discussion of any real improvement effected by this replacement there should be some consideration of limiting the choice of the second medium. With XLD as the medium of first choice, the H_2S negative and lactose and saccharose positive strains will not be detected as suspected *Salmonella* colonies. Thus the second medium should be one that can cover at least one or, better, all of these shortcomings.

It must be considered that all these changes promoted by the Standards Organisations only have the effect of 'polishing the Gold Standard' and are lacking innovative inspiration.

In the future it is likely that most of the changes will take place in the pre-enrichment phase. This is because all current non-cultural techniques require some form of enrichment to fulfil the internal detection limits of these types of test kits (Van der Zee and Huis in't Veld, 2000).

References

Andrews, W.H. (1996) Evolution of methods for the detection of *Salmonella* in foods. J. AOAC Int. 79, 4–12.

Anon. (2002) Microbiology of food and animal feeding stuffs – horizontal method for the detection of *Salmonella spp.* (EN ISO 6579:2002) International Organisation for Standardisation Geneva.

Anon. (1993) Microbiology – General guidance on methods for the detection of *Salmonella*. Int. Standard ISO 6579. International Organisation for Standardisation Geneva.

Arroyo, G. and Arroyo, J.A. (1995) Selective action of inhibitors used in different culture media on the competitive micro flora of *Salmonella*. J. Appl. Bacteriol. 75, 281–289.

Bailey, J.S. and Cox, N.A. (1992) Universal Preenrichment Broth for the Simultaneous Detection of *Salmonella* and *Listeria* in Foods. J. Food Protect. 44, 256–259.

Busse, M. (1995) Media for *Salmonella*. In: J.E.L. Corry , G.D.W. Curtis and R.M. Baird (Eds,) Culture Media for Food Micobiology. Elsevier Amsterdam, P.187–201.

Carlson, V.L. and Snoeyenbos, G.H. (1974) Comparative efficiencies of selenite and tetrathionate enrichment broth's for the isolation of *Salmonella* serotypes. Am. J. Vet. Res. 35, 711–718.

Cooke, V.M., Miles, R.J., Price, R.G. and Richardson, A.C. (1999) A novel chromogenic ester agar medium for detection of salmonellae. Appl. Environ. Microbiol. 65, 807–812.

Curtis, G.D.W and Clarke, L.A. (1994) Comparison of the MSRV method with an in-house conventional method for the detection of *Salmonella* in various high and low moisture foods (Letter). Lett. Appl. Microbiol. 18, 239–240.

Davies, R.H. and Wray, C. (1994a) Evaluation of SMID agar for identification of *Salmonella* in naturally contaminated veterinary samples. Lett. Appl. Microbiol. 18, 15–17.

Davies, R.H. and Wray, C. (1994b) Evaluation of a rapid cultural method for identification of salmonellas in naturally contaminated veterinary samples. J. Appl. Bacteriol. 77, 237–241.

De Smedt, J., Bolderdijk, R., Rappold, H. and Lautenschlaeger, D. (1986) Rapid *Salmonella* detection in foods by motility enrichment on a modified semi-solid Rappaport-Vassiliadis medium. J. Food Protect. 49, 510–514.

Dolzinsky, B. and Kruse, K. (1992) Zur Isolierung von Salmonellen unter Verwendung Ammoniumeisen(III)-citrat-haltiger Medien. Arch. Lebensmittelhyg. 43, 124–125.

Dusch, H. and Altwegg, M. (1993) Comparison of Rambach agar, SM-ID medium and Hektoen Enteric agar for primary isolation of non-typhi salmonellae from stool samples. J. Clin. Microbiol. 31, 410–412.

Entis, P. and Boleszczuk, P. (1991) Rapid detection of *Salmonella* in foods using EF-18 agar in conjunction with the Hydrophobic Grid Membrane Filter. J. Food Protect. 54, 930–934.

Entis, P. (1996) Validation of the ISO-GRID 2-day rapid Screening method for detection of *Salmonella* spp. in egg products. J. Food Protect. 59, 555–558.

Fagerburg, D.J. and Avens, J.S. (1976) Enrichment and plating methodology for *Salmonella* detection in food. A review. J. Milk Food Technol. 39, 628–646.

Freydiere, A.M., and Gille, Y. (1991) Detection of *Salmonella*e by using Rambach agar and by a C8 esterase spot test. J. Clin. Microbiol. 29, 2357–2359.

Fricker, C.R. (1987) The isolation of salmonellas and campylobacters. J. Appl. Bacteriol. 63, 99–116.

Fung, D.Y.C., Yu, L., Niroomand, F. and Tuitemwong, K. (1993) Novel methods to stimulate growth of food pathogens by oxyrase and related membrane fractions. *In:* Spencer, Wright and Newsom (eds.) Rapid methods and automation in Microbiology and Immunology. Intercept Limited: 313–318.

Gaillot, O., Di Cammillo, P., Berche, P., Courcol, R. and Savage, C. (1999) Comparison of CHROMagar *Salmonella* medium and Hektoen Enteric Agar for isolation of *Salmonella*e from stool samples. J. Clin. Microbiol. 37, 762–765.

Gast, R. K. (1993) Recovery of *Salmonella* enteritidis from inoculated pools of egg contents. J. Food Protect. 56, 21–24.

Gast, R. K. and Holt, P. S (1995) Iron Supplementation to Enhance the Recovery of *Salmonella enteritidis* from Pools of egg Contents. J. Food Protect. 58, 268–272.

Gruenewald, R., Henderson, R.W. and Yappow, S. (1991) Use of rambach propylene glycol containing agar for identification of *Salmonella* spp. J. Clin. Microbiol. 29, 2354–2356.

Hajna, A.A. (1965) A new enrichment broth for Gram negative organisms of the intestinal group. Pub. Hlth. Lab. 13, 83–89.

Hammack, T.S., Amaguana, R.M., June, G.A., Sherrod, P.S. and Andrews, W.H. (1999) Relative effectiveness of selenite cystine broth, tetrathionate broth and Rappaport-Vassiliadis medium for the recovery of *Salmonella* spp. from foods with a low microbial load. J. Food Protect. 62, 16–21.

Holbrook, R., Andersen, J.M., Baird-Parker, A.G., Doods, L.M., Sawhney, A., Stuchbury, S.H. and Swaine, D. (1989) Rapid detection of *Salmonella* in foods, a convenient two-day procedure. Lett. Appl. Microbiol. 8, 139–142.

Jeffries, L. (1959) Novobiocin-tetrathionate broth: a medium of improved selectivity for the isolation of *Salmonella*e from faeces. J. Clin. Pathol. 12, 568–571.

Jiang, J., Larkin, C., Steele, M., Poppe, C. and Odumeru, J.A. (1998) Evaluation of Universal Preenrichment Broth for the recovery of food borne pathogens from milk and cheese. J. Dairy Sci. 81, 2798–2803.

June, G.A., Sherrod, P.S., Hammack, T.S., Amagua, R.M. and Andrews, W.H. (1996) Relative effectiveness of selnite cystine broth, tetrathionate broth and Rappaport-Vassiliadis medium for recovery of *Salmonella* spp. from raw flesh, highly contaminated foods, and poultry feed: Collaborative study. J. AOAC Int. 79, 1307–1322.

Kauffmann, F. (1935) Weitere Erfahrungen mit dem kombinierten Anreicherungsverfahren fuer *Salmonella* Bazillen. Zeitschrift fuer Hygiene und Infectionskrankheiten 117, 26–32. *Cited by* Fricker, C.R. 1987. The isolation of *Salmonellas* and campylobacters. J. Appl. Bacteriol. 63, 99–116.

Kühn, H., Wonde, B, Rabsch, W. and Reissbrodt, R. (1994) Evaluation of Rambach agar for detection of *Salmonella* Subspecies I to VI. Appl. Environ. Microbiol. 60, 749–751.

Leifson, E. (1936) New selenite enrichment media for the isolation of typhoid and paratyphoid (*Salmonella*) bacilli. American Journal of Hygiene 24, 423–432. *Cited by* Fricker, C.R. (1987) The isolation of *Salmonellas* and campylobacters. J. Appl. Bacteriol. 63, 99–116.

Lucore, L.A., Cullison, M.A. and Jaykus, L. (2000) Immobilization with Metal Hydroxides as a means to concentrate food-borne bacteria for detection by cultural and molecular methods. Appl. Env. Microb. 66, 1769–1776.

Miller, R.G., Tate, C.R., Mallinson, E.T. and Scherrer, J.A. (1991) Xylose-Lysine-Tergitol 4: an improved selective agar medium for the isolation of *Salmonella*. Poult. Sci. 70, 2429–2432.

Miller, R.G., Tate, C.R., and Mallinson, E.T. (1995) Improved XLT4 agar: Small addition of peptone to promote stronger production of hydrogen-sulfide by *Salmonellae*. J. Food Protect. 58, 115–119.

Mueller, L. (1923) Un nouveau milieu d'enrichissement pour la recherche du bacile typhique et des paratiphique. Comptes Rendus des Séances de la Societé de Biologie et de ses Filiales 89, 434–437. *Cited by* Fricker, C.R. 1987. The isolation of salmonellas and campylobacters. J. Appl. Bacteriol. 63, 99–116.

Patel, P. (1997) Recent advances in microbiological methods in food control laboratories. *In:* The second Symposium on Food Safety, organised by Ministry of Public Health, Preventive Health dept., Doha, Quatar, 29 April–1 May 1997.

Perales, I. and Erkiaga, E. (1991) Comparison between semisolid Rappaport and modified semisolid Rappaport-Vassiliadis media for the isolation of *Salmonella* species from foods and feed. Int. J. Food Microbiol. 14, 51–57.

Petrz, M., Wiberg, C. and Norberg, P. (1989) The effect of incubation temperature and magnesium chloride concentration on growth of *Salmonella* in home made and in commercially available dehydrated Rappaport-Vassiliadis broths. J. Appl. Bacteriol. 66, 523–528.

Perry, J.D., Ford, M., Taylor, J., Jones, A.L., Freeman, R and Gould, F.K. (1999) ABC Medium, a new chromogenic agar for selective isolation of *Salmonella* spp. J. Clin. Microbiol. 37, 766–768.

Poppe, C., Johnson, R.P., Forsberg, C.M. and Irwin, R.J. (1992) *Salmonella* enteritidis and other *Salmonella* in laying hens and eggs from flocks with *Salmonella* in their environment. Can. J. Vet. Res. 56, 226–232.

Rambach, A. (1990) New plating medium for facilitated differentiation of *Salmonella* spp. from *Proteus* spp. and other enteric bacteria. Appl. Environ. Microbiol. 56, 127–130.

Rappaport, F., Konforti, N. and Navon, B. (1956) A new enrichment medium for certain *Salmonellae*. J. Clin. Pathol. 9, 261–266.

Reissbrodt, R.L. and Rabsch, W. (1993) Selective pre-enrichment of *Salmonella* from eggs by siderophore supplements. Zbl. Bakteriol. 279, 344–353.

Robertson, D.S. (1970) Selenium, a possible teratogen. Lancet, 1 (7645); 518–519.

Schwab, A.H., Leininger, H.V. and Powers, E.M. (1984) Media, reagents and stains. In: Speck, M.L (Ed.) Compendium of methods for the microbiological examination of foods. p. 810 Am. Publ.

Health Assoc.

Svastova, A., Skalka, B. and Smola, J. (1984) A modified medium for *Salmonella* isolation by the selective motility test. Zbl. Vet. Med. (Serie B) 31, 396–399.

Van der Zee, H. and van Netten. P. (1992) Diagnostic selective semi-solid media based on Rappaport-Vassiliadis broth for the detection of *Salmonella* spp. and *Salmonella* enteritidis in foods. Proc. Symposium *Salmonella* and Salmonellosis, Ploufragan: Reports and Communications: 69–77.

Van der Zee, H. (1994) Conventional methods for the detection and isolation of *Salmonella* enteritidis. Int. J. Food Microbiol. 21, 41–46.

Van der Zee, H. and. Huis in't Veld, J.H.J. (2000) Rapid methods for the detection of and isolation of *Salmonella. In*: Wray (ed.) *Salmonella* in domestic animals. CAB International: 373–391.

Van Netten, P., van der Zee, H. and van Moosdijk, A. (1991) The use of a diagnostic semisolid medium for the isolation of *Salmonella* enteritidis from poultry. *In:* R.W.A.W. Mulder (ed.) Quality of poultry products. III Safety and marketing aspects. Spelderholt Jubilee Symposia, Doorwerth. pp 59–66.

Van Schothorst, M. and Renaud, A.M. (1983) Dynamics of salmonella isolation with modified Rappaports's medium (R10). J. Appl. Microbiol. 54, 209–215

Van Schothorst, M., Renaud, A. and van Beek, C. (1987) *Salmonella* isolation using RVS broth and MLCB agar. Food Microbiol. 4, 11–18.

Van Velzen, H. and Verberkt, P.E.J. (1999) Gebruik van CHROMagar bij het reinstrijken van *Salmonella* verdachte groeizones van DIASALM- en MSRV-medium. De Ware(n)-Chemicus 29, 137–139.

Vassiliadis, P., Kalapothaki, V., Trichopoulos, D., Mavrommati, Ch. and Sérié, Ch. (1981) Improved isolation of *Salmonellae* from naturally contaminated meat products by use Rappaport-Vassiliadis enrichment broth. Appl. Environ. Microbiol. 42, 615–618.

Waltman, D. (1998) Isolation of *Salmonella* from poultry environments. Proc. Int. Symp. on foodborne *Salmonella* in poultry, Baltimore: 133–153.

Warburton, D.W, Arling, V, Worobec, S., Mackenzie, J., Todd, E.C.D., Lacasse, P., Lamontagne, G., Plante, R., Shaw, S., Bowen, B., and Konkle, A. (1994) A comparison study of the EF-18 agar/ Hydrophobic Grid Membrane Filter (HGMF) method and the enzyme linked antibody (ELA) / HGMF method to the HPB standard method in the isolation of *Salmonella.* Int J. Food Microbiol. 23, 89–98.

Wiberg,C. and Norberg, P. (1996) Comparison between a cultural procedure using Rappaport-Vassiliadis broth and Motility enrichment on modified semisolid Rappaport-Vassiliadis medium for *Salmonella* detection from food and feed. Int. J. Food Microbiol. 29, 353–360.

Wray, C. and Davies, R.H. (Eds; 1994) Guidelines on detection and monitoring of *Salmonella* infected poultry flocks with particular reference to *Salmonella* enteritidis. Report of a WHO consultation on strategies for detection and monitoring of *Salmonella* infected poultry flocks. WHO/Zoon./94.173.

Chapter 14

Media for the isolation of *Shigella* spp.

H. van der Zee

Inspectorate for Health Protection and Veterinary Public Health, De Stoven 2, Postbus 202, 7200 AE, Zutphen, The Netherlands

Media used for isolation of *Shigella* spp. are not very specific or sensitive and are also in use for isolation of other Enterobacteriaceae. Media for enrichment and isolation from foods are available and in use in reference methods developed by official organisations. In general these methods use one or two liquid enrichment media and up to three isolation media. These media, as well as some modifications, are reviewed in this paper.

Introduction

Shigella species were classically regarded as waterborne pathogens, but nowadays they are recognised also as foodborne pathogens, restricted primarily to higher primates, including humans.

The genus consists of four species: *S. dysenteriae* with 10 serovars, *S. flexneri* with 8 serovars and 9 sub serovars, *S. boydii* with 15 serovars and *S. sonnei* with 1 serovar which can have two "phases" I and II.

S. dysenteriae, *S. boydii* and, to a lesser extend, *S. flexneri* and are confined to the developing countries, whereas *S. sonnei* is encountered in the industrialised nations. The isolation of *Shigella* spp. from foods is considered to be more difficult than that of other pathogenic Enterobacteriaceae, which may have resulted in under reporting of foodborne shigellosis.

Enrichment and isolation procedures for food and water

Conventional techniques and media used for the isolation and detection of shigellae are neither very specific nor very sensitive. This gives rise to a number of problems.

Since the minimum infective dose of shigellae is small, the occurrence of the organisms in food, milk and water may be significant even when only a small number of organisms is present, and isolation methods must be capable of detecting low numbers. Thus an enrichment procedure is necessary. Also shigellae are easily overgrown by other bacteria present in the food (Beckers and Soentoro, 1989) and acids produced from carbohydrates in enrichment media by other Enterobacteriaceae are toxic for shigellae (Mehlmann et al., 1985). No specific media for shigellae are available and there is no absolutely reliable and effective single enrichment method. Therefore combinations of enrichment media must be used in an attempt to overcome the effect of overgrowth of shigellae by other Enterobacteriaceae after a short incubation time (Reusse, 1984). Compared to other Enterobacteriaceae shigellae are relatively inactive biochemically. This lack of biochemical activity results in weak electivity on most currently used media and therefore in difficulties in differentiating shigellae from other colonies on solid media. *Shigella* spp. lack enzyme activity such as lysine decarboxylase, phenylalanine deaminase and urease, do not use gluconate as a carbon source and do not liquefy gelatine and are Voges-Proskauer negative. Most media for the isolation of *Shigella* spp. are also in use for the selective isolation of other Enterobacteriaceae. They may contain bile salts or deoxycholate which may inhibit repair of sub-lethally injured cells giving rise to false-negative results.

For foods, procedures generally include homogenisation of 25 g food sample in 225 ml of one or two enrichment broths and, after incubation, streaking on one or more selective agars. The proportion of sample in the enrichment broth should not exceed the ratio 1: 9, since addition of more sample material can lead to formation of too much acid from carbohydrates during incubation, which can be harmful to shigellae.

Procedures for analysis of water recommend concentration of the organisms from a 1–10 litre sample onto a membrane, followed by liquid enrichment in one or two enrichment media and streaking out on selective agar(s).

Enrichment media

It has been demonstrated (Nakamura and Dawson, 1962) that freeze-thaw injured *S. sonnei* have more demanding nutritional requirements, that chlorine-injured *S. dysenteriae*, *S. flexneri* and *S. sonnei* do not grow in presence of deoxycholate (LeChavallier et al., 1985) and heat-injured *S. flexneri* do not repair in the presence of deoxycholate or bile salts (Tollison and Johnson, 1985). Therefore, if the presence of injured cells is suspected, pre-enrichment in a non-selective medium such as tryptone soya broth with yeast extract is essential. For selective enrichment it is preferable to use media which contain little or no carbohydrate to avoid the toxic effect on shigellae of acidification of the medium. Enrichment media commonly used by official organisations for this purpose are:

- *Shigella* broth+ novobiocin (0.5µg/ml and/or 3µg/ml)
- Gram Negative (GN) broth according to Hajna
- Selenite Cystine (SC) broth

Shigella broth is a phosphate-buffered tryptone glucose broth containing 1 g/l of the surface active agent Tween 80 and various concentrations of the antibiotic novobiocin as selective agents. This medium allows complete repair of heat-stressed *S. flexneri* cells (Smith and Dell, 1990). The medium is used by the FDA with novobiocin added at concentrations of 0.5 µg/ml (for *S. sonnei*) or 3 µg/ml (for other *Shigella* spp.) at pH 7.0 ± 0.2 and incubated under anaerobic conditions at different temperatures. For the detection of *S. sonnei* 44°C is used, and 42°C for detection of the other shigellae. This

Table 1
Enrichment media specified by Standard Organisations.

Organisation	Enrichment Media	Incubation	Selective agent(s)
Nordic Committee on Food Analysis	GN broth	37°C	Sodium deoxycholate 0.5 g/l Sodium citrate 5.0 g/l
FDA/BAM	*Shigella* broth + Novobiocin	44°C and 42°C	Novobiocin 0.5 μg/ml and 3.0 μg/ml
APHA	SC broth GN broth	35–37°C 35–37°C	Sodium selenite 4.0 g/l Sodium deoxycholate 0.5 g/l Sodium citrate 5.0 g/l
ISO (draft)	*Shigella* broth + novobiocin	41.5±1°C	Novobiocin 0.5 μg/ml

medium was proposed for the isolation of *Shigella* spp. with the addition of 0.5 μg/ml novobiocin at pH 7.0 ± 0.2 and incubation at 41.5 ± 1°C for 16–20 h under anaerobic conditions in an ISO draft.

Gram Negative (GN) broth is a medium originally formulated for general detection of Gram-negative pathogens in faeces, blood and urine. GN broth can also be used for examining food samples for *Shigella* spp. To avoid overgrowth by other Gram negative bacteria the pH is adjusted to 6.0–7.0 after the food sample is added (Morris, 1984). Dipotassium hydrogen phosphate and potassium dihydrogen phosphate buffer the broth and prevent premature over-acidification by acidic metabolic products. The medium contains deoxycholate but the presence of tryptose and glucose suppresses the inhibitory effects on injured shigellae. The higher concentration of mannitol over glucose is intended to stimulate the growth of *Shigella* species over mannitol non-fermenting *Proteus* species. Modification by the addition of DL-serine is specified in a procedure for examination of water (Anon., 1994). The Nordic Committee on Food Analysis uses GN broth as sole enrichment broth for all *Shigella* species (Anon., 1995).

The selenite cystine broth used in isolation of shigellae is identical to the formulation which is also used for enrichment of *Salmonella* spp. The medium contains tryptone as a source of carbon, nitrogen, vitamins and minerals, and lactose as carbohydrate (Anon., 1998). Sodium bi-selenite is the selective agent and L-cystine is a reducing agent which mitigates the toxic effect of selenite to bacteria. Use of the medium is recommended in addition to GN broth by the American Public Health Association (Morris, 1984).

The organisations and the enrichment media recommended are shown in Table 1.

Isolation media

Because different species of *Shigella* vary in their ability to grow on any particular medium, it is advisable to streak aliquots of the enrichment broth(s) onto agars of different selectivity (Table 2).

Agars of the low selectivity group allow isolation of more fragile species such as *S.*

Table 2
Properties of isolation media for *Shigella*.

Medium	Colour medium	Colour *Shigella* colony	Inhibitors
Low selectivity			
MacConkey agar	Red-brown	Translucent, reddish-Brown	Bile salts No. 3 0.15% Crystal violet 0.0001%
Tergitol-7 agar	Yellow-green	Opalescent, yellowish-green with blue zone	Tergitol-7 0.01%
Moderate selectivity			
DCA	Orange-red	Reddish	Sodium deoxycholate 0.1%
XLD	Red	Translucent, reddish	Sodium deoxycholate 0.25% Sodium thiosulphate 0.5%
High selectivity			
Hektoen Enteric agar	Green opalescent	Greenish blue	Bile salts 0.9% Brom thymol blue 0.0065% Acid fuchsin 0.01%
Salmonella Shigella agar	Yellow	Translucent	Bile salts No. 3 0.85% Sodium thiosulphate 0.85% Brilliant green 0.000033%

dysenteriae, the intermediate media are suitable for most species and a high selectivity medium should be used when background flora of the product is expected to be high. Using an agar from each of these groups will improve the chances of isolating *Shigella* species from a food sample. Details of the media and the colonial appearance of *Shigella* spp. on each are shown in Table 2.

An outline of the enrichment and isolation media used in the methods developed or proposed by some Standard Organisations is given in Table 3.

Modifications of isolation media

Injured cells are always a possibility in processed foods and shigellae are no exception. Incorporation of an H_2O_2 scavenger such as pyruvate (1%), even in low selective media like MCA and Tergitol-7 agar, will improve recovery of injured organisms (Smith and Dell, 1990).

When used for direct culturing of stool specimens the addition of 1% xylose to MacConkey agar gave better specificity and at least equal sensitivity than the conventional agar (Altweg et al., 1996).

A further modification of MacConkey agar was suggested by Munshi et al. (1997), who added potassium tellurite to the original medium at a concentration of 1 μg/ml. The modified medium increased the overall isolation rate of *Shigella* spp. in stool samples, compared to MAC and SS, but appeared unsuitable for isolation of *S. sonnei*.

Xylose-Galactosidase medium contains peptone, yeast extract and beef extract for growth support, novobiocin and bile salts as selective agents and D-xylose, 5-bromo-4-chloro-3-indolyl-β-D galactopyranoside and iso-propyl-β-D-thiogalactopyranoside as biochemical markers. This results in green colonies for *S. sonnei* (xylose negative

Table 3
Methods used by Standard Organisations.

Organisation	Enrichment Media	Isolation media
Nordic Committee on Food Analysis	GN broth	At least one medium (own choice) of – low selectivity – medium selectivity – high selectivity
FDA/BAM	*Shigella* broth + novobiocin 0.5μg/ml and *Shigella* broth + novobiocin 3.0μg/ml	MacConkey agar
APHA	SC broth GN broth	At least one medium (own choice) of – low selectivity – medium selectivity – high selectivity
ISO (draft)	*Shigella* broth + novobiocin	– MacConkey agar – XLD agar – Hektoen agar

and β-galactosidase positive) and colourless colonies for other species of *Shigella* (xylose and β-galactosidase negative). The medium produces better results for specificity when compared with XLD, MAC and HEA, but no statistically significant improvements for sensitivity (Garcia-Aguayo et al., 1999).

Conclusions

A wide range of non-specific media is in use for isolation of *Shigella* spp. from foods.

At the international level there is no agreement about the choice of enrichment media for *Shigella* spp., nor for the procedure concerning incubation time and/or temperature. As isolation is concerned, at least some agreement is achieved in a procedure outlining the use of agars of different selectivity. Due to the fact that there are at least three media available for use at each selectivity level, the diversity in methods will persist in the near future. Much work is still going on to modify and optimise media and methods for detection of shigellae from foods.

References

Altweg, M., Buser, J. and Von Graevenitz. (1996) Stool cultures for *Shigella* spp: Improved specificity by using McConkey agar with xylose. Diagn. Microbiol. Infect. Dis. 24, 121–124.

Anonymous (1994) The Microbiology of Water. Part 1, Drinking Water. Methods for the Examination of Water and Associated Materials. HMSO London.

Anonymous (1995a) *Shigella* bacteria. Detection in foods. Nordic Committee on food analysis. No.

151.

Anonymous (1995b). FDA Bacteriological Analytical Manual, 8th edition, Chapter 6 AOAC International

Anonymous (1998) The Difco Manual, 11th Edn. Difco Laboratories, Division of Becton Dickinson and Company, Sparks MD. pp. 455–456.

Anonymous (2000) Microbiological examination of foods and animal feeding stuffs – Horizontal method for the detection of *Shigella* species. ISO TC 34/SC 9N (draft).

Beckers, H.J. and Soentoro, P.S.S. (1989) Method for the detection of *Shigella* in Foods. Zbl. Bakt. Hyg. B187, 261–265.

Garcia-Aguayo, J.M., Ubeda, P. and Gobernado, M. (1999) Evaluation of Xylose-Galactosidase Medium, a new plate for the isolation of *Salmonella*, *Shigella*, *Yersinia* and *Aeromonas* species. Eur. J. Clin. Microbiol. Infect. Dis. 18, 77–78.

Hajna, A.A. (1955) A new enrichment broth medium for gram-negative organisms of the intestinal group. Pub. Hlth. Lab. 13, 83–89.

LeChavallier, M.W., Singh, A., Shiemann, D.A. and Mc Feters, G.A. (1985) Changes in virulence of waterborne enteropathogens with chlorine injury. Appl. Environ. Microbiol. 50, 412–419.

Mehlmann, I.J., Romero, A. and Wentz, B.A. (1985) Improved enrichment for recovery of *Shigella sonnei* from foods. J. Assoc.Off. Anal. Chem. 68, 552–555.

Morris, G.K. (1984) *Shigella. In :* Compendium of Methods for the Microbiological Examination of Foods, 2nd edition. APHA, Washington DC. pp. 343–350.

Munshii, M.M., Morshed, M.G., Ansaruzzaman, M., Alam, K., Kay, A., Aziz, K.M. and Rahaman, M.M. (1997) J. Trop. Pediatr. 43, 307–310.

Nakamura, M and Dawson, D.A. (1962) Role of suspending and recovery media on the survival of frozen *Shigella sonnei*. Appl. Micobiol. 10, 40–43.

Post, D.E. (1998) Food borne pathogens, Monograph number 5. Escherichia coli-*Shigella* species. Oxoid Limited, Basingstoke, UK.

Reusse, U. (1984) Untersuchugen zur Isolierung van Shigellen aus Lebensmitteln tierischer Herunft. Archiv Lebensmittelhyg. 35, 138–141.

Smith, J.L. and Dell, B.J. (1990) Capability of selective media to detect heat injured *Shigella flexneri*. J. Food Protect. 53, 141–144.

Smith, J.L. (1987) Shigella as a foodborne pathogen. J. Food Protect. 50, 788–801.

Smith, J.L. and Buchanan, R.L. (1992) *Shigella. In:* Compendium of methods for the microbiological examination of foods, chapter 26. APHA, Washington, DC.

Tollison, S.B. and Johnson, M.G. (1985) Sensitivity to bile salts of *Shigella flexneri* sublethally heat stressed in buffer or broth. Appl. Environ. Microbiol. 50, 337–341.

Chapter 15

Isolation of *Yersinia enterocolitica* from foods

E. de Boer

Inspectorate for Health Protection, PO Box 202, 7200 AE Zutphen, The Netherlands

Many selective enrichment and plating media for the isolation of *Yersinia enterocolitica* from foods have been described. Use of many of these results in the isolation of non-pathogenic *Yersinia* strains. At present no single isolation procedure is available for the recovery of all pathogenic strains of *Y. enterocolitica*. Cold enrichment in phosphate-buffered saline plus 1% sorbitol and 0.15% bile salts (PBSSB) and two-step enrichment with tryptone soy broth (TSB) and bile oxalate sorbose (BOS) broth are very efficient methods for the recovery of a wide spectrum of *Y. enterocolitica* serotypes. Enrichment in irgasan ticarcillin chlorate (ITC) broth is the most efficient method for recovery of strains of serotype O:3, the most prevalent clinical serotype of *Y. enterocolitica* in Europe. Post-enrichment alkali treatment often results in higher isolation rates. Cefsulodin irgasan novobiocin (CIN) agar and Salmonella-Shigella deoxycholate calcium chloride (SSDC) agar are the most frequently used plating media. For the recovery of serotype O:8 strains, the common clinical isolates in North America, enrichment in BOS and plating on CIN agar seems the most efficient procedure. Selection of the proper isolation procedure will depend on the bio/serogroups of *Yersinia* spp. sought and on the type of food to be examined. Use of more than one medium for both enrichment and plating will result in higher recovery rates of *Yersinia* spp. from foods. Serotyping, biotyping and virulence testing is essential for differentiation between pathogenic and environmental *Yersinia* strains. The International Standard Organization method for the detection of presumptive pathogenic *Y. enterocolitica* includes parallel use of the following two isolation procedures: (1) Enrichment in peptone, sorbitol and bile salts (PSB) broth for 2–3 days at 22–25°C with agitation or 5 days without agitation; plating on CIN agar directly and after alkaline treatment and incubation for 24 h at 30°C. (2) Enrichment in ITC for 2 days at 24°C; plating on SSDC agar and incubation for 2 days at 30°C.

Introduction

Yersinia enterocolitica is a Gram-negative, facultatively anaerobic coccobacillus belonging to the genus *Yersinia* in the Enterobacteriaceae family. The organism is recognised as a foodborne pathogen and some large food-associated outbreaks of yersiniosis have been reported (Bottone, 1997). In developed countries, *Y. enterocolitica* can be isolated from 1–2% of all human cases of acute enteritis (Kapperud, 1991). Yersiniosis in the United States has been characterized by foodborne outbreaks, whereas in Europe and Japan it is endemic (Schiemann and Wauters, 1992). The disease can range in severity from self-limiting gastroenteritis to pseudoappendicitis, septicaemia in neonates and immunodeficient patients and postinfectious complications like reactive arthritis and myocarditis (Bottone, 1997). Though *Y. enterocolitica* was considered an emerging pathogen after the large outbreaks in the seventies and eighties, epidemiological data do not indicate an increase in cases and outbreaks of yersiniosis in the past decade.

Yersinia enterocolitica and related species have been isolated from many types of both raw and processed foods (De Boer et al., 1986; De Boer, 1995). The majority of these food isolates differ in biochemical and serological characteristics from typical clinical strains and are usually classified as 'non-pathogenic'or 'environmental' *Yersinia* strains. These strains, which include the 'related species' *Y. frederiksenii, Y. kristensenii, Y. intermedia, Y. aldovae, Y. rohdei, Y. mollaretti* and *Y. bercovieri*, are ubiquitous and probably have no clinical significance with the exception of a few atypical cases (Kapperud, 1991). Pathogenic *Yersinia* strains harbour a virulence plasmid, as well as chromogenic virulence genes, such as the enterotoxin gene *yst* and the invasion-associated gene *ail* (Burnens et al., 1996). Strains associated with human disease mainly belong to the serogroups O:3, O:5,27, O:8 and O:9 of *Y. enterocolitica* sensu stricto (Stolk-Engelaar and Hoogkamp-Korstanje, 1996).

The epidemiology of *Y. enterocolitica* infections is, for the greater part, not understood. There is an association with consumption of contaminated foods, especially pork. The oral cavity and the intestinal tract of healthy pigs have been found to be important reservoirs of pathogenic serotypes of *Y. enterocolitica* and reduction of the contamination of pig carcasses at the slaughterhouse and, consequently, of raw pork and pork products has been suggested as a means of prevention of foodborne yersiniosis (De Boer and Nouws, 1991; De Boer et al., 1998; Verhaegen et al., 1998).

The increasing interest in *Y. enterocolitica* infections and the role of foods in some outbreaks of yersiniosis has led to the development of improved procedures for the isolation of this organism from foods during the last 15–20 years.

As the numbers of *Y. enterocolitica* organisms in foods are usually low and there is often a great variety of background flora, direct isolation on selective plating media is seldom successful. Isolation methods usually involve enrichment of the sample followed by plating onto selective agar media, confirmation of typical colonies and testing for virulence properties of isolated strains.

Enrichment

As a psychrotrophic organism *Y. enterocolitica* is able to multiply at 4ºC and enrichment at this temperature for 2–4 weeks is widely used. At this temperature, the growth rate of competitive bacteria is slowed sufficiently to enable *Y. enterocolitica* to multiply to numbers necessary for isolation on plating media.

Media used for this 'cold enrichment' include simple buffers like phosphate-buffered saline (PBS), PBS modified by addition of 1% sorbitol and 0.15% bile salts (PBSSB) (Mehlman et al., 1978), PBS supplemented with 1% mannitol (Schiemann, 1979a), PBS with 0.5% peptone (Weagant and Kaysner, 1983), PBS with peptone and cycloheximide (Vidon and Delmas, 1981), tryptone soy broth (Van Pee and Stragier, 1979), trypticase soy broth (Schiemann, 1983a), tryptic soy broth plus polymyxin and novobiocin (TSPN) (Landgraf et al., 1993) and tris-buffered peptone water, pH 8.0 (Greenwood and Hooper, 1989).

The long period required for cold enrichment is often unacceptable for quality assurance of foods. Schiemann and Olson (1984) showed than incubation at 15°C for 2 days was as efficient as enrichment at 4°C for some weeks. Doyle and Hugdahl (1983) incubated PBS for 1–3 days at 25°C, while tris-buffered peptone water was incubated at 9°C for 11–14 days (Greenwood and Hooper, 1989) or at 21°C with subculturing after 4–7 days and 11–14 days of incubation (Greenwood, 1993). Landgraf et al (1993) recommended incubation at 18°C for 3 days.

Several other enrichment procedures involving incubation at higher temperature for shorter periods and using selective media have been proposed. Modified Rappaport broth (Wauters, 1973) has been used for many years as the first–choice medium for the isolation of the major pathogenic serotypes of *Y. enterocolitica* in Europe. This medium has been shown to inhibit the common North American serotype O:8 strains of *Y. enterocolitica* (Schiemann, 1983a; Walker and Gilmour, 1986). Carbenicillin in modified Rappaport broth was shown to inhibit the growth of certain serotype O:3 strains (Schiemann, 1982). However, Wauters et al. (1988a) stated that serogroup O:3 strains are not inhibited by carbenicillin and that omission of this antibiotic results in a decrease in selectivity of the enrichment medium.

Lee et al. (1980) described two modified selenite media that effectively recovered certain strains of *Y. enterocolitica* from meats. They found it critical to limit the sample size of the blended meat suspension to 0.2 g per 100 ml enrichment medium to restrict the growth of competitive bacteria. Otherwise the slower growing *Y. enterocolitica* would be overgrown by the faster growing normal bacterial flora of the meat.

Schiemann (1982) developed a two-step enrichment procedure for recovery of *Y. enterocolitica* from food. In this procedure pre-enrichment for 9 days at 4°C in yeast extract rose bengal broth was followed by selective enrichment with bile oxalate sorbose (BOS) broth at 22°C for 5 days. As the pre-enrichment medium was less selective, it allowed multiplication of small inocula and repair of injured cells. BOS broth was found especially useful for the isolation of serotype O:8 strains, but strains of serotype O:5,27 were more difficult to recover (Schiemann, 1983a).

Wauters et al. (1988a) developed a new enrichment broth, named ITC broth, derived from modified Rappaport broth and based on the selective agents irgasan, ticarcillin and potassium chlorate. In comparative studies ITC broth was especially effective for the recovery of *Y. enterocolitica* serotype O:3 from pork and procine tonsils, while cold and two-step enrichments yielded better results for non-pathogenic strains (Wauters et al., 1988a; Kwaga et al., 1990; De Boer and Nouws, 1991). However, it was found that ITC broth had some short-comings for the enrichment of *Y. enterocolitica* serotype O:9 from meat, because of the sensitivity of this serotype to chlorate. De Zutter et al. (1994) suggested omitting chlorate in ITC broth and reducing the concentrations of magnesium chloride and malachite green to 80% of the original concentrations for optimal recovery of serotype O:9 strains. For the inoculation of ITC broth a filtrate or supernatant of a meat homogenate should preferably be used, allowing the examination of 1 g of meat (De Zutter et al., 1995). A study by Toora et al. (1994) indicated that the addition of ticarcillin and chlorate to ITC did not improve its efficiency and that irgasan alone functioned as a better selective supplement.

Alkaline treatment

Aulisio et al. (1980) found that strains of *Y. enterocolitica* were more tolerant of alkaline solutions than other Gram-negative bacteria. By treating food enrichments with potassium hydroxide (KOH) solutions before plating, the background flora was markedly reduced, thus improving separation of target colonies from similar colonies on the isolation medium. However, Schiemann (1983b) found that various factors, including medium, temperature and growth phase, influence tolerance of alkaline conditions and reduce the effectiveness of this treatment. Weagant and Kaysner (1983) concluded that a specific treatment time cannot be recommended. They found that streaking three to four successive plates from the KOH rinse at 10-s intervals enhanced the probability of obtaining isolated colonies of *Y. enterocolitica* even when growing in the presence of numerous organisms.

Direct KOH treatment of meat samples proved to be a valuable rapid method for direct isolation of *Yersinia* from meat contaminated with more than 10^2 cells per g (Fukushima, 1985). Based on results of comparative experiments the use of 0.125% KOH with an exposure time of 5 min was recommended for the direct detection of *Y. enterocolitica* in foods (Schraft and Untermann, 1989). Direct plating after KOH treatment was only suitable for strongly positive material (Wauters et al., 1988a).

Plating media

Different plating media have been used to isolate *Y. enterocolitica* from clinical specimens and food (Table 1). Initially media like MacConkey, Salmonella-Shigella, desoxycholate citrate and bismuth sulphite agars, designed for the isolation of enteropathogens, were used. Since *Y. enterocolitica* ferments lactose slowly, colonies are colourless on media such as MacConkey agar, which contains lactose as an indicator substrate. Lee (1977) modified MacConkey agar by addition of Tween 80 to improve differentiation of *Yersinia* colonies from other lactose-negative colonies. However, lipolytic *Yersinia* strains, easily recognized on this medium as white wrinkled colonies surrounded by a sheen, are usually non-pathogenic (De Boer and Seldam, 1987).

Salmonella-Shigella agar was made more selective for *Y. enterocolitica* by addition of sodium deoxycholate and $CaCl_2$ (SSDC) (Wauters, 1973; Wauters et al., 1988a). Colonies of *Y. enterocolitica* on this medium are small, round and colourless. Some species of *Morganella*, *Proteus*, *Serratia* and *Aeromonas* are also able to develop on SSDC and differentiation of *Yersinia* from these competing organisms may be difficult.

Schiemann (1979b) developed cefsulodin-irgasan-novobiocin (CIN) agar, a selective and differential agar medium for *Y. enterocolitica*. Organisms capable of fermenting mannitol, including yersiniae, produce red coloured ‘bullseye’ colonies on this medium. CIN agar was found inhibitory to *Pseudomonas aeruginosa*, *Escherichia*

Table 1
Selective agents in enrichment and plating media for *Yersinia enterocolitica*.

Medium[a]	Selective agents
Enrichment	
MRB	magnesium chloride (2.8%), malachite green (0.0013%), carbenicillin (0.00025%)
Selenite media	sodium selenite (0.15 or 0.25%), malachite green (0.002%), carbenicillin (0.001%)
PBSSB	bile salts (0.15%)
BOS	sodium oxalate (0.5%), bile salts (0.2%), irgasan (0.0004%), sodium furadantin (0.001%)
ITC	magnesium chloride (6%), malachite green (0.001%), irgasan (0.00001%), ticarcillin (0.00001%), potassium chlorate (0.1%)
Plating	
CIN	sodium deoxycholate (0.05%), crystal violet (0.0001%), irgasan (0.0004%), cefsulodin (0.0015%), novobiocin (0.00025%)
VYE	sodium deoxycholate (0.1%), crystal violet (0.0001%), irgasan (0.0004%), cefsulodin (0.0004%), oleandomycin (0.001%), josamycin (0.002%)
SSDC	sodium deoxycholate (0.85%)
MacConkey	bile salts (0.15%), crystal violet (0.0001%)

[a] MRB, modified Rappaport broth (Wauters, 1973); Selenite media (Lee et al., 1980); PBSSB, phosphate-buffered saline with sorbitol and bile salts (Mehlman et al., 1978); BOS, bile oxalate sorbose broth (Schiemann, 1982; Part 2 of this volume); ITC, irgasan ticarcillin chlorate (Wauters et al., 1988a; Part 2 of this volume); CIN, cefsulodin irgasan novobiocin agar (Schiemann, 1979b; Part 2 of this volume); VYE, virulent *Yersinia enterocolitica* agar (Fukushima, 1987); SSDC, Salmonella-Shigella deoxycholate calcium chloride agar (Wauters et al., 1988a; Part 2 of this volume); MacConkey agar no.3 (Oxoid CM115).

coli, *Klebsiella pneumoniae* and *Proteus mirabilis*, but some *Enterobacter*, *Aeromonas* and *Proteus* strains showed a colonial appearance similar to that of *Yersinia* (De Boer and Seldam, 1987). The addition of 1 µg/ml of streptomycin improved the selectivity of CIN agar, but resulted in smaller *Yersinia* colonies (Schiemann, 1987). As the recovery rate and the colony size of *Y. enterocolitica* diminishes during storage of the medium, it is recommended to use CIN medium within 14 days of preparation (Petersen, 1985).

On CIN agar colonies of pathogenic and environmental *Yersinia* strains appear similar. Fukushima (1987) developed a selective agar medium for isolation of pathogenic (virulent) *Y. enterocolitica* (VYE agar). Pathogenic strains form red colonies on this medium, the result of mannitol fermentation and aesculin nonhydrolysis, while most environmental *Yersinia* strains form dark red colonies with a dark peripheral zone as a result of mannitol fermentation and aesculin hydrolysis. Another medium for the direct detection and isolation of plasmid-bearing virulent serotypes of *Y. enterocolitica* was described by Bhaduri and Cottrell (1997). This medium, low-calcium – Congo red – Brain Heart Infusion – agarose (CR-BHO), is incubated at 37°C for 24 h and plasmid-bearing virulent *Y. enterocolitica* strains appear as red pinpoint colo-

Table 2
Differential characteristics of *Yersinia* and related genera.

Characteristics	*Y. enterocolitica*	*Hafnia*	*Serratia*	*Citrobacter*	*Enterobacter*	*Escherichia*	*Klebsiella*	*Proteus*
Urease	+[a]	–	d	d	d	–	d	+
Motility at 25ºC	+	+	+	+	+	+	–	+
Motility at 37ºC	–	+	+	+	+	d	–	+
Arginine dihydrolase	–	–	–	+	d	–	–	–
Lysine decarboxylase	–	+	d	–	d	+	d	–
Phenylalanine deaminase	–	–	–	–	–	–	–	+
H_2S production	–	–	+	d	–	–	–	d

[a] +, positive; –, negative; d, different reactions.

Table 3
Biochemical differentiation within the genus *Yersinia* (from Wauters et al., 1988b).

Test	*Y. enterocolitica*	*Y. intermedia*	*Y. frederiksenii*	*Y. kristensenii*	*Y. aldovae*	*Y. rhodei*	*Y. mollaretii*	*Y. bercovieri*	*Y. pseudotuberculosis*
Indole	d[a]	+	+	d	–	–	–	–	–
Voges–Proskauer	d	+	d	–	+	–	–	–	–
Citrate (Simmons)	–	+	d	–	d	+	–	–	–
L-Ornithine	+	+	+	+	+	+	+	+	–
Mucate, acid	–	d	d	–	d	–	+	+	–
Pyrazinamidase	d	+	+	+	+	+	+	+	–
Sucrose	+	+	+	–	–	+	+	+	–
Cellobiose	+	+	+	+	–	+	+	+	–
L-Rhamnose	–	+	+	–	+	–	–	–	+
Melibiose	–	+	–	–	–	d	–	–	+
L-Sorbose	d	+	+	+	–	ND	+	–	–
L-Fucose	d	d	+	d	d	ND	–	+	–

[a] d, different reactions; +, positive; –, negative; ND, not determined.

nies.

Identification

For the differentiation of *Yersinia* from related genera the tests listed in Table 2 may be used. Table 3 shows the characteristics differentiating the foodborne species within

Table 4
Biotypes of *Yersinia enterocolitica* (Wauters et al., 1987).

Test	Biogroups					
	1A	1B	2	3	4	5
Lipase (Tween-esterase)	+[a]	+	–	–	–	–
Aesculin hydrolysis (24 h)	+	–	–	–	–	–
Salicin (acid production in 24 h)	+	–	–	–	–	–
Indole production	+	+	d	–	–	–
Xylose (acid production)	+	+	+	+	–	–
Trehalose (acid production)	+	+	+	+	+	–
Nitrate reduction	+	+	+	+	+	–
Pyrazinamidase	+	–	–	–	–	–

[a] +, positive; –, negative; d, different reactions.

the genus *Yersinia*. Some biochemical activities (cellobiose, raffinose, indole, ONPG hydrolysis, ornithine decarboxylase, Voges-Proskauer) of *Yersinia* strains are temperature-dependent (Bercovier and Mollaret, 1984). These test are preferably incubated at 25 or 30°C, rather than at 37°C, as yersiniae are more active biochemically at these lower temperatures. Miniaturised identification kits like the API 20E (bioMérieux) and the Crystal E/NF system (Becton Dickinson) have proven to be valuable for rapid identification of *Y. enterocolitica* strains (Peele et al., 1997; Neubauer et al., 1998; Linde et al., 1999), though for definitive species identification additional testing is often necessary.

Serotyping, biotyping and virulence testing are essential for differentiation between pathogenic and environmental *Yersinia* strains. The serotyping scheme of *Y. enterocolitica* and related *Yersinia* species now involves 67 major O factors and 44 H factors (Wauters et al., 1991). Table 4 shows the biotyping scheme of *Y. enterocolitica* as proposed by Wauters et al. (1987). In Europe, *Y. enterocolitica* strains of serotypes O:3 (biogroup 4), O:9 (biogroup 2) and O:5,27 (biogroup 2) are the most frequently isolated human pathogenic strains. In North America, strains causing human yersiniosis usually belong to biogroup 1B (serotypes O:4, O:8, O:13a,13b, O:18, O:20). Though biotype 1A strains are considered to be non-pathogenic, some of these strains may be human-adapted and therefore potentially pathogenic (Burnens et al., 1996; Grant et al., 1998).

Several *in vitro* tests have been described to determine the potential virulence of *Yersinia* isolates and many of these tests are easy to perform in routine laboratories (Farmer et al., 1992) (Table 5). Virulence plasmid expression in *Yersinia* mostly occurs at temperatures greater than 30°C. As most of these tests are plasmid-dependent, it must be realized that plasmids may easily be lost during subculture. DNA-based tests for the identification of potentially pathogenic *Y. enterocolitica* strains include polymerase chain reaction (PCR) tests for detection of the heat stable enterotoxin gene (*yst*)(Ibrahim et al., 1997), the attachment and invasion gene plasmid (*ail*)(Blais and Phillippe, 1995), the invasin gene locus (*inv*)(Rasmussen et al., 1994) and the virulence plasmid (*yadA*)(Blais and Phillippe, 1995).

Table 5
Pathogenicity testing of *Yersinia enterocolitica* strains.

Test	Reference
Aesculin hydrolysis, 24h, 25°C (negative)	Schiemann and Devenish, 1982
Salicin fermentation, 24h, 37°C (negative)	Schiemann and Devenish, 1982
Calcium-dependent growth at 37°C on magnesium oxalate agar	Gemski et al., 1980
Congo red uptake at 25 or 37°	Prpic et al., 1983
Pyrazinamidase activity at 25° (absent)	Kandolo and Wauters, 1985
Autoagglutination at 37°C	Laird and Cavanaugh, 1980
Crystal violet binding at 37°C	Bhaduri et al., 1987

In studies on the molecular epidemiology of pathogenic *Y. enterocolitica* strains a number of DNA-based techniques have been utilized, of which pulsed-field gel electrophoresis (PFGE) seems very efficient for differentiation of bioserotype 4/O:3 strains (Fredriksson-Ahomaa et al., 1999).

Comparative studies

Reporting a comparison of pre-enrichment media, Schiemann (1983a) proposed tryptone soy broth as the medium of choice. Comparison of the two-step enrichment against cold enrichment and modified Rappaport broth showed improved recovery of human pathogenic strains of *Y. enterocolitica* from inoculated foods using pre-enrichment in tryptone soy broth followed by enrichment in BOS broth (Schiemann, 1982; 1983a; Walker and Gilmour, 1986).

In a study on the occurrence of *Y. enterocolitica* in foods Delmas and Vidon (1985) evaluated seven enrichment procedures. Most positive samples obtained within the shortest time were found using enrichment in PBSSB and BOS, with alkali treatment before plating onto CIN agar. However, only environmental serogroups of *Y. enterocolitica* were isolated.

The enrichment medium ITC was found superior to enrichment in BOS for isolation of pathogenic *Y. enterocolitica* from pork products, although enrichment in BOS isolated more non-pathogenic strains (Wauters et al., 1988a; Kwaga et al., 1990; De Boer and Nouws, 1991).

In different studies cold enrichment in PBSSB gave high isolation rates for *Yersinia* spp., which proved to be mainly non-pathogenic environmental strains (De Boer et al., 1982; 1986; De Boer and Seldam, 1987). In these studies MRB was found a suitable medium for the isolation of serogroup O:3 and O:9 strains, but much less efficient for the isolation of environmental strains.

Alkali treatment often resulted in a marked increase in the number of *Yersinia* isolations. This occurred especially when high numbers of competing organisms were present, as was usually the case for PBSSB, and when a plating medium with low selectivity, such as MacConkey agar, was used (De Boer et al., 1982; De Boer and Seldam, 1987).

MacConkey agar proved to be a very productive medium for *Y. enterocolitica* (Mehlman et al., 1978; Schiemann, 1979b; De Boer and Seldam, 1987). However, *Yersinia* colonies are hard to recognize on this medium because of its low selectivity.

In several comparative studies CIN agar was found to be the most selective isolation medium for *Yersinia* spp. (Aldova et al., 1990; Cox et al., 1990 ; Schiemann, 1983a ; Walker and Gilmour, 1986). Fukushima (1987) found that biotype 3B, serogroup O: 3 strains were inhibited in CIN agar. Differentiation of *Yersinia* colonies from other colonies on CIN agar is not always easy (Fukushima, 1987; De Boer and Seldam, 1987). Growth kinetics of *Y. enterocolitica* in CIN broth were influenced by selective agents, incubation temperature and virulence plasmid carriage (Logue et al., 2000).

Plating ITC enrichments onto SSDC isolated more serogroup O:3 strains than onto CIN agar (Wauters et al., 1988a; De Boer and Nouws, 1991). SSDC proved to be unsuitable after alkali treatment (Wauters et al., 1988a).

In a recent study on the occurrence of pathogenic *Y. enterocolitica* in faeces and tonsils of pigs at slaughterhouses the enrichment media peptone, sorbitol and bile salts (PSB) broth, ITC broth and Yersinia Selective Enrichment broth (YSEB)(Merck) and the plating media CIN agar and SSDC agar were compared. ITC broth was found to be the most productive enrichment medium, whereas much lower isolation percentages were obtained with YSEB and PSB broths. Results with CIN agar were slightly better than those obtained with SSDC agar (De Boer et al., 1998).

Discussion

Addition of selective agents including magnesium chloride, malachite green, bile salts, irgasan, and the antibiotics carbenicillin, ticarcillin and cefsulodin to *Yersinia* isolation media results in growth inhibition of Gram-positive and some Gram-negative flora (Table 1). The colonial appearance of *Yersinia* spp. on currently used isolation media is not always characteristic, thus confirmation of presumptive colonies is always necessary.

At present, no single isolation procedure is available for the recovery of all pathogenic strains of *Y. enterocolitica* from foods. Cold enrichment in PBSSB, two-step enrichment with PBS and BOS, and enrichment in ITC are the most commonly used enrichment procedures. CIN and SSDC agars are the most frequently used plating media and are also commercially available. These enrichment and plating media are not particularly selective for *Y. enterocolitica* as they support the growth of several other members of the Enterobacteriaceae. This makes the isolation of low numbers of *Yersinia* in products containing many other contaminants rather difficult. Moreover, non-pathogenic environmental *Yersinia* strains are very common in many raw foods and may greatly hinder the isolation of pathogenic *Yersinia* strains from these products.

Cold enrichment and two-step enrichment are very efficient methods for the recovery of a wide spectrum of *Y. enterocolitica* serotypes. However, usually only pathogenic serotypes are of interest. Since the colonial appearance of pathogenic and environmental strains on CIN and SSDC agar is similar, selection of pathogenic strains on these media requires much effort. This highlights the need for the development and evaluation of agar media selecting for pathogenic serotypes, like VYE agar (Fukushima, 1985).

Enrichment in ITC was found to be the most efficient method for the recovery of strains of serotype O:3, which is the most common clinical serotype of *Y. enterocolitica* in Europe; however for other pathogenic serotypes this method was less efficient, as was also the case for SSDC agar.

For the recovery of the North American serotype O:8 strains, enrichment in BOS and plating on CIN agar seems the most efficient procedure.

Selection of a suitable enrichment procedure will depend on the bio/serotypes of *Yersinia* spp. sought and on the type of food to be examined. The use of more than one medium for both enrichment and plating will obviously result in higher recovery rates of *Yersinia* spp. from foods. The International Standard Organization method for the detection of presumptive pathogenic *Y. enterocolitica* (ISO, 1994) includes parallel use of the following two isolation procedures:

(1) Enrichment in peptone, sorbitol and bile salts (PSB) broth for 2–3 days at 22–25°C with agitation or 5 days without agitation; plating on CIN agar directly and after alkaline treatment and incubation for 24 h at 30°C.

(2) Enrichment in ITC for 2 days at 24°C; plating on SSDC agar and incubation for 2 days at 30°C.

Higher sensitivity and specificity for the detection of pathogenic *Y. enterocolitica* is obtained by using PCR methods directed at virulence genes (Harnett et al., 1996; Nilsson et al., 1998; Fredriksson-Ahomaa et al., 2000).

References

Aldova, E., Svandova, E., Votypka, J. and Sourek, J. (1990) Comparative study of culture methods to detect *Yersinia enterocolitica* serogroup O:3 on swine tongues. Zbl. Bakteriol. 272, 306–312.

Aulisio, C.C.G., Mehlman, I.J. and Sanders, C. (1980) Alkali method for rapid recovery of *Yersinia enterocolitica* and *Yersinia pseudotuberculosis* from foods. Appl. Environ. Microbiol. 39, 135–140.

Bercovier, H. and Mollaret, H.H. (1984) *Yersinia*. In: Krieg, N.R. and Holt, J.G. (Eds.), Bergey's Manual of Systematic Bacteriology, Vol. 1, Williams and Wilkins, Baltimore and London.

Bhaduri, S., Conway, L.K. and Lachica, R.V. (1987) Assay of crystal violet binding for rapid identification of virulent plasmid-bearing clones of *Yersinia enterocolitica*. J. Clin. Microbiol 25, 1039–1042.

Bhaduri, S. and Cottrell, B. (1997) Direct detection and isolation of plasmid-bearing virulent serotypes of *Yersinia enterocolitica* from various foods. Appl. Environ. Microbiol. 63, 4952–4955.

Blais, B.W. and Phillippe, L.M. (1995) Comparative analysis of *yad*A and *ail* polymerase chain reaction methods for virulent *Yersinia enterocolitica*. Food Control 4, 211–214.

Bottone, E.J. (1997) *Yersinia enterocolitica*: The charisma continues. Clin. Microbiol. Rev. 10, 257–276.

Burnens, A.P., Frey, A. and Nicolet, J. (1996) Association between clinical presentation, biogroups and virulence attributes of *Yersinia enterocolitica* strains in human diarrhoeal disease. Epidemiol. Infect. 116, 27–34.

Cox, N.A., Bailey, J.S., Del Corral, F., Shotts, E.B. (1990) Comparison of enrichment and plating media for isolation of *Yersinia*. Poultry Sci. 69, 686–693.

De Boer, E., Hartog, B.J. and Oosterom, J. (1982) Occurrence of *Yersinia enterocolitica* in poultry products. J. Food Prot. 45, 322–325.

De Boer, E., Seldam, W.M. and Oosterom, J. (1986) Characterization of *Yersinia enterocolitica* and related species isolated from foods and porcine tonsils in the Netherlands. Int. J. Food Microbiol. 3, 217–224.

De Boer, E. and Seldam, W.M. (1987) Comparison of methods for the isolation of *Yersinia enterocolitica* from porcine tonsils and pork. Int. J. Food Microbiol. 5, 95–101.

De Boer, E. and Nouws, J.F.M. (1991) Slaughter pigs and pork as a source of human pathogenic *Yersinia enterocolitica*. Int. J. Food Microbiol. 12, 375–378.

De Boer, E. (1995) Isolation of *Yersinia enterocolitica* from foods. Contrib. Microbiol. Immunol. 13, 71–73.

De Boer, E., Hulleman, A. and Kleverwal, M. (1998) Comparison of culture media for the isolation of *Yersinia enterocolitica* from porcine faeces and tonsils. Ned. Tijdschrift voor Medische Microbiologie 6, S46–47 (Abstracts 7th International Congress on Yersinia, Nijmegen, The Netherlands).

De Zutter, L., Le Mort, L., Janssens, M. and Wauters, G. (1994) Short-comings of irgasan ticarcillin chlorate broth for the enrichment of *Yersinia enterocolitica* biotype 2, serotype 9 from meat. Int. J. Food Microbiol. 23, 231–237.

De Zutter, L., Janssens, M. and Wauters, G. (1995) Detection of *Yersinia enterocolitica* serogroup O:3 using different inoculation methods of the enrichment medium Irgasan Ticarcillin Chlorate. Contrib. Microbiol. Immunol. 13, 123–125.

Delmas, C.L. and Vidon, D.J.M. (1985) Isolation of *Yersinia enterocolitica* and related species from foods in France. Appl. Environ. Microbiol. 50, 767–771.

Doyle, M.P. and Hugdahl, M.B. (1983) Improved procedure for recovery of *Yersinia enterocolitica* from meats. Appl. Environ. Microbiol. 45, 127–135.

Farmer, J.J., Carter, G.P., Miller, V.L., Falkow, S. and Wachsmuth, I.K. (1992) Pyrazinamidase, CR-MOX agar, salicin fermentation-esculin hydrolysis, and D-xylose fermentation for identifying pathogenic serotypes of *Yersinia enterocolitica*. J. Clin. Microbiol. 30, 2589–2594.

Fredriksson-Ahomaa, M., Autio, T. and Korkeala, H. (1999) Efficient subtyping of *Yersinia enterocolitica* bioserotype 4/O:3 with pulsed-field gel electrophoresis. Lett. Appl. Microbiol. 29, 308–312.

Fredriksson-Ahomaa, M., Korte, T. and Korkeala, H. (2000) Contamination of carcasses, offals, and the environment with *yadA*-positive *Yersinia enterocolitica* in a pig slaughterhouse. J. Food Protect. 63, 31–35.

Fukushima, H. (1985) Direct isolation of *Yersinia enterocolitica* and *Yersinia pseudotuberculosis* from meat. Appl. Environ. Microbiol. 50, 710–712.

Fukushima, H. (1987) New selective agar medium for isolation of virulent *Yersinia enterocolitica*. J. Clin. Microbiol. 25, 1068–1073.

Gemski, P., Lazere, J.R. and Casey, T. (1980) Plasmid associated with pathogenicity and calcium dependency of *Yersinia enterocolitica*. Infect. Immun. 27, 682–685.

Grant, T., Bennett-Wood, V. and Robins-Browne, R.M. (1998) Identification of virulence-associated characteristics in clinical isolates of *Yersinia enterocolitica* lacking classical virulence markers. Infect. Immun. 66, 1113–1120.

Greenwood, M.H. and Hooper, W.L. (1989) Improved methods for the isolation of *Yersinia* species from milk and foods. Food Microbiol. 6, 99–104.

Greenwood, M.H. (1993) Comparison of enrichment at 9°C and 21°C for recovery of *Yersinia* species from food and milk. Food Microbiology 10, 23–30.

Harnett, N., Lin, Y.P. and Krishnan, C. (1996) Detection of pathogenic *Yersinia enterocolitica* using the multiplex polymerase chain reaction. Epidemiol. Infect. 117, 59–67.

Ibrahim, A., Liesack, W., Griffiths, M.W. and Robins-Browne, R.M. (1997) Development of a highly specific assay for rapid identification of pathogenic strains of *Yersinia enterocolitica* based on PCR amplification of the *Yersinia* heat-stable enterotoxin gene (*yst*). J. Clin. Microbiol. 35, 1636–1638.

ISO. (1994) Microbiology – General guidance for the detection of presumptive pathogenic *Yersinia enterocolitica*. ISO 10273. International Standard Organization, Geneva.

Kandolo, K. and Wauters, G. (1985) Pyrazinamidase activity in *Yersinia enterocolitica* and related organisms. J. Clin. Microbiol. 21, 980–982.

Kapperud, G. (1991) *Yersinia enterocolitica* in food hygiene. Int. J. Food Microbiol. 12, 53–66.

Kwaga, J., Iversen, J.O. and Saunders, J.R. (1990) Comparison of two enrichment protocols for the detection of *Yersinia* in slaughtered pigs and pork products. J. Food Protect. 53, 1047–1049.

Laird, W.J. and Cavanaugh, D.C. (1980) Correlation of autoagglutination and virulence of yersinae. J. Clin. Microbiol. 11, 430–432.

Landgraf, M., Iaria, S.T. and Falcão, D.P. (1993) An improved enrichment procedure for the isolation of *Yersinia enterocolitica* and related species from milk. J. Food Protect. 56, 447–450.

Lee, W.H. (1977) Two plating media modified with Tween 80 for isolating *Yersinia enterocolitica*. Appl. Environ. Microbiol. 33, 215–216.

Lee, W.H., Harris, M.E., McClain, D., Smith, R.E. and Johnston, R.W. (1980) Two modified selenite media for recovery of *Yersinia enterocolitica* from meats. Appl. Environ. Microbiol. 39, 205–209.

Linde, H.J., Neubauer, H., Meyer, H., Aleksic, S. and Lehn, N. (1999) Identification of *Yersinia* species by the VITEK GNI card. J. Clin. Microbiol. 37, 211–214.

Logue, C.M., Sheridan, J.J., McDowell, D.A., Blair, I.S. and Hegarty, T. (2000) The effect of temperature and selective agents on the growth of *Yersinia enterocolitica* serotype O:3 in pure culture. J. Appl. Microbiol. 88, 1001–1008.

Mehlman, I.J., Aulisio, C.C.G. and Sanders, A.C. (1978) Problems in the recovery and identification of *Yersinia* from food. J. Assoc. Off. Anal. Chem. 61, 761–771.

Neubauer, H., Sauer, T., Becker, H., Aleksic, S. and Meyer, H. (1998) Comparison of systems for identification and differentiation of species within the genus *Yersinia*. J. Clin. Microbiol. 36, 3366–3368.

Nilsson, A., Lambertz, S.T., Stålhandske, P., Norberg, P. and Danielsson-Tham, M.-L. (1998) Detection of *Yersinia enterocolitica* in food by PCR amplification. Lett. Appl. Microbiol. 26, 140–144.

Peele, D., Bradfield, J., Pryor, W. and Vore, S. (1997) Comparison of identifications of human and animal source Gram-negative bacteria by API 20E and Crystal E/NF systems. J. Clin. Microbiol. 35, 213–216.

Petersen, T. (1985) Keeping quality of cefsulodin-irgasan-novobiocin (CIN) medium for detection and enumeration of *Yersinia enterocolitica*. Int. J. Food Microbiol. 2, 49–54.

Prpic, J.K., Robins-Browne, R.M. and Davey, R.B. (1983) Differentiation between virulent and avirulent *Yersinia enterocolitica* isolates by using Congo red agar. J. Clin. Microbiol. 18, 486–490.

Rasmussen, H.N., Rasmussen, O.F., Andersen, J.K. and Olsen, J.E. (1994) Specific detection of pathogenic *Yersinia enterocolitica* by two-step PCR using hot-start and DMSO. Molecular and Cellular Probes 8, 99–108.

Schiemann, D.A. (1979a) Enrichment methods for recovery of *Yersinia enterocolitica* from foods and raw milk. Contr. Microbiol. Immunol. 5, 212–217.

Schiemann, D.A. (1979b) Synthesis of a selective agar medium for *Yersinia enterocolitica*. Can. J. Microbiol. 25, 1298–1304.

Schiemann, D.A. (1982) Development of a two-step enrichment procedure for recovery of *Yersinia enterocolitica* from food. Appl. Environ. Microbiol.43, 14–27.

Schiemann, D.A. and Devenish, J.A. (1982) Relationship of HeLa cell infectivity to biochemical, serological and virulence characteristics of *Yersinia enterocolitica*. Infect. Immun. 35, 497–506.

Schiemann, D.A. (1983a) Comparison of enrichment and plating media for recovery of virulent strains of *Yersinia enterocolitica* from inoculated beef stew. J. Food Protect. 46, 957–964.

Schiemann, D.A. (1983b) Alkalotolerance of *Yersinia enterocolitica* as a basis for selective isolation from food enrichments. Appl. Environ. Microbiol. 46, 22–27.

Schiemann, D.A. and Olson, S.A. (1984) Antagonism by Gram-negative bacteria to growth of *Yersinia enterocolitica* in mixed cultures. Appl. Environ. Microbiol. 48, 539–544.

Schiemann, D.A. (1987) *Yersinia enterocolitica* in milk and dairy products. J. Dairy Sci. 70, 383–391.

Schiemann, D.A. and Wauters, G. (1992) *Yersinia*. In: Vanderzant, C. and Splitttoesser (eds.), Compendium of methods for the microbiological examination of foods, p. 433–450. American Public Health Association, Washington, D.C.

Schraft, H. and Untermann, F. (1989) Use of KOH treatment for direct detection of *Yersinia enterocolitica* in foods. 10th WAVFH Int. Symp., Stockholm, Abstracts, p.31.

Stolk-Engelaar, V.M.M. and Hoogkamp-Korstanje, J.A.A. (1996) Clinical presentation and diagnosis of gastrointestinal infections by *Yersinia enterocolitica* in 261 Dutch patients. Scand. J. Infect. Dis. 28, 571–575.

Toora, S., Budu-Amoako, E., Ablett, R.F. and Smith, J. (1994) Evaluation of different antimicrobial agents used as selective supplements for isolation and recovery of *Yersinia enterocolitica*. J. Appl. Bacteriol. 77, 67–72.

Van Pee, W. and Stragier, J. (1979) Evaluation of some cold enrichment and isolation media for the recovery of *Yersinia enterocolitica*. Antonie van Leeuwenhoek J. Microbiol. Serol. 45, 465–477.

Verhaegen, J., Charlier, J., Lemmens, P., Delmée, M., Van Noyen, R., Verblist, L. and Wauters, G. (1998) Surveillance of human *Yersinia enterocolitica* infections in Belgium: 1967–1996. Clin. Infect. Dis. 27, 59–64.

Vidon, D.J.M. and Delmas, C.L. (1981) Incidence of *Yersinia enterocolitica* in raw milk in Eastern France. Appl. Environ. Microbiol. 41, 355–359.

Walker, S.J. and Gilmour, A. (1986) A comparison of media and methods for the recovery of *Yersinia enterocolitica* and *Yersinia enterocolitica*-like bacteria from milk containing simulated raw milk microfloras. J. Appl.. Bacteriol. 60, 175–183.

Wauters, G. (1973) Improved methods for the isolation and the recognition of *Yersinia enterocolitica*. Contr. Microbiol. Immunol. 2, 68–70.

Wauters, G., Kandolo, K. and Janssens, M. (1987) Revised biogrouping scheme of *Yersinia enterocolitica*. Contr. Microbiol. Immunol. 9, 14–21.

Wauters, G., Goossens, V., Janssens, M., and Vandepitte, J. (1988a) New enrichment method for isolation of pathogenic *Yersinia enterocolitica* serogroup O :3 from pork. Appl. Environ. Microbiol. 54, 851–854.

Wauters, G., Janssens, M., Steigerwalt, A.G. and Brenner, D.J. (1988b) *Yersinia mollaretii* sp.nov. and *Yersinia bercovieri* sp.nov., formerly called *Yersinia enterocolitica* biogroups 3A and 3B. Int. J. Syst. Bacteriol. 38, 424–429.

Wauters, G., Aleksic, S., Charlier, J. and Schulze, G. (1991) Somatic and flagellar antigens of *Yersinia enterocolitica* and related species. Contrib. Microbiol. Immunol. 12, 239–243.

Weagant, S.D. and Kaysner, C.A. (1983) Modified enrichment broth for isolation of *Yersinia enterocolitica* from nonfood sources. Appl. Environ. Microbiol. 45, 468–471.

Chapter 16

Review of media for the isolation of diarrhoeagenic *Escherichia coli*

Annet E. Heuvelink

Inspectorate for Health Protection, P.O. Box 202, 7200 AE Zutphen, The Netherlands

The species *Escherichia coli* contains both diarrhoeagenic and non-diarrhoeagenic strains and it is very important to have methods available which can differentiate between them. Adequate culture methods have been developed for the isolation of enterohaemorrhagic *E. coli* (EHEC) of serogroup O157 from foods. However, at present no single isolation procedure is available for the recovery of all EHEC. Additionally, there are still no simple sensitive procedures available for the direct cultivation of strains of the other groups of diarrhoeagenic *E. coli*. The isolation of these organisms will best be accomplished by a combination of culture and molecular biological methods.

In this paper some comparative studies of the media described for EHEC, especially EHEC O157, are noted and the difficulties associated with the isolation and enumeration of these organisms considered. Modified trypticase soya broth supplemented with novobiocin or modified *E. coli* broth supplemented with novobiocin and incubated at 41–42°C are the most appropriate selective enrichments. Injured EHEC O157 cells require pre-enrichment in a non-selective broth. Methods for the isolation of EHEC O157 should include sorbitol MacConkey agar supplemented with cefixime and potassium tellurite as the most effective isolation medium for typical sorbitol-non-fermenting EHEC O157 and a second isolation medium not based on the fermentation of sorbitol but, for instance, on ß-D-glucuronidase activity. Where the background flora is low, washed sheep blood agar supplemented with calcium ("EHEC agar") may be used.

Introduction

Escherichia coli is the predominant bacterium of the normal facultative anaerobic microflora of the human intestine. These commensal strains of *E. coli* usually remain harmlessly confined to the intestinal lumen and have been shown to play an important role in maintaining intestinal physiology. However, given the right opportunities any *E. coli* strain can probably cause invasive disease, and *E. coli* has therefore aptly been called an opportunistic pathogen. On the other hand, several highly adapted *E. coli* clones have evolved the ability to cause a broad spectrum of human diseases. Some strains of *E. coli* are well equipped for extra-intestinal multiplication, causing urinary tract infections, neonatal meningitis, wound infections, peritonitis, and septicaemia; however, inherently pathogenic strains of *E. coli* are generally considered as a cause of diarrhoeal diseases. Presently, six major groups of diarrhoeagenic *E. coli* strains have been defined based primarily on pathogenic mechanisms (Nataro and

Table 1
Foods involved in outbreaks of infection with diarrhoeagenic *E. coli.*

Type of *E. coli*	Known or presumed food vehicles of outbreaks
EPEC	Coffee substitute, cold pork, meat pie, weaning foods or infant formula, water
EIEC	Brie and Camembert cheese, canned salmon, potato salad, guacamole, water
ETEC	Brie cheese, curried turkey mayonnaise, tuna pasta (school lunch), fresh fruits, salads containing raw vegetables, especially lettuce (airline flight, mountain lodge), prepared food (restaurant, cafeteria, cruise ship), water
EHEC	Undercooked ground beef, unpasteurised cow's milk, pasteurised cow's milk, goat's milk, unpasteurised cream, unpasteurised yogurt, unpasteurised cheese, roast beef, salami, meat from poultry, sheep and deer, fresh potatoes, lettuce, radish and alfalfa sprouts, unpasteurised apple juice and cider
EAEC	?
DAEC	?

Kaper, 1998): enteropathogenic *E. coli* (EPEC), enteroinvasive *E. coli* (EIEC), enterotoxigenic *E. coli* (ETEC), enterohaemorrhagic *E. coli* (EHEC), enteroaggregative *E. coli* (EAEC or EAggEC), and diffusely adherent *E. coli* (DAEC).

Importance of diarrhoeagenic *E. coli* as foodborne pathogens

Infection with diarrhoeagenic *E. coli* is usually acquired by ingestion of the organisms via contaminated food or water. Different types of food including water have been associated with outbreaks of diarrhoea due to these *E. coli* (Table 1). The source of infection is usually not investigated in sporadic cases. In addition to the food associated with outbreaks of *E. coli*-induced diarrhoea, several investigations of food and water as a potential vehicle have demonstrated the presence of pathogenic *E. coli* strains. Humans, symptomatic or symptom-free carriers, are presumed to be the principal reservoir of EPEC, EIEC, and ETEC strains that cause human illness. These bacteria are present in the intestinal tract of carriers and are excreted in their faeces. Infected foodhandlers with poor personal hygiene or water contaminated by human sewage are sources of food contamination. The principal reservoir of EHEC is the intestinal tract of cattle. Other animals used in the production of food have also been shown to carry the organisms, although less frequently. Animals carrying EHEC are usually asymptomatic. Raw foods of animal origin may be contaminated with the organisms via faecal contact during slaughter or milking procedures. Most EHEC infections are caused by ingestion of undercooked ground beef and cow's milk. In the past few years, however, fruits and vegetables have accounted for a growing number of recognised EHEC outbreaks. While contamination of fresh produce may be due to cross-contamination from meat products, contact with faeces from domestic or wild animals at some stage during cultivation or handling of fresh produce is another presumed route of contamination. Person-to-person contact has been shown to be an important cause of outbreaks in day-care and chronic-care facilities. Less is known about the reservoirs and routes of transmission of EAEC and DAEC. It has been suggested that the ability to cause disease is a property of only certain EAEC strains. Most reports have implicated EAEC in sporadic endemic diarrhoea but the source of infection was frequently unclear. In addition to the uncertainty about the contribution of DAEC to the human disease burden, the mode of acquisition of DAEC infection is also as yet undetermined.

Isolation of diarrhoeagenic *E. coli* from foods

E. coli consists of both pathogenic and non-pathogenic strains and, as the latter constitute part of the normal intestinal flora, differentiation between the two is very important. The detection and isolation of pathogenic *E. coli* from foods is quite difficult. Serological testing of *E. coli* isolates has limited value in identifying pathogenic strains. Traditional cultivation from food samples using selective enrichment broth at elevated temperatures (44–45.5°C) has been shown to favour non-pathogenic, environmental strains compared to pathogenic strains. In addition, some pathogenic strains do not have the typical characteristics of *E. coli*, such as fermentation of lactose and gas production. Adequate culture methods have been developed for EHEC, especially EHEC of serogroup O157. *E. coli* strains of serotype O157:H7, and the non-motile (NM) variant *E. coli* O157:NM, are the most common EHEC strains in many parts of the world. However, there are still no simple sensitive procedures available for the direct cultivation of the other diarrhoeagenic *E. coli* from foods. For the isolation of these organisms resuscitation in brain heart infusion broth for 3 h at 35°C, followed by enrichment in tryptone phosphate broth for 20 h at 44°C and plating on Levine's eosin-methylene blue agar and MacConkey agar is advised. Both typical (lactose-fermenting) and non-typical (non-lactose-fermenting) colonies have to be picked and characterised biochemically, serologically and on the basis of virulence properties (US Food and Drug Administration, 1995). A phenotypic approach to test for virulence properties requires the use of cell cultures and sometimes fluorescence microscopy, and a genotypic method requires the use of DNA hybridisation or the polymerase chain reaction (PCR). This means that from one sample numerous isolates must be assayed.

Enrichment

The low infectious dose of EHEC (~100–200 organisms for EHEC O157) necessitates the detection of low numbers in foods. Owing to the lack of sensitivity of direct plating, enrichment media have been developed to allow target cells to multiply to detectable levels. The recovery of low numbers of EHEC during the enrichment procedure is dependent upon the competitive flora present, both the total number of bacteria and more specifically the number of Enterobacteriaceae. Selective enrichment media for EHEC are listed in Table 2. Bile salts inhibit many non-enteric bacteria. Novobiocin, acriflavine and vancomycin are incorporated to inhibit Gram-positive bacteria, and cefsulodin and cefixime suppress the growth of *Aeromonas* spp. and *Proteus* spp., respectively.

Incubation conditions during selective enrichment should suppress the growth of

Table 2
Selective enrichment media for EHEC.

Medium	Acronym	Selective agents	Antibiotics	Reference
Modified trypticase soy broth	mTSB	bile salts no. 3 (1.5 g/l)	novobiocin (20 mg/l)	Doyle and Schoeni, 1987
Double-modified trypticase soy broth containing casamino acids (10 g/l)	dmTSB or dmTSB-CA	bile salts no. 3 (1.5 g/l)	acriflavine (10 mg/l)	Padhye and Doyle, 1991
Trypticase soy broth with cefixime, tellurite and vancomycin	TSB-CTV	bile salts no. 3 (1.5 g/l)	cefixime (0.05 mg/l) potassium tellurite (2.5 mg/l) vancomycin (40 mg/l)	Kudva et al., 1996
EHEC enrichment broth (=mTSB base (without novobiocin) with vancomycin, cefixime and cefsulodin),	EEB	bile salts no. 3 (1.5 g/l)	vancomycin (8 mg/l) cefixime (0.05 mg/l) cefsulodin (10 mg/l)	Weagant et al., 1995
Buffered peptone water with vancomycin, cefixime and cefsulodin	BPW-VCC	–	vancomycin (8 mg/l) cefixime (0.05 mg/l) cefsulodin (10 mg/l)	Chapman et al., 1993
Modified *E. coli* (EC) broth with novobiocin	mEC+n	bile salts no. 3 (1.12 g/l)	novobiocin (20 mg/l)	Okrend et al., 1990a
Brilliant green bile lactose broth	BRILA	brilliant green (0.0133 g/l) ox bile (20.0 g/l)	–	Heckötter et al., 1997

competing microflora and maximise the outgrowth of EHEC. As many other micro-organisms in foods show optimal growth at 37°C, enrichment of EHEC at 41–42°C is preferred. The main advantage is the inhibition of other non-sorbitol-fermenting organisms such as *Hafnia alvei*, which can be a problem when testing raw meats. However, precise temperature control is critical as poor growth of EHEC O157 strains has been observed at temperatures above 42°C (Raghubeer and Matches, 1990). The incubation period required will depend on the anticipated competing microflora and the subsequent isolation procedure.

Immunomagnetic separation (IMS) may be applied at this stage. Para-magnetic beads coated with antibodies specific to the target organism are mixed with enrichment cultures capturing the target organism onto the beads which are then removed from food debris and background microorganisms by application of a magnetic field. Following washing to remove non-specifically bound bacteria and sample particles, the beads are resuspended in a reduced volume and cultured on solid media. Commercially available IMS systems for the isolation of *E. coli* O157 are Dynabeads™ anti-*E. coli* O157 (Dynal), Captivate™ O157 (Lab M) and *E. coli* O157-IMS 'Seiken' (Denka Seiken). Immunomagnetic beads with anti-O26 and anti-O111 antibodies are also available commercially. Several studies have shown that the use of IMS is more sensitive compared to direct subculture from food and bovine faeces enrichment cultures (Fratamico et al., 1992; Chapman et al., 1994; Wright et al., 1994; Bennett et al., 1995; 1996; Heckötter et al., 1997; Heuvelink et al., 1998). IMS increases the sensitivity of the detection system by concentrating the target organism relative to the background microflora, which may mask or mimic target cells on selective/differential solid media. Non-specific binding of organisms in samples with a high background flora can pose a problem which can be resolved by reducing the enrichment time, plating onto a highly selective agar or using a low-ionic-strength washing solution (Tomoyasu, 1998). Another application of immunocapture is the automated Vitek Immuno Diagnostic Assay System for immunoconcentration of *E. coli* O157 (VIDAS-ICE) of BioMérieux (Vernozy-Rozand et al., 1998; 2000). A second approach to separate EHEC O157 and other EHEC from background flora present in food enrichment cultures is treatment with hydrochloric acid before streaking organisms onto the selective plating medium, usually sorbitol MacConkey agar supplemented with cefixime and tellurite (Fukushima and Gomyoda, 1999a, 1999b; Fukushima et al., 2000). This method is based on the acid resistance of *E. coli* and the tellurite resistance of many EHEC strains associated with severe disease in humans. Acid treatment of IMS concentrates may further improve isolation rates of EHEC from food samples (Fukushima and Gomyoda, 1999a, 1999b). Often an incubation period of 6 h is used before immunocapture, acidic treatment or direct plating from enrichment cultures followed by a further 12 to 18 h to retest samples negative after 6 h of incubation. An enrichment period of only 6 h has the advantage that the results will be available the following day. Though incubation with shaking is common, the advantages over static incubation are not well documented. However, for short enrichment periods (< 6 h) the use of pre-warmed enrichment broth and a shaking water bath is recommended to obtain sufficient growth of the target organism.

In several studies the efficacy of different enrichment procedures has been evaluated, but the results vary. During an investigation of a food-borne outbreak attributed to the consumption of undercooked hamburger patties, Johnson et al. (1995) analysed food and environmental samples and found it more difficult to recover viable *E. coli* O157:H7 from dmTSB than from mEC+n enrichments after screening with commercial immunoassays. They suggested that dmTSB promotes good expression of the somatic 157 antigens but adversely affects cell viability after a certain incubation time. Bennett et al. (1996) showed that the isolation rate of *E. coli* O157 from inoculated minced beef by the use of IMS was increased when samples were enriched in mEC+n compared with enrichment in BPW-VCC. Equivalent results were achieved by Fratamico et al. (1992) with enrichments in lauryl tryptose broth and EC broth containing novobiocin (20 mg/l) when isolating *E. coli* O157:H7 from artificially inoculated raw ground beef by using IMS. EEB was found to be inhibitory to pure cultures of EHEC O157, and cefixime appeared to be the most critical factor in the enrichment broth (Heuvelink et al., 1997). Heckötter et al. (1997) recommended the simultaneous use of the enrichment media TSB and BRILA for the isolation of *E. coli* O157 from foods. The best recovery rates of *E. coli* O157:H7 inoculated into ground beef samples were recorded using both enrichment media for 6 and 24 h followed by IMS and subcultivation on two different selective media. Seo et al. (1998) compared modified BPW with casamino acids (mBPW), first described by Restaino et al. (1996), BPW, and mEC+n and found mBPW most effective for enrichment of *E. coli* O157:H7 in ground beef in 6 h. Both mBPW and dmTSB were effective for increasing the concentration of *E. coli* O157:H7 in raw milk in 6 h (Seo et al., 1998). Fukushima and Gomyoda (1999b) studied the recovery of *E. coli* O157:H7 from artificially contaminated ground beef samples enriched in TSB, TSB-CTV and mEC+n and incubated at both 37 and 42°C for 18 h with agitation and standing. Enrichments were plated onto selective agar after hydrochloric acid treatments. *E. coli* O157:H7 strains were most frequently recovered from samples incubated in TSB and TSB-CTV at 42°C without agitation. The growth of the *E. coli* O157:H7 strains in mEC+n was suppressed regardless of culture conditions. Additional studies were done to compare the recovery after TSB, TSB-CTV and mEC+n enrichments at 42°C for 6 and 18 h. It was concluded that TSB culture at 42°C for 6 h without agitation was the most effective procedure to enrich *E. coli* O157:H7 in food samples prior to hydrochloric acid treatment and subsequent streaking of organisms onto selective plating medium.

Hara-Kudo et al. (1999) evaluated enrichment conditions for isolation of *E. coli* O157:H7 from inoculated ground beef and radish sprouts using the enrichment broths TSB, mTSB, mEC+n, EEB and TSB-CTV. After static incubation for 6 or 18 h at 37 or 42°C, enrichment cultures were streaked onto selective agar both directly and after IMS. Incubation at 37°C tended to be less effective than at 42°C for isolation of *E. coli* O157:H7 from both ground beef and radish sprouts. The most effective enrichment condition was incubation in mTSB or mEC+n at 42°C for 18 h for ground beef, and in mEC+n at 42°C for 18 h for radish sprouts, both in combination with the use of IMS. Similarly, Blais et al. (1997) demonstrated that the recovery and detection of *E. coli* O157:H7 in ground meats (beef, pork, turkey) by static enrichment in mEC+n for 24 h was best accomplished by incubation at 42°C, rather than at 37°C. They suggested that *E. coli* O157:H7 cells may be (physiologically) less sensitive to the selective agents at the higher temperature. In addition to the effect of temperature, Blais et al. (1997) studied the effect of agitation and concluded that incubation at 42°C without shaking effectively suppressed competing non-target micro-organisms (meat microflora) while allowing good growth of *E. coli* O157:H7 cells. On the other hand, in one of their studies Bennett et al. (1995) demonstrated the successful use of shaken mEC+n enrichment culture as opposed to static enrichment for the recovery of *E. coli* O157 inoculated into minced beef followed by enrichment at 37°C, but the success appeared to be dependent upon the detection agar. In agreement with the observation of Hara-Kudo et al. (1999), Bolton et al. (1996) reported that the combination of enrichment in mTSB at 42°C followed by IMS gave successful recovery of *E. coli* O157 from ground beef, though the incubation time giving the best recovery differed between the two studies. Hara-Kudo et al. (1999) recommended selective enrichment culture for 18 h, while Bolton et al. (1996) obtained good results after 6 h of enrichment. Tilburg et al. (1999) evaluated the efficacy of mEC+n (37°C), mTSB (37 and 41°C), BPW-VCC (37°C) and lactose bile brilliant green broth (LBBG) (37°C) by examining pure bacterial cultures and artificially inoculated minced beef samples by the use of IMS and VIDAS-ICE. The following procedures were found the most effective for the isolation of EHEC O157 from samples of minced beef: (i) selective enrichment in mTSB (6 h, 37 or 41°C) followed by the Dynabeads anti-*E. coli* O157 procedure and (ii) selective enrichment in mTSB (24 h, 41°C) followed by the Dynabeads anti-*E. coli* O157 procedure or VIDAS-ICE.

For EHEC strains of other serogroups than O157 few detection and isolation methods have been evaluated. Hara-Kudo et al. (2000) compared various enrichment procedures that are effective for isolating *E. coli* O157 for their efficiency in isolating *E. coli* O26 from inoculated broth containing ground-beef or radish-sprout extract, or from artificially contaminated ground beef or radish sprouts. The enrichment broths used were TSB, mTSB, mEC+n, EEB, and TSB-CTV. They concluded that enrichment in mEC+n at 42°C for 18 h in combination with IMS using anti-O26 coated magnetic beads was the most efficient enrichment procedure for isolating *E. coli* O26 from foods. Enrichment in TSB at 37°C for 6 h was also effective in recovering the bacterium from ground beef. Concordant with the results of Hara-Kudo et al. (2000), Peitz et al. (2000) reported that TSB supplemented with VCC was not suitable for enrichment of EHEC non-O157. Fukushima and Gomyoda (1999b) determined that TSB culture at 42°C for 6 h without agitation followed by hydrochloric acid treatment and subsequent streaking of organisms onto selective plating medium was an effective procedure not only to isolate EHEC O157, but also to recover EHEC O26 and EHEC O111.

Plating media

Table 3 lists plating media commonly used to isolate EHEC strains, especially EHEC O157. Typical strains of EHEC O157, unlike the majority of *E. coli*, do not ferment sorbitol within 24 h and are ß-D-glucuronidase-negative, which differential characteristics are applied in several media for their isolation. The most widely accepted method of screening for EHEC O157 involves observation of sorbitol MacConkey (SMAC) agar

Table 3
Selective plating media for EHEC.

Medium	Acronym	Diagnostic system	EHEC serogroup	Reference or manufacturer
Sorbitol MacConkey agar	SMAC	Sorbitol fermentation	O157	March and Ratnam, 1986
Sorbitol MacConkey agar supplemented with cefixime and rhamnose	CR-SMAC	Sorbitol fermentation	O157	Chapman et al., 1991
Sorbitol MacConkey agar supplemented with cefixime and potassium tellurite	CT-SMAC	Sorbitol fermentation	O157	Zadik et al., 1993
Haemorrhagic-colitis agar	HC agar	Sorbitol fermentation, β-D-glucuronidase, indole reaction	O157	Szabo et al., 1986
Phenol red sorbitol MacConkey agar supplemented with 4-methylumbelli-feryl-ß-D-glucuronide	PRS-MUG	Sorbitol fermentation, β-D-glucuronidase	O157	Okrend et al., 1990a
Sorbitol MacConkey agar supplemented with cefixime, potassium tellurite and 8-hydroxyquinoline-β-D-glucuronide	CT-SMAC-HQG	Sorbitol fermentation, β-D-glucuronidase	O157	Reinders et al., 2000
SD-39 agar	–	Sorbitol fermentation, β-D-glucuronidase, lysine decarboxylase	O157	Entis, 1998
Fluorocult™ *E. coli* O157:H7 agar	–	Sorbitol fermentation, β-D-glucuronidase	O157	Merck

CHROMagar® O157	–	Break down of chromogenic substrates	O157	CHROMagar
Rainbow™ Agar O157	–	β-D-galactosidase, β-D-glucuronidase	O157 and some non-O157	Biolog
Rainbow™ Agar O157 supplemented with potassium tellurite and novobiocin	–	β-D-galactosidase, β-D-glucuronidase	O157	Stein and Bochner, 1998
O157:H7 ID medium	–	Not given	O157	bioMérieux
Biosynth Culture Medium O157:H7	–	Not given	O157	Biosynth
Enterohaemolysin or enterohaemorrhagic *E. coli* agar (syn. washed sheep blood agar supplemented with calcium)	EHEC agar (syn. WSBA-Ca)	Enterohaemolysin production after 24 h	O157 and non-O157	Beutin et al., 1989

Other modifications of SMAC include SMAC supplemented with the fluorogenic compound 4-methylumbelliferyl-ß-D-glucuronide (MUG) or the chromogenic compound 5-bromo-4-chloro-3-indolyl-ß-D-glucuronide (BCIG); the selectivity of the commercially available chromogenic media and EHEC agar may be improved by the incorporation of antibiotics.

for colourless (sorbitol-negative) colonies (March and Ratnam, 1986). The sensitivity of SMAC is limited by the capacity to recognise non-sorbitol-fermenting colonies against the background of other organisms and the possible presence of other non-sorbitol-fermenters, such as *Proteus* spp., *Aeromonas* spp. and some other *E. coli* which makes it necessary to test many colonies for confirmation. Chapman et al. (1991) improved the isolation rate of EHEC O157 by supplementing SMAC with cefixime and rhamnose (CR-SMAC). Cefixime inhibits *Proteus* spp. at a concentration not inhibitory to *E. coli*, and rhamnose is fermented by most non-sorbitol-fermenting *E. coli* strains other than those of serogroup O157. The inclusion of these components appeared to reduce the number of colourless colonies from human faecal samples by approximately 60%. Zadik et al. (1993) reported a further improvement in EHEC O157 isolation rates by using SMAC supplemented with cefixime and potassium tellurite (CT-SMAC). EHEC O157 strains are generally less susceptible to tellurite than are many other non-sorbitol-fermenters such as *Aeromonas* spp., *Pleisomonas* spp., *Morganella* spp., *Providencia* spp., and most other *E. coli* strains.

Beta-D-glucuronidase activity can be detected by the addition of the fluorogenic compound 4-methylumbelliferyl-ß-D-glucuronide (MUG) (Szabo et al., 1986; Okrend et al., 1990a) or the chromogenic compound 5-bromo-4-chloro-3-indolyl-ß-D-glucoronide (BCIG) (Okrend et al., 1990b) to SMAC plates. Haemorrhagic colitis (HC) agar contains both sorbitol and MUG to distinguish phenotypes based on reactions to these reagents (Szabo et al., 1986). Entis (1998) developed SD-39 agar, for use with the ISO-GRID hydrophobic grid membrane filter. This medium relies on three differential biochemical reactions, lysine decarboxylase (positive for typical EHEC O157 strains), sorbitol fermentation, and ß-D-glucuronidase, which are read simultaneously. Selectivity is achieved through use of monensin to inhibit Gram-positive bacteria, incubation at 44.0 to 44.5°C to inhibit many Gram-negative bacteria, including *Hafnia alvei*, a common competitor that would otherwise mimic the appearance of *E. coli* O157:H7, and novobiocin to slow the growth of some of the fast-growing temperature-tolerant Gram-negative bacteria, such as *Klebsiella* spp. At the specified incubation temperature, presumptive *E. coli* O157:H7 appear pink; other *E. coli* are green, and most other organisms are either unable to initiate growth, or develop yellow colonies. The ISO-GRID method using SD-39 agar has been validated for the direct presumptive enumeration of *E. coli* O157:H7 in a broad range of foods (pasteurised apple cider, pasteurised 2% milk, cottage cheese, cooked ground pork, raw ground beef, and frozen whole egg). Recently, Reinders et al. (2000) studied the use of an alternative substrate, 8-hydroxyquinoline-ß-D-glucuronide (HQG), for presumptive identification of EHEC O157 on SMAC agars. HQG is less expensive than BCIG, is visible in normal daylight and does not diffuse into the agar like MUG. On CT-SMAC agar with HQG, EHEC O157 strains grew as cream-coloured colonies and could be easily distinguished from strains utilising sorbitol and/or producing ß-D-glucuronidase that formed purple or black colonies. The medium showed a high sensitivity (100%) and specificity (99.7%) for typical EHEC O157.

CHROMagar® O157 (CHROMagar), a novel commercial chromogenic medium of unknown composition has also been reported to be a sensitive and specific medium for EHEC O157 (Wallace and Jones, 1996). EHEC O157 can readily be recognised by typical pink colonies (Bettelheim, 1998a). Some other EHEC strains, including many

EHEC O111, also produced pink colonies on CHROMagar®. However, most other EHEC strains consistently grew as blue colonies, while commensal *E. coli* formed blue or, rarely, colourless colonies. Another new medium that recently has been introduced on the market is Rainbow™ Agar O157 (Biolog). This medium contains chromogenic substrates that are specific for two enzymes usually associated with *E. coli*, ß-D-galactosidase and ß-D-glucuronidase. The substrate for ß-D-galactosidase is blue-black and that for ß-D-glucuronidase is red. Typical EHEC O157 strains, being ß-D-galactosidase-positive and ß-D-glucuronidase-negative, form characteristic charcoal grey or steel black colonies, whereas most other *E. coli* strains produce both enzymes and appear as violet or red colonies on this medium. It has been claimed that some typical EHEC non-O157 strains overproduce ß-D-galactosidase relative to ß-D-glucuronidase on this medium, giving the colonies a distinctive intermediate blue or purple colour (Bettelheim, 1998b). The addition of tellurite and novobiocin has been shown to enhance the recovery of *E. coli* O157:H7 considerably on Rainbow™ Agar O157 (Stein and Bochner, 1998). Other chromogenic media may also need the addition of inhibiting compounds like cefixime, novobiocin and tellurite to obtain sufficient selectivity. Variations in the incubation temperature and the use of agar plates stored for some weeks may cause fluctuations in typical colony colours on chromogenic media.

Selective plating media for EHEC O157 have been compared in several studies. A brief survey of the literature reveals the following. In a comparison of the selective plating media SMAC, CT-SMAC, and Fluorocult™ *E. coli* O157:H7 agar (Merck) using pure bacterial cultures, CT-SMAC proved to be the most selective and Fluorocult™ *E. coli* O157:H7 agar, containing sorbitol and MUG, the least selective (Heuvelink et al., 1997). In a similar study the electivity and productivity of CT-SMAC, Rainbow™ Agar O157 (Biolog), O157:H7 ID medium (BioMérieux) and two types of CHROMagar O157 (from CHROMaga and BBL), with and without the incorporation of selective agents, were compared (Tilburg et al., 1999). Overall, best results were obtained with CHROMagar O157 from CHROMaga. Bennett et al. (1995) reported that replacing SMAC agar with CT-SMAC agar clearly increased the rate and ease of isolation of *E. coli* O157 from inoculated minced beef, both when plating directly after 24 h enrichment and when plating after performing IMS. Applying IMS for isolating *E. coli* O157 from artificially contaminated ground beef, Heckötter et al. (1997) found that on HC agar better results were observed after a 6 h enrichment and on CT-SMAC agar after a 24 h enrichment. The best results were obtained by applying both incubation times and both selective plating media. Hammack et al. (1997) studied the relative efficacies of HC agar and several formulations of SMAC agar, with and without MUG, in recovering unstressed and heat-stressed *E. coli* O157:H7 from different dairy products. HC agar and the SMAC agar formulations did not differ significantly in their ability to recover both stressed and unstressed *E. coli* O157:H7 from ice cream and whole milk, but HC agar was superior to the SMAC agars for recovery of unstressed *E. coli* O157:H7 from Brie cheeses (Hammack et al., 1997).

It is important to realise that by using culture methods based on sorbitol-fermentation and ß-D-glucuronidase-activity, several EHEC strains will be missed. Sorbitol-fermenting, ß-D-glucuronidase-positive EHEC strains of serotype O157:NM, for example, have been isolated from humans in several European countries (Gunzer et al., 1992; Feng, 1995; Herpay et al., 1997; Bielaszewska et al., 1998; Keskimaki et al., 1998). These strains will not be detected by using SMAC. Moreover, Hayes et al. (1995) reported the isolation from a patient with bloody diarrhoea of an atypical *E. coli* O157:H7 strain that did not ferment sorbitol but produced an active ß-D-glucuronidase. Some strains of sorbitol-fermenting *E. coli* O157:NM are very sensitive to tellurite and do not grow on CT-SMAC. Furthermore, in recent years an increasing number of outbreaks and sporadic cases related to EHEC, other than EHEC O157, have been reported in many countries (Goldwater and Bettelheim, 1996; Johnson et al., 1996), the majority of these strains being sorbitol-positive and having variable ß-D-glucuronidase reactions (Krishnan et al., 1987).

The occurrence of phenotypic variants of EHEC O157 and of other EHEC serotypes necessitates the use of techniques that detect virulence characteristics of EHEC, such as production of Shiga toxin (Stx), Stx encoding genes, or the *E. coli* attaching-and-effacing (*eae*) gene. The *eae* gene encodes intimin which is produced by all enteric pathogens that induce the characteristic attaching-and-effacing histopathology. However, a non-selective, but differential plating medium called enterohaemolysin or enterohaemorrhagic *E. coli* (EHEC) agar has been described, which is a variation of sheep blood agar (SBA) and covers more EHEC strains than EHEC O157 (Beutin et al., 1989). Sheep red blood cells are washed three times in phosphate buffered saline (PBS), resuspended to the original volume and added at 5% (v/v) to trypticase soy agar (TSA) supplemented with 10 mmol/l (final concentration) calcium chloride, hence these plates are also named WSBA-Ca. The plates are inoculated by streaking and incubated at 37°C. Nearly all (approx. 90%) EHEC O157 strains and a significant proportion of other EHEC strains (approx. 70%) produce enterohaemolysin (Beutin et al., 1989; Bettelheim, 1995). Enterohaemolytic *E. coli* are characterised on EHEC agar by small turbid zones of haemolysis around the colonies occurring after 18 to 24 h incubation at 37°C. Alpha-haemolytic *E. coli* form large, clear zones of haemolysis after only 3 to 6 h incubation.

The use of EHEC agar for rapidly detecting EHEC in human stool samples has been reported by Beutin et al. (1996). Hudson et al. (2000) recently evaluated this medium for its ability to recover EHEC strains of serotypes O157:H7, O26 and O113:H21 from foods following selective enrichment for 24 h. It was concluded that the use of selective enrichment and plating onto EHEC agar represents a useful addition to the methods currently available for detecting EHEC non-O157. The method worked best on cooked meat products and pasteurised milk where there was a low background flora. Selective enrichment followed by streaking onto EHEC agar was also useful for salami, but possibly less sensitive than for cooked meat products. The procedure was least useful for raw minced meat products. EHEC agar contains no selective agents, allowing overgrowth of EHEC by concomitant flora which may also produce haemolysins. The authors suggested that testing additional suspect colonies would increase the sensitivity of the method. It should also be possible to incorporate agents, such as novobiocin and cefsulodin, into EHEC agar that would assist in inhibiting the background flora that interferes with use of the medium as it is currently formulated. Lehmacher et al. (1998) recommended the use of blood agar supplemented with vancomycin, cefixime, and cefsulodin (BVCC) for isolation and presumptive identification of EHEC. They showed this medium to be superior to enterohaemolysin agar for the detection of haemolysis by EHEC and suggested that vancomycin facilitated the secretion of haemolysin through an increase in the permeability of the cell wall. Sixteen EHEC strains of the seven predominant serovars in Germany were added separately to samples of ground beef and raw milk and after 6 h enrichment in mTSB (containing only 1.12 g/l bile salts no. 3 and 10 mg/l novobiocin) the EHEC strains were detected by culturing on BVCC. Even the smallest amounts of 8 of EHEC added (6.2 CFU per g) of ground beef and per ml of raw milk were detected. However, detection of EHEC was hampered in ground beef samples by haemolysis caused by *Serratia* spp. Enterohaemolysin agar supplemented with vancomycin has been successfully used for the isolation of EHEC from beef, pork meat and raw sausages (Pozzi et al, 1996; Peitz et al., 2000), but concentrations higher than 8 mg/l vancomycin gave rise to false-positive haemolysis with some bacteria (Peitz et al., 2000).

Resuscitation

The occurrence of heat-, freeze-, acid- or salt-stressed EHEC in foods makes it important to be able to detect cells that are in a stressed state, since EHEC generally have a very low infectious dose and injured cells mostly retain their pathogenic properties. The use of selective broths or agars for the direct enrichment or enumeration of *E. coli* O157:H7 in foods may lead to a reduced recovery and subsequently reduced multiplication of injured cells.

Many enrichment broths contain bile salts as a selective agent. These have been reported to be inhibitory to the growth of injured *E. coli* O157:H7 cells (Weagant et al., 1994; McCarthy et al., 1998; Stephens and Joynson, 1998; Blackburn and McCarthy, 2000). Blackburn and McCarthy (2000) evaluated several rapid methods for the detection of *E. coli* O157:H7 in food using beefburgers, parsley and fermented meat artificially contaminated with injured cells. Poor performance was observed from methods that used direct enrichment into broth containing selective agents, including bile salts, with or without an elevated incubation temperature. The incorporation of a non-selective pre-enrichment medium improved the detection rates of these assays by up to tenfold. Commercial enrichment systems are being developed that include a release of selective compounds after a repair period.

Similarly, SMAC agar has been shown to perform poorly in recovering heat-, freeze-, acid- and salt-stressed *E. coli* O157:H7 (Conner and Hall, 1994; Ahmed and Conner, 1995; McCarthy et al., 1998; Sage and Ingham, 1998; Blackburn and McCarthy, 2000; Riordan et al., 2000). Where levels of competitive bacteria are low, complete media such as TSA or phenol red sorbitol agar with MUG can be used (Conner and Hall, 1994). Czechowicz et al. (1996) recommended plate count agar supplemented with 1% sodium pyruvate for the recovery of thermally-stressed *E. coli* O157:H7, finding that the spread plate method gave greater recovery of heat-stressed cells than the pour plate method. Levine's eosin methylene blue agar modified by the addition of sorbitol and novobiocin (MEMB) was found a suitable alternative to SMAC agar for the recovery of *E. coli* O157:H7 from heated and dried meat samples (Harrison et al., 1998). Sage and Ingham (1998) showed that the hydrophobic grid membrane filter method using SD-39 agar was preferable to spread plating onto SMAC for enumerating *E. coli* O157:H7 cells that have survived freezing and thawing in apple cider. McCleery and Rowe (1995) found improved recoveries of injured cells after a resuscitation period (2 h at 25°C) on TSA before overlay with SMAC agar supplemented with MUG. They obtained maximum recovery by adding catalase to the TSA. Similarly, Riordan et al. (2000) reported that the use of TSA (2 h at 37°C)-SMAC allowed the recovery and enumeration of uninjured and sublethally injured *E. coli* O157:H7 cells heated in pepperoni. Blackburn and McCarthy (2000) found improved recoveries using a resuscitation stage on TSA (4 h at 37°C) followed by membrane transfer to SMAC. The membrane method was used to monitor the numbers of artificially contaminated *E. coli* O157:H7 during the fermentation of a meat product and demonstrated better survival when compared to counts on SMAC. In a previous study they demonstrated that the choice of membrane material was critical for maintaining the positive sorbitol colour change used to differentiate non-*E. coli* O157:H7 (McCarthy et al., 1998). Track-etched polycarbonate membranes allowed the typical colour reactions to be visualised, whereas cellulose acetate did not.

Probably the best approach for the effective recovery of stressed EHEC is non-selective pre-enrichment for at least 18 to 24 h, similar to the classical *Salmonella* detection method (Stephens and Joynson, 1998). As a pre-enrichment medium BPW, modified BPW with casamino acids (Restaino et al., 1996) or Universal Pre-enrichment Broth (Jiang et al., 1998) may be used.

Identification

Presumptive EHEC must be biochemically confirmed as *E. coli* by the conventional IMViC (indole, methyl red, Voges-Proskauer and citrate) tests or by commercial identification kits (API, Minitek, Enterotube). Presumptive identification of *E. coli* can be done by inoculation onto Levine's eosin methylene blue agar and observation for colonies with a metallic sheen. *Escherichia hermannii* is, like EHEC O157, sorbitol- and ß-D-glucuronidase-negative, but can be distinguished from EHEC O157 by three additional tests; growth in the presence of potassium cyanide, fermentation of cellobiose, and production of a yellow pigment. *E. coli* is negative and *E. hermannii* positive for these tests (Borczyk et al., 1987).

Several latex agglutination assays are commercially available for the rapid presumptive identification of *E. coli* O157. These tests consist of latex beads coated with antibodies that agglutinate O157 antigens and form an antigen-antibody complex visible as a precipitate. Determination of the H7 antigen is not essential for the presumptive identification of EHEC O157 strains. Some *E. coli* O157 strains possess the H7 flagellin but are non-motile and negative in H-serology. Nevertheless, these strains frequently produce Stx and are very similar to *E. coli* O157:H7 (Strockbine et al., 1998). Other strains possess H7 antigens but are of serogroups other than O157; these strains include both Stx-producing *E. coli* and *E. coli* strains not producing Stxs.

Most EHEC produce enterohaemolysin, a useful epidemiological marker for EHEC strains, which can easily be detected using washed sheep blood agar supplemented with calcium (Beutin et al., 1996).

Since strains of several species cross-react with O157 antiserum and some disease-causing EHEC strains may not demonstrate the enterohaemolytic phenotype, isolates should additionally be tested for Stx production or the presence of *stx* genes for which several ELISA tests have been developed and commercialised. The Verotox F™ test (Denka Seiken) is a rapid microplate latex agglutination test for detecting and characterisation of Stxs. Results with this test were obtained after 48 h and were consistent with PCR results, which were obtained after 26 to 28 h (Heuvelink et al., 1997).

Strains identified as EHEC (O157) can be further characterised by several methods, including both subtyping and fingerprinting techniques (Strockbine et al., 1998).

Concluding remarks

Cultural methods for the enrichment, isolation and confirmation of EHEC O157 are still evolving. Several selective enrichment media have been described of which mTSB and mEC+n seem to be the most appropriate. These minimally selective broths give a rather limited differential specificity favouring isolation of EHEC O157 as opposed to other Gram-negative bacteria. Incubation at 41–42°C further enhances selectivity. For the isolation of stressed EHEC O157 pre-enrichment in a non-selective broth is necessary. CT-SMAC agar is the most commonly used isolation medium for typical sorbitol-non-fermenting EHEC strains of serogroup O157. As some EHEC O157 are sensitive to tellurite and/or sorbitol-fermenting, the use of a second isolation medium such as one of the newer chromogenic media is recommended. Immunological detection systems have been developed which give a significant reduction of analysis time. These methods include latex agglutination tests, ELISAs, colony immunoblot assays, direct immunofluorescent filter techniques and immunocapture techniques. Both polyclonal and monoclonal antibodies specific for the O157 and H7 antigen are used in these methods. Many of these test systems are able to detect less than one EHEC O157 cell per g of raw meat after overnight enrichment. Presumptive results are available after just one day, but need to be confirmed with the isolation of the organisms. The primary use of these procedures is therefore to identify food samples that possibly contain EHEC O157. IMS following enrichment and spread plating of the concentrated target cells onto CT-SMAC agar appears to be the most sensitive and cost-effective method for the isolation of *E. coli* O157 from raw foods.

EHEC isolates from humans now comprise more than 100 different serovars. Because of the biochemical and serological diversity, detection of the major virulence factor Stx by cytotoxicity assays with Vero cells, Stx ELISA, or *stx* PCR is the method of choice for identifying EHEC. For simple detection of EHEC by culturing, EHEC (syn. WSBA-Ca) agar, which is based on the relationship between Stx and enterohaemolysin production may be used.

As yet, no culture methods have evolved for the isolation of strains of the other five groups of diarrhoeagenic *E. coli*. Isolation and differentiation of these organisms will best be accomplished by a combination of culture and molecular biological methods.

References

Ahmed, N.M. and Conner, D.E. (1995) Evaluation of various media for recovery of thermally-injured *Escherichia coli* O157:H7. J. Food Protect. 58, 357–360.

Bennett, A.R., MacPhee, S. and Betts, R.P. (1995) Evaluation of methods for the isolation and detection of *Escherichia coli* O157 in minced beef. Lett. Appl. Microbiol. 20, 375–379.

Bennett, A.R., MacPhee, S., and Betts, R.P. (1996) The isolation and detection of *Escherichia coli* O157 by use of immunomagnetic separation and immunoassay procedures. Lett. Appl. Microbiol. 22, 237–243.

Bettelheim, K.A. (1995) Identification of enterohaemorrhagic *Escherichia coli* by means of their production of enterohaemolysin. J. Appl. Bacteriol. 79, 178–180.

Bettelheim, K.A. (1998a) Reliability of CHROMagar® O157 for the detection of enterohaemorrhagic *Escherichia coli* (EHEC) O157 but not EHEC belonging to other serogroups. J. Appl. Microbiol. 85, 425–428.

Bettelheim, K.A. (1998b) Studies of *Escherichia coli* cultured on Rainbow™ Agar O157 with particular reference to enterohaemorrhagic *Escherichia coli* (EHEC). Microbiol. Immunol. 42, 265–269.

Beutin, L., Montenegro, M.A., Orskov, I., Orskov, F., Prada, J., Zimmermann, S. and Stephan, R. (1989) Close association of verotoxin (Shiga-like toxin) production with enterohemolysin production in strains of *Escherichia coli*. J. Clin. Microbiol. 27, 2559–2564.

Beutin, L., Zimmerman, S. and Gleier, K. (1996) Rapid detection and isolation of Shiga-like toxin (verocytotoxin)-producing *Escherichia coli* by direct testing of individual enterohemolytic colonies from washed sheep blood agar plates in the VTEC-RPLA assay. J. Clin. Microbiol. 34, 2812–2814.

Bielaszewska, M., Schmidt, H., Karmali, M.A., Khakhria, R., Janda, J., Blahova, K and Karch, H. (1998) Isolation and characterization of sorbitol-fermenting Shiga toxin (Verocytotoxin)-producing *Escherichia coli* O157:H- strains in the Czech Republic. J. Clin. Microbiol. 36, 2135–2137.

Blackburn, C.W. and McCarthy, J.D. (2000) Modifications to methods for the enumeration and detection of injured *Escherichia coli* O157:H7 in foods. Int. J. Food Microbiol. 55, 285–290.

Blais, B.W., Booth, R.A., Phillippe, L.M. and Yamazaki, H. (1997) Effect of temperature and agitation on enrichment of *Escherichia coli* O157:H7 in ground beef using modified EC broth with novobiocin. Int. J. Food Microbiol. 36, 221–225.

Bolton, F.J., Crozier, L. and Williamson, J.K. (1996) Isolation of *Escherichia coli* O157 from raw meat products. Lett. Appl. Microbiol. 23, 317–321.

Borczyk, A.A., Lior, H. and Ciebin, B. (1987) False-positive identification of *Escherichia coli* O157 in foods. Int. J. Food Microbiol. 4, 347–349.

Chapman, P.A., Siddons, C.A., Wright, D.J., Norman, P., Fox, J. and Crick, E. (1993) Cattle as a possible source of verocytotoxin-producing *Escherichia coli* O157 infections in man. Epidemiol. Infect. 111, 439–447.

Chapman, P.A., Siddons, C.A., Zadik, P.M. and Jewes, L. (1991) An improved selective medium for the isolation of *Escherichia coli* O157. J. Med. Microbiol. 35, 107–110.

Chapman, P.A., Wright, D.J. and Siddons, C.A. (1994) A comparison of immunomagnetic separation and direct culture for the isolation of verocytotoxin-producing *Escherichia coli* O157 from bovine faeces. J. Med. Microbiol. 40, 424–427.

Conner, D.E. and Hall, G.S. (1994) Efficacy of selected media for recovery of *Escherichia coli* O157:H7 from frozen chicken meat containing sodium chloride, sodium lactate or polyphosphate. Food Microbiol. 11, 337–344.

Czechowicz, S.M., Santos, O. and Zottola, E.A. (1996) Recovery of thermally-stressed *Escherichia coli* O157:H7 by media supplemented with pyruvate. Int. J. Food Microbiol. 33, 275–284.

Doyle, M.P. and Schoeni, J.L. (1987) Isolation of *Escherichia coli* O157:H7 from retail fresh meats and poultry. Appl. Environ. Microbiol. 53, 2394–2396.

Entis, P. (1998) Direct 24-hour presumptive enumeration of *Escherichia coli* O157:H7 in foods using Hydrophobic Grid Membrane Filter followed by serological confirmation: collaborative study. J. AOAC Int. 81, 403–418.

Feng, P. (1995) *Escherichia coli* serotype O157:H7: novel vehicles of infection and emergence of phenotypic variants. Emerg. Infect. Dis. 1, 47–52.

Fratamico, P.M., Schultz, F.J. and Buchanan, R.L. (1992) Rapid isolation of *Escherichia coli* O157:H7 from enrichment cultures of foods using an immunomagnetic separation method. Food Microbiol. 9, 105–113.

Fukushima, H. and Gomyoda, M. (1999a) Hydrochloric acid treatment for rapid recovery of Shiga toxin-producing *Escherichia coli* O26, O111 and O157 from faeces, food and environmental samples. Zbl. Bakteriol. 289, 285–299.

Fukushima, H. and Gomyoda, M. (1999b) An effective, rapid and simple method for isolation of Shiga toxin-producing *Escherichia coli* O26, O111 and O157 from faeces and food samples. Zbl. Bakteriol. 289, 415–428.

Fukushima, H., Hoshina, K. and Gomyoda, M. (2000) Selective isolation of *eae*-positive strains of Shiga toxin-producing *Escherichia coli*. J. Clin. Microbiol. 38, 1684–1687.

Goldwater, P.N. and Bettelheim, K.A. (1996) An outbreak of hemolytic uremic syndrome due to *Escherichia coli* O157:H-: Or was it? Emerg. Inf. Dis. 2, 153–154.

Gunzer, F., Böhm, H., Rüssmann, H., Bitzan, M., Aleksic, S. and Karch, H. (1992) Molecular detection of sorbitol-fermenting *Escherichia coli* O157 in patients with hemolytic-uremic syndrome. J. Clin. Microbiol. 30, 1807–1810.

Hammack, T.S., Feng, P., Amaguaña, R.M., June, G.A., Sherrod, P.S. and Andrews, W.H. (1997) Comparison of sorbitol MacConkey and hemorrhagic coli agars for recovery of *Escherichia coli* O157:H7 from brie, ice cream, and whole milk. J. AOAC Int. 80, 335–340.

Hara-Kudo, Y., Konuma, H., Nakagawa, H. and Kumagai, S. (2000) *Escherichia coli* O26 detection from foods using an enrichment procedure and an immunomagnetic separation method. Lett. Appl. Microbiol. 30, 151–154.

Hara-Kudo, Y., Onoue, Y., Konuma, H., Nakagawa, H. and Kumagai, S. (1999) Comparison of enrichment procedures for isolation of *Escherichia coli* O157:H7 from ground beef and radish sprouts. Int. J. Food Microbiol. 50, 211–214.

Harrison, J.A., Harrison, M.A. and Rose, R.A. (1998) Survival of *Escherichia coli* O157:H7 in ground beef jerky assessed on two plating media. J. Food Protect. 61, 11–13.

Hayes, P.S., Blom, K., Feng, P., Lewis, J., Strockbine, N.A. and Swaminathan, B. (1995) Isolation and characterization of a ß-D-glucuronidase-producing strain of *Escherichia coli* serotype O157: H7 in the United States. J. Clin. Microbiol. 33, 3347–3348.

Heckötter, S., Bülte, M. and Lücker, E. (1997) Detection of *Escherichia coli* serogroup O157 in foods by immunomagnetic separation (IMS). Arch. Lebensmittelhyg. 48, 85–87.

Herpay, M., Czirok, E., Gado, I. and Milch, H. (1997) Detection of verocytotoxin-producing *Escherichia coli* in Hungary, abstract V39/VI. *In*: 3rd International symposium and workshop on Shiga toxin (verotoxin)-producing *Escherichia coli* infections. Melville, N.Y., USA: Lois Joy Galler Foundation for Hemolytic-Uremic Syndrome Inc., 1997:95.

Heuvelink, A.E., van den Biggelaar, F.L., de Boer, E., Herbes, R.G., Melchers, W.J., Huis in 't Veld, J.H. and Monnens, L.A.H. (1998) Isolation and characterization of verocytotoxin-producing *Escherichia coli* O157 strains from Dutch cattle and sheep. J. Clin. Microbiol. 36, 878–882.

Heuvelink, A.E., Zwartkruis-Nahuis, J.T.M. and de Boer, E. (1997) Evaluation of media and test kits for the detection and isolation of *Escherichia coli* O157 from minced beef. J. Food Protect. 60, 817–824.

Hudson, J.A., Nicol, C., Capill, J. and Bennett, J. (2000) Isolation of Shiga toxin-producing *Escherichia coli* (STEC) from foods using EHEC agar. Lett. Appl. Microbiol. 30, 109–113.

Jiang, J., Larkin, C., Steele, M., Poppe, C. and Odumeru, J.A. (1998) Evaluation of universal preenrichment broth for the recovery of foodborne pathogens from milk and cheese. J. Dairy Sci. 81, 2798–2803.

Johnson, R.P., Clarke, R.C., Wilson, J.B., Read, S.C., Rahn, K., Renwick, S.A., Sandhu, K.A., Alves, D., Karmali, M.A., Lior, H., McEwan, S.A., Spika, J.S. and Gyles, C.L. (1996) Growing concerns and recent outbreaks involving non-O157:H7 serotypes of verotoxigenic *Escherichia coli*. J. Food Protect. 59, 1112–1122.

Johnson, J.L., Rose, B.E., Sharar, A.K., Ransom, G.M., Lattuada, C.P. and McNamara, A.M. (1995) Methods used for detection and recovery of *Escherichia coli* O157:H7 associated with a food-borne disease outbreak. J. Food Protect. 58, 597–603.

Keskimaki, M., Saari, M., Heiskanen, T. and Siitonen, A. (1998) Shiga toxin-producing *Escherichia coli* in Finland from 1990 through 1997: prevalence and characteristics of isolates. J. Clin. Microbiol. 36, 3641–3646.

Krishnan, C., Fitzgerald, V.A., Dakin, S.J. and Behme, R.J. (1987) Laboratory investigation of outbreak of hemorrhagic colitis caused by *Escherichia coli* O157:H7. J. Clin. Microbiol. 25, 1043–1047.

Kudva, I.T., Hatfield, P.G. and Hovde, C.J. (1996) *Escherichia coli* O157:H7 in microbial flora of sheep. J. Clin. Microbiol. 34, 431–433.

Lehmacher, A., Meier, H., Aleksic, S. and Bockemühl, J. (1998) Detection of hemolysin variants of Shiga toxin-producing *Escherichia coli* by PCR and culture on vancomycin-cefixime-cefsulodin blood agar. Appl. Environ. Microbiol. 64, 2449–2453.

March, S.B. and Ratnam, S. (1986) Sorbitol-MacConkey medium for detection of *Escherichia coli* O157:H7 associated with hemorrhagic colitis. J. Clin. Microbiol. 23, 869–872.

McCarthy, J., Holbrook, R. and Stephens, P.J. (1998) An improved direct plate method for the enumeration of stressed *Escherichia coli* O157:H7 from food. J. Food Protect. 61, 1093–1097.

McCleery, D.R. and Rowe, M.T. (1995) Development of a selective plating technique for the recovery of *Escherichia coli* O157:H7 after heat stress. Lett. Appl. Microbiol. 21, 252–256.

Nataro, J.P. and Kaper, J.B. (1998) Diarrheagenic *Escherichia coli* [published erratum in Clin Microbiol Rev 1998 Apr;11(2):403]. Clin. Microbiol. Rev. 11, 142–201.

Okrend, A.J.G., Rose, B.E. and Bennett, B. (1990a) A screening method for the isolation of *Escherichia coli* O157:H7 from ground beef. J. Food Protect. 53, 249–252.

Okrend, A.J.G., Rose, B.E. and Lattuada, C.P. (1990b) Use of 5-bromo-4-chloro-3-indoxyl-beta-D-glucuronide in MacConkey sorbitol agar to aid in the isolation of *Escherichia coli* O157:H7 from ground beef. J. Food Protect. 53, 941–943.

Padhye, N.V. and Doyle, M.P. (1991) Rapid procedure for detecting enterohemorrhagic *Escherichia coli* O157:H7 in food. Appl. Environ. Microbiol. 57, 2693–2698.

Peitz, R., Weber, H., Gleier, K., Zimmermann, S. and Beutin, L. (2000) Nachweis von enterohämorrhagischen *Escherichia coli* (EHEC) in fleischproben und rohwürsten. Fleischwirtsch. 3, 71–74.

Pozzi, W., Beutin, L. and Weber, H. (1996) Überleben und nachweis von enterohäemorrhagischen *Escherichia coli* in streichfähiger rohwurst. Fleischwirtsch. 76, 1300–1311.

Raghubeer, E.V. and Matches, J.R. (1990) Temperature range for growth of *Escherichia coli* serotype O157:H7 and selected coliforms in *E. coli* medium. J. Clin. Microbiol. 28, 803–805.

Reinders, R.D., Bijker, P.G.H., Huis in 't Veld, J.H.J. and van Knapen, F. (2000) Use of 8-hydroxyquinoline-ß-D-glucuronide for presumptive identification of Shiga toxin-producing *Escherichia coli* O157. Lett. Appl. Microbiol. 30, 411–414.

Restaino, L., Castillo, H.J., Stewart, D. and Tortorello, M.L. (1996) Antibody-direct epifluorescent filter technique and immunomagnetic separation for 10-h screening and 24-h confirmation of *Escherichia coli* O157:H7 in beef. J. Food Protect. 59, 1072–1075.

Riordan, D.C.R., Duffy, G., Sheridan, J.J., Whiting, R.C., Blair, I.S. and McDowell, D.A. (2000) Effects of acid adaptation, product pH, and heating on survival of *Escherichia coli* O157:H7 in pepperoni. Appl. Environ. Microbiol. 66, 1726–1729.

Sage, J.R. and Ingham, S.C. (1998) Evaluating survival of *Escherichia coli* O157:H7 in frozen and thawed apple cider: potential use of a hydrophobic grid membrane filter–SD-39 agar method. J. Food Protect. 61, 490–494.

Seo, K.H., Brackett, R.E. and Frank, J.F. (1998) Rapid detection of *Escherichia coli* O157:H7 using immunomagnetic flow cytometry in ground beef, apple juice, and milk. Int. J. Food Microbiol. 44, 115–123.

Stein, K.O. and Bochner, B.R. (1998) Tellurite and novobiocin improve recovery of *E. coli* O157 on Rainbow® Agar O157. 98th Annual Meeting of the American Society for Microbiology, P-80.

Stephens, P.J. and Joynson, J.A. (1998) Direct inoculation into media containing bile salts and antibiotics is unsuitable for the detection of acid/salt stressed *Escherichia coli* O157:H7. Lett. Appl. Microbiol. 27, 147–151.

Strockbine, N.A., Wells, J.G., Bopp, C.A. and Barrett, T.J. (1998) Overview of detection and subtyping methods. *In: Escherichia coli* O157:H7 and other Shiga toxin-producing *E. coli.* Kaper, J.B. and O'Brien A.D. (Eds.). American Society for Microbiology, Washington, D.C. pp. 331–356.

Szabo, R.A., Todd, E.C.D. and Jean, A. (1986) Method to isolate *Escherichia coli* O157:H7 from food. J. Food Protect. 49, 768–772.

Tilburg, J.J.H.C., Creemers, O., van der Zee, H., Heuvelink, A.E. and de Boer, E. (1999) Comparison of methods for the detection and isolation of *Escherichia coli* O157 from foods. Abstract. *In:* Food microbiology and food safety into the next millennium; proceedings of the 17th international conference of the International Committee on Food Microbiology and Hygiene, Veldhoven, The Netherlands, 13–17 September, 1999. Tuijtelaars, A.C.J., Samson, R.A., Rombouts, F.M. and Notermans, S. eds. Ponsen & Looyen, Wageningen, The Netherlands. pp. 601.

Tomoyasu, T. (1998) Improvement of the immunomagnetic separation method selective for *Escherichia coli* O157 strains. Appl. Environ. Microbiol. 64, 376–382.

US Food and Drug Administration (1995) Bacteriological Analytical Manual (BAM), 8th edition, AOAC International, Gaithersburg, MD.

Vernozy-Rozand, C., Feng, P., Montet, M.P., Ray-Gueniot, S., Villard, L., Bavai, C., Meyrand, A., Mazuy, C. and Atrache, V. (2000) Detection of *Escherichia coli* O157:H7 in heifers' faecal samples using an automated immunoconcentration. Lett. Appl. Microbiol. 30, 217–222.

Vernozy-Rozand, C., Mazuy, C., Ray-Gueniot, S., Boutrand-Loeï, S., Meyrand, A. and Richard, Y. (1998) Evaluation of the VIDAS methodology for detection of *Escherichia coli* O157 in food samples. J. Food Protect. 61, 917–920.

Wallace, J.S. and Jones, K. (1996) The use of selective and differential agars in the isolation of *Escherichia coli* O157 from dairy herds. J. Appl. Bacteriol. 81, 663–668.

Weagant, S.D., Bryant, J.L. and Bark, D.H. (1994) Survival of *Escherichia coli* O157:H7 in mayonnaise and mayonnaise-based sauces at room and refrigerated temperatures. J. Food Protect. 57, 629–631.

Weagant, S.D., Bryant, J.L. and Jinneman, K.G. (1995) An improved rapid technique for isolation of *Escherichia coli* O157:H7 from foods. J. Food Protect. 58, 7–12.

Wright, D.J., Chapman, P.A. and Siddons, C.A. (1994) Immunomagnetic separation as a sensitive method for isolating *Escherichia coli* O157 from food samples. Epidemiol. Infect. 113, 31–39.

Zadik, P.M., Chapman, P.A. and Siddons, C.A. (1993) Use of tellurite for the selection of verocytotoxigenic *Escherichia coli* O157. J. Med. Microbiol. 39, 155–158.

Handbook of Culture Media for Food Microbiology, J.E.L. Corry et al. (Eds.)

Chapter 17

Culture media for the isolation and enumeration of pathogenic *Vibrio* species in foods and environmental samples

James D. Oliver

Department of Biology, University of North Carolina at Charlotte, Charlotte, NC 28223, USA

The genus *Vibrio* contains over 20 described species, of which 12 are known human pathogens, and of these, eight are food-associated. The vibrios are normal microflora in estuarine waters, and thus occur in high numbers in seafood. *V. cholerae, V. parahaemolyticus,* and *V. vulnificus* are the most important vibrios worldwide, causing diseases ranging from mild gastroenteritis to fatal infections. A large number of media have been developed over the last 40 years, both for the selective enrichment and for the isolation of these pathogens. Of these, alkaline peptone water (APW) and thiosulphate citrate bile salts sucrose agar are the most widely employed for the enrichment and isolation of *V. cholerae* and *V. parahaemolyticus*. For *V. vulnificus*, cellobiose polymyxin colistin agar is the most widely used selective medium. These media, numerous others which have been described, their modes of action, their use, and typical results, are the topic of this review.

Introduction

At least twelve *Vibrio* spp. have been described which are human pathogens. Of these, eight are known to be directly food-associated. These include *V. alginolyticus, V. fluvialis, V. furnissii, V. hollisae* and *V. mimicus*. Of greatest significance, however, are *V. cholerae, V. parahaemolyticus,* and *V. vulnificus*, and culture media employed for these three pathogens are emphasized in this chapter.

The occurrence of pathogenic vibrios in foods

Vibrios exist as a major component of the normal microflora of marine and estuarine waters, and thus occur in high numbers in seafood. *V. cholerae* is also found with some vegetables, including cooked rice, that become contaminated with sewage-contaminated water. As a result, not only foods but the potential sources (typically seawater and contaminated freshwater) of these pathogens need to be examined for their presence. The occurrence, identification, epidemiology, incidence of infections, susceptibility to physical and chemical treatments, and virulence mechanisms of the eight food-associated vibrios have recently been reviewed (Oliver and Kaper, 1998).

A few studies on the occurrence of pathogenic vibrios in foods are summarized here to illustrate the significance of this genus. Lowry et al. (1989) reported 100% of the raw oysters from Louisiana they tested contained *V. parahaemolyticus* and 67% harboured *V. vulnificus*. In a study of frozen raw shrimp from Mexico, China and Ecuador, 63% were reported to contain *Vibrio* spp., including *V. vulnificus* and *V. parahaemolyticus* (Berry et al., 1994). In a comprehensive study of pathogenic vibrios in tropical oysters, Matté et al. (1994) reported the presence of *V. alginolyticus* (81% of samples), *V. parahaemolyticus* (77%), *V. cholerae* non-01 (31%), *V. fluvialis* (27%), *V. furnissii* (19%), *V. mimicus* (12%), and *V. vulnificus* (12%). As vibrios typically prefer warm waters as their natural reservoir, their occurrence in waters in northern Europe is less common, and this probably accounts for the decreased incidence of the pathogenic vibrios in foods in these countries (see e.g. Dalsgaard et al., 1996 and references therein).

V. cholerae

V. cholerae 01 is the causative agent of cholera, and one of the few food-borne pathogens capable of producing epidemic and pandemic outbreaks. While the majority of cases involve mild diarrhoea or are even asymptomatic, approximately 11% of patients develop the classic infection involving explosive diarrhoea. The resulting massive loss of fluid (500–1000 ml/h) leads to dehydration, tachycardia, hypotension, vascular collapse, and ultimately death. Essential to infection is the production of the cholera enterotoxin. While there are exceptions, production of the cholera toxin is limited to cells of the O1 and (more recently) the O139 serogroups; non-O1 serogroups generally do not produce the cholera toxin. The biology of this species has been recently reviewed by Kaper et al. (1995), and the role of food in cholera transmission by Kaysner and Hill (1994). Foods implicated in the spread of *V. cholerae* include seafood (crabs, shrimp, raw fish, mussels, cockles, squid, oysters, clams), rice, raw pork, street vendor food, frozen coconut milk, and raw fruits and vegetables (Oliver and Kaper, 1998), with greater survival occurring in cooked foods (Mintz et al., 1994).

V. parahaemolyticus

Numerous food-borne outbreaks of *V. parahaemolyticus* have been reported worldwide since this pathogen was first described in 1950. In Japan, up to 70% of all bacterial food-borne disease is caused by this species. Gastroenteritis is exclusively associated with seafood that is consumed raw or undercooked. Primarily implicated are raw fish (in Japan), crab, shrimp, lobster, and oysters. The largest outbreak in the United States occurred in 1978 and ultimately affected 1133 people. Boiled shrimp that had been returned into the original shipping boxes after cooking, and held for over 7 h in an unrefrigerated truck were implicated (Oliver and Kaper, 1998). Symptoms (typically beginning 16–24 h after ingestion and lasting 3–7 days) include diarrhoea, abdominal cramps, nausea and vomiting, and fever. As in *V. cholerae*, only certain strains of *V. parahaemolyticus* are involved in disease production, and these are cells that produce the so-called 'Kanagawa haemolysin'. Interestingly, whereas virtually all clinical isolates produce this toxin, only about 1% of environmental strains are Kanagawa positive (KP^+). It is believed that selection of KP^+ cells occurs in the intestinal tract, and that this accounts for the predominance of this type in stool samples (Oliver and Kaper, 1998).

V. vulnificus

While accounting for only about 1% of all food-borne infections in the United States, this species has the second highest rates of hospitalization and highest case fatality of all food-borne pathogens in that country (Mead et al., 1999). Indeed, this single species is responsible for 95% of all seafood-borne deaths in the United States, with a fatality rate exceeding 60% (Oliver and Kaper, 1998). *V. vulnificus* occurs as part of the normal microflora of warm estuarine waters worldwide, but occurs in especially high numbers in filter-feeding bivalve molluscs (oysters, clams, and mussels). When consumed raw or undercooked, these are the primary source of the infection. Unlike *V. cholerae* and *V. parahaemolyticus*, no outbreaks of *V. vulnificus* have been reported. This is probably because cases of this disease are largely restricted to persons who have an underlying chronic disease (Oliver, 1989), the most common being liver-related disorders such as alcohol-induced cirrhosis. An unusual aspect of the at-risk group is that over 80% of cases occur in males whose average age exceeds 50 years. Symptoms appear a median of 26 h after infection, with fever, chills, nausea, and hypotension being the most common (Oliver and Kaper, 1998). An unusual symptom occurring in most cases is the development of secondary lesions on the extremities. In fatal infections, death typically ensues within a few days. *V. vulnificus* is also able to produce potentially fatal infections when introduced into a wound (typically acquired when removing shrimp shells, stepping on a crab or oyster, or via a finfish puncture). Such infections generally occur in healthy persons, and carry a fatality rate of about 25% (Oliver, 1989).

Other vibrios

Along with the species listed above, seafood-associated infections due to *V. mimicus, V. fluvialis, V. furnissii, V. hollisae*, and *V. alginolyticus* have also been reported. All of these vibrios cause disease whose primary symptoms are similar to those produced by *V. cholerae*, although of a milder nature. For a recent review, see Oliver and Kaper (1998).

The need for selective media for the enrichment and isolation of vibrios

While vibrios typically comprise the major bacterial genus in estuarine waters (the source of many of the seafoods harvested worldwide), there are numerous competing bacteria which must be selected against in order to adequately characterize the vibrio species present in a sample. In addition, while vibrios such as *V. parahaemolyticus*, *V. alginolyticus*, and *V. vulnificus* may constitute a large population in seafoods, those

Table 1
Modes of action of selective agents in media for isolation of vibrios[a].

Agent	Mode of Action
Bile salts (including taurocholate)	inhibit Gram-positive bacteria, some Gram-negative other than the enterics
Tellurite	inhibits some Gram-positive bacteria
Polymyxins (including colistin)	bactericidal towards many Gram-negative bacteria
pH	high pH values select for vibrios (optimum 8.4–8.6)
Salts	NaCl (0.5–1%) required by most vibrios; higher levels inhibit many Gram-negative bacteria
Sodium lauryl (dodecyl) sulphate	membrane solubilizer; bactericidal for many bacteria

[a]Adapted and extended from Donovan and van Netten (1995).

such as *V. cholerae* generally occur at low numbers, and their presence is not easily determined unless the cells are allowed to grow preferentially over other naturally occurring bacterial groups. For this reason, enrichment and plating media have been described which take advantage of the resistance typically exhibited by vibrios to compounds such as bile salts, tellurite and certain antibiotics, as well as to elevated salt concentrations and pH values. The latter considerations are based on the fact that the pathogenic vibrios have a preference for alkaline conditions and, with the exception of *V. cholerae*, are all halophilic. These attributes are the basis for their frequent employment in media for the selection of vibrios, whether obtained from clinical, environmental, or food samples.

Many media, both enrichment broths and plating agars, have been described over the last 40 years. For the most part, these have not been accepted by clinical or research investigators, the food industry, or government officials. This lack of acceptance is largely due to a lack of success of the various media for selectively enriching the various *Vibrio* spp. from the competing microflora, and the phenotypic similarity between many *Vibrio* spp. In the following sections, I have tried to mention most of these media, although this is primarily of historical interest only. Those few media that are widely accepted by the scientific community are emphasized. The reader is encouraged to review the previous edition of this chapter (Donovan and van Netten, 1995) for historical aspects and discussions of the principles on which the various media are based.

Mode of action of agents employed in vibrio-selective media

The reagents employed in many of the media designed to be selective for *Vibrio* spp., along with their modes of action, are shown in Table 1.

Enrichment broths for vibrios

Whereas quantitative methods (MPN and direct plating) for the isolation of vibrios are often used, many investigators prefer a qualitative enrichment step prior to plating to various selective media. Of these, only alkaline peptone water (APW) has been widely accepted. This broth, which has been employed since 1887 (Donovan and van Netten, 1995) takes advantage of the fact that vibrios thrive in moderately alkaline environments. While many variations have been suggested during that time (Donovan and van Netten, 1995, and see below), APW (1% tryptone peptone and 1% NaCl at a final pH of 8.6) has been, and continues to be, most commonly employed for the isolation of vibrios from foods and other natural sources. In one study, Sloan et al. (1992) compared the effectiveness of APW with four other enrichment broths, and found APW to provide the greatest detection of *V. vulnificus* (see below).

Some investigators have employed additions to this basic broth in attempts to enrich various *Vibrio* spp. (Table 2). Among these, NaCl additions have been the most common. For example, Arias et al. (1998) recommended the use of APW with 3% NaCl (APWS) for the isolation of *V. vulnificus*. Recently, Høi et al. (1998a) proposed the addition of 2×10^4 U/liter polymyxin B (APWP) for the enrichment of *V. vulnificus*. While such additions may be of value in the isolation of certain species, other laboratories have yet to confirm these modifications. None of the suggested additions, however, have proven to be of value (or to be adopted) in the case of *V. cholerae* or *V. parahaemolyticus*.

V. cholerae

The American Public Health Association (APHA) analytical method (Clesceri et al., 1998) for *V. cholerae* recommends a 6–8 h enrichment in APW at 35°C followed by isolation on TCBS agar (see also Madden et al., 1989 for a detailed description). This simple enrichment is used worldwide, and while several modifications or alternate enrichments have been proposed, no other enrichment medium has found acceptance among investigators or government authorities.

Some proposed enrichments, such as gelatin-phosphate-saline (GPS; Madden et al., 1989) have been shown to be of no value for the selective enrichment of *V. cholerae* (Spira, 1984). Others, such as Monsur's taurocholate tellurite peptone (TTP), have been found by some investigators to be as good or better than APW in supporting growth of *V. cholerae* (Furniss et al., 1978), but have not been adopted by investigators. Similarly, starch gelatin polymyxin broth (SGP), developed by Kitaura et al. (1983) for the isolation of *V. cholerae*, has not been accepted by the research community. However, one simple modification of the APW enrichment appears to warrant consideration. An elevated temperature (42°C) enrichment method for recovery of *V. cholerae* from oysters was found by DePaola et al. (1988) to give a significantly ($p<0.05$) higher recovery and a greater specificity ($P<0.01$) than the standard enrichment protocol at 35°C.

Table 2
Enrichment broths described for vibrios[a].

	Vibrio sp.[b]	NaCl	pH	Carbohydrate	Selective agent(s)	Reference
Alkaline peptone water (APW)	all	1%	8.6	–	pH	Furniss et al., 1978
Taurocholate tellurite peptone (TTP)	Vc	1%	9.2	–	taurocholate, tellurite, pH	Monsur, 1963
Gelatin phosphate saline (GPS)	Vc	1%	ND[c]	–	–	Madden et al., 1989
Salt polymyxin broth (SPB)	Vp	2%	7.4	–	polymyxin	Sakazaki, 1973 (see also Karunasagar et al., 1986)
Salt colistin broth (SCB)	Vp	2%	7.4	–	colistin	Sakazaki, 1973
Glucose salt teepol broth (GSTB)	Vp	3%	9.4	glucose	methyl violet, teepol, salt, pH	Akiyama et al., 1963.
Horie's arabinose ethyl violet broth (HAE)	Vp	3%	9.0	arabinose	ethyl violet, pH	Horie et al., 1964
Alkaline peptone water salt (APWS)	Vv	3%	8.6	–	pH, salt	Arias et al., 1998
Alkaline peptone water polymyxin (APWP)	Vv	1%	8.6	–	polymyxin	Høi et al., 1998a
Starch gelatin polymyxin (SGP)	Vv, Vc	2%[d]	7.6	starch	polymyxin	Kitaura et al., 1983
Peptone saline cellobiose (PNC)	Vv	1.0	8.0	cellobiose	pH	Hsu et al., 1998
Vibrio fluvialis medium (FEM)	Vf	4.0	8.5	–	novobiocin	Nishibuchi et al., 1983

[a] Adapted and expanded from Donovan and van Netten (1995).
[b] Vc = *V. cholerae*, Vp = *V. parahaemolyticus*, Vv = *V. vulnificus*, Vf = *V. fluvialis*.
[c] Not described.
[d] Reduced to 0.5% for *V. cholerae*.

V. parahaemolyticus

As with *V. cholerae*, a number of enrichment broths, including glucose salt teepol broth (GSTB), salt polymyxin broth (SPB), and salt colistin broth (SCB), have been proposed for the enrichment of this pathogen (see Joseph et al., 1982, Karunasagar et al., 1986, and Twedt, 1989 for reviews). To date, however, none has been found to be preferable over APW. While SPB and SCB were reported by Nakanishi and Murase (1974) to be of value in the enrichment of *V. parahaemolyticus* from raw fish, Karunasagar et al. (1986) found direct plating to TCBS to yield better recovery than the MPN technique using GSTB, SPB, or two other marine broths. Further, Hagen et al. (1994) compared SPB and APW for the isolation of *V. parahaemolyticus* from crab, oysters, shrimp, lobster, and shark, and found APW to be significantly more efficient than SPB. Similarly, APW was found by Eyles et al. (1985) to be the most effective for isolating this species from oysters and prawns. Hagen et al. (1994) also found APW to be more effective for the isolation of *V. parahaemolyticus* from seafood samples which had been refrigerated at 2–4°C for up to 7 days, or frozen at –15°C for up to 28 days.

An interesting method for enriching and quantifying *V. parahaemolyticus* in foods has been described by Miyamoto et al. (1990). They employed an arabinose glucuronate enrichment medium in which they incubated samples at 37°C overnight. They then examined the trypsin-like activity of the bacteria, measured by fluorescence with the fluorogenic substrate benzoyl-L-arginine-7-aminomethyl-coumarin. They reported that, even in the presence of $>10^5$ cells of *V. alginolyticus*, 20 cells of *V. parahaemolyticus* could be detected after 6 h. In a study of 14 seafoods (50 different samples), they found a 0.95 correlation after a 6 h incubation when comparing this method with a conventional assay (bromothymol blue teepol agar and the MPN method). The presence of 10 cells of *V. parahaemolyticus* could be detected in the seafoods after a 10 h detection period. No fluorogenic activity was detected with eight other *Vibrio* spp., or members of *Pseudomonas, Bacillus, Staphylococcus, Aeromonas*, or a variety of Enterobacteriaceae.

V. vulnificus

Until recently, the United States FDA recommended the use of glucose salt teepol broth (GST) as an enrichment for *V. vulnificus* in an MPN procedure, with positive tubes being streaked onto TCBS agar. However, teepol is no longer commercially available, and other enrichment broths, primarily APW, have been proposed.

Along with their study on *V. parahaemolyticus*, Hagen et al. (1994) compared SPB and APW for the isolation of *V. vulnificus* from a variety of fish and shellfish, and found APW to be significantly more efficient than SPB. As they also observed with *V. parahaemolyticus*, APW was also found to be superior when the seafood samples had been cold stressed. In a more comprehensive study, Sloan et al. (1992) compared five selective enrichment broths for the isolation of *V. vulnificus* from seeded oysters. APW, Marine broth (Difco), Horie's arabinose-ethyl violet broth (HAE; Horie et al., 1964), Monsur's TTP broth (Monsur, 1963; Kaysner et al., 1987), and glucose salt teepol broth (GSTB; U.S. Food and Drug Administration, 1984) were examined. They found that APW and marine broths yielded significantly higher recovery than the others, with APW being the most successful. Similarly, Kaysner et al. (1989) found that an 18 h MPN enrichment step in APW gave higher recovery levels of *V. vulnificus* from two species of oysters than did direct plating to TCBS or CPC (see below for a discussion of this medium).

Hsu et al. (1998) recently described PNC enrichment broth for *V. vulnificus*, containing 5% peptone, 1% NaCl, and 0.08% cellobiose, the concentrations of which were said to be optimized for this species. A further modification of this broth (PNCC) included 1.0–4.1 U of colistin methanesulfonate per ml, which the authors found increased the growth of low levels of *V. vulnificus* while suppressing non-target bacteria. To date, this medium has not been tested outside the laboratory.

In a study examining vibrios in shellfish from coastal waters of Spain, Arias et al. (1998) concluded that the best combination of methods was enrichment in APWS (APW with 3% NaCl) for three hours at 40°C followed by plating onto CPC agar.

Dalsgaard and Høi (1997) and Høi et al. (1998 a,b) described the addition of polymyxin B (2.0×10^4 U) to APW to give APWP. Following incubation at 37°C for 18–24 h, a loopful of the surface pellicle was streaked onto CPC agar (or one of its modified versions; see below) which was then incubated for an additional 18–24 h at 40°C. These authors have employed this modified enrichment for the isolation of *V. vulnificus* from Danish mussels and from shrimp products imported into Denmark, noting that this amendment appears to be of value in the pre-enrichment of samples that contain large numbers of bacteria capable of growing under alkaline conditions.

Other vibrios

Nishibuchi et al. (1983) described FEM (*Vibrio fluvialis* enrichment medium), containing novobiocin as an inhibitor, for the enrichment and enumeration of this species from environmental sources. In a test of 177 samples (including crabs), they reported that FEM was more effective than APW in enriching this species, particularly from samples taken from waters of low (<6%) salinity.

Plating media for vibrios

A variety of plating media has been proposed for the isolation of vibrios from clinical and environmental sources (Table 3), and these are reviewed in Donovan and van Netten (1995). Of these, however, only thiosulphate citrate bile salts sucrose (TCBS; Kobayashi et al., 1963) is routinely employed. All of the pathogenic vibrios, with the exception of *V. hollisae*, grow on this medium, which depends on ox bile and an alkaline pH to suppress growth of other bacteria. *V. damsela* does not grow on TCBS when incubated at 37°C, but will grow at lower temperatures (e.g. 28°C; see e.g. Song et al., 1993). Differentiation between *Vibrio* spp. depends on their ability to ferment sucrose and turn the bromothymol blue indicator from green to yellow (Table 4). The medium should not be autoclaved, requiring only boiling for its preparation, and is available

Table 3
Selective agars for vibrios[a].

Medium	*Vibrio* sp.[b]	NaCl	pH	Carbohydrate	Selective agent(s)	Reference
Thiosulphate citrate bile salts sucrose (TCBS)	all except *V. hollisae*	1%	8.6	sucrose	bile salts	Kobayashi et al., 1963
Thiosulphate chloride iodide (TCI)	pathogenic vibrios	0.5	ND[c]	–	potassium iodide	Beazley and Palmer, 1992
Gelatin taurocholate tellurite agar (GTT)	Vc,Vp	1%	8.5	–	taurocholate, tellurite	Monsur, 1961 (modified by O'Brien and Colwell, 1985)
Polymyxin mannose tellurite (PMT)	Vc (O1/non-O1)	1%	8.4	mannose	polymyxin, tellurite, SDS	Shimada et al., 1990
Sucrose teepol tellurite (STT)	Vc	0%	8.0	sucrose	teepol, tellurite, lack of salt	Chatterjee et al., 1977
Bromothymol blue teepol agar (BTBT)	Vp	4%	7.8	sucrose	teepol, salt	Sakasaki, R., 1973
Vibrio parahaemolyticus (VP)	Vp	2%	8.6	sucrose	taurocholate, SDS	De et al., 1977
Trypticase soya agar triphenyltetrazolium (TSAT)	Vp,Va	2.5%	7.1	sucrose	bile salt, triphenyltetrazolium	Kourany, 1983
SDS polymyxin sucrose (SPS)	Vv,Vc	2%	7.6	sucrose	SDS, polymyxin	Kitaura et al., 1983
Cellobiose polymyxin B colistin (CPC)	Vv,Vc	2%	7.6	cellobiose	colisitin, polymyxin B,	Massad and Oliver, 1987 (modifications suggested by Tamplin et al., 1991, Høi et al., 1998b; Cerdà-Cuéllar et al., 2000)
Vibrio vulnificus agar (VV)	Vv	1%	8.6	salicin	tellurite, ox gall, crystal violet	Brayton et al., 1983
Vibrio vulnificus enumeration (VVE)	Vv	2%	8.5	cellobiose	ox gall, lactose, X-gal, taurocholate, tellurite	Micelli et al., 1993

[a] Adapted and extended from Donovan and van Netten (1995).
[b] Vc = *V. cholerae*, Vp = *V. parahaemolyticus*, Vv = *V. vulnificus*, Va = *V. alginolyticus*.
[c] Not described.

Table 4
Sucrose reactions of the pathogenic vibrios.

Sucrose +	*V. cholerae*	Sucrose -	*V. mimicus*
	V. metschnikovii[a]		*V. hollisae*
	V. cincinnatiensis[a]		*V. damsela*[b]
	V. fluvialis		*V. parahaemolyticus*
	V. furnissii		*V. vulnificus*
	V. alginolyticus		
	V. carchariae[b]		

[a] The route of infection of these pathogens has not been established.
[b] Infections appear to result solely from wound infections.

from commercial suppliers. As with all fermentable carbohydrate-containing media, one drawback of TCBS is the inability to perform the oxidase test (crucial for differentiating vibrios from the Enterobacteriaceae) directly on colonies. Variations in effectiveness of TCBS obtained from different suppliers has also been documented by several investigators (McCormack et al., 1974; Morris, 1982). The major disadvantage of this medium, however, is its lack of selectivity and differentiation among vibrios. A number of non-vibrios are known to grow on TCBS (McCormack et al., 1974; Morris et al., 1976; West et al., 1982; Lotz et al., 1983; Spira, 1984). Indeed, Lotz et al. (1983) found that 177 of 188 strains from 15 genera, including three Gram-positive genera, were able to grow on TCBS. These included *Acinetobacter, Aeromonas* (47 of 47 strains tested), *Alcaligenes, Enterobacter*, *Escherichia coli* (5 of 7 strains), *Pasteurella, Pseudomonas* spp. (12 of 12 strains), *Salmonella*, and *Proteus*. Some non-vibrios produce black colonies (due to FeS precipitation as a result of H_2S production from thiols).

On the other hand, TCBS appears to be the best medium currently available for the isolation of vibrios (Morris et al., 1979; Rennels et al., 1980; Karunasager et al., 1986). In a study of several media designed for the isolation of vibrios, Bolinches et al. (1988) found TSAT (trypticase soy agar sucrose bile triphenyltetrazolium; Kourany, 1983), TTGA (modified gelatin taurocholate tellurite agar; O'Brien and Colwell, 1985), GS (glucose-salts; Simidu and Tsukamoto, 1980), and GSTC (glucose salt tellurite crystal violet; Bolinches et al., 1988) agars recovered only 55, 13, 12, and 0%, respectively, of the vibrios present in three estuarine water samples. In contrast, their study reported TCBS provided an 83% recovery of vibrios from these same natural samples. To cite just two examples of its usage, Barbieri et al. (1999) employed TCBS (following enrichment in APW) to examine estuarine waters along the Italian Adriatic coast for the presence of pathogenic vibrios. They readily isolated *V. cholerae*, *V. parahaemolyticus*, *V. vulnificus*, and *V. alginolyticus*. Similarly, Hanharan et al. (1995) employed TCBS to characterize the microbial flora of mussels and oysters from Canada. They isolated *V. parahaemolyticus*, *V. alginolyticus*, *V. vulnificus*, *V. damsela*, *V. cholerae* (non-01), and *V. metschnikovii*, along with eight additional species of non-pathogenic vibrios. Such studies clearly indicate that this medium is capable of cultivating a variety of *Vibrio* spp. from natural waters and shellfish. Indeed, the medium has been employed extensively in characterizing the vibrio flora of seafood (e.g. see Kelly, 1982; Oliver et al., 1982; Oliver et al., 1983; West et al., 1982; Karunasagar et al., 1986).

Thiosulphate chloride iodide (TCI) agar, which employs potassium iodide in lieu of bile as an inhibitory agent, was described for the isolation of pathogenic *Vibrio* spp. by Beazley and Palmer (1992). These authors reported that the plating efficiency of five *Vibrio* spp. was superior to that of TCBS, and because it contains no fermentable carbohydrate, colonies can be directly tested for oxidase activity. In a more extensive examination of TCI, Abbott et al. (1993) reported that 102 strains of vibrios grew well on this medium. Of interest is that none of the *Aeromonas* spp. (14 strains), *Plesiomonas shigelloides* (10 strains), or three species of non-pathogenic vibrios grew on TCI agar. On the negative side, extremely poor (<1%) plating efficiencies were reported for *V. vulnificus* and *V. damsela*, and most of the pathogenic vibrios developed into relatively small (typically <3 mm) colonies. We (Pfeffer and Oliver, unpublished) recently compared this medium to TCBS in isolating vibrios from estuarine water samples. We found that 61% of the colonies developing on TCBS could be presumptively identified as being Vibrio spp., whereas only 46% were so identified on TCI agar. Further, over twice as many presumptive vibrio colonies developed on TCBS compared to TCI. Thus, this medium does not appear to have great potential in isolating vibrios from environmental samples.

Plating media for V. cholerae

The primary plating medium of choice for *V. cholerae* continues to be TCBS, and this medium has been recommended by the World Health Organization for the identification of this species (Morris, 1982) from stool samples. While few organisms in clinical samples, other than vibrios, grow on this medium, numerous genera found in foodstuffs are capable of growth, and this has been discussed above. Being sucrose-positive, *V. cholerae* produces large yellow colonies on TCBS (Table 5).

A variety of solid media has been proposed for the cultivation of *V. cholerae*. Monsur (1961) developed gelatin taurocholate tellurite (GTT) agar, taking advantage of the resistance to tellurite, taurocholate, and high pH of this species. *V. cholerae* produces small, translucent colonies surrounded by a cloudy zone of gelatinase activity on this light brown medium (Morris, 1982; Donovan and van Netten, 1995). Morris et al. (1979) found this medium to be comparable to TCBS for the isolation and identification of this species.

Sucrose tellurite teepol (STT) agar (Chatterjee et al., 1977) also employs tellurite as a selective agent, but substitutes teepol for the bile salts found in TCBS. The latter are said to be the cause of the variation that is often reported in TCBS from different vendors (Donovan and van Netten, 1995). STT is also salt-free, which allows the preferential growth of *V. cholerae* over other vibrios which are more halophilic. Morris et al. (1979), however, found this medium to be the poorest in performance among the four plating media they tested.

Polymyxin mannose tellurite (PMT) agar was developed specifically to differentiate *V. cholerae* O1 from non-O1 strains (Shimada et al., 1990). The medium owes its selectivity to tellurite and polymyxin B. Its differentiating abilities depend on O1

Table 5
Colony size and appearance of the major pathogenic vibrios[a].

Medium	*Vibrio* spp.	Size (mm)	Appearance
TCBS[b]	*V. cholerae*	>5	yellow
	V. parahaemolyticus	>5	green
	V. vulnificus	3–5	green
	V. alginolyticus	>5	yellow
TCI	*V. cholerae*	<1–1.5	ND[c]
	V. parahaemolyticus	2.5–3.5	ND
	V. vulnificus	2	ND
	V. alginolyticus	3	ND
GTT	*V. cholerae*	2–3	transparent with surrounding halo and faint black centre
	V. parahaemolyticus	2–5	opaque grey with surrounding halo and black centre
	V. vulnificus	ND	
	V. alginolyticus	2–5	opaque grey with surrounding halo and black centre
PMT	*V. cholerae*	2–3	yellow or dark violet[d]
	V. parahaemolyticus	3–4	yellow
	V. vulnificus	1–2	yellow
	V. alginolyticus	ND	
SPS	*V. cholerae*	1–3	yellow with halo
	V. parahaemolyticus	2–3	purple-green
	V. vulnificus	1–3	purple green with halo
	V. alginolyticus	1–2	yellow
CPC	*V. cholerae*	2–3	purple surrounded by blue zone
	V. parahaemolyticus	NG[e]	
	V. vulnificus	2–3	flat, yellow with darker centre
	V. alginolyticus	NG	
	other *Vibrio* spp.	NG	
STT	*V. cholerae*	3–5	yellow
	V. parahaemolyticus	3-5	blue
	V. vulnificus	ND	
	V. alginolyticus	3-5	yellow
TSAT	*V. cholerae*	ND	
	V. parahaemolyticus	2–4	red
	V. vulnificus	ND	
	V. alginolyticus	1–3	white (possible pink centre)
VV	*V. cholerae*	NG	
	V. parahaemolyticus	V[f]	
	V. vulnificus	2–4	light grey, translucent with black centre
	V. alginolyticus	ND	
VVE	*V. vulnificus*	ND	blue to greenish-blue
	other *Vibrio* spp.	NG	

[a] Adapted and expanded from Donovan and van Netten (1995).
[b] For a more complete list of the various pathogenic vibrios on TCBS agar, see Table 4.
[c] No description.
[d] Mannose-fermenting strains are yellow; mannose non-fermenting strains are violet.
[e] No growth.
[f] Variations in size and appearance.

strains fermenting the mannose in the medium, resulting in yellow colonies, whereas non-O1 strains yield dark violet colonies. However, mannose fermentation has proven to be a poor indicator of serogroups, and this medium has not found acceptance for the isolation of *V. cholerae*.

VP agar, originally developed by De et al. (1977) for the isolation of *V. parahaemolyticus*, contains the selective agents sodium taurocholate and sodium lauryl sulphate. The medium was reported to have value in isolating *V. cholerae*, but Morris et al. (1979), comparing this medium to several others for its ability to isolate *V. cholerae* from patients with cholera symptoms, reported that TCBS and GTT were superior to both VP agar and STT agar, based on the total number of positive specimens obtained.

Sodium dodecylsulphate polymyxin sucrose (SPS) agar, developed by Kitaura et al. (1983) and cellobiose polymyxin colistin (CPC) agar, developed by Massad and Oliver (1987), were both said to be of value in the isolation of *V. cholerae* as well as *V. vulnificus*. To date, however, these media have not been examined outside the laboratory for this purpose.

Plating media for V. parahaemolyticus

Many solid media have been proposed for the isolation of *V. parahaemolyticus*, but none has proven to be effective. Furniss et al. (1978) reported that Monsur's GTT, originally developed for the isolation of *V. cholerae*, was of value in isolating *V. parahaemolyticus*, but this has not been corroborated.

Trypticase soy agar triphenyltetrazolium (TSAT), developed by Kourany (1983), employs sucrose and triphenyltetrazolium to differentiate *V. parahaemolyticus* from *V. alginolyticus*. *V. parahaemolyticus* colonies are red on this medium, as compared to white colonies for the latter. Unfortunately, the medium does not appear to be very selective, and *Proteus* spp. grow with colonies similar to *V. parahaemolyticus*. *Escherichia* spp. and other Enterobacteriaceae also grow on the medium.

VP agar was developed by De et al. (1977) for the isolation of *V. parahaemolyticus*. However, when this medium was tested with samples of seawater, oysters and clams by Cleland et al. (1985), only 22% of the colonies developing on this medium could be identified as vibrios. Further, whereas 19.2% of the colonies developing on TCBS from these samples were subsequently identified as *V. parahaemolyticus*, only 4% of those on VP medium were identified as this species.

Plating media for V. vulnificus

V. vulnificus is the only *Vibrio* species for which there exists a widely adopted, selective medium. While many media have been proposed for the isolation of *V. vulnificus* (Table 3, and see Donovan and van Netten, 1995, for a historic treatment), the cellobiose polymyxin B colistin (CPC) agar described by Massad and Oliver (1987) is the only medium generally used for this purpose. The medium has also been adopted by the United States FDA for isolation of this bacterium. This medium takes advantage of the colistin and polymyxin B resistance of *V. vulnificus* to eliminate most other pathogenic vibrios, high temperature incubation (40°C) to eliminate many marine bacteria, and the fermentation of cellobiose as a differential agent. In the initial study (Massad and Oliver, 1987), 136 strains representing 19 species of *Vibrio*, as well as marine isolates of three other genera, were tested for growth on this medium. CPC agar was found to be highly selective for *V. vulnificus* (which ferments cellobiose and produces yellow colonies) and for *V. cholerae* (which does not ferment the sugar and produces purple colonies). Of the 79 strains of the remaining 17 species of *Vibrio*, only one out of nine *V. parahaemolyticus* strains tested grew on CPC agar. Similarly, no growth was observed for 17 strains of *Photobacterium, Pseudomonas,* or *Flavobacterium*. Subsequent field testing of this medium, involving direct plating of oyster and clam homogenates (followed by gene probe and monoclonal antibody technologies), confirmed the value of CPC agar, finding it superior to SPS agar and TCBS (Oliver et al., 1992). A later study (Sun and Oliver, 1995) compared CPC agar and VVE agar for isolation of *V. vulnificus* from 224 oysters, with gene probe hybridization used to confirm identification. Of over 3,500 cellobiose-positive colonies tested, 28.7% of those on CPC were identified as *V. vulnificus* on the basis of the probe whereas only 2.8% of 19,000 colonies developing on VVE agar could be identified as this species. When colony morphology (flat, with a darker central area) as well as colony colour was considered, 81.6% of over 1000 colonies developing on CPC agar proved to be *V. vulnificus*. Using these same criteria, Sloan et al. (1992) found 81% of the 'typical' *V. vulnificus* colonies on CPC to be identified as this species. The observation that only yellow colonies with a flat, darker yellow central area represent *V. vulnificus* is clearly very important in the use of this medium. Kaysner et al. (1989) examined *V. vulnificus* in shellstock and shucked oysters, and concluded that CPC gave the best recovery (following APW enrichment), and that the background microflora level on CPC agar was much lower than other selective media, and that in most instances *V. vulnificus* was in pure culture on CPC agar.

Several modifications of CPC agar have been proposed, primarily involving reductions or deletions of the antibiotics. Tamplin et al. (1991) described mCPC, with the colistin level lowered to 400,000U/l. This medium has been employed by DePaola et al. (1994) to isolate *V. vulnificus* from the intestines of fish from the U.S. Gulf Coast, by Parker et al. (1994) for examining the levels of *V. vulnificus* in frozen and vacuum-packaged oysters, and by Høi et al. (1998a) in a study on the occurrence of *V. vulnificus* in Danish mussels and fish. CC agar, recently described by Høi et al. (1998b), was reported to give a significantly higher isolation rate of *V. vulnificus* from water and sediment samples than did mCPC, and to have a statistically higher plating efficiency than TCBS agar. CC agar has the same composition as CPC agar, but with no polymyxin B and a slight reduction in colistin concentration. Cerdà-Cuéllar et al. (2000) recently described another medium, termed VVM agar, for the isolation of *V. vulnificus*. This medium, which contains a number of minor modifications from the original CPC, was combined with a 16S rDNA probe to isolate and identify this species. Like CPC, the medium appears to be selective for *V. vulnificus*, but several other *Vibrio* spp. (*V. campbellii, V. carchariae*, and *V. navarrensis*) also produced yellow colonies. Several other vibrios produce yellow colonies, as does *P. aeruginosa*. The medium has not been tested in the field or by other laboratories. Indeed, with all of these media, more studies are required to determine if any show sufficient improvement on the original to justify replacing CPC as the recommended plating medium for *V. vulnificus*.

Other media have been developed for the isolation of *V. vulnificus*. VV agar (Brayton et al., 1983) was tested by Dinnuzzo et al. (1984) and Cleland et al. (1985) who reported that only 9–10% of the colonies could be identified as being *V. vulnificus*. In another field study, Tilton and Ryan (1987) also found VV agar to be inadequate for the isolation of *V. vulnificus*. SPS (sodium dodecylsulphate polymyxin B sucrose), developed by Kitaura et al. (1983) is selective for *V. vulnificus* through its polymyxin B resistance, and is differential through the sulphatase activity of *V. vulnificus*. The latter, in the presence of sodium dodecylsulphate, results in haloes around colonies of this species. However, growth and halo production also occurs with *V. cholerae* and *V. anguillarum*. The medium was found by Kitaura et al. (1983) to isolate *V. vulnificus* from a variety of shellfish and Bryant et al. (1987) found it to be of value for the direct isolation of *V. vulnificus* from shellfish. Oliver et al. (1992), however, in comparing CPC to SDS and TCBS for their ability to select and differentiate *V. vulnificus* from background vibrios in shellfish, reported CPC to be superior to both these media. Using monoclonal antibody and gene probe technology, as well as classic taxonomic methods, to verify the identity of presumptive *V. vulnificus* colonies, these authors reported that approximately twice as many colonies on CPC agar could be identified as this species when compared to TCBS, with none of the sulphatase-positive colonies taken from SDS agar being so identified.

VVA agar, described by Wright et al. (1993), was designed to be non-selective, but to be used in combination with a *V. vulnificus*-specific gene probe ('VVAP'). However, in a study examining seawater, sediment, plankton and oysters for the presence of *V. vulnificus*, they reported that LB agar yielded higher counts than did TCBS agar, which yielded higher levels than did VVA.

VVE medium, described by Micelli et al. (1993) for the direct isolation of *V. vulnificus*, contains lactose as well as cellobiose, and several inhibitors (oxgall, sodium cholate, sodium taurocholate, and tellurite). A field test using VVE to determine the levels of *V. vulnificus* in oysters found 10^4 to $>10^6$ cells/100g oyster. A study by Sun and Oliver (1995) on 224 oysters, however, reported only 2.8% of 19,000 colonies developing on VVE agar could be identified as this species. The use of this medium has not been reported by other laboratories.

Plating media for other vibrios

TSAT was designed by Kourany (1983) to differentiate *V. parahaemolyticus* from *V. alginolyticus*, employing sucrose, bile salts, and triphenyltetrazolium chloride as selective and differential agents. The medium has not been used routinely for the isolation of *V. alginolyticus*.

Recommended culture media

The only medium that has proven of value for the culture of *Vibrio* spp. from clinical, environmental and food sources, continues to be TCBS. Despite its limitations, this medium remains the medium of choice by investigators worldwide.

V. cholerae

A combination of alkaline peptone water (APW) enrichment (with incubation at 35–37°C for 18 h) and subsequent plating to TCBS agar as the selective/differential medium remains the most commonly employed (and recommended) method for isolation of *V. cholerae*. When large numbers of background bacterial populations exist, it is advisable to subculture APW at 2 h as well as 18 h.

SPS and CPC agars may have value in the isolation of *V. cholerae*, especially following APW enrichment, but further testing is required to establish these media for this purpose.

V. parahaemolyticus

Despite advocacy for numerous alternate enrichment and plating media for this species, APW coupled with plating onto TCBS agar is the only generally accepted method for the isolation of this pathogen.

V. vulnificus

Unlike the case with *V. cholerae* and *V. parahaemolyticus*, a medium exists which is highly selective for this pathogenic species, and which has found wide acceptance for use. The general recommendation is a 16 h enrichment at 35–37°C in APW followed by plating a loopful from the top cm of the broth to CPC agar. For quantitative determinations, direct plating of samples onto CPC agar or one of its derivatives (e.g. mCPC) is recommended. Enrichment in APW with colistin (Høi et al., 1998b) may have merit, and should be examined further.

Identification of vibrios

Because of the phenotypic variation exhibited by the vibrios, and the need to determine which isolates are toxin producers, it has become commonplace for investigators to supplement isolation media with a variety of molecular techniques for identification of presumptive vibrio isolates. The reader is referred to studies by Koch et al. (1993) and Popovic et al. (1994) for identification of cholera toxin (*ctx*)-producing strains of *V. cholerae* in foods. Similarly, since only Kanagawa haemolysin-positive (*tdh* +) strains of *V. parahaemolyticus* are pathogenic, some investigators now use gene probes or PCR primers against these genes to differentiate K^+ and K^- strains (DePaola et al., 1990; Beasley et al., 1994; McCarthy et al., 1999). Identification of *V. vulnificus* presents special problems, as significant phenotypic variation exists among isolates. Further, and as is the case with *V. cholerae* and *V. parahaemolyticus*, not all strains of *V. vulnificus* may be of concern in food safety (Warner and Oliver, 1999). For these reasons, a probe (VVAP) developed by Wright et al. (1993) against the haemolysin/cytotoxin of *V. vulnificus* has proved to be of great value in identifying this species. The probe was shown in laboratory studies to have 100% specificity against this species (Morris et al., 1987), and has subsequently been used in numerous studies (e.g. Oliver et al., 1992; Kaspar and Tamplin, 1993; Dalsgaard and Høi, 1997; DePaola et al., 1997).

Acknowledgement

I am deeply indebted to Drs. Donovan and van Netten for their contributions to the previous edition of this chapter.

References

Abbott, S.L., Cheung, W.W.K.W. and Janda, J.M. (1993) Evaluation of a new selective agar, thiosulfate-chloride-iodide (TCI), for the growth of pathogenic *Vibrio* species. Med. Microbiol. Lett. 2, 362–370.

Akiyama, S., Takizawa, K., Ichinoe, H., Enomto, S., Kobayashi, T. and Sakazaki, R. (1963) Application of teepol to isolation of *Vibrio parahaemolyticus*. Jpn. J. Bacteriol. 18, 255–256.

Arias, C.R., Aznar, R., Pujalte, M.J. and Garay, E. (1998) A comparison of strategies for the detection and recovery of *Vibrio vulnificus* from marine samples of the Western Mediterranean coast. Systm. Appl. Microbiol. 21, 28–134.

Barbieri, E., Falzano, L., Fiorentini, C., Pianetti, A., Baffone, W., Fabbri, A., Matarrese, P., Casiere, A., Katouli, M., Kühn, I., Möllby, R., Bruscolini, F. and Donelli, G. (1999) Occurrence, diversity, and pathogenicity of halophilic *Vibrio* spp. and non-O1 *Vibrio cholerae* from estuarine waters along the Italian Adriatic Coast. Appl. Environ. Microbiol. 65, 2748–2753.

Beasley, L., Jones, D.D. and Bej, A.K. (1994) A rapid method for detection and differentiation of KP+ and KP- *Vibrio parahaemolyticus* in artificially contaminated shellfish by in vitro DNA amplification and gene probe hybridization methods. Abstr. Ann. Meet. Amer. Soc. Microbiol. P100, p. 386.

Beazley, W.A. and Palmer, G.G. (1992) TCI – a new bile free medium for the isolation of *Vibrio* species. Austr. J. Med. Sci. 13, 25–27.

Berry, T.M., Park, D.L. and Lightner, D.V. (1994). Comparison of the microbial quality of raw shrimp from China, Ecuador, or Mexico at both wholesale and retail levels. J. Food. Protect. 57, 150–153.

Bolinches, J., Romalde, J.L. and Toranzo, A.E. (1988) Evaluation of selective media for isolation and enumeration of vibrios from estuarine waters. J. Microbiol. Meth. 8, 151–160.

Brayton, P.R., West, P.A., Russek, E. and Colwell, R.R. (1983) New selective plating medium for isolation of *V. vulnificus* biogroup 1. J. Clin. Microbiol. 17, 1039–1044.

Bryant, R.G., Jarvis, J. and Janda, J.M. (1987) Use of sodium dodecyl sulfate-polymyxin B-sucrose medium for isolation of *Vibrio vulnificus* from shellfish. Appl. Environ. Microbiol. 53, 1556–1559.

Cerdà-Cuéllar, M., Jofre, J. and Blanch, A.R. (2000) A selective medium and a specific probe for detection of *Vibrio vulnificus*. Appl. Environ. Microbiol. 66, 855–859.

Chatterjee, B.D., De, P.K. and Sen, T. (1977) Sucrose teepol tellurite agar: a new selective indicator medium for isolation of *Vibrio* species. J. Infect. Dis. 135, 654–658. (Erratum: J. Infect. Dis. 135, 716).

Cleland, D., Thomas, M.B., Strickland, D. and Oliver, J.D. (1985) A comparison of media for the isolation of *Vibrio* spp. from environmental sources. Abstr. Annu. Meet. Amer. Soc. Microbiol. N16, p. 220.

Clesceri, L.S., Greenberg, A.E. and Eaton, A.D. (ed). (1998) Standard Methods for the Examination of Water and Wastewater, 20th ed. Amer. Publ. Health Assoc., Washington, D.C.

Dalsgaard, A. and Høi, L. (1997) Prevalence and characterization of *Vibrio vulnificus* isolated from shrimp products imported into Denmark. J. Food Protect. 60, 1132–1135.

Dalsgaard, A., Frimodt-Møller, N., Bruun, B, Høi, L. and Larsen, J.L. (1996) Clinical manifestations and molecular epidemiology of *Vibrio vulnificus* infections in Denmark. Eur. J. Clin. Microbiol. Infect. Dis. 15, 227–232.

De, S.P., Sen, D., De, P.C., Ghosh, A. and Pal., S.C. (1977) A simple selective medium for isolation of Vibrios with particular reference to *Vibrio parahaemolyticus*. Ind. J. Med. Res. 66, 398–399.

DePaola, A., Capers, G.M. and Alexander, D. (1994) Densities of *Vibrio vulnificus* in the intestines of fish from the U.S. Gulf Coast. Appl. Environ. Microbiol. 60, 984–988.

DePaola, A., Motes, M.L. and McPherson, R.M. (1988) Comparison of APHA and elevated enrichment methods for recovery of *Vibrio cholerae* from oysters: Collaborative study. J. Assoc. Off. Anal. Chem. 71, 584–589.

DePaola, A., Hopkins, L.H., Peeler, J.T., Wentz, B. and McPherson, R.M. (1990) Incidence of *Vibrio parahaemolyticus* in U.S. coastal waters and oysters. Appl. Environ. Microbiol. 56, 2299–2302.

DePaola, A., Motes, M.L., Cook, D.W., Veazey, J., Cartright, W.E. and Blodgett, R. (1997) Evaluation of alkaline phosphatase-labeled DNA probe for enumeration of *Vibrio vulnificus* in Gulf Coast oysters. J. Microbiol. Meth. 29, 115–120.

Dinnuzzo, A.R., Kelly, M.T. and Tacquard, E.C. (1984) Evaluation of selective *Vibrio* media for the isolation of *Vibrio vulnificus* in environmental sampling. Abstr. Annu. Meet. Amer. Soc. Microbiol. Q9, p. 206.

Donovan, T.J. and van Netten, P. (1995) Culture media for the isolation and enumeration of pathogenic *Vibrio* species in foods and environmental samples. In: Culture Media for Food Microbiology edited by J.E.L. Corry, G.D.W. Curtis, and R.M. Baird, Progress in Industrial Microbiology series No. 34. Elsevier, Amsterdam, pp. 203–217.

Eyles, M.J., Davey, G.R., Arnold, G. and Ware., H.M. (1985) Evaluation of methods for the enumeration and identification of *Vibrio parahaemolyticus* in oysters. Food Technol. Aust. 37, 302–304.

Furniss, A.L., Lee, J.V. and Donovan, T.J. (1978) The Vibrios. Monograph Series, Public Health Laboratory Service, HMSO, London.

Hagen, C.J., Sloan, E., Lancette, G.A., Peeler, J.T. and Sofos, J.N. (1994) Enumeration of *Vibrio parahaemolyticus* and *Vibrio vulnificus* in various seafoods with two enrichment broths. J. Food Protect. 5:403–409.

Hanharan, H., Giles, J.S., Heaney, S.B., Arsenault, G., McNair, N. and Rainnie, D.J. (1995) Bacteriological studies of mussels and oysters from six river systems in Prince Edward Island, Canada. J. Shellfish Res. 14, 527–532.

Høi, L., Larsen, J.L., Dalsgaard, I. and Dalsgaard, A. (1998a). Occurrence of *Vibrio vulnificus* biotypes in Danish marine environments. Appl. Environ. Microbiol. 64, 7–13.

Høi, L., Dalsgaard, I. and Dalsgaard, A. (1998b). Improved isolation of *Vibrio vulnificus* from seawater and sediment with cellobiose-colistin agar. Appl. Environ. Microbiol. 64, 1721–1724.

Horie, S., Saheki, K., Kozima, T., Nara, M. and Sekine, Y. (1964) Distribution of *Vibrio parahaemolyticus* in plankton and fish in the open sea. Bull. Jpn. Soc. Sci. fish. 30, 786–791.

Hsu, W.-Y., Wei, C.-I. and Tamplin, M.L. (1998) Enhanced broth media for selective growth of

Vibrio vulnificus. Appl. Environ. Microbiol. 64, 2701–2704.
Joseph, S.W., Colwell, R.R. and Kaper, J.B. (1982). *Vibrio parahaemolyticus* and related halophilic vibrios. Crit. Rev. Microbiol. 10, 77–124.
Kaper, J.B., Morris, J.G. Jr. and Levine, M.M. (1995). Cholera. Clin. Microbiol. Rev. 8, 48–86.
Karunasager, I., Venugopal, M.N., Karunasager, I. and Segar, K. (1986) Evaluation of methods for enumeration of *Vibrio parahaemolyticus* from seafood. Appl. Environ. Microbiol. 52, 583–585.
Kaspar, C.W. and Tamplin, M.L. (1993) Effects of temperature and salinity on the survival of *Vibrio vulnificus* in seawater and shellfish. Appl. Environ. Microbiol. 59, 2425–2429.
Kaysner, C.A. and Hill, W.E. (1994). Toxigenic *Vibrio cholerae* O1 in food and water. pp. 27–39 in Wachsmuth, I.K, Blake, P.A., and Olsvik, Ø. (Eds.). *Vibrio cholerae* and cholera: Molecular to Global Perspectives. ASM Press, Washington, D.C.
Kaysner, C.A., Abeyta, C., Jr., Wekell, M.M., DePaola, A., Jr., Stott, R.F. and Leitch, J.M. (1987) Incidence of *Vibrio cholerae* from estuaries of the United States West Coast. Appl. Environ. Microbiol. 53, 1344–1348.
Kaysner, C.A., Tamplin, M.L., Wekell, M.M., Stott, R.F. and Colburn, K.G. (1989) Survival of *Vibrio vulnificus* in shellstock and shucked oysters (*Crassostrea gigas* and *Crassostrea virginica*) and effects of isolation medium on recovery. Appl. Environ. Microbiol. 55, 3072–3079.
Kelly, M.T. (1982) Effect of temperature and salinity on *Vibrio (Beneckea) vulnificus* occurrence in a Gulf Coast environment. Appl. Environ. Microbiol. 44, 820–824.
Kitaura, T., Doke, S., Azuma, I., Miyano, K., Harada, K. and Yabuuchi, E. (1983) Halo production by sulphatase activity in *V. vulnificus* and *V. cholerae* O:1 on a new selective sodium dodecyl sulphate containing medium: a screening marker in environmental surveillance. FEMS Microbiol. Lett. 17, 205–209.
Kobayashi, T., Enomoto, S., Sakazaki, R. and Kuwahara, S. (1963) A new selective isolation medium for vibrio group on a modified Nakanishi's medium (TCBS agar medium). Jpn. J. Bacteriol. 18, 387–392.
Koch, W.H., Payne, W.L., Wentz, B.A. and Cebula, T.A. (1993) Rapid polymerase chain reaction method for detection of *Vibrio cholerae* in foods. Appl. Environ. Microbiol. 59, 556–560.
Kourany, M. (1983) Medium for isolation and differentiation of *Vibrio parahaemolyticus* and *Vibrio alginolyticus*. Appl. Environ. Microbiol. 45, 310–312.
Lotz, M.J., Tamplin, M.L. and Rodrick, G.E. (1983) Thiosulfate-citrate-bile salts-sucrose agar and its selectivity for clinical and marine vibrio organisms. Ann. Clin. Lab. Sci. 13, 45–48.
Lowry, P.W., McFarland, L.M., Peltier, B.H., Roberts, N.C., Bradford, H.B., Herndon J.L., Stroup, D.F., Mathison, J.B., Blake, P.A. and Gunn, R.A. (1989) *Vibrio* gastroenteritis in Louisiana: a prospective study among attendees of a scientific congress in New Orleans. J. Infect. Dis. 154,730–731.
Madden, J.M., McCardell, B. and Morris, J.G., Jr. (1989) Vibrio cholerae. *In*: Foodborne Bacterial Pathogens, M.P. Doyle (Ed.). Marcel Dekker, Inc., NY.
Massad, G. and J.D. Oliver. (1987) New selective and differential medium for *Vibrio cholerae* and *V. vulnificus*. Appl. Environ. Microbiol. 53, 2262–2264.
Matté, G.R., Matté, M.H., Rivera, I.G. and Martins, M.T. (1994) Distribution of pathogenic vibrios in oysters from a tropical region. J. Food Protect. 57, 870–873.
McCarthy, S.A., DePaola, A., Cook, D.W., Kaysneer, C.A. and Hill, W.E. (1999) Evaluation of alkaline phosphatase- and digoxigenin-labelled probes for detection of the thermolabile hemolysin (*tlh*) gene of *Vibrio parahaemolyticus*. Lett. Appl. Microbiol. 28, 66–70.
McCormack, W.M., Dewitt, W.E., Bailey, P.E., Morris, G.K., Socharjono, P. and Gangarosa, R.J. (1974) Evaluation of thiosulfate-citrate-bile salts sucrose agar, a selective medium for the isolation of *Vibrio cholerae* and other pathogenic vibrios. J. Infect. Dis. 129, 497–500.
Mead, P.S., Slutsker, L., Dietz, V., McCaig, L.F., Bresee, J.S., Shapiro, C., Griffin, P.M. and Tauxe, R.V. (1999) Food-related illness and death in the United States. Emerging Infect. Dis. vol. 5. Centers for Disease Control and Prevention Web journal: (http://www.dcd.gov/ncidod/eid/

vol5no5/mead.htm). Atlanta, Georgia, USA

Micelli, G.A., Watkins, W.D. and Rippey, S.R. (1993) Direct plating procedure for enumerating *Vibrio vulnificus* in oysters (*Crassostrea virginica*). Appl. Environ. Microbiol. 59, 3519–3524.

Mintz, E.D., Popovic, T. and Blake, P.A. (1994) Transmission of Vibrio cholerae O1. In: *Vibrio cholerae* and cholera: Molecular to Global Perspectives. I.K. Wachsmuth, P.A. Blake, and Ø. Olsvik (Eds.) ASM Press, Washington, D.C., pp 345–356.

Miyamoto, T., Miwa, H. and Hatano, S. (1990) Improved fluorogenic assay for rapid detection of *Vibrio parahaemolyticus* in foods. Appl. Environ. Microbiol. 56, 1480–1484.

Monsur, K.A. (1961) A highly selective gelatin-taurocholate-tellurite medium for the isolation of *Vibrio cholerae*. Trans. R. Soc. Trop. Med. Hygiene 55, 440–442.

Monsur, K.A. (1963) Bacteriological diagnosis of cholera under field conditions. Bull. W.H.O. 28, 387–389.

Morris, G.K. (1982) Media for *Vibrio* species. *In*: Quality Assurance and Quality Control of Microbiological Culture Media. J.E.L. Corry (Ed.) G.I.T. Verlag, Darmstadt, Germany, pp. 169–174.

Morris, G.K., Dewitt, W.E., Gangarosa, E.J. and McCormack, W.M. (1976) Enhancement by sodium chloride of the selectivity of thiosulfate citrate bile salts sucrose agar for isolating *Vibrio cholerae* biotype El Tor. J. Clin. Microbiol. 4, 133–136.

Morris, G.K., Merson, M.H., Huq, I, Kibrya, A.K.M.G. and Black, R. (1979) Comparison of four plating media for isolating *Vibrio cholerae*. J. Clin. Microbiol. 9, 79–83.

Morris, J.G., Jr., Wright, A.C., Roberts, D.M., Wood, P.K., Simpson, L.M. and Oliver, J.D. (1987) Identification of environmental *Vibrio vulnificus* isolates with a DNA probe for the cytotoxin-hemolysin gene. Appl. Environ. Microbiol. 53, 193–195.

Nakanishi, H. and Murase, M. (1974) Enumeration of *Vibrio parahaemolyticus* in raw fish meat. *In*: International Symposium on *Vibrio parahaemolyticus*. R. Fujino, R. Sakagachi, R. Sakazaki, and Y. Takeda (Eds.) Saikon Publishing Co., Tokyo.

Nishibuchi, M., Roberts, N.C., Bradford, H.B., Jr. and Seidler, R.J. (1983) Broth medium for enrichment of *Vibrio fluvialis* from the environment. Appl. Environ. Microbiol. 46, 425–429.

O'Brien, M. and Colwell, R.R. (1985) Modified taurocholate-tellurite-gelatin agar for improved differentiation of *Vibrio* species. J. Clin. Microbiol. 22, 1011–1013.

Oliver, J.D. (1989) *Vibrio vulnificus*. *In*: Foodborne Bacterial Pathogens . M.P. Doyle (Ed.) Marcel Dekker, New York. pp. 569–600.

Oliver, J.D. and Kaper, J.P. (1998) *Vibrio* species. *In*: Food Microbiology, Fundamentals and Frontiers. M.P. Doyle, L.R. Beuchat, and T.J. Montville (Eds.) ASM Press, Washington, D.C. pp. 228–264.

Oliver, J.D., Warner, R.A. and Cleland, D.R. (1982) Distribution and ecology of *Vibrio vulnificus* and other lactose-fermenting marine vibrios in coastal waters of the southeastern United States. Appl. Environ. Microbiol. 4, 1404–1414.

Oliver, J.D., Warner, R.A. and Cleland, D.R. (1983) Distribution of *Vibrio vulnificus* and other lactose-fermenting vibrios in the marine environment. Appl. Environ. Microbiol. 45, 985–998.

Oliver, J.D., Guthrie, K., Preyer, J., Wright, A., Simpson, L.M., Siebeling, R. and Morris, J.G., Jr. (1992) Use of colistin-polymyxin B-cellobiose agar for isolation of *Vibrio vulnificus* from the environment. Appl. Environ. Microbiol. 58, 737–739.

Parker, R.W., Maurer, E.M., Childers, A.B. and Lewis, D.H. (1994) Effect of frozen storage and vacuum-packaging on survival of *Vibrio vulnificus* in Gulf Coast oysters (Crassostrea virginica). J. Food. Protect. 57, 604–606.

Popovic, T., Fields, P.I. and Olsvik, Ø. (1994) Detection of cholera toxin genes. *In*: *Vibrio cholerae* and cholera: Molecular to Global Perspectives. I.K. Wachsmuth, P.A. Blake, and Ø. Olsvik (Eds.) ASM Press, Washington, D.C. pp. 41–52.

Rennels, M.B., Levin, M.M., Kaya, V., Angle, P. and Young, C. (1980) Selective vs nonselective media and direct plating vs enrichment technique in isolation of *Vibrio cholerae*: recommendations for clinical laboratories. J. Infect. Dis. 142, 328–331.

Sakazaki, R. 1973. *In*: B.C. Hobbs and J.H.B. Christian, The Microbiological Safety of Food. Academic Press, NY.

Shimada, T., Sakazaki, R., Fujimura, S., Niwano, K., Mishina, M. and Takizawa, K. (1990) A new selective, differential agar medium for isolation of *Vibrio cholerae* O1: PMT (polymyxin-mannose-tellurite) agar. Jap. J. Med. Sci. Biol. 43:37–41.

Simidu, U. and Tsukamoto, K. (1980) A method of the selective isolation and enumeration of marine Vibionaceae. Microb. Ecol. 6, 181–184.

Sloan, E.M., Hagen, C.J., Lancette, G.A., Peeler, J.T. and Sofos, J.N. (1992) Comparison of five selective enrichment broths and two selective agars for recovery of *Vibrio vulnificus* from oysters. J. Food Protect. 55, 356–359.

Song, Y.-L., Cheng, W. and Wang, C.-H. (1993) Isolation and characterization of *Vibrio damsela* infectious for cultured shrimp in Taiwan. J. Invert. Pathol. 61, 24–31.

Spira, W.M. (1984) Tactics for detecting pathogenic vibrios in the environment. Chapter 17 *In*: Vibrios in the Environment. R.R. Colwell (Ed.) Wiley, New York, NY. pp. 251–268.

Sun, Y. and Oliver, J.D. (1995) Value of cellobiose-polymyxin B-colistin agar for isolation of *Vibrio vulnificus* from oysters. J. Food Protect. 58, 439–440.

Tamplin, M.L., Martin, A.L., Ruple, A.D., Cook, D.W. and Kaspar, C.W. (1991) Enzyme immunoassay for identification of *Vibrio vulnificus* in seawater, sediment, and oysters. Appl. Environ. Microbiol. 57, 1235–1240.

Tilton, R.C. and Ryan, R.W. (1987) Clinical and ecological characteristics of *Vibrio vulnificus* in the Northeastern United States. Diagn. Microbiol. Infect. Dis. 6, 109–117.

Twedt, R.M. (1989) *Vibrio parahaemolyticus*. *In*: Foodborne Bacterial Pathogens. M.P. Doyle (Ed.) Marcel Dekker, New York, NY. pp. 543–569.

U.S. Food and Drug Administration (1984) Bacteriological Analytical Manual, 6th ed. Assoc. Off. Analyt. Chem., Arlington, VA.

Warner, J.M. and Oliver, J.D. (1999) Randomly amplified polymorphic DNA analysis of clinical and environmental isolates of *Vibrio vulnificus* and other *Vibrio* species. Appl. Environ. Microbiol. 65, 1141–1144.

West, P.A., Russek, E., Brayton, P.R. and Colwell, R.R. (1982) Statistical evaluation of a quality control method for isolation of pathogenic *Vibrio* species on selected thiosulfate-citrate-bile salts-sucrose agars. J. Clin. Microbiol. 16, 1110–1116.

Wright, A.C., Micelli, G.A., Landry, W.L., Christy, J.B., Watkins, W.D. and Morris, J.G., Jr. (1993) Rapid identification of *Vibrio vulnificus* on nonselective media with an alkaline phosphatase-labeled oligonucleotide probe. Appl. Environ. Microbiol. 59, 541–546.

Chapter 18

Culture media for the isolation of campylobacters, helicobacters and arcobacters

Janet E.L. Corry[a], H. Ibrahim Atabay[b], Stephen J. Forsythe[c] and Lucielle P. Mansfield[c]

[a]*Department of Clinical Veterinary Medicine, University of Bristol, Langford, Bristol, BS40 5DU, UK*

[b]*Department of Microbiology, Veterinary Faculty, University of Kafkas, Kars, Turkey*

[c]*Department of Life Sciences, The Nottingham Trent University, Clifton Lane, Nottingham, NG11 8NS, UK*

The history of the development of selective media for isolation of campylobacters, including the rationale for choice of selective agents is described. Developments have included modifications to allow incubation at 37°C instead of 42 or 43°C and changes in the types and concentrations of antibiotics in order not to inhibit organisms such as *Campylobacter upsaliensis*, *C. jejuni* subsp. *doylei* and some strains of *C. coli* and *C. lari*. When examining foods, plating media originally developed for isolation from faeces are normally used, sometimes after liquid enrichment. Most of the media include ingredients intended to protect campylobacters from the toxic effect of oxygen derivatives. Most commonly used are lysed or defibrinated blood, charcoal, a combination of ferrous sulphate, sodium metabisulphite and sodium pyruvate (FBP); and haemin or haematin.

The manner in which liquid enrichment media are used has been modified for food samples to avoid inhibitory effects on sublethally damaged cells by toxic components in the formula. This is done by a preliminary period of incubation at reduced temperature and sometimes by delayed addition of antibiotics. Expensive and time-consuming methods have been proposed to achieve a microaerobic atmosphere while using liquid enrichment media.

To date there is no generally accepted "standard" method of isolating campylobacters from food, although Bolton broth and modified charcoal, cefoperazone deoxycholate agar have been proposed in a number of standard methods.

Media for isolating arcobacters are similar to those for campylobacters, except that lower temperatures and sometimes aerobic atmosphere are used for incubation. Some strains of *Arcobacter cryaerophilus* and *A. skirrowii* are sensitive to 32 mg/l cefoperazone and all *Arcobacter* sp. are sensitive to colistin used in some media.

Media for the human pathogen *Helicobacter pylori* have been developed, although, with one exception, using immunomagnetic beads, they have not been successful in isolating the organism from foods or the environment. Many other *Helicobacter*-like organisms, seen in gastric or intestinal tissue samples from a variety of animals have not been successfully cultivated.

Introduction

Campylobacters were originally classified within the genus *Vibrio*, but differ from vibrios in a number of respects, particularly in their DNA base composition and their ability to grow under conditions of reduced oxygen tension. They are Gram negative, oxidase positive, curved or spiral rod-shaped bacteria, 0.2–0.5mm wide and 0.5–8mm long. They possess one polar flagellum, which gives them a very characteristic "cork-screw" motility. Survival of campylobacters is poor under most conditions and particularly in dry, relatively warm and aerobic situations. As cultures lose their viability the rod-shape changes to a coccoid form which has been described as "viable but non-culturable". This is because, although these forms cannot be grown using normal media, there have been reports that they can infect animals (e.g. Jones et al., 1991; Cappelier et al., 1999a and b). However, others have suggested that this could be due to regrowth of a small proportion of cells which never lost culturability (e.g. Bovill and Mackey, 1997; Kell et al., 1998).

The genus *Campylobacter* was subdivided 10–15 years ago into three related genera: *Campylobacter, Arcobacter* and *Helicobacter.* The most important species of *Campylobacter* for this review are the thermophilic species: *C. jejuni* ssp. *jejuni, C. coli* and *C. lari* (formerly known as "nalidixic acid resistant thermophilic campylobacters - NARTC"). UPTC (urease positive thermophilic campylobacters) are variants of *C. lari* found in the aquatic environment which do not appear to be pathogenic (Megraud et al., 1998; Jones, 2001; On, 2001). Other species which sometimes cause diarrhoea are *C. upsaliensis, C. fetus* ssp. *fetus* (some of which are thermophilic) and *C. jejuni* ssp. *doylei.* Other non-thermophilic species include *C. hyointestinalis, C. mucosalis, C. concisus, C. sputorum, C. helveticus, C. rectus, C. showae, C. hominis, C, curvus and C. gracilis* (On, 2001). *Helicobacter* contains *H. pylori*, (previously *C. pyloridis*, responsible for gastritis, gastric and duodenal ulcers and gastric cancer in humans – Jones et al., 2001)) *H. cinaedi, H. fennelliae, H. mustelae* and *H. pullorum,* and a number of other species, many of which can be found in the stomachs of a variety of animals (On, 2001). The genus *Arcobacter* comprises *A. cryaerophilus, A. butzleri, A. skirrowii* and *A. nitrofigilis* (Vandamme et al., 1991, 1992a; Vandamme and Goossens, 1992). *Arcobacter* spp. (except for *A. nitrofigilis)* are increasingly being implicated as important causes of human gastro-intestinal disease (Vandamme et al. 1992b; Lerner et al., 1994; On et al., 1995; Lauwers et al., 1996; Lastovica and Skirrow, 2000). Arcobacters and helicobacters have many similarities with campylobacters. All grow in a microaerobic atmosphere (containing about 6% oxygen, and 10% carbon dioxide), some also requiring the presence of hydrogen. Their morphology is similar to campylobacters, except that helicobacters often have multiple polar flagella, and their flagella are sheathed. However, helicobacters usually have an optimum growth temperature of about 37°C, while the thermophilic campylobacters can grow at 42–43°C but not below about 31°C , and arcobacters can grow at 25°C or below. Arcobacters are also more aerotolerant than campylobacters or helicobacters, being able to grow in air even on first isolation. Table 1 lists the most important species of *Campylobacter* and *Arcobacter*, which are associated with human and/or animal diarrhoea together with their important characteristics.

Table 1a
Diagnostic features of *Campylobacter* species causing illness in man.

	C. *jejuni*		C. *coli*	C. *fetus*	C. *lari*	C. *upsaliensis*	C. *hyointestinalis*	*Arcobacter* spp
	subsp. *jejuni*	subsp. *doylei*						
Growth aerobically								
At 30°C or 36°C	–	–	–	–	–	–	–	+
Growth microaerobically								
25°C	–	–	–	+	–	–	D	+
37°C	+	+	+	+	+	+	+	D
42°C	+	V	+	W	+	D	D	D
Catalase	+	D	+	+	+	W	+	W or +
Oxidase	+	+	+	+	+	+	+	+
Nitrate reduction	+	–	+	+	+	+	+	+ or –
Cephalothin 30 μg	R	S	R	S	R	S	S	R/S
Nalidixic acid 30 μg	S	S	S	R	R	S	R	S
Hydrogen sulphide								
production in TSI	–	–	W	–	–	–	+	–
Indoxyl acetate hydrolysis	+	+	+	–	–	+	–	+/W
Hippurate hydrolysis	+	+	–	–	–	–	–	–
Urea hydrolysis	–	–	–	–	–/+	–	–	–/+
Growth in 1% glycine	+	+	+	W	+	+	+	+/–
Growth on MacConkey agar	+	?	+	+	+	+	+	D/+
Growth on Campy–BAP`								
agar (BBL)	+	?	+	?	+	?	?	–
Growth on Campy–CVA								
minus polymyxin)	+	?	+	?	+	?	?	+

V, 50% positive; W, negative or weakly positive; D, some +, some –; R, resistant; S, sensitive; TSI, triple sugar iron medium.
From: Barrett et al., 1988; Lastovica et al., 1989; Patton et al., 1989: Warmsley and Karmali, 1989; Boudreau et al., 1991; Kiehlbauch et al., 1991; Tenover and Fennell, 1992; Vandamme et al., 1991, 1992; Vandamme and Goossens, 1992.

Table 1b
Differentiating characteristics of *Arcobacter* spp. Numbers in columns indicate percentage positive (modified from Mansfield and Forsythe 2000).

Test[a]	*A. butzleri*	*A. cryaerophilus*	*A. skirrowii*	*A. nitrofigilis*	*C. jejuni*
Growth 15°C	Positive	Positive	Not determined	Positive	Negative
25°C	Positive	Positive	Positive	Positive	Negative
30°C	Positive	Positive	Positive	Positive	90
37°C, aerobic	Positive	50	Positive	50	Negative
37°C, microaerophilic	Positive	58	Positive	Negative	Positive
37°C, anaerobic	Positive	16–95	Positive	Positive	Negative
42°C, microaerophilic	25–67[b]	0–18[b]	11–33[b]	Negative	Positive
Anaerobic + TMAO	Positive	11	Negative	Positive	Negative
Nutrient agar	Positive	Positive	Positive	Positive	Positive
Buffered charcoal yeast medium	Positive	Positive	Positive	Negative	Positive
Campylobacter charcoal-deoxycholate medium	Positive	95	Positive	Negative	Positive
Campylobacter minimal medium	Positive	Negative	Negative	Negative	10
MacConkey agar	83–100[b]	16–43[b]	Negative	Negative	5
Lecithin	75	16	11	Negative	35
Growth in presence of glycine 1%	58	9–23[b]	78	Not tested	Growth
NaCl 2%	92	84	Positive	Positive	Negative
NaCl 3.5%	42–67	0–33[b]	61	Positive	Negative
NaCl 4.0%	Negative	Negative	Positive	Positive	Negative
Bile 2%	Positive	79	Positive	Negative	80
Glucose 8%	Positive	45–48	50	Not determined	Not determined
Pteridine 0/129 vibriostat	Positive	Positive	Not determined	Positive	Not determined
Cadmium chloride (2.5 μg disc)	Resistant	Sensitive	Not determined	Not determined	Sensitive
Oxidase	Positive	Positive	Positive	Positive	Positive
Catalase	33–100[b]	Positive	Positive	Positive	Positive
Alpha haemolysis	Negative[f]	Negative	Positive	Negative	95
H_2S production (lead acetate paper)	0–25	0–11	Negative	Negative	Positive
Nitrate reduction	Positive	30–36	Positive	Positive	Positive
Glucose utilization	Negative	Negative	Not determined	Not determined	Negative
Hippurate hydrolysis	Negative	Negative	Negative	Negative	Positive
Urea hydrolysis	Negative[d]	Negative	Negative	Positive	Negative

DNase activity	0–92	0–72	22–100	100	25–33[b]
Selenite reduction	Negative	Negative	11	50	70
Alkaline phosphatase	Negative	Negative	Negative	Negative	Negative
Triphenyl-tetrazolium chloride reduction	Positive	95	77	Negative	90
Indoxyl acetase	Positive	Positive	Positive[e]	Positive	Positive
Pyrazinamidase activity	Negative	Negative	Not determined	Not determined	Positive
Resistance to nalidixic acid (30μg disc)	14–25[b]	0–17[b]	Negative	Negative	Negative
Resistance to cephalothin (30μg disc)	83–100[b]	72–100[b]	Positive	0–22[b]	95
Metronidazole resistance[c] (4 mg/L)	92	Positive	Positive	Negative	70
Carbenicillin resistance[c] (32 mg/L)	Positive	Positive	Positive	Negative	65
Cefoperazone resistance[c] (64 mg/L)	Positive	Positive	Positive	Negative	Positive
5-fluorouracil resistance[c] (100 mg/L)	Positive	Positive	Positive	Negative	90
G+C content (mol%)	29–31	28–29	29[9]	28–29	30–33

Sources: On et al. (1996), Schroeder-Tucker et al. (1996), Atabay et al. (1998a and b), Jacob et al. (1993), Harrass et al. (1998), Vandamme and De Ley (1991). It should be noted that numbers of strains tested varied between researchers. Therefore % values are not necessarily comparable.

[a] Tests for *Arcobacter* and *Campylobacter* spp. were at 25°C and 37°C respectively unless otherwise stated.

[b] Variation between sources.

[c] Tested on blood agar medium

[d] Some poultry isolates are positive for urease (Atabay, unpublished).

[e] Some strains of *A. skirrowii* are negative for indoxyl acetate hydrolysis (Atabay et al., 1998a; Atabay et al. 2001).

[f] Alpha-haemolysis (positive) strains of *A. butzleri* exist (Atabay et al., 1998a; Atabay et al., 2001).

[9] Type strain.

Selective media for thermophilic campylobacters

Selective media were originally designed to isolate *C. jejuni* from faeces, by use of a cocktail of antibiotics in a rich basal medium and exploiting the ability of this organism to grow at 42 or 43°C (Butzler and Skirrow, 1979). Later, *C. coli* and *C. lari* were distinguished, although many workers either failed to differentiate between these three species when monitoring the functioning of selective media or only tested for *C. jejuni* (e.g. Park et al., 1981; Rosef, 1981; Acuff et al., 1982a & b; Bolton et al., 1982; Goossens et al., 1983; Martin et al., 1983; Wesley et al., 1983; Barot and Bokkenheuser, 1984; Stern et al., 1984; Beuchat et al., 1985; Furanetto et al. 1991; Holler 1991; Jacob and Stelzer, 1992; Stern et al., 1992; Stern and Line, 1992; Scotter et al., 1993).

There is evidence that some strains of *C. coli* and even a few strains of *C. jejuni* are likely to have been missed due to their sensitivity to cephalothin (Brooks et al., 1986; Ng et al., 1985, 1988; Endtz et al. 1991). Further work has indicated that other species of *Campylobacter* of importance in human intestinal disease will not be isolated by the usual selective media because of their susceptibility to the antibiotics used in most media e.g. *C. upsaliensis* (Warmsley and Karmali, 1989; Goossens et al., 1990a). If isolation temperatures of 42 or 43°C are used, *Arcobacter* species, and some strains of *C. fetus* and *C. upsaliensis* are unlikely to be detected as causes of human (or animal) diarrhoea, neither would they be detected in routine examination of human or animal faeces, foods, water or environmental specimens unless membrane filtration and non-selective media at 37°C are used. However, a modification to blood-free selective agar has been described which enables *C. upsaliensis* to be isolated, thus removing the practical difficulties of the filtration method (Aspinall et al., 1993). Between 1989 and 1991 only 15% of thermophilic campylobacters reported as causes of human diarrhoea in England and Wales were identified to species level. Of these 89–93% were *C. jejuni* and 7–10% were *C. coli* or *C. lari,* with 0.2–0.4% other species of campylobacter (Pearson and Healing, 1992). More recent data (1999) from the UK Public Health Laboratory Service Campylobacter Reference Laboratory gives a similar picture, with 93% *C. jejuni,* 6.5% *C. coli,* 0.5% *C. lari* and a few cases each of *C. jejuni* subsp. *doylei, C. fetus* subsp. *fetus* and *C. upsaliensis.* However, Goossens et al. (1990b) reported *C. upsaliensis* to be the cause of 13% of campylobacter diarrhoea cases in Belgium. *C. jejuni* subsp. *doylei* is another strain implicated as a cause of diarrhoea that frequently does not grow at 42° C and is sensitive to cephalothin (Taylor et al., 1991; Bolton et al., 1992). These considerations should be borne in mind when reading this review, which will be concerned mostly with the thermophilic campylobacters.

The media used for isolating campylobacters from foods and water have been derived from those first developed for the isolation of campylobacters from faeces. In some cases the same plating media are still used for both purposes. In line with techniques developed for isolating other pathogens, such as salmonellas, from foods and other environments, liquid (enrichment) media have been developed, and also pre-enrichment media, intended to aid recovery of sublethally damaged campylobacters. Incubation has usually been carried out at 42 or 43°C, but more recently there has been a trend to reduce the incubation temperature to 37°C.

Besides a variety of selective agents, almost all of which are antibiotics, media for campylobacters usually contain ingredients to neutralise the toxic effects of substances that form in the presence of oxygen and light. In addition, almost all workers have found it necessary to incubate plates in an atmosphere of about 5–7% oxygen, 10% carbon dioxide and 80% nitrogen and/or hydrogen. This can be achieved by using sachets designed for microaerophiles, or a bottled gas mixture of those proportions, or by replacing two thirds of the atmosphere with a mixture of 5–15% carbon dioxide plus nitrogen and/or hydrogen (Skirrow et al., 1982). Alternatively a sachet intended for use for anaerobes, but without a catalyst, which liberates hydrogen and carbon dioxide, is reported to be effective, even though the oxygen level must be higher than 7% (Dr David Wareing, personal communication). Skirrow et al. (1991) reported that the presence of hydrogen at not less than 7% improved the primary isolation of *C. jejuni* from faeces. Candle jars have been reported to be successful for incubation of plates of Butzler's medium Virion, provided 37°C is used (Goossens et al., 1983). Ribeiro et al. (1985) compared growth of *C. jejuni*, *C. coli* and *C. lari* in atmospheres generated by burning ethanol (methylated spirit) rather than a candle (paraffin wax) and found higher colony counts and larger colonies using ethanol. Colony counts for *C. lari* were particularly low when using candles. Similar comparisons by Skirrow et al. (1987) confirmed these results. Best results were obtained when the spirit was burned in a 90mm diameter Petri dish with the jar not more than half full with dishes. Pennie et al. (1984) found that a satisfactory atmosphere could be obtained by using a mixture of grade 0 steel wool, previously soaked in cupric sulphate solution and an Alka-Seltzer (sodium bicarbonate) tablet in water, all placed in a plastic bag! Jacob and Stelzer (1992) reported that this method worked well. In our experience, it is advisable to use microaerobic atmospheres with hydrogen whenever isolating any of the Campylobacteraceae, since some (e.g. *H. pullorum*) have a requirement for hydrogen, and many seem to grow better in the presence of hydrogen (e.g. thermophilic campylobacters and *Arcobacter* spp.).

Jones et al. (1993) have shown that *C. jejuni* can adapt to grow in a normal air atmosphere. The implications of this observation are not yet clear, but may explain to some extent the "viable but non-culturable" phenomenon. Fraser et al. (1992) reported that, provided the relative humidity was 99%, normal air containing 10% carbon dioxide gave larger colonies of *C. jejuni* and *C. coli* than the conventional gas mixture. However at 95% RH growth was poor. Other workers have reported that *C. jejuni*, as evidenced by its colonial morphology (drier plates yield smaller colonies), is sensitive to the moisture level in plating media (Buck and Kelly, 1981; Corry, unpublished observations), although changes in colonial morphology could also be due to inhibitory oxygen derivatives formed during storage and drying of plates.

Experienced workers generally claim to be able to recognise the organism by the typical appearance of colonies - flat, glossy and effuse, thinly spreading if the agar is moist. However, colonies are sometimes atypical, especially if plates are rather dry, so Gram staining of oxidase positive colonies and examination by microscope for the characteristic morphology is advisable. If they have been isolated at 37°C, the possibility that the isolates are *Arcobacter* spp. can be checked by testing for the ability to

Table 2
Most widely used isolation media for thermophilic campylobacters.

Plating Media
Skirrow (Skirrow, 1977 - modified by addition of amphotericin)
Campy BAP (Blaser et al., 1978)
Preston agar (Bolton and Robertson, 1982)
Campylosel (BioMerieux)
mCCD agar (Bolton et al., 1984 modified according to Hutchinson and Bolton, 1984; see also Baird et al.,1987)
Karmali agar (Karmali et al., 1986)
Enrichment Media
Preston broth (Bolton et al., 1982)
Park and Sanders broth (1991)
Hunt and Radle broth (Hunt, 1992)
Exeter broth (De Boer and Humphrey, 1991)
Bolton broth (Bolton, personal communication, 1995; Hunt et al.,1998).

grow aerobically at 25°C. Indicator systems, which produce colonies with a specific colour (e.g. pH indicator plus a sugar) are difficult to develop because Campylobacteraceae do not ferment carbohydrates. However, triphenyltetrazolium chloride has been used as indicator for a selective medium for thermophilic campylobacters (Line, 2001; Table 3) and also for *H. pylori* (Glupczinski et al., 1988; Table 5).

The number of formulations proposed for the isolation of thermophilic campylobacters probably exceeds that for any other group of bacteria, especially if one considers that all have been published since 1972 and almost all since 1977. Table 3 summarises many of these. Before the medium of Dekeyser et al. (1972) was available isolation of "related Vibrio" (as campylobacters were then known) depended on the use of membrane filtration followed by subculture onto nutritionally rich blood agar. The method of Dekeyser and co-workers used a combination of centrifugation, filtration through a 0.65 μm membrane filter and plating of the filtrate onto a selective agar.

Even if all the campylobacter media formulae had been published with a full description of the rationale used in their development, there would be insufficient space to consider them in detail. This review will examine the basic principles that the media have in common, and then concentrate on the details of the most widely used ones. The current most commonly used plating and liquid media are listed in Table 2. Filtration methods have frequently been used to concentrate campylobacters in samples of water (Bolton et al., 1982, 1987; Ribeiro and Price, 1984; Rosef et al., 1987; Stelzer et al., 1988; Brennhovd et al., 1992). Filtration is sometimes followed by liquid enrichment, sometimes by direct plating onto selective agar. Centrifugation has also been used when examining food samples (Lovett et al., 1983; Hunt et al., 1998). Media for isolation of *Arcobacter* spp. and *Helicobacter pylori* will also be considered (Tables 4 and 5).

Basal media

Few publications provide an explanation for the choice of basal medium and many have been used. Although campylobacters will grow in relatively simple media such as nutrient agar, most workers have used basal media developed for other fastidious capnophilic or anaerobic pathogens such as brucella or thioglycollate medium, Columbia or blood agar base and Müller Hinton broth. However, Bolton and Robertson (1982) chose nutrient broth no. 2, an unsophisticated medium, as the basis for their Preston media because it contained less thymidine, an antagonist to the activity of the selective agent trimethroprim, used in most selective solid and liquid media (see Table 3).

Blood

Most solid media incorporate blood at levels between 5 and 15%. Some media use defibrinated blood from various animals and others lysed horse blood. Only one uses lysed sheep blood (Stelzer and Jacob, 1992). Skirrow et al. (1982) stated that lysed horse blood was needed to neutralise trimethoprim antagonists which are present in most media. Many media which incorporate trimethoprim do not contain lysed horse blood although blood of some type is usually present. The media of Martin et al. (1983) and Goossens et al. (1989), neither of which contain blood, have the highest concentrations of trimethoprim. Presumably this is necessary to counteract the thymidine. Blood is also thought to be active in neutralising toxic oxygen derivatives (Juven and Rosenthal, 1985; Weinrich et al., 1990).

FBP and other bloodless supplements

A combination of ferrous sulphate ($FeSO_4 7H_20$), sodium metabisulphite and sodium pyruvate (FBP), each at 0.25 or 0.5g per litre was suggested by George et al. (1978) and Hoffman et al. (1979a) as an addition to campylobacter media to counteract the toxic effect of oxygen (Hoffman et al., 1979b). Many selective media contain some or all of these compounds but concentrations vary. A few media contain both FBP supplement and blood (e.g. Gilchrist et al., 1981; Weber et al., 1987; Stern et al., 1992) but in most cases media contain either blood or FBP. Alternatives to blood or FBP include haematin (Razi and Park, 1979; Wesley et al., 1983; Karmali et al., 1986) and charcoal (Bolton et al., 1984b; Karmali et al., 1986). Lignite-derived humic acids plus 0.05% $FeSO_4$ $7H_2O$ have been proposed as an alternative system (Weinrich et al., 1990). The mode of action of these supplements is not clear, but they may help to neutralize hydrogen peroxide, singlet oxygen and/or superoxide ions.

Storage of Campylobacter plating media

Hoffman et al. (1979a) found that growth of campylobacters was substantially reduced when plates of nutrient medium were stored in the presence of light and, in particular, air. Similar results were obtained by Bolton et al. (1984a) and Juven and Rosenthal (1985). The incorporation of aerotolerant supplements such as FBP or blood into campylobacter media reduces this effect but does not completely eliminate it (Fricker, 1985; Weinrich et al., 1990). Preston agar stored aerobically at room temperature inhibits the growth of *C. lari* to a greater extent than *C. jejuni* or *C. coli* (Fricker, 1985). Loss of moisture is probably also a factor (Buck and Kelly, 1981; Fraser et al., 1992). Campylobacter plates prepared in the laboratory should therefore either be used immediately or after storage in the dark anaerobically at room temperature, or aerobically in the refrigerator with precautions to prevent dehydration. A maximum storage time of 5 days for laboratory prepared plates is probably advisable.

pH

The pH of many of the isolation media is not specified, but presumably approximates to that of the basal medium used, normally near neutrality. Some liquid enrichment media have a pH ca. 8.0 (Wesley et al., 1983) although Park et al. (1983) found that their medium functioned best at pH 7.0. Liquid media with pH ca. 8.0 could be used to counteract acid production by competitive flora. The ability of competitive flora to acidify the enrichment media may depend on the food under examination (Humphrey, 1986b).

Choice of antibiotics

Antibiotics in selective media developed for campylobacters were chosen on the basis of those to which test strains were resistant and those most effective in inhibiting competitive flora. Various studies of antibiotic resistance have been published (Plastridge et al., 1964; Vanhoof et al., 1978; Walder, 1979; Karmali et al., 1980,1981; Ahonkai et al., 1981; Michel et al., 1983; Ng et al., 1985, 1988; Gebhart et al., 1985; Patton et al., 1989; Kiehlbauch et al., 1992). Neither Butzler (Dekeyser et al., 1972; Butzler et al., 1973; Lauwers et al., 1978), Skirrow (Skirrow, 1977; Butzler and Skirrow, 1979) nor Blaser (Blaser et al., 1979) published the full rationale for the development of their media. Probably these authors used data from Plastridge et al. (1964), Butzler et al. (1974) and Vanhoof et al. (1978). Later workers will have had access to other studies listed above. Plastridge et al. (1964) examined 57 strains of "related vibrio" bacteria that probably belonged to the *C. jejuni/C. coli* group, finding them resistant to bacitracin, novobiocin and polymyxin B. Butzler et al. (1974) found over 90% of 114 strains of related vibrios resistant to cephalothin and all were resistant to vancomycin and rifampicin. With few exceptions (Ng et al., 1985 and presumably Plastridge et al., 1964) the strains tested had been isolated using antibiotic-containing selective media and so the possibility of missing strains sensitive to the antibiotic used in the selective media will have been perpetuated.

Antibiotic activity

Polymyxin is generally active only against Gram-negative bacteria, and *Proteus* spp. are sometimes resistant. Trimethoprim usually inhibits *Proteus* spp. as well as other Gram-negative bacteria. Colistin is closely related to polymyxin, has a similar spectrum of activity and is used in some media instead. Vancomycin and rifampicin are both used, but not usually together. They are effective against Gram-positive bacteria. Rifampicin is also active against Gram-negative organisms. Butzler's media have generally incorporated bacitracin instead of vancomycin. The cephalosporins (cephalothin, cefaperazone) have a wide spectrum of activity against Gram-positive bacteria. Cefoperazone is now usually preferred because some campylobacters, especially *C. coli*, are sensitive to cephalothin (Brooks et al., 1986; Ng et al., 1985; 1988; Burnens and Nicolet, 1992). Cycloheximide (actidione), amphotericin B or nystatin are present in many media to inhibit yeasts and moulds, enabling the media to be incubated at 37°C rather than 42° or 43°C. Recently concern has been expressed over the toxicity of cycloheximide towards humans, and some media containing cycloheximide have been reformulated with amphotericin B. It appears that this is unlikely to adversely affect their performance (Martin et al., 2002).

Development of Plating Media (Table 3)

Bolton and co-workers gave the most detailed rationale for their formulation of Preston and CCD (charcoal, cefazolin (later replaced by cefoperazone) deoxycholate) media. When developing Preston medium, Bolton and Robertson (1982) carried out a survey of the MIC's of 104 strains of *Campylobacter* spp. and a variety of competitive organisms against four antibiotics (polymyxin, rifampicin, vancomycin and trimethoprim). They chose polymyxin because of its activity against Gram-negative bacteria (except for *Proteus* spp., which were suppressed by the trimethoprim). Since rifampicin showed a wide spectrum of activity against Gram-positive and Gram-negative bacteria it was chosen in preference to vancomycin, which has limited activity against Gram-negative organisms. Amphotericin B was added to inhibit yeasts. The level of polymyxin was set at 5,000 iu per litre on the basis of VanHoof's (1978) results, which showed that some campylobacters were sensitive to 10,000 iu per litre. When developing their CCD medium Bolton and co-workers determined that campylobacters grow best on solidified nutrient broth no. 2, after comparison with other media such as Columbia blood agar base, veal or brain heart infusion and diagnostic sensitivity test agar. They then carried out a systematic survey of alternatives to blood for neutralising oxygen toxicity by comparison with the growth of campylobacter obtained on nutrient agar plus 5% lysed horse blood (Bolton and Coates, 1983; Bolton et al., 1984a). A combination of 0.4% charcoal, 0.25% ferrous sulphate and 0.25%

Table 3
Formulation of liquid and solid selective media for thermophilic campylobacters (concentrations in mg per litre unless otherwise stated).

Medium name/ reference	Basal medium Other details	Antioxygen system	Cephalo-sporins	Trimetho-prim	Polymyxin B or Colistin (C) in iu	Vanco-mycin or Teico-planin (T)	Rifam-picin	Novo-biocin	5-Fluorouracil or Na-deoxy-cholate (D)	Bacitracin in iu	Anti-fungals
Dekeyser A [1]	TA	15%S			10000			5		25000	50 CY
Skirrow A [2,a]	BAB	7% LH		5	2500	10					
Blaser A [3]	BA	10%S		5	2500	10		2			
Butzler A [4]	TB	10%S	15 cthin		10 000 (C)			5		25000	50 CY
CAMPY-BAP A [5]	BA	10%S	15 cthin	5	2500	10					2 AM
Blaser-Wang A [6]	BAB	7% LH	15 cthin	5	2500	10					2 AM
C-2 A [7]	BA	5% H/FBP	5 cthin	5	8000	10					
C-3 A [7]	BA	5% H/FBP	1 cthin	5	2500	10					
Park (1981) B [8]	BB	–		4	8000	8					
BU40 B [9]	TB	10%S	15 cthin		40 000 (C)			5		25000	50 CY
Rosef B [10,b]	NB	–		5	2500	10					
Preston A [11]	NB	5% LH		10	5000		10				100 CY
Preston B [12]	NB	5% LH/FBP		10	5000		10				100 CY
Doyle-Roman B [13,c]	BB	7% LH		5	20000	15					50 CY
Christopher B + A [14]	BB	0.5g P	15 cthin	5	2500	10					2 AM
FBP-AM B/A [15]	BB/A	FBP (B) 5% DH (A)	15 cthin	5	2500	10					2 AM
Lander B [16]	V1	5g C		10	50000	40			100		100 CY
Park-Stankiewicz B [17,d]	BB	2g F, 0.25g B, 0.5g P	30 cthin	4	3 mg (C)	8					
Butzler medium Virion A [18]	CA	5–7%S	15 czone		10 000 (C)		10				2 AM
Butzler medium Oxoid A [19]	CA	5–7%S	15 czolin		10 000 (C)	5				25000	50 CY
Modified Park (1981) B [20]	BB	FBP		7.5	5000	15					
CEB B [21]	BB	–	32 czone	32					333		
VTP-FBP B [22]	BB	FBP, 5% LH		10	5000	20					

CAK A[23]	GA	7% H	15 cthin						1 AM
Rosef-Kapperud A[24]	GA, IVx	–	15 cthin		10 000 (C)				25 000 iu NY
ATB[25,e]	T; YE, NaCI	FBP, HT	6.25 codin		20000		25		
CCD A[26,f]	NB	4g C, 0.25g F, 0.25g P	10 czolin					1000 (D)	
CCD B[26,f]	NB	4g C, 0.25g FBP	10 czolin					1000 (D)	
mCCD A[27,f]	NB	4g C,0.25g F 0.25g P	32 czone					1000 (D)	10 AM
Doyle-Roman B[28,g]	BB	1.0%S + FBP		5	2.9 mg (C)	10			
Waterman A[29]	BAB	5% DH		5	2500		10		100 CY
Waterman B[29]	TB	5% DH		20	10000		10		100 CY
Virion[30]	CA	5–7%S	30 czone				10		
Exeter A + B[31], b	NB	5% LH, 0.5g F, 0.2g B, 0.2g P	15 czone	10	4 mg (C)	10[X]	10[Y]		
Karmali A[32,i]	CA	4g C, 0.32g HT, 0.1g P	32 czone			20			
CAR B/A[33]	BA/BB	FBP, 3% LH	32 czone				10		2.5 AM
Exeter A + B[34]	NB	FBP, 5% LH	15 czone		4 mg		10		2 AM
SSM B+A[35]	MHB + 4 g/1 A	–	30 czone	50					
Bolton[36 j]		5% LH, 0.1g HE, 0.5g P, 0.5g B	20 czone	20		20			50 CY
Park-Sanders B[37]	BB	0.25 P, 1 g Na citrate, 5% LH	32 czone**				10		100 CY or 2 AM
Hunt-Radle B[38]	NB,YE	FBP, 5% LH	15 czone***	12.5			10		100 CY
Campy Cefex A[39]	BA	0.5 g F, 0.2 g B, 0.5 g P, 5% LH	33 czone						200 CY
CAT A[40,f]	NB	4g C	8 czone			4 (T)		1000 (D)	10 AM
Line A[41,j]	BA,YE	0.5g F, 0.2g B, 0.5g P, 10g HE or 5% LH	33 czone	5	3.5mg	10			100 CY

cont.

Table 3. Continued.

Key: COLUMN 1. *References:* [1]Dekeyser et al., 1972; [2]Skirrow, 1977; [3]Blaser et al., 1978; [4]Lauwers et al., 1978; [5]Blaser et al., 1979; [6]Blaser et al., 1980; [7]Gilchrist et al., 1981; [8]Park et al., 1981; [9]Patton et al., 1981; [10]Rosef, 1981; [11]Bolton and Robertson, 1982; [12]Bolton et al., 1982; [13]Doyle and Roman, 1982; [14]Christopher et al., 1982; [15]Ehlers et al., 1982; [16]Lander, 1982; [17]Park and Stankiewicz, 1982; [18]Butzler et al., 1983; [19]Goossens et al., 1983; [20]Lovett et al., 1983; [21]Martin et al., 1983; [22]Park et al., 1983; [23]Rosef et al., 1983a; [24]Roseff and Kapperud, 1983; [25]Wesley et al., 1983; [26]Bolton et al., 1984b; [27]Hutchinson and Bolton, 1984; [28]Ray and Johnson, 1984b; [29]Waterman et al., 1984; [30]Goossens et al., 1986; [31]Humphrey, 1986b; [32]Karmali et al.,1986; [33]Weber et al., 1987; [34]De Boer and Humphrey, 1989; [35]Goossens et al., 1989; [36]Bolton, personal communication; [37]Parl and Sanders, 1981; [38]Hunt and Radle, 1992 (cited by Stern and Line, 1992); [39]Stern et al., 1992; [40]Aspinall et al., 1993; [41]Line, 2001.

A, agar; B, broth; SSM, semi-solid selective motility medium; ATB, alkaline tryptone broth.

Other details (pH, growth factors, other additives) [a] pH 7.3-7.4 ; [b] yeast extract, resazurin; [c] Na succinate, cysteine HCl; [d] 3% calf serum; [e] 10 g bicine, pH 8.0; [f] casein hydrolysate; [g].modified protocol: add polymyxin after 6 h at 37C for freeze-stressed campylobacters; [h] antibiotics added afte 4 h at 37'C; [i] pH 7.4, 1 g Na deoxycholate; [j] α-ketoglutaric acid, $NaC0_3$.

COLUMN 2. Basal medium: TA, thioglycollate agar; TB, thioglycollate broth base; BA, brucella agar base; BB, brucella broth base; NB, nutrient broth; BAB, blood agar base; VI, veal infusion; CA, Columbia agar base; GA, gonococcus agar base; T, tryptose; MHB, Mueller-Hinton broth; YE yeast extract; IVx, IsoVitalex (vitamin and nutrient mixture (BBL)); A, 10 g meat peptone, 5 g lactoalbumin hydrolysates, 5 g yeast extract, 5 g NaCl, 0.6 g $NaCO_3$.

COLUMN 3. Antioxygen system: S, sheep blood; H, horse blood; LH, lysed horse blood; DH, defibrinated horse blood; C, charcoal; F, ferrous sulphate ($FeSO_4.7H_2O$); B, sodium metabisulphite ($NaHSO_3$); P, sodium pyruvate at 0.5 g per litre unless otherwise stated: HT, haematin; HE, haemin.

COLUMN 4. Cephalosporins; cthin, cephalothin; czolin, cephazolin; czone, cefoperazone., codin, cefsulodin. * 15 mg cefalexin instead of cephalothin in recipe of Butzler and Skirrow, 1979 (used without cephalosporins by ButzIer et al., 1973). ** added later. *** + 15mg later. [x] broth only. [Y] agar only.

COLUMN 12. Antifungals: CY, cycloheximide (actidione); AM, amphotericin B; NY, nystatin.

sodium pyruvate was best. A further study (Bolton et al., 1984b) surveyed the effect of 11 dyes, 17 chemical compounds and 14 chemotherapeutic agents on one strain each of *C. jejuni* biotype 1, *C. jejuni* biotype 2, *C. coli* and a NARTC (*C. lari*) and a selection of Gram-positive and Gram-negative competitive bacteria. No details were given of either the competitive test strains or the inhibitors examined, but deoxycholate and cefazolin were chosen as the most effective inhibitory agents. Casein hydrolysate was found necessary to stimulate the growth of environmental NARTC (*C. lari*) strains. CCDA is the only medium except for CAT agar (see below) to use deoxycholate and apart from Karmali's medium is the only widely used plating medium not using blood. This is useful, particularly for laboratories specialising in food microbiology in which the use of blood in media is unusual. Blood is expensive, has a short shelf life and is easily contaminated. Later Hutchinson and Bolton (1984) replaced cefazolin (10mg/l) with cefoperazone (32mg/l). This allowed fewer contaminants to grow, and permitted the modified medium (mCCDA) to be used at 37°C. However, amphotericin B was needed to prevent overgrowth by yeasts able to grow at 37°C but not at 42°C.

No rationale is provided by Blaser et al. (1979) for their "Campy-BAP" medium, which has a similar formulation to Skirrow (1977) agar, but with added cephalothin, as used by Lauwers et al. (1978).

In 1983 Butzler and co-workers produced a second selective plating medium called Butzler medium Virion (Goossens et al., 1983). This was said to be superior to a slightly modified version of their 1978 medium (Lauwers et al., 1978), "Butzler medium Oxoid", which had a different basal medium, less blood and cephazolin instead of cephalothin. Butzler medium Virion substituted cefoperazone for cefazolin, rifampicin for vancomycin, colistin for polymyxin and trimethroprim and amphotericin for cycloheximide. The Virion medium was said to suppress competitive flora more effectively, especially pseudomonads and Enterobacteriaceae. Campylobacter colonies were easier to recognise. Incubation temperature was reduced from 42°C to 37°C (Butzler et al., 1983). Goossens et al. (1986) modified the Virion medium, omitting the colistin and doubling the concentration of cefoperazone in order to detect colistin-sensitive campylobacters, including some strains of *C. coli*, reported by Ng et al. (1985) to be sensitive to colistin and polymyxin. Goossens and co-workers found no significant improvement over their Butzler Virion agar. However, they recommended its use in preference to the earlier medium, presumably because they felt it would be able to isolate colistin and polymyxin sensitive campylobacters. Burnens and Nicolet (1992) were able to isolate *C. upsaliensis* after changing to the medium of Goossens et al. (1986).

The medium of Karmali et al. (1986) is a variation of modified CCD medium (Hutchinson and Bolton, 1984) using haematin rather than ferrous sulphate, vancomycin instead of deoxycholate and cycloheximide instead of amphotericin B. Vancomycin at 20mg per litre rather than the more common 10mg per litre was chosen for better suppression of Gram positive competitors (*Bacillus* spp. and enterococci). Cycloheximide was chosen because of superior yeast suppression. In agreement with Hutchinson and Bolton (1984) cefoperazone was found superior to cephalothin. The most numerous contaminants were found to be Enterobacteriaceae which are resistant to cefoperazone when present in high numbers, especially *Klebsiella oxytoca*. The rationale for the modification to the oxygen quenching system was not stated.

The semi-solid medium of Goossens et al. (1989) relies upon the ability of campylobacters to swarm, analogous to the system used in semi-solid Rappaport Vassiliadis media for salmonellas (Busse, 1995). The use of a semi-solid medium, where growth and swarming is mostly below the surface apparently dispenses with the need for blood, charcoal or other anti-oxygen system other than a microaerobic atmosphere. The medium has to be stored in the dark at 4°C, and prepared twice weekly. Cefoperazone (30mg/l) and a high level of trimethoprim (50mg/l) are used as selective agents. Three strains of *C. jejuni* and seven strains of *C. coli* with MICs to cefoperazone <100μg/ml (screened in a previous study of 200 inhibitors) were tested for their ability to swarm in the new medium; all gave satisfactory results. Eleven different campylobacter strains were tested for their ability to initiate swarming in the medium. A range of 1–15 cells was necessary to initiate growth, with a mean of 6.7. The medium functioned very well by comparison with other plating media. Neat faecal samples were inoculated using a loop at the edge of 50mm diameter Petri dishes of the medium. Incubation was at 42°C for 42h in a microaerobic atmosphere or a candle jar. Positive campylobacter cultures were recognised by the characteristic swarming growth. Presumably isolation required subculture to selective or non-selective plates to obtain a pure growth. Polymyxin or other Gram-positive inhibitors, as well as anti-fungal antibiotics were not necessary because competitor bacteria did not swarm. The medium had a number of advantages: cheapness because of the small volumes needed, no necessity for blood, only two antibiotics in the formula, easy interpretation of swarming, low incidence of contaminants, and the ability to fit large numbers of 50mm plates into gas jars. A disadvantage would be the unsuitability of semi-solid agar for quantitative estimates of campylobacters, unless MPN techniques were used.

In 1993 Aspinall et al. developed a modification of mCCD agar designed for use at 37°C to isolate *C. upsaliensis* as well as the other thermophilic campylobacters. Of 51 strains of C. *upsaliensis* tested, 47 were resistant to 8 mg per litre of cefoperazone in CCD agar base and all were resistant to 64 mg per litre of teicoplanin. The medium contains 8 mg/l cefoperazone and 4 mg/l teicoplanin, replacing 32 mg/l cefoperazone in mCCD agar (Table 3). Teicoplanin has an antimicrobial spectrum similar to that of vancomycin, active mainly against Gram-positive bacteria. By comparison with mCDD agar the final formulation isolated equivalent numbers of *Campylobacter* spp. other than *C. upsaliensis* from faeces and was superior to mCCD agar for *C. upsaliensis*, with slightly more growth of competitors. These results were confirmed using faeces artificially inoculated with C. *upsaliensis.* Aspinall et al. (1996) found CAT agar superior to mCCDA for isolating *C. upsaliensis* from animal faeces, and this medium is used routinely in Preston Public Health Laboratory for the examination of faeces (Dr David Wareing, personal communication). Line (2001) proposed a medium for isolation of campylobacters from poultry, which is a modification of Campy-Cefex agar, and includes triphenyltetrazolium chloride (TTC) as indicator, as well as trimethoprim, polymyxin and vancomycin, in addition to the 33mg/l cefaperazone used in Campy-Cefex (Table 3). . Colonies of campylobacters are a deep red (cf. use of TTC in the medium of Glupczinski et al. (1988) which yields golden-yellow colonies of *H. pylori* – Table 5). This medium apparently performs well (Line et al., 2001).

Membrane filter method of Steele and McDermott (1984)

This method is a simplification of the filtration method of Dekeyser et al. (1972). It employed 47 mm 0.45 μm pore size cellulose triacetate membrane filters, placed centrally on a 90 mm deep 6% blood agar plate. Ten to twelve drops of a 1 in 10 suspension of faeces in physiological saline solution were placed onto the membrane filter using a Pasteur pipette and taking care not to let the drops spill over the edge of the membrane. The membrane was removed after 30 min, by which time the fluid (and presumably sufficient campylobacters) had passed through. It appears that some of the campylobacters are able to move through the membrane pores. The plates were then incubated micro-aerobically at 41°C for 3–5 days, and examined daily. Cellulose nitrate filters were unsatisfactory (a result confirmed by Megraud and Gavinet, 1987) and 0.8 μm pore size gave unacceptable levels of contamination; 0.65 μm filters were not tested. Comparison of the filter method with a selective medium containing sheep blood, FBP supplement, trimethoprim, vancomycin and colistin gave 45 positive with the selective medium and 50 with the filter method out of a total of 56 positive samples from the 1000 tested. Six strains of *C. jejuni* were not isolated by the filter method and 11 *Campylobacter* spp. were not isolated with the selective medium. Six of these were *C. jejuni*. None of the non-*C. jejuni* strains grew in the selective agar because of the colistin present (they were also sensitive to polymyxin). These authors found that the filter method usually gave either campylobacter colonies or no colonies at all. However, only about 10% of campylobacters in the faecal suspension actually passed through the filter.

This method or a similar one (e.g. using 0.65 μm filters) has been used by various workers, often in parallel with selective agar media or after selective enrichment (Megraud and Gavinet, 1987; Goossens et al., 1990b; De Boer and Reitsma, 1991; Moreno et al., 1993; Scotter et al., 1993; Van Etterijck et al., 1996; Engberg et al., 2000). Shanker et al. (1991), working with 0.45 μm cellulose acetate or mixed cellulose ester filters and artificially contaminated faeces, found that 10^4/g of most strains of campylobacter could be detected. In their procedure the filter was removed 10 min after applying the faecal suspension. We have found this method very useful for direct plating or for plating after enrichment, particularly when groups other than thermophilic campylobacters are sought. We normally use 0.65 μm pore size cellulose acetate filters (Atabay and Corry, 1997). The study by Van Etterijck et al. (1996) found a combination of a 0.45μm or 0.65μm pore membrane filter plating, plus plating on a selective agar medium the most effective method of isolating *C. concisus* from faeces. Aspinall et al. (1996) found that 0.45μm pore filters caused 150-fold reduction in recovery, and 0.65 μm pore filters 50-fold reduction compared with direct plating of pure cultures onto CAT agar. De Boer and Reitsma (1991) found that there was little difference between the performance of cellulose acetate or nitrate filters of 0.65 μm pore with respect to penetration of campylobacters, but that competitors were least able to penetrate the 0.65 μm pore cellulose nitrate filters.

An interesting modification of this technique was devised by Baggerman and Koster (1992). These workers enriched 20 g of raw poultry samples in 100 ml of mCCD enrichment broth for 8 h at 42°C, and then allowed the culture to filter by gravity through a 0.45 μm pore cellulose nitrate filter unit. Incubation of the mCCD broth was continued microaerobically and campylobacters detected by use of a latex agglutination kit or plating. This method was found to be as effective as the original Steele and McDermott method or conventional enrichment and plating, but more rapid and more sensitive because a larger volume of sample passes through the filter.

Enrichment methods for isolation from food

The Preston and mCCD plating media of Bolton and co-workers have been utilised as enrichment media using the same inhibitors, and incubation at 37 or 42°C. Liquid Preston medium contains 0.25 g/l FBP in addition to 5% horse blood (Bolton et al., 1982; Baird et al. 1987). Both media have been found useful for isolating campylobacters from samples in which campylobacter numbers are comparatively low (e.g. faeces of normal animals, food, water and drain swabs - Bolton et al., 1982; 1983).

Park et al. (1981) developed an enrichment broth based on Brucella broth, containing vancomycin, trimethroprim and polymyxin B (VTP). It was later modified (Park et al., 1983) by adding FBP supplement and lysed horse blood, reducing the amount of the polymyxin B and increasing that of vancomycin and trimethoprim (VTP-FBP) - see Table 3. The media were used to examine raw chickens. Carcasses were rinsed with nutrient broth, which was then filtered through cheesecloth and centrifuged. The sediment was resuspended in Brucella broth and then plated either directly onto selective agar or added to the enrichment broth. Incubation was carried out at 37°C under microaerobic conditions. VPT broth was found to detect 0.2 campylobacters per g of sediment in the presence of 10^4–10^6 competitors per g (Park et al., 1981). Enrichment showed 62% of chickens to be positive for campylobacters, compared with 32% by direct plating. VPT-FBP broth functioned better at 37°C than 42°C and 48h incubation was better than 24 h or 72 h. pH 7.0 was optimal.

Christopher et al. (1982) used a liquid version of Campy-BAP agar (see Table 3) with added pyruvate. This medium contained cephalothin and amphotericin B in addition to vancomycin, polymyxin and trimethroprim. Ehlers et al. (1982) modified this medium by adding FBP. The broth was incubated at 42°C in a microaerobic atmosphere. The limit of detection in cheese artificially contaminated with *C. jejuni* and using an MPN method was about 0.3 campylobacters per g. Acuff et al. (1982) published a similar modification containing FBP, but without blood. Doyle and Roman (1982) developed a broth similar to that of Park and co-workers, but with a very high level of polymyxin (20,000 iu per litre), cycloheximide and an oxygen quenching system of 7% lysed horse blood, as well as added succinate and cysteine HCl. Incubation was microaerobic at 42°C, for 16–18 h. The medium was evaluated using a variety of foods artificially inoculated with 46 strains of *C. jejuni,* two of *C. coli* and two of *C. lari*. Ten or 25 g of food were homogenised directly in 90 or 100 ml of enrichment broth. Campy-BAP agar was used as the plating medium. All strains in raw hamburger and raw milk were detected at 1–4 cells per g and most at 0.1 to 0.4 cells per g. The broth was less effective for chicken, probably due to the numbers and types of competitors.

Wesley et al. (1983) used a medium (alkaline tryptone broth) containing rifampicin, polymyxin at 20,000 iu per litre and 6.25 mg/l of cefsulodin. The levels of polymyxin and cefsulodin chosen were necessary to inhibit *Pseudomonas aeruginosa* contaminants. FBP and haematin were included as oxygen quenchers and the optimum pH level was 8.0. The broth (100 ml) was inoculated with 10 ml of carcass rinse fluid. Incubation was microaerobic at 42°C for 48 h. Comparison of this medium with the medium and sampling method of Park et al. (1981) for isolation of campylobacters from naturally contaminated chicken gave a much higher isolation rate (25 versus 6 out of 50 chickens using the Wesley or Park medium respectively).

Lovett et al. (1983) modified the medium of Park et al. (1981). They added FBP, and, as a result of testing growth of *C. jejuni* strains and competitors, reduced the level of polymyxin and increased the levels of the other antibiotics (Table 3). In the Park formulation polymyxin at 8,000 iu/l was found to be inhibitory to many strains of *C. jejuni*.

Effect of damage

Ray and Johnson (1984a and b) observed that viability of freeze-injured *C. jejuni* was reduced when cells were incubated in selective broth or (to a lesser extent) on selective agar at 42°C. They used the liquid medium of Acuff et al. (1982) and Campy-BAP agar. The basal media without antibiotics were also inhibitory, an effect attributed partly to the high incubation temperature. Blood reduced the toxicity, as also did succinate and cysteine. Polymyxin was toxic to the injured cells but the other antibiotics were not. As a result of these studies a medium was proposed using a modification of the broth of Ehlers et al. (1982), which contained succinate and cysteine but no FBP supplement. Incubation was at 37°C for the first 6 h to resuscitate injured cells before adding the polymyxin and raising the incubation temperature to 42°C.

Humphrey and Cruickshank (1985) investigated the effect of a variety of inhibitors used in selective media on six serotypes of *C. jejuni* uninjured or injured by freezing, heating or EDTA. Bacitracin, trimethoprim, cefoperazone, colistin, novobiocin, vancomycin and deoxycholate were tested in blood agar, all at levels used in selective media. The growth rate and colony size of undamaged strains were reduced by rifampicin. Injured cells were more affected. Deoxycholate was toxic for some but not all injured cells. The toxic effect of rifampicin and deoxycholate in complete medium was confirmed for one strain when freeze-damaged. Counts on Preston medium were reduced compared to Skirrow, Campy-BAP or blood agar for one of the six strains. Further work (Humphrey 1986a and b) confirmed the observations of Ray and Johnson (1984b) that damaged *C. jejuni* recovered better in media at 37°C than 42 or 43°C, even in the absence of inhibitors. An enrichment medium and method was subsequently proposed which had nutrient broth as the base, to which 5% lysed horse blood and FBP were added. Antibiotics were omitted for the initial period of incubation. Food samples were incubated in the basal broth for 2 h at 37°C before addition of the antibiotics and incubation was continued at 43°C for up to 48 h. The antibiotics were trimethoprim, cefoperazone, colistin, amphotericin B and either vancomycin or rifampicin (see Table 3). Some inhibition of *C. jejuni* was observed even after the 2 h resuscitation period at 37°C. Vancomycin was less inhibitory than rifampicin. No information was supplied about the incubation atmosphere. The plating medium used following enrichment had a similar formulation, but substituted rifampicin for vancomycin in order to suppress competing flora. In a later paper, Humphrey (1989) suggested that pre-enrichment at 37°C should be continued for 4 h and that addition of all antibiotics should be delayed until the 4h pre-enrichment had been completed. The rifampicin-containing enrichment medium was recommended in parallel with a similar plating medium (Humphrey 1986b; Table 3). Pre-enrichment at 37°C for 4 h was recommended for all types of sample, even those heavily contaminated, such as sewage and chicken skin, although for these selective medium, rather than the basal enrichment broth, was recommended. Later the broth was modified to contain polymyxin instead of colistin, because of difficulty in obtaining colistin (De Boer and Humphrey, 1991; Humphrey personal communication). The latest version of Exeter broth contains (mg per l) cefoperazone (15), trimethoprim (10). polymyxin (2.500 iu), rifampicin (5), amphotericin B (2) (Humphrey 2001, personal communication).

A similar enrichment medium for use with food was suggested by Bolton (Table 3; personal communication; Anon, undated). It has a rich basal medium to aid resuscitation of sublethally damaged campylobacters. As with Humphrey's Exeter enrichment medium, preliminary incubation of the medium complete with antibiotics, for 4 h at 37°C was recommended to aid resuscitation of injured organisms, followed by 42°C for 14–48 h. The combination of some of the FBP ingredients with lysed horse blood and haemin, together with a small headspace in the culture vessel was designed to avoid the need for a microaerobic atmosphere. Sodium carbonate was added, presumably to prevent acidification and/or provide a source of carbon dioxide. The broth is used in a 1:4 (w/v) ratio of food to medium, with a 1.5 cm headspace in a screw-capped bottle (with the cap presumably tightened). It can also be used to enrich faeces (1 ml of 10% suspension added to 5 ml broth). Recently the medium has been modified by omitting the lysed horse blood and adding FBP supplement (Bolton, 2000; Bolton et al., 2002). The incubation recommendations have also changed; 37°C for 24 h, followed by a further 24 h at 42°C is recommended (Bolton et al., 2002).

Another enrichment method employing similar principles was published in abstract only by Park and Sanders (1991) (see Table 3 for formulation). No rationale was supplied for the choice of antibiotics. The method involved incubation of the broth containing vancomycin and trimethoprim for 4 h at 31–32°C, followed by addition of cefoperazone and cycloheximide. Incubation temperature was raised to 37°C for 2 h and then 42°C for 40–42h. All incubation was microaerobic; static at 30–32° and 37°, and shaking at 42°C.

Yet another enrichment medium, a variation of Park and Sanders broth, also employing a period of incubation at reduced temperature, and delayed addition of antibiotics was devised by Hunt and Radle (cit. Stern and Line, 1992; Hunt, 1992). This medium contained 0.25 g/l FBP and 5% lysed horse blood, cefoperazone, trimethoprim, vancomycin, and amphotericin B or cycloheximide (see Table 3). Pre-enrichment was carried out in the medium containing all the antibiotics (except for 15 mg/l cefoperazone of the final total of 30 mg/l) at 32°C in a microaerobic atmosphere in flasks sealed in plastic bags and shaken in a water bath. After 3 h the remainder of the cefoperazone was added and the temperature of the water bath raised to 37°C. After a further 2 h at 37°C, the bath temperature was raised to 42°C.

The FDA BAM (Bacteriological Analytical Manual – FDA, 1992) recommended use of Hunt and Radle broth for most types of food, modified Exeter broth for water and environmental samples and Hunt and Radle broth and modified Preston agar for dairy products. Frozen products or products stored chilled for 10 days or more were pre-enriched according to the Hunt and Radle protocol or the Humphrey (1989) protocol. Foods not chill-stored were examined by the Hunt and Radle protocol, but adding all antibiotics at the beginning of incubation. Shaking water baths were recommended for incubation, with continuous gas flow (except during incubation at 30°C if it was static). Plating was on mCCDA and Campy-Cefex directly and after enrichment. In general 25g food was added to 100ml of enrichment broth. Liquid products such as milk and ice cream were centrifuged. Chickens were rinsed, the rinse fluid centrifuged and the pellet added to the enrichment broth.

The 1998 version of the BAM recommends only Bolton broth for enrichment followed by mCCDA (plus added yeast extract) or Abeyta-Hunt-Bark agar, which is similar to the CAR medium of Weber et al. (1987), and has (mg/l) cefoperazone (32), rifampicin (10) and amphotericin B (2.5) in a base of heart infusion agar with yeast extract and FBP. Liquids such as milk and ice cream are centrifuged to concentrate the microflora, foods such as vegetables and meat are rinsed with enrichment broth. Samples with large numbers of competitors are diluted at 1:100 as well as 1:10 for enrichment. Enrichment broths can be incubated in gassed pouches or jars, as well as using continuous gassing.

Mason et al. (1999) examined naturally contaminated river water and frozen chicken to see whether delayed addition of selective agents improved isolation rates. A delay of 4–8 h before adding antibiotics to broth significantly increased the isolation rate from water, compared with direct culture in selective broth. With the chicken, however, significantly better results were obtained with selective broth as the primary medium. This was in agreement with results obtained by Uyttendale and Debevere (1996), who concluded that non-selective pre-enrichment was not beneficial when isolating campylobacters from poultry products using Preston broth. They recommended direct enrichment at 42°C for 24 h. They also found no benefit in including blood in Preston enrichment medium, which already contains FBP supplement.

Isolation from surfaces Campylobacters do not survive well on dry surfaces, and Humphrey et al. (1995) found that they are most likely to be found on moist surfaces. Comparison of survival rates in various dilutents and media indicated that it is best to put swabs straight into enrichment medium rather than diluent, that addition of FBP supplement to diluents does not improve survival, and that Fastidious Anaerobe Broth is better than buffered peptone water or peptone saline as a transport medium (Humphrey et al., 1995).

Atmosphere during incubation of enrichment media

No systematic investigation of the effect of atmosphere during incubation of liquid media has been carried out. Some earlier workers used static incubation in normal air in screw capped bottles, sometimes with a small airspace, sometimes with the airspace undefined (Bolton et al., 1982; Fricker et al., 1983; Martin et al., 1983; Waterman et al., 1984). Others used static incubation in bottles or tubes in a microaerobic atmosphere (Acuff et al., 1982; Ehlers et al., 1982; Megraud, 1987; De Boer and Reitsma, 1991). Doyle and Roman (1984) used 250ml conical flasks possessing side-arms and replaced the atmosphere with the microaerobic gas mixture. These were evacuated and filled three times. Park et al. (1981; 1983) and Lovett et al. (1983) bubbled microaerobic gas mixture through their enrichment broths. Park and Sanders (1991) used static incubation of plastic storage bags flushed with microaerobic atmosphere. The report of Ribeiro and Price (1984) who used Preston broth containing FBP supplement (George et al., 1978) to isolate campylobacters from water indicated that a microaerobic atmosphere during incubation gave a higher isolation rate. Wesley et al. (1983) reported that better results were obtained with a high surface area of broth in a flask compared to a lower surface area in tubes, in both cases using a microaerobic atmosphere. Comparison of Doyle and Roman's evacuation-replacement procedure with a continuous gas flow procedure using Doyle and Roman's enrichment broth for examining inoculated hamburgers, indicated that the continuous gas flow gave counts 1–2 log cycles higher for six out of eight strains *of C. jejuni* for both tests (Heisick et al., 1984). There appears to be no published justification for the FDA recommendation of continuous gas flow with agitation (Hunt, 1992; Hunt et al. 1998).

A preparation of membranes from *Escherichia coli* ("Oxyrase Enzyme System", Oxyrase Inc., Manfield, Ohio, USA) has been found effective at 0.15 or 0.6 units per ml of liquid medium instead of a microaerobic atmosphere for cultivation of thermophilic camplyobacters (Adler and Spady, 1995; Tran 1995; Wonglumsom et al., 2001). Wonglumsom et al. (2000) found that Oxyrase could compensate for a high headspace to medium ratio in containers, without the need to use modified atmosphere. However, with low headspace to medium ratio there was little advantage in using Oxyrase, over aerobic incubation of sealed containers.

Comparisons of media for isolation from faeces

Comparisons of media (Morris et al., 1982; Bolton et al., 1983; Fricker, 1983; Merino et al., 1986; Gun-Munro et al., 1987; Rosef et al., 1987; Albert et al., 1992; Moreno et al., 1993) using pure cultures, faeces or water, reviewed in the previous edition of this review (Corry et al., 1995) will not be covered in detail here. Bolton et al. (1983) and Fricker (1983) found that Butzler medium grew *C. coli* and *C. lari* poorly compared with other plating media. Studies using the membrane filter method of Steele and McDermott (1984) usually found that it was productive, especially in isolating a wider range of Campylobacteraceae than direct plating onto selective agar, and use of more than one plating medium/method increases the number of samples found positive (Megraud and Gavinet, 1987; Albert et al., 1992; Moreno et al., 1993; Persimoni et al., 1995; Engberg et al., 2000). Membrane filters were used in the conventional way by Rosef et al. (1987) to examine water. Filters of 0.45 μm pore size were found more useful than 0.22 μm pore size, incubating face up on selective agar media, or putting into Preston enrichment broth. More recent comparisons, which have included the charcoal-containing selective media (mCCDA and/or Karmali) have found these to be more productive and selective for faecal samples than other selective media (Endtz et al., 1991; Albert et al., 1992; Piersimoni et al., 1995).

When large numbers of campylobacters and competitive flora are likely to be present (e.g. in chicken caecal contents), direct plating may be more productive than enrichment followed by plating (Musgrove et al., 2001).

Comparison of media for isolation from foods

Beuchat (1985) compared Campy-BAP, CCDA (unmodified - Table 3 - Bolton and Coates (1983)) and Butzler agar (as in Table 3: with cephalothin), plus five enrichment media (by MPN) for the isolation of campylobacters from chill-stored chicken. The chicken had been artificially inoculated with five strains of *C. jejuni*. Recovery was poor with all the enrichment/plating combinations, but the enrichment media of Christopher et al. (1982) and Park et al. (1983) combined with Butzler's or CCDA media gave best results. The medium of Martin et al. (1983) performed particularly poorly. Preston broth and the medium of Rosef and Kapperud (1983) gave intermediate results. Direct plating of chicken samples diluted in 0.1% peptone water gave better recovery than enrichment, with CCDA and Campy-BAP yielding higher counts than Butzler agar. Preliminary study of Doyle and Roman's enrichment broth indicated that it was superior to other media.

Fricker (1984) undertook a comparison of three enrichment media: Preston, Lander and Doyle and Roman all incubated at 42°C, one transport/storage medium stored at 4°C (Campy-thio: Blaser et al. 1979) and two plating media - Preston and Campy-BAP. All media had 0.05% FBP added and were incubated aerobically in 6 ml quantities in quarter oz bottles (ca. 7.5 ml). Naturally contaminated giblets from frozen chickens were examined, of which 177/198 were positive for campylobacters. Doyle and Roman and Preston media both detected 176 positive samples while Lander medium detected 159 and the Campy-thio medium 69. Samples were not plated directly, but the result using Campy-thio, in which multiplication of campylobacters would not have occurred, was probably similar to the result that would have been found by direct plating. Preston agar isolated campylobacters more frequently than Campy-BAP (577 v. 562 times). Forty-eight hours was the optimal time for incubation at 42°C. Enrichment in either Doyle and Roman or Preston broth followed by plating onto Preston agar was recommended.

A comparison of the Doyle and Roman enrichment method with that of Lovett et al. (1983) for isolation from milk was carried out by Hunt et al. (1985). The Lovett method was significantly better. Heisick et al. (1984) and Heisick (1985) also compared the Doyle and Roman broth with the Lovett et al. medium, using artificially inoculated beefburgers or milk-puddings and various strains of *C. jejuni*. The two methods gave similar results, although higher numbers of *C. jejuni* were sometimes achieved in the Lovett medium, especially when added in the absence of food.

Peterz (1991) conducted a collaborative trial amongst six laboratories in Scandinavia, testing chicken liver artificially inoculated with two strains of *C. jejuni* at levels between 0.3 and 59/g. Samples of 2.5g were enriched in 50 ml of Preston broth, without FBP but incubated microaerobically, at 42°C for 24h, followed by plating onto mCCDA and Preston agar. A detection limit of about one campylobacter per g of liver was determined. The two plating media gave similar isolation rates, although mCCDA contained fewer contaminants.

De Boer and Humphrey (1991) compared Park and Sanders method to Preston broth and mCCD broth for chilled or frozen chicken and found Park and Sanders superior, but gave no details. Exeter medium (modified from that described by Humphrey (1986b) see Table 3) was reported to give good results when used with an 18 h pre-enrichment step at 37°C.

De Boer and Reitsma (1991) compared the effectiveness of Preston broth (without FBP) and mCCD broth in combination with plating onto Skirrow, Preston, mCCDA and filtration through an 0.65μm pore cellulose nitrate membrane used with blood agar (modified Steele and McDermott, 1984). All incubation was microaerobic and naturally contaminated chilled and frozen chickens were examined. The two enrichment broths gave similar results, most positives occurring with mCCD broth plated via the membrane filter. Direct plating of frozen chicken gave much lower numbers of positive results than when samples were enriched (e.g. 8% positive versus 27% on mCCDA with or without enrichment in Preston broth respectively). With chilled chicken, direct plating on mCCDA gave 32% positive, compared with 47% positive after enrichment. Direct plating using the membrane filter method was completely ineffective.

The study of Furanetto et al. (1991), also examining naturally contaminated chilled chicken carcasses, compared direct plating on Campy-BAP or VTP-FBP agar with plating on the same media after enrichment in Doyle and Roman or Preston broth. All incubation was static in a micro-aerobic atmosphere. The combination of Doyle and Roman enrichment with Campy-BAP gave the highest number (16/42) of positive samples, although the results were not significantly different from direct plating on either Campy-BAP (13/42) or VTP-FBP agar (11/42). The combination of Doyle and Roman broth with VTP-FBP agar gave particularly poor results (1/42).

Stern and Line (1992) compared the enrichment broths of Doyle and Roman, Park and Sanders and Hunt and Radle (plating onto Campy-BAP, mCCDA and Campy-Cefex) for examining whole chilled, naturally contaminated chickens. Inocula were prepared by rinsing the carcasses with 200 ml buffered peptone, filtering the rinse through cheesecloth, centrifuging, discarding the supernatant, streaking some of the pellet onto the selective agars, resuspending the rest of the pellet in 5 ml buffered peptone water and inoculating 1 ml of the resuspension to 100 ml of enrichment broth. Microaerobic atmosphere was obtained by replacing the air in polythene bags with the appropriate gas mixture (flushing 3 times). All bags were incubated in water baths with shaking. The Doyle and Roman broth was incubated at 42°C. The Park and Sanders and Hunt and Radle broths were incubated at 32°C for 3.5 h, 37°C for 2.5 h and then at 42°C. Cefoperazone supplement was added to the Hunt and Radle broth and cefoperazone and cycloheximide to the Park and Sanders broth before raising the temperature from 32 to 37°C. Re-gassing of the enrichment broths was carried out after adding the supplements. Ten out of 50 chicken carcasses were positive for campylobacter by direct plating. Forty-nine out of 50 were detected by all the enrichment/plating methods combined Using Doyle and Roman, Park and Sanders and Hunt and Radle broths, 23, 40 and 43 positive chickens were detected respectively. mCCD agar gave most positive samples overall, and with the Park and Sanders method, although all three selective media detected similar numbers of positive samples from Hunt and Radle broth.

The study of Scotter et al. (1993) used frozen samples of naturally contaminated chicken skins, in which the campylobacters were more likely to be damaged, and therefore require pre-enrichment. Twelve laboratories in the UK participated in the trial. Tests were carried out using two methods defined in a draft International Standards Organisation (ISO) Method. The three enrichment methods used 10g quantities of food in 90ml of enrichment medium, under a small headspace. They were as follows: (1) Preston broth without FBP supplement, incubated in micro-aerobic atmosphere at 42°C for 48 h; (2) Park and Sanders broth incubated aerobically, with antibiotic supplement and at times and temperatures prescribed by Park and Sanders (1991); (3) Preston broth containing FBP supplement, aerobically, but according to the temperatures and times prescribed by Humphrey (1989) – 37°C for 4 h and then 42°C for 44 h. All broths were plated directly onto Skirrow agar. Method 1 used any selective medium chosen by the participant, method 2 used a 0.65 μm pore membrane on non-selective blood agar and method 3 used Preston agar in addition. All plates were incubated at 42°C. At 10 cells per 10 g chicken skin all three methods were equally effective. At about 2 cells per 10 g, method 2 gave significantly higher isolation rates. Skirrow agar was frequently overgrown with pseudomonads and proteus. The membrane plus non-selective agar was as effective as selective agars such as mCCDA. No advantage was gained by pre-enriching at 37°C for 4 h (method 3). However, pre-enrichment at 37°C for 4 h was intended by Humphrey (1989) for use with his (Exeter) medium (Humphrey, 1986b), not Preston broth.

De Boer et al. (1998) found Preston broth to be superior to mCCD broth or CAT broth, and Karmali, mCCDA and CAT agar better than SSM for isolation of campylobacters from chicken meat.

Aquino et al. (1996) compared isolation of campylobacters from chicken meat directly onto Blaser or Skirrow agar and after enrichment in Doyle and Roman broth, modified by use of Blaser's enrichment supplement. Direct plating detected 38 out of 40 samples positive, while enrichment found only 19 out of 40. A more recent study by Line et al. (2001) found that recovery of campylobacters from chicken rinse samples was no better after enrichment in Hunt and Radle broth and plating onto mCCDA than direct plating onto Campy-Cefex. Our experience (Corry, unpublished) using Exeter broth and mCCDA indicates that enrichment and plating of carcass rinse samples is more sensitive than direct plating, but that this is not true when examining caecal contents.

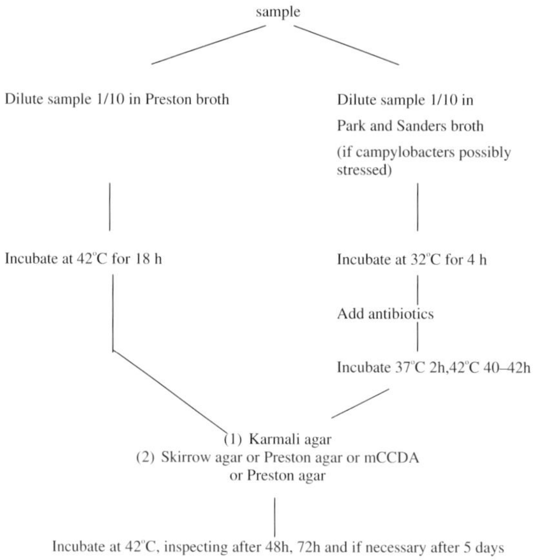

Fig. 1. International Standards Organisation method for the detection of thermophilic campylobacter in foods (ISO 10272, 1995).

Published comparisons of Bolton broth with other enrichment media are rare, in spite of its popularity. Musgrove et al. (1997) and Baylis et al. (2001) found Bolton broth better than Hunt or Preston broth for isolation from naturally contaminated poultry litter and food respectively.

Standard methods for detection of thermophilic campylobacters in foods

Fig. 1 shows the ISO method (1995). It recommended enrichment in Preston broth if the campylobacters were not likely to be stressed, or Park and Sanders broth if stressed. A surprisingly short incubation time was recommended for Preston broth of only 18 h, which may explain why Federighi et al. (1999) found Park and Sanders method (48 h) to be superior to Preston (18 h) for isolating campylobacters from food. The ISO method is currently being revised, and is likely to be changed as follows: Preston broth will be incubated at 41.5°C for 24 and 48 h, instead of 42°C for 18 h; Bolton broth will be used instead of Park and Sanders (incubating Bolton broth at 37°C for 4 h and 41.5°C for 42–48 h). mCCDA is retained as first choice plating medium, with a second medium of own choice (De Boer, personal communication). The incubation temperature of 41.5°C for both enrichment and plating media matches the temperature recommended in the standard for salmonella enrichment. The FDA standard method has also adopted Bolton broth (see p. 291).

Isolation methods for *Arcobacter* species

The selective agents used for arcobacters are similar to those used for thermophilic campylobacters except for the use of 5-fluorouracil (5-FU: see Table 4). This agent was found by Johnson and Rogers (1964) to be useful for selecting Leptospirae and used by Ellis et al. (1977) in a *Leptospira* semisolid enrichment medium, EMJH P-80 (Ellinghausen-McCullough-Johnson-Harris Polysorbate-80) supplemented with 100 μg/ml of 5-FU. These authors unexpectedly isolated a spiral-shaped organism from aborted bovine foetuses which was almost certainly an *Arcobacter* (Neill et al. 1985). 5FU has also been used in an enrichment medium for *Campylobacter* spp., especially *C. fetus* subsp. *venerealis* (Lander, 1982, 1990; Lander and Gill, 1985). Since then, various enrichment and plating media and isolation protocols have been devised for the recovery of *Arcobacter* spp. and are summarised in Table 4. Methods used in order to avoid isolating campylobacters include incubation at 30°C or below, omitting protective agents such as blood, FBP or charcoal, and aerobic incubation.

Lammerding *et al.* (1996) used an enrichment broth, first used by Rosef (1981) for *C. jejuni*, to recover *A. butzleri* from fresh poultry and poultry products. The medium was composed of (per l): peptone (10 g), Lab Lemco Powder (8 g), yeast extract (1 g), sodium chloride (5 g), 0.0025% resazurin (16 ml) and cefoperazone as selective agent. After enrichment, the samples were filtered through a 0.45 μm pore size low-protein-binding syringe filter fitted to a 10 ml syringe onto modified CCDA plates as single drops (approx. 0.45 ml of the enrichment broth was inoculated). The enrichment broths were incubated microaerobically and the plates aerobically at 30°C. Subsequently, Atabay and Corry (1997) used CAT broth, which consisted of campylobacter enrichment basal medium (Lab M), 5% (v/v) laked horse blood and CAT (cefoperazone, amphotericin B and teicoplanin) selective supplement, and compared it with the Lammerding enrichment broth (1997) to isolate *Arcobacter* spp. from fresh chicken carcasses. After enrichment in both broths, these workers used the Steele and McDermott membrane filtration technique (see page 287), in parallel with the Lammerding filtration technique onto CAT, mCCDA and blood agar. These techniques yielded three *Arcobacter* spp.: *A. butzleri*, *A. cryaerophilus* and *A. skirrowii*. *A. skirrowii* was only isolated on blood agar, possibly due to a synergistic effect of cefoperazone and sodium deoxycholate. Corry and Atabay (1997) compared the productivity of CAT agar and mCCDA, both of which are used for the isolation of thermophilic campylobacters, for various strains of *Arcobacter* and found that CAT agar performed better than mCCDA, in particular for *A. butzleri* strains and for one of two strains of *A. cryaerophilus* tested. However, recently seven strains of *A. skirrowii* isolated from ducks with CAT broth and blood agar were found not to grow on either CAT agar or mCCDA (Atabay et al. 2001). De Boer et al. (1996) developed Arcobacter Selective Enrichment Broth (ASB) and a semisolid Arcobacter Selective Plating Medium (ASM) for the isolation of *Arcobacter* spp. from meats under aerobic conditions. These media contained cefoperazone, piperacillin, trimethoprim and cycloheximide as selective agents. The media were incubated at 24°C aerobically for at least 48 h for each step and the isolation was based on the swarming of arcobacters on semisolid medium. *A. butzleri* or *A. butzleri*-like strains only were isolated. Collins *et al.* (1996) used EMJH

Table 4
Selective agents and oxygen quenching agents used in media for the isolation of *Arcobacter* spp. (concentrations in mg per litre).

Medium[a](incubation temp. and atmosphere)	Oxygen quenching agents	Cephalo-sporins	Trimetho-prim	Colistin	5-fluoro-uracil	Other	Vancomycin or Teicoplanin	Antifungals
Ellis *et al.*, 1977 EMJH P80 - broth (30°C A)					100			
Lammerding *et al.* (1996) (30°C MA)		32 czone						
Collins *et al.* (1996) - CVA-agar (30°C MA)	10% S	20 cthin					10 V	5 AM
Collins *et al.* (1996) - mCIN - agar (30°C MA)					200			
De Boer *et al.* (1996) – ASM - agar and broth (24°C A)	5% H (broth only)	32 czone	20			75 PIP		100 CY
Atabay and Corry (1997) CAT broth (30° and 37°C MA)	haemin 5% LH	8 czone					4 T	10 AM
Corry and Atabay (1997) mCCDA (30 and 37°C MA)	4% C, 0.25% F, 0.25% P	32 czone						10 AM
Atabay and Corry (1998) Oxoid broth (25°C A)		8 czone				1% BS	4 T	10 AM
Johnson and Murano (1999a) - agar (30°C A)	5% S; 0.5% P; 0.05% TG	32 czone						
Johnson and Murano (1999b) - broth (30°C A)	0.5% P; 0.05% TG; 3%C	32 czone			200	0.25%BS		
Marinescu *et al.* (1996b) broth (25°C A)		7.5		5000 iu	100			1000 AM
Marinescu *et al.* (1996b) Karmali agar (25°C MA)	4% C, haematin	32 czone					20V	
Houf et al. (2001b) Oxoid Arcobacter broth/agar (28°C MA)	5% LH; 0.5%P; 0.5% TA	16 czone	64		100	32N		10AM

czone, cefoperazone; cthin, cephalotin; V, vancomycin; T, teicoplanin; AM, amphotericin B; CY, cycloheximide; PIP, piperacillin; BS, bile salts; N, novobiocin; P, pyruvate; C, charcoal; TA, thioglycollic acid; H, horse blood; LH lysed horse blood; S sheep blood; A aerobic atmosphere; MA microaerobic atmosphere.

P-80 (see above) as enrichment medium in combination with two selective plating media to recover arcobacters from ground pork. Their isolates were not identified to the species level. The first plating medium consisted of CIN (*Yersinia* selective cefsulodin-irgasan-novobiocin) base agar with various detoxifying and growth-promoting agents and 5-FU at a concentration of 200 mg/l as the only selective agent. The other plating medium was CVA (cephalothin-vancomycin-amphotericin B) agar, which contains brain heart infusion agar supplemented with 10% bovine blood and cephalothin, vancomycin and amphotericin B. Atabay and Corry (1998) described a new (Oxoid) Arcobacter Enrichment Broth which contained (g/l) peptone (18), yeast extract (1) and sodium chloride (5) and incorporated CAT selective supplement. The basal medium (CM965, Oxoid) and selective agent (SR174E) are now commercially available. The productivity of this medium was compared with two campylobacter enrichment media, Preston broth and Lab M Bolton enrichment broth, for the recovery of various *Arcobacter* strains, and the new broth was found to be better than the two campylobacter broths tested for the isolation of a wide variety of arcobacters. Johnson and Murano (1999a and b) examined several common components used in media intended for the isolation of *Campylobacter*, *Helicobacter* and other Gram-negative rods and, developed a new enrichment broth and plating medium for the isolation of *Arcobacter* spp. from poultry. Both media contained cefoperazone, bile salts and thioglycollate. The enrichment broth (Johnson and Murano 1999b) additionally contained 5-FU, and an aerobic atmosphere was used throughout. Arcobacter colonies were coloured red on the plating medium, apparently due to the thioglycollate. Comparison of these media with the media of De Boer et al. (1996) and enrichment in the EMJH P-80 followed by plating onto CVA agar, indicated that their new media were superior (Johnson and Murano 1999b). However, the strains isolated were not speciated, and the ability of *A. skirrowii* to grow on their plating medium was not tested.

Houf et al. (2001a) studied the susceptibility of 101 strains of *Arcobacter* spp. (*A. butzleri, A. cryaerophilus* and *A. skirrowii*) to a variety of selective agents. All were susceptible to colistin (MIC < 4 mg/l) and rifampicin, and some strains of *A. cryaerophilus* and *A. skirrowii* were susceptible 32 mg/l cefoperazone. The authors concluded that media with CAT supplement and EMJH P-80 medium supplemented with 5-FU are suitable for isolation of all three *Arcobacter* spp., although none of these media is sufficiently selective to satisfactorily inhibit competitors. They also concluded that *A. skirrowii* is the most sensitive species, and that suitable selective agents for this species include 5-FU, novobiocin, trimethoprim and cefoperazone. In their subsequent paper, Houf et al. (2001b) described a selective liquid and solid medium for arcobacters containing the four selective agents above, plus amphotericin B (Table 4). The basal medium was Oxoid arcobacter broth. *A. butzleri* and *A. cryaerophilus* grew better than *A. skirrowii* in these media.

In our experience it is important to incubate plating media for up to seven days for recovering strains of *A. cryaerophilus* and *A. skirrowii*, which may in part account for the rarity with which their isolation is reported.

The major way in which arcobacters are considered to differ from campylobacters is in their tolerance to oxygen, although they grow well in microaerobic atmosphere. In addition they tend to have lower growth temperatures. Methods of isolation are often aerobic and frequently use an incubation temperature of 30°C, partly in order to avoid isolating (thermophilic) campylobacters. Little has been done to investigate whether special conditions are needed to isolate sublethally damaged arcobacters (e.g late addition of selective agents or use of lower incubation temperature), or to investigate whether initial isolation is better with aerobic or microaerophilic atmosphere. However, a study by Liu et al. (1995) found that Oxyrase (see p. 292) aided growth of arcobacters in non-selective broth incubated aerobically at 37°C, but not at 30°C.

Detection methods for *Helicobacter* species

The most well-known *Helicobacter* species is *H. pylori,* which was first reported by Marshall and Warren (1984) to colonise the mucosa of the human stomach. It causes gastritis, ulcers and gastric carcinoma. This organism was first designated '*Campylobacter pyloridis*', later '*C. pylori*', and finally assigned to a separate genus. At least 17 other species have now been described (On, 2001), and still more probable new species have been observed, but attempts to grow them in culture media have not succeeded. Many colonise the stomachs of various animals including cats, dogs, monkeys, pigs, cattle and rodents. Others colonise the lower part of the intestine, and may be a cause of hepatitis and hepatic carcinoma (Fox, 1997; Fox et al., 1997, 2000). *H. pullorum* is of particular interest because it has been found in chickens and chicken meat, and has also been implicated in chicken and human disease (see below).

The morphology of helicobacters such as *H. pylori* or *H. pullorum* is similar to campylobacters and arcobacters, except that helicobacters usually have sheathed flagella (*H. pullorum* does not have sheathed flagella, however.) Some, especially those more difficult to cultivate, have a much more marked spiral appearance, which earned them the epithet "gastrospirillum". All require a microaerobic environment for growth, and added hydrogen has been recommended (Fox, 1997). Extended incubation (4–7 days) at 37°C has been found useful for *H. pylori* and other *Helicobacter* species (Drumm et al., 1987; Krajden et al., 1987; Fox and Lee, 1997), and the surface of solid media should be moist.

Selective media for *H. pylori*

Solid media have been developed for isolating this bacterium from gastric biopsies. Table 5 summarises the selective agents used. Skirrow agar has been widely used, but the first medium devised specifically for *H. pylori* seems to be that of Goodwin et al. (1985). These workers found that a combination of nalidixic acid, vancomycin and amphotericin B was most productive. They also observed that growth was enhanced if hydrogen was included in the gas mixture for incubation, and that whole blood was better than lysed blood in the medium. Dent and McNulty (1988) found that 14% of 97 strains of *H. pylori* were sensitive to the level of nialdixic acid used in Goodwin's medium, and devised a combination of cefsulodin, trimethoprim and vancomycin instead. The selective supplement of Dent and McNulty (1988) is avail-

Table 5
Media used for isolating for *H. pylori* (concentrations in mg per litre unless otherwise stated).

Reference	basal medium	oxygen-quenching agents	indicators	cefsu-lodin	trimetho-prim	polymyxin B or colistin (C)	vanco-mycin	nalidixic acid	sulfameth-oxazole	ampho-tericin
Skirrow, 1977	Brucella agar	10% horse blood			5	2,500iu	10			
Goodwin et al. 1985	BHI + Isovitalex	7% horse blood					6	20		2 (A)
Dent and McNulty 1988	Columbia agar	7% sheep blood		5	5		10			5(A)
Glupczinski et al. 1988	BHI + YE +10% FCS	2% charcoal	40 TTC	5	5		10			10(A)
Stevenson *et al.* (2000b)	HPSPA	pyruvate		10	40	62,000iu	10		20	5(A)

Key: BHI: brain heart infusion; FCS: foetal calf serum; YE yeast extract; TTC: triphenyltetrazolium chloride; HPSPA (g/l): 10 special peptone, 5 YE, 5 NaCl, 5 beef extract.

able commercially. Quieroz et al. (1987) developed an indicator system using 2,3,5 triphenyl tetrazolium chloride (TTC) in a medium containing the selective supplements of Goodwin et al. (1985) and 10% sheep blood. *H. pylori* produced tiny convex non-haemolytic colonies with entire edges and a brilliant golden pigment. TTC was also incorporated into the medium of Glupczinski et al. (1988), who used the selective agents of Dent and McNulty, but with a different basal medium and a higher concentration of amphotericin (10 rather than 5 mg per l). Most media contain blood as oxygen-detoxifying agent, rather than FBP, following the observation of Goodwin et al. (1985) that some strains of *H. pylori* are inhibited by sodium metabisulphite in this mixture. Stevenson et al. (2000a) examined the productivity of a variety of basal liquid and solid media for two strains of *H. pylori.* They found that the bacteria grew similarly in brain heart infusion, Mueller Hinton and a modification of Johnson and Murano's (1999a) broth for arcobacters. Growth on Columbia agar, Mueller Hinton agar, basal media for Johnson and Murano medium and Glupczynski's medium, and HPSPA, a similar modification of Johnson and Murano's basal medium ("*H. pylori* special peptone agar" with no phosphate, and 3% activated charcoal replaced by 5% defibrinated sheep blood) measured in terms of numbers of cfu and size of colonies, showed that there were no significant differences between the numbers of cfu, but that colonies were significantly larger on HPSPA. Extensive trials (Stevenson et al. 2000b) using HPSPA with various combinations of selective agents resulted in a medium containing six antibiotics (see Table 5) which was reported to be effective for isolating *H. pylori* from artificially contaminated samples of cattle rumen and abomasum contents. The reason for using cattle samples for this purpose was not clear, but this medium might be useful for detecting *H. pylori* in food, human faeces or the environment. TTC was excluded from HPSPA because it reduced colony size significantly. To date there appear to be no reports of successful isolation of *H. pylori* from food or environmental sources using conventional methods, with the exception of the report by Lu et al. (2002) who isolated the organism from waste water using immunomagnetic separation and Columbia blood agar plates (see p. 303).

Helicobacter pullorum

Helicobacter pullorum was first isolated from the caeca of healthy broilers as well as from the livers and intestinal contents of laying hens with hepatic lesions (Stanley et al., 1994; Burnens et al., 1996) and later from asymptomatic broiler chickens (Atabay and Corry, 1997; Atabay et al., 1998b). It has also been isolated from both immunocompetent and immunocompromised human patients suffering from gastroenteritis (Burnens et al., 1994; Stanley et al., 1994; Steinbrueckner et al., 1997) and detected in the blood of human patients with chronic liver diseases (Ananieva et al., 2001). This may indicate the possibility of transmission of *H. pullorum* from poultry to humans, and its pathogenicity for the human host. However, there is limited information available on the epidemiology of this species, and its clinical importance is probably underestimated since it does not always grow on popular *Campylobacter* isolation media, and requires a microaerobic atmosphere including hydrogen (Burnens et al., 1994; Atabay et al., 1998b). It could also be mistaken for *C. coli* (Burnens et al., 1994; Atabay et al., 1998b; Fox et al., 2000).

H. pullorum was first isolated on Columbia agar plates with 5% sheep blood and the following antimicrobials (mg/l): cefoperazone 32, vancomycin 10 and amphotericin B 3 (Campylosel; BioMériux, France). Later isolations of *H. pullorum* from poultry were made by the use of the membrane filter method of Steele and McDermott (1984) on a non-selective blood agar plate (Atabay et al., 1998b). The selective agents used in CAT (cefoperazone amphotericin teicoplanin) agar and mCCDA have inhibitory effects on some isolates of *H. pullorum* (Corry and Atabay, 1997).

H. pullorum is catalase and oxidase positive, urease negative and does not hydrolyse indoxyl acetate. Inability to hydrolyse indoxyl acetate differentiates it from *C. coli*. *H. pullorum* reduces nitrate and grows at both 37 and 42°C under microaerobic conditions with hydrogen. Most strains are susceptible to nalidixic acid and are resistant to cephalothin and polymyxin B (Stanley et al., 1994; Burnens et al., 1996; Steinbrueckner et al., 1997; Atabay et al., 1998b; Fox et al., 2000).

Immunomagnetic separation methods

An alternative method of pre-enrichment for separating target from non-target organisms is immunomagnetic separation (IMS). This technique uses paramagnetic particles coated with target specific antibodies to capture the target cell in the presence of a mixed flora (Safarik et al., 1995). Although this technique has been applied successfully for the isolation of Salmonella (Mansfield and Forsythe, 2000) and *E. coli* O157 (Chapman and Siddons, 1996) it has not been extensively applied to Campylobacteraceae. Yu et al. (2001) demonstrated that labelled Dynabeads could be used to detect 10^4 campylobacter cfu/g, without pre-enrichment, from ground poultry meats. Lamoureux et al. (1997) described an IMS-hybridisation assay for capturing and detecting thermophilic *Campylobacter* cells from chicken meat after enrichment in Rosef broth. Genomic DNA was extracted from IMS captured bacteria and *Campylobacter* detected by nucleic acid hybridisation methods. Che et al. (2001) tested eight types of beads (three sizes and functional groups) and found that Streptavidin-labeled beads (2.8µm diameter) were the most effective at capturing *C. jejuni* from poultry samples. By combining the IMS method with a biosensor the total detection time was 1.5 hours with a detection limit of 10^6 cfu/ml, or 2.5h with a detection limit of $2x10^4$ cfu/ml. *C. hominis* was first isolated from the human intestinal tract using IMS to select *Campylobacter* from a mixed microbial flora after filtration onto non-selective agar (Lawson et al., 2000) followed by 16S rDNA sequence analysis. *H. pylori* has been isolated from faecal samples of experimentally inoculated mice (Osaki et al., 1998), and naturally contaminated environmental samples (Lu et al., 2002) using IMS followed by plating.

Conclusions

Media for isolating campylobacters and related bacteria from faeces and other environments such as food and water are not optimal. A standard method is being drafted for isolation of thermophilic campylobacters, but there is so far no consensus concerning the best media and methods for arcobacters or helicobacters. PCR-based methods of detection are increasingly being used for all these genera, along with genetic methods for identification. However, while being slower, cultural methods are usually cheaper and simpler. The viability of the organisms detected culturally in food or water is not in doubt. Cultural methods also yield isolates that can be further characterised and typed. Whether 'viable but not culturable' campylobacters occur in food and other menstrua and whether or not they can infect humans and other animals is not clear. Neither is it known whether *Helicobacter* spp. are transmited via food or water.

Of the plating media listed in Table 2, mCCD and Karmali agars have performed best in comparative studies using faecal samples. Other media, for example, the semi-solid medium of Goossens et al. (1989) and the medium of Aspinall et al. (1993), have yet to be tested in comparative studies. If cefaperazone- and polymyxin- or colistin-sensitive campylobacters are sought and *C. jejuni* subsp. *doylei*, which does not grow at 42ºC or 43ºC, media such as mCCD, those of Goossens et al. (1986) and Aspinall et al. (1993) and the semi-solid medium of Goossens et al. (1989) incubated at 37ºC, might be useful. Modifications of the membrane filter method of Steele and McDermott (1984) in combination with non-selective blood agar are useful for this group, but only if relatively high numbers of campylobacters are present in the faecal suspension or enrichment culture. Use of this method in combination with selective media often yields more positive samples.

Development of liquid enrichment media has so far been confined to the isolation of thermophilic campylobacters, predominantly *C. jejuni* subsp. *jejuni*. Future work may be directed towards modifying these media and current methods in order to detect the mesophilic and antibiotic-sensitive types mentioned above. This may, in turn, depend on whether those types are perceived to be an important cause of diarrhoea, and whether contaminated food is shown to be a vector. Comparative studies are needed on faeces, food and environmental samples using the Arcobacter enrichment medium of Houf et al. (2001b) and the *H. pylori* enrichment medium of Stevenson et al. (2000b). Although the medium of Bolton has been accepted in several standard methods for isolation of thermophilic campylobacters from food, published comparisons are few.

More investigation is also needed concerning the optimal atmospheric conditions necessary for thermophilic campylobacters, particularly for growth in liquid enrichment media.

References

Acuff, G.R., Vanderzant, C., Gardner, F.A. and Golan, F.A. (1982) Evaluation of an enrichment plating procedure for the recovery of *Campylobacter jejuni* from turkey eggs and meat. J. Food Protect. 45, 1276–1278.

Adler, H. and Spady, G. (1997) The use of microbial membranes to achieve anaerobiosis. J. Rapid Meth. Automat. Microbiol. 5, 1–12.

Ahonkhai, V.I., Cherubin, C.E., Sierra, M.F., Bokkenheuser, V.D., Shulman, M.A. and Mosenthal, A.C. (1981) In vitro susceptibility of *Campylobacter fetus* subsp. *jejuni* to N-formimidoyl thienamycin, rosaramicin, cefoperazone and other antimicrobial agents. Antimicrob. Agents Chemother. 20, 850–851.

Albert, M.J., Tee, W., Leach, A.S., Asche, V. and Penner, J.L. (1992) Comparison of a blood-free medium and a filtration technique for the isolation of *Campylobacter* spp. from diarrhoeal stools of hospitalised patients in central Australia. J. Med. Microbiol. 37, 176–179.

Ananieva, O, Nilsson, I., Vorobjova, T., Prukk, T., Wadstrom, T., Uibo, R., 2001. A comparative study of immune response to *H. pullorum, H. hepaticus, H. bilis* and *H. pylori* in patients with chronic liver diseases, a randomized Estonian population group and healthy blood donors. Abstracts of 11th International Workshop on Campylobacter, Helicobacter and related organisms. Int. J. Med. Microbiol. 291 (suppl. 31), 58 (abstract no. G-06).

Anon. (undated – ca. 1995) "Lab-M Culture Media" pp. 36 and 88. Lab-M, Bury BL9 6AU, UK.

Aquino, M.H.C., Carvalho, J.C.P., Tibana, A. and Franko, R.M. (1996) *Campylobacter jejuni/coli* methodology of isolation and possible interfering facors in primary culture. J. Food Protect. 59, 429–432.

Aspinall, S.T., Wareing, D.R.A., Hayward, P.G. and Hutchinson, D.N. (1993) Selective medium for thermophilic campylobacters including *Campylobacter upsaliensis*. J. Clin. Pathol. 46, 829–831.

Aspinall, S.T., Warcing, D.R.A., Hayward, P.G. and Hutchinson, D.N. (1996) A comparison of a new campylobacter selective medium (CAT) with membrane filtration for the isolation of thermophilic campylobacters including *Campylobacter upsaliensis*. J. Appl. Microbiol., 80, 645–650.

Atabay, H.I. and Corry, J.E.L. (1997) The prevalence of campylobacters and arcobacters in broiler chickens. J. Appl. Microbiol. 83: 619–626.

Atabay, H.I. and Corry, J.E.L. (1998) Evaluation of a new arcobacter enrichment medium and comparison with two media developed for enrichment of *Campylobacter* spp. Int. J. Food Microbiol. 41 53–58.

Atabay, H.I., On, S.L.W and Corry, J.E.L. (1998a) Diversity and prevalence of *Arcobacter* spp. in broiler chickens J. Appl. Microbiol. 84, 1007–1016.

Atabay, H.I., Corry, J.E.L. and On, S.L.W (1998b) Identification of unusual *Campylobacter*-like isolates from poultry products as *Helicobacter pullorum*. J. Appl. Microbiol. 84, 1017–1024.

Atabay, H.I., Waino, M. and Madsen, M. (2001) Comparison of PCR and conventional isolation methods for detection of arcobacters in Danish poultry. Identification of isolates using multiplex-PCR and phenotypic tests. 11th International Workshop on Campylobacter, Helicobacter and related organism, September 1–5, 2001, Freiburg, Germany (Abstract and poster presentation also published in Int. J. Med. Microbiol, 291, p. 139).

Baggerman, W.I. and Koster, T. (1992) A comparison of enrichment and membrane filtration methods for the isolation of *Campylobacter* from fresh and frozen foods. Food Microbiol. 9, 87–94.

Barot, M.S. and Bokkenheuser V.D. (1984) Systematic investigation of enrichment media for wild-type *Campylobacter jejuni* strains. J. Clin. Microbiol. 20, 77–80.

Baird, R.M., Corry, J.E.L. and Curtis, G.D.. (eds) (1987) Pharmacopoeia of culture media for food microbiology. Int. J. Food Microbiol. 5, 212–213 (mCCD agar), 250–253 (Preston media)

268–269 (Skirrow agar).

Baylis, C.L., MacPhee, S., Martin, K.W., Humphrey, T.J. & Betts, R.P. (2000) Comparison of three enrichment media for the isolation of *Campylobacter* spp. from foods. J. Appl. Microbiol. 89, 884–891.

Beuchat, L.R. (1985) Efficacy of media and methods for detecting and enumerating *Campylobacter jejuni* in refrigerated chicken meat. Appl. Environ. Microbiol. 50, 934–939.

Blaser, M.J., Cravens, J., Powers, R. and Wang, W.L. (1978) Campylobacter enteritis associated with canine infection. Lancet 2, 979–981.

Blaser, M.J., Berkowicz, I.D., Laforce, F.M., Cravens, J., Reller, L.B. and Wang, W.L. (1979) Campylobacter enteritis: clinical and epidemiological features. Ann. Intern. Med. 91, 179–185.

Blaser, M.J., Hardesty, H.L., Powers, B. and Wang, W.L.L. (1980) Survival of *Campylobacter fetus* subsp. *jejuni* in biological milieus. J. Clin. Microbiol. 11, 309–313.

Bolton, F.J., Sails, A.D., Fox, A.J., Wareing, D.R.A. and Greenaway, D.L.A. (2002) Detection of *Campylobacter jejuni* and *Campylobacter coli* in foods by enrichment culture and polymerase chain reaction enzyme-linked immunosorbent assay. J. Food Protect. 65, 760–767.

Bolton, F.J. (2000) Methods for isolation of campylobacters from humans, animals, food and water. In: "The Increasing Incidence of Human Campylobacteriosis" Report and Proceedings of a WHO Consultation of Experts. Copenhagen, Denmark, 21–25 November 2000, WHO/CDS/CSR/APH 2001.7 pp. 87–93.

Bolton, F.J. and Coates, D. (1983) Development of a blood-free campylobacter medium: screening tests on basal media and supplements, and the ability of selected supplements to facilitate aerotolerance. J. Appl. Bacteriol. 54, 115–125.

Bolton, F.J. and Robertson, L. (1982) A selective medium for isolating *Campylobacter jejuni/coli*. J. Clin. Pathol. 35, 462–467.

Bolton, F.J., Hinchcliffe, P.M., Coates, D. and Robertson, L. (1982) A most probable number method for estimating small numbers of campylobacters in water. J. Hyg. Camb. 89, 185–190.

Bolton, F.J., Coates, D., Hinchcliffe, P.M. and Robertson, L. (1983) Comparison of selective media for isolation of *Campylobacter jejuni/coli*. J. Clin. Pathol. 36, 78–83.

Bolton, F.J., Coates, D. and Hutchinson, D.N. (1984a) The ability of *Campylobacter* media supplements to neutralise photochemically induced toxicity and hydrogen peroxide. J. Appl. Bacteriol. 56, 151–157.

Bolton, F.J., Hutchinson, D.N. and Coates, D. (1984b) A blood-free selective medium for the isolation of *Campylobacter jejuni* from faeces. J. Clin. Microbiol. 19, 169–171.

Bolton, F.J., Coates, D. Hutchinson, D.N. and Godfree, A.F. (1987) A study of thermophilic campylobacters in a river system. J. Appl. Bacteriol. 62, 167–176.

Bolton, F.J., Wareing, D.R.A., Skirrow, M.B. and Hutchinson, D.N. (1992) Identification and biotyping of campylobacters. In: Board, R.D., Jones, D. and Skinner, F.A. (eds) Identification Methods in Applied and Environmental Microbiology. Society for Applied Bacteriology Technical Series no. 29, pp151–161. Oxford: Blackwell.

Bovill R.A. and Mackey B.M. (1997) Resuscitation of 'non-culturable' cells from aged cultures of *Campylobacter jejuni*. Microbiology, U.K. 143, 1575–1581.

Boudreau, M., Higgins, R and Mittal, K.R. (1991) Biochemical and serological characterisation of *Campylobacter cryaerophila*. J. Clin. Microbiol. 29, 54–58.

Brennhovd, O. and Kapperud, G. (1991) Comparison of three methods of isolation of thermotolerant *Campylobacter* spp. from water (in Norwegian) In: Brennhovd, O. "Termotolerante *Campylobacter* spp. og *Yersinia* spp. i noem norske vannforekomster" Ph.D. thesis, Norwegian College of Veterinary Medicine, Oslo.

Brennhovd, O., Kapperud, G. and Langeland, G. (1992) Survey of thermotolerant *Campylobacter* spp. and *Yersinia* spp. in three surface water sources in Norway. Int. J. Food Microbiol. 15, 327–338.

Brooks, B.W., Garcia, M.M., Frazer, A.D.E., Lior, H., Stewart, R.B. and Lammerding, A.M. (1986)

Isolation and characterisation of cephalothin-susceptible *Campylobacter coli* from slaughtered cattle. J. Clin. Microbiol. 24, 591–595.

Buck, G.E. and Kelly, M.T. (1981) Effect of moisture content of the medium on colony morphology of *Campylobacter fetus* subsp. *jejuni*. J. Clin. Microbiol. 14, 585–586.

Burnens, A.P. and Nicolet, J. (1992) Detection of *Campylobacter upsaliensis* in diarrheic dogs and cats, using a selective medium with cephoperazone. Am. J. Vet. Res. 53, 48–51.

Burnens, A.P, Stanley, J., Morgenstern, R.and Nicolet, J. (1994) Gastroenteritis associated with *Helicobacter pullorum*. Lancet, 344, 1569–1570.

Burnens, A.P., Stanley, J. and Nicolet, J. (1996) Possible association of *H. pullorum* with lesions of vibrionic hepatitis in poultry. *In Campylobacters, Helicobacters and Related Organisms. Proceedings of the 8th International Workshop on Campylobacters, Helicobacters and Related Organisms* ed. Newell, D.G., Ketley, J.M. and Feldman, R.A. pp. 291–293. New York: Plenum Publishing Corporation.

Busse, M. (1995) Media for salmonellae. Int. J. Food Microbiol. 26, 117–131.

Butzler, J.P., Dekeyser, P. and Detrain, M. (1973) Related vibrio in stools. J. Pediatr. 82, 493–495.

Butzler, J.P., Dekeyser, P. and Lafontaine, T. (1974) Susceptibility of related vibrios and *Vibrio fetus* to twelve antibiotics. Antimicrob. Agents Chemother. 5, 86–89.

Butzler, J.P. and Skirrow, M.B. (1979) Campylobacter enteritis. Clin. Gastroenterol. 8, 737–765.

Butzler, J.P., DeBueck, M. and Goossens, H. (1983) New selective medium for isolation of *Campylobacter jejuni* from faecal specimens. Lancet i, 818.

Cappelier, J.M., Magras C., Jouve, J.L. and Federighi, M. (1999a) Recovery of viable but non-culturable *Campylobacter jejuni* cells in two animal models. Food Microbiol. 16, 375–383.

Cappelier, J.M., Minet, J., Magras C., Colwell, R.R. and Federighi, M. (1999b) Recovery in embryonated eggs of viable but nonculturable *Campylobacter jejuni* cells and maintenance of ability to adhere to HeLa cells after resuscitation. Appl. Environm. Microbiol. 65, 5154–5157.

Chapman, P.A. and Siddons, C.A. (1996) A comparison of immunomagnetic separation and direct culture for the isolation of verocytotoxin-producing *Escherichia coli* O157 from cases of bloody diarrhoea, nonbloody diarrhoea and asymptomatic contact. J. Med. Microbiol. 44, 267–271.

Che, Y.H., Li, Y.B. and Slavik, M. (2001) Detection of *Campylobacter jejuni* in poultry samples using an enzyme-linked immunoassay coupled with an enzyme electrode. Biosens. Bioelectron. 16, 791–797.

Christopher, F.M., Smith, G.C. and Vanderzant, C. (1982) Examination of poultry giblets, raw milk and meat for *Campylobacter fetus* subsp. *jejuni*. J. Food Protect. 45, 260–262.

Collins, C.I., Wesley, I.V. and Murano, E.A. (1996) Detection of *Arcobacter* spp. in ground pork by modified plating methods. J. Food Protect. 59, 448–452.

Corry, J.E.L. and Atabay, H.I. (1997) Comparison of the productivity of cefoperazone amphotericin teicoplanin (CAT) agar and modified charcoal cefoperazone deoxycholate (mCCD) agar for various strains of *Campylobacter, Arcobacter* and *Helicobacter pullorum*. Int. J. Food Microbiol., **38**, 201–209.

Corry, J.E.L., Post, D.E., Colin, P. & Laisney, M.J. 1995 Culture media for the isolation of campylobacters. Int. J. Food Microbiol., 26, 43–76.

De Boer, E. and Humphrey, T.J. (1991) Comparison of methods for the isolation of thermophilic campylobacters from chicken products. Microb. Ecol. Health Dis. 4, (Special Issue) S43.

De Boer, E. and Reitsma, W. (1991) Isolation of *Campylobacter* from poultry products using membrane filtration. In: Ruiz-Palacios G.M., Calva, F. and Ruiz-Palacios, B.R. (eds). "Campylobacter V", Proceedings of the 5th International Workshop on Campylobacter Infection, pp113–115. Instituto Nacional de la Nutricion, Vasco de Quiroga 15, Mexico 1400, DF.

De Boer, E., Tilburg, J.J.H.C., Woodward, D.L., Lior, H. and Johnson, W.M. (1996) A selective medium for the isolation of *Arcobacter* from meats. Lett. Appl. Microbiol. 23, 64–66.

De Boer, E., van Beek, P. and Pelgrom, K. (1998) Comparison of culture media for the isolation of campylobacters from chicken meat. In: Lastovica, A.J., Newell, D.G. and Lastovica, E.E. (Eds)

"Campylobacter, Helicobacter and Related Organisms" Proc. 9th International Workshop; Cape Town 1997, University of Cape town, pp. 370–372.

Dekeyser, P., Gossuin-Detrain, M., Butzler, J.P. and Sternon, J. (1972) Acute enteritis due to related vibrio: first positive stool cultures. J. Infect. Dis. 125, 390–392.

Dent, J.C. and McNulty, C.A. (1988) Evaluation of a new selective medium for *Campylobacter pylori*. Eur. J. Clin. Microbiol. Infect. Dis. 7, 555–558.

Doyle, M.P. and Roman, D.J. (1982) Recovery of *Campylobacter jejuni* and *Campylobacter coli* from inoculated foods by selective enrichment. Appl. Environ. Microbiol. 43, 1343–1353.

Drumm, B., Sherman, P., Cutz, E. and Karmali, M. (1987) Association of *Campylobacter pylori* on the gastric mucosa with antral gastritis in children. New Eng. J. Med. 316, 1557–1561.

Ehlers, J.G., Chapparo-Serrano, M., Richter, R.L. and Vanderzant, C. (1982) Survival of *Campylobacter fetus* subsp. *jejuni* in cheddar and cottage cheese. J. Food Protect. 45, 1018–1021.

Ellis, W.A., Neill, S.D., O'Brien, J.J., Ferguson, H.W. and Hanna, J. (1977) Isolation of *Spirillum/Vibrio*-like organisms from bovine fetuses. Vet. Record 100, 451–452.

Endtz, H.P., Ruijs, G.J.H.M., Zwinderman, A.H., Van der Reijden, T., Biever, M. and Mouton, R.P. (1991) Comparison of six media, including a semisolid agar, for the isolation of various *Campylobacter* species from stool specimens. J. Clin. Microbiol. 29, 1007–1010.

Engberg, J., On, S.L.W. , Harrington, C.S. and Gerner-Smidt, P. (2000) Prevalence of *Campylobacter, Arcobacter, Helicobacter, Sutterella* spp. in human fecal samples as estimated by a reevaluation of isolation methods for Campylobacters. J. Clin. Microbiol. 38, 286–291.

FDA (Food and Drug Administration) Bacteriological Analytical Manual (1992) 7th edition. Arlington, Virginia, USA: AOAC International.

Federighi, M., Magras, C., Pilet, M.F., Woodward, D. Johnson, W., Jugiau, F. and Jouve, J.L. (1999) Incidence of thermotolerant campylobacter in foods assessed by NF ISO 10272 standard: results of a two-year study. Food Microbiol. 16, 195–204.

Fox, J. G. (1997) Helicobacters: the next generation. Baillières Clin. Infect. Dis.4, 449–471.

Fox, J. G. and Lee, A. (1997) The role of *Helicobacter* species in newly recognised gastrointestinal tract diseases of animals. Lab. Animal Sci., 47, 222–255.

Fox, J.G., Chien, C.C., Dewhurst, F.E., Paster, B.J., Shen, Z., Melito, P.L., Woodward, D.L. and Rogers, F. G. 2000 *Helicobacter canadensis* sp. nov. isolated from humans with diarrhea as an example of an emerging pathogen. J. Clin. Microbiol. 38, 2546–2549.

Fraser, A.D.E., Chandan, V., Yamazaki, H., Brooks, B.W. and Garcia, M.M. (1992) Simple and economical culture of *Campylobacter jejuni* and *Campylobacter coli* in CO_2 in moist air. Int. J. Food Microbiol. 15, 377–382.

Fricker, C.R. (1984) Procedures for the isolation of *Campylobacter jejuni* and *Campylobacter coli* from poultry. Int. J. Food Microbiol. 1, 149–154.

Fricker, C.R. (1985) A note on the effect of different storage procedures on the ability of Preston medium to recover campylobacters. J. Appl. Bacteriol. 58, 57–62.

Fricker, C.R., Girdwood, R.W.A. and Munro, D. (1983) A comparison of procedures for the isolation of campylobacters from seagull faeces. J. Hyg. Camb. 91, 445–450.

Furanetto, S.M.P., Nascimento, D.D., Cerqueira-Campos, M.L. and Iaria, S.T. (1991) Efficacy of direct plating and selective enrichment media for detecting *Campylobacter jejuni* in fresh eviscerated whole market chickens - Sao Paulo, Brazil. Rev. Microbiol., Sao Paulo 22, 303–307.

Gebhart, C.J., Ward, G.E. and Kurtz, H.J. (1985) In vitro activities of 47 antimicrobial agents against three *Campylobacter* spp. from pigs. Antimicrob. Agents Chemother. 27, 55–59.

George, H.A., Hoffman, P.S., Smibert, R.M. and Krieg, N.R. (1978) Improved media for growth and aerotolerance of *Campylobacter fetus*. J. Clin. Microbiol. 8, 36–41.

Gilchrist, M.J.R., Grewell, C.M. and Washington, J.A. (1981) Evaluation of media for isolation of *Campylobacter fetus* subsp. *jejuni* from faecal specimens. J. Clin. Microbiol. 14, 393–395.

Glupczinski, Y., Labbe, M. and Thibaymont, F. (1988) Comparative evaluation of a new selective culture medium for improved isolation of *Campylobacter pylori* from gastric biopsy specimens. Excerpta Medica, Amsterdam pp.3–9.

Goodwin, C.S., Blincow, E.D., Warren, J.R., Waters, T.E., Samderson, C.R. and Easton, L. (1985) Evaluation of cultural techniques for isolating *Campylobacter pyloridis* from endoscopic biopsies of gastric mucosa. J. Clin. Pathol. 38, 1127–1131.

Goossens, H., De Boeck, M. and Butzler, J.P. (1983) A new selective medium for the isolation of *Campylobacter jejuni* from human faeces. Europ. J. Clin. Microbiol. 2, 389–394.

Goossens, H. De Boeck, M., Coignau, H., Vlaes, L., Van Den Borre, C. and Butzler, J.P. (1986) Modified selective medium for isolation of *Campylobacter* spp. from faeces: comparison with Preston medium and blood free medium and a filtration system. J. Clin. Microbiol. 24, 840–845.

Goossens, H., Vlaes, L., Galand, I., Van Den Borre, C. and Butzler, J.P. (1989) Semisolid blood-free selective motility medium for the isolation of campylobacters from stool specimens. J. Clin. Microbiol. 27, 1077–1080.

Goossens, H., Pot, B., Vlaes, L. et al. (1990a) Characterisation and description of "*Campylobacter upsaliensis*" isolated from human faeces. J. Clin. Microbiol. 28, 1039–1046.

Goossens, H., Vlaes, L., DeBoeck, M. et al., (1990b) Is "*Campylobacter upsaliensis*" an unrecognised cause of human diarrhoea? Lancet 335, 584–586.

Gun-Munro, J., Rennie, R.P., Thornley, J.H., Richardson, H.L., Hodge, D. and Lynch, J. (1987) Laboratory and clinical evaluation of isolation media for *Campylobacter jejuni*. J. Clin. Microbiol. 25, 2274–2277.

Hald B; Knudsen K; Lind P; Madsen M. (2001) Study of the infectivity of saline-stored *Campylobacter jejuni* for day-old chicks. Appl. Environ. Microbiol. 67, 2388–2392.

Harrass, B., Schwarz, S. and Wenzel, S. (1998) Identification and characterisation of *Arcobacter* isolated from broilers by biochemical tests, antimicrobial resistance patterns and plasmid analysis. J. Vet. Med. Series B-Infect Dis Vet. Publ Health 45, 87–94.

Heisick, J. (1985) Comparison of enrichment broths for isolation of *Campylobacter jejuni*. Appl. Environ. Microbiol. 50, 1313–1314.

Heisick, J., Lanier, J. and Peeler, J. (1984) Comparison of enrichment methods and atmosphere modification procedures for isolating *Campylobacter jejuni* from foods. Appl. Environ. Microbiol. 48, 1254–1255.

Hoffman, P., Krieg, N.R. and Smibert, R.M. (1979a) Studies of the microaerophilic nature of *Campylobacter fetus* subsp. *jejuni*. I. Physiological aspects of enhanced aerotolerance. Can. J. Microbiol. 25, 1–7.

Hoffman, P.S., George, H.A., Krieg, N.R. and Smibert, R.M. (1979b) Studies of the microaerophilic nature of *Campylobacter fetus* subsp. *jejuni*. II. Role of exogenous superoxide anions and hydrogen peroxide. Can. J. Microbiol. 25, 8–16.

Höller, C. (1991) A note on the comparative efficacy of three selective media for isolation of *Campylobacter* species from environmental samples. Zntbl. Hyg. Unwelmedizin 192, 116–123.

Houf, K., Devriese, L.A., De Zutter, L., Van Hoof, J. and Vandamme, P. (2001a) Susceptibility of *Arcobacter butzleri, Arcobacter cryaerophilus,* and *Arcobacter skirrowii* to antimicrobial agents used in selective media. J. Clin. Microbiol. 39, 1654 – 1656.

Houf, K., Devriese, L.A., De Zutter, L., Van Hoof, J. and Vandamme, P (2001b) Development of a new protocol for the isolation and quantification of *Arcobacter* species from poultry products. Int. J. Food Microbiol. 71, 189–196.

Humphrey, T.J. (1986a) Injury and recovery in freeze-or-heat damaged *Campylobacter jejuni*. Letts. Appl. Microbiol. 3, 81–84.

Humphrey, T.J. (1986b) Techniques for optimum recovery of cold-injured *Campylobacter jejuni* from milk and water. J. Appl. Bacteriol. 61, 125–132.

Humphrey, T.J. (1989) An appraisal of the efficacy of pre-enrichment for the isolation of

Campylobacter jejuni from water and food. J. Appl. Bacteriol. 66, 119–126.

Humphrey, T.J. and Cruikshank, J.G. (1985) Antibiotic and deoxycholate resistance in *Campylobacter jejuni* following freezing or heating. J. Appl. Bacteriol. 59, 65–71.

Humphrey, T., Mason, M., Martin, K. (1995) The isolation of *Campylobacter jejuni* from contaminated surfaces and its survival in diluents. Int. J. Food Microbiol. 26, 295–303.

Hunt, J.M. (1992) Campylobacter. In: Food and Drug Administration Bacteriological Analytical Manual, pp77–94, 7th edition, Arlington, Virginia, USA: AOAC.

Hunt, J.M., Abeyta, C. and Tran, T (1998) Campylobacter. In: F.D.A. Bacteriological Analytical Manual, 8th Edition, (Revision A) 7.01–7.27. AOAC, Arlington Va, USA 7.01–7.27.

Hunt, J.M., Francis, D.W., Peeler, J.T. and Lovett, J. (1985) Comparison of methods for isolating *Campylobacter jejuni* from raw milk. Appl. Environ. Microbiol. 50, 535–536.

Hutchinson, D.N. and Bolton, F.J. (1984) Improved blood-free selective medium for the isolation *of Campylobacter jejuni* from faecal specimens. J. Clin. Pathol. 37, 956–957.

ISO (International Standards Organisation) 1995 Microbiology of food and animal feeding stuffs – horizontal method for detection of thermotolerant *Campylobacter.* ISO 10272:1995(E) (including technical corrigenda 1: 1996 and 2: 1997).

Jacob, J., Lior, H. and Feuerpfeil, I. (1993) Isolation of *Arcobacter butzleri* from a drinking water reservoir in Eastern Germany. Zentralbl. Hyg. 193, 557–562

Jacob, J. and Stelzer, W. (1992) Comparison of two media for the isolation of thermophilic campylobacters from waste waters of different quality. Zentbl. Mikrobiol. 147, 42–44.

Johnson, R.C. and Rogers, P. (1964) 5-Fluorouracil as a selective agent for growth of Leptospirae. J. Bacteriol. 87, 422–426.

Johnson, L.G. and Murano, E.A. (1999a) Development of a new medium for the isolation of *Arcobacter* spp. J. Food Protect. 62, 456–462.

Johnson, L.G. and Murano, E.A. (1999b) Comparison of three protocols for the isolation of arcobacter from poultry. J. Food Protect., 62, 610–614.

Jones, D.M., Sutcliffe, E.M. and Curry, A. (1991) Recovery of viable but non-culturable *Campylobacter jejuni*. J. Gen. Microbiol. 137, 2477–2482.

Jones, D.M., Sutcliffe, E.M., Rios, R., Fox, A.J. and Curry, A. (1993*) Campylobacter jejuni* adapts to aerobic metabolism in the environment. J. Med. Microbiol. 38, 145–150.

Jones, K. (2001) Campylobacters in water, sewage and the environment. J. Appl. Microbiol. 90, 68S–79S.

Jones, RG; Trowbridge, DB; Go, MF (2001) *Helicobacter pylori* infection in peptic ulcer disease and gastric malignancy. Front. Biosci. 6, E213–E226.

Juven, B.J. and Rosenthal, I. (1985) Effect of free-radical and oxygen scavengers on photochemically generated oxygen toxicity and on the aerotolerance of *Campylobacter jejuni*. J. Appl. Bacteriol. 59, 413–419.

Karmali, M.A., De Grandis, S. and Fleming, P.C. (1980) Antimicrobial susceptibility of *Campylobacter jejuni* and *Campylobacter fetus* subsp. *fetus* to eight cephalosporins with special reference to species differentiation. Antimicrob. Agents Chemother. 18, 948–951.

Karmali, M.A., De Grandis, S. and Fleming, P.C. (1981) Antimicrobial susceptibility of *Campylobacter jejuni* with special reference to resistance patterns of Canadian isolates. Antimicrob. Agents Chemother. 19, 593–597.

Karmali, M.A., Simor, A.E., Roscoe, M., Fleming, P.C., Smith, S.S. and Lane, J. (1986) Evaluation of a blood-free, charcoal-based, selective medium for the isolation of *Campylobacter* organisms from faeces. J. Clin. Microbiol. 23, 456–459.

Kell, D.B., Kaprelyants, A.S., Weichart, D.H. Harwood, C.R. and Barer, M.R. (1998) Viability and activity in readily culturable bacteria: a review and discussion of the practical issues. Antonie Van Leeuwenhoek, 73, 169–187.

Kiehlbauch, J.A. Brenner, D.J., Nicholson, M.A., Baker, C.N., Patton, C.M., Steigerwalt, A.B. and Wachsmuth, I.K. (1991) *Campylobacter butzleri* sp. nov. isolated from humans and animals with diarrhoeal illness. J. Clin. Microbiol. 29, 376–385.

Kiehlbauch, J.A., Baker, C.N. and Wachsmuth, I.K. (1992) In vitro susceptibilities of aerotolerant *Campylobacter* isolates to 22 antimicrobial agents. Antimicrob. Agents Chemother. 36, 717–722.

Krajden, S., Bohnen, J., Anderson, J., Kempston, J., Fuksa, J., Matlow, A., Marcon, N., Haber, G., Kortan, P., Karmali, M., Corey, P., Petrea, C., Babida, C. and Hayman, S. (1987) Comparison of selective and non-selective media for recovery of *Campylobacter pylori* from antral biopsies. J. Clin. Microbiol. 25, 1117–1118.

Lammerding, A.M., Harris, J.E., Lior, H., Woodward, D.E., Cole, L. and Muckle, C.A. (1996) Isolation method for recovery of *Arcobacter butzleri* from fresh poultry and poultry products. In: Newell, D.G. and Ketley, J. (eds) Campylobacter VIII. Plenum, New York, p.329–333.

Lamoureux, M., MacKay, A., Messier, S., Fliss, I., Blais, B.W., Holley, R.A. and Simard, R.E. (1997). Detection of *Campylobacter jejuni* in food and poultry viscera using immunomagnetic separation and microtitre hybridisation. J. Appl. Microbiol. 83, 641–651.

Lander, K.P. (1982) A selective, enrichment and transport medium for campylobacters. In: D.G. Newell (ed) Campylobacter, Epidemiology, Pathogenesis and Biochemistry. pp77–78. MTP Press, Lancaster.

Lander, K.P. (1990) The development of a transport and enrichment medium for *Campylobacter fetus*. Brit. Vet. J. 146, 327–333.

Lander, K.P. and Gill, K.P.W. (1985) Campylobacters. In: Isolation and Identification of Micro-organisms of Medical and Veterinary Importance. eds Collins, C. H. and Grange, J.H. Academic Press, London, pp. 123–142.

Lastovica, A.J., Le Roux, E. and Penner, J.L. (1989) "*Campylobacter upsaliensis*" isolated from blood cultures of paediatric patients. J. Clin. Microbiol. 276, 657–659.

Lastovica, A.J. and Skirrow, M.B., 2000. Clinical significance of *Campylobacter* and related species other than *Campylobacter jejuni* and *C. coli*. In: Nachamkin, I., Blaser, M. J. (Eds.), Campylobacter. American Society for Microbiology, Washington DC, pp. 89–120.

Lauwers, S. De Boeck, M. and Butzler, J.P. (1978) Campylobacter enteritis in Brussels. Lancet i, 604–605.

Lauwers, S., Breynaert, J., Van Etterijck, R., Revets, H., Mets, T., 1996 *Arcobacter butzleri* in the elderly in Belgium. In: Newell, D.G., Ketley, J., Feldman, R.A. (Eds.), Campylobacters, Helicobacters and Related Organisms. Proc. 8th International Workshop on Campylobacters, Helicobacters and Related Organisms. Plenum Publishing Corporation, New York, pp. 515–518.

Lawson, A.J., On, S.L.W., Logan, J.M.J. and Stanley, J. (2000) *Campylobacter hominis* sp. nov., from the human gastrointestinal tract. Int. J. System. Evol. Microbiol. 35, 5.

Lerner, J., Brumberger, V., Preac-Mursic, V., 1994. Severe diarrhoea associated with *A. butzleri*. Europ. J. Clin. Microbiol. Infect. Dis. 13, 660–662.

Line, J.E. (2001) Development of a selective differential agar for isolation and enumeration of *Campylobacter* spp. J. Food Protect. 64, 1711–1715.

Line, J.E., Stern, N.J., Lattuada, C.P. and Benson, S.T. (2001) Comparison of methods for recovery and enumeration of *Campylobacter* from freshly processed broilers. J. Food Protect. 64, 982–986.

Liu, X., Phebus, R.K., Fung, D.Y.C. and Kastner, C.L. (1995) Evaluation of culture protocols and Oxyrase supplementation for *Arcobacter* spp. J. Rapid Meth. Autom. Microbiol. 4, 115–126.

Lovett, J., Francis, D.W. and Hunt, J.M. (1983) Isolation of *Campylobacter jejuni* from raw milk. Appl. Environ. Microbiol. 46, 459–462.

Lu, Y., Fredlonger, T.E., Avitia, R., Galindo, A. and Goodman, K. (2002) Isolation and genotyping of *Helicobacter pylori* from untreated municipal wastewater. Appl. Environm. Microbiol. 68, 1436–1439.

Mansfield, L.P. and Forsythe, S.J. (2000) *Arcobacter butzleri, A. skirrowii* and *A. cryaerophilus*

– potential emerging human pathogens. Rev. Med. Microbiol. 11, 161–170.

Marinescu, M., Collignon, A., Squinazi, F., Derimay, R., Woodward, D.L., Lior, H., 1996a. Two cases of persistent diarrhoea associated with *Arcobacter* spp. In: Newell, D.G., Ketley, J., Feldman, R.A. (Eds.), Campylobacters, Helicobacters and Related Organisms. Proc. 8th International Workshop on Campylobacters, Helicobacters and Related Organisms. Plenum Publishing Corporation, New York, pp. 521–523.

Marinescu, M., Collignon, A., Squinazi, F., Woodward, D.L., Lior, H., 1996b. Biotypes and serogroups of poultry strains of *Arcobacter* spp. isolated in France. In: Newell, D.G., Ketley, J., Feldman, R.A. (Eds.), Campylobacters, Helicobacters and Related Organisms. Proc. 8th International Workshop on Campylobacters, Helicobacters and Related Organisms. Plenum Publishing Corporation, New York, pp. 519–520.

Marshall B.J. and Warren, J.R. (1984) Unidentified curved bacilli in the stomach of patients with gastritis and peptic-ulceration. Lancet, 1, 1311–1315.

Martin, W.H., Patton, C.M., Morris, G.K., Potter, M.E. and Puhr, N.D. (1983) Selective enrichment broth medium for isolation of *Campylobacter jejuni*. J. Clin. Microbiol. 17, 853–855.

Martin, K.W., Mattick, K.L., Harrison, M. and Humphrey, T.J. (2002) Evaluation of selective media for *Campylobacter* isolation when cycloheximide is replaced with amphotericin B. Lett. Appl. Microbiol. 34, 124–129.

Mason, M.J., Humphrey, T.J. & Martin, K.W. (1989) Isolation of sublethally injured campylobacters from poultry and water sources. Br. J. Biomed. Sci. 56, 2–5.

Megraud, F. (1987) Isolation of *Campylobacter* spp. from pigeon faeces by a combined enrichment-filtration technique. Appl. Environ. Microbiol. 53, 1394–95.

Megraud, F. and Gavinet, A.M. (1987) Comparison of three protocols to isolate *Campylobacter* species from human faeces. In: Kayser, B and Falsen, E. (Eds.) "Campylobacter IV" Proceedings of the 4th International Workshop on Campylobacter Infections pp 90–91, Gôteborg, Sweden.

Megraud, F., Chevrier, D., Desplaces, N., Sedallian, A. and Guesdon, JL. (1988) Urease-positive thermophilic campylobacter (*Campylobacter laridis* variant) isolated from an appendix and from human feces. J. Clin. Microbiol., 26, 1050–1051.

Merino, F.J., Agulla, A., Villasante, P.A., D¡az, A., Saz, J.V. and Velasco, A.C. (1986) Comparative efficacy of seven selective media for isolating *Campylobacter jejuni*. J. Clin. Microbiol. 24, 451–452.

Michel, J., Rogol, M. and Dickman, D. (1983) Susceptibility of clinical isolates of *Campylobacter jejuni* to sixteen antimicrobial agents. Antimicrob. Agents Chemother. 23, 796–797.

Moreno, G.S., Griffiths, P.L., Connerton, I.F. and Park, R.W.A. (1993) Occurrence of campylobacters in small domestic and laboratory animals. J. Appl. Bacteriol. 75, 49–54.

Morris, G.K., Bopp, C.A., Patton, C.M. and Wells, J.G. (1982) Media for isolating campylobacter. Arch. Lebensmittelhyg. 33, 151–153.

Musgrove, M.T., Stern, N.J. and Bailey, J.S. (1997) A comparison of enrichment methods for the recovery of *Campylobacter* spp. in broiler litter samples. Poult. Sci. 76, Supplement, p.30.

Musgrove, M.T., Berrang, M.E., Byrd, J.A., Stern, N.J. and Cox, N.A. (2001) Detection of *Campylobacter* spp. in ceca and crops with and without enrichment. Poult. Sci. 80, 825–828.

Neill, S.D., Campbell, J.N., O'Brien, J.J., Weatherup, S.T.C. and Ellis, W. A. (1985) Taxonomic position of *Campylobacter cryaerophila* sp. nov. Int. J. Syst. Bacteriol. 35, 342–356.

Ng, L.K., Stiles, M.E. and Taylor, D.E. (1985) Inhibition of *Campylobacter coli* and *Campylobacter jejuni* by antibiotics used in selective growth media. J. Clin. Microbiol. 22, 510–514.

Ng, L.K., Taylor, D.E. and Stiles, M.E. (1988) Characterisation of freshly isolated *Campylobacter coli* strains and suitability of selective media for their growth. J. Clin Microbiol. 26, 518–523.

On, S.L.W. (1996) Identification methods for campylobacters, helicobacters and related organisms. Clin. Microbiol. Rev. 9, 405–422.

On, S.L.W. (2001) Taxonomy of *Campylobacter, Arcobacter, Helicobacter* and related bacteria: currnt status, furure prospects and immediate concerns. J. appl. Microbiol. 90, 1S–15S.

On, S.L.W., Holmes, B. and Sackin, M.J. (1996) A probability matrix for the identification of campylobacters, helicobacters and allied taxa. J. Appl. Bacteriol. 81, 425–432.

On, S.L.W., Stacey, A. and Smyth, J., (1995) Isolation of *Arcobacter butzleri* from a neonate with bacteraemia. J. Infect. 31, 225–227.

Osaki, T., Taguchi, H., Yamaguchi., H. and Kamiya, S. (1998) Detection of *Helicobacter pylori* in fecal samples of gnotobiotic mice infected with *H. pylori* by an immunomagnetic bead separation technique. J. Clin. Microbiol. 36, 321–323.

Park, C.E., Stankiewicz, Z.K., Lovett, J. and Hunt, J. (1981) Incidence of *Campylobacter jejuni* in fresh eviscerated whole market chickens. Can. J. Microbiol. 27, 841–842.

Park, C.E. and Stankiewicz, Z.K. (1982) A selective enrichment procedure for the isolation of *Campylobacter jejuni* from foods. In: Newell, D.G. (ed) "Campylobacter, Epidemiology, Pathogenesis and Biochemistry". Proceedings of International Workshop on Campylobacter Infections p79. Lancaster, UK: MTP Press Ltd.

Park, C.E., Stankiewicz, Z.K., Lovett, J., Hunt, J., and Francis, D.W. (1983) Effect of temperature, duration of incubation and pH of enrichment culture on the recovery of *Campylobacter jejuni* from eviscerated market chickens. Can. J. Microbiol. 29, 803–806.

Park, C.E. and Sanders, G.W. (1991) A sensitive enrichment procedure for the isolation of *Campylobacter jejuni* from frozen foods. In: Riuz-Palacios, G.M., Calva, F. and Ruiz-Palacios, B.R. (Eds). "Campylobacter V", "Proceedings of 5th International Workshop on Campylobacter Infections p102. Puerto Vallarta, Mexico: National Institute of Nutrition.

Patton, C.M., Mitchell, S.W., Potter, M.E. and Kaufman, A.F. (1981) Comparison of selective media for primary isolation of *Campylobacter fetus* subsp. *jejuni*. J. Clin. Microbiol. 13, 326–330.

Patton, C.M., Shaffer, N., Edmonds, P. et al. (1989) Human disease associated with *Campylobacter upsaliensis* (catalase-negative or weakly positive *Campylobacter* species) in the United States. J. Clin. Microbiol. 27, 66–73.

Pearson, A.D. and Healing, T.D. (1992) The surveillance and control of campylobacter infection. Communicable Disease Report Review 2, R133–140. London: Public Health Laboratory Service.

Pennie, R.A., Zurino, J.N., Rose, C.E. and Guerrant, R.L. (1984) Economical simple method for production of the gaseous environment required for cultivation of *Campylobacter jejuni*. J. Clin. Microbiol. 20, 320–322.

Piersimoni, C., Bornigia, S., Curzi, L. and De Sio, G. (1995) Comparison of two selective media and a membrane filter technique for isolation of *Campylobacter* species from diarrhoeal stools. Eur. J. Clin. Microbiol. Infect. Dis., 14, 539–542.

Peterz, M. (1991) Comparison of Preston agar and a blood-free selective medium for detection of *Campylobacter jejuni* in food. J. Assoc. Off. Anal. Chem. 74, 651–654.

Plastridge, W.N., Williams, L.F. and Trowbridge, D.G. (1964) Antibiotic sensitivity of physiologic groups of microaerophilic vibrios. Am. J. Vet. Res. 25, 1295–1299.

Quieroz, D.M.M., Mendes, E.N. and Rocha, G.A. (1987) Indicator medium for isolation of *Campylobacter pylori*. J. clin. Microbiol. 25, 2378–2379.

Ray, B. and Johnson, C. (1984a) Sensitivity of cold-stressed *Campylobacter jejuni* to solid and liquid enrichments. Food Microbiol. 1, 173–176.

Ray, B. and Johnson, C. (1984b) Survival and growth of freeze-stressed *Campylobacter jejuni* cells in selective media. J. Food Safety 6, 183–195.

Razi, M.H.H. and Park, R.W.A. (1979) Studies of media for the isolation and storage of *Campylobacter* spp. J. Appl. Bacteriol. 47, x.

Ribeiro, C.D. and Price, T.H. (1984) The use of Preston enrichment broth for the isolation of "thermophilic" campylobacters from water. J. Hyg., Camb. 92, 45–51.

Ribeiro, C.D., Marks, J., and Grimshaw, A.D. (1985) Economic cultivation of "thermophilic" *Campylobacter* spp. J. Clin. Pathol. 38, 1311–1313.

Rosef, O. (1981) Isolation of *Campylobacter fetus* subsp. *jejuni* from the gall bladder of normal slaughter pigs, using an enrichment procedure. Acta Vet. Scand. 22, 149–151.

Rosef, O. and Kapperud, G. (1983) House flies (*Musca domestica*) as possible vectors of *Campylobacter fetus* subsp. *jejuni*. Appl. Environ. Microbiol. 45, 381–383.

Rosef, O., Gondrosen, B., Kapperud, G. and Underdal, B. (1983) Isolation and characterisation of *Campylobacter jejuni* and *Campylobacter coli* from domestic animals and wild animals in Norway. Appl. Environ. Microbiol. 46, 855–859.

Rosef, O., Kapperud, G. and Skjerve, E. (1987) Comparison of media and filtration procedures for qualitative recovery of thermotolerant *Campylobacter* spp. from naturally contamined surface water. Int. J. Food Microbiol. 5, 29–39.

Safarik, I., Safarikova, M. and Forsythe, S. (1995) The application of magnetic separation in applied microbiology. J. Appl. Bacteriol. 78, 575–585.

Schroeder-Tucker, L., Wesley, I.V., Kiehlbauch, J.A., Larson, D.J., Thomas, L-A. and Erickson, G.A. (1996) Phenotypic and ribosomal RNA characterisation of *Arcobacter* species isolated from procine aborted fetuses. J. Vet. Diag. Invest. 8, 186–195.

Scotter, S.L., Humphrey, T.J. and Henley, A. (1993) Methods for the detection of thermotolerant campylobacters in foods: results of an inter-laboratory study. J. Appl. Bacteriol. 74, 155–163.

Shanker, S., Gordon, S.W., Fuller, H. and Gilbert, G.L. (1991) Sensitivity of a filtration method for the detection of *Campylobacter/Helicobacter* spp. in faeces. Microb. Ecol. Hlth Dis. 4, S47.

Skirrow, M.B. (1977) Campylobacter enteritis: a new disease. Brit. Med. J. 2, 9–11.

Skirrow, M.B., Benjamin, J., Razi, M.H.H. and Waterman, S. (1982) Isolation, cultivation and identification of *Campylobacter jejuni* and *Campylobacter coli*. In: Corry, J.E.L., Roberts, D. and Skinner, F.A. Eds. "Isolation and Identification Methods for Food Poisoning Organisms". London. Academic Press. pp313–328.

Skirrow, M.B., Vickery, C.R., Lovelock, M.R. and Evans, J.J. (1987) Cheap microaerobic atmospheres: modification of the methylated spirit burn method. In: Kaijser, B. and Falsen, E. (Eds) Campylobacter IV. Proceedings of the 4th International Workshop on Campylobacter Infections. Sweden: University of Göteborg.

Skirrow, M.B., Merry, T.J. and Vickery, C.R. (1991) Hydrogen facilitates the isolation of *Campylobacter jejuni*. In: Ruiz-Palacios, G.M., Calva, F. and Ruiz-Palacios, B.R. (eds) "Campylobacter V". Proceedings of the 5th International Workshop on Campylobacter Infection, p109. Instituto Nacional de la Nutricion, Vasco de Quiroga 15, Mexico 1400, DF.

Stanley, J., Linton, D., Burnens, A.P., Dewhirst, F.E., On, S.L.W., Porter, A., Owen, R.J. and Costas, M. (1994) *Helicobacter pullorum* sp. nov. - genotype and phenotype of a new species isolated from poultry and human patients with gastroenteritis. Microbiology 140, 3441–3449.

Steele, T.W. and McDermott, S.W. (1984) The use of membrane filters applied directly to the surface of agar plates for the isolation of *Campylobacter jejuni* from faeces. Pathol. 16, 263–265.

Steinbrueckner, B., Haerter, G., Pelz, K., Weiner, S., Rump, J.A., Deissler, W., Bereswill, S. and Kist, M. (1997) Isolation of *Helicobacter pullorum* from patients with enteritis. Scand. J. Infect. Dis., 29, 315–318.

Stelzer, W. and Jacob, J. (1992) Das Vorkommen von Campylobacter in einem Mittelgebirgsbach. Zbl. Mikrobiol. 147, 45–50.

Stelzer, W., Mochmann, H.P., Richter, U. and Dobberkau, H.J. (1988) Characterisation of *Campylobacter jejuni* and *Campylobacter coli* isolated from waste water. Zbl. Bakteriol. Hyg. A 269, 188–196.

Stern, N.J., Green, S., Thaker, N., Krout, D. and Chju, H. (1984) Recovery of *Campylobacter jejuni* from fresh and frozen meat and poultry collected at slaughter. J. Food Protect. 47, 372–374.

Stern, N.J. and Line, J.E. (1992) Comparison of three methods for recovery of *Campylobacter* spp. from broiler carcasses. J. Food Protect. 55, 663–666.

Stern, N.J., Wojton, B. and Kwiatek, K. (1992) A differential-selective medium and dry ice-generated atmosphere for recovery of *Campylobacter jejuni*. J. Food Protect. 55, 514–517.

Stevenson, T.H., Castillo, A., Lucia, L.M. and Acuff, G.R. (2000a) Growth of *Helicobacter pylori* in various liquid and plating media. Lett. Appl. Microbiol. 30, 192–196.

Stevenson, T.H., Lucia, L.M. and Acuff, G.R. (2000b) Development of a selective medium for isolation of *Helicobacter pylori* from cattle and beef samples. Appl. Environm. Microbiol. 66, 723–727.

Taylor, D.N., Kiehlbauch, J.A., Tee, W., Pitarangsi, C. and Echeverria, P. (1991) Isolation of group 2 aerotolerant *Campylobacter* species from Thai children with diarrhoea. J. Infect. Dis. 163, 1062–1067.

Tenova, F.C. and Fennell, C.L. (1992) The genera *Campylobacter* and *Helicobacter*. In: Balows, A., Truper, H.G., Dworkin, M., Hurder, W. and Schleifer, K.H. (eds) The Prokaryotes. 2nd edition vol. IV pp. 3488–3511. New York: Springer-Verlag.

Tran, T.T. (1998) A blood-free enrichment medium for growing *Campylobacter* spp. under aerobic conditions. Lett. Appl. Microbiol. 26, 145–148.

Uyttendaele, M. and Debevere, J. (1996) Evaluation of Preston medium for detection of *Campylobacter jejuni* in vitro and in artificially and naturally contaminated poultry products. Food Microbiol., 13, 115–122.

Van Etterijck, R., Breynaert, J., Revets, H., Devreker, T., Vandenplas, Y., Vandamme, P. and Lauwers, S. (1996) Isolation of *Campylobacter concisus* from feces of children with and without darrhea. J. Clin. Microbiol. 34, 2304–2306.

Vandamme, P. and De Ley, J. (1991) Proposal for a new family, *Campylobacteraceae*. Int. J. Sytem. Bacteriol. 41, 451–455.

Vandamme, P. and Goossens, H. (1992) Taxonomy of campylobacter, arcobacter and heliobacter. Zbl. Bakteriol. 276, 447–472.

Vandamme, P., Falsen, E., Rossau, R., Hoste, B., Segers, P., Tytgat, R. et al. (1991) Revision of *Campylobacter, Helicobacter* and *Wolinella* taxonomy: emendation of generic descriptions and proposal of *Arcobacter* gen. nov. Int. J. Syst. Bacteriol. 41, 88–103.

Vandamme, P., Vancanneyt, M., Pot, B., Mels, L., Hoste, B., Dewettinck, D. et al. (1992a) Polybasic taxonomic study of the emended genus *Arcobacter* with *Arcobacter butzleri* comb. nov. and *Arcobacter skirrowi* sp. nov., an aerotolerant bacterium isolated from veterinary specimens. Int. J. Syst. Bacteriol. 42, 344–356.

Vandamme, P., Pugina, P., Benzi, G., Van Etterijck, R., Vlaes, L., Kersters, K. et al. (1992b) Outbreak of recurrent abdominal cramps associated with *Arcobacter butzleri* in an Italian school. J. Clin. Microbiol. 30, 2335–2337.

Vanhoof, R., Vanderlinden, M.P., Diereckx, R., Lauwers, S., Yourassowsky, E. and Butzler, J.P. (1978) Susceptibility of *Campylobacter fetus* subsp. *jejuni* to twenty nine antimicrobial agents. Antimicrob. Agents Chemother. 14, 553–556.

Walder, M. (1979) Susceptibility of *Campylobacter fetus* subsp. *jejuni* to twenty antimicrobial agents. Antimicrob. Agents Chemother. 16, 37–39.

Warmsley, S.L. and Karmali, M.A. (1989) Direct isolation of atypical thermophilic *Campylobacter* species from human faeces on selective agar medium. J. Clin. Microbiol. 27, 668–670.

Waterman, S.C., Park, R.W.A. and Bramley, A.J. (1984) A search for the source of *Campylobacter jejuni* in milk. J. Hyg., Camb. 93, 333–337.

Weber, G., Manafi, M. and Reisinger, H. (1987) Die Beduntung von *Yersinia enterocolitica* und thermophilen Campylobactern für die Wasserhygiene. Zbl. Bakteriol. Hyg. B 184, 501–514.

Weinrich, K., Winkler, K. and Heberer, E. (1990) Untersuchungen zur Eignung ausgewhlter Huminsaurenchargen als Nahbodenzusatz für Medien sur Isolierung thermophiler Campylobacterspecies. Deusche Tierarztliche Wochenschrift 97, 511–515.

Wesley, R.D., Swaminathan, B. and Stadelman, W.J. (1983) Isolation and enumeration of *Campylobacter jejuni* from poultry products by a selective enrichment method. Appl. Environ. Microbiol. 46, 1097–1102.

Wonglumsom, W., Vishnubhatla, A. and Fung, D.Y.C. (2000) Effect of volume of liquid enrichment medium containing Oxyrase on growth of *Campylobacter jejuni.* J. Rapid Meth. Automat.

Microbiol. 8, 111–139:
Wonglumsom, W., Vishnubhatla, A., Kim, J.M. and Fung, D.Y.C. (2001) Enrichment media for isolation of *Campylobacter jejuni* from inoculated ground beef and chicken skin under normal atmosphere. J. Food Protect. 64, 630–634.
Yu, L.S.L., Uknalis, J. and Tu, S.I. (2001) Immunomagnetic separation methods for the isolation of *Campylobacter jejuni* from ground poultry meats. J. Immunol. Meth. 256, 11–18.

Chapter 19

Culture media for *Aeromonas* spp. and *Plesiomonas shigelloides*

Ildefonso Perales

Laboratorio Normativo de Salud Pública, Departamento de Sanidad, Gobierno Vasco, María Díaz de Haro 60, 48010 Bilbao, Spain.

The genera *Aeromonas* and *Plesiomonas* are included in the family Vibrionaceae and are primarily aquatic inhabitants. *Aeromonas* taxonomy has changed very much in recent years and now at least 16 genospecies have been defined. Although identification to the species level of the most commonly found *Aeromonas* can be done on the basis of a limited number of biochemical tests, identification of some species is more difficult. *Plesiomonas shigelloides* is the only species of the genus *Plesiomonas*.

A review of the culture media components used for *Aeromonas* and *Plesiomonas* isolation is made. This paper also reports on the media used for the isolation of *Aeromonas* and *Plesiomonas* from foods and comparative studies.

In the case of *Aeromonas* spp. the most promising media are Starch Ampicillin Agar (SAA) and Bile Salts Irgasan Brilliant Green agar (BSIBG) as plating media and Alkaline Peptone Water (APW) or Trypticase Soy Broth Ampicillin (TSBA) as enrichment broths. With regard to *P. shigelloides*, it is more difficult to make recommendations, but enrichment in APW or tetrathionate broth without iodine and plating on Inositol Brilliant green Bile salts agar (IBB) and *Plesiomonas* agar (PL) can be useful.

Taxonomy

The genera *Aeromonas* and *Plesiomonas* are included in the family Vibrionaceae and are primarily aquatic. They are Gram-negative rods, motile by polar flagella or nonmotile, oxidase positive and facultatively anaerobic (Holt et al., 1994). Although it was proposed that *Aeromonas* should be placed in the new family Aeromonadaceae (Colwell et al., 1986) and *Plesiomonas* be transferred to the family Enterobacteriaceae (Martínez-Murcia et al., 1992) in the 9th edition of *Bergey´s Manual of Determinative Bacteriology* both genera are retained in the family Vibrionaceae (Holt et al., 1994).

The genus *Plesiomonas* only includes one species, *P. shigelloides* (Holt et al. 1994). Aeromonad taxonomy has changed very much in the last years with regard to species identification at both the biochemical and the molecular levels. In 1984 four species in two separate groups were included in the genus *Aeromonas* (Popoff, 1984), one encompassed the psychrophilic and nonmotile aeromonads into one species *Aerom-*

Table 1
Current genospecies and phenospecies within the genus *Aeromonas* and their occurrence in clinical specimens.

Hybridization group (HG)	Genospecies	Phenospecies	Subspecies	Important in human disease
1	*A. hydrophila*	*A. hydrophila*		Yes [++++]
2	*A. bestiarum*	*A. hydrophila*-like		No **
3*	*A. salmonicida*	*A. salmonicida*	subsp. *achromogenes* subsp. *masoucida* subsp. *salmonicida* subsp. *smithia*	No**
4	*A. caviae*	*A. caviae*		Yes [++++]
5	*A. media*	*A. media*		No **
6	*A. eucrenophila*	*A. eucrenophila*		No
7	*A. sobria*	*A. sobria*		No
8/10	*A. veronii* biovar sobria	*A. sobria*		Yes [++++]
10/8	*A. veronii* biovar veronii	*A. veronii*		Yes [++]
9	*A. jandaei*	*A. jandaei*		Yes [++]
11	*A. encheleia*	*A. encheleia*		No
12	*A. schubertii*	*A. schubertii*		Yes [++]
13	Unnamed	*Aeromonas* Group 501		No
14	*A. trota*	*A. trota*		Yes [++]
15	*A. allosaccharophila*	*A. allosaccharophila*		No
16	*A. popoffii*	*A. popoffii*		No

* HG3 contains both non-motile psychrotrophic *A. salmonicida* organisms as well as motile mesophilic *A. hydrophila*-like.
** Rarely isolated from human clinical samples.
[++++] Frequently isolated from human clinical samples.
[++] Sometimes isolated from human clinical samples.

onas salmonicida which included three subspecies. The second group included the mesophilic and motile aeromonads and comprised three species that were phenotypically and genetically differentiated: *Aeromonas hydrophila, A. caviae* and *A. sobria*.

In 2000 the taxonomy is still very confusing (Altwegg, 1999) and there are at least 16 species that represent clearly differentiated DNA hybridization groups (HG) and most of these have been named as phenospecies (Janda, 1991; Esteve et al., 1995; Huys et al., 1997a, b; Altwegg, 1999; Martinez-Murcia, 1999). Because of the lack of phenotypic characteristics that correlate with the genetic species, some of the DNA hybridization groups have not yet been named. The geno- and phenospecies of *Aeromonas* are shown in Table 1.

A. hydrophila (HG1) is frequently isolated from human clinical samples. The biochemically similar HG2 is now named *A. bestiarum* (Ali et al., 1996) and is very rare in humans (Kuijper et al., 1989; Janda et al., 1996).

A. salmonicida (HG3) is usually psychrophilic and nonmotile and now includes four subspecies. It is a strict parasite under natural conditions and produces furunculosis in salmonid fish (Popoff, 1984; Joseph et al., 1988; Jeppesen, 1995). In this HG3 genospecies some mesophilic and motile isolates are included and, although rarely, they can be isolated from humans (Janda et al., 1996).

A. caviae (HG4) is the aeromonad most often isolated from human faeces in many countries (Kuijper et al., 1989). The biochemically similar *A. media* (HG5) and *A. eucrenophila* (HG6) are isolated rarely; the former, but never the latter, from clinical samples (Janda et al., 1996; Huys et al., 1997a).

HG8 *(A. sobria)* and HG10 (*A. veronii*) were independently established and are biochemically dissimilar, but have been shown to be genetically identical (Kuijper et al., 1989). Since HG 10 has precedence in the literature, HG8 is considered a biovar of *A. veronii*, even though it is much more frequently recovered from clinical material than group 10 (Janda, 1991). Therefore, essentially all clinical isolates identified as *A. sobria* are actually *A. veronii* and they are called *A. veronii* biovar sobria. On the other hand *A. sobria sensu stricto* (HG7) has not yet been recovered from clinical specimens (Hickman-Brenner et al., 1987; Janda, 1991; Janda et al., 1996).

A. schubertii (HG12) can cause bacteraemia and wound infections in humans (Hickman-Brenner et al., 1988; Kirov, 1993; Abbot et al., 1994, Janda et al., 1994). Other *Aeromonas* species such as *A. jandaei* (HG9) and *A. trota* (HG14) can be isolated from cases of bacteraemia in humans and also from faecal samples (Carnahan et al., 1991a; Carnahan et al., 1991b; Abbot et al., 1994; Janda et al., 1994).

A. enchelia (HG11) and *A. allosaccharophila* (HG15) have never been reported in human clinical samples (Esteve et al., 1995; Janda, 1991, Huys et al., 1996). *A. popoffi* has been described recently and was isolated from water (Huys et al., 1997b)

Other species such as *A. enteropelogenes* and *A. ichthiosmia* are of questionable validity and they are probably identical to the earlier proposed species *A. trota* and *A. veronii* biovar sobria, respectively (Collins et al., 1993; Huys et al., 1996).

Identification of *Aeromonas* spp. and *Plesiomonas shigelloides* in the laboratory.

The genus *Aeromonas* (Hänninen 1994; Holt et al., 1994; Altwegg, 1999) has the general characteristics of Vibrionaceae: Gram-negative rods, usually motile by a single polar flagellum, but peritrichous flagella may be found on solid media in young cultures. Nonmotile species are *A. salmonicida* and *A. media*. Aeromonads are facultatively anaerobic, oxidase and catalase positive, reduce nitrate to nitrite, and ferment D-glucose with or without production of gas. They grow over a wide temperature range (0 to 45°C). Human (mesophilic) strains grow between 10 and 42°C, whereas the non-motile, psychrophilic species have a maximum growth temperature of 37°C (e.g. *A. media*) or lower (e.g. *A. salmonicida*). Usually they produce many exoenzymes including gelatinase, DNase, amylase, esterases, peptidases and arylamidases. They are resistant to 150 μg disks of the vibriostatic agent 2,4-diamino-6,7-diisopropylpteridine (O/129).

P. shigelloides (Schubert, 1984; Holt et al., 1994; Altwegg, 1999) is a Gram-negative rod usually motile with two to five lophotrichous flagella. *Plesiomonas* is faculta-

Table 2
Differentiation of the most important genera of the family Vibrionaceae, found in foods.

Test	Genera		
	Vibrio	*Aeromonas*	*Plesiomonas*
Na^+ required for or stimulates growth	+	–	–
Sensitivity to vibriostatic compound 0/129	+	–	+
Sucrose	(+)	+*	–
D-Mannitol	(+)	+**	–
myo-Inositol	(+)	–	+
Gelatinase	+	+	–
Lipase	–	(+)	–

Symbols: +, All species positive; (+), Most species positive; –, All species negative;
*Except *A. salmonicida* subsp. *salmonicida*; **Except *A. salmonicida* subsp. *Achromogenes.*

tively anaerobic, oxidase and catalase positive, reduces nitrate to nitrite, and ferments D-glucose without the production of gas. Most strains are sensitive to 150 μg disks of O/129. It is positive for lysine and ornithine decarboxylase, and arginine dihydrolase. Its temperature range is 8 to 45ºC.

Differentiation between the three most important genera of the family Vibrionaceae, *Vibrio, Aeromonas* and *Plesiomonas*, can be made by a few tests (Holt et al., 1994; Palumbo et al., 2000) (Table 2).

Differential characteristics of *Aeromonas* species are listed in Table 3. Most microbiology laboratories dealing with food, water or clinical samples still report *Aeromonas* isolates as *A. hydrophila, A. sobria,* or *A. caviae* because it is difficult to differentiate the new *Aeromonas* species from the better known species. This is now being done with the understanding that these phenospecies are a collection of several genetically distinct groups that are biochemically similar but could not unambiguously be separated from one another by phenotypic methods (Abbott et al., 1992). For this reason it is more appropriate to report the most frequent isolates as members of three Aeromonas groups: (i) *Aeromonas hydrophila* group that includes: *Aeromonas hydrophila* and *Aeromonas salmonicida* (motile biogroup); (ii) the *Aeromonas caviae* group: *Aeromonas caviae, Aeromonas eucrenophila, Aeromonas media* and (iii) the *Aeromonas sobria* complex: *Aeromonas veronii* biovar sobria, *Aeromonas veronii* biovar veronii and *Aeromonas schubertii* (Janda, 1991; Holt et al. 1994).

However the identification of most *Aeromonas* isolates to the species level can be done on the basis of a limited number of biochemical tests. From the public health point of view a dichotomous key is applicable that uses seven tests for identifying clinically important aeromonads: *A. hydrophila, A. caviae, A. veronii* biovar veronii, *A. schubertii, A. jandaei, A. trota* and *A. veronii* biovar sobria (Carnahan et al. 1991c). Also, many of the recently named DNA groups appear to have unique biochemical properties that are not commonly found in other members of this genus, and this will permit an easier identification. Recently, it has been shown that a battery of 18 biochemical tests allows identification of 97 % of strains from diverse sources to genospecies level (Janda et al., 1996). Also commercial identification kits are available

(Joseph et al., 1988) which allow correct identification of between 80 and 97.5 % of isolates (Ogden et al., 1994; Vivas et al., 2000).

For the differentiation of the phenospecies of *Aeromonas*, biochemical tests (including kits such as API 20E) must be incubated at 30ºC, because 37ºC can give false negative results (Hänninen, 1994).

Some studies have been made for the identification of *Aeromonas* strains by PCR (Cascón et al., 1996; Borrell, et al., 1997; Kingombe et al., 1999).

Significance of *Aeromonas* spp. in foods

Aeromonas spp. are commonly found in fresh water and in foods (Abeyta and Wekell, 1988; Archer and Kvenberg, 1988). Although their isolation from stools of persons with diarrhoea has implicated them as a cause of diarrhoeal disease, these microbes have failed to produce disease, even at very high doses, in volunteer human feeding studies (Archer and Kvenberg, 1988). Also, there is no readily available animal model that reproduces *Aeromonas*-associated diarrhoea (Janda, 1991; Kirov, 1993) and no well-documented foodborne outbreaks have been reported in the literature (Abeyta and Wekell, 1988; Janda, 1991).

Most aeromonads produce a variety of extracellular enzymes: proteases, DNase, RNAse, elastase, lecithinase, amylase, lipases, gelatinase, and chitinases; some of them are now confirmed as toxins: the cytotoxic/cytolytic enterotoxin, three different haemolysins and cytotonic enterotoxins (Janda, 1991; Kirov et al., 1993; Merino et al., 1999). It is not yet possible to identify virulent strains definitively because *Aeromonas* virulence can be multifactorial (Kirov et al., 1999). Some serogroups are more associated with infection, for example serogroups O:11, O:34 and O:16 predominate clinically (Janda et al., 1996).

A. hydrophila, A. caviae and *A. veronii* biovar sobria predominate in human clinical infections (Moyer, 1987; Kirov, 1993; Rautelin et al., 1995; Janda et al., 1996). There is evidence to suggest that strains of *A. veronii* biovar sobria are most often virulent, more commonly invasive than *A. hydrophila* and more frequently associated with bacteraemia (Kirov, 1993). *A. caviae* may be more common in paediatric diarrhoea (Kirov, 1993).

In a study dealing with strains isolated from water, foods and human clinical samples, Hänninen and Shtonen (1995) found that the majority of *Aeromonas* spp. occurring in water were adapted to water, with clinically important species present as a minority. By contrast chicken and ground beef commonly contained *Aeromonas* spp. also found in human diarrhoeal and non diarrhoeal faecal samples.

It is therefore difficult to recognise potentially significant strains in foods and water and this poses a dilemma for public health authorities (Kirov, 1993) making regulatory limits based on health concerns seem insupportable (Archer and Kvenberg, 1988). However the presence of large numbers of aeromonads in foods should be regarded as a potential health threat, particularly for immunocompromised individuals (Kirov, 1993).

Culture media components

The first isolations of *Aeromonas* spp. were made from faeces with 'enteric' agars used for the isolation of pathogens such as *Salmonella* and *Shigella* and containing bile salts and lactose. Because aeromonads are usually lactose-negative (Popoff, 1984) and the response to selective agents is similar to some Enterobacteriaceae, some gastroenteritis-related aeromonads were detected and described in the literature (von Graevenitz and Mensch, 1968). Specific media for the isolation of aeromonads from faeces, foods and environmental samples were developed subsequently.

Basal media

Aeromonas spp. have no specific nutritional requirements and like most of the Enterobacteriaceae and Vibrionaceae they grow in relatively simple media such as nutrient agar or trypticase soya agar (Popoff, 1984). As with 'enteric' media, the specifically developed media for aeromonads usually contain either peptones or extracts of meat or yeast, alone or in combination. The amounts of peptone incorporated in the media are usually between 5 and 30 g/l, in the case of meat extracts between 1 and 5 and between 2 and 5 in the case of yeast extracts. Most media contain a total of nutrients between 10 and 20 g/l. However some media include as little as 2.5 g of yeast extract or 5 g of peptone as single nutrients (Rippey and Cabelli, 1979; Myers et al., 1982).

Selective systems

Most of the selective agents used are similar to the ones incorporated in media for the isolation of Enterobacteriaceae and *Vibrio* spp. These include dyes, surfactants, bile salts, antiseptics, antibiotics and vibriostatics.

Dyes

In the case of dyes, fuchsin (Schubert, 1967) has been used together with sodium sulphite to inhibit Gram-positive bacteria. This is based on the formulation of Endo agar used for coliform isolation, but because of this, coliforms can grow easily (von Graevenitz and Zinterhofer, 1970). Crystal violet is also active against Gram-positive bacteria and is included in media based on MacConkey agar (Kaper et al., 1981). Toluidine blue is used to show the DNase activity of *Aeromonas* spp. in differential media, but at the concentration of 50 mg/l is also inhibitory for Gram-positive cocci. Brilliant green is included in many media for Enterobacteriaceae. In the case of aeromonads it has generally been used at concentrations of 0.5 to 5 mg/l (Millership and Chattopadhyay, 1984; Hunt et al. 1987; Want and Millership, 1990). Millership and Chattopadhyay (1984) observed that below 5 mg/l, colony counts were the same as on blood agar, but at 1 mg/l or above the colonies were much smaller. Hence, 0.5

mg/l was chosen as the optimum concentration. In general the concentrations of brilliant green used in media for aeromonads are higher than those used for the isolation of coliforms and lower than those used for the isolation of *Salmonella* spp.

Surfactants

Sodium lauryl sulphate has been used at concentrations between 0.1 and 0.8 g/l in order to make media inhibitory for Gram-positive bacteria (McCoy and Pilcher, 1974; Holmes and Sartory, 1993). Also 'Pril', a quaternary ammonium detergent has been added at 200 mg/l for inhibition of *Proteus* swarming. (Rogol et al., 1979).

Antibiotics

Ampicillin is the antibiotic most used in aeromonad culture media as most strains of *A. hydrophila* were found to be resistant (von Graevenitz and Mensch, 1968). It was used by von Graevenitz and Zinterhofer (1970) at 30 mg/l in order to inhibit the growth of coliforms. However, although *A. hydrophila* is usually resistant, some strains of *A. sobria* and *A. caviae* are inhibited (von Graevenitz and Bucher, 1983; Singh and Sanyal, 1994), and *A. trota* is always susceptible to ampicillin (Carnahan et al., 1991b). Because of this, some authors use lower concentrations, but in general the amount of ampicillin incorporated is between 10 and 30 mg/l (Rippey and Cabelli, 1979; Rogol et al., 1979; Millership and Chattopadhyay, 1984; Agger et al., 1985; Palumbo et al., 1985; Havelaar et al., 1987; Okrend et al., 1987; Stern et al., 1987; Cunliffe and Adcock, 1989; Huguet and Ribas, 1991; Holmes and Sartory, 1993; Mattick and Donovan, 1998a). One medium prepared with 100 mg/l (Kelly et al., 1988) of ampicillin showed poor productivity, probably because of the high concentration of the antibiotic.

Penicillin was used as a selective agent before ampicillin (Kielwein, 1969), but has been little used because of its narrower spectrum. Some authors combined penicillin with ampicillin because they detected a possible synergism against *Pseudomonas* (Huguet and Ribas, 1991).

Because some Enterobacteriaceae are sensitive to novobiocin and aeromonads are resistant, this antibiotic was included in the composition of some media at concentrations between 2.5 and 5 mg/l (Shotts and Rimler, 1973; Altorfer et al., 1985). Altorfer et al. (1985) used noboviocin at a concentration of 2 mg/l and found that no *Aeromonas* were inhibited. Even at 8 mg/l only 4% of the strains tested were found to be sensitive. However other authors found that a concentration of 5 mg/l decreased recovery rates for *A. hydrophila* (Palumbo et al., 1985).

Altorfer et al. (1985) observed than some aeromonads could grow on CIN agar, a medium for the isolation of *Yersinia* spp. containing cefsulodin and they used this antibiotic at a concentration reduced to 4 mg/l for the isolation of *Aeromonas* spp. from faeces. However, some strains are susceptible to cefsulodin (Altorfer et al., 1985; Moyer et al., 1991; Brenner et al., 1993; Alonso et al., 1996) and although it might be useful for isolation from faeces when numbers are very high (Altorfer et al., 1985; Moyer et al., 1991), cefsulodin is not appropriate for isolation of aeromonads from food and environmental samples.

Antiseptics

The antiseptic Irgasan that is also used in CIN agar can inhibit the growth of some Gram-negative bacteria that interfere with the isolation of *Aeromonas* spp. and has been used at concentrations of 4–5 mg/l for the isolation of *Aeromonas* spp. (Altorfer et al., 1985; Hunt et al., 1987; Moyer et al., 1991). Ethanol has been included in one medium in order to inhibit the growth of *Klebsiella* spp. (Rippey and Cabelli, 1979).

Bile salts

Bile salts or sodium deoxycholate have been included in some formulations for the inhibition of Gram-positive and, at the higher concentrations, of some Gram-negative microorganisms. The amounts of bile salts used are between 1 and 8.5 g/l of the final medium (Kaper et al. 1981; Millership and Chattopadhyay, 1984; Hunt et al., 1987; Cunliffe and Adcock 1989). Bile salts of some manufacturers are more inhibitory than others and concentrations of 2.5 g/l can be inhibitory (Want and Millership, 1990). Sodium deoxycholate is usually used at concentrations between 0.1 and 1 g/l (Shotts and Rimler, 1973; Rippey and Cabelli, 1979; Havelaar et al., 1987).

Vibriostatics

Species of the genus *Vibrio* can be present in some samples and interfere with the detection of *Aeromonas*. In order to inhibit their growth some authors have added the compound O/129 at concentrations of 50 mg/l (Havelaar et al., 1987; Alonso and Garay, 1989). Other workers have used media without NaCl for the same purpose (Kielwein, 1969).

Diagnostic systems

Various products have been included to differentiate aeromonads from background flora. Carbohydrates are the most common but blood, DNA, Tween 80, and amino acids have also been used.

Carbohydrates

Most of the 'enteric' media contain lactose to differentiate the intestinal pathogens from the competitive flora. The colonies of *Aeromonas* spp. are usually lactose negative, but a few strains may develop lactose-fermenting colonies. For this reason most of the media specifically developed for the isolation of *Aeromonas* spp. are based on other diagnostic systems.

Some media use a fermentable carbohydrate and a pH indicator to detect acid production. The carbohydrates most often used are: dextrin (Schubert, 1967; Havelaar et al., 1987), trehalose (Rippey and Cabelli, 1979; Kaper et al., 1981), maltose (Shotts and Rimler, 1973) glycogen (McCoy and Pilcher, 1974) and mannitol (Altorfer et al., 1985; Okrend et al., 1987).

On the other hand some authors prefer to use non-fermentable carbohydrates. The reason for this is that for differentiation between *Aeromonas* spp. and Enterobacteriaceae, some workers perform the oxidase test directly on the plate, or on colonies taken from the plates (von Graevenitz and Zinterhofer, 1970). If the pH of the medium is 5.1 or lower, the oxidase reaction can give a false-negative reaction (Hunt et al., 1981; von Graevenitz and Bucher, 1983) and colonies from media with fermentable carbohydrates must be subcultured to a carbohydrate-free medium before testing for oxidase. Some authors therefore prefer media with carbohydrates which are not fermented by aeromonads, which allow the oxidase test to be done directly. The most frequently used carbohydrates are xylose (Roland 1977; Rogol et al., 1979; Hunt et al., 1987; Okrend et al., 1987; Cunliffe and Adcock, 1989; Holmes and Sartory, 1993) and m-inositol (Schubert, 1977; Cunliffe and Adcock, 1989; Anonymous, 1990)

Starch is included, usually at concentrations between 10 and 20 g/l, in some differential media (Kielwein, 1969, Roland, 1977; Palumbo et al., 1985; Huguet and Ribas, 1991) because aeromonads are amylase-positive. A pH indicator shows acid production while amylase activity is demonstrated by flooding the plates with Lugol's iodine solution, resulting in a black medium except around amylase positive colonies, that are surrounded by a clear zone (Palumbo et al., 1985; Palumbo et al., 1992). Glycogen is another polysaccharide that has occasionally been used (McCoy and Pilcher, 1974).

Pectin (sodium polygalacturonate) can be digested by *A. hydrophila* and a few other bacteria. Its inclusion in a medium for the detection of *Yersinia* showed a high number of samples containing *A. hydrophila* (Myers et al., 1982) and the authors proposed pectin as a differential agent.

β-*haemolysis*

β-haemolysis is exhibited by most strains on sheep and horse blood agar (von Graevenitz and Zinterhofer, 1970) but non-haemolytic aeromonads are not uncommon (von Graevenitz and Bucher, 1983). Agger et al. (1985) found 75% of isolates were haemolytic. Some species such as *A. caviae* are usually β-haemolysin-negative, and this can be used to distinguish between *Aeromonas* species, such as *A. hydrophila* (positive) and *A. caviae* (Janda et al., 1984), but this also makes the differentiation of the background flora difficult. Media with blood as a differential agent have been used particularly with clinical samples (Millership et al., 1983) and, because they do not contain carbohydrates, allow the oxidase reaction to be done directly (Agger et al., 1985).

DNA

Aeromonads usually produce a DNase, and this activity can be detected by adding toluidine blue to a DNA-containing medium (von Graevenitz and Zinterhofer, 1970;

Table 3
Biochemical properties to differentiate *Aeromonas* phenospecies.

					Acid produced from						Use of.				Hydrolysis		
Species	HG	Mo-tility	In-dole	Gas from D-glu-cose	Sali-cin	Cello biose	LDC	ODC	Utili-zation of DL-lactate	Glu-conate	Argi-nine	Gluta-mine	Arabi -nose	Growth at 42°C	Ela-stin	Aesc-ulin	Arbu-tin
A. hydrophila	1	+	+	+	+	d	+	–	+	+	+	d	+	d	+	+	+
A. bestiarum	2	+	ND	d	–	ND	+	–	–	ND	ND	ND	ND	ND	–	+	ND
A. salmonicida	3	–	d	d	d	–	–	–	–	ND	ND	ND	d	–	ND	d	ND
A. caviae	4	+	+	–	+	d	–	–	d	+	+	+	+	d	–	+	+
A. media	5	–	+	–	+	+	–	–	+	+	+	+	d	d	–	+	+
A. eucrenophila	6	+	+	+	+	+	–	–	–	+	+	+	+	–	+	+	–
A. sobria	7	+	+	+	–	d	+	+	d	+	–	–	–	d	d	d	–
A. veronii bv. *sobria*	8	+	+	+	–	d	+	+	–	ND	ND	ND	–	+	ND	–	–
A. veronii bv. *Veronii*	10	d	+	+	+	d	+	+	ND	+	+	+	–	+	–	+	d
A. jandaei	9	+	+	+	–	–	+	–	+	d	d	d	–	+	d	–	–
A. schubertii	12	+	–	–	–	–	+	–	d	+	+	ND	–	+	+	–	–
A. trota	14	+	+	+	–	+	+	–	ND	+	+	+	–	+	–	–	d
A. encheleia	11	+	+	d	+	–	–	–	–	–	–	–	–	–	–	+	+
A. allosaccha-rophila	15	+	+	+	–	+	+	d	ND	+	+	d	+	+	–	d	–
A. popoffii	16	+	d	+	–	–	–	–	+	+	d	+	ND	ND	ND	–	ND

+, 85 % or more of the strains are positive; **-,** 85 % or more of the strains are negative; d, 16 to 84 % are positive; ND, not data found.
Data from: Carnahan et al., 1991a, b; Abbot et al., 1992; Hasan et al., 1992; Holt et al., 1994; Esteve et al., 1995; Janda et al., 1996; Huys et al., 1997a, b.

von Graevenitz and Bucher 1983). DNase activity is shown by a bright rose-pink zone around the colonies.

Tween 80

This surface active compound is hydrolyzed by aeromonads, and positive colonies show a halo of precipitate (Kelly et al., 1988).

Other differential compounds

Because some media are derived from those used for isolation of salmonella, they contain indicator systems for amino acid decarboxylation and for H_2S production (Shotts and Rimler, 1973; Anonymous, 1990). The purpose of the latter is to differentiate of *Aeromonas* spp. from H_2S producing microorganisms. However, the use of lysine decarboxylation is of minor interest because this test is variable in the *Aeromonas* genus (Table 3).

Incubation temperature

Although the optimal growth temperature of the genus *Aeromonas* is 22–28°C, and the use of 28–30°C is recommended for biochemical tests (Palumbo et al., 1992), most aeromonads grow well at 35–37°C (Popoff, 1984; Holt et al., 1994). Many *Aeromonas* spp. can grow in a variety of fresh foods stored at 5°C (Palumbo et al., 2000)

The incubation temperature of the culture media for aeromonads depends on the samples analysed and the medium used. In general media for the isolation of *Aeromonas* spp. from clinical samples are incubated at 35–37°C. In some studies incubation at 37°C produced more isolates that at 28°C (Wilcox et al., 1992). In others the results were dependent on the media. Whereas Blood Ampicillin Agar (BAA) gave better recovery when incubated at 37°C, some media, such as the CIN II agar gave better results when incubated at 25°C than at 37°C (Altorfer et al., 1985), probably because the *Aeromonas* are more sensitive to the antibiotics included in the media when incubated at the higher temperature. Clinical strains are generally able to grow at higher temperatures than those from drinking water.

When using enrichment in Alkaline Peptone Water (APW) for the isolation of aeromonads from faeces, incubation at 25°C or at room temperature gave better results than at 37°C (Millership and Chattopadhyay, 1984; Moyer et al., 1991).

Media specifically developed for isolation from water have in general a recommended incubation temperature of 25–30°C (Holmes and Sartory 1993; Bernagozzi et al., 1994; Jeppesen, 1995; Villari et al., 1999). In a study with river water incubation at 28°C gave better results than at 37°C (Huguet and Ribas, 1991).

In the case of foods, both 28–30°C (Palumbo, 1991; Hänninen, 1993; Mattick and Donovan, 1998a) and 35–37°C (Fricker and Tompsett, 1989) have been used, but the former is much more common.

Table 4
Plating media for the isolation of *Aeromonas* spp.

Medium	Original purpose	Diagnostic system	Selective system	Reference
Ampicillin-dextrin agar (**ADA**)	Water	Dextrin 10 g/l Bromothymol blue	Ampicillin Sodium deoxycholate (O/129 for sea water)	Havelaar et al., 1987
Bile salts Brilliant Green (**BBG**)	Faeces	Oxidase on the plate	Bile salts Brilliant green	Millership and Chattopadhyay, 1984
Bile salts brilliant green starch agar (**BBGS**)	Food and water	Starch + Lugol's iodine solution	Bile salts Brilliant green	Nishikawa, and Kishi, 1987
Bile salts Irgasan Brilliant Green agar **(BSIBG)**	Faeces	Xylose Neutral red Sodium thiosulphate	Bile salts Irgasan Brilliant green	Hunt et al., 1987
Blood Agar Ampicillin (**BAA**)	Faeces	Erythrocytes	Ampicillin	Millership et al., 1983
Cefsulodin Irgasan agars (**CIN I** & **CIN II**)	Faeces	Mannitol	Cefsulodin Irgasan Novobiocin Bile salts Crystal violet	Altorfer et al., 1985
Dextrinfuchsin sulphite (**DFS**)	Water	Dextrin Bromothymol blue	Sodium sulphite Fuchsin	Schubert, 1967
DNase Ampicillin (**DNTA**)	Faeces	DNA Toluidine blue	Ampicillin Toluidine blue	Von Graevenitz and Zinterhofer, 1970
Glutamate Starch Penicillin (**GSP**)	Water	Starch Phenol red	Penicillin	Kielwein, 1969

Inositol Brilliant green bile salts agar (**IBB**)*	Faeces	Inositol	Brilliant green Bile salts	Schubert, 1977
mA	Water	Trehalose	Sodium deoxycholate Ethanol Ampicillin	Rippey and Cabelli, 1979
MacConkey Mannitol Ampicillin (**MMA**)	Food	Mannitol	Bile Salts Ampicillin	Okrend et al., 1987
MacConkey Trehalose (non lactose) (**MT**)	Water	Trehalose	Bile salts Crystal violet	Kaper et al., 1981
MacConkey Xylose Ampicillin (**MXA**)	Food	Xylose	Bile Salts Ampicillin	Okrend et al., 1987
MacConkey Ampicillin (**MA**)	Faeces	Lactose	Bile Salts Ampicillin	Stern et al., 1987
MacConkey-Tween 80-Ampicillin (**MAT**)	Faeces	Lactose Tween 80	Bile salts Crystal violet Ampicillin	Kelly et al., 1988
MIX agar	Water	Xylose *Meso*-inositol Bromothymol blue	Bile salts Ampicillin	Cunliffe and Adcock, 1989
Pectin agar (**PA**)	Food	Pectin (polygalacturonic acid)	No	Myers et al., 1982
Peptone beef extract glycogen (**PBG**)	Multipurpose	Glycogen Bromothymol blue	Sodium lauryl sulphate	McCoy and Pilcher, 1974
Pril xylose ampicillin (**PXA**)	Faeces	Xylose Phenol red	Pril Ampicillin	Rogol et al,. 1979
Rimler Shotts (**RS**)	Multipurpose	Lysine Ornithine Maltose Bromothymol blue Sodium thiosulphate Ferric ammonium citrate	Sodium deoxycholate Novobiocin	Shotts and Rimler, 1973

Table 4. Continued.

Medium	Original purpose	Diagnostic system	Selective system	Reference
Ryan	Multipurpose	Lysine, Arginine, Inositol, lactose, sorbose, xylose, ferric ammonium citrate + sodium thiosulphate bromothymol blue	Bile salts Ampicillin	Anonymous, 1990
Salt Starch Xylose Lysine Deoxycholate (**SSXLD**)**	Faeces	Lysine Starch Xylose	Sodium deoxycholate Citrate NaCl	Roland, 1977
SGAP-10C	Water	Soluble starch Phenol red	Sodium G penicillin Ampicillin	Huguet and Ribas, 1991
Starch Bile Salts agar (**SB**)	Clinical	Starch	Bile salts	Hansen and Bonde, 1973
Starch Ampicillin Agar (**SAA**)	Food	Starch + Lugol's iodine solution Phenol red	Ampicillin	Palumbo et al., 1985
Xylose Ampicillin Agar (**XAA**)	Water	Xylose Phenol red	Sodium lauryl sulphate Ampicillin	Holmes and Sartory, 1993
Xylose sodium deoxycholate-citrate (**XDC**)	Faeces	Xylose Neutral red Sodium thiosulphate Ferric ammonium citrate	Sodium deoxycholate Sodium citrate	Shread et al., 1981

* Originally developed for the isolation of *Plesiomonas shigelloides*.
** Originally developed for the isolation of *Vibrio* spp.

Atmosphere

Aeromonas media are generally incubated in air. In a study using water samples (Cunliffe and Adcock, 1989), anaerobic incubation for 18 h at 30ºC (without CO_2), followed by aerobic incubation at 35ºC for 24 h, reduced the growth of competitors and improved the confirmation rate of presumptive *Aeromonas* spp.

Plating media

About 30 different media have been suggested for isolation of *Aeromonas* spp. from faeces, water or food samples (Table 4). Few of these were developed primarily for the examination of foodstuffs. More often they were developed for other types of samples.

Glutamate Starch Penicillin agar (GSP) was described by Kielwein (1969). This medium contained penicillin as the selective agent and, because it did not contain NaCl, inhibited the growth of halophilic vibrios. The diagnostic system was starch and phenol red as pH indicator. Incubation was at 25ºC for 2–3 days. Aeromonads produced yellow colonies on this medium surrounded by a transparent halo (Encinas et al., 1999) and were easily distinguished from the background flora (Huguet and Ribas, 1991). A modification of this medium (SGAP-10C) was proposed by Huguet and Ribas (1991). They added 20 mg/l of ampicillin and 25 µg/l of D-glucose (10µg/l of carbon) in order to inhibit *Pseudomonas* spp. and improve the resuscitation of stressed cells of aeromonads respectively.

Shotts and Rimler (1973) described a medium called Rimler Shotts (RS). This includes maltose, lysine and ornithine with bromothymol blue and ferric ammonium citrate as diagnostic components. The selective system was 1 g/l of sodium deoxycholate and 5 mg/l of novobiocin. They incubated the plates at 37ºC for 20–24 h. *A. hydrophila* produced yellow colonies on this medium. However other authors found that it failed to recognise lysine-positive colonies (von Graevenitz and Bucher, 1983). This medium was used by Abeyta et al. (1986) to isolate *A. hydrophila* from oysters.

Peptone Beef extract Glycogen agar (PBG), described by McCoy and Pilcher (1974), used glycogen and bromothymol blue as the diagnostic system and 0.1 g/l of sodium lauryl sulphate to provide selectivity. Aeromonads produced yellow colonies. For *A. hydrophila* isolation the authors recommended incubation at 37ºC for 24 h.

Inositol Brilliant green Bile salts agar (IBB) was proposed by Schubert (1977) for *Plesiomonas* isolation, but has also been used for *Aeromonas* spp. detection (von Graevenitz and Bucher, 1983). The medium contained inositol as differential agent and brilliant green and bile salts as selective system. Aeromonad colonies were colourless. The recommended incubation temperature was 37ºC (von Graevenitz and Bucher, 1983).

Myers et al. (1982) reported that on Pectin agar (PA), *Aeromonas* spp. produced 3–5 mm diameter colonies with pectionolyzed zones beyond the outer edge of the colonies. This characteristic was also common to *Yersinia* spp. and *Klebsiella oxytoca*, but no other bacteria from vacuum-packaged pork had pectinolytic activity. They incubated the medium at 25ºC for 12–18 h.

Blood agar with ampicillin (BAA) was proposed for the isolation of *Aeromonas* spp. by Millership et al. (1983). The differentiation between aeromonads and other organisms was based on β-haemolysis and ampicillin inhibits part of the background flora. Oxidase testing could be performed directly on the plate. The incubation temperature was 35–37ºC for 24 h.

Palumbo et al. (1985) described Starch Ampicillin Agar (SAA). Phenol red agar base was chosen as the basal medium, and 10 g/l of starch and 10 mg/l of ampicillin were added as the differential and selective agents respectively. Incubation was at 28ºC for 24 h. Typical *A. hydrophila* colonies were 3 to 5 mm in diameter and yellow to honey coloured. After incubation, the plates were flooded (ca. 5 ml) with Lugol's iodine solution, and amylase-positive colonies (those having a clear zone surrounding the colony) scored as presumptive *A. hydrophila* (Palumbo et al, 1985. 1992). This medium has been one of the most widely used in food microbiology for detection of aeromonads (Callister and Agger, 1987; Berrang et al., 1989; Majeed et al., 1989; Hudson and De Lacy, 1991; Sierra et al., 1995; Melas et al., 1999; Wang and Silva, 1999).

A commercial medium named Ryan (Anonymous, 1990) was a modification of XLD medium and the diagnostic system was based on it. The selectivity of the medium for *Aeromonas hydrophila* is achieved by the addition of 5 mg/l of ampicillin. After incubation at 30–35ºC for 24 h, aeromonad colonies were dark green, opaque with a diameter between 0.5 and 1.5 mm.

Okrend et al. (1987) modified MacConkey agar by using xylose or mannitol instead of lactose, to form MacConkey Xylose Agar (MXA) and MacConkey Mannitol Agar (MMA) They also added ampicillin to increase the selectivity of the media. Colonies were colourless on MXA and red on MMA after incubation for 18–24 h at 28ºC.

Ampicillin Dextrin Agar (ADA) was formulated by Havelaar et al. (1987) and Havelaar and Vonk (1988). Most aeromonads fermented dextrin which was incorporated at 10 g/l with bromothymol blue to form the diagnostic system. Ampicillin (10 mg/l) provided selectivity. Incubation was at 30ºC for 24 h. Aeromonads produced yellow colonies from 1 to 2 mm in diameter. This medium was used in surveys for aeromonads in ground meat, chicken and fish products (Hänninen, 1993; Hänninen et al., 1997).

Nishikawa and Kishi (1987) described Bile salts Brilliant Green Starch agar (BBGS) that was a modification of a previously published medium by Millership and Chattopadhyay (1984). BBGS was intended to isolate aeromonads from food and water. Bile salts and brilliant green constituted the selective system and starch was the differential agent. After incubation at 30ºC for 24–48 h, plates were flooded with Lugol's iodine solution and amylase-positive colonies showed a clear surrounding zone (Nishikawa and Kishi, 1987, 1988; Palumbo et al., 1992).

Hunt et al. (1987) proposed Bile salts Irgasan Brilliant Green agar (BSIBG), which contained Irgasan, bile salts and a high concentration of brilliant green (5 mg/l) as the selective system. The diagnostic system was formed by xylose that is not fermented by aeromonads, and neutral red and allowed oxidase testing of colonies picked directly from the medium. Incubation was overnight at 37ºC. *Aeromonas* spp. colonies were

Table 5
Selective enrichment media for the isolation of *Aeromonas* spp.

Medium	Original purpose	Selective system	Reference
Alkaline Peptone Water (**APW**)	Faeces	pH 8.6	Shread et al., 1981
Ampicillin Dextrin Broth (**ADB**)	Water	Ampicillin Sodium deoxycholate	Havelaar et al., 1987
Modified Rimler Shotts-Broth (**MRSB**)	Water	Novobiocin Bile salts	Kaper et al., 1981
Trypticase Soy Broth Ampicillin (**TSBA**)	Faeces	Ampicillin	von Graevenitz and Zinterhofer, 1970
Trypticase Soy Broth Ampicillin Irgasan (**TSBAI**)	Food	Ampicillin Irgasan	Mattick and Donovan, 1998a

translucent and 1–2 mm in diameter. This medium has been used in surveys of aeromonads in ready-to-eat salads (Mattick and Donovan, 1998b)

Enrichment media

When small numbers of aeromonads are anticipated or when injured cells are suspected an enrichment broth must be used (Palumbo et al., 1992). Two hundred and twenty five ml of the enrichment medium can be inoculated with a 25 g sample of food and homogenized. Quantification can be achieved by a most probable number (MPN) procedure using the enrichment medium in a 3 or 5 tube method. The most important media used for the enrichment of aeromonads are showed in Table 5.

Trypticase soy broth supplemented with 30 mg/l of ampicillin (TSBA) was developed by von Graevenitz and Zinterhofer (1970) for *A. hydrophila* enrichment. The broth was incubated at 37°C for 16 h before plating. TSBA incubated at 30°C for 24 h has been used in different surveys for aeromonads in various foods (Hänninen, 1993; Hänninen and Shtonen, 1995; Hänninen et al., 1997). In a single survey TSBA was incubated at 7°C for 72 h before plating (Melas et al., 1999)

Kaper et al. (1981) modified RS medium to make a liquid enrichment medium Modified Rimler Shotts Broth (MRSB). Like the original RS, MRSB contained novobiocin and bile salts as selective agents. Incubation temperature was 37°C for 24 h. These authors did not find the broth entirely satisfactory because it was not very selective in the presence of large numbers of competing organisms (Kaper et al., 1981; Abeyta et al., 1989).

Alkaline Peptone Water (APW), widely used for vibrio enrichment, has also been used for aeromonads (Shread et al., 1981). Incubation of APW at room temperature gave better results than 37°C (Millership and Chattopadhyay, 1984). Other authors used APW as enrichment with various foods (Majeed et al., 1989; Hudson and De Lacy, 1991) incubating at 28°C for 18–24 h.

Ampicillin Dextrin Broth (ADB), ADA medium without agar, was described by

Havelaar et al. (1987) Ampicillin (10 mg/l) and sodium deoxycholate were the selective agents. Incubation was at 30°C for 20–24 h.

Mattick and Donovan (1998a) designed an enrichment medium, TSBAI that consisted of tryptone soy broth supplemented with ampicillin (30 mg/l) and Irgasan (40 mg/l). Incubation was at 30°C for 24 h.

Comparisons of media for isolation from foods

There are many comparisons of media for isolation of *Aeromonas* spp. from water (Arcos et al., 1988; Monfort and Baleux, 1988; Alonso and Garay, 1989; Knøchel, 1989; Ribas et al., 1991; Holmes and Sartory, 1993; Bernagozzi et al., 1994; Gavriel and Lamb, 1995; Jenkins and Taylor, 1995; Handfield et al., 1996; Kersters et al., 1996; Vilari et al., 1999) and faeces (von Graevenitz and Bucher, 1983; Millership and Chattopadhyay, 1984; Stern et al., 1987; Kelly et al., 1988; Moyer et al., 1991). Discussion here will be confined to comparisons of media for isolation from food either using pure cultures, artificially inoculated foods or naturally contaminated foods.

Okrend et al. (1987) compared five enrichment broths and five selective and differential plating media for isolating *Aeromonas* spp. using five pure cultures (three *A. hydrophila* and two *A. caviae* strains) and 30 meat samples from chicken, beef and pork. Enrichment media were selective and non-selective: APW, TSB, TSBA (TSB with 10 mg/l of ampicillin), TSBA with 2% extra NaCl and Tryptone Broth, incubated at 28°C for 6 and/or 24 h. Plating media were SAA, MMA, MXA, BAA (10 mg/l ampicillin) and PBG. All agar media except PBG were incubated at 28°C for 18–24 h. After inoculation, PBG was overlaid with 10–15 ml of sterile non-nutrient agar and incubated at 37°C for 24 h. When samples were plated directly, recovery of *Aeromonas* spp. was better on SAA than on MMA. However both media were equivalent when samples were first enriched in TSBA. BAA proved to be unsatisfactory because not all strains of aeromonads were haemolytic. MXA was not suitable because of the many xylose-negative, non-aeromonads that grew on this medium. PBG also proved to be unsuitable because the non-nutrient agar overlay made picking colonies difficult. Of the enrichment broths TSBA, performed slightly better than APW and both were better than Tryptone Broth, TSB or TSBA plus 2% extra NaCl. Overnight (18–24 h) enrichment at 28°C gave better recovery of aeromonads than did 6 h enrichment.

Fricker and Tompsett (1989) examined 563 samples of various foodstuffs to compare three plating media, MacConkey agar, BAA (with 10 mg/l of ampicillin) and BSIBG (Difco). Samples (10 g) were enriched in APW incubated at 37°C for 24 h then one loopful of the enrichment was plated on each of the three media and incubated at 37°C for 24 h. The results showed that BSIBG was the most efficient with 48.0 % of samples positive. BAA showed 43.3 % of samples positive and MacConkey 31.2 %. The MacConkey plates were frequently overgrown with Enterobacteriaceae and occasionally the BAA plates were overgrown with swarming *Proteus* spp.

Pin et al. (1994) used pure cultures of four strains of *Aeromonas* spp. (two *A. hydrophila*, one *A. caviae* and one *A. veronii* biovar sobria) and artificially inoculated food samples. Three agar media were evaluated for quantitative recovery: SAA, Ryan and GSP. Ryan gave the best overall recovery of 94.1 % compared with GSP and SAA that gave 79.7 and 79.5 % respectively. However, this was because of much lower counts with the single strain of *A. veronii* biovar sobria on GSP and SAA.

Gobat and Jemmi (1995) evaluated two enrichment media and seven selective agar media for the isolation of mesophilic *Aeromonas* species. For comparisons they used 32 pure cultures of aeromonads, artificially contaminated foods and naturally contaminated meat, fish and shellfish samples. The enrichment broths evaluated were APW and TSBA (with 10 mg/ of ampicillin) and they found slightly better results with the APW that with the TSBA. In both cases incubation at 28°C gave better results than 35°C. The selective agars evaluated were CIN II, BSIBG, Starch Ampicillin DNA Agar (a commercial agar by Biolife), Ryan and BAA with 10, 20 or 30 mg/l of ampicillin, incubating at 35°C for 18–24 h. Among the seven media evaluated, BSIBG was the most selective but inhibited some *A. caviae* strains. For the naturally contaminated foods they used BSIBG and BAA with 30 mg/l of ampicillin and APW as enrichment. BSIBG isolated more *A. veronii* biovar sobria and BAA more *A. caviae* strains. The authors recommended the use of more than one agar for optimal recovery of mesophilic *Aeromonas* spp.

Tsai and Chen (1996) evaluated three selective media, SAA, Ryan and BAA with 10 mg/l ampicillin for the isolation of *A. hydrophila* from seafood. The incubation temperatures were 28°C for SAA and 35°C for Ryan and BAA. BAA gave the highest recovery rate of *A. hydrophila*. The low detection rates of SAA and Ryan agar in this study may be due to the fact that the workers looked only for *A. hydrophila* and other species as *A. caviae* and *A. veronii* biovar sobria could have overgrown *A. hydrophila*.

Singh (1997) compared GSP, Ryan and SAA for their ability to recover *Aeromonas* spp. from pure culture, raw ground beef and spiked autoclaved ground beef samples. In all instances SAA medium proved to be superior. Selectivity with SAA and GSP was better than with Ryan. SAA was incubated at 37°C for 24 h.

Mattick and Donovan (1998a) compared two enrichment broths: APW and TSBAI and five selective agars: BSIBG, SAA, Ryan, BAA and TCBS, in a study with naturally contaminated ready-to-eat salads. They found that 30°C was the optimal temperature for recovery of mesophilic strains of aeromonads. TSBAI was more selective than APW. Of the plating media, BSIBG had good selectivity and SAA gave easy differentiation and they recommended using both to detect most strains of aeromonads. Ryan medium was also useful but lacked selectivity.

Recommended culture media

At present there are neither ISO nor AOAC standards for the examination of foods for *Aeromonas* spp. or *P. shigelloides*. For *Aeromonas* spp. the American Public Health Association (APHA) suggests SAA and SAA modified by Lachica as plating media (Palumbo et al., 1992). BBGS is also used if samples with large numbers of *Proteus* spp. are likely to be encountered. The plates are incubated at 28°C overnight (24 h maximum). After incubation the plates are flooded with Lugol's iodine solution.

As enrichment media they suggest APW and TSBA (30 mg/l of ampicillin), incubating at 28°C for 24 h.

The Nordic Committee on Food Analysis (NMKL; 1995) proposes SAA as plating medium incubating at 37°C for 24 h.

The Public Health Laboratory Service (Roberts et al., 1995) recommends BSIBG or Ryan as plating medium and APW for enrichment. All media are incubated at 30°C for 18–24 h.

In conclusion, after reviewing the literature, especially the media comparisons, the most promising plating media for isolating *Aeromonas* spp. from foods are SAA and BSIBG with APW or TSBA (30 mg/l of ampicillin) for enrichment, all incubated at 28°C for 18–24 h.

Culture media for *Plesiomonas shigelloides*

Plesiomonas shigelloides is an opportunistic pathogen (Koburger and Wei, 1992). It has been associated with epidemics of diarrhoea (Tsukamoto et al., 1978; Rutala et al., 1982) and causes mild, self-limited diarrhoeal diseases in previously healthy adults and severe extraintestinal disease in immunocompromised individuals (Perales et al., 1983; Ingram et al., 1987; Olsvik et al., 1990; Rautelin et al., 1995). However, its role as enteric pathogen is still controversial (Abbott et al., 1991).

P. shigelloides can be isolated from water, animals and humans (Arai et al., 1980) and has been isolated from freshwater fishes, oysters and crabs (Arai et al., 1980; Van Damme and Vandepitte, 1980; Freund et al., 1988b; Marshall et al., 1996).

Historically, *P. shigelloides* has been isolated using routine clinical media for Enterobacteriaceae such as MacConkey, Salmonella Shigella, Hektoen Enteric, XLD and other agars (Richard et al., 1978; Miller and Koburger, 1985; Koburger and Wei, 1992). There are few studies comparing isolation rates on enteric agars but Van Damme and Vandepitte (1980) found that direct plating on MacConkey agar yielded better isolation rates than XLD. They also found that enrichment in selenite broth and tetrathionate with iodine was unproductive.

Plating media specifically described for isolation of *P. shigelloides* are shown in Table 6. Schubert (1977; 1982) described the first selective medium for isolation of *P. shigelloides*, called Inositol Brilliant green Bile salts agar (IBB). The differential properties were based on inositol fermentation and the selective properties on brilliant green and bile salts. In IBB plesiomonad colonies were whitish to pink (von Graevenitz and Bucher, 1983). Millership and Chattopadhyay (1984), modified IBB by removing inositol because they thought that inositol delayed the oxidase reaction done directly on the plate. They named this medium BBG. Miller and Koburger (1985) developed a new medium, *Plesiomonas* agar (PL). The medium's differential properties were analogous to XLD agar: incorporation of lysine and nonfermentable carbohydrates allowed for the selection of lysine descarboxylase-positive, non-fermenting organisms such as Plesiomonas (Koburger, 1988). The inhibitory properties were due to bile salts present in low concentrations.

Few comparative studies have been made with *Plesiomonas* media. Von Graeven-

Table 6
Plating media for *Plesiomonas shigelloides*.

Medium	Original purpose	Diagnostic system	Selective system	Reference
Inositol Brilliant green Bile salts agar (**IBB**)	Faeces	Inositol	Brilliant green Bile salts	Schubert, 1977
Bile salts Brilliant Green (**BBG**)	Faeces	Oxidase on the plate	Brilliant green Bile salts	Millership and Chattopadhyay 1984
Plesiomonas agar (**PL**)	Food and water	Mannitol Arabinose Lysine Phenol Red	Bile salts	Miller and Koburger, 1985

itz and Bucher (1983) evaluated different media for isolation of *P. shigelloides* from human faeces. They used nine plating media and two enrichment media and recommended IBB as plating medium and APW as enrichment medium for *P. shigelloides*.

Miller and Koburger (1986) compared PL and IBB with water samples. They recommended using both media because recovery of plesiomonads was higher using IBB, but in laboratory controlled studies the PL agar was less inhibitory for injured organisms.

Freund et al. (1988a, b) compared five enrichment broths (Gram-negative broth, APW, tetrathionate broth without iodine and two *Plesiomonas* broths) using fresh water and oyster samples. They found that tetrathionate broth consistently gave greater recovery of *P. shigelloides* than the other four tested. They also observed that incubation of enrichment broths at 40°C resulted in higher recovery than 35°C.

Marshall et al., (1996) compared three agar media, MacConkey, Salmonella Shigella and IBB for isolation of *P. shigelloides* from blue crab. Of these, IBB was most satisfactory for distinguishing between *P. shigelloides* and *A. hydrophila*.

Since there are very few studies of media for the isolation of *P. shigelloides* from foods it is difficult to make recommendations. With other types of sample, enrichment in APW or tetrathionate broth without iodine and plating on IBB and PL can be useful. All the media are incubated for 24 h at 35–37°C except tetrathionate broth, which must be incubated at 40°C.

References

Abbott, S.L., Cheung, W.K.W., Kroske-Briston, S., Malekzadeh, T. and Janda, M.J. (1992) Identification of *Aeromonas* strains to the genospecies level in the clinical laboratory. J. Clin. Microbiol. 30, 1262–1266.

Abbott, S.L., Kokka, R.P. and Janda, J.M. (1991). Laboratory investigations on the low pathogenic potential of *Plesiomonas shigelloides*. J. Clin. Microbiol. 29, 148–153.

Abbott, S.L., Serve, H. and Janda, J.M. (1994) Case of *Aeromonas veronii* (DNA group 10)

bacteremia. J. Clin. Microbiol. 32, 3091–3092.
Abeyta Jr, C., Kaysner, C.A., Wekell, M.M., Sullivan, J.J. and Stelma, G.N. (1986) Recovery of *Aeromonas hydrophila* from oysters implicated in an outbreak of foodborne illness. J. Food Protect. 49, 643–646.
Abeyta Jr, C., Weagant, S.D., Kaysner, C.A., Wekell, M.M., Stott, R.F., Krane, M.H. and Peeler, J.T. (1989) *Aeromonas hydrophila* in shellfish growing waters: Incidence and media evaluation. J. Food Protect. 52, 7–12.
Abeyta Jr, C. and Wekell, M.M. (1988) Potential sources of *Aeromonas hydrophila*. J. Food Safety 9, 11–22.
Agger, W.A., McCormick, J.D. and Gurwith, M.J. (1985) Clinical and microbiological features of *Aeromonas hydrophila*-associated diarrhea. J. Clin. Microbiol. 21, 909–913.
Ali, A., Carnahan, A.M., Altwegg, M., Lüthy-Hottenstein, J. and Joseph, S.W. (1996) *Aeromonas bestiarum* sp. nov. (formerly genomospecies DNA group 2 *A. hydrophila*), a new species isolated from non-human sources. Med. Microbiol. Lett. 5, 156–165.
Alonso J.L. and Garay, E. (1989) Two membrane filter media (mADA/O129 and mSA/O129 agars) for enumeration of motile *Aeromonas* in sea water. Zbl. Hyg. 189, 14–19.
Alonso J.L., Amoros, I. and Alonso, M.L. (1996) Differential susceptibility of aeromonads and coliforms to cefsulodin. Appl. Environ. Microbiol. 62, 1885–1888.
Altorfer, R., Altwegg, M., Zollinger-Iten, J. and von Graevenitz, A. (1985) Growth of *Aeromonas* spp. on cefsulodin-Irgasan-novobiocin agar selective for *Yersinia enterocolitica*. J. Clin. Microbiol. 22, 478–480.
Altwegg, M. (1999) *Aeromonas* and *Plesiomonas*. *In*: Manual of Clinical Microbiology, 7th edition, American Society for Microbiology, Washington. pp. 507–516.
Anonymous (1990) Aeromonas Medium Base (Ryan), *In*: Bridson E.Y. (editor) The Oxoid Manual, 6th edition. Unipath Ltd., Wade Road, Basingstoke, UK, pp. 2.32–2.33.
Arai, T., Ikejima, N., Itoh, T., Sakai, S., Shimada, T. and Sakazaki, R. (1980) A survey of *Plesiomonas shigelloides* from aquatic environments, domestic animals, pets and humans. J. Hyg. 84, 203–211.
Archer, D.L. and Kvenberg, J.E. (1988) Regulatory significance of *Aeromonas* in foods. J. Food Safety 9, 53–58.
Arcos, M.L., de Vincente, A., Moriñigo, M.A., Romero, P. and Borrego, J.J. (1988) Evaluation of several selective media for recovery of *Aeromonas hydrophila* from polluted waters. Appl. Environ. Microbiol. 54, 2786–2792.
Berrang, M.E., Brackett, R.E. and Beuchat, L.R. (1989) Growth of *Aeromonas hydrophila* on fresh vegetables stored under a controlled atmosphere. Appl. Environ. Microbiol. 55, 2167–2171.
Bernagozzi, M., Bianucci, F., Scerre, E. and Sacchetti, R. (1994) Assessment of some selective media for the recovery of *Aeromonas hydrophila* from surface waters. Zbl. Hyg. 195, 121–134.
Borrell, N., Acinas, S. G., Figueras, M.A., and Martínez-Murcia, A. J. (1997) Identification of *Aeromonas* clinical isolates by restriction fragment length polymorphism of PCR-amplified 16S rRNA genes. J. Clin. Microbiol. 35, 1671–1674.
Brenner, K.P., Rankin, C.C., Roybal, Y.B., Stelma Jr. G.N., Scarpino, P.V. and Dufour, A.P. (1993) New medium for the simultaneous detection of total coliforms and *Escherichia coli* in water. Appl. Env. Microbiol. 59, 3534–3544.
Callister, S.M. and Agger, W.A. (1987) Enumeration and characterization of *Aeromonas hydrophila* and *Aeromonas caviae* isolated from grocery store produce. Appl. Environ. Microbiol. 53, 249–253.
Carnahan, A.M., Fanning, G.R., and Joseph, S.W. (1991a) *Aeromonas jandaei* (formerly genospecies DNA group 9 *A. sobria*), a new sucrose-negative species isolated from clinical specimens. J. Clin. Microbiol. 29, 560–564.
Carnahan, A.M., Chakkraborty, T., Fanning, G.R., Verma, D., Ali, A., Janda, J.M. and Joseph, S.W. (1991b) *Aeromonas trota* sp. nov., an ampicillin-susceptible species isolated from clinical

specimens. J. Clin. Microbiol. 29, 1206–1210.

Carnahan, A.M., Behram, S. and Joseph, S.W. (1991c) Aerokey II: a flexible key for identifying clinical *Aeromonas* species. J. Clin. Microbiol. 29, 2843–2849.

Cascón, A., Anguita, J., Hernanz, C., Sánchez, M., Fernández, M. y Haharro, G. (1996) Identification of *Aeromonas hydrophila* hybridization group 1 by PCR assays. Appl. Environ. Microbiol. 62, 1167–1170.

Collins. M.D., Martínez-Murcia, A.J., and Cai, J. (1993) *Aeromonas enteropelogenes* and *Aeromonas ichthiosmia* are identical to *Aeromonas trota* and *Aeromonas veronii*, respectively, as revealed by small-subunit rRNA sequence analysis. Int. J. Syst. Bacteriol. 43, 855–856.

Colwell, R.R., MacDonnell, M.R., and DeLey, J. (1986) Proposal to recognize the family *Aeromonadaceae* fam. nov. Int. J. Syst. Bacteriol. 36, 473–477.

Cunliffe, D.A. and Adcock, P. (1989) Isolation of *Aeromonas* spp. from water by using anaerobic incubation. Appl. Environ. Microbiol. 55, 2138–2140.

Encinas, J.P., González, C., García-López, M.L. and Otero, A. (1999) Numbers and species of motile aeromonads during the manufacture of naturally contaminated Spanish fermented sausages (longaniza and chorizo). J. Food Protect. 62, 1045–1049.

Esteve, C., Gutiérrez, M.C. and Ventosa, A. (1995) *Aeromonas encheleia* sp. nov., isolated from European eels. Int. J. Syst. Bacteriol. 45, 462–466.

Freund, S.M., Koburger, J.A. and Wei, C.I. (1988a) Enhanced recovery of *Plesiomonas shigelloides* following an enrichment technique. J. Food Protect. 51, 110–112.

Freund, S.M., Koburger, J.A. and Wei, C.I. (1988b) Isolation of *Plesiomonas shigelloides* from oysters using tetrathionate broth enrichment. J. Food Protect. 51, 925–929.

Fricker, C.R. and Tompsett, S. (1989) *Aeromonas* spp. in foods: A significant cause of food poisoning?. Int. J. Food Microbiol. 9, 17–23.

Gavriel, A. and Lamb, A.J. (1995) Assessment of media used for selective isolation of *Aeromonas* spp. Lett. Appl. Microbiol. 21, 313–315.

Gobat, P.F. and Jemmi, T. (1995) Comparison of seven selective media for the isolation of mesophilic *Aeromonas* species in fish and meat. Int. J. Food Microbiol, 24, 375–384.

Handfield, M., Simard, P. and Letarte, R. (1996) Differential media for quantitative recovery of waterborne *Aeromonas hydrophila*. Appl. Environ. Microbiol. 62, 3544–3547.

Hänninen, M.L. (1993) Occurrence of *Aeromonas* spp. in samples of ground meat and chicken. Int. J. Food Microbiol. 18, 339–342.

Hänninen, M.L. (1994) Phenotypic characteristics of the three hybridization groups of *Aeromonas hydrophila* complex. J. Appl. Bacteriol. 76, 455–462.

Hänninen, M.L., Oivanen, P. and Hirvelä-Koski, V. (1997) *Aeromonas* species in fish, fish-eggs, shrimp and freshwater. Int. J. Food Microbiol. 34, 17–26.

Hänninen, M.L. and Shtonen, A. (1995) Distribution of *Aeromonas* phenospecies and genospecies among strains isolated from water, foods or from human clinical samples. Epidemiol. Infect. 115, 39–50.

Hasan, J.A.K., Macaluso, P., Carnahan, A. M. and Joseph, S.W. (1992) Elastolytic activity among *Aeromonas* spp. using a modified bilayer plate assay. Diagn. Microbiol. Infect. Dis. 15, 201–206.

Hansen, J.C. and Bonde, G.J. (1973) *Aeromonas* (*S. liquefaciens*) som mulig arsag til furunkulose has al. Nord. Veterinaermed. 25, 121–130.

Havelaar, A.H., During, M. and Versteegh, J.FM. (1987) Ampicillin-dextrin agar medium for the enumeration of *Aeromonas* species in water by membrane filtration. J. Appl. Bacteriol. 62, 279–287.

Havelaar, A.H. and Vonk, M. (1988) The preparation of ampicillin dextrin agar for the enumeration of *Aeromonas* in water. Lett. Appl. Microbiol. 7, 169–171.

Hickman-Brenner, F.W., Fanning, G.R., Arduino, M.J., Brenner, D.J. and Farmer III J.J. (1988) *Aeromonas schubertii*, a new mannitol-negative species found in human clinical specimens. J.

Clin. Microbiol. 26, 1561–1564.

Hickman-Brenner, F.W., MacDonald, K.L., Steigerwalt, A.G., Fanning, G.R., Brenner, D.J. and Farmer III J.J. (1987) *Aeromonas veronii*, a new ornithine decarboxylase-positive species that may cause diarrhea. J. Clin. Microbiol. 25, 900–906.

Holmes, P. and Sartory, D.P. (1993) An evaluation of media for the membrane filtration enumeration of *Aeromonas* from drinking water. Lett. Appl. Microbiol. 17, 58–60.

Holt, J.G., Krieg, N.R., Sneath, P.H.A., Staley, J.T. and Williams, S.T. (1994) Bergey's Manual of Determinative Bacteriology. Williams & Wilkins, Baltimore, USA.

Hudson, J.A. and De Lacy, K.M. (1991) Incidence of motile aeromonads in New Zealand retail foods. J. Food Protect. 54, 696–699.

Huguet, J.M. and Ribas, F. (1991) SGAP-10C agar for the isolation and quantification of *Aeromonas* from water. J. Appl. Bacteriol. 70, 81–88.

Hunt, G.H., Price, E.H., Patel, U., Messenger, L., Stow, P. and Salter, P. (1987) Isolation of *Aeromonas* sp. from faecal specimens. J. Clin. Pathol. 40, 1382–1384.

Hunt, L.K., Overman, T.L. and Otero, R.B. (1981) Role of oxidase variability of *Aeromonas hydrophila*. J. Clin. Microbiol. 13, 1054–1059.

Huys, G., Coopman, R., Janssen, P. and Kersters, K. (1996) High-resolution genotypic analysis of the genus *Aeromonas* by AFLP fingerprinting. Int. J. Syst. Bacteriol 46, 572–580.

Huys, G., Kämpfer P., Altwegg M., Coopman, R., Janssen, P., Gillis, M. and Kersters, K. (1997a) Inclusion of *Aeromonas* Hybridation Group 11 in *Aeromonas enchelia* and extended descriptions of the species *Aeromonas eucrenophila* and *A. enchelia*. Int. J. Syst. Bacteriol. 47, 1157–1164.

Huys, G., Kämpfer, P., Altwegg, M., Kersters, I., Lamb, A., Coopman, R., Lüthy-Hottenstein, J., Vancanneyt, M., Janssen, P. and Kersters, K. (1997b) *Aeromonas popoffii* sp. nov., a mesophilic bacterium isolated from drinking water production plants and reservoirs. Int. J. Syst. Bacteriol. 47, 1165–1171.

Ingram, C.W., Morrison, Jr. A.J. and Levitz, R.E. (1987) Gastroenteritis, sepsis, and osteomyelitis caused by *Plesiomonas shigelloides* in an immunocompetent host: Case report and review of the literature. J. Clin. Microbiol. 25, 1791–1793.

Janda, J.M. (1991) Recent advances in the study of the taxonomy, pathogenicity and infectious syndromes associated with the genus *Aeromonas*. Clin. Microbiol. Rev. 4, 397–410.

Janda, J.M., Abbott, S.L., Khashe, S., Kellogg, G.H. and Shimada, T. (1996) Further studies on biochemical characteristics and serologic properties of the genus *Aeromonas*. J. Clin. Microbiol. 34, 1930–1933.

Janda, J.M., Dixon, A., Raucher, B., Clark, R.B. and Bottone, E.J. (1984) Value of blood agar for primary plating and clinical implication of simultaneous isolation of *Aeromonas hydrophila* and *Aeromonas caviae* from a patient with gastroenteritis. J. Clin. Microbiol. 20, 1221–1222.

Janda, J.M., Guthertz, L.S., Kokka, R.P. and Shimada, T. (1994) *Aeromonas* species in septicaemia: laboratory characteristics and clinical observations. Clin. Infect. Dis. 19, 77–83.

Jenkins, J.A. and Taylor P.W. (1995) An alternative bacteriological medium for the isolation of *Aeromonas* spp. J. Wildl. Dis. 31, 272–275.

Jeppesen C. (1995) Media for *Aeromonas* spp., *Plesiomonas shigelloides* and *Pseudomonas* spp. from food and environment. Int. J. Food Microbiol. 26, 25–41.

Joseph, S. W., Janda, M. and Carnahan, A. (1988) Isolation, enumeration and identification of *Aeromonas* sp. J. Food Safety 9, 23–35.

Kaper, J.B., Lockman, U. and Colwell, R.R. (1981) *Aeromonas hydrophila*: ecology and toxigenicity of isolates from an estuary. J. Appl. Bacteriol. 50, 359–377.

Kelly, M. T., Dan Stroh, E.M. and Jessop, J. (1988) Comparison of blood agar, ampicillin blood agar, MacConkey-Novobiocin agar for isolation of *Aeromonas* spp, from stool specimens. J. Clin. Microbiol. 26, 1738–1740.

Kersters, I., Smeyers, N. and Verstraete, W. (1996) Comparison of different media for the enumeration of *Aeromonas* sp. in freshwaters. J. Appl. Bacteriol. 81, 257–261.

Kielwein, G. (1969) Ein Nährboden zur selektiven Züchtung von Pseudomonaden und Aeromonaden. Arch. Lebensmittelhyg. 20, 131–133.
Kingombe C. I. B., Huys, G., Tonolla, M., Albert, M.J., Swings, J., Peduzi, R. and Jemmi, T. (1999) PCR detection, characterization, and distribution of virulence genes in *Aeromonas* spp. Appl. Environ. Microbiol. 65, 5293–5302.
Kirov, S.M. (1993) The public health significance of *Aeromonas* spp. in foods. Int. J. Food Microbiol. 20, 179–198.
Kirov, S.M., O'Donovan, L.A. and Sanderson, K. (1999) Functional characterization of type IV pili expressed on diarrhea-associated isolates of *Aeromonas* species. Infect. Immun. 67, 5447–5454.
Kirov, S.M., Ardestani, E.K. and Hayward, L.J. (1993) The growth and expression of virulence factors at refrigeration temperature by *Aeromonas* strains isolated from foods. Int. J. Food Microbiol. 20, 159–168.
Knøchel, S. (1989) The suitability of four media for enumerating *Aeromonas* spp. from environmental samples. Lett. Appl. Microbiol. 9, 67–69.
Koburger, J. A. (1988) *Plesiomonas shigelloides. In*: Foodborne Bacterial Pathogens edited by M.P. Doyle, Marcel Decker, Inc, New York, pp. 311–325.
Koburger, J.A. and Wei, C.I. (1992) *Plesiomonas shigelloides. In*: Compendium of Methods for the Microbiological Examination of Foods. American Public Health Association, Washington, DC, pp. 517–522.
Kuijper, E.J., Steigerwalt, A.G., Schoenmakers B.S.C.I.M., Peeters, M.F., Zanen. H.C. and Brenner D.L. (1989) Phenotypic characterization and DNA relatedness in human faecal isolates of *Aeromonas* spp. J. Clin. Microbiol. 27,132–138.
Majeed, K.N., Egan, A.F. and Mac Rae, I.C. (1989) Incidence of aeromonads in samples from an abattoir processing lambs. J. Appl. Bacteriol. 67, 597–604.
Marshall, D.L., Kim, J.J. and Donnelly, S.P. (1996) Antimicrobial susceptibility and plasmid-mediated streptomycin resistance of *Plesiomonas shigelloides* isolated from blue crab. J. Appl. Bacteriol. 81, 195–200.
Martínez-Murcia, A.J., Benlloch, S. and Collins M.D. (1992) Phylogenetic interrelationships of members of the genera *Aeromonas* and *Plesiomonas* as determined by 16S ribosomal DNA sequencing: lack of congruence with results of DNA-DNA hybridations. Int. J. Syst. Bacteriol. 42, 412–421.
Martínez-Murcia, A.J. (1999) Phylogenetic positions of *Aeromonas enchelia, Aeromonas popoffii, Aeromonas* DNA hybridization Group 11 and *Aeromonas* Group 501. Int. J. Syst. Bacteriol. 49, 1403–1408.
Mattick, K.L. and Donovan, T.J. (1998a) Optimisation of the protocol for detection of *Aeromonas* species in ready-to-eat salads, and its use to speciate isolates and establish their prevalence. Commun. Dis. Public Health 1, 263–266.
Mattick, K.L. and Donovan, T.J. (1998b) The risk to public health of aeromonads in ready-to-eat salad products. Commun. Dis. Public Health 1, 267–270.
McCoy, R.H. and Pilcher, K.S. (1974) Peptone beef extract glycogen agar, a selective and differential *Aeromonas* medium. J. Fish. Res. Board Can. 31, 1553–1555.
Melas, D.S., Papageorgiou, D.K. and Mantis, A.L. (1999) Enumeration and confirmation of *Aeromonas hydrophila, Aeromonas caviae* and *Aeromonas sobria* isolated from raw milk and other milk products in northern Greece. J. Food Protect. 62, 463–466.
Merino, S., Aguilar, A., Nogueras, M.M., Regue, M., Swift, S. and Tomás, J.M. (1999). Cloning, sequencing, and role in virulence of two phospholipases (A1 and C) from mesophilic *Aeromonas* sp. serogroup O:34. Infect. Immun. 67, 4008–4013.
Miller, M.L. and Koburger, J.A. (1985) *Plesiomonas shigelloides*: An opportunistic food and waterborne pathogen. J. Food Protect. 48, 449–457.
Miller, M.L. and Koburger, J.A. (1986) Evaluation of Inositol Brilliant Green Bile Salts and Plesiomonas agars for recovery of *Plesiomonas shigelloides* from aquatic samples in a seasonal

survey on the Suwannee river estuary. J. Food Protect. 49, 274–278.

Millership, S.E., Curnow, S.R. and Chattopadhyay, B. (1983) Faecal carriage rate of *Aeromonas hydrophila.* J. Clin. Pathol. 36, 920–923.

Millership, S.E. and Chattopadhyay, B. (1984) Methods for the isolation of *Aeromonas hydrophila* and *Plesiomonas shigelloides* from faeces. J. Hyg. 92, 145–152.

Monfort, P. and Baleux, B. (1988) Évaluation de deux milieux de culture pour l'isolement le dénombrement des aeromonas mobiles dans différents types d'eaux. C. R. Acad. Sci. Paris. 307, 523–527.

Moyer, N.P. (1987) Clinical significance of *Aeromonas* species isolated from patients with diarrhea. J. Clin. Microbiol. 25, 2044–2048.

Moyer, N.P., Geiss, H.K., Marinescu, M., Rigby, A., Robinson, J. and M. Altwegg. (1991) Media and methods for isolation of aeromonads from faecal specimens. A multilaboratory study. Experientia 47, 409–412.

Myers, B.R., Marshall, R.T., Edmondson, J.E. and Stringer, W.C. (1982) Isolation of pectinolytic *Aeromonas hydrophila* and *Yersinia enterocolitica* from vacuum-packaged pork. J. Food Protect. 45, 33–37

Nishikawa, Y. and Kishi, T. (1987) A modification of bile salts brilliant green agar for isolation of motile *Aeromonas* from foods and environmental specimens. Epidemiol. Infect. 98, 331–336.

Nishikawa, Y. and Kishi, T. (1988) Isolation and characterization of motile *Aeromonas* from human, food and environmental specimens. Epidemiol. Infect. 101, 213–223.

Nordic Committee on Food Analysis. (1995) *Aeromonas* species, motile. Determination in foods. NMKL Proposed Method No. 150.

Ogden, I.D., Millar, I.G., Watt, A.J. and Wood, L. (1994) A comparison of three identification kits for the confirmation of *Aeromonas* spp. Lett. Appl. Microbiol. 18, 97–99.

Okrend, A.J.G., Rose, B.E. and Bennett, B. (1987) Incidence and toxigenicity of *Aeromonas* species in retail poultry, beef and pork. J. Food Protect. 50, 509–513.

Olsvik, O., Waschmuth, K., Kay, B., Birkness, K.A., Yi, A. and Sack, B. (1990) Laboratory observations on *Plesiomonas shigelloides* strains isolated from children with diarrhea in Peru. J. Clin. Microbiol. 28, 886–889.

Palumbo, S.A., Maxino, F., Williams, A.C., Buchanan, R.L. and Thayer, D.W. (1985) Starch-ampicillin agar for the quantitative detection of *Aeromonas hydrophila.* Appl. Environ. Microbiol. 50, 1027–1030.

Palumbo, S.A. (1991) A review of methods for detection of the psychrotrophic foodborne pathogens *Listeria monocytogenes* and *Aeromonas hydrophila*. J. Food Safety 11, 105–122.

Palumbo, S., Abeyta, C. and Stelma Jr., G. (1992) The *Aeromonas hydrophila* group. *In*: Compendium of Methods for the Microbiological Examination of Foods. American Public Health Association, Washington, DC, pp. 497–515.

Palumbo, S., Stelma, Jr. G.N. and Abeyta, C. (2000) The *Aeromonas hydrophila In*: The Microbiological safety and quality of food. B.M. Lund, T.C. Baird-Parker and G.W. Gould (ed.) Aspen Publishers, Inc, Maryland pp. 1011–1028.

Perales, I., García, F., Michaus, L., Blanco, S. and Lantero, M. (1983) Aislamiento de *Plesiomonas shigelloides* en un caso de gastroenteritis. Rev. Clin. Esp. 171, 115–117.

Pin, C., Marín, M.L., García, M.L., Tormo, J. and Casas, C. (1994) Comparison of different media for the isolation and enumeration of *Aeromonas* spp. in foods. Lett. Appl. Microbiol. 18, 190.192.

Popoff, M. (1984) Genus III. *Aeromonas In*: N.R. Krieg and J.G. Holt (Eds.) Bergey's Manual of Systematic Bacteriology, Vol. 1, pp. 545–548. Williams & Wilkins, Baltimore, USA.

Rautelin, H., Sivonen, A., Kuikka, A., Renkonen, O.V., Valtonen, V. and Kosunen, T.U. (1995) Enteric *Plesiomonas shigelloides* in Finnish patients. Scand. J. Infect. Dis. 27, 49–498.

Ribas, F., Araujo, R., Frias, J., Huguet, J.M., Ribas, F.R. and Lucena, F. (1991) Comparison of different media for the identification and quantification of *Aeromonas* spp. in water. Antonie van Leeuwenhoek 59, 225–228.

Richard, C., Lhuillier, M. and Laurent, B. (1978) *Plesiomonas shigelloides*: une vibrionacée entéropathogène exotique. Bull. Int. Pasteur, 76, 187–200.
Rippey, S.R. and Cabelli, V.J. (1979) Membrane filter procedure for enumeration of *Aeromonas hydrophila* in fresh waters. Appl. Environ. Microbiol. 38, 108–113.
Roberts, D., Hooper, W. and Greenwood, M. (1995) Practical Food Microbiology. Methods for the examination of food micro-organisms of public health significance. Public Health Laboratory Service, London
Rogol, M., Sechter, I., Grinberg, L. and Gerichter, C.B. (1979) Pril-xylose-ampicillin agar, a new selective medium for the isolation of *Aeromonas hydrophila.* J. Med. Microbiol. 12, 229–231.
Roland, F. P. (1977) Salt-starch xylose lysine deoxycholate agar. A single medium for the isolation of sodium and non-sodium dependent enteric Gram-negative bacilli. Med. Microbiol. Immunol. 163, 241–249.
Rutala, W.W., Sarubbi Jr. F.A., Finch, C.S. and Steinkraus, G.E. (1982) Oyster-associated outbreak of diarrhoeal disease possibly caused by *Plesiomonas shigelloides.* Lancet 1, 739.
Schubert, R.H.W. (1967) Das Vorkommen der Aeromonaden in oberirdischen Gewässern. Arch. Hyg. 150, 688–708.
Schubert, R.H.W. (1977) Ueber den Nachweis von *Plesiomonas shigelloides* Habs und Schubert, 1962, und ein Elektivmedium, den Inositol-Brillantgrum-Gallesalz-Agar. Ernst-Rodenwaldt-Arch. 4, 97–103.
Schubert, R.H.W. (1982) On the problems of isolation of *Plesiomonas shigelloides* and on a new elective medium: inositol brilliant green bile salts agar. *In:* J.E.L. Corry (Ed.) Quality Assurance and Quality Control of Microbiological Culture Media, pp. 163–167. G.I.T. Verlag – Ernst Giebeler, Darmstadt.
Schubert, R.H.W. (1984) Genus IV. *Plesiomonas In*: N.R. Krieg and J.G. Holt (Eds.) Bergey's Manual of Systematic Bacteriology, Vol. 1, pp. 548–550. Williams & Wilkins, Baltimore, USA.
Shotts Jr. E.B., and Rimler, R. (1973) Medium for the isolation of *Aeromonas hydrophila.* Appl. Microbiol. 26, 550–553.
Shread, P., Donovan, T.J. and Lee, J.V. (1981) A survey of the incidence of *Aeromonas* in human faeces. Soc. Gen. Microbiol. Q. 8, 184.
Sierra, M.L., González-Fandos, E., García-López, M.L., García Fernández, M.C. and Prieto, M. (1995) Prevalence of *Salmonella, Yersinia, Aeromonas, Campylobacter*, and cold-growing *Escherichia coli* on freshly dressed lamb carcasses. J. Food Protect. 58, 1183–1185.
Singh, U. (1997) Isolation and identification of *Aeromonas* spp. from ground meats in eastern Canada. J. Food Protect. 60, 125–130.
Singh, D.V. and Sanyal, S.C. (1994) Antibiotic resistance in clinical and environmental isolates of *Aeromonas* spp. J. Antimicrob. Chemother. 33, 368–368.
Stern, N. J., Drazek, E.S. and Joseph, S.W. (1987) Low incidence of *Aeromonas* sp. in livestock faeces. J. Food Protect. 50, 66–69.
Tsai, G.J. and Chen, T.H. (1996) Incidence and toxigenicity of *Aeromonas hydrophila* in seafood. Int. J. Food Microbiol. 31, 121–131.
Tsukamoto, T., Kinoshita, Y., Shimada, T. and Sakazaki, R. (1978) Two epidemics of diarrhoeal disease possibly caused by *Plesiomonas shigelloides.* J. Hyg. 80, 275–280.
Van Damme, L.R. and Vandepitte, J. (1980) Frequent isolation of *Edwarsiella tarda* and *Plesiomonas shigelloides* from healthy Zairese freshwater fish: a possible source of sporadic diarrhea in tropics. Appl. Environ. Microbiol. 39, 475–479.
Villari, P., Pucino, A., Santagata, N. and Torre, I. (1999) A comparison of different culture media for the membrane filter quantification of *Aeromonas* in water. Lett. Appl. Microbiol. 29, 253–257.
Vivas, J., Sáa, A.I., Tinajas, A., Barbeyto, L. and Rodríguez, L.A. (2000) Identification of motile *Aeromonas* strains with the MicroScan WalkAway system in conjunction with the Combo Negative Type 1S panels. Appl. Environ. Microbiol. 66, 1764–1766.
von Graevenitz, A. and Zinterhofer, L. (1970) The detection of *Aeromonas hydrophila* in stool

specimens. Health Lab. Sci. 7, 124–127.
von Graevenitz, A. and Mensch, A.H. (1968) The genus *Aeromonas* in human bacteriology. Report of 30 cases and review of the literature. New Eng. J. Med. 278, 245–249.
von Graevenitz, A. and Bucher, C. (1983) Evaluation of differential and selective media for isolation of *Aeromonas* and *Plesiomonas* spp. from human feces. J. Clin. Microbiol. 17, 16–21.
Wang, C. and Silva, J. L. (1999) Prevalence and characteristics of *Aeromonas* species isolated from processed channel catfish. J. Food Protect. 62, 30–34.
Want, S.V. and Millership, S.E. (1990) Effects of incorporating ampicillin, bile salts and carbohydrates in media on the recognition and selection of *Aeromonas* spp. from faeces. J. Med. Microbiol. 32, 49–54.
Wilcox, M.H., Cook, A.M., Eley, A. and Spencer, R.C. (1992) *Aeromonas* spp. as a potential cause of diarrhoea in children. J. Clin. Pathol. 45, 959–963.

Chapter 20

Media for *Pseudomonas* spp. and related genera from food and environmental samples.

Vibeke From Jeppesen[a], Claus Jeppesen[b]

[a] *Eurofins Denmark Ltd., Frydendalsvej, 30, 1809 Frederiksberg C., Denmark*
[b] *Regional Food Control, Fjeldhammervej 15, 2610 Rødovre, Denmark*

Despite changes in taxonomy in recent years media for the isolation and enumeration of this group of organisms show little change.

Pseudomonas spp. belong to a group of organisms of great importance as pathogens and as spoilage organisms. Methods described for isolation and enumeration are primarily selective, since non-selective pre-enrichment is not possible due to competing flora. The selective principles are based on antibiotics and the indicative principles on production of fluorescence, pigmentation and hydrolysis of casein. As with all selective media, the recovery of stressed cells is sometimes prevented and the competing flora is not always completely inhibited so that confirmatory tests need to be made on presumptive positive colonies.

1. Introduction

The genus *Pseudomonas* has undergone taxonomic changes in recent years, with species being removed and placed in the genera *Xanthomonas*, *Stenotrophomonas*, *Burkholderia*, *Brevundimonas*, *Comamonas*, *Sphingomonas* and *Chryseomonas* among others (www, 2000). However, even though the taxonomy has changed, the bacteria and the media for isolation, enumeration and identification have not.
In this chapter the term "*Pseudomonas* spp." is used not only for *Pseudomonas* spp. but also for related species in other genera as mentioned above.

Pseudomonas spp. are ubiquitous and have been isolated from a variety of sources including drinking water, animals, humans, plants and therefore also from various foods. They are Gram-negative rods, catalase positive, oxidase positive or negative and typically motile. They are aerobic, although nitrate in some cases can be used as an alternative electron acceptor, allowing growth to occur anaerobically, e.g. in vacuum-packed meat-products. The predominant spoilage organisms in proteinaceous foods with high a_w belong to the genus *Pseudomonas*.

Pseudomonas spp. grow well on non-selective media (e.g. tryptose soya agar (TSA), plate count agar and blood agar) and most of them grow on routine primary isolation media (e.g. MacConkey agar and Eosin methylene blue agar) when incubated at a temperature suitable for their growth. Depending on the purpose of the examination, the following factors will influence the choice of medium for isolation and enumeration: (1) selectivity, i.e. the ability to suppress background flora; (2) ability to differentiate between the organisms sought and the competing flora on the basis of colonial morphology; (3) quantitative recovery of the desired organism, including recovery of injured cells.

In food and environmental samples *Pseudomonas* spp. may only be present in low concentrations and are often accompanied by a competing microflora which is physiologically similar. A problem associated with the use of media developed for isolation of *Pseudomonas* spp. from foods is the considerable interference from non-pseudomonads (Brant, 1975). Furthermore the organisms may often be injured, e.g., due to refrigeration, heating, chemical preservation or disinfectants. Food and water microbiologists are concerned not only about the presence of certain organisms, but also about their numbers since these determine not only the risk of contamination but also of spoilage. Even though *Pseudomonas* spp. may only represent a minority of the total microflora at the beginning of shelf-life, they may subsequently become the dominating microflora under the right conditions, i.e. high a_w, neutral pH, high oxygen tension and suitable temperature.

The aim of this paper is to review media proposed in the literature for isolation, enumeration and identification of *Pseudomonas* and related species in the new taxonomic genera. The paper is a revision of Jeppesen (1995).

2. Enrichment procedures

2.1. Non-selective enrichment

The European Pharmacopoeia (Ph.Eur., 2001) as well as the United States Pharmacopoeia (USP, 2000) prescribes non-selective enrichment of pharmaceutical products in TSB (casein soya bean digest broth) before streaking on Cetrimide agar. A non-selective enrichment broth can be used where there are very low numbers of micro-organisms in a product, i.e. low content of competing flora.

2.2. Selective enrichment

The EU-directive on natural mineral water and spring water prescribes absence of *P. aeruginosa* in 250 ml water without specifying any method of analysis, i.e. there is free choice of method. If *P. aeruginosa* should be present in these products, they are expected to be in very low numbers and the use of the ISO-method (1988a) for bottled water with enrichment in liquid medium is recommended. Enrichment is made in "Drake's medium 10" (also called asparagine broth) with ethanol at 37°C for 48 h. The ISO-standard proposes incubation at 38°C or 39°C if the samples are expected to contain large numbers of other bacteria. However, the improved selectivity for *P. aeruginosa* brought about by this temperature increase is doubtful. Instead the use of 42°C, as proposed in the ISO membrane filtration method (1988b) and as prescribed for confirmatory tests, would be more logical. All tubes of enrichment broth with growth and/or fluorescence should be subcultured to milk agar with cetrimide (0.03%) for confirmation (growth, pigment production and casein hydrolysis) and if necessary further biochemical tests.

3. Plating media

3.1. Media for foods

The only current standard on *Pseudomonas* spp. in foods in general (except for bottled water) is ISO (1995) for meat and meat products which recommends the use of Cephaloridine Fucidin Cetrimide agar (CFC) with verification by oxidase reaction and aerobic sugar metabolism (Kligler's agar). CFC was developed by Mead and Adams (1977) by adding cephaloridine, fucidin and cetrimide as selective agents to Difco heart infusion agar. This medium allows the growth of pigmented as well as non-pigmented pseudomonads. Mead and Adams (1977) found that CFC-agar effectively suppressed Gram-positive bacteria and supported good growth of *Pseudomonas* spp. whilst inhibiting other Gram-negative bacteria, although *Shewanella putrefaciens* (formerly *Alteromonas putrefaciens* and *Pseudomonas putrefaciens*) was only partly inhibited.

Mead (1985) pointed out that CFC medium can be used for enumeration of *Pseudomonas* spp. in various types of food. The medium is inoculated by surface plating and incubated for 48 h at 25°C (ISO, 1995). CFC aims to suppress virtually all unwanted organisms. There are two types of colonies: pigmented and non-pigmented, 2 – 5 mm in diameter. Confirmatory tests are usually not required but in case of doubt the oxidase test may be used, although it might be a problem in the case of *Stenotrophomonas maltophilia*, *Burkholderia mallei* and *Chryseomonas luteola* which are oxidase-negative and therefore need further identification.

Mead and Adams (1977) investigated the selectivity and productivity of four existing media for isolation of *Pseudomonas* spp. (ALVC containing crystal violet, MGV containing erythromycin and chloramphenicol, Difco heart infusion agar with diamide and CETCH containing cetrimide and 2-hydroxy-2',4,4'-trichlorodiphenyloxide). They found that ALVC medium (Gyllenberg et al., 1960a,b, c.f. Mead and Adams, 1977), CETCH medium (Solberg et al., 1972 c.f. Mead and Adams, 1977) and MGV medium (Masurovsky et al., 1963 c.f. Mead and Adams, 1977) all gave adequate recoveries of 12 pure cultures of pseudomonads isolated from poultry. These media effectively suppressed the growth of Gram-positive organisms, but there were marked differences in their ability to suppress unwanted Gram-negative micro-organisms. ALVC medium permitted growth of most strains of Gram-negative bacteria, whereas *Serratia liquefaciens* could grow on MGV and *Shewanella putrefaceins, Aeromonas* spp. and *S. liquefaciens* grew on CETCH. On MGV the majority of *Pseudomonas* spp. produced only small colonies and were difficult to count. The Difco medium (Verbovsky and Collins, 1973) was the least successful, inhibiting most of the non-pigmented *Pseudomonas* spp. whilst supporting the growth of certain yeasts.

Table 1
Media for isolation, enumeration and identification of *Pseudomonas* spp.

Medium	Purpose	Selective agents	Differential agents	Incubation	Typical growth	Reference
Arginine Broth acc. to Schuberg	Enrichment broth for chlorine damaged *P. aeruginosa*	Brilliant green	Bromothymol blue Cresol red	37°C, 24–48 h.	Colour change from grey-green to blue-violet.	Precribed in German Standard Method according to Merck (1996).
CFC	Enumeration of *Pseudomonas* spp. in foods	Cephaloridine Fucidin Cetrimide		25°C, 48 h	Pigmented or non-pigmented. All colonies are counted as presumptive *Pseudomonas* spp.	ISO (1995)
Cetrimide agar	Enumeration (DS, 1990), subculturing and verification (Ph.Eur, 2001; USP, 2000)	Cetrimide DS also includes: Nalidixic acid (0.015 ‰)	Magnesium chloride[1)] Potassium sulphate[2)]	35–37°C, 18–72 h.	Gram-negative rods	Ph.Eur. (2001)
				30–35°C, 24–48 h.	Greenish pigmented colonies with greenish fluorescence	USP (2000)
				42±0.5°C, 48 h.		DS (1990)
Drake's medium 10	Detection of *P. aeruginosa* in bottled water and disinfected water.	Ethanol Asparagine[6] Proline[6]	Potassium phosphate[3)] Magnesium sulphate[4)] Potassium sulphate[2)]	37°C, 48 h. May be increased to 38-39°C	May be fluorescent.	ISO (1988a)
Drake's medium 19	Detection and enumeration of *P. aeruginosa* in environmental samples (except from bottled water and swimming pool water).	Cetrimide Ethanol 42°C if used.	Potassium sulphate[2)] Magnesium chloride[1)]	37±1°C, 48 h or 42±0.5°C, 48 h if large number of competing flora is expected	Blue-green, greenish-brown or fluorescent colonies	ISO (1988b)

GSP Agar	Detection of *Pseudomonas* and *Aeromonas* in food and environment	Glutamate[6] Starch[6] Penicillin (Pimaricin)	Phenol red	25°C, 72 h.	Large, blue-violet colonies surrounded by a red violet zone	Merck (1996); Kielwein (1969).
King's B	*Pseudomonas* spp. in environmental samples		Magnesium sulphate[4] Phosphate[3] Glycerol[5]	21°C, 72 h.	Fluorescence may be seen, especially when surface-inoculated. Can not be recommended for enumeration of fluorescent pseudomonads.	King et al. (1954) DS (1983)
Malachite-green Broth	Selective enrichment of *P. aeruginosa* in water	Malachite-green	None Growth has to be subcultured and verified.	36±1°C, 24–48 h.	–	Prescribed in German Standard Method according to Merck (1996)
Milk agar without cetrimide	Verification of *P. aeruginosa*	None	Casein	42±0.5°C, 24 h.	Casein hydrolysis (clearing of the medium), pigmentation (brown, green or blue) and fluorescence.	DS (1990)
Milk agar with cetrimide	Verification of *P. aeruginosa*	Cetrimide	Casein	42±0.5°C	Casein hydrolysis (clearing of the medium), pigmented (blue-green) and fluorescent	ISO (1988a,b)
PA-F	Verification of *P. aeruginosa*, pharmaceutical isolates	None	Phosphate[3] Magnesium sulphate[4]	35±2°C, minimum three days	Colourless to yellowish colonies with yellowish fluorescence	USP (2000)

Table 1. Continued.

Medium	Purpose	Selective agents	Differential agents	Incubation	Typical growth	Reference
PA-P	Verification of *P. aeruginosa*, pharmaceutical isolates	None	Magnesium chloride[1)] Potassium sulphate[2)]	35±2°C, minimum three days	Greenish colonies with blue fluorescence	USP (2000)
TSB	Non-selective enrichment of *P. aeruginosa* in pharmaceutical samples	None. Subculturing on Cetrimide-agar.	–	35–37°C, 18–48 h. 30-35°C, 24-48h.	–	Ph.Eur. (2001) USP (2000)

1) Magnesium chloride enhances pigment production (pyocyanin).
2) Potassium sulphate enhances pigment production (pyocyanin).
3) Phosphate stimulates fluorescein production, inhibits pyocyanin.
4) Magnesium sulphate activates fluorescein production.
5) Glycerol helps to promote pyocyanin production.
6) Cannot be metabolised by the dominating part of accompanying flora.

None of these media seems to have gained general acceptance.

The examination of bottled water for the presence of *P. aeruginosa* according to ISO has been described earlier (see 2.2). In general drinking water, including bottled water, has a microflora dominated by *Pseudomonas* spp. As this flora can grow under the right conditions to high numbers and as the pseudomonads are potentially pathogenic and may be multiresistant to antibiotics (Rosenberg and Hernandez Duquino, 1989; Payment et al., 1991; Jeppesen and Beck, 1992), counts should be performed and the results compared with standards set by the authorities. For this purpose it might be relevant to use non-selective agars such as TSA or yeast extract agar, as prescribed in the European reference method (EN, 1999), for colony counts in drinking water.

3.2. Media for environmental samples

A medium that has frequently been used for the isolation of fluorescent *Pseudomonas* spp. from environmental samples is King's B agar (King et al., 1954), which enhances pyoverdin production and at the same time suppresses the production of the pigment pyocyanin. The colonies can be identified on plates by a characteristic diffusible pigment under UV-light. King's A medium acts in the opposite way by enhancing pyocyanin, while suppressing pyoverdin.

The Danish Standard for examination of drinking water (DS, 1983) used King's B agar for colony counts as well as for enumeration of fluorescent pseudomonads, but an interlaboratory trial later revealed that the reproducibility for the fluorescent count was so low that King's B agar was only acceptable for general colony counting (Bagge, 2000). The low reproducibility may be due to the use of a pour-plate method, although even surface-plating is expected to give problems due to difficulties in separating fluorescent from non-fluorescent colonies when counting.

Where relatively high levels of residual disinfectant, especially chlorine, are expected, the ISO method of enrichment in liquid medium (ISO, 1988a) may be used (see 2.2). For all other environmental samples ISO prescribes the membrane filtration method (ISO, 1988b) provided that the sample is suitable for filtration. This method uses "Drake's medium 19" as a selective fluid in which to soak a sterile filter pad, before placing the membrane on the filter pad in a Petri dish and incubating at 37°C for 48 h. Counting and confirmation of typical colonies is carried out on milk agar with cetrimide (growth, pigment production and casein hydrolysis) and if necessary by biochemical tests. Incubation at 42°C is mentioned as a possibility if a high level of competing flora is expected, but the higher selective pressure on *P. aeruginosa* should be taken into consideration (ISO, 1988b).

In the Danish Standard for examination of swimming pool water for *P. aeruginosa* (DS, 1990) the water samples are membrane filtered, the filter placed on Cetrimide Nalidixine agar and incubated at 42°C for 48 h. Verification is made on milk agar (without cetrimide) at 42°C for 24 h with casein-hydrolysing isolates tested for oxidase, catalase, O/F, fluorescence and pigment production.

Many commercial producers of dehydrated media exist, e.g. Oxoid, Difco and Merck. Most of their media are composed in accordance with the different international standards.

Table 2
Important tests differentiating *Pseudomonas* spp. from other organisms in the same environment.

Pseudomonas spp.	Distinguishes *Pseudomonas* spp. from:
Oxidase-positive and **O/F** non-fermentative	*Enterobacteriaceae*
O/F non-fermentative	*Vibrio* spp., *Aeromonas* spp. and *Plesiomonas* spp.[1)]
No reduction of **TMAO**[2)] and no production of $\mathbf{H_2S}$	*Shewanella putrefaciens*

1) For differentiation between *Vibrio, Aeromonas* and *Plesiomonas*, see Jeppesen (1995).
2) Trimethylamineoxide.

Table 1 gives an overview of media proposed for isolation, enumeration and verification of *Pseudomonas* spp. The table includes media from the standards described in this paper as well as some commercially produced media proposed for *Pseudomonas* spp.

4. Identification

Identification of microorganisms is more and more commonly made by genetically based methods, which have also resulted in the taxonomic changes mentioned earlier. However there is still a need for biochemical characterisation to compare new isolates using established methods and diagnostic tables, as in Bergey's Systematic Manual (Krieg and Holt, 1984) and in Cowan and Steel (Barrow and Feltham, 1993). These biochemical criteria can be carried out in a traditional way by testing the specific criteria mentioned in the methods and in the diagnostic tables, or they can be performed using commercial diagnostic kits with matching databases as, for example API 20 NE (BioMerieux, France) or Biolog MicroStation™.

Table 2 gives an overview of the most important phenotypical criteria to differentiate *Pseudomonas* from other genera.

5. Conclusion

Several media have been proposed for the detection of *Pseudomonas* spp.: some are for general purposes and others specifically for isolation of certain species or for isolation from clinical, environmental or food samples.

Standards exist for the examination of pseudomonads, e.g. *P. aeruginosa* in pharmaceutical samples (Ph. Eur., 2001; USP, 2000), *Pseudomonas* spp. in meat and meat products (ISO, 1995) and *P. aeruginosa* in water and environmental samples (ISO, 1988a,b., DS, 1983; DS, 1990).

The selective media generally give a good recovery but, as with all selective media, problems concerning the recovery of injured cells and inhibition of competing flora may be considerable. The media vary in sensitivity and selectivity and their suitability may vary according to the source of the samples and whether qualitative detection or quantitative recovery is required. The final choice of media and incubation temperature will thus depend on the intended purpose of the investigation. Achieving an increase in one regard may result in reduction in another.

References

Bagge, L. (2000) Personal communication. Danish Environmental Protection Agency. Strandgade 29, 1401 Kbh. K., Denmark.

Barrow, G.I. and Feltham, R.K.A. (1993) Cowan and Steel's Manual for the Identification of Medical Bacteria. Third Edition, Cambridge University Press, UK.

Brant, A.W. (1975) Selective media for psychrotrophic spoilage bacteria of fresh poultry. 2nd Eur. Symp. Poult. Meat, Oosterbeek, 31.

DIN (1982) DIN 38411, Teil 8, Nachweis von Pseudomonas aeruginosa. Deutsches Institut für Normung. Germany.

DS (1983) Determination of the number of bacteria and fluorescent bacteria at 21°C in Kings agar B. First edition DS 2252. Danish Standard, Copenhagen, Denmark.

DS (1990) Water quality – Determination of Pseudomonas aeruginosa in water – Membrane filtering. First edition DS 268. Danish Standard, Copenhagen, Denmark.

EN (1999) EN/ISO 6222, Water quality – Enumeration of culturable micro-organisms – Colony count by inoculation in a nutrient agar culture medium. CEN, Brussels.

Gyllenberg, H., Eklund, E., Antila, M. and Vartiovaara, U. (1960a) Contamination and deterioration of market milk. II Significance of pseudomonads as deteriorating bacteria. Acta Agric. Scand. 10, 50–64.

Gyllenberg, H., Eklund, E., Antila, M. and Vartiovaara, U. (1960b) Contamination and deterioration of market milk. III. A selective plating test for the demonstration of significant numbers of pseudomonads. Acta Agric. Scand. 10, 65–73.

ISO (1988a) Water quality – Detection and enumeration of *Pseudomonas aeruginosa*. Part 1: Method by enrichment in liquid medium. First edition ISO 8360-1 (E). International Organization for Standardization.

ISO (1988b) Water quality – Detection and enumeration of *Pseudomonas aeruginosa*. Part 2: Membrane filtration method. First edition ISO 8360-2 (E). International Organization for Standardization.

ISO (1995) Meat and meat products – Enumeration of *Pseudomonas* spp. First edition ISO 13720 (E). International Organization for Standardization.

Jeppesen, C. (1995) Media for *Aeromonas* spp., *Plesiomonas shigelloides* and *Pseudomonas* spp. from food and environment. In: Culture Media for Food Microbiology edited by J.E.L Corry, G.D.W. Curtis, and R.M. Baird, Progress in Industrial Microbiology series No. 34. Elsevier, Amsterdam, pp. 111–127.

Jeppesen, C. and Beck, J. (1992) Spring water coolers and natural mineral water (In Danish). Vandteknik, 60, 7–11.

Kielwein, G. (1969) Ein Nährboden zur selektiven Züchtung von Pseudomonaden und Aeromonaden. Arch. Lebensmittelhyg. 20, 131–138.

King, E.O., Ward, M.K. and Raney, D.E. (1954) Two simple media for the demonstration of pyocyanin and fluorescin. J. Lab. Cin. Med. 44, 301–307.

Krieg, N.R. and Holt, J.G. (1984) Bergey's Manual of Systematic Bacteriology, volume 1. Williams & Wilkins, USA.

Masurovsky, E.B., Goldblith, S.A. and Voss, J. (1963) Differential medium for selection and enumeration of the genus *Pseudomonas*. J. Bacteriol. 85, 722–723.

Mead, G.C. (1985) Enumeration of pseudomonads using cephaloridine-fucidin-cetrimide agar

(CFC). Int. J. Food Microbiol. 2, 21–26.

Mead, G.C. and Adams, B.W. (1977) A selective medium for the rapid isolation of pseudomonads associated with poultry meat spoilage. Br. Poult. Sci. 18, 661–670.

Merck (1996) Microbiology Manual. Darmstadt, Germany.

Payment, P. Franco, E., Richardson, L. and Siemiatycki, J. (1991) Gastrointestinal Health Effects Associated with the consumption of Drinking Water Produced by Point-of-Use Domestic Reverse-Osmosis Filtration Units. Appl. Environ. Microbiol., 57, 945–948.

Ph. Eur. (2001) European Pharmacopoeia, third edition, supplement 2001. 2.6.13. Council of Europe, Strasbourg.

Rosenberg, F.A. and Hernandez Duquino, H. (1989) Antibiotic resistance of Pseudomonas from German mineral waters. Toxicity Assessment, 4, 281–294.

Solberg, M., O'Leary, V.S. and Riha, W.E. (1972) New medium for the isolation and enumeration of pseudomonads. Appl. Microbiol. 24, 544–550.

USP (2000) U.S. Pharmacopeia. The National Formulary, 24 <61>. United States Pharmacopeial Convention,USA.

Verbovsky, P.D. and Collins, E.B. (1973) Effect of diamide on growth of selected bacteria. J. Dairy Sci. 56, 1180–1182.

www (2000) www-sv.cict.fr/bacterio/. 2000.06.24.

Chapter 21

Culture Media for genera in the family Flavobacteriaceae

Celia J. Hugo[a] and P.J. Jooste[b]

[a]*Department of Food Science, University of the Free State, PO Box 339, Bloemfontein 9300, South Africa*

[b]*Animal Nutrition and Products Institute, Private Bag X2, Irene 0062, South Africa*

The taxonomy of the flavobacteria has undergone many changes over the past 10 years. This fact may lead to confusion when seeking culture media for this group of bacteria. This chapter is organised in such a way that it orientates the reader in terms of the taxonomy of the family and then deals with the isolation, cultivation and maintenance media that can be used for members of the Flavobacteriaceae family that are of significance for the food microbiologist.

1. Introduction

Flavobacteria are widely distributed in the natural and clinical environment (Owen and Snell, 1973), soil (Hayes, 1977), fresh and marine waters (Shewan and McMeekin, 1983) and poultry and meat (McMeekin et al., 1971; Hayes, 1977). As part of the Gram-negative rod, psychrotrophic, proteolytic group, they are considered to be spoilers of food (Jooste and Britz, 1986; Jooste et al., 1986a,b; Fischer et al., 1987; Welthagen and Jooste, 1992; Engelbrecht et al., 1996a,b). They have been associated with a variety of defects such as surface taint and apple odour in butter (Wolochow et al., 1942), 'sweaty feet' odour in skim milk (Jooste et al., 1986a), 'thinning' in creamed rice (Everton et al., 1968), odour production and proteolysis in fresh and frozen fish (Engelbrecht et al., 1996a,b) and bitterness due to the production of phospholipase C (Fox et al., 1976).

The taxonomy of the flavobacteria, however, was in disarray from its inception (Bergey et al., 1923). In 1985, the family name, Flavobacteriaceae, was suggested by Jooste (1985) and formally applied by Reichenbach (1989). It provided affiliation for a large number of genera including *Flavobacterium, Cytophaga* and *Flexibacter.* Bernardet et al. (1996) emended the description of this family on the basis of 16S rRNA oligonucleotide catalogue and sequence data (Paster et al., 1985; Woese et al., 1985; Woese et al., 1990; Gherna and Woese, 1992; Kath and Reichenbach, 1993; Nakagawa and Yamasato, 1993; Olsen et al., 1994) as well as DNA-rRNA hybridization data (Segers et al., 1993; Vandamme et al., 1994a). As a result of these studies, the following genera were included in the Flavobacteriaceae family: *Bergeyella, Capnocytophaga, Chryseobacterium, Empedobacter, Flavobacterium, Myroides, Ornithobacterium, Riemerella, Weeksella* and *[Flexibacter] maritimus* (the brackets indicate generically misclassified organisms). Of these genera, only *Bergeyella, Chryseobacterium, Empedobacter, Myroides* and *Weeksella* have so far been found to be of relevance to the food microbiologist.

The taxonomy of the latter members of the Flavobacteriaceae family are well documented in reviews by Holmes et al. (1984a), Holmes (1992), Vandamme et al. (1994a), Bernardet et al. (1996) and Jooste and Hugo (1999).

The aim of this chapter will, therefore, be to discuss routine and selective isolation procedures, growth conditions, maintenance of cultures, some differential characteristics of genera and species and the use of specific cultivation media for fatty acid methyl ester analysis of the genera of Flavobacteriaceae which are relevant to the food microbiologist.

2. General culture media

The genera in the Flavobacteriaceae family have been isolated from a vast range of habitats (terrestrial, aquatic and diseased humans or animals; Jooste and Hugo, 1999). They are, however, not difficult to isolate and cultivate. They are chemo-organotrophic (Bernardet et al., 1996) and most of them do not require enrichment or special isolation procedures (Holmes, 1992).

McCurdy agar (Oyaizu and Komagata, 1981), medium M1 (Weeks, 1955), nutrient agar (Hayes, 1977, Holmes et al., 1977; 1978), 0.1% peptone agar (Oyaizu and Komagata, 1981), trypticase soy agar (Segers et al., 1993) and yeast extract-malt extract agar (Oyaizu and Komagata, 1981) are all suitable and have been used as general growth media for the flavobacteria. Food related flavobacteria may be isolated and grown on food *Flavobacterium* medium (McMeekin et al., 1971) or plate count agar (Jooste et al., 1986b).

General selective isolation media for Gram-negative bacteria and flavobacteria that have been used include brilliant green phenol red agar (Engelbrecht, 1992), nutrient or plate count agar containing 0.0002% (w/v) crystal violet (Jooste et al., 1986b; Engelbrecht et al., 1996a,b) and, except for *Weeksella* and *Bergeyella* that are susceptible to penicillin (Jooste and Hugo, 1999), plate count agar supplemented with 160 μg/ml penicillin G (Engelbrecht, 1992). Other methods for the selective isolation of specific species or genera in the Flavobacteriaceae family will be discussed where applicable.

General maintenance media that can be used for most of the flavobacteria are nutrient agar slopes kept at 4°C and subcultured every 6–12 weeks (according to the duration of their viability) (Hayes, 1977). Maintenance of the members of the Flavobacteriaceae will be discussed in more detail under each specific genus. General maintenance media are considered to be Dorset egg medium (Cowan, 1974), medium M1 (Weeks, 1955), nutrient agar (Hayes, 1977; Holmes et al., 1977, 1978), peptonized milk yeast extract agar (PMYA II; Christensen and Cook, 1972) and yeast extract-peptone agar (Oyaizu and Komagata, 1981). Cultures can also be frozen in liquid nitrogen or freeze-dried by standard bacteriological procedures. The freeze-drying technique is often preferred because of its convenience for the storage and shipping of samples.

3. Culture media for food related genera of the Flavobacteriaceae family

In this section, enrichment, isolation, cultivation and maintenance media of the food related genera in the Flavobacteriaceae family will be discussed. The cultivation media proposed by two international culture collections (the American Type Culture Collection [ATCC] and the Belgian Co-ordinated Collections of Micro-organisms, Laboratory for Microbiology, University of Gent [BCCM-LMG]) will also be referred to. A summary of culture media that are used for food related Flavobacteriaceae, are given in Table 1.

3.1 Bergeyella spp.

This genus was formerly known as *Flavobacterium* CDC Group IIj (Tatum et al., 1974) and later as *Weeksella* (Holmes et al., 1986b). Strains in this genus are frequently isolated from the upper respiratory tracts of dogs, from human wounds caused by dog bites (Rubin et al., 1985; Holmes et al., 1986b; Reina and Borrell, 1992) and from the food environment (Botha et al., 1989; 1998a,b; Engelbrecht et al., 1996a,b). After genotypic (DNA-rRNA), chemotaxonomic and phenotypic studies by Vandamme et al. (1994a), this genus was placed in a new genus, *Bergeyella*, consisting of only one species, namely *B. zoohelcum*. Studies by Botha et al. (1989; 1998a,b), however, suggested that there may be other species present in the food environment. Engelbrecht et al. (1996b) found that these organisms produced stale or pungent odours in fresh or frozen fish.

This organism is an obligate aerobe and will grow at room temperature and 37°C, but not at 5 or 42°C (Holmes et al., 1986b). It will give moderate growth on blood agar plates after 24 h of incubation (Tatum et al., 1974) and has been successfully cultivated on 5% horse blood agar (Holmes et al., 1986b), brain-heart infusion medium (Difco; ATCC, 2000), trypticase soy medium (BBL; BCCM-LMG, 1998), plate count agar containing 0.5% NaCl (Engelbrecht et al., 1996a,b) and nutrient agar (Oxoid; Botha et al., 1989, 1998a,b). This organism will not, however, grow in nutrient broth with 6% sodium chloride, on MacConkey, Salmonella Shigella (SS; Oxoid CM99), cetrimide, Simmon's citrate and β-hydroxybutyrate media (Tatum et al., 1974; Holmes et al., 1986b) or on penicillin agar (Jooste and Hugo, 1999).

No medium for the selective isolation of *Bergeyella* is known. Freeze-drying techniques are used for the long-term preservation of this organism (BCCM-LMG, 1998; ATCC, 2000).

Table 1
Media used for the cultivation of food related Flavobacteriaceae genera.

Species	Isolation and cultivation media	Incubation temperature	Reference
Bergeyella spp.	5% Horse blood agar	25–30°C	Holmes et al., 1986b
	Brain-heart infusion medium		ATCC, 2000
	Trypticase soy medium		BCCM-LMG, 1998
	Nutrient agar		Botha et al., 1989, 1998a,b
	Plate count agar + 0.5% NaCl		Engelbrecht et al., 1996a,b
Chryseobacterium balustinum	Plate count agar + 0.5% NaCl	20–30°C	Engelbrecht et al., 1996a,b
	Nutrient agar		Harrison, 1929; Holmes et al., 1984a; BCCM-LMG, 1998; ATCC, 2000
Ch. gleum	Nutrient agar, trypticase soy agar	25–30°C	Holmes et al., 1984b; BCCM-LMG, 1998; ATCC, 2000
Ch. indologenes	Heart infusion agar, nutrient agar, trypticase soy agar	25–30°C	Hugh and Gilardi, 1974; Yabuuchi et al., 1983; BCCM-LMG, 1998
Empedobacter brevis	Nutrient agar	25–30°C	Holmes et al., 1978; BCCM-LMG, 1998
	5% Horse blood agar		Holmes et al., 1978
	Heart infusion medium + 5% rabbit blood		Dees et al., 1986
	Brain heart infusion medium		ATCC, 2000
	Plate count agar + 0.5% NaCl		Engelbrecht et al., 1996a,b
Myroides spp.	Plate count agar + 0.5% NaCl	25–30°C	Engelbrecht et al., 1996a,b
	MacConkey agar and Trypticase soy agar		Vancanneyt et al., 1996; BCCM-LMG, 1998; ATCC, 2000
	Modified Shieh agar		Bernardet et al., 1996
	Heart infusion medium + 5% rabbit blood		Dees et al., 1986
Weeksella spp.	5% Horse blood agar	25–30°C	Holmes et al., 1986a
	Brain-heart infusion medium		ATCC, 2000
	Trypticase soy medium		BCCM-LMG, 1998
	Nutrient agar		Botha et al., 1989, 1998a,b
	Plate count agar + 0.5% NaCl		Engelbrecht et al., 1996a,b

3.2 Chryseobacterium spp.

This genus consists of six species, namely *Ch. indologenes* (type species), *Ch. gleum, Ch. balustinum, Ch. indoltheticum* (isolated from the marine environment and degrades chitin; Campbell and Williams, 1951), *Ch. scophthalmum* (isolated from diseased gill tissue of turbot; Mudarris and Austin, 1989) and *Ch. meningosepticum* (isolated from the hospital environment; King, 1959; Rubin et al., 1985). All these species previously belonged to the genus *Flavobacterium,* formerly also known as CDC Group IIb strains (Yabuuchi et al., 1983; Holmes et al., 1984a,b). Of the six species, only *Ch. balustinum, Ch. indologenes* and *Ch. gleum* are regularly isolated from the food environment where they may play a role as proteolytic, psychrotrophic spoilers of food (Jooste et al., 1985; Welthagen, 1991; Hugo et al., 1999). A new species (*C. joosteii*), isolated from raw milk, has recently been described (Hugo, 1997; Hugo et al., 1999). All the species are aerobic and chemoorganotrophic and will grow at 30°C (Table 1; Vandamme et al., 1994a). There are no selective isolation media known for the species in this genus apart from the general selective media referred to earlier. General maintenance media (e.g. nutrient agar slants) are used for this group of organisms unless otherwise specified.

Ch. balustinum

Ch. balustinum has been isolated from marine fish and may be associated with spoilage (odours, proteolysis and discolouration) of fresh and frozen fish (Harrison, 1929; Bergey and Breed, 1948; Engelbrecht et al., 1996a,b). The species is aerobic and will grow well on media such as nutrient agar or plate count agar containing 0.5% NaCl at 20 - 30°C (Table 1; Harrison, 1929; Holmes et al., 1984a; Engelbrecht et al., 1996a,b; BCCM-LMG, 1998; ATCC, 2000). According to Harrison (1929) best growth was observed in media containing 2% proteose peptone (Difco). *Ch. balustinum* will also grow on MacConkey agar and β-hydroxybutyrate agar but not on cetrimide agar or Simmon's citrate agar (Holmes et al., 1984a).

Ch. gleum

Ch. gleum was first described by Holmes et al. (1984b) and was isolated from the hospital environment (Holmes et al., 1984b; Yabuuchi et al., 1990). This organism is, however, regularly isolated from the dairy environment (Jooste et al., 1985; Welthagen, 1991; Hugo and Jooste., 1997; Hugo et al., 1999). Growth is aerobic and it will grow at room temperature and also at 30 and 37°C within 24 h but not at 5°C. This organism grows well on ordinary peptone media such as nutrient agar and trypticase soy agar (Table 1; Holmes et al., 1984b; BCCM-LMG, 1998; ATCC, 2000). It will also grow on MacConkey agar and β-hydroxybutyrate agar but not on cetrimide agar or Simmon's citrate agar (Holmes et al., 1984a).

Ch. indologenes

Ch. indologenes, was first described as a new species in 1983 (Yabuuchi et al., 1983). This organism occurs in the clinical (Yabuuchi et al., 1983; Rubin et al., 1985) and food environment (Tatum et al., 1974; Jooste et al., 1985; Welthagen and Jooste,

1992). The metabolism of the species is respiratory and not fermentative. It can, however, grow under anaerobic conditions in the presence of fumarate (Yabuuchi et al., 1983). Luxurious growth is produced on ordinary peptone media (e.g. heart infusion agar, nutrient agar and trypticase soy agar; Table 1), but it does not grow in standard mineral base medium (Hugh and Gilardi, 1974) supplemented with glucose and ammonium sulphate as the sole sources of carbon and nitrogen (Yabuuchi et al., 1983; BCCM-LMG, 1998). According to Yabuuchi et al. (1983), some of the strains (almost 50%) are inhibited on MacConkey agar.

3.3 Empedobacter brevis

This genus consists of only one species (Vandamme et al., 1994a) formerly known as *Flavobacterium breve* (Bergey et al., 1923; Holmes et al., 1978). Strains have been isolated from canal water, eye swabs, bronchial secretions, blood and urine (Holmes et al., 1978) and the food environment (Jooste et al., 1985; Welthagen, 1991; Engelbrecht et al., 1996b).

This genus is aerobic and has an optimum growth temperature of 30°C (Table 1). Visible growth at this temperature appears within 18–24 h. Most strains grow at 37°C (Vandamme et al., 1994a). The strains are isolated and propagated on plate count agar containing 0.5% NaCl (Engelbrecht, 1996b), nutrient agar (Holmes et al., 1978; BCCM-LMG, 1998), 5% (v/v) horse blood agar (Holmes et al., 1978), heart infusion medium with 5% (v/v) rabbit blood (Dees et al., 1986) or brain heart infusion medium (ATCC, 2000). This organism will grow on MacConkey agar, triphenyl tetrazolium chloride agar and β-hydroxybutyrate agar but not on cetrimide agar, Simmon's citrate agar or Salmonella-Shigella agar (Holmes et al., 1978). The general maintenance media recommended for flavobacteria are suitable for maintaining this genus.

3.4 Myroides spp.

This genus was previously known as *Flavobacterium odoratum* (Holmes et al., 1979; Vancanneyt et al., 1996). As a result of further 16S rRNA sequencing data, *Fa. odoratum* was excluded from the emended genus *Flavobacterium* (Gherna and Woese, 1992; Nakagawa and Yamasato, 1993; Vandamme et al., 1994b; Bernardet et al., 1996). Vancanneyt et al. (1996) described the new genus *Myroides* with *M. odoratus* as the type species together with a new species, *M. odoratimimus*.

Strains have been isolated from human clinical specimens and hospital environments (Holmes et al., 1979; Holmes et al., 1984b; Vancanneyt et al., 1996) as well as from marine fish where they can cause spoilage (odours and proteolysis) of fresh and frozen fish (Engelbrecht et al., 1996b). Growth is strictly aerobic at room temperature (18–22°C) and 37°C. No growth occurs at 5 and 42°C (Vancanneyt et al., 1996). The strains may also be grown at the optimum temperature in nutrient broth No. 2 (Oxoid) with stirring at 120 rpm to provide aeration (Owen and Holmes, 1978).

Isolation of the marine strains can commence in 0.1% tryptone water (Merck) followed by plating onto the surface of plate count agar containing 0.5% NaCl (Liston, 1980; Hobbs, 1983; Simmonds and Lamprecht, 1985; Engelbrecht et al., 1996b).

Good growth occurs on nutrient and MacConkey agar (Table 1; Vancanneyt et al., 1996; BCCM-LMG, 1998; ATCC, 2000). Other media that have also been used are: trypticase soy agar (Vancanneyt et al., 1996; ATCC, 2000), modified Shieh agar (Bernardet et al., 1996) and heart infusion medium with 5% rabbit blood (Dees et al., 1986).

When examining the proteolytic spoilage characteristics of fish isolates, standard methods caseinate agar provided more sensitive detection of proteolysis than 10% skim milk agar (Engelbrecht et al., 1996b).

3.5 Weeksella

This genus was formerly known as *Flavobacterium* CDC Group IIf (Tatum et al., 1974). Holmes et al. (1986a) described the new genus, *Weeksella,* consisting of only one species, *W. virosa*. Strains in this genus are frequently isolated from the female genital tract (Holmes et al., 1986a) and from the food environment (Botha et al., 1989; 1998a,b; Engelbrecht et al., 1996b). Studies by Botha et al. (1989; 1998a,b), however, suggested that other species of this genus may be present in the food environment. Engelbrecht et al. (1996b) found that these organisms produced stale or pungent odours in fresh or frozen fish.

W. virosa will grow as a strict aerobe at temperatures from 18 to 42°C but will not grow at 5°C (Holmes et al., 1986a). It will grow well on blood agar medium within 18 to 24 h (Tatum et al., 1974).

Weeksella spp. may be isolated and cultivated on any of the following media (Table 1): 5% horse blood agar (Holmes et al., 1986a), brain-heart infusion medium (Difco; ATCC, 2000), heart infusion medium with 5% rabbit blood (Dees et al., 1986), trypticase soy medium (BBL; BCCM-LMG, 1998; ATCC, 2000), plate count agar containing 0.5% NaCl (Engelbrecht et al., 1996a,b) and nutrient agar (Oxoid; Botha et al., 1989; 1998a,b). It will also grow on β-hydroxybutyrate media and on MacConkey agar (Holmes et al., 1986a), but not in nutrient broth with 6% sodium chloride, on Salmonella Shigella agar (SS; Oxoid CM99), cetrimide and Simmon's citrate media (Tatum et al., 1974; Holmes et al., 1986a) or penicillin agar (Jooste and Hugo, 1999).

No selective isolation medium is known for *Weeksella* spp. and freeze-drying is used for the long-term preservation of strains of this genus (BCCM-LMG, 1998; ATCC, 2000).

4. Differentiation of members of the Flavobacteriaceae

Phenotypic tests

The characteristics that are used to differentiate the food-related genera in the Flavobacteriaceae family are: pigment production; resistance to penicillin G; production of indole; catalase, DNase and urease activity; degradation of aesculin and gelatin; growth at 37°C and 42°C; growth on MacConkey and β-hydroxybutyrate agar; acid production from glucose and sucrose; G + C content; type of respiratory quinone and habitat (Bernardet et al., 1996; Jooste and Hugo, 1999).

Of the above-mentioned characteristics, pigment production is one of the most important. Medium composition, incubation temperature and incubation period may, however, influence this characteristic. The media, incubation times and temperatures will be the same as given under each genus discussed above, except where other media, incubation temperatures and times are recommended.

Pigment production

According to Holmes et al. (1984a) the degree of pigmentation may be more pronounced at lower temperatures (15–20°C) and daylight may be required for maximum pigmentation development. Casein, milk and starch agar may also enhance pigment development (Holmes et al., 1984a).

B. zoohelcum is non-pigmented on nutrient agar and does not produce a brown, diffusible pigment on tyrosine agar (Holmes et al., 1986b). *E. brevis* produces a light yellow pigment on nutrient agar but no pigmentation occurs on tyrosine agar (Holmes et al., 1978). The hue on nutrient agar does not change with variation in temperature (Holmes et al., 1984a).

Ch. balustinum produces a cadmium-yellow pigment on agar media with a deep orange pigment on 2% proteose media (Harrison, 1929). *Ch. gleum* colonies on nutrient agar are bright yellow and produce a water-soluble dark brown pigment on tyrosine agar (Holmes et al., 1984b) while *Ch. indologenes* colonies on any medium are deep orange (Yabuuchi et al., 1983).

E. brevis produces a light yellow pigment on nutrient agar but no pigment on tyrosine agar (Holmes et al., 1978). The hue on nutrient agar does not change with variation of medium or temperature (Holmes et al., 1984a).

Myroides colonies are yellow pigmented on trypticase soy agar (Vancanneyt et al., 1996) while *W. virosa* is non-pigmented on nutrient agar (Holmes et al., 1986a).

Media for the determination of fatty acid methyl esters

The determination of the fatty acid methyl ester (FAME) profiles is a useful tool in the differentiation of many of the genera and species belonging to the family Flavobacteriaceae. The preliminary cultivation of the organisms for FAME is, however, just as important especially when the genera are being compared with each other using a computer programme (e.g. Microbial Identification (MIDI) System software package, Microbial ID Inc., Newark, Del.). The media that have been used for the cultivation of the genera for FAME analysis by gas-liquid chromatography are presented in Table 2. From this it is clear that trypticase soy agar and heart infusion agar supplemented with 5% rabbit blood are most widely used for the determination of FAMEs.

In order to compare the results of fatty acid analysis and even the protein profiles of the genera in this family, Bernardet (personal communication) recommends the use of Difco marine 2216E agar for marine strains and modified Shieh medium (Song et al., 1988) for the other strains. Bernardet also recommends a further modification of the modified Shieh medium (i.e. 1 g/l yeast extract instead of 0.5 g/l).

Table 2
Cultivation media, incubation time and temperature of *Flavobacteriaceae* family members for the determination of fatty acid methyl esters.

Genus (and species)	Medium	Incubation Time	Incubation temperature	Reference
Bergeyella zoohelcum	Trypticase soy agar	48 h	30°C	Segers et al., 1993
Chryseobacterium	Trypticase soy agar	48 h	30°C	Segers et al., 1993
	Trypticase soy agar	24 h	28°C	Hugo et al., 1999
Ch. balustinum	Trypticase soy agar	48 h	28°C	Segers et al., 1993
Empedobacter brevis	Heart infusion agar + 5% rabbit blood	24 h	35°C	Dees et al., 1986
	Trypticase soy agar	24 h	28°C	Hugo et al., 1999
Myroides	Trypticase soy agar	24 h	28°C	Vancanneyt et al., 1996
	Heart infusion agar + 5% rabbit blood	24 h	35°C	Dees et al., 1986
Weeksella virosa	Heart infusion agar + 5% rabbit blood	24 h	35°C	Dees et al., 1986

5. Conclusions

Culture media enabling satisfactory cultivation of food-borne members of the Flavobacteriaceae are available as described in this chapter, but media for the selective isolation of most of the genera and species are lacking. The reason for this is that these genera share many of the same characteristics and require more or less the same substrates. Nevertheless, the yellow or orange pigmentation of the *Chryseobacterium, Empedobacter* and *Myroides* genera is a strongly differentiating characteristic and is always useful in detecting colonies of these genera on agar plates containing a mixed culture. The *Bergeyella* and *Weeksella* genera are not pigmented, but their penicillin sensitivity is very uncommon amongst Gram-negative rod-shaped bacteria, and can serve as a useful pointer in screening for these organisms.

References

American Type Culture Collection (2000) Catalogue of strains. http://www.atcc.org.

BCCM-LMG (1998) Belgian Co-ordinated Collections of Micro-organisms, Catalogue 1998. Universiteit Gent (RUG), Laboratorium voor Microbiologie.

Bergey, D.H., Harrison, F.C., Breed, R.S., Hammer, B.W. and Huntoon, F.M. (1923) Bergey's Manual of Determinative Bacteriology, 1st edition, Williams and Wilkins, Co., Baltimore, pp 97–117.

Bergey, D.H. and Breed, R.S. (1948) Genus III. *Flavobacterium* Bergey et al. In: Bergey's Manual of Determinative Bacteriology, 6th edition, edited by R.S. Breed, E.G.D. Murray, and A.P. Hitchens. Baillière, Tindall and Cox, London, pp 427–442.

Bernardet, J.-F., Segers, P., Vancanneyt, M., Berthe, F., Kersters, K. and Vandamme, P. (1996) Cutting a Gordian knot: emended classification and description of the genus *Flavobacteria*, emended description of the family Flavobacteriaceae, and proposal of *Flavobacterium hydatis* nom. nov. (basonym, *Cytophaga aquatilis* Strohl and Tait 1978). Int. J. Syst. Bacteriol. 46, 128–148.

Botha, W.C., Jooste, P.J. and Britz, T.J. (1989) The taxonomic relationship of certain environmental flavobacteria to the genus *Weeksella*. J. Appl. Bacteriol. 67, 551–559.

Botha, W.C., Jooste, P.J. and Hugo, C.J. (1998a) The incidence of *Weeksella*- and *Bergeyella*-like bacteria in the food environment. J. Appl. Microbiol. 84, 349–356.

Botha, W.C., Jooste, P.J. and Hugo, C.J. (1998b) Taxonomic interrelationship of *Weeksella*- and *Bergeyella*-like strains from dairy sources using electrophoretic protein profiles, DNA base composition and non-polar fatty acid composition. Food Microbiol. 15, 479–489.

Campbell, L.L. and Williams, O.B. (1951) A study of chitin-decomposing micro-organisms of marine origin. J. Gen. Microbiol. 5, 894–905.

Christensen, P.J. and Cook, F.D. (1972) The isolation and enumeration of cytophagas. Can. J. Microbiol. 18, 1933–1940.

Cowan, S.T. (1974) Cowan and Steel's Manual for the Identification of Medical Bacteria, 2nd edn., Cambridge University Press, Cambridge.

Dees, S.B., Moss, C.W., Hollis, D.G. and Weaver, R.E. (1986) Chemical characterization of *Flavobacterium odoratum, Flavobacterium breve*, and *Flavobacterium*-like groups IIe, IIh, and IIf. J. Clin. Microbiol. 23, 267–273.

Engelbrecht, K. (1992) Spoilage bacteria associated with Cape hake (*Merluccius capensis* & *Merluccius paradoxus*) and other edible South Atlantic species. M.Sc. thesis, Department of Microbiology and Biochemistry, University of the Orange Free State, Bloemfontein, South Africa.

Engelbrecht, K., Jooste, P.J. and Prior, B.A. (1996a) Quantitative and qualitative determination of the aerobic bacterial populations of Cape marine fish. S. Afr. J. Food Sci. Nutr. 8, 60–65.

Engelbrecht, K., Jooste, P.J. and Prior, B.A. (1996b) Spoilage characteristics of Gram-negative genera and species isolated from Cape marine fish. S. Af. J. Food Sci. Nutr. 8, 66–71.

Everton, J.R., Bean, P.G. and Bashford, T.E. (1968) Spoilage of canned milk products by flavobacteria. J. Food Technol. 3, 241–247.

Fischer, P.L., Jooste, P.J. and Novello, J.C. (1987) The seasonal distribution of psychrotrophic bacteria in Bloemfontein raw milk supplies. S. Af. J. Dairy Sci. 19, 73–76.

Fox, C.W., Chrisope, G.L. and Marshall, R.T. (1976) Incidence and identification of phospholipase C-producing bacteria in fresh and spoiled homogenized milk. J. Dairy Sci. 59, 1857–1864.

Gherna, R. and Woese, C.R. (1992) A partial phylogenetic analysis of the "flavobacter-bacteroides" phylum: basis for taxonomic restructuring. Syst. Appl. Microbiol. 15, 513–521.

Harrison, F.C. (1929) The discoloration of halibut. Can. J. Res. 1, 214–239.

Hayes, P.R. (1977) A taxonomic study of flavobacteria and related Gram negative yellow pigmented rods. J. Appl. Bacteriol. 43, 345–367.

Hobbs, G. (1983) Microbial spoilage of fish. In: Food Microbiology, Advances and Prospects, edited by T.A. Roberts and F.A. Skinner. Academic Press, London, pp 217–229.

Holmes, B. (1992) The genera *Flavobacterium, Sphingobacterium* and *Weeksella*. In: The Prokaryotes, Volume IV, 2nd edition, edited by A. Balows, H.G. Trüper, M. Dworkin, W. Harder, and K.-H. Schleifer. Springer Verlag, New York, pp 3620–3630.

Holmes, B., Snell, J.J.S. and Lapage, S.P. (1977) Revised description, from clinical isolates, of *Flavobacterium odoratum* Stutzer and Kwaschnina 1929, and designation of the neotype strain. Int. J. Syst. Bacteriol. 27, 330–336.

Holmes, B., Snell, J.J.S. and Lapage, S.P. (1978) Revised description, from clinical strains, of *Flavobacterium breve* (Lustig) Bergey et al. 1923 and proposal of the neotype strain. Int. J. Syst. Bacteriol. 28, 201–208.

Holmes, B., Snell, J.J.S. and Lapage, S.P. (1979) *Flavobacterium odoratum*: a species resistant to a wide range of antimicrobial agents. J. Clin. Pathol. 32, 73–77.

Holmes, B., Owen, R.J. and McMeekin, T.A. (1984a) Genus *Flavobacterium* Bergey, Harrison, Breed, Hammer and Huntoon 1923. In: Bergey's Manual of Systematic Bacteriology, Volume 1 edited by N.R. Krieg and J.G. Holt. Williams and Wilkins, Co., Baltimore, pp 353–361.

Holmes, B., Owen, R.J., Steigerwalt, A.G. and Brenner, D.J. (1984b) *Flavobacterium gleum*, a new species found in human clinical specimens. Int. J. Syst. Bacteriol. 34, 21–25.

Holmes, B., Steigerwalt, A.G., Weaver, R.E. and Brenner, D.J. (1986a) *Weeksella virosa* gen. nov., sp. nov. (formerly Group IIf), found in human clinical specimens. Syst. Appl. Microbiol. 8, 185–190.

Holmes, B., Steigerwalt, A.G., Weaver, R.E. and Brenner, D.J. (1986b) *Weeksella zoohelcum* sp. nov. (formerly Group IIj), from human clinical specimens. Syst. Appl. Microbiol. 8, 191–196.

Hugh, R. and Gilardi, G.L. (1974) *Pseudomonas*. In: Manual of Clinical Microbiology, 2nd edition, edited by E.-H. Lennette, E.H. Spaulding and J.P. Truant. American Society for Microbiology, Washington, D.C., pp 250–269.

Hugo, C.J. (1997) A taxonomic study of the genus *Chryseobacterium* from food and environmental sources. Ph.D. thesis. University of the Orange Free State, Bloemfontein, South Africa.

Hugo, C.J. and Jooste, P.J. (1997) Preliminary differentiation of food strains of *Chryseobacterium* and *Empedobacter* using multilocus enzyme electrophoresis. Food Microbiol. 14, 133–142.

Hugo, C.J., Jooste, P.J., Segers, P., Vancanneyt, M. and Kersters, K. (1999) A polyphasic taxonomic study of *Chryseobacterium* strains isolated from dairy sources. System. Appl. Microbiol. 22, 586–595.

Jooste, P.J. (1985) The taxonomy and significance of *Flavobacterium-Cytophaga* strains from dairy sources. Ph.D. study, Department of Microbiology, University of the Orange Free State, Bloemfontein, South Africa.

Jooste, P.J. and Britz, T.J. (1986) The significance of flavobacteria as proteolytic psychrotrophs in milk. Milchwissenschaft 41, 618–621.

Jooste, P.J., Britz, T.J. and De Haast, J. (1985) A numerical taxonomic study of *Flavobacterium-Cytophaga* strains from dairy sources. J. Appl. Bacteriol. 59, 311–323.

Jooste, P.J., Britz, T.J. and Lategan, P.M. (1986a) The prevalence and significance of *Flavobacterium* strains in commercial salted butter. Milchwissenschaft 41, 69–73.

Jooste, P.J., Britz, T.J. and Lategan, P.M. (1986b) Screening for the presence of *Flavobacterium* strains in dairy sources. S. Afr. J. Dairy Sci. 18, 45–50.

Jooste, P.J. and Hugo, C.J. (1999) The taxonomy, ecology and cultivation of bacterial genera belonging to the family *Flavobacteriaceae*. Int. J. Food Microbiol. 53, 81–94.

Kath, T. and Reichenbach, H. (1993) A study into the taxonomy of cellulose-degrading *Cytophaga* and *Sporocytophaga*. In: Advances in the Taxonomy and Significance of *Flavobacterium, Cytophaga* and Related Bacteria, edited by P.J. Jooste. University of the Orange Free State Press, Bloemfontein, South Africa, pp 171–181.

King, E.O. (1959) Studies on a group of previously unclassified bacteria associated with meningitis in infants. Am. J. Clin. Pathol. 31, 241–247.

Liston, J. (1980) Microbiology in fishery science. In: Advances in Fish Science and Technology, edited by J.J. Connell. Fishing News (Books).

McMeekin, T.A., Patterson, J.T. and Murray, J.G. (1971) An initial approach to the taxonomy of some Gram-negative yellow pigmented rods. J. Appl. Bacteriol. 34, 699–716.

Mudarris, M. and Austin, B. (1989) Systemic disease in turbot *Scophthalmus maximus* caused by a previous unrecognised *Cytophaga*-like bacterium. Dis. Aquat. Org. 6, 161–166.

Nakagawa, Y. and Yamasato, K. (1993) Phylogenetic diversity of the genus Cytophaga and revealed by 16S rRNA sequencing and menaquinone analysis. J. Gen. Microbiol. 139, 1155–1161.

Olsen, G.J., Woese, C.R. and Overbeek, R. (1994) The winds of (evolutionary) change: breathing new life into microbiology. J. Bacteriol. 176, 1–6.

Owen, R.J. and Holmes, B. (1978) Heterogeneity in the characteristics of deoxyribonucleic acid from *Flavobacterium odoratum*. FEMS Microbiol. Lett. 48, 41–46.

Owen, R.J. and Snell, J.J.S. (1973) Comparison of group IIf with *Flavobacterium* and *Moraxella*. A. van Leeuw. 39, 473–480.

Oyaizu, H. and Komagata, K. (1981) Chemotaxonomic and phenotypic characterization of the strains of species in the *Flavobacterium-Cytophaga* complex. J. Gen. Appl. Microbiol. 27, 57–107.

Paster, B.J., Ludwig, W., Weisburg, W.G., Stackebrandt, E., Hespell, R.B., Hahn, C.M., Reichenbach, H., Stetter, K.O. and Woese, C.R. (1985) A phylogenetic grouping of the bacteroides, cytophagas and certain flavobacteria. Syst. Appl. Microbiol. 6, 34–42.

Reichenbach, H. (1989) Order I. Cytophagales. In: Bergey's Manual of Determinative Bacteriology, Volume 3, edited by J.T. Staley, M.P. Bryant, N. Pfennig and J.G. Holt. Williams and Wilkins, Co., Baltimore, pp 2010–2082.

Reina, J. and Borrell, N. (1992) Leg abscess caused by *Weeksella zoohelcum* following a dog bite. Clin. Infect. Dis. 14(5): 1162–1163.

Rubin, S.J., Granato, P.A. and Wasilauskas, B.L. (1985) Glucose-nonfermenting Gram-negative bacteria. In: Manual of Clinical Microbiology, 4th edition, edited by E.H. Lennette, A. Balows, W.J. Hausler Jr., and H.J. Shadony. Washington DC, American Society for Microbiology, pp 330–349.

Segers, P., Mannheim, W., Vancanneyt, M., De Brandt, K., Hinz, K.-H., Kersters, K. and Vandamme, P. (1993) *Riemerella anatipestifer* gen. nov., comb. nov., the causative agent of septicemia anserum exsudativa, and its phylogenetic affiliation within the *Flavobacterium-Cytophaga* rRNA homology group. Int. J. Syst. Bacteriol. 43, 768–776.

Shewan, J.M. and McMeekin, T.A. (1983) Taxonomy and ecology of *Flavobacterium* and related genera. Ann. Rev. Microbiol. 37, 233–252.

Simmonds, C.K. and Lamprecht, E.C. (1985) Microbiology of frozen fish and related products. In: Microbiology of Frozen Foods, edited by R.K. Robinson. Elsevier Applied Science Publishers, London, pp 169–208.

Song, Y.L., Fryer, J.L. and Rohovec, J.S. (1988) Comparison of six media for the cultivation of *Flexibacter columnaris*. Fish Pathol. 23, 91–94.

Tatum, H.W., Ewing, W.H. and Weaver, R.E. (1974) Miscellaneous Gram-negative nacteria. In: Manual of Clinical Microbiology, 2nd edition, edited by E.H. Lenette, E.H. Spaulding, and J.P. Truant. Washington D.C., American Society for Microbiology, pp 270–294.

Vancanneyt, M., Segers, P., Torck, U., Hoste, B., Bernardet, J.-F., Vandamme, P. and Kersters, K. (1996) Reclassification of *Flavobacterium odoratum* (Stutzer 1929) strains to a new genus, *Myroides,* as *Myroides odoratus* comb. nov. and *Myroides odoratimimus* sp. nov. Int. J. Syst. Bacteriol. 46, 926–932.

Vandamme, P., Bernardet, J.-F., Segers, P., Kersters, K. and Holmes, B. (1994a) New perspectives in the classification of the flavobacteria: description of *Chryseobacterium* gen. nov., *Bergeyella* gen. nov., and *Empedobacter* nom. rev. Int. J. Syst. Bacteriol. 44, 827–831.

Vandamme, P., Segers, P., Vancanneyt, M., Van Hove, K., Mutters, R., Hommez, J., Dewhirst, F., Paster, B., Kersters, K., Falsen, E., Devriese, L.A., Bisgaard, M., Hinz, K.-H. and Mannheim, W. (1994b) *Ornithobacterium rhinotracheale* gen. nov., sp. nov., isolated from the avian respiratory tract. Int. J. Syst. Bacteriol. 44, 24–37.

Weeks, O.B. (1955) *Flavobacterium aquatile* (Frankland and Frankland) Bergey et al., type species of the genus *Flavobacterium*. J. Bacteriol. 69, 649–658.

Welthagen, J.J. (1991) Evaluering van 'n verkorte diagnostiese skema vir die roetine identifisering van psigrotrofe flavobakterieë uit roumelk. M.Sc. (Agric) thesis. University of the Orange Free State, Bloemfontein, South Africa.

Welthagen, J.J. and Jooste, P.J. (1992) Isolation and characterization of pigmented psychrotrophic bacteria from refrigerated raw milk. S. Afr. J. Dairy Technol. 24, 47–52.

Woese, C.R., Mandelco, L., Yang, D., Gherna, R. and Madigan, M.T. (1990) The case for relationship of the flavobacteria and their relatives to the green sulfur bacteria. Syst. Appl. Microbiol. 13, 258–262.

Woese, C.R., Stackebrandt, E., Macke, T.J. and Fox, G.E. (1985) A phylogenetic definition of the major eubacterial taxa. Syst. Appl. Microbiol. 6, 143–151.

Wolochow, H., Thornton, H.R. and Hood, E.G. (1942) Studies on surface taint butter. VI. Other bacterial species as casual agents. *Flavobacterium maloloris* (n. sp.). Sci. Agric. 22, 637–644.

Yabuuchi, E., Kaneko, T., Yano, I., Moss, C.W. and Miyoshi, N. (1983) *Sphingobacterium* gen. nov., *Sphingobacterium spiritivorum* comb. nov., *Sphingobacterium multivorum* comb. nov., *Sphingobacterium mizutae* sp. nov., and *Flavobacterium indologenes* sp. nov.: Glucose-nonfermenting Gram-negative rods in CDC Groups IIk-2 and IIb. Int. J. Syst. Bacteriol. 33, 580–598.

Yabuuchi, E., Hashimoto, Y., Ezaki, T., Ido, Y. and Takeuchi, N. (1990) Genotypic and phenotypic differentiation of *Flavobacterium indologenes* Yabuuchi et al 1983 from *Flavobacterium gleum* Holmes et al 1984. Microbiol. Immunol. 34, 73–76.

Chapter 22

Media for detecting and enumerating yeasts and moulds

Larry R. Beuchat

Center for Food Safety, Department of Food Science and Technology, University of Georgia, 1109 Experiment Street, Griffin, GA 30223-1797, USA

Dilution plating techniques are designed to determine populations of viable fungal propagules per unit weight or volume of food. Direct plating techniques, on the other hand, are designed to assess the internal mycoflora of individual pieces of foods, e.g., seeds or dried fruits, and results are expressed as a percentage of infected pieces. Both techniques are used by industry and regulatory agencies to monitor fungal contamination at various stages of food handling, storing, processing and marketing. Peptone (0.1%) water is commonly used as a diluent for samples to be homogenized, pummelled or blended. Buffered diluents containing up to 30% glycerol, 40% glucose, or 60% sucrose are recommended for enumerating xerophiles. No one medium is satisfactory for detection or enumeration of all yeasts and moulds in all foods. Antibiotic-supplemented media are superior to acidified media for enumeration of yeasts and moulds. Dichloran 18% glycerol agar performs well for enumerating moderately xerophilic yeasts and moulds. Fastidious xerophiles require media containing high concentrations of sugars and/or sodium chloride. Media have been formulated to detect potentially aflatoxigenic aspergilli and mycotoxigenic strains of penicillia, fusaria and other moulds but increased selectivity and specificity of media for detecting mycotoxigenic moulds are needed. Heat-resistant mould ascospores often require heat treatment prior to plating in order to activate the germination process. The spread-plate technique is strongly preferred over the pour-plate technique for enumerating yeasts and moulds. The recommended incubation temperature is 25°C, but incubation time between plating and counting colonies ranges from 5 days for determination of general populations of mycoflora to 4 weeks or more for fastidious xerophiles. There is a need for new and improved media for selectively isolating various groups, genera, species and/or strains of fungi capable of growing only under specific environmental conditions, e.g., low a_w, low pH or, in the case of sublethally injured cells, under conditions which facilitate resuscitation. Improved media are needed which accurately detect moulds producing specific mycotoxins in a wide range of food types.

Introduction

Fungi, i.e., yeasts and moulds, are distributed widely in decayed plant materials, soils, water and air. Consequently, unprocessed materials of both plant and animal origin are contaminated with fungi at the time they reach the food manufacturer. Processing schemes can either render the finished food product free of fungi or merely reduce populations. Given enough time, survivors may grow and eventually spoil the product. The detection and enumeration of viable yeasts and moulds in unprocessed and processed foods is an integral part of total quality management programs, and can be used to monitor the effectiveness of sanitation practices at each step during post-harvest and post-slaughter handling, processing and distribution of foods.

Dilution plating techniques are designed to determine populations of viable fungal propagules per unit weight or volume of food. Direct plating techniques, on the other hand, are designed to assess the internal mycoflora of individual pieces of foods, e.g., seeds or dried fruits, and results are expressed as a percentage of infected pieces. The direct plating technique is qualitative, then, rather than quantitative. Both techniques are used by industry as well as regulatory agencies to monitor relative levels of fungal contamination at various stages of food handling, storing, processing and marketing. Procedures for analysis of any given product may differ from laboratory to laboratory. Certainly, procedures differ for analyzing various types of foods. There are reviews and descriptions of dilution and direct plating procedures in the literature (Corry, 1982a; Jarvis et al., 1983; Jarvis and Williams, 1987; Beuchat and Hocking, 1990; Beuchat, 1992; 1993; 1998; Deak and Beuchat, 1996; Beuchat and Cousin, 2001) and the reader is referred to these publications. For detailed descriptions of common foodborne fungi, see Pitt and Hocking (1997) and Samson et al. (1985). The following text provides essentially only an introduction to procedures for determining the mycological quality of foods.

Sampling procedure

Procedures for selecting samples of foods for mycological analyses differ depending upon the consistency or physical state of the food. Replicate samples representative of the lot to be examined must be obtained. The size of the sample for dilution plating may range from 5 to 250 g (ml). Samples selected for direct plating should consist of at least 100 pieces, six to ten pieces per 9-cm plate. All foods, whether solid or liquid, should be thoroughly mixed before samples are taken. Dry foods such as nuts and cereal grains are usually ground to a powder if they are to be analyzed using the dilution plating technique. Sampling procedures are outlined by the International Commission on Microbiological Specifications for Foods (ICMSF, 1986).

Preparation of samples

Preparation of samples for dilution plating may consist of manually shaking with a known volume of diluent, direct blending with a diluent in a blender or pummelling with a diluent in a Stomacher. In some instances, rinsing of the entire sample with a diluent may be necessary. Jarvis et al. (1983) reported that higher populations are usually obtained by stomaching or blending than by shaking techniques. Hastings et al. (1986) observed that there were no significant differences between mould populations detected in samples prepared by blending or stomaching. Methods used to homogenize samples prior to secondary dilution are described by Jarvis and Williams (1987).

Soaking foods in diluent prior to mixing and diluting can have at least three effects (Seiler, 1986). First, soaking may allow resuscitation of sublethally damaged cells from dried, intermediate moisture or acidic foods. Soaking may also facilitate better release of cells which are present within tissues. Thirdly, with hard or sharp materials such as cereals and nuts, the softening effect of soaking will help prevent damage to stomacher bags used for preparing primary dilutions. After homogenizing, samples should not be allowed to stand for more than a few minutes before removing a portion for subsequent dilution and plating, since fungal propagules may settle to the bottom of the container, resulting in under- or overestimation of populations, depending upon the location in the mixture from which the portion is withdrawn.

To assess the internal mycoflora of cereal grains, nuts and dried fruits and vegetables by direct plating, various chemicals are used for surface disinfection. These include 3% hydrogen peroxide, 2% potassium permanganate, 75% ethanol, 0.001% mercuric chloride and 0.35% sodium or calcium hypochlorite (Booth, 1971a). Contact time ranges from 1 to 5 min, followed by rinsing with sterile distilled water. Andrews (1986) suggested that most food particles should be effectively surface disinfected by immersion in a 0.35% (w/v) sodium hypochlorite solution for 2 min. In a later study (Andrews, 1996), an ethanol pre-rinse of seeds and nuts prior to chlorine treatment was recommended to optimize surface disinfection. Rinsing with sterile water after treatment is not essential, provided samples are adequately drained after disinfection.

Diluent

Peptone (0.1%) water and 0.05–0.1 M potassium phosphate buffer (pH 7.0) are among the most commonly used diluents for yeast and mould enumeration. However, other diluents may be employed, depending upon the type of food under investigation. For example, it is important to use a diluent containing a sufficient amount of solute to minimize osmotic shock to fungal cells in high-sugar or high-salt foods when serial dilutions are made prior to plating. Buffered diluents containing up to 30% glycerol, 40% glucose, or 60% sucrose have been used for this purpose. Death of halophilic fungi is minimized by incorporating 6–10% sodium chloride into diluents.

Surface active agents may be added to diluents to reduce clumping of mould spores and conidia. For a discussion of diluents suitable for various foods, see Beuchat and Hocking (1990), Hernadez and Beuchat (1994), Abdul-Raouf et al. (1995), Deak and Beuchat (1996), Mian et al. (1997), and Samson et al. (1985).

Culture media

General purpose media

Because foods differ widely in composition and the environments to which they have been exposed, the types and populations of mycoflora also differ. Pitt (1986) characterized the ideal enumeration medium as having the following attributes. It should suppress bacterial growth completely, without affecting growth of food fungi. It should be nutritionally adequate and support the growth of relatively fastidious fungi. The radial growth of mould colonies should be constrained but spore germination should not be inhibited.

Unfortunately, no one medium is satisfactory for detection or enumeration of all yeasts and moulds in all foods. Traditionally, acidified potato dextrose agar (PDA) has been used for enumeration, but this medium is not an exceptionally good nutrient source and it may inhibit resuscitation of injured cells due its low pH (3.5). Antibiotic supplemented media such as oxytetracycline glucose extract (OGY) agar (Mossel et al., 1962, 1970), rose bengal chlortetracycline (RBC) agar (Jarvis, 1973) and dichloran rose bengal chloramphenicol (DRBC) agar (King et al., 1979) are superior to acidified media for both dilution and direct plating techniques. Monographs have been published for all these media in the ICFMH Pharmacopoeia (see this volume). Antibiotic supplemented media are less inhibitory to injured yeasts and moulds, more effective in inhibiting bacterial growth and less likely to cause precipitation of food particles because of their higher pH (5–6). Basal media to which chloramphenicol, chlortetracycline, oxytetracycline, gentamicin, kanamycin or streptomycin is added at concentrations up to 100 μg/ml include OGY, DRBC, plate count agar, mycophil agar, malt agar and tryptone glucose yeast extract agar. The effectiveness of most antibiotics is diminished at alkaline pH, so if highly alkaline foods are being analyzed, adjustment of the medium pH to less than 8.0 may be necessary to minimize bacterial growth. DRBC agar has replaced traditional mycological media as a recommended medium for enumerating yeasts and moulds in foods and beverages (Taniwaki et al., 1999; Beuchat and Cousin, 2001).

Rose bengal (Jarvis, 1973; King et al., 1979) and dichloran (2,6-dichloran-4-nitroaniline) have been successfully used to control growth of spreading moulds. Caution should be taken to avoid exposure of rose bengal supplemented media to light, since cytotoxic breakdown products (Banks et al., 1985; Gogna et al., 1992; Chilvers et al., 1999) may result in underestimation of mycoflora in samples.

Bartschi et al. (1991) developed a selective medium for enumerating *Mucor* species. The medium supports the growth and reproduction of *Mucor* but inhibits the development of a wide range of hyphomycetes. Various dyes have been used in media to control the growth of spreading moulds (Bragulat et al., 1995). Twomey and Mackie (1985), on the other hand, developed a medium to support rapid mould growth. The medium, apple pulp gelatin, contains an optical brightener.

Two techniques have been developed for general enumeration of yeasts and moulds which are a clear departure from traditional methods. The Hydrophobic Grid Membrane Filter (HGMF) technique, using an automated ISO-GRID™ sample processor counting system, was developed by Brodsky et al. (1982). The HGMF is essentially a membrane base upon which is applied a hydrophobic grid. The food sample is passed through the filter before transferring it to the surface of YM-11 agar (Entis and Lerner, 1996). Plates are incubated 48±4 h at 25°C before colonies are counted.

A culture film (Petrifilm™ YM) method has been developed for enumerating yeasts and moulds in foods. A dry selective indicator medium on a special support film is rehydrated upon application of 1 ml of diluted sample. The sample is then covered with a second film to prevent drying during incubation at 25°C for 4 or 5 days. This system has been demonstrated to perform well for enumerating yeasts and moulds in a wide range of foods (Beuchat et al., 1990, 1991).

Xerophilic fungi

While many yeasts and moulds capable of growing at low a_w can grow at high a_w, some species require low a_w in recovery media to enhance growth and colony development. Reduction of a_w is achieved by supplementing basal media with sodium chloride or solutes such as glycerol, glucose, fructose or sucrose. Dichloran 18% glycerol (DG18) agar (Hocking and Pitt, 1980; Baird et al., 1987), glucose yeast extract sucrose agar, yeast extract agar, malt salt agar (King et al., 1986), 25% glycerol nitrate agar (Pitt, 1979) and malt yeast 5% salt 12% glucose agar (Pitt and Hocking, 1985) are but a few of the mycological media formulated to select for fungi capable of growing at low a_w. Others are described by Beuchat and Hocking (1990). Czapek casein 50% glucose agar can be used to distinguish differences among various species of *Chrysosporium* (Kinderlerer, 1995).

DG18 agar (a_w 0.95) was developed for enumeration of moderately xerophilic moulds and osmophilic yeasts in products such as grains, flours, nuts and spices. The medium supports growth of the *Aspergillus restrictus* series, *Wallemia sebi, Zygosaccharomyces rouxii, Debaryomyces hansenii* and many penicillia and other aspergilli (Beuchat and Hocking, 1990). Unfortunately, some *Eurotium* species grow too rapidly on DG18 agar, obscuring the growth of other xerophiles. The use of surfactants and fungicides to control spreading of *Eurotium* colonies on DG18 agar has been recommended (Beuchat and Tapia de Daza, 1992; Tapia de Daza and Beuchat, 1992). A collaborative study has shown variability in commercial sources of DG18 agar to control radial growth of moulds (Frandberg and Olsen, 1999).

Differences exist in tolerance of xerophilic moulds to various solutes and hydrogen ion concentration. In an investigation of germination and growth of six xerophilic moulds, Pitt and Hocking (1977) concluded that a universal isolation medium for enumeration could be based on glycerol or glucose/fructose but not on sodium chloride as the a_w-limiting solute.

Apart from *Xeromyces bisporus, Wallemia sebi, Chrysosporium* species and some *Eurotium* and *Scopulariopsis* species, the vast majority of moulds are not able to grow in the a_w range of 0.61–0.70. Likewise, only a few species of yeasts will grow in this a_w range. Those that can grow have been described as osmophilic and can cause spoilage of high-sugar products. Particularly troublesome is *Zygosaccharomyces rouxii*. Raw sugar cane, maple, chocolate and fruit syrups, honey, confectionery products, jams and jellies, fruit concentrates and dried fruits are most vulnerable to degradation by *Z. rouxii*. Spoilage of these products may be due to growth of *Z. bailii*, but this yeast is not particularly tolerant to low a_w and is more likely to be implicated in spoilage of acidic foods such as salad dressings, mayonnaise and pickles with substantially higher a_w.

As with xerophilic moulds, plating media with reduced a_w are required for recovering osmophilic yeasts from foods. Potato dextrose agar supplemented to contain 60%

sucrose (Restaino et al., 1985), a yeast extract medium containing 50% glucose (Jermini et al., 1987) and a wide range of other sugar-supplemented media (Scarr, 1959; King et al., 1986; Beuchat and Hocking, 1990; Deak and Beuchat, 1996) are suitable for isolating and enumerating osmophilic yeasts. Often, xerophilic moulds will also develop on these media. Tryptone 10% glucose yeast extract (TGY) agar is most suitable for enumerating *Z. rouxii* in a wide range of reduced-a_w foods (Andrews et al., 1997; Beuchat et al., 1998). Incubation of plates at 30°C for 3 days has been recommended (Beuchat, 1999). Braendlin (1996) recommended the use of DG18 agar for enumerating xerophilic yeasts in the presence of xerophilic moulds.

Heat-resistant moulds

Spoilage of thermally processed fruits and fruit products by heat-resistant strains of *Byssochlamys fulva, B. nivea, Neosartorya fischeri, Talaromyces flavus, Talaromyces bacillisporus* and *Eupenicillium brifeldianum* is an endemic problem (Beuchat and Rice, 1979; Beuchat and Pitt, 2001). Ascospores of these moulds show high heat resistance, in some instances comparable to bacterial spores. Because of their low incidence in fruit products (usually less than 10 per 100 g or ml), relatively large samples must be analyzed. Centrifugation may be used to concentrate ascospores in liquid fruit products, the force and time necessary being influenced by volume, viscosity and specific gravity of the sample.

Heat-resistant ascospores may require heat activation before germination and growth will occur (Splittstoesser et al., 1972; Katan, 1985; Beuchat, 1987). Samples are heated at 75–85°C for 1 h before 50-ml quantities are combined with 100 ml of 1.5× strength potato dextrose agar in 150-mm Petri plates (Beuchat and Pitt, 2001). Plates are incubated at 30°C for up to 30 days before colonies are counted. Most viable ascospores will germinate and form visible colonies within 10 days; however, heat-injured and other debilitated ascospores may require additional time to form colonies. An alternative method which avoids error due to aerial contamination involves heating samples in flat-sided bottles followed by incubation on their sides. Colonies develop on the surface of the fruit product. This system is suitable for pulps and homogenates.

Mycotoxigenic moulds

Several media have been formulated for use in recovering specific genera or groups of yeasts and moulds based on metabolic activities. Among these are media suitable for enumeration of moulds capable of producing mycotoxins. These media are of value to any quality assurance program concerned with mycological quality of foods from a public health standpoint. The presence of mycotoxin producers in a food does not necessarily mean that mycotoxins are present. Conversely, the absence of mycotoxin producers is not evidence that mycotoxins are not present in the food, since growth followed by death of mycotoxigenic moulds may have occurred at some point prior to analysis. Nevertheless, media formulated to select for mycotoxin producers are of great interest to food mycologists and considerable effort is being expended in several laboratories to develop these media for routine use.

Detection of *Aspergillus flavus* and *Aspergillus parasiticus* in foods indicates the possibility of aflatoxin contamination. In connection with their studies on aflatoxin contamination of grains, Bothast and Fennell (1974) developed a diagnostic medium which could be used to rapidly differentiate between members of the *A. flavus* group and other common storage fungi. The medium contains ferric citrate which promotes the formation of a persistent bright yellow-orange reverse pigmentation of colonies. The medium (*Aspergillus* differential medium, ADM) also effectively distinguishes between members of the *A. flavus* group and other *Aspergillus* species of interest in medical mycology (Salkin and Gordon, 1975). Noting that supplementation or rose bengal streptomycin agar with botran (2,6-dichlo-4-nitroaniline) facilitated the isolation and enumeration of *A. flavus* from peanuts (Bell and Crawford, 1967), ADM was modified by Hamsa and Ayres (1977) to yield a medium with greater selectivity.

Hara et al. (1974) developed a medium to simplify the screening of large numbers of *A. flavus* isolates for aflatoxin production. Detection of aflatoxin-positive strains utilized ultraviolet-induced fluorescence of aflatoxin produced in a Czapek's solution agar containing corn steep liquor, $HgCl_2$ and $(NH_4)H_2PO_4$ instead of $NaNO_3$. Pitt et al. (1983) developed a selective medium for enumerating *A. flavus* and *A. parasiticus* based on the medium of Bothast and Fennell (1974). Results can be obtained after incubating plates 42 h at 30°C, making the medium suitable for use in industrial quality control laboratories. Results were reproducible and comparable with those obtained using standard fungal enumeration media incubated for much longer periods. A very low percentage of false positives or negatives was reported. *Aspergillus flavus* and *parasiticus* agar (AFPA) is included in this volume.

Coconut cream agar (CCA) was developed for detecting aflatoxin production by *A. flavus* and related species (Dyer and McCammon, 1994). The medium consists of 50% coconut cream and 1.5% agar. Fluorescence colouring of colonies can be used to differentiate *A. flavus* from *A. parasiticus* and *A. nomius*. In addition, conidial colour of *A. flavus* and *A. nomius* is distinct from that of *A. parasiticus*. Coconut cream agar can be used to screen *Aspergillus carbonarius* and *Aspergillus niger* for ochratoxin A production (Heenan et al., 1998). Fluorescence of colony reverses under long wave UV light is used to distinguished ochratoxin producers from non-producers.

A simple method for screening aflatoxin-producing *A. flavus* and *A. parasiticus* using ultraviolet photography has been developed (Yabe et al., 1987). In ultraviolet photographs of colonies, aflatoxin producers appear as grey or black colonies, whereas non-producers appear as white colonies. The technique may be useful as a simple, safe and rapid method of screening aflatoxin-producing moulds.

A selective medium was developed by Frisvad (1983, 1986) to screen for *Penicillium viridicatum* and *Penicillium verruculosum* as an aid in the examination of stored cereal products for toxic metabolites. The medium, pentachloronitrobenzene rose bengal yeast extract sucrose (PRYES) agar, contains 15% sucrose, chloramphenicol and chlortetracycline (50 mg/1), as selective agents. Members of the Mucorales are completely inhibited, allowing important storage moulds to grow. Ochratoxin A and citrinin producers in the *P. viridicatum* group II appear as violet brown reverse on PRYES agar, whereas producers of xanthomegnin and viomellein (*P. viridicatum* group I and *P. aurantiogriseum*) are indicated by their yellow reverse and obverse colours. The medium appears to be quite suitable for screening for nephrotoxin-producing fungi in cereals.

Dichloran rose bengal yeast extract sucrose (DRYES) agar was originally formulated as a screening medium to detect *Penicillium verrucosum*, but may be used for a more general determination of toxigenic *Penicillium* and *Aspergillus* species (Frisvad et al., 1990a). They recommend the use of both DG18 agar and DRYES agar if populations and identity of moulds and their mycotoxin production is important in the examination of dried products.

Three media have been developed for selective isolation and direct identification of various groups of moulds associated with cereal products, meat and cheese (Frisvad et al., 1990b). Acetic acid dichloran yeast extract sucrose (ADYES) agar is useful for selective and indicative isolation of moulds spoiling acid-preserved bread. Roquefortine C and PR-toxin produced by *Penicillium roqueforti* var. *roqueforti* on ADYES agar can be detected by an agar plug method. Dichloran creatine (DC) agar is an effective screening medium for penicillia associated with meat, cheese, nuts and other lipid and/or protein foods. The third medium, dichloran yeast extract sucrose (DYES) agar, was developed for qualitative determination of toxigenic penicillia and aspergilli.

Creatine sucrose agar, containing creatine as a sole nitrogen source and bromocresol purple as a pH indicator, is reported to be a good medium to differentiate subgenera of penicillia and to subdivide taxa and subgroups based on mycotoxin production into approximately two equal groups (Frisvad, 1985). Pitt (1993) modified creatine sucrose agar by varying sucrose and creatine concentrations and pH. Termed neutral creatine sucrose agar, the medium produces eight different reactions among twenty species in *Penicillium* subgenus *Penicillium*, making it a useful taxonomic medium.

Andrews and Pitt (1986) developed dichloran chloramphenicol peptone agar (DCPA), a selective medium for isolating *Fusarium* species and some dematiaceous hyphomycetes (e.g., *Alternaria* species) from cereals. The medium was shown to select against species of *Aspergillus, Penicillium, Cladosporium* and mucoraceous fungi. *Fusarium* species produced well-formed colonies with good conidial production, permitting rapid identification. The medium can also be used for the identification of *Fusarium* species (Hocking and Andrews, 1987). Conner (1990) reported that DCPA and DCPA supplemented with 0.5 μg/ml crystal violet were effective in suppressing *Aspergillus* and *Penicillium* while allowing selective growth of *Fusarium* species.

The possibility of using the growth response of *Fusarium* species on tannin sucrose agar as an aid to identification was investigated by Thrane (1986). Of the eleven *Fusarium* species investigated, eight were able to grow on tannin sucrose agar. The difference in growth response proved to be an additional character usable for identification of species. A selective medium for *Fusarium* species was developed using Czapek-Dox agar containing iprodione [3-(3-5-dichlorophenyl)-*N*-(1-methyl-ethyl)-2,4-dioxo-1-imidazolidine-carboxamide] (3 mg/l) and dichloran (2 mg/l) (Abildgren et al., 1987). This medium (CZID agar) is selective against numerous species of *Alternaria, Epicoccum, Penicillium* and mucoraceous fungi. Fusaria produce large, easily recognizable colonies on CZID agar, facilitating isolation and subculture. Thrane (1996) compared CZID agar, DCPA and pentachloronitrobenzene peptone agar for their ability to enumerate *Fusarium* species. The three media performed equally in supporting the numbers of colonies developed but CZID agar was recommended because species could be differentiated by pigmentation of colonies.

Andrews (1990) evaluated media for differentiation of potentially toxigenic *Alternaria alternata* from the *Alternaria* state of *Pleospora infectoria*. The two species can be differentiated on dichloran chloramphenicol peptone agar (DCPA) but are more readily differentiated on dichloran chloramphenicol malt agar (DCMA).

Selective media for yeasts

Relatively few media are available specifically for enumerating or at least facilitating the growth of yeasts at the expense of moulds and bacteria. However, since yeasts may be the predominant microflora in some foods, e.g., fruit juice concentrates, special media for enumeration of yeasts are desirable. Davenport (1980) outlined a guide to media and methods for studying yeasts in foods.

Oxytetracycline glucose yeast extract (OGYE) agar (Mossel et al., 1970; Baird et al., 1987) has been widely used for many years, especially in Europe, as a general medium for enumerating yeasts. The medium loses its bacteriostatic effect if incubated at temperatures greater than 25°C. Moulds will also develop colonies on OGYE agar. For selective recovery of psychrotrophic yeasts from chilled proteinaceous foods, oxytetracycline and gentamicin seem to be the bacteriostatic combination of choice (Dijkmann et al., 1979). Tryptone glucose yeast extract (TGY) agar supplemented with antibiotics can also be used successfully for products containing high populations of yeasts.

Acid-resistant yeasts, particularly *Zygosaccharomyces bailii* and, to a lesser extent, *Z. rouxii*, cause spoilage of foods containing benzoic and sorbic acids. Some strains of *Schizosaccharomyces pombé* and *Pichia membranaefaciens* also exhibit unusual tolerance to acids. These yeasts can be enumerated with some success on malt extract agar supplemented to contain 0.5% acetic acid (Samson et al., 1992). Other media used to enumerate acid-resistant yeasts are acidified (0.5% acetic acid) TGY agar (Deak and Beuchat, 1996; Hocking, 1996) and *Zygosaccharomyces bailii* agar (Erickson, 1993). Modifications of TGY agar have enhanced its performance (Makdesi and Beuchat, 1996a; 1996b; 1996c); however, a highly selective medium for enumerating acid-resistant yeasts has yet to be developed.

Media have also been developed for enumerating yeasts in specific food products. For example, Kish et al. (1983) formulated a selective medium for wine yeasts. The medium, containing 150 mg/l bisulphite and 12% (by volume) ethanol, was suitable for the enumeration of wine yeasts when present at low populations in the natural microflora during early stages of grape juice fermentation. In the brewing industry, lysine agar (Walters and Thiselton, 1953; Lin, 1975), Schwarz differential medium, and other selective media (Lin, 1973) are used to detect wild yeasts. Heard and Fleet (1986) evaluated selective media for enumeration of yeasts during wine fermentation. Colonies of *Saccharomyces cerevisiae* dominated on malt extract agar and sometimes masked the presence of other genera. Lysine agar suppressed the growth of *S. cerevisiae* and enabled the enumeration of non-*Saccharomyces* species such as *Kloeckera apiculata, Candida stellata* and *Saccharomycodes ludwigii*. Growth of non-*Saccharomyces* yeasts on ethanol sulphite agar was variable.

Molybdate agar fortified with 0.125% calcium propionate was successfully used by Rale and Vakil (1984) to selectively isolate and identify yeasts from tropical fruits. This medium requires a longer incubation period for full expression of the natural pigmentation of yeasts. A yoghurt whey medium developed for enumerating yeasts in foods is reported to compare favorably with acidified PDA but not as well as antibiotic-supplemented plate count agar (Yamani, 1993). A differential medium for isolating *Kluyveromyces marxianus* and *Kluyveromyces lactis* from dairy products (Valderrama et al., 1999) has been described. Carreira and Loureiro (1998) developed a differential medium to detect *Yarrowia lipolytica* within 24 h.

Plating technique

The spread-plate technique is strongly preferred to the pour-plate technique for enumeration of yeasts and moulds in foods using dilution plating. Spread plating avoids any risk of thermal inactivation of fungal propagules which may be associated with the pour-plate technique and facilitates maximum exposure of cells to atmospheric oxygen, thus allowing sporulation to proceed unencumbered in those instances when identification of mycoflora is desired.

Media should be prepared and poured (ca. 20 ml per plate) 16–40 h in advance of use to facilitate 'drying' of the surface. Generally, 0.1 ml of appropriately diluted sample is deposited in duplicate or triplicate on the surface of media and then spread evenly over the surface using a sterile bent glass rod. Rods should not exceed 2 mm in diameter in order to minimize the amount of sample adhering to them at the end of the spreading procedure. Larger samples, e.g., up to 0.33 ml of undiluted of sample, can be spread on each plate to facilitate enumeration of low populations of yeasts and moulds.

Incubation

Media used for general enumeration of yeasts and moulds should be incubated at 25°C for 4 or, preferably, 5 days before colonies are counted. Higher temperatures, e.g., 40–45°C, must be used for thermophiles, whereas incubation at 5–15°C is needed for psychrotrophic fungi. The incubation time necessary before counting colonies of specific genera, species or groups of fungi varies. For example, accurate enumeration of moderate xerophiles on DG18 agar may require incubation for 6–8 days and mycelial development by fastidious xerophiles may require 4 weeks or longer.

Petri plates should be incubated in an upright position and should not be disturbed until colonies are ready to count. Movement of plates can result in release of mould conidia or spores and subsequent development of satellite colonies which would give an overestimation of population in the test sample. If spreading moulds are a problem, one may be forced to count colonies after 3 days and again after 5 days of incubation.

No attempt should be made to select plates containing 30–300 colonies for counting. Rather, select dilutions giving 10–100 colonies on 9-cm plates or 20–200 colonies on 14-cm plates. Several colonies should be picked from countable plates periodically and examined microscopically by wet mount to ensure that bacteria are not growing. Development of bacterial colonies is more likely to occur on acidified media than on antibiotic-supplemented media.

Non-linearity of counts from dilution plating often occurs, i.e., 10-fold dilutions of samples may not result in 10-fold reductions in numbers of colonies recovered on plating media. This has been attributed to fragmentation of mycelia and breaking of spores clumps during dilution in addition to competitive inhibition when large numbers of colonies are present on plates (Jarvis et al., 1983).

Standard methods for determining numbers of yeasts and moulds in foods.

There are two current ISO (International Standards Organisation) methods. ISO 7954 (1987) uses yeast extract glucose agar with either chloramphenicol or chlortetracycline as a pour-plate, incubating at 25°C and counting colonies after 2, 4 and 5 days. ISO 13681 (1995) applies specifically to meat and meat products, and acknowledges that the ISO 7954 medium does not inhibit Gram negative bacteria in meat sufficiently, and therefore suggests using a combination of gentamicin and chlortetracycline in yeast extract glucose agar. Pour plates are again suggested, but with the option of using surface plating so that yeasts and moulds can be distinguished, and also to avoid damaging heat-sensitive fungi. This would also assist resuscitation of heat-damaged cells, although this was not mentioned.

The latest Nordic Committee on Food Analysis method (NMKL 1995) suggests surface-plating onto DRBC agar for fresh foods, including fruit, vegetables meat and milk products, and surface-plating onto DG18 agar for raw materials or foods with a_w <0.95, incubating at 25°C for 5–7 days. The new draft ISO method (ISO, 2002) specifies DG18 agar for all samples, by pour plate, but again with an option for use of surface plates. This is contrary to interlaboratory studies recommending that DG18 agar not be used as a general purpose medium for enumerating yeasts in foods (Beuchat et al., 2001; Deak et al. 2001).

Interpretation of data

Unprocessed fruits, vegetables, grains and nuts may harbour large populations of yeasts and/or moulds without apparent spoilage. Counts exceeding 10^6 CFU/g are not uncommon. The same populations in processed ingredients or finished products may indicate spoilage. The significance of high populations in either unprocessed or processed foods and food ingredients can only be assessed by reference to the history of the product and its intended treatment before it reaches the consumer.

Results from direct plating reveal the extent of internal infection of samples and therefore may give an indication of the environmental conditions under which the product was stored prior to analysis. The major genera of moulds detected using direct plating techniques may also give some indication concerning time at which infection occurred, i.e., in the field or during storage. The presence of potentially mycotoxigenic moulds in the internal areas of grains or nuts, however, is not evidence that mycotoxins are present. Conversely, the absence of low incidence of mycotoxigenic moulds is not evidence that the product does not contain mycotoxins.

Precautions

Mycological media should be monitored for performance by comparing recovery of yeasts and moulds as well as unwanted bacterial colonies (Curtis and Beuchat, 1998). Seiler (1985) concluded that the Miles-Misra method (Corry, 1982b) can be used to assess recovery of certain yeasts and for ensuring that bacteria are adequately inhibited. The method is not suitable for use with the more rapidly spreading yeasts and moulds which require a stab inoculation method and measurement of rate of colony diameter increase with incubation time. Strains of microorganisms selected for monitoring performance of media should exhibit stable characteristics which do not change after repeated subculturing. A range of microorganisms representing rapid- and slow-growing yeasts, moulds and bacteria should be included. Microorganisms likely to be present in the food under study should be selected for the evaluation.

A proportion of mould and yeast cells in any given foods being analyzed may be metabolically or structurally injured as a result of physical and chemical stress (Beuchat, 1984). Such cells require optimum nutrient, pH, osmotic and temperature conditions for recovery if subsequent repair and colony formation are to occur on enumeration media. If stressed cells are likely to be present food samples, these cells should be considered when the enumeration medium is selected. Methods for quality control of mycological media should also consider the presence of stressed cells in foods (Curtis and Beuchat, 1998).

Infections, intoxications and respiratory problems as a result of inhaling mould spores can occur in laboratory personnel working with moulds. It is important to not remove lids from Petri plates containing colonies of moulds unless absolutely necessary for purposes of identification. All plates on which colonies have developed should be properly enclosed and heat sterilized before discarding. Anyone who has mishandled moulds in a laboratory will be remembered by those who work in that laboratory for many years to come.

References

Abdul-Raouf, U. M., Hwang, C. A. and Beuchat, L. R. (1994) Comparison of combinations of diluents and media for enumerating *Zygosaccharomyces rouxii* in intermediate water activity foods. Lett. Appl. Microbiol. 19, 28–31.

Abildgren, M. P., Lund, F., Thrane, U. and Elmholt, S. (1987) Czapez-Dox agar containing iprodione and dichloran as a selective medium for the isolation of *Fusarium* species. Lett. Appl. Microbiol. 5, 83–36.

Andrews, S. (1986) Optimization of conditions for the surface disinfection of sorghum and sultanas using sodium hypochlorite solutions. In: A. D. King, J. I. Pitt, L. R. Beuchat and J. E. L. Corry (Eds.), Methods for Mycological Examination of Foods. NATO ASI Series A: Life Sciences, Vol. 122. Plenum Press, New York. pp. 28–32.

Andrews, S. (1990) Differentiation of *Alternaria* species isolated from cereals on a dichloran malt extract agar. International Workshop on Standardization on Methods for the Mycological Examination of Foods, Progr. and Abstracts. Baarn, The Netherlands, p. 25.

Andrews, S. (1996) Evaluation of surface disinfection procedures for enumerating fungi in foods: a collaborative study. Int. J. Food Microbiol. 29, 177–184.

Andrews, S., de Graaf, H. and Stamation, H. (1997) Optimization of methodology for enumeration of xerophilic yeasts from foods. Int. J. Food Microbiol. 35, 109–116.

Andrews, S. and Pitt, J. I. (1986) Selective medium for isolation of *Fusarium* species and dematiaceous hyphomycetes from cereals. Appl. Environ. Microbiol. 51, 1235–1238.

Baird, R. M., Corry, J. E. L. and Curtis, G. D. W. (Eds.) (1987) Pharmacopoeia of Culture Media for Food Microbiology. Int. J. Food Microbiol. 5, 187–300.

Bandler, R., Stack, M. E., Koch, H. A., Tournas, V. and Mislivec, P. B. (1995) Yeasts, molds and mycotoxins. In: Bacteriological Analytical Manual, 8th Edn. U.S. Food and Drug Administration. Washington, DC.

Banks, J. G., Board, R. G., Carger, J. and Dodge, A. D. (1985) The cytotoxic and photodynamic inactivation of microorganisms by rose bengal. J. Appl. Bacteriol. 58, 392–400.

Bartschi, C., Berthier, J. Guiguettaz, C. and Valla, G. (1991) A selective medium for the isolation and enumeration of *Mucor* species. Mycol. Res. 95, 373–374.

Bell, D. K. and Crawford, J. L. (1967) A botran-amended medium for isolating *Aspergillus flavus* from peanuts and soil. Phytopathology. 57, 939–941.

Beuchat, L. R. (1984) Injury and repair of yeasts and moulds. In: Revival of Injured Microbes, M. H. E. Andrews and A. D. Russell (Eds.), Academic Press, London. pp 293–308.

Beuchat, L. R. (1987) Influence or organic acids on the heat resistance characteristics of *Talaromyces flavus* ascospores. Int. J. Food Microbiol. 6, 97–105.

Beuchat, L. R. (1992) Media for detecting and enumerating yeasts and moulds. Int. J. Food Microbiol. 17, 145–158.

Beuchat, L. R. (1993) Selective media for detecting and enumerating foodborne yeasts. Int. J. Food Microbiol. 19, 1–14.

Beuchat, L. R. (1998) Influence of pH and temperature on performance of tryptone yeast extract 10% glucose (TY10G) agar in supporting colony development by *Zygosaccharomyces rouxii*. J. Food Mycol. 1, 181–186.

Beuchat, L. R. (1998) Progress in conventional methods for detection and enumeration of foodborne yeasts. Food Technol. Biotechnol. 36, 267–272.

Beuchat, L. R. and Cousin, M. A. (2001) Yeasts and molds. In: F. Pouch-Downes and K. Ito (Eds.), Compendium of Methods for the Microbiological Examination of Foods, Am. Public Health Assoc. Washington, DC, pp. 209–215.

Beuchat, L. R., Frandberg, E., Deak, T., Alzamora, S.M., Chen, J., Guerrero, S., Lopez-Malo, A., Ohlsson, I., Olson, M., Pienado, J.M., Schneuerer, J., de Siloniz, M.I., and Tornai-Lehoczki, J. Performance of mycological media for enumerating desiccated food-spoilage yeasts: an interlaboratory study. Int. J. Food Microbiol. 70, 89–96.

Beuchat, L. R. and Hocking, A. D. (1990) Some considerations when analyzing foods for the presence of xerophilic fungi. J. Food Prot. 53, 984–989.

Beuchat, L. R., Jung, Y., Deak, T., Keffler., T., Golden, D. A., Peinado, J. M., Gonzolo, P., de Silonez,

M. I. and Valderrama, M. J. (1998) An interlaboratory study on the suitability of diluents and recovery media for enumeration of *Zygosaccharomyces rouxii* in high sugar foods. J. Food Mycol. 1, 117–130.

Beuchat, L. R., Nail, B. V., Brackett, R. E. and Fox, T. L. (1990) Evaluation of a culture film (Petrifilm™ YM) methods for enumerating yeasts and molds in dairy and high-acid foods. J. Food Prot. 53, 864, 869–874.

Beuchat, L. R., Nail, B. V., Brackett, R. E. (1991) Comparison of the Petrifilm™ yeast and mold culture film method to conventional methods for enumerating yeasts and molds in foods. J. Food Prot. 54, 443–447.

Beuchat, L. R., Pitt, J. I. (2001) Detection and enumeration of heat resistant molds. In: F. Pouch-Downes and K. Ito (Eds.), Compendium of Methods for the Microbiological Examination of Foods. Am. Public Health Assoc., Washington, DC, pp. 217–222.

Beuchat, L. R. and Rice, S. L. (1979) *Byssochlamys* spp. and their importance in processed fruits. Advan. Food Res. 25, 237–288.

Beuchat, L. R. and Tapia de Daza, M. S. (1992) Evaluation of chemicals for restricting colony spreading by a xerophilic mold, *Eurotium amstelodami*, on dichloran-18% glycerol agar. Appl. Environ. Microbiol. 58, 2093–2095.

Booth, C. (1971a) Fungal culture media. In: C. Booth, (Ed.) Methods in Microbiology, Vol. 4, Academic Press, London. pp. 50–93.

Booth, C. (1971b) Introduction to general methods. In: C. Booth (Ed.), Methods in Microbiology, Vol. 4, Academic Press, London.

Bothast, R. J. and Fennell, D. I. (1974) A medium for rapid identification and enumeration of *Aspergillus flavus* and related organisms. Mycologia. 66, 365–369.

Braendlin, N. (1996) Enumeration of xerophilic yeasts in the presence of xerophilic moulds: a collaborative study. Int. J. Food Microbiol. 29, 185–192.

Bragulat, M. R., Abarca, M. L. and Castella, G. and Cabanes, F. J. (1995) Dyes as fungal inhibitors: effect on colony enumeration. J. Appl. Bacteriol. 79, 578–582.

Brodsky, M. N., Entis, P., Entis, M. P., Sharpe, A. N. and Jarvis, G. A. (1982) Determination of aerobic plate and yeast mold counts in foods using an automated hydrophobic grid-membrane filter technique. J. Food Prot. 45, 301–304.

Carriera, A. and Loureiro, V. (1998) A differential medium to detect *Yarrowia lipolytica* within 24 hours. J. Food Mycol. 1, 3–12.

Chilvers, K. F., Reed, R. H. and Perry, J. D. (1999) Phototoxicity of rose bengal in mycological media - implications for laboratory practice. Lett. Appl. Microbiol. 28, 103–107.

Conner, D. E. (1990) Evaluation of methods for selective enumeration of *Fusarium* spp. in foodstuffs. International Workshop on Standardization of Methods for the Mycological Examination of Foods, Progr. and Abstracts. Baarn, The Netherlands. p. 26.

Corry, J. E. L. (1982a) Assessment of the selectivity and productivity of media used in analytical mycology. Arch. Lebensm. Hyg. 33, 160–164.

Corry, J. E. L. (1982b) Quality assessment of culture media by the Miles-Misra method. In: J. E. L. Corry (Ed.), Quality Assurance and Quality Control in Microbiological Culture Media. G. I. T.-Verlag, Darmstadt, pp. 21–37.

Curtis, G. D. W. and Beuchat, L. R. 1998. Quality control of culture media – perspectives and problems. Int. J. Food Microbiol. 45, 3–6.

Curtis, G. D. W. and Baird, R. M. (Eds.) (1993) Pharmacopoeia of cultive media for food microbiology: additional monographs (II). Int. J. Food Microbiol. 17, 201–268.

Davenport, R. R. (1980) An outline to media and methods for studying yeasts and yeast-like organisms. In: F. A. Skinner, S. M. Passmore and R. R. Davenport (Eds.), Biology and Activities of Yeasts. Soc. Appl. Bacteriol. Series No. 9, Academic Press, London, pp. 261–278.

Deak, T. and Beuchat. L. R. (1996) Handbook of Food Spoilage Yeasts. CRC Press, Boca Raton, FL.

Deak, T., Chen, J., Tapia, M.S., Tornai-Lehoczki, J., Viljoen, B.C., Wyder, M.T. and Beuchat, L.R. (2001) Comparison of dichloran glycerol (DG18) agar with general purpose mycological media for enumerating food spoilage yeasts. Int. J. Food Microbiol. 657, 49–53.

Golden, D.A., Dijkmann, K. E., Koopmans, J. and Mossel, D. A. A. (1979) The recovery and identification of psychrotrophic yeasts from chilled and frozen comminuted fresh meats. J. Appl. Bacteriol. 47. ix.

Dyer, S. K. and McCammon, S. (1994) Detection of toxigenic isolates of *Aspergillus flavus* and related species on coconut cream agar. J. Appl. Bacteriol. 76, 75–78.

Entis, P. and Lerner, I. (1996) Two-day yeast and mold enumeration using the ISO-GRID membrane filtration system in conjunction with YM-11 agar. J. Food Prot. 59, 416–419.

Erickson, J. P. (1993) Hydrophobic membrane filtration method for the selective recovery and differentiation of *Zygosaccharomyces bailii* in acidified ingredients. J. Food Prot. 56, 234–238.

Frandberg, E. and Olsen, M. (1999) Performance of DG18 media, a collaborative study. J. Food Mycol. 2, 239–249.

Frisvad, J. C. (1983) A selective and indicative medium for groups of *Penicillium viridicatum* producing different mycotoxins in cereals. J. Appl. Bacteriol. 54, 409–416.

Frisvad, J. C. (1983) Creatine sucrose agar, a differential medium for mycotoxin producing terverticillate *Penicillium* species. Lett. Appl. Microbiol. 1, 109–113.

Frisvad, J. C. (1986) Selective medium for *Penicillium viridicatum* in cereals. In: A. D. King, J. I. Pitt, L. R. Beuchat and J. E. L. Corry (Eds.), Methods for the Mycological Examination of Food, NATO ASI Series A: Life Sciences, Vol. 122. Plenum Publ. Co., New York, pp. 132–135.

Frisvad, J. C., Filtenborg, O. and Thrane, U. (1990a) Collaborative study on media for detecting and enumerating toxigenic fungi. 2nd International Workshop on the Standardization of Methods for the Mycological Examination of Foods, Progr. and Abstracts. Baarn, The Netherlands, p. 27.

Frisvad, J. C., Filtenborg, O. and Thrane, U. (1990b) Selective media for the detection of the associated toxigenic mycoflora of cereal products, meat and cheese. 2nd International Workshop on the Standardization of Methods for the Mycological Examination of Foods, Progr. and Abstracts. Baarn, The Netherlands, pp. 27–28.

Gogna, E., Vohra, R. and Sharma. P. (1992) Biodegradation of rose bengal by *Phanerochaete chryosporium*. Lett. Appl. Microbiol. 14, 58–60.

Hasma, T. A. P. and Ayres, J. C. (1977) A differential medium for the isolation of *Aspergillus flavus* from cottonseed. J. Food Sci. 42, 449–453.

Hara, S., Fennell, D. I. and Hesseltine, C. W. (1974) Aflatoxin-producing strains of *Aspergillus flavus* detected by fluorescence of agar medium under ultraviolet light. Appl. Microbiol. 27, 1118–1123.

Hastings, J. W., Tsai, W. Y. J. and Bullerman, L. B. (1986) Comparison of stomaching versus blending in sample preparation for mold enumeration. In: A. D. King, J. I. Pitt, L. R. Beuchat and J. E. L. Corry (Eds.), Methods for Mycological Examination of Foods, NATO ASI Series A: Life Sciences, Vol. 122, Plenum Press, New York, pp. 7–9.

Heard, G. M. and Fleet, G. H. (1986) Evaluation of selective media for enumeration of yeasts during wine fermentation. J. Appl. Bacteriol. 60, 477–481.

Heenan, C. N., Shaw, K. J. and Pitt, J. I. (1998) Ochratoxin A production by *Aspergillus carbonarius* and *A. niger* isolates and detection using coconut cream agar. J. Food Mycol. 1, 67–72.

Hernandez, P. and Beuchat, L. R. (1995) Evaluation of diluents and media for enumerating *Zygosaccharomyces rouxii* in blueberry syrup. Int. J. Food Microbiol. 25, 11–18.

Hocking, A. D. (1996) Media for preservative resistant yeasts: a collaborative study. Int. J. Food Microbiol. 29, 167–175.

Hocking, A. D. and Andrews, S. (1987) Dichloran chloramphenicol peptone agar as in identification medium for *Fusarium* species and some dematiaceous hyphomycetes. Trans. Br. Mycol. Soc. 89, 239–244.

Hocking, A. D. and Pitt, J. I. (1980) Dichloran glycerol medium for enumeration of xerophilic fungi from low-moisture foods. Appl. Environ. Microbiol. 39, 488–492.

International Commission on Microbiological Specification for Foods. (1986). Microorganisms in Foods. 2. Sampling for Microbiological Analysis: Principles and Specific Applications. University of Toronto Press, Toronto.

ISO (1987) ISO 7954 Microbiology – General guidance for enumeration of yeasts and moulds – colony count technique at 25°C.(http://www.iso.org/iso/en/ISOOnline.frontpage)

ISO (1995) ISO 13681: 1995E Meat and meat products – enumeration of yeasts and moulds – colony count technique. (http://www.iso.org/iso/en/ISOOnline.frontpage)

ISO Draft (2002) ISO 21527 "Microbiology of food and animal feeding stuffs – Horizontal method for the enumeration of yeasts and moulds – Colony count technique". In preparation.

Jarvis, B. (1973) Comparison of an improved rose-bengal-chlortetracycline agar with other media for the selective isolation and enumeration of moulds and yeasts in food. J. Appl. Bacteriol. 36, 723–727.

Jarvis, B. Seiler, D. A. L., Ould, A. J. L. and Williams, A. P. (1983) Observations on the enumeration of moulds in food and feeding stuffs. J. Appl. Bacteriol. 55, 325–336.

Jarvis, B. and Williams, A. P. (1987) Methods for detecting fungi in foods and beverages. In: L. R. Beuchat (Ed.), Food and Beverage Mycology, Van Nostrand Reinhold, New York, pp. 599–636.

Jermini, M. F. G., Geiges, O. and Schmidt-Lorenz, W. (1987) Detection isolation and identification of osmotolerant yeasts from high-sugar products. J. Food Prot. 50, 468–472.

Katan, T. (1985) Heat activation of dormant ascospores of *Talaromyces flavus*. Trans. Br. Mycol. Soc. 84, 748–750.

Kinderlerer, J. L. (1995) Czapek casein 50% glucose (CZC50G): a new medium for the identification of foodborne *Chrysosporium* spp. Lett. Appl. Microbiol. 21, 131–136.

King, A. D., Hocking, A. D. and Pitt, J. I. (1979) Dichloran rose bengal medium for enumeration isolation of molds from foods. Appl. Environ. Microbiol. 37, 959–964.

King, A. D., Pitt, J. I., Beuchat, L. R. and Corry, J. E. L. (Eds.) (1986) Methods for the Mycological Examination of Food. NATO ASI Series A, Life Sciences, Vol. 122. Plenum Press, New York.

Kish, S., Sharf, R. and Margalith, P. (1983) A note on a selective medium for the wine yeasts. J. Appl. Bacteriol. 55, 177–179.

Lin, Y. (1973) Detection of wild yeasts in the brewery: a new criterion. Brewers Digest. 48, 60–69.

Lin, Y. (1975) Detection of wild yeasts in the brewery: efficiency of differential media. J. Inst. Brew. 81, 410–417.

Makdesi, A. K. and Beuchat, L. R. (1996a) Evaluation of media for enumerating heat-stressed, benzoate-resistant *Zygosaccharomyces bailii*. Int. J. Food Microbiol. 33, 169–181.

Makdesi, A. K. and Beuchat, L. R. (1996b) Improved selective medium for enumerating of benzoate-resistant, heat-stressed *Zygosaccharomyces bailii*. Food Microbiol. 13, 281–290.

Makdesi, A. K. and Beuchat, L. R. (1996c) Performance of selective media for enumerating *Zygosaccharomyces bailii* in acidic foods and beverages. J. Food Prot. 59, 652–656.

Mian, M. A., Fleet, G. H. and Hocking, A. D. (1997) Effect of diluent type on viability of yeasts enumerated from foods or pure culture. Int. J. Food Microbiol. 35,103–107.

Mossel, D. A. A., Visser, M. and Mengerink, W. H. J. (1962) A comparison of media for the enumeration of moulds and yeasts for foods and vegetables. Lab. Prac. 11, 109–112.

Mossel, D. A. A., Kleynen-Semmeling, A. M. C., Vincentie, H. M., Beerens, H. and Catsavas, M. (1970) Oxytetracycline glucose yeast extract agar for selective enumeration of moulds and yeasts in food and clinical material. J. Appl. Bacteriol. 33, 454–457.

NMKL (1995) Proposed method. Mould and yeasts. Determination in foods. No. 98, 3rd edition 1995. (http://www.nmkl.org)

Pitt, J. I. (1979) The Genus *Penicillium* and its Teleomophic States *Eupenicillium* and *Talaromyces*. Academic Press, London.

Pitt, J. I. (1986) Properties of the ideal enumeration medium. In: A. D. King, J. I. Pitt, L. R. Beuchat and J. E. L. Corry (Eds.), Methods for the Mycological Examination of Foods, NATO ASI Series

A: Life Sciences, Vol. 122. Plenum Publ. Co., New York, pp. 132–132.
Pitt, J. I. (1993) A modified creatine sucrose medium for differentiation of species in *Penicillium* subspecies *Penicillium*. J. Appl. Bacteriol. 75, 559–563.
Pitt, J. I. and Hocking, A. D. (1977) Influence of solute and hydrogen ion concentration on the water relations of some xerophilic fungi. J. Gen. Microbiol. 101, 35–40.
Pitt, J. I. and Hocking, A. D. (1997) Fungi and Food Spoilage Second Eds. Blackie Academic and Professional, London.
Pitt, J. I. and Hocking, A. D. (1985) New species of fungi from Indonesian dried fish. Mycotaxon. 22, 197–208.
Pitt, J. I. Hocking, A. D. and Glenn, D. R. (1983) An improved medium for the detection of *Aspergillus flavus* and *Aspergillus parasiticus*. J. Appl. Bacteriol. 54, 109–114.
Rale, V. B. and Vakil, J. R. (1984) A note on an improved molybdate agar for the selective isolation of yeasts from tropical fruits. J. Appl. Bacteriol. 56, 409–413.
Restaino, L., Bills, S. and Lenovich, L. M. (1985) Growth response of an osmotolerant, sorbate-resistant *Saccharomyces rouxii* strain: evaluation of plating media. J. Food Prot. 48, 207–209.
Salkin, I. F. and Gordon, M. A. (1975) Evaluation of *Aspergillus* differential medium. J. Clin. Microbiol. 2, 74–75.
Samson, R. A., Hocking, A. D., Pitt, J. I. and King, A. D. (1992) Modern Methods in Food Mycology. Elsevier, London.
Samson, R. A., Hoetstra, E. S., Frisvad, J. C. and Filtenborg, O. (1985) Methods for the detection and isolation of food-borne fungi. In: R. A. Samson, E. S. Hoekstra, J. C. Frisvad and O. Filtenborg (Eds.), Introduction to Food-borne Fungi. Centraalblureau voor Schimmelcultures, Baarn, The Netherlands, pp. 235–242.
Scarr, M. P. (1959) Selective media used in the microbiological examination of sugar products. J. Sci. Food Agric. 10, 678–681.
Seiler, D. A. L. (1985) Monitoring mycological media. Int. J. Food Microbiol. 2, 123–131.
Seiler, D. A. L. (1986) Effect of presoaking on recovery of fungi from cereals and cereal products. In: A. D. King, J. I. Pitt, L. R. Beuchat and J. E. L. Corry (Eds.), Methods for the Mycological Examination of Food. NATO ASI Series A., Life Sciences, Vol. 122. Plenum Press, New York, pp. 26–28.
Splittstoesser, D. F., Wilkison, M. and Harrison, N. (1972) Heat activation of *Byssochlamys fulva* ascospores. J. Milk Food Technol. 35, 399–401.
Taniwaki, M. H., Iamanaka, B. T. and Banhe, A. A. (1999) Comparison of culture media to recover fungi from flour and tropical fruit pulps. J. Food Mycol. 2, 291–302.
Thrane, U. (1996) Comparison of three selective media for detecting *Fusarium* species in foods: a collaborative study. Int. J. Food Microbiol. 29, 149–156.
Twomey, D. G. and Mackie, D. L. (1985) A medium for the rapid growth and detection of moulds. Lett. Appl. Microbiol. 1, 105–107.
Valderrama. M.-J., de Siloniz, M. I., Gonzalo, P. and Peinado, J. M. (1999) A differential medium for the isolation of *Kluyveromyces marxianus* and *Kluyveromyces lactis* from dairy products. J. Food Prot. 62, 189–193.
Walters, L. S. and Thiselton, M. A. (1953) Utilization of lysine of yeasts. J. Inst. Brew. 59, 401–404.
Yabe, K., Ando, Y., Ito, M. and Terakado, N. (1987) Simple method for screening aflatoxin-producing molds by UV photography. Appl. Environ. Microbiol. 53, 230–234.
Yamani, M. I. (1993) Yoghurt whey medium for food-borne yeasts. Int. J. Food Microbiol. 28, 111–116.

Part 2

Pharmacopoeia of Culture Media

Notes on the use of the monographs

Composition. Under this heading will be found all the ingredients of the medium in the quantities in which they are present in the finished product. Where salts are available in both anhydrous and hydrated forms the quantities given refer to the anhydrous form unless otherwise stated. The quantity of agar specified assumes the use of a product that produces a satisfactory gel in nutrient broth when used at a concentration of 15g/l. Local circumstances may require variations in this quantity depending on the gelling properties of the agar used. In media such as semi-solid agars where the gel strength is critical, the specification of the agar should be followed exactly.

Bile and its derivatives are subject to considerable variation and where these substances are specified in these monographs they should be tested in trial batches of media before general use.

Preparation. For optimal performance media should be prepared as directed. With some media it is possible to prepare and sterilize basal ingredients for subsequent re-melting and addition of other ingredients to produce the final product. Whilst this practice may be convenient, particularly in small laboratories, in general the re-heating of media should be avoided.

Autoclaving is sometimes undesirable and often unnecessary in the preparation of media. The total heat input is difficult to standardize in general purpose laboratory autoclaves and more consistent results are obtained by the use of a purpose-built media sterilizer (Corry et al., 1986).

Physical properties. pH values given are for readings made at 20–25°C unless otherwise stated.

No values have been given for gel strength measurements. Machines are available which can be used to measure this property but none is in general use. Costin (1978) gave details of a method using the Gelomat^R machine (Heinrich Bareiss Apparatebau Company, 7931 Oberdischingen, Germany) and stated that satisfactory batches of media should have a gel rigidity of between 50 and 80 g. Users of the Van der Bijl penetrometer regard a penetration of less than 2 mm within 10 min as satisfactory but whilst this ensures a minimum strength of gel it provides no indication of an excessively stiff matrix. The LFRA Texture Analyser (C.S. Stevens & Son Ltd., Unit 4, Executive Park, Hatfield Road, St. Albans, Herts, AL1 4TA, U.K.) may also be used.

Shelf life. Dehydrated media and dry ingredients should be stored according to the manufacturer's instructions and with regard to the stated shelf life or expiry date. Some ingredients, particularly dyes, undergo changes when exposed to light which result in the formation of substances toxic to microorganisms. For this reason the shelf life given for ready to use media applies to poured plates or plugged tubes stored in the dark. Shelf life can be maximised by packing poured plates in sealed bags and tubed media in screw capped containers to reduce the rate of water loss due to evaporation.

Resuscitation of stressed organisms. Test strains will not normally have been stressed (i.e. exposed to sublethal heating, drying, chilling, acid) but when using these media for samples where stressing may have occurred appropriate resuscitation procedures should be carried out. For a review of this problem see Mossel and van Netten (1984) and Ray (1986).

Drying of poured plates. Facilities for drying vary from laboratory to laboratory. It is not possible therefore to recommend a single procedure to obtain a uniform dryness of agar surfaces. Where local production methods are consistent it should be possible to adopt a standardized method to give uniform results. Over-drying should be avoided.

Incubation temperatures. Strict adherence to the stated incubation temperature is recommended as variations in temperature can affect the selectivity of many media. Monitoring of incubator temperatures, preferably by means of a thermograph with a sensor in the medium, will provide warnings of inconsistencies in performance. Loading with many plates in high stacks or the use of high sided non-perforated trays can result in extremely long delays in achieving the desired temperature.

Reading of results and interpretation. Many of the media described in these monographs are designed for the isolation of a single genus or species. Very few however are entirely specific and the identity of characteristic colonies should be regarded as 'presumptive' until confirmatory tests for identity have been carried out.

Quality assessment. Test strains. For each medium a minimum set of test strains is given which should be used on all batches of media prepared in the laboratory. Where other strains are known to perform satisfactorily with a given medium these are also listed under the heading of Supplementary Strains. These extra strains can be used for more extensive testing undertaken when assessing new batches of dehydrate from manufacturers or when investigating problems. The occasional use of freshly isolated 'wild' strains may also prove valuable.

Where strains of *Proteus* sp. are used for quantitative testing of agars, a reference medium on which the organism does not swarm should be used. Alternatively the ability of a test medium to prevent swarming of *Proteus* sp. may be checked qualitatively and another organism (e.g. *Morganella morganii*) used in quantitative tests.

Strains are identified by their numbers in the National Collection of Industrial and Marine Bacteria (NCIMB) Quality Control catalogue. Cross references to numbers in other collections will be found in Appendix II. Those strains which are not available from NCIMB are listed in each monograph with the appropriate number(s) in other collections.

Methods. Recommended methods are given in Appendix I.

Criteria. Unless otherwise stated in individual monographs the criteria for assessment of prepared media are as follows. For the modified Miles-Misra and dilution to extinction methods, recovery of all strains on non-selective media should be within 0.7 $\log_{10}$ of the recovery of the same strains on the reference medium. When ecometry is used the Absolute Growth Index (AGI) should be at least 3 and an equivalent score should be obtained with the streaking method.

For unwanted strains on selective media recovery by modified Miles-Misra or dilution to extinction methods should be more than 5 $\log_{10}$ below the recovery with the reference medium. For wanted strains recovery should be less than 1 $\log_{10}$ below that with the reference medium. Ecometry should yield AGIs not greater than 2.0 for unwanted strains and not less than 2.5 for wanted strains. Equivalent scores should be obtained with the streaking method. The reference medium, unless otherwise stated, is tryptone soya agar or broth. Standards for mycological media are given in individual monographs.

The quality assessment procedures detailed in these monographs relate primarily to productivity, selectivity and colonial appearance. With some media other methods may also be appropriate and these have been reviewed by Curtis (1985).

Status. Since the publication of the Pharmacopoeia of Culture Media these monographs have been incorporated as source documents in laboratory Quality Manuals. In order to improve the usefulness of the monographs the WPCM decided at its meeting in Budapest in 1996 that each should be placed in one of three categories. A *draft* monograph has been prepared by an expert for circulation to interested parties prior to discussion at a WPCM meeting. Once such a monograph has been agreed as suitable for assessment its status is raised to *proposed.* An assessed monograph which a WPCM meeting has agreed to be satisfactory is classed as *approved.* A statement on the formal system of approval for monographs together with the rationale for the choice of test strains can be found in Curtis et al. (1998).

References

Corry, J.E.L., Leclerc, M., Mossel, D.A.A., Skovgaard, N.P., Terplan, G. and van Netten, P. (1986) An investigation into the quality of media prepared and poured by an automatic system. Int. J. Food Microbiol. 3, 109–120.

Costin, I.D. (1978) A semiautomatic instrument for the determination of gel rigidity in microbiological nutrients and gelling agents. *In:* Mechanizing microbiology. A.N. Sharpe and D.S. Clark (Eds.), Charles C. Thomas, Springfield, IL, pp. 170–196.

Curtis, G.D.W. (1985) A review of methods for quality control of culture media. Int. J. Food Microbiol. 2, 13–20.

Curtis, G.D.W., Baird, R.M., Skovgaard, N.P. and Corry, J.E.L. (1998) A formal system of approval for monographs in the pharmacopoeia of culture media – statement from the IUMS-ICFMH working party on culture media. Int. J. Food Microbiol. 45, 59–63.

Mossel, D.A.A. and van Netten, P. (1985) Harmful effects of selective media on stressed microorganisms: nature and remedies. *In:* The revival of injured microorganisms. M.H.E. Andrew and A.D. Russell (Eds.), Academic Press, London, pp. 329–369.

Ray, B. (1986) Impact of bacterial injury and repair in food microbiology: its past, present and future. J. Food Protect. 49, 651–655.

Summary of organisms and recommended media

Aeromonas spp.	Bile salts irgasan brilliant green agar Starch ampicillin agar
Bacillus cereus	Mannitol egg yolk polymyxin agar Polymyxin pyruvate egg yolk mannitol bromothymol blue agar
Brochothrix thermosphacta	Streptomycin thallous acetate actidione agar
Campylobacter spp.	Charcoal cefoperazone deoxycholate agar Charcoal cefoperazone deoxycholate broth Preston campylobacter selective agar Preston enrichment broth Skirrow campylobacter selective agar
Carnobacterium spp.	Cresol red thallium acetate sucrose agar
Clostridium perfringens	Differential clostridial agar Iron sulphite agar Oleandomycin polymyxin sulphadiazine perfringens agar Rapid perfringens medium Sulphite cycloserine azide agar Tryptose sulphite cycloserine agar
Clostridium thermosaccharolyticum	Iron sulphite agar
Coli-aerogenes group (see also Enterobacteriaceae)	Brilliant green bile broth Laurel sulphate MUG X-gal broth Lauryl tryptose broth
Desulfotomaculum nigrificans	Iron sulphite agar
Enterobacteriaceae	Enterobacteriaceae enrichment broth Violet red bile glucose agar Violet red bile agar
Enterococcus spp.	Citrate azide tween carbonate agar Enterococcosel agar/broth Kanamycin aesculin azide agar M-enterococcus agar Thallous acetate tetrazolium glucose agar

Escherichia coli	Brilliant green bile broth Enterobacteriaceae enrichment broth Lauryl sulphate MUG X-gal broth Lauryl tryptose broth Tryptone bile agar Violet red bile glucose agar Violet red bile agar
Escherichia coli O157:H7	Cefixime tellurite sorbitol MacConkey agar Haemorrhagic colitis agar
Lactic acid bacteria	All purpose tween agar-modified Briggs agar L-S differential agar M-17 agar de Man, Rogosa and Sharpe agar de Man, Rogosa and Sharpe agar with sorbic acid (pH 6.2)
Lactobacilli (see also Lactic acid bacteria)	Briggs agar Lactobacillus sorbic acid agar de Man, Rogosa and Sharpe agar Rogosa agar
Lactococcus lactis ssp. *lactis*	Briggs agar de Man, Rogosa and Sharpe agar de Man, Rogosa and Sharpe agar with sorbic acid Rogosa agar (pH 6.2)
Leuconostocs	see Lactic acid bacteria
Listeria spp.	FDA listeria enrichment broth (1995) Fraser broth Half Fraser broth Lithium chloride phenylethanol moxalactam agar Oxford agar Oxford agar-modified Polymyxin acriflavine lithium chloride ceftazidime aesculin mannitol agar Polymyxin acriflavine lithium chloride ceftazidime aesculin mannitol egg yolk broth University of Vermont broths I & II
Moulds	*Aspergillus flavus* and *parasiticus* agar Dichloran glycerol (DG-18) agar Dichloran rose bengal chloramphenicol agar Oxytetracycline glucose yeast extract agar Rose bengal chloramphenicol agar
Pediococci	Briggs agar de Man, Rogosa and Sharpe agar de Man, Rogosa and Sharpe agar with sorbic acid Rogosa agar

Pseudomonas spp.	Cephaloridine fucidin cetrimide agar
Salmonella spp. (See also Enterobacteriaceae)	Bismuth sulphite agar Diagnostic salmonella selective semisolid medium Hektoen enteric agar Mannitol lysine crystal violet brilliant green agar Muller Kauffmann tetrathionate broth Phenol red brilliant green agar Rambach agar Rappaport-Vassiliadis broth Rappaport-Vassiliadis medium-semisolid modification Selenite cystine broth SM ID medium Xylose lysine deoxycholate agar Xylose lysine tergitol 4 agar
Shigella spp. (See also Enterobacteriaceae)	Violet red bile agar Xylose lysine deoxycholate agar
Staphylococcus aureus	Baird-Parker agar Baird-Parker liquid medium Giolitti and Cantoni broth Rabbit plasma fibrinogen agar Tryptone soya broth with 10% NaCl & 1% sodium pyruvate
Streptococcus bovis	Kanamycin aesculin azide agar M-enterococcus agar
Streptococcus thermophilus	Briggs agar LS differential agar M 17 agar
Vibrio spp.	Cellobiose polymyxin B colistin agar Thiosulphite citrate bile salts agar
Yersinia spp.(see also Enterobacteriaceae)	Bile oxalate sorbose broth Cefsulodin irgasan novobiocin agar Irgasan ticarcillin chlorate broth Salmonella shigella deoxycholate calcium agar

All Purpose Tween (APT) agar – modified (for H_2O_2 detection)

This monograph has been reviewed by members of the IUMS-ICFMH Working Party on Culture Media and given 'Proposed' status.

Description and history

This medium was devised by Evans and Niven (1951) and a modification developed by Shipp (1964) for the rapid non-selective enumeration of bacteria capable of elaborating H_2O_2 during growth under aerobic conditions. Such reaction on cooked cured meats results in a "green" discoloration. This principle was also used by Whittenbury (1964) and Reuter (1970).

Composition (grams)

Tryptic digest of casein	12.5
Yeast extract	7.5
Glucose	10.0
Sodium chloride	5.0
Tri-sodium citrate	5.0
Di-potassium hydrogen orthophosphate	5.0
Sorbitan monooleate (Tween 80)	0.2
Magnesium sulphate $\cdot 7H_2O$	0.8
Manganese (II) chloride $\cdot 4H_2O$	0.14
Iron (II) sulphate $\cdot 7H_2O$	0.04
Thiamine hydrochloride	0.001
Agar	13.5
Distilled or deionized water	1000.0

Preparation

Double layer plates are prepared.
Bottom layer: APT agar (15 ml).
Top layer: 20 g of MnO_2 are added to 200 ml APT **broth**, dispensed in 10 ml amounts and sterilized by autoclaving at 121°C for 15 min. To each 100 ml amount of sterile, cooled (45°C) APT agar is added 10 ml of MnO_2 suspension. When the bottom layer is firmly set a thin layer of APT agar with MnO_2 is added.

Physical properties

Appearance	Bottom layer: light brown, clear. Top layer: black, with the suspended MnO_2.
pH	6.7 ± 0.2

Shelf life

Ready to use medium	At least 7 days at 4 ± 2°C.

Inoculation method for samples

Sample dilutions (in 0.1% peptone + 0.85% NaCl) are surface plated to obtain discrete colonies.

Incubation method

At 30°C for a minimum of 5 days, in air, or 25°C for 5 days or 7°C for 10 days.

Reading of results and interpretation

H_2O_2 forms soluble compounds with suspended MnO_2. Colonies surrounded by a clear zone are regarded as potentially positive. These may be inoculated onto the surface of a sterile cooked, cured meat and after incubation in the dark at 20°C for 24–72 h, green discoloration should be apparent for a positive result.

Quality assessment

(i) *Positive reaction*

Test strains	*Lactobacillus viridescens* NCIMB 50057 *Lactobacillus brevis* NCIMB 50043
Inoculation method	Broth cultures streaked for discrete colonies.
Criteria	Growth with clear zones around colonies.

(ii) *Negative reaction*

Test strain	*Lactobacillus sakei* ssp. *sakei* NCIMB 50056
Inoculation method	Broth cultures streaked for discrete colonies.
Criteria	Growth without clear zones around colonies.

(iii) *Characteristic appearance of colonies*

All H_2O_2 producing colonies are surrounded by a clear zone on the medium.

References

Evans, J.B. and Niven, C.F. (1951) Nutrition of the heterofermentative lactobacilli that cause greening of cured meat products. J. Bacteriol. 62, 599–603.
Reuter, G. (1970) Laktobazillen und eng verwandte Mikroorganismen in Fleisch und Fleischerzeugnissen. 2.Mitteilung: Die Charakterisierung der isolierten Laktobazillenstamme. Fleischwirtschaft 50, 954–962.
Shipp, H.L. (1964) The green discolouration of cooked cured meats of bacterial origin. Technical circular No. 266, British Food Manufacturing Industries Research Association, Leatherhead, U.K.
Whittenbury, R. (1964) Hydrogen peroxide formation and catalase activity in the lactic acid bacteria. J. Gen. Microbiol. 35, 13–26.

Aspergillus flavus and *parasiticus* agar (AFPA)

This monograph has been reviewed by members of the IUMS-ICFMH Working Party on Culture Media and given 'Proposed' status.

Description and history

This is a selective medium for the enumeration in foods of the mycotoxin producing fungi *Aspergillus flavus* and *Aspergillus parasiticus*. Bothast and Fennell (1974) developed *Aspergillus* differential medium (ADM), containing 1.0% yeast extract, 1.5% tryptone and 0.5% ferric chloride, recommending incubation at 28°C for 3 days. Hamsa and Ayres (1977) incorporated streptomycin and dichloran into ADM and recommended incubation at 28°C for 5 days. Both media relied upon the formation of a bright orange-yellow reverse pigment by *Aspergillus flavus* and related species. Pitt et al. (1983) further refined the formulation, producing *Aspergillus flavus* and *parasiticus* agar (AFPA), which gives sufficient colour development to enable recognition of *Aspergillus flavus* (or *Aspergillus parasiticus*) colonies within 42–48 h at 30°C.

Composition (grams)

Yeast extract	20.0
Peptone (bacteriological)	10.0
Iron (III) ammonium citrate	0.5
Dichloran (2,6-dichloro-4-nitroaniline)	0.002
Chloramphenicol	0.1
Agar	15.0
Distilled or deionised water	1000.0

Preparation

Dissolve yeast extract, peptone, iron (III) ammonium citrate and agar in the water by heating. Add 1.0 ml of a 0.2% (w/v) ethanolic solution of dichloran and 10 ml of a 1% ethanolic solution of chloramphenicol. Adjust to pH 6.2 ± 0.2 and sterilize by autoclaving at 121°C for 15 min. Cool to 50°C and dispense 15 ml amounts into sterile 9 cm diameter Petri dishes. Dry and use immediately or store at 4 ± 2°C in the dark for up to 4 weeks before using.

Physical properties

Appearance	Amber, clear.
pH	6.2 ± 0.2

Shelf life

Ready to use medium	4 weeks at 4 ± 2°C.

Inoculation method for samples

Surface spread 0.1 or 0.2 ml of diluted sample per 9 cm diameter plate.

Incubation

At 30°C for 42–48 h.

Reading of results and interpretation

Aspergillus flavus and *Aspergillus parasiticus* produce an orange-yellow (chrome yellow) reverse colony pigmentation. *Aspergillus niger* sometimes produces colonies with a light yellow reverse, but is readily distinguished from *Aspergillus flavus* after further 24–48 h incubation by the production of black conidial heads. *Aspergillus ochraceus* produces an orange-yellow reverse, but only forms colonies after prolonged incubation.

Quality assessment

Use a stab inoculation procedure. Examine reverse of colonies after 42–48 h incubation at 30°C for typical pigmentation (see above). If doubt exists over *Aspergillus niger*, continue incubation for a further 24–48 h. Prolonged incubation is not recommended.

Test strains	CSIRO[1]	A-NRRL[2]
Aspergillus flavus	3084	3251
Aspergillus parasiticus	2744	2999
Aspergillus niger	2522	3361

[1] Commonweatlh Scientific and Industrial Research Organisation, PO Box 52, North Ride, New South Wales, 2231, Australia.

[2] USDA Northern Utilization Research and Development Division, Peoria, Illinois, USA.

References

Bothast, R.J. and Fennell, D.I. (1974) A medium for rapid identification and enumeration of *Aspergillus flavus* and related organisms. Mycologia 66, 365–369.

Hamsa, T.A. and Ayres, J.C. (1977) A differential medium for the isolation of *Aspergillus flavus* from cottonseed. J. Food Sci. 42, 449–453.

Pitt, J.I., Hocking, A.D. and Glenn, D.R. (1983) An improved medium for the detection of *Aspergillus flavus* and *A. parasiticus*. J. Appl. Bacteriol. 54, 109–114.

Baird-Parker agar

This monograph has been assessed by members of the IUMS-ICFMH Working Party on Culture Media and given 'Approved' status.

Description and history

This is a selective medium for the enumeration of *Staphylococcus aureus* in foods which was first reported by Baird-Parker (1962). It is now widely recommended by national and international bodies. Selectivity is attained with potassium tellurite and lithium chloride. Addition of sulphamezathine (Baird-Parker, 1969) is advised only if *Proteus* species are suspected in the test sample. Sodium pyruvate is a critical component, essential to both recovery of damaged *Staphylococcus aureus* cells and their subsequent growth (Baird-Parker and Davenport, 1965). On this medium *Staphylococcus aureus* forms grey to black or brown-grey colonies due to tellurite reduction (ISO, 1999). *Staphylococcus aureus* colonies with egg yolk factor are surrounded by an opaque zone, frequently with an outer clear zone (Bennett and Lancette, 1995).

Composition (grams)

Note: From the point of quality assurance the use of commercially available medium conforming to the formulation given below is recommended. Comercially available egg yolk or egg yolk-tellurite emulsion are used as supplements to base medium.

Base medium:

Pancreatic digest of casein	10.0
Yeast extract	1.0
Meat extract	5.0
Sodium pyruvate	10.0
L-Glycine	12.0
Lithium chloride	5.0
Agar	12 to 20.0
Distilled or deionized water	950.0

Complete medium:

Base medium	95.0 ml
Potassium tellurite solution	1.0 ml

Egg yolk emulsion	5.0 ml
Sulphamezathine solution (if necessary)	2.5 ml

Preparation

Dissolve the base medium components or the dehydrated base in the water by boiling. If necessary, adjust the pH so that after sterilization it is 7.2 ± 0.2 at 25°C. Transfer the medium in quantities of 95 ml to flasks or bottles and sterilize in an autoclave for 15 min at 121°C.

Prepare membrane-filtered (0.22 μm) aqueous solutions of potassium tellurite 1 % w/v and, if necessary, sulphamezathine (sulphamezathine, sulphadimidine) 0.2 % w/v. Performance of this medium is greatly dependant on the potassium tellurite. It is important to use a satisfactory brand, e.g. Merck or Sigma, and to buy in small quantities as, once opened, the salt begins to deteriorate. The salt must be completely dissolved prior to filtration. Solutions with white precipitate must be discarded. The sulphamezathine solution is prepared by dissolving 0.2 g of pure sulphamezathine in 10 ml of 0.1 N NaOH and diluting to 100 ml with water. Both solutions may be stored for a maximum of one month at 4 ± 2°C.

To prepare the complete medium melt and cool the base medium to appoximately 47°C by means of a water bath. To each of the 95 ml add 1 ml of potassium tellurite solution and 5 ml egg yolk. Commercially available egg yolk should be used. Alternatively egg yolk-tellurite emulsion is also commercially available. If necessary, add 2.5 ml sulphamezathine solution. Each solution should be previously warmed in a water bath at 45–47°C, mixing well after each addition.

Preparation of agar plates:

Transfer the complete medium in quantities of 12–15 ml to 9 cm diameter Petri dishes or 28–30 ml to 14 cm diameter dishes. The plates may be stored, prior to drying, at 4 ± 2°C. Before use, dry the plates, preferably with the lids off and the agar surface downwards, in an oven or incubator at 45°C for 30 min, or at 37°C for 1 h. The plates can also be dried in a laminar flow cabinet for 30 min with half-open lids.

Physical properties

Appearance	Complete medium opaque, cream/pale fawn in colour.
pH	Base medium 7.2 ± 0.2 at 25°C.

Shelf life

Complete medium	14 days at 4 ± 2°C.
Solutions	Sulphamezathine and potassium tellurite 1 month at 4 ± 2°C. Discard tellurite if a white precipitate forms. Egg yolk according to manufacturer's instructions.

Inoculation method for samples

Spread plate technique using 0.1 ml inocula on 9 cm diameter dishes or 1 ml on 14 cm diameter dishes in order to raise the detection limit by a factor of 10, if necessary. It is also possible to inoculate 1 ml fractionated in 0.3, 0.3 and 0.4 ml to the surface of three 9 cm diameter plates. Carefully spread the inoculum as quickly as possible over the surface of the plates using a sterile spreader. Allow the plates to dry with their lids on for about 15 min at room temperature.

Incubation method

At 37°C for 24–48 h, in air.

Reading of results and interpretation

After incubation for 24 h, mark on the bottom of the plates the positions of any typical colonies. Reincubate all plates for a further 24 h and mark any newly developed typical colonies; also mark any atypical colonies.

Typical colonies are black to grey, shining and convex (1–1.5 mm diameter after 24 h and 1.5–2.5 mm diameter after 48 h) with egg yolk reaction (clear halo and/or opaque zone around colony). Atypical colonies are black to grey without egg yolk reaction. These are formed mainly by strains of coagulase-positive staphylococci contaminating e.g. dairy products, shrimps and giblets (ISO, 1999).

If several types of colonies are observed which appear to be *Staphylococcus aureus*, count the number of colonies of each type and record counts separately. Select five colonies of each type counted and test for coagulase production using commercially available rabbit plasma with EDTA.

Quality assessment

(i) *Productivity*

Test strains	*Staphylococcus aureus* NCIMB 50081 *Staphylococcus aureus* NCIMB 50080
Inoculation method	Modified Miles-Misra or spiral plating onto Baird-Parker agar and Heart Infusion agar.
Criteria	Counts on Baird-Parker agar should be within 0.5 $\log_{10}$ of the counts on the non-selective medium. 50081 will show a reduction in count (>0.5 $\log_{10}$) if the Baird-Parker agar containing sulphamezathine is deficient in pyruvate.

(ii) *Selectivity*

Test strains	*Escherichia coli* NCIMB 50034

	Proteus mirabilis (ATCC 29906/CECT 4168/NCTC 11938)
Inoculation method	As above.
Criteria	Recovery on Baird-Parker agar should be 3.0 $\log_{10}$ below the recovery on Heart Infusion agar.

(iii) *Characteristic appearance of* Staphylococcus aureus

Staphylococcus aureus produces grey to jet-black colonies up to 3 mm diameter (at 48 h) frequently with a colourless margin. They are typically surrounded by an opaque zone, often with an outer clear zone. From various foods and dairy products non-lipolytic strains of similar appearance may be encountered, except that surrounding opaque and clear zones are absent.

References

Baird-Parker, A.C. (1962) An improved diagnostic and selective medium for isolating coagulase positive Staphylococci. J. Appl. Bacteriol. 25, 12–19.

Baird-Parker, A.C. (1969) Isolation methods for microbiologists, edited by D.A. Shapton and G.W. Gould. SAB Technical Series No. 3. Academic Press, London, pp. 1–8.

Baird-Parker, A.C. and Davenport, E. (1965) The effect of recovery medium on the isolation of *Staphylococcus aureus* after heat treatment and after storage of frozen or dried cells. J. Appl. Bacteriol. 28, 390–402.

Bennett, R.W. and Lancette, G.A. (1995) Chapter 12. *Staphylococcus aureus*. *In:* FDA Bacteriological Analytical Manual, 8th edition. AOAC International, Gaithersburg, MD pp. 12.01–12.05.

Chopin, A., Malcolm, S., Jarvis, G., Asperger, H., Beckers, H.J., Bertona, A.M., Cominazzini, C., Carini, S., Lodi, R., Hahn, G., Heeschen, W., Jans, J.A., Jervis, D.I., Lanier, J.M., O'Connor, F., Rea, M., Rossi, J., Seligmann, R., Tesone, S., Waes, G., Mocquot, G., and Pivnick, H. (1985) ICMSF Methods studies XV. Comparison of four media and methods for enumerating *Staphylococcus aureus* in powdered milk. J. Food Protect. 48, 21–27.

Harvey, J. and Gilmour, A. (1985) Application of current methods for isolation and identification of staphylococci in raw bovine milk. J. Appl. Bacteriol. 59, 207–221.

IDF (1997) Milk and milk-based products – Enumeration of coagulase-positive staphylococci – Colony count technique at 37°C. Provisional IDF Standard 145A: 1997. International Dairy Federation, Brussels.

ISO (1999) Microbiology of food and animal feeding stuffs – Horizontal method for the enumeration of coagulase-positive staphylococci (*Staphylococcus aureus* and other species) – Part 1: Technique using Baird-Parker agar medium. ISO 6888–1: 1999. International Organisation for Standardization, Geneva.

Baird-Parker liquid (LBP) medium

This monograph has been reviewed by members of the IUMS-ICFMH Working Party on Culture Media and given 'Proposed' status.

Description and history

This modification of Baird-Parker's agar allows the selective enrichment of small numbers of injured cells of *Staphylococcus aureus*. Selectivity is attained through addition of potassium tellurite and lithium chloride, whilst sodium pyruvate enhances productivity, especially the recovery of stressed cells. Anaerobic incubation of the medium reduces the likelihood of false positive results from micrococci, a common isolate on Baird-Parker agar. The medium has been used for the isolation of *Staphylococcus aureus* from pharmaceutical products, where small populations of stressed cells may be of significance (Baird and van Doorne, 1982).

Composition (grams)

Peptone	8.0
Pancreatic digest of casein	2.0
Beef extract	5.0
Yeast extract	1.0
Sodium pyruvate	10.0
Glycine	12.0
Lithium chloride	5.0
Potassium tellurite	0.1
Distilled or deionized water	1000.0

Preparation

Dissolve all the ingredients except the potassium tellurite in the water with the aid of heat. Cool to room temperature and adjust the pH to 6.6. Dispense into test tubes and sterilize at 121°C for 15 min. When cooled to approximately 50°C, add filter sterilized potassium tellurite solution to give a final concentration of 100 μg/ml.

Physical properties

Appearance	Clear, amber.

pH	6.6 ± 0.2

Shelf life

Basal medium without potassium tellurite	1 month at 4°C.
Complete medium	Use the same day.

Before use, stored basal medium should be heated to 100°C for 15 min to remove oxygen, cooled to approximately 47°C and the potassium tellurite solution added.

Inoculation method for samples

Use food macerates or decimal dilutions and inoculate broths in the proportion 1: 10. MPN procedures need at least three tubes for at least three dilution steps. If no anaerobic jar is available, overlay the broth with agar or molten paraffin (melting point 56°C).

Incubation method

At 37°C for 48 h anaerobically.

Reading of results and interpretation

After removal of agar or paraffin plug subculture all tubes showing growth, whether or not they show a black precipitate, by streaking a loopful of culture to Baird-Parker agar. Incubate the agar plates and examine as detailed in the monograph on Baird-Parker agar.

Quality assessment

(i) *Productivity*

Test strain	*Staphylococcus aureus* NCIMB 50080
Inoculation method	Dilution to extinction.
Criteria	Recovery in LBP should be within 1 titre unit of the recovery in TSB after 48 h at 37°C.

(ii) *Selectivity*

Test strain	*Micrococcus luteus* NCIMB 50063 *Escherichia coli* NCIMB 50034
Inoculation method	Dilution to extinction.

Criteria Recovery in LBP should be less than 5 titre units of recovery in TSB after 48 h at 37°C.

References

Baird, R.M. and van Doorne, H. (1982) Enrichment techniques for *Staphylococcus aureus*. Arch. Lebensmittelhyg. 33, 146–150.

Deutsche Norm (1988) Bestimmung Koagulase-positiver Staphylokokken in Trockenmilcherzeugnissen und Schmelzkäse – Verfahren mit selektiver Anreicherung, DIN 10178, Teil 1.

Bile Oxalate Sorbose (BOS) broth

This monograph has been reviewed by members of the IUMS-ICFMH Working Party on Culture Media and given 'Proposed' status.

Description and history

A two-step enrichment procedure for the recovery of *Yersinia enterolitica* from foods was developed by Schiemann (1982). In this procedure pre-enrichment in yeast extract-rose bengal broth or in tryptone soya broth is followed by selective enrichment in BOS broth. The selective agents in BOS broth are sodium oxalate, bile salts, irgasan and sodium furadantin. BOS broth was found especially useful for the isolation of serotype 0:8 strains (Schiemann, 1983).

Composition (grams)

Di-sodium hydrogen orthophosphate $\cdot 7H_2O$	17.25
Sodium oxalate	5.0
Bile salts No. 3	2.0
Sodium chloride	1.0
Magnesium sulphate $\cdot 7H_2O$	0.01
Sorbose*	10.0
Asparagine*	1.0
Methionine*	1.0
Metanil yellow*	0.025
Yeast extract*	0.025
Sodium pyruvate*	0.05
2,3,4'-Trichloro-2'-hydroxyl diphenyl ether* (Irgasan, Ciba-Ceigy)	0.004
Sodium furadantin*	0.01
Distilled or deionized water	659.0

Preparation

Dissolve ingredients except those marked * in water. Adjust pH to 7.6. Autoclave at 121°C for 15 min. Cool to 50°C and add the following filter sterilized solutions:

Sorbose (10%)	100.0 ml
Asparagine (1.0%)	100.0 ml

Methionine (1.0%)	100.0 ml
Metanil yellow (2.5 mg/ml)	10.0 ml
Yeast extract (2.5 mg/ml)	10.0 ml
Sodium pyruvate (0.5%)	10.0 ml
Irgasan (0.4% in 95% ethanol)	1.0 ml

Adjust pH to 7.6. On day of use add the following solution:

Sodium furadantin (1.0 mg/ml)	10.0 ml

Distribute aseptically in the required volumes in flasks.

Physical properties

Appearance	Yellow, clear.
pH	7.6 ± 0.1

Shelf life

Prepared medium (without sodium furadantin)	7 days at 4 ± 2°C.

Inoculation method for samples

Add food sample to pre-enrichment medium in proportion 1:10. Incubate for 9 days at 4°C or for 20–24 h at 24 ± 2°C. Inoculate pre-enriched medium in BOS broth in proportion 1:10.

Incubation method

At 24 ± 2°C for 5 days, in air.

Reading of results and interpretation

Growth of yersiniae usually results in obvious turbidity. The medium should always be cultured by streaking one loopful on CIN agar.

Quality assessment

(i) *Productivity*

Test strains	*Yersinia enterocolitica*, biotype 4 serotype 0:3 NCIMB 50087 *Yersinia enterocolitica*, biotype 1 serotype 0:8 NCIMB 50085

Inoculation method	Dilution to extinction.
Criteria	Growth should be within 2 titre units of the growth in tryptone soya broth after 5 days at 24°C.

(ii) *Selectivity*

Test strains	*Proteus mirabilis* (ATCC 29906/CECT 4168/NCTC 11938) *Pseudomonas aeruginosa* NCIMB 50067
Inoculation method	Dilution to extinction.
Criteria	Difference in growth should be equal to or less than 5 titre units of the growth in tryptone soya broth after 5 days at 24°C.

References

Kwaga, J., Iversen, J.O. and Saunders, J.R. (1990) Comparison of two enrichment protocols for the detection of *Yersinia* in slaughtered pigs and pork products. J. Food Protect. 53, 1047–1049.

Schiemann, D.A. (1982) Development of a two-step enrichment procedure for recovery of *Yersinia enterocolitica* from food. Appl. Environ. Microbiol. 43, 14–27.

Schiemann, D.A. (1983) Comparison of enrichment and plating media for recovery of virulent strains of *Yersinia enterocolitica* from inoculated beef stew. J. Food Protect. 46, 957–964.

Bile Salts Irgasan Brilliant Green (BSIBG) agar

This monograph has been assessed by members of the IUMS-ICFMH Working Party on Culture Media and given 'Approved' status.

Description and history

This medium was originally designed for the selective isolation of *Aeromonas* spp. from faeces (Hunt et al., 1987). Gram positive organisms are inhibited by bile salts and brilliant green and Gram negative organisms which possess a type A nitratase are inhibited by Irgasan. Organisms which survive the selective process are differentiated by their ability to attack xylose. *Aeromonas* spp. do not ferment xylose and so oxidase tests may be performed on colonies that do not produce acid.

The absence of antibiotics from this medium allows the isolation of ampicillin sensitive strains (Rahim et al., 1984) and the performance of the current formulation has been shown to be superior to that of ampicillin-containing media for the isolation of *Aeromonas* spp. from foods (Fricker and Tomsett, 1989; Gobat and Jemmi, 1995).

Composition (grams)

Beef extract	5.0
Proteose peptone	5.0
Xylose	10.0
Bile salts No. 3	8.5
Sodium thiosulphate	5.44
2,3,4'-Trichloro-2'-hydroxyldiphenyl ether (Irgasan, Ciba-Geigy)	0.005
Brilliant green	0.005
Neutral red	0.025
Agar	11.5
Distilled or deionized water	1000.0

Preparation

Add the ingredients, including 10 ml of Irgasan solution (50 mg/100 ml in ethanol), to the water, bring gently to the boil to dissolve completely and hold for one minute at boiling point. Allow to cool and pour into Petri dishes.

Physical properties

Appearance	Purple, transparent.
pH	7.0 ± 0.2

Shelf life

Ready to use medium	4 weeks at 4 ± 2°C.

Inoculation method for samples

1. Direct counting:
   surface spreading of dilutions of food homogenates over the whole plate.

2. Subculturing after enrichment:
   a loopful of incubated alkaline peptone water is steaked onto the medium to obtain isolated colonies.

Incubation method

At 35–37°C for 18–24 h, in air.

Reading results and interpretation

Mesophilic aeromonads will appear as translucent colonies 1–2 mm in diameter which must be distinguished from *Pseudomonas* spp. The majority of contaminating flora will be inhibited on the selective medium or will appear as mauve/green opaque colonies often surrounded by a zone of precipitated bile salts. Confirmation of presumptive aeromonad colonies is necessary.

Quality assessment

(i) *Productivity*

Test strains	*Aeromonas hydrophila* NCIMB 50013 *Aeromonas sobria* (ATCC 35993/NCTC 11215)
Inoculation method	Modified Miles-Misra or spiral plating.

(ii) *Selectivity*

Test strains	*Escherichia coli* NCIMB 50034 *Proteus mirabilis* (ATCC 29906/CECT 4168/NCTC 11938)
Inoculation method	Modified Miles-Misra or spiral plating.

(iii) *Characteristic appearance of colonies*
See above.

References

Fricker, C.R. and Tompsett, S. (1989) *Aeromonas* spp. in foods: A significant cause of food poisoning? Int. J. Food Microbiol. 9, 17–23.

Gobat, P.F. and Jemmi, T. (1995) Comparison of seven selective media for the isolation of mesophilic *Aeromonas* species in fish and meats. Int. J. Food Microbiol. 24, 375–384.

Hunt, G.H., Price, E.H., Patel, U., Messenger, L., Stow, P. and Salter, P. (1987) Isolation of *Aeromonas* sp. from faecal specimens. J. Clin. Pathol. 40, 1382–1384.

Rahim, Z., Sanyal, S.C., Aziz, K.M.S., Huq, M.I. and Chowdhury, A.A. (1984) Isolation of enterotoxigenic, hemolytic, and antibiotic-resistant *Aeromonas hydrophila* strains from infected fish in Bangladesh. Appl. Env. Microbiol. 48, 865–867.

Bismuth sulphite agar

This monograph has been assessed by members of the IUMS-ICFMH Working Party on Culture Media and given 'Approved' status.

Description and history

This is a modification of the original Wilson and Blair (1927) selective diagnostic medium used for the isolation of *Salmonella typhi* and other salmonellae from food and other materials e.g. sewage, water, etc. Bismuth sulphite and brilliant green are the selective agents. The medium also contains indicators for sulphide production. The use of this medium is advocated by several authorities (Andrews et al., 1995; Flowers et al., 1992) and it may be particularly useful when lactose fermenting strains of salmonellae are sought.

Composition (grams)

Beef extract	5.0
Peptone	10.0
Glucose	5.0
di-Sodium hydrogen orthophosphate $\cdot 12H_2O$	4.0
Iron (II) sulphate $\cdot 7H_2O$	0.3
Bismuth ammonium citrate	1.85
Sodium sulphite	6.15
Brilliant green	0.016
Agar	20.0
Distilled or deionized water	1000.0

* concentration varies with formulation and activity of the dye

Preparation

Suspend the ingredients in the water. Mix and heat very carefully to boiling, with frequent agitation, to dissolve soluble material. An insoluble precipitate is formed. Cool to 45–50°C. Adjust pH to 7.6 ± 0.2. Keeping the precipitate in suspension, pour 15–20 ml quantities into 9 cm diameter Petri dishes or about 40 ml into 14 cm diameter Petri dishes.

Physical properties

Appearance	Pale green/straw, opaque with flocculent precipitate which must be uniformly dispersed in the liquid medium.
pH	7.6 ± 0.2

Shelf life

Ready to use medium	5 days at 4°C dependent on formulation and specific application. Manufacturers of some dehydrated preparations recommend storage of the plated medium for a minimum period of one to three days, others recommend use on the day of preparation. In general the medium becomes less inhibitory on storage.

Inoculation method for samples

Streak inoculation on pre-dried plates (see above) from selective enrichment broths by method specified by van Leusden et al. (1982). Other streaking techniques which result in well-isolated colonies may be satisfactory.

Incubation method

At 37 ± 1°C for 24–48 h, in air.

Reading of results and interpretation

After 24 h well-isolated colonies of most salmonellae are small, black, flat or raised, with blackening of the medium and characteristic metallic sheen. These reactions are intensified on a further 24 h incubation. Non-H_2S producing strains of salmonellae grow as green colonies with no blackening. Because of this and other variations in characteristics of salmonellae on bismuth sulphite agar, almost any growth on the medium should be subject to further tests.

Quality assessment

(i) *Productivity*

Test strains	*Salmonella enteritidis* NCIMB 50073 *Salmonella virchow* NCIMB 50077
Supplementary strain	*Salmonella saintpaul* NCIMB 50075
Inoculation method	Modified Miles-Misra, spiral plating or streaking/ecometry.

(ii) *Selectivity*

Test strains	*Escherichia coli* NCIMB 50034 *Proteus mirabilis* (ATCC 29906/CECT 4168/NCTC 11938) – qualitative test for control of swarming.
Inoculation method	Modified Miles-Misra, spiral plating or streaking/ ecometry.

(iii) *Characteristic appearance of colonies*
See above.

References

Andrews, W.H., June, G.A., Sherrod, P.S., Hammack, T.S. and Amaguana, R.M. (1995) FDA Bacteriological Analytical Manual, 8th edn., Chapter 5, Salmonella.

Flowers, R.S., D'Aoust, J-Y, Andrews, W.H. and Bailey, J.S. (1992) Salmonella. *In:* Compendium of Methods for the Microbiological Examination of Foods. 3rd edn., American Public Health Association, Washington, DC.

van Leusden, F.M., van Schothorst, M. and Beckers, H.J. (1982) The standard *Salmonella* isolation method. *In:* Isolation and identification methods for food poisoning organisms, edited by J.E.L. Corry, D. Roberts and F.A. Skinner. SAB Technical Series No. 17. Academic Press, London, pp. 35–49.

Wilson, W.J. and Blair, E.M.M'V. (1927) Use of a glucose bismuth sulphite iron medium for the isolation of *B. typhosa* and *B. proteus*. J. Hyg. 26, 374–391.

Handbook of Culture Media for Food Microbiology, J.E.L. Corry et al. (Eds.)

Briggs agar

This monograph has been reviewed by members of the IUMS-ICFMH Working Party on Culture Media and given 'Proposed' status.

Description and history

Briggs tomato juice agar (Briggs, 1953) was developed primarily for the cultivation of lactobacilli from milk and dairy products and has since been successfully used for strains from other sources (Cox and Briggs, 1954; Lerche and Reuter, 1960; Reuter, 1964). It may be used for the non-selective enumeration of the whole group of lactic acid bacteria and has better productivity for some strains of *Lactobacillus delbrueckii* ssp. *bulgaricus*, *Streptococcus thermophilus* and *Lactococcus lactis* ssp. *lactis* than MRS agar.

Composition (grams)

Tryptic digest of casein	8.0
Peptic digest of meat	8.0
Yeast extract	6.0
Glucose	20.0
Starch, soluble	0.5
Sorbitan monooleate (Tween 80)	1.0
Sodium chloride	5.0
Agar	20.0

Distilled or deionized water 600.0, 800.0 or 900.0 (depending on tomato juice used, a, b or c)

additionally:

a) *Tomato juice dilution* — One part of the commercial product without salt and preserving agents is diluted with two parts water, boiled and filtered. 400 ml of this dilution is added to 600 ml basal agar.

b) *Fresh tomato juice* — Filter the juice of freshly pulped tomatoes to remove seeds and coarse particles. Autoclave and add 200 ml to 800 ml of the basal agar.

c) *Tomato extract dilution* Mix 40 g tomato extract (triple), 5.75 g Na_2HPO_4, 0.8 g KH_2PO_4 in 500 ml water, filter and autoclave at 121°C for 15 min. Add 100 ml to 900 ml basal agar. The remainder may be stored for future use.

Preparation

Suspend the ingredients in the appropriate amount of water and add the diluted tomato juice or tomato extract dilution. Heat to boiling to dissolve completely. Adjust pH to 6.8 with 10% NaOH. Sterilize in the autoclave for 15 min at 121°C. Cool to 50°C and distribute into sterile Petri dishes. Avoid overheating this medium which hydrolyses the agar, resulting in a soft medium.
N.B. The tomato juice must be free of Cu contamination.

Physical properties

Appearance	Medium to dark amber, slightly opalescent, may have a slight precipitate.
pH	6.8 ± 0.2

Shelf life

Tomato extract dilution	4–6 weeks at 4 ± 2°C.
Ready to use medium	14 days at 4 ± 2°C.

Inoculation method for samples

Surface spreading over whole plate or modified Miles-Misra.

Incubation method

This depends on the particular habitat of the organisms to be cultivated. Dairy strains should be incubated at 30°C for 2 days followed by 1 day at 22°C, intestinal or yoghurt strains for 2 days at either 37 or 42°C. All incubations should be performed under anaerobic or microaerobic (6% 0_2: 10% CO_2 in N_2) conditions.

Reading of results and interpretation

Briggs agar is an elective medium that gives good colony counts and a similar colony size and appearance for all lactic acid bacteria. Other microorganisms may be distinguished by pigments and extraordinary colonial morphology (micrococci, Gram negative species, yeasts). Confirmation tests e.g. Gram stain, catalase reaction, are necessary to distinguish mixed microflora. A selective medium for lactobacilli or lactic acid bacteria should be used in parallel where very mixed microflora are present.

Quality assessment

(i) *Productivity*

Test strains	*Lactobacillus delbrueckii* ssp. *bulgaricus* NCIMB 50050 *Streptococcus thermophilus* NCIMB 50083 *Lactococcus lactis* ssp. *lactis* NCIMB 50058
Inoculation method	Modified Miles-Misra, spiral plating or streaking/ ecometry.
Criteria	Recovery on Briggs agar should be within 1 $\log_{10}$ of recovery on MRS agar.

(ii) *Characteristic appearance of colonies*

Small greyish-white colonies, 1–3 mm diameter, flat or raised, smooth, rough or intermediate. Non-lactic acid bacteria e.g. yeasts and Gram-negative species, may be distinguished by colony size and appearance after further incubation at room temperature.

References

Briggs, M.J. (1953) An improved medium for lactobacilli. J. Dairy Res. 20, 36–40.

Cox, C.P. and Briggs, M.J. (1954) Experiments on growth media for lactobacilli. J. Appl. Bacteriol. 17, 18–26.

Lerche, M. and Reuter, G. (1960) Beitrag zur Methodik der Isolierung und Differenzierung von aerob wachsenden Laktobazillen. Zbl. Bakteriol. I. Orig. 179, 354–370.

Reuter, G. (1964) Vergleichsuntersuchungen an 90 thermophilen Lactobazillen Stammen verschiedener Herkunft. Zbl. Bakteriol. I. Orig. 193, 454–466.

Brilliant Green Bile (BGB) broth

This monograph has been assessed by members of the IUMS-ICFMH Working Party on Culture Media and given 'Approved' status.

Description and history

This broth is a modification of MacConkey's liquid medium for the isolation of Enterobacteriaceae, formulated by Dunham and Schoenlein in 1926 to attain maximum recovery of bacteria of the coli-aerogenes group, while inhibiting most Gram positive organisms which might hinder the development of the bacteria sought. It contains brilliant green and bile as the inhibitory agents for Gram positive organisms and lactose as the carbon source, which is dissimilated rapidly by the coli-aerogenes group, mostly by a heterofermentative pathway, leading to gas formation. Mackenzie et al. (1948) found it markedly superior to MacConkey broth for confirmation of *Escherichia coli* at 44°C and it is now common practice to carry out preliminary MPN tests using a less selective medium such as lauryl tryptose broth or minerals modified glutamate medium and confirm any tubes showing a positive reaction by subculture to BGB broth.

Composition (grams)

Peptone	10.0
Lactose	10.0
Ox bile	20.0
Brilliant green	0.0133
Distilled or deionized water	1000.0

Preparation

Dissolve the ingredients in water, distribute in the required volumes in flasks or tubes and heat at 100°C for 30 min. Although lactose media are markedly less sensitive to heat damage than those containing glucose, the performance of the medium, with respect to both selectivity and productivity is much more consistent when it is decontaminated by standardised pasteurization as recommended.

Physical properties

Appearance	Green, clear.

pH	7.4 ± 0.2

Shelf life

Ready to use medium	1 month at 4 ± 2°C in screw capped containers.

Inoculation method for samples

Use food macerates or decimal dilutions and inoculate the broth in the proportion 1:9. BGB broth can be used at double strength. In this instance, it must not be autoclaved. Equal volumes of food macerates, or decimal dilutions (or fluid samples) are added to double strength BGB broth.

Incubation method

To detect members of the coli-aerogenes group incubate at 35 or 37°C for 24–48 h. For dairy purposes and where it is desired to detect the full range of coliform bacteria, incubate at 30°C for 24–48 h. For *Escherichia coli* the temperature of 44 ± 0.1°C for 18 h is specifically recommended. Psychrotrophic coliforms can be detected by incubation at 4°C for 10 days.

Reading of results and interpretation

Turbidity and often change of colour of the medium towards yellowish-green provides presumptive evidence of the presence of bacteria of the coli-aerogenes group, particularly when accompanied by copious gas formation. This should be confirmed by isolation on violet red bile glucose agar and subsequent study of the mode of attack on glucose and negative oxidase reaction, preferably by the technique described in the monograph on VRB agar.

Quality assessment

(i) *Productivity*

Test strains	*Escherichia coli* NCIMB 50034 *Citrobacter freundii* NCIMB 50025
Inoculation method	Dilution to extinction.
Criteria	Recovery in BGB broth should be within one titre unit of the recovery in tryptone soya broth and copious gas formation should occur with *Escherichia coli* and *Citrobacter freundii* within 18–24 h at 30°C; the same should apply for *Escherichia coli* only after incubation for 24 h at 44 ± 1°C.

(ii) *Selectivity*

Test strains	*Enterococcus faecalis* NCIMB 50030 *Staphylococcus aureus* NCIMB 50080
Supplementary strains	*Bacillus cereus* NCIMB 50014 *Lactococcus lactis* ssp. *lactis* NCIMB 50058
Inoculation method	Dilution to extinction.
Criteria	Recovery in BGB should be less than 5 titre units of the recovery in tryptone soya broth after 18–24 h at 30°C. At 44 $\pm$ 0.1°C the medium should be even more selective.

References

Dunham, H.G. and Schoenlein, H.W. (1926) Brilliant green bile media. Stain Technol. 1, 129–134.
MacKenzie, E.F.W., Taylor, E.W. and Gilbert, W.E. (1948) Recent experiences in the rapid identification of *Bacterium coli*, type 1. J. Gen. Microbiol 2, 197–204.

Cefixime Tellurite Sorbitol MacConkey (CT-SMAC) agar

This monograph has been assessed by members of the IUMS-ICFMH Working Party on Culture Media and given 'Approved' status.

Description and history

CT-SMAC agar was developed by Zadik et al. (1993) as a modification of Sorbitol MacConkey (SMAC) agar (March and Ratnam, 1986) for the isolation of verocytotoxigenic *Escherichia coli* O157 from faeces and foods. Sorbitol was initially added to MacConkey agar in place of lactose to differentiate *Escherichia coli* O157 strains, most of which do not ferment sorbitol within 24 h, from other *Escherichia coli* strains, which are predominantly fast sorbitol fermenters. *Escherichia coli* O157 produces colourless colonies on SMAC agar, while other *Escherichia coli* strains form pink colonies.

Addition of cefixime and tellurite to SMAC inhibits *Proteus* spp. and *Aeromonas* spp. which are often sorbitol negative, non-O157 *E. coli* and other unwanted organisms.

Composition (grams)

Peptone	20.0
Sorbitol	10.0
Bile salts No.3	1.5
Sodium chloride	5.0
Neutral red	0.03
Crystal violet	0.001
Cefixime*	0.00005
Potassium tellurite*	0.0025
Agar	15.0
Distilled or deionized water	1000.0

Preparation

Suspend the ingredients except those marked * in the water and heat to boiling to dissolve completely. Sterilize by autoclaving at 121°C for 15 min. Add to the cooled (50°C) medium 1.0 ml of a solution of cefixime in ethanol (0.05 mg/ml) and 1.0 ml of a filter sterilized aqueous solution of potassium tellurite (2.5 mg/ml).

Physical properties

Appearance	Pale violet.
pH	7.1 ± 0.2 at 45°C.

Shelf life

Ready to use medium	2 weeks at 10–15°C.

Inoculation method for samples

Surface streaking of dry plates for isolation from enrichment broth or directly from dilutions of food homogenates.

Incubation method

At 37°C for 18–20 h, in air.

Reading of results and interpretation

Colourless or brownish colonies are considered to be presumptive *Escherichia coli* O157:H7. Colonies confirmed as belonging to the species *Escherichia coli* and which are negative for β-glucuronidase and positive in an *Escherichia coli* O157 latex test, should be further investigated for verocytotoxin production and other virulence characteristics.

Quality assessment

(i) *Productivity*

Test strain	*Escherichia coli* O157:H7 NCIMB 50139 – this strain is non-toxigenic.
Inoculation method	Modified Miles-Misra, spiral plating or streaking/ecometry.

(ii) *Selectivity*

Test strains	*Escherichia coli* NCIMB 50034 *Hafnia alvei* NCIMB 50037 *Enterococcus faecalis* NCIMB 50030 *Proteus mirabilis* (ATCC 29906/CECT 4168/NCTC 11938)
Supplementary strain	*Escherichia coli* NCIMB 50109

Inoculation method	Modified Miles-Misra, spiral plating or streaking/ ecometry.

(iii) *Characteristic appearance of colonies*
Colourless or brownish colonies of more than 1 mm diameter.

References

March, S.B. and Ratnam, S. (1986) Sorbitol MacConkey medium for the detection of *Escherichia coli* O157:H7 associated with hemorrhagic colitis. J. Clin. Microbiol. 23, 869–872.

Zadik, P.M., Chapman, P.A. and Siddons, C.A. (1993) Use of tellurite for the selection of verocytotoxigenic *Escherichia coli* O157. J. Med. Microbiol. 39, 155–158.

Cefoperazone Amphotericin Teicoplanin (CAT) agar

This monograph has been reviewed by members of the ICFMH-IUMS Working Party on Culture Media and given 'Proposed' status.

Description and history

This employs the blood free basal medium described by Hutchinson and Bolton in 1984 (Modified CCD agar) which was an improved version of the original CCDA agar (Bolton et al., 1984). Cefoperazone is incorporated, as with Modified CCD agar but at a reduced concentration, together with amphotericin B and teicoplanin as selective agents. The medium was developed for the isolation of *Campylobacter* spp. including *C. upsaliensis* from human and animal faeces (Aspinall et al., 1993). Subsequently it has proved to be successful in the isolation of thermophilic *Campylobacter* spp. including *C. upsaliensis* without the need for membrane filtration (Aspinall et al., 1996).

Composition (grams)

Nutrient broth No 2	25.0
Bacteriological charcoal	4.0
Casein hydrolysate	3.0
Sodium deoxycholate	1.0
Iron (II) sulphate	0.25
Sodium pyruvate	0.25
Cefoperazone*	0.008
Amphotericin*	0.010
Teicoplanin*	0.004
Agar	12.0
Distilled or deionized water	1000.0

Preparation

Suspend the above ingredients except those marked * in the water and bring to the boil to dissolve completely. Sterilize by autoclaving at 121°C for 15 min. Cool to 50°C and aseptically add the cefoperazone, amphotericin and teicoplanin to give the final concentrations stated.

Physical properties

Appearance	Black, opaque.
pH	7.4 ± 0.2

Shelf life

Ready to use medium	10 days at 4 ± 2°C.

Inoculation method for samples

Samples can be inoculated either directly onto the medium or using a 1 in 10 dilution of the food in 0.1% bacteriological peptone.

Incubation method

At 37°C for 48 h microaerobically in an atmosphere containing approximately 5% O_2, 10% CO_2 and 85% N_2 or H_2.

Reading of results and interpretation

Campylobacter spp. produce grey, moist, flat and occasionally spreading growth which may be accompanied by a green hue and/or a metallic sheen. *C. upsaliensis* strains often produce slightly smaller colonies than *C. jejuni*.

Quality assessment

(i) *Productivity*

Test strains	*Campylobacter jejuni* NCIMB 50091 *Campylobacter coli* NCIMB 50092 *Campylobacter upsaliensis* (ATCC 43953/NCTC 11540)
Inoculation method	Modified Miles-Misra, spiral plating or streaking/ecometry.

(ii) *Selectivity*

Test strains	*Proteus mirabilis* (ATCC 29906/CECT 4168/NCTC 11938) *Pseudomonas aeruginosa* NCIMB 50067 *Escherichia coli* NCIMB 50034
Inoculation method	Modified Miles-Misra, spiral plating or streaking/ecometry.

(iii) *Characteristic appearance of colonies*
See above.

References

Aspinall, S.T., Wareing, D.R.A., Hayward, P.G. and Hutchinson, D.N. (1993) A selective medium for thermophilic campylobacters including *Campylobacter upsaliensis*. J. Clin. Pathol. 46, 829–831.

Aspinall, S.T., Wareing, D.R.A., Hayward, P.G. and Hutchinson, D.N. (1996) A comparison of a new campylobacter selective medium (CAT) with membrane filtration for the isolation of thermophilic campylobacters including *Campylobacter upsaliensis*. J. Appl. Bacteriol. 80, 645–650.

Bolton, F.J., Hutchinson, D.N. and Coates, D. (1984) Blood-free selective medium for isolation of *Campylobacter jejuni* from faeces. J. Clin. Microbiol. 19, 169–171.

Hutchinson, D.N. and Bolton, F.J. (1984) An improved blood-free selective medium for the isolation of *Campylobacter jejuni* from faecal specimens. J. Clin. Pathol. 37, 959–957.

Cefsulodin Irgasan Novobiocin (CIN) agar

This monograph has been assessed by members of the IUMS-ICFMH Working Party on Culture Media and given 'Approved' status.

Description and history

The medium was originally formulated by Schiemann (1979) for detection of *Yersinia enterocolitica*. He subsequently (1982) revised it by substituting sodium deoxycholate for bile salts and reducing the novobiocin content. It relies on the use of the selective inhibitory components sodium deoxycholate, crystal violet, cefsulodin, irgasan and novobiocin. The indicative principle is fermentation of mannitol with localised pH reduction which forms a red colony due to the neutral red, and a zone of precipitation due to the deoxycholate.

Composition (grams)

Peptone	20.0
Yeast extract	2.0
Mannitol	20.0
Sodium pyruvate	2.0
Sodium chloride	1.0
Magnesium sulphate $\cdot 7H_2O$	0.01
Sodium deoxycholate	0.5
Neutral red	0.03
Crystal violet	0.001
Cefsulodin*	0.015
2,4,4'-Trichloro-2'-hydroxyl diphenyl ether*	
(Irgasan, Ciba-Geigy)	0.004
Novobiocin*	0.0025
Agar	15.0
Distilled or deionized water	1000.0

Preparation

Suspend the ingredients (except those marked *) in 970 ml water. Bring to the boil to dissolve completely. Allow to cool to 50°C and aseptically add the following filter-sterilized solutions: (i) 10 ml of cefsulodin (150 mg/100 ml ethanol/water 1:1); (ii) 10 ml of irgasan (40 mg/100 ml ethanol); (iii) 10 ml of novobiocin (25 mg/100 ml water).

Mix gently and pour into sterile Petri dishes.

Physical properties

Appearance	Pink, clear.
pH	7.4 ± 0.2

Shelf life

Ready to use medium	7 days at 4 ± 2°C.

Inoculation method for samples

Surface spreading over whole plate using 0.1 ml per 9 cm diameter pre-dried plate.

Incubation method

At 30°C for 18–24 h or at 22°C for 48 h, in air.

Reading of results and interpretation

Yersinia enterocolitica appears as round pink colonies about 2 mm diameter, with dark pink centres surrounded by a zone of precipitated bile. Confirmatory tests are required.

Quality assessment

(i) *Productivity*

Test strains	*Yersinia enterocolitica*, biotype 4 serotype 0:3 NCIMB 50087 *Yersinia enterocolitica*, biotype 1 serotype 0:8 NCIMB 50085
Inoculation method	Modified Miles-Misra, spiral plating or streaking/ ecometry.

(ii) *Selectivity*

Test strains	*Escherichia coli* NCIMB 50034 *Pseudomonas aeruginosa* NCIMB 50067
Supplementary strains	*Proteus mirabilis* (ATCC 29906/CECT 4168/NCTC 19938) *Staphylococcus aureus* NCIMB 50080

Inoculation method	Modified Miles-Misra, spiral plating or streaking/ ecometry.

(iii) *Characteristic appearance of colonies*
Round pink, about 2 mm in diameter with a dark pink centre and surrounded by a precipitation zone.

References

Schiemann, D.A. (1979) Synthesis of a selective agar medium for *Yersinia enterocolitica*. Can. J. Microbiol. 25, 1298–1304.

Schiemann, D.A. (1982) Development of a two step enrichment procedure for recovery of *Yersinia enterocolitica* from food. Appl. Environ. Microbiol. 43, 14–27.

Cellobiose Polymyxin B Colistin (CPC) agar

This monograph has been reviewed by members of the IUMS-ICFMH Working Party on Culture Media and given 'Proposed' status.

Description and history

The medium was developed for the isolation of *Vibrio vulnificus* and *Vibrio cholerae* from environmental sources (Massad and Oliver, 1987). It was shown to be superior to other media in its ability to select and differentiate *Vibrio vulnificus* from other species found in the environment (Oliver et al., 1992). Its use in the examination of oysters was reported by Kaysner et al. (1989) and Tamplin et al. (1991). The medium is less inhibitory to *Vibrio vulnificus* than is TCBS, the medium used for isolation of most other pathogenic *Vibrio* spp. The selective components of CPC agar are the antibiotics colistin and polymyxin B, and a high incubation temperature (40°C). Differentiation is based on cellobiose fermentation which is detected by the pH indicators.

Composition (grams)

Solution 1	
Peptone	10.0
Beef extract	5.0
Sodium chloride	20.0
Bromothymol blue	0.04
Cresol red	0.04
Agar	15.0
Distilled or deionized water	900.0
Solution 2	
Cellobiose	15.0
Colistin methanesulphonate (i.u.)	1 360 000.0
Polymyxin B (i.u)	100 000.0
Distilled or deionized water	100.0

Preparation

Suspend the ingredients for Solution 1 in water. Adjust to pH 7.6. Autoclave at 121°C for 15 min. Cool to 55°C. For solution 2, dissolve cellobiose in water by heating gently. Cool. Add antibiotics. Filter sterilize solution 2. Add to cooled solution 1. Pour plates.

mCPC Modification

CPC agar has been used in a modified form to make it less inhibitory to *Vibrio vulnificus*. Modified CPC (mCPC) is used by the U.S. FDA. The modifications are a reduction of cellobiose concentration to 10 g/l and colistin methanesulphonate to 400 000 i.u./l. Solution 1 may be boiled rather than autoclaved and filter sterilization of Solution 2 is not necessary.

CC Modification

This modification was proposed by Høi et al. (1998) for the isolation of *Vibrio vulnificus* from environmental sources. Polymyxin is omitted, cellobiose content reduced to 10.0 g/l and colistin to 400 000 i.u./l.

Physical properties

Appearance	Olive green to light brown-purple.
pH	7.6 ± 0.2

Shelf life

Use freshly prepared medium. Old plates of CPC are more inhibitory, especially to *Vibrio cholerae*, than is freshly prepared medium.

Inoculation method for samples

Surface spreading or streaking over plate from sample, homogenate or alkaline peptone water enrichment broth culture.

Incubation method

At 39–40°C for 24 to 48 h, in air. Lower temperature incubation decreases the inhibitory nature of the medium and allows growth of some other species of *Vibrio*.

Reading of results and interpretation

The medium is very selective and most other *Vibrio* spp. will not grow on it. *Vibrio vulnificus* produces flat yellow (cellobiose fermenting) colonies with opaque centres and translucent peripheries, about 2 mm in diameter. *Vibrio cholerae* El Tor produces raised purple or green colonies. Classical *Vibrio cholerae* are sensitive to polymyxin B and will not grow on the medium.

Quality assessment

(i) *Productivity*

Test strains	*Vibrio vulnificus* NCIMB 50104 *Vibrio cholerae* (NCTC 11348)
Inoculation method	Modified Miles-Misra, spiral plating or streaking/ ecometry.

(ii) *Selectivity*

Test strains	*Vibrio fluvialis* NCIMB 50084 *Vibrio parahaemolyticus* (NCTC 11344) *Photobacterium leiognathi* NCIMB 50103 *Pseudomonas aeruginosa* NCIMB 50067
Inoculation method	Modified Miles-Misra, spiral plating or streaking/ ecometry.

(iii) *Characteristic appearance of colonies*
See above.

References

Høi, L., Dalsgaard, I. and Dalsgaard, A. (1998) Improved isolation of *Vibrio vulnificus* from seawater and sediment using cellobiose-colistin agar. Appl. Environ. Microbiol. 64, 1721–1724.

Kaysner, C.A., Tamplin, M.L., Wekell, M.M., Stott, R.F. and Colburn, K.G. (1989) Survival of *Vibrio vulnificus* in shellstock and shucked oysters (*Crassostrea gigas* and *Crassostrea virginica*) and effects of isolation medium on recovery. Appl. Environ. Microbiol. 55, 3072–3079.

Massad, G. and Oliver, J.D. (1987) New selective and differential medium for *Vibrio cholerae* and *Vibrio vulnificus*. Appl. Environ. Microbiol. 53, 2262–2264.

Oliver, J.D., Guthrie, K., Preyer, J., Wright, A., Simpson, L.M., Siebeling, R. and Morris, Jr., J.G. (1992) Use of colistin-polymyxin B-cellobiose agar for isolation of *Vibrio vulnificus* from the environment. Appl. Environ. Microbiol. 58, 737–739.

Tamplin, M.L., Martin, A.L., Ruple, A.D., Cook, D.W. and Kaspar, C.W. (1991) Enzyme immunoassay for identification of *Vibrio vulnificus* in seawater, sediment, and oysters. Appl. Environ. Microbiol. 57, 1235–1240.

Cephaloridine Fucidin Cetrimide (CFC) agar

This monograph has been assessed by members of the IUMS-ICFMH Working Party on Culture Media and given 'Approved' status.

Description and history

The selectivity of CFC agar depends upon a specific combination of antibacterial compounds: cephaloridine, fucidin and cetrimide (Mead and Adams, 1977). The medium has no differential properties and aims to suppress virtually all unwanted organisms. Although developed for use in the microbiological examination of poultry-meat products, CFC agar can also be applied to other foods for isolating *Pseudomonas* spp. (Gardner, 1980; Banks and Board, 1983).

Composition (grams)

Heart infusion agar	40.0
Cephaloridine (Ceporin)	0.05
Sodium fusidate (Fucidin)	0.01
Cetyltrimethyl ammonium bromide (Cetrimide)	0.01
Distilled or deionized water	1000.0

Preparation

Prepare 1 litre of heart infusion agar. Dispense into bottles (or final containers) and sterilize by autoclaving at 121°C for 15 min. Cool the sterile basal medium to 50–52°C and add the required amounts of the following filter-sterilized aqueous solutions: (i) 5 ml of 1% cephaloridine; (ii) 1 ml of 1% sodium fusidate; (iii) 1 ml of 1% cetyltrimethyl ammonium bromide.

Physical properties

Appearance	Straw coloured.
pH	7.4 ± 0.2

Shelf life

Ready to use medium	At least 1 month at 20 ± 2°C.

Inoculation method for samples

Surface spreading over whole plate using 0.1 ml per 9 cm diameter pre-dried plate. Ensure adequate ventilation of plates during incubation.

Incubation method

At 25°C for 48 h, in air.

Reading of results and interpretation

There are two types of colonies, pigmented and non-pigmented. Round, white or cream colonies or greenish, fluorescent colonies, 2–5 mm diameter are considered to be pseudomonads. Confirmatory tests are not usually required but in cases of doubt plates can be flooded with oxidase reagent (1% tetramethyl-*p*-phenylenediamine dihydrochloride). Colonies showing an outer dark-purple ring within 10 seconds are counted as pseudomonads.

Quality assessment

(i) *Productivity*

Test strains	*Pseudomonas fluorescens* NCIMB 50068 *Pseudomonas fragi* NCIMB 50069
Supplementary strain	*Pseudomonas putida* NCIMB 50070
Inoculation method	Modified Miles-Misra, spiral plating or streaking/ ecometry.

(ii) *Selectivity*

Test strains	*Acinetobacter* sp. NCIMB 50011 *Escherichia coli* NCIMB 50034
Inoculation method	Modified Miles-Misra, spiral plating or streaking/ ecometry.

(iii) *Characteristic appearance of colonies*

Round, white or cream colonies or greenish, fluorescent colonies, 2–5 mm diameter.

References

Banks, J.G. and Board, R.G. (1983) The classification of pseudomonads and other obligately aerobic Gram-negative bacteria from British pork sausage and other ingredients. Syst. Appl. Microbiol. 4, 424–438.

Gardner, G.A. (1980) The occurrence and significance of *Pseudomonas* in Wiltshire bacon brines. J. Appl. Bacteriol. 48, 69–74.

Mead, G.C. and Adams, B.W. (1977) A selective medium for the rapid isolation of pseudomonads associated with poultry meat spoilage. Brit. Poult. Sci. 18, 661–670.

Charcoal Cefoperazone Deoxycholate (CCD) agar – modified

This monograph has been assessed by members of the IUMS-ICFMH Working Party on Culture Media and given 'Approved' status.

Description and history

This blood-free medium is an improved version of the original CCD agar (Bolton et al., 1984). Modified CCD agar (Hutchinson and Bolton, 1984) was developed for the direct isolation of campylobacters from human and animal faeces and its major constituents are charcoal, cefoperazone and sodium deoxycholate. It has subsequently proved to be a very successful subculture medium when used in conjunction with Preston enrichment broth (Bolton et al., 1986).

Composition (grams)

Beef extract	10.0
Peptone	10.0
Sodium chloride	5.0
Charcoal	4.0
Casein hydrolysates	3.0
Iron (II) sulphate	0.25
Sodium pyruvate	0.25
Sodium deoxycholate	1.0
Cefoperazone*	0.032
Amphotericin*	0.01
Agar	15.0
Distilled or deionized water	1000.0

Preparation

Suspend the above ingredients except those marked * in the water and bring to the boil to dissolve completely. Sterilize by autoclaving at 121°C for 15 min. Cool to 50°C and aseptically add the cefoperazone, and amphotericin if required, to give the final concentration(s) stated.

Physical properties

Appearance	Black, opaque.
pH	7.4 ± 0.2

Shelf life

Ready to use medium	10 days at 4 ± 2°C.

Inoculation method for samples

Plates should not be overdried but preferably left overnight at room temperature. This medium can be inoculated as a surface spread plate or conventionally to produce discrete colonies.

Incubation method

At 42°C for 48 h microaerobically in a jar with an atmosphere containing approximately 5% O_2, 10% CO_2 and 85% N_2 or H_2.

Reading of results and interpretation

Campylobacter jejuni strains produce grey, moist, flat and occasionally spreading growth which may be accompanied with a green hue and/or a metallic sheen. *Campylobacter coli* strains tend to be creamy-grey in colour, moist and often produce a more discrete type of colony. *Campylobacter lari* strains are more varied and produce both types of colonial morphology.

Quality assessment

(i) *Productivity*

Test strains	*Campylobacter jejuni* NCIMB 50091 *Campylobacter coli* NCIMB 50092
Inoculation method	Modified Miles-Misra, spiral plating or streaking/ecometry.

(ii) *Selectivity*

Test strains	*Proteus mirabilis* (ATCC 29906, CECT 4168/NCTC 11938) *Pseudomonas aeruginosa* NCIMB 50067
Supplementary strains	*Bacillus cereus* NCIMB 50014 *Escherichia coli* NCIMB 50034 *Staphylococcus aureus* NCIMB 50080

Inoculation method	Modified Miles-Misra, spiral plating or streaking/ ecometry.

(iii) *Characteristic appearance of colonies*
See above. Occasionally contaminating organisms may grow on this medium. These include cefoperazone resistant *Pseudomonas* spp. and Enterobacteriaceae, some streptococci and yeasts.

References

Bolton, F.J., Hutchinson, D.N. and Coates, D. (1984) Blood-free selective medium for isolation of *Campylobacter jejuni* from faeces. J. Clin. Microbiol. 19, 167–171.
Bolton, F.J., Hutchinson, D.N. and Coates, D. (1986) Comparison of three selective agars for the isolation of campylobacters. Eur. J. Clin. Microbiol. 5, 466–468.
Hutchinson, D.N. and Bolton, F.J. (1984) An improved blood-free selective medium for the isolation of *Campylobacter jejuni* from faecal specimens. J. Clin. Pathol. 37, 956–957.

 Handbook of Culture Media for Food Microbiology, J.E.L. Corry et al. (Eds.)

Charcoal Cefoperazone Deoxycholate (CCD) broth

This monograph has been reviewed by members of the IUMS-ICFMH Working Party on Culture Media and given 'Proposed' status.

Description and history

This blood-free broth is the liquid version of modified CCD agar (Bolton et al., 1984; Hutchinson and Bolton, 1986). Charcoal, ferrous sulphate and sodium pyruvate replace blood, while sodium deoxycholate and cefoperazone are the selective agents in this enrichment broth. It is used in conjunction with modified CCD agar or Skirrow's agar for the isolation of low numbers of themophilic campylobacters from food and environmental samples (Bolton et al., 1986; Korhonen and Martikainen, 1990).

Composition (grams)

Beef extract	10.0
Peptone	10.0
Sodium chloride	5.0
Charcoal	4.0
Casein hydrolysate	3.0
Iron (II) sulphate	0.25
Sodium pyruvate	0.25
Sodium deoxycholate	1.0
Cefoperazone*	0.032
Amphotericin*	0.01
Distilled or deionized water	1000.0

Preparation

Suspend the ingredients except those marked * in the water and bring to the boil. Sterilize by autoclaving at 121°C for 15 min. Cool to 50°C and aseptically add the required amount of cefoperazone and amphotericin to give the final concentrations stated. Distribute aseptically in the required volumes in flasks or tubes.

Physical properties

Appearance	Black, opaque.
pH	7.4 ± 0.2

Shelf life

Ready to use medium	14 days at 4 ± 2°C.

Inoculation method for samples

Use macerates of foods or environmental samples and inoculate broth in proportion 1:10.

Incubation method

At 42°C for 24–48 h microaerobically in a jar with an atmosphere containing approximately 5% O_2, 10% CO_2 and 85% N_2 or H_2.

Reading of results and interpretation

Growth of campylobacters in the broth is usually not visible, but can be detected by microscopic examination. The medium should always be subcultured by streaking one loopful on CCD agar or Skirrow's medium.

Quality assessment

(i) *Productivity*

Test strains	*Campylobacter jejuni* NCIMB 50091 *Campylobacter coli* NCIMB 50092
Inoculation method	Dilution to extinction.
Criteria	Growth should be within 2 titre units of the growth in tryptone soya broth after 48 h at 42°C.

(ii) *Selectivity*

Test strains	*Proteus mirabilis* (ATCC 29906/CECT 4168/NCTC 11938) *Pseudomonas aeruginosa* NCIMB 50067
Inoculation method	Dilution to extinction.
Criteria	Difference in growth should be equal to or less than 5 titre units of the growth in tryptone soya broth.

References

Bolton, F.J., Hutchinson, D.N. and Coates, D. (1984) Blood-free selective medium for isolation of *Campylobacter jejuni* from faeces. J. Clin. Microbiol. 19, 167–171.

Bolton, F.J., Hutchinson, D.N. and Coates, D. (1986) Comparison of three selective agars for the isolation of campylobacters. Eur. J. Clin. Microbiol. 5, 466–468.

Hutchinson, D.N. and Bolton, F.J. (1984) An improved blood-free selective medium for the isolation of *Campylobacter jejuni* from faecal specimens. J. Clin. Pathol. 37, 956–957.

Korhonen, L.K. and Martikainen, P.J. (1990) Comparison of some enrichment broths and growth media for the isolation of thermophilic campylobacters from surface water samples. J. Appl. Bacteriol. 68, 593–599.

Citrate Azide Tween Carbonate (CATC) agar

This monograph has been assessed by members of the IUMS-ICFMH Working Party on Culture Media and given 'Approved' status.

Description and history

CATC agar is a selective medium for the isolation and cultivation of enterococci. High concentrations of citrate and azide achieve almost complete suppression of other bacterial flora and allow good growth of enterococci (*Enterococcus faecalis* ssp. and *Enterococcus faecium*). (Reuter, 1968; 1978; 1985).

Composition (grams)

Tryptic digest of casein	15.0
Yeast extract	5.0
Potassium di-hydrogen orthophosphate	5.0
Sodium citrate	15.0
Sorbitan monooleate (Tween 80)	1.0
Sodium carbonate*	2.0
2,3,5,-Triphenyltetrazolium chloride*	0.1
Sodium azide*	0.4
Agar	15.0
Distilled or deionized water	1000.0

Preparation

Suspend the ingredients except those marked * in the water and heat to boiling to dissolve completely. Sterilize in the autoclave for 15 min at 121°C. Cool to about 50°C and add:

20 ml of 10% solution of sodium carbonate (prepared in sterile water without heating)

10 ml of 1% solution of 2,3,5,-triphenyltetrazolium chloride

4 ml of 10% solution of sodium azide.

Adjust pH to 7.0 ± 0.1 at 50°C. Distribute into sterile Petri dishes.

Note: High selectivity and productivity can be achieved only by strictly following the method of preparation detailed here. Heating of sodium carbonate, triphenyltetrazolium chloride and sodium azide must be avoided.

Physical properties

Appearance	Pale pink.
pH	7.0 ± 0.1 at 37–40°C.

Shelf life

Ready to use medium	7 days at 4 ± 2°C.

Inoculation method for samples

Surface spreading over whole plate or modified Miles-Misra.

Incubation method

24 h at 37–40°C, in air. If plates show no growth or unusually small colonies, then incubation for a further 24 h is advisable.

Reading of results and interpretation

Distinctly red well-developed colonies can be considered as *Enterococcus faecalis* whilst pink colonies or colonies with a red centre are likely to be *Enterococcus faecium*. Since, especially after prolonged incubation, other enterococci or streptococci may appear as pink colonies though smaller in size (<0.5 mm diameter), confirmation tests should be carried out.

Quality assessment

(i) *Productivity*

Test strains	*Enterococcus faecalis* NCIMB 50030 *Enterococcus faecium* NCIMB 50032
Supplementary strain	*Enterococcus hirae* NCIMB 50031
Inoculation method	Modified Miles-Misra, spiral plating or streaking/ecometry.

(ii) *Selectivity*

Test strains	*Escherichia coli* NCIMB 50034 *Lactococcus lactis* ssp. *lactis* NCIMB 50058

Supplementary strains	*Lactobacillus casei* ssp. *rhamnosus* NCIMB 50045 *Pediococcus damnosus* NCIMB 50065 *Staphylococcus aureus* NCIMB 50080
Inoculation method	Modified Miles-Misra, spiral plating or streaking/ecometry.

(iii) *Characteristic appearance of colonies*
Red, mostly smooth colonies: *Enterococcus faecalis*
Pink, partly rough colonies, or red centre: *Enterococcus faecium*
Small colourless or pink colonies (< 0.5 mm diameter): other streptococci or enterococci, which might grow at a slower rate and after prolonged incubation.

References

Anon. (1991) DIN 10106: Bestimmung von *Enterococcus faecalis* und *Enterococcus faecium*, Normenausschuss lebensmittel und landwirtschafte Produkte (NAL) im DIN (Deutsches Institut für Normung e. V). Berlin, 1991.

Reuter, G. (1968) Erfahrungen mit Nahrboden für die selektive mikrobiologische Analyse von Fleischerzeugnissen. Arch. Lebensmittelhyg. 19, 53–57 u. 84–89.

Reuter, G. (1978) Selektive Kultivierung von "Enterokokken" aus Lebensmitteln tierischer Herkunft. Arch. Lebensmittelhyg. 29, 84–91.

Reuter, G. (1985) Selective media for group D streptococci. Int. J. Food Microbiol. 2, 103–114.

Cresol red Thallium Acetate Sucrose (CTAS) agar

This monograph has been assessed by members of the IUMS-ICFMH Working Party on Culture Media and given 'Approved' status.

Description and history

This medium was devised for the detection of carnobacteria (syn. *Lactobacillus divergens*) which fail to grow well on conventional selective media for lactobacilli (Holzapfel and Gerber, 1983). Investigations showed pH values below 6.0 to be inhibitory to carnobacteria especially in the presence of acetate. Growth of carnobacteria was found to take place at pH values of up to 11.0 (Holzapfel and Long, 1984; unpublished results) compared to a maximum pH of 7.2 for growth of practically all lactobacilli. Also, relatively high concentrations of between 0.4 and 1% manganese sulphate have been found to stimulate growth of carnobacteria more than that of enterococci (Bosch and Holzapfel, 1985; unpublished results). In order to minimise Maillard reactions during heat sterilization of such alkaline media, glucose was substituted by sucrose. Thus far, all carnobacteria have been found to ferment sucrose. The selectivity of the CTAS medium is based on its high pH (8.5 to 9.0) and the presence of thallium acetate, nalidixic acid and a relatively high concentration of sodium citrate.

Composition (grams)

Peptone from casein	10.0
Yeast extract	10.0
Sucrose	20.0
Sorbitan monooleate (Tween 80)	1.0
Sodium citrate	15.0
Manganese (II) sulphate $\cdot 4H_2O$	4.0
Di-potassium hydrogen orthophosphate	2.0
Thallium (I) acetate	1.0
Nalidixic acid	0.04
Cresol red	0.004
2,3,5-Triphenyl-tetrazolium chloride	0.01
Agar	15.0
Distilled or deionized water	1000.0

Preparation

Add all the components except triphenyl-tetrazolium chloride to 990 ml of water and bring to the boil to dissolve completely. Cool to 55°C and adjust the pH of the medium to 9.1 with 1N NaOH. Autoclave at 121°C for 10 min, cool to 55°C and add 10 ml of a 10% solution of triphenyl-tetrazolium chloride. Pour about 15 ml of medium per Petri dish.

Physical properties

Appearance	Red to purple-red with precipitate.
pH	9.0 ± 0.2

Shelf life

Ready to use medium	Freshly prepared plates are preferred but the medium may be stored for 1–2 weeks at 4 ± 2°C in sealed plastic bags.

Inoculation method for samples

Surface spreading or direct plating of dilution series.

Incubation method

At 30°C for 24–48 h or at 25°C for 3–4 days, in air.

Reading of results and interpretation

Strains of *Carnobacterium piscicola* (syn. *Carnobacterium carnis*) appear as small, bronze-metallic shiny yellowish to pinkish colonies, causing a yellow colour change of the medium and clearance of the precipitate. The metallic shine is best observed on the periphery of growth. Growth of *Carnobacterium divergens* strains is more sparse resulting mainly in pin-point colonies, often without colour change of the medium, and showing a bronze-metallic shine. Good growth of most *Enterococcus* spp. is observed with yellow colour change and clearance of precipitate in the medium. These small to pin-point (0.5 to 0.1 mm) colonies are convex, shiny and pinkish to yellowish, and without metallic shine. *Listeria* spp. grow sparsely, if at all, with a bronze metallic shine resembling *Carnobacterium divergens*. Phase contrast microscopy for determination of cell morphology is a useful tool to distinguish rods (carnobacteria) from cocci (enterococci).

Quality assessment

(i) *Productivity*

Test strains	*Carnobacterium piscicola* NCIMB 50020

	Carnobacterium divergens NCIMB 50021
Inoculation method	Modified Miles-Misra, spiral plating or streaking/ ecometry.

(ii) *Selectivity*

Test strains	*Lactobacillus sakei* ssp. *sakei* NCIMB 50056
Inoculation method	Modified Miles-Misra, spiral plating or streaking/ ecometry.
Criteria	No growth after 48–72 h at 30°C.

(iii) *Characteristic appearance of colonies*
See above.

Note:
Tests on more than 60 different antibiotics showed carnobacteria to have similar resistance spectra as enterococci; the same was found for heavy metal salts (Bosch and Holzapfel, 1985; unpublished results). However, substitution of sucrose by 2% inulin may be used to distinguish colonies of 'non-divergens' carnobacteria from enterococci. Unlike enterococci these carnobacteria (e.g. *Carnobacterium piscicola*) ferment inulin causing a change in colour of the medium from red to yellow.

Reference

Holzapfel, W.H. and Gerber, E.S. (1983) *Lactobacillus divergens* sp. nov., a new heterofermentative *Lactobacillus* species producing L(+)-lactate. System. Appl. Microbiol. 4, 522–534.

Diagnostic Salmonella Selective Semisolid Medium (DIASALM)

This monograph has been reviewed by members of the IUMS-ICFMH Working Party on Culture Media and given 'Proposed' status.

Description and history

DIASALM is a diagnostic selective motility agar to be used in Petri dishes for the isolation of *Salmonella* spp. from food and environmental samples (van Netten and van der Zee, 1991). The basal medium is a semi-solid indole motility medium (SIM). The selective system exploits the resistance of *Salmonella* spp. as compared to other Enterobacteriaceae to relatively high osmotic pressures, relatively low pH, malachite green and novobiocin. The diagnostic system of DIASALM consists of sucrose, lactose and bromocresol purple. This differentiates salmonellae not only from the motile non-pathogenic lactose fermenters but also from many non-pathogens which ferment lactose or sucrose.

The semi-solid approach simultaneously enriches salmonellae and separates motile salmonellae from most competitive organisms resistant to the selective system. As a result salmonellae are mostly isolated in pure culture but sometimes occur mixed with *Proteus* spp., *Hafnia* spp. or *Enterobacter* spp. as interfering motile Enterobacteriaceae.

Composition (grams)

Pancreatic digest of casein*	20.0
Peptic digest of animal tissue*	6.1
Iron (II) ammonium sulphate*	0.2
Sodium thiosulphate*	5.0
Sucrose	7.5
Lactose	0.5
Potassium di-hydrogen orthophosphate	1.47
Malachite green oxalate	0.037
Magnesium chloride $\cdot 6H_2O$	23.34
Bromocresol purple	0.08
Sodium novobiocin	0.010
Agar*	3.5
Distilled or deionized water	1000.0

Note: Commercial formulations may vary from the above composition.

Preparation

Solution A
Suspend 30 g SIM-medium (ingredients marked*), 7.5 g sucrose, 0.5 g lactose, 1.47 g potassium di-hydrogen orthophosphate and 23.34 g magnesium chloride in 980 ml water. This suspension is made on the day of preparation of DIASALM.

Solution B
Dissolve 3.7 g of analytically pure malachite green oxalate in 100 ml of water. Solution B can be kept for at least 8 months at room temperature in a dark bottle.

Solution C
Dissolve 0.8 g bromocresol purple in 18.5 ml 0.01N NaOH and 82.5 ml water. Store in refrigerator in dark bottle.

Solution D
Dissolve 100 mg of sodium novobiocin in 100 ml of water. Sterilize by filtration. Store in refrigerator in an airtight dark bottle.

The medium is prepared by adding to 990 ml of solution A, 1 ml of solution B and 10 ml of solution C. Heat the suspension in a steamer or waterbath to about 80°C until the ingredients are completely dissolved. Cool to 50°C and add 10 ml of solution D. Mix and pour 20 ml in 9 cm diameter Petri dishes. Before inoculation plates are dried at 55°C for 5 min to remove surface moisture.

Physical properties

Appearance	Dark green-blue, transparent.
pH	5.5 ± 0.1

Shelf life

Ready to use medium	5 days at room temperature in the dark.

Inoculation method for samples

The plate is inoculated either with three drops (3×0.033 ml) in separate spots or with 0.1 ml in one spot in the middle.

Incubation

18–24 h at 42°C.

Reading of results and interpretation

After incubation plates with a purple and/or black migration zone require further investigation for the presence of motile salmonellae. The edge of the migration zone and the purple zone are not always identical. To obtain the purest subculture it is imperative to take a loopful of culture from the edge of the migration zone onto a salmonella selective diagnostic agar inoculated in such a way as to obtain well-isolated colonies. The identity of suspected salmonella colonies is confirmed by appropriate biochemical and serological tests.

Quality assessment

(i) *Productivity*

Test strains	*Salmonella enteritidis* NCIMB 50073 *Salmonella virchow* NCIMB 50077 *Salmonella dublin* NCIMB 50072
Inoculation method	Miles-Misra (without spreading of drop).
Criteria	Typical appearance and motility (zone should extend at least 2 cm from point of inoculation).

(ii) *Selectivity*

Test strains	*Enterobacter cloacae* (ATCC 23355) *Proteus mirabilis* (ATCC 29906/CECT 4168/NCTC 11938) *Citrobacter freundii* NCIMB 50025 *Pseudomonas aeruginosa* NCIMB 50067
Inoculation method	Miles-Misra (without spreading of drop).
Criteria	No growth or growth confined to centre of plate.

(iii) *Characteristic appearance of growth*
See above.

References

Van der Zee, H. and van Netten, P. (1992) Diagnostic selective semi-solid media based on Rappaport-Vassiliadis broth for the detection of *Salmonella* spp. and *Salmonella enteritidis* in foods. Proc. Symposium on Salmonella and Salmonellosis, Ploufragan, France: Reports and Communications, pp 69–77.

Van der Zee, H. (1994) Conventional methods for the detection and isolation of *Salmonella enteritidis*. Int. J. Food Microbiol. 21, 41–46.

Van Netten, P. and van der Zee, H. (1991) The use of a diagnostic semi-solid medium for the isolation of *Salmonella enteritidis* from poultry-meat. Quality of Poultry Products, Safety and Marketing aspects. Proceedings of the 10th Symposium on the quality of poultry-meat, Doorwerth, May 12–17 1991. Spelderholt, Beekbergen.

Dichloran Glycerol (DG18) agar

This monograph has been reviewed by members of the ICFMH-IUMS Working Party on Culture Media and given 'Proposed' status.

Description and history

This reduced water activity medium (a_w 0.95) described by Hocking and Pitt (1980) is suitable for the enumeration and isolation of moderately xerophilic moulds in foods. Dichloran inhibits the rapid spreading of mucoraceous fungi and restricts the colony size of most other genera, thus enabling more accurate counting of colonies. The addition of chloramphenicol and the reduced a_w prevent the growth of bacteria. Although specifically developed to enumerate xerophilic fungi in dried and semi-dried foods it has proved to be a useful general purpose medium for counting yeasts and moulds in a range of foodstuffs (Beckers et al., 1982). The addition of Triton X-301 (Tapia de Daza and Beuchat, 1992) enables easier enumeration of xerophiles when *Eurotium* spp. are present (Beuchat and Hwang, 1995).

Composition (grams)

Mycological peptone	5.0
Glucose	10.0
Potassium di-hydrogen orthophosphate	1.0
Magnesium sulphate $\cdot 7H_2O$	0.5
Dichloran (2,6-dichloro-4-nitroaniline)	0.002
Chloramphenicol	0.1
*Triton X-301™	0.1
Glycerol	220.0
Agar	15.0
Distilled or deionized water	1000.0

*Rohm and Haas Co., Philadelphia, PA, USA. Can be included in formula to aid in enumerating xerophiles when eurotia are present.

Preparation

Suspend all the ingredients except chloramphenicol, Triton X-301 and glycerol in about 800 ml of water and heat to dissolve completely. Add 10 ml of a 1% ethanolic solution of chloramphenicol and bring to 1000 ml with water. Add glycerol, mix and sterilize by autoclaving at 121°C for 15 min. Cool to 50°C, add Triton X-301 if desired, mix well and dispense 15 ml amounts into sterile Petri dishes. Dry and use immediately or store dried plates at 4 ± 2°C for up to 7 days (Olsen et al., 1997).

Physical properties

Appearance	Medium amber, slightly opalescent.
pH	5.6 ± 0.2

Shelf life

Ready to use medium	7 days at 4 ± 2°C.

Inoculation method for samples

Spread 0.1 or 0.2 ml of dilutions of the food sample over the whole of the surface of 9 cm diameter plates.

Incubation

At 25°C for 5 days in the dark with the lid uppermost. Where identification is required, prolong incubation until characteristic colonies are formed.

Reading of results and interpretation

Where separate counts of moulds and yeasts are required, identify by morphological appearance and, where necessary, microscopic examination of the two groups of microorganisms.

Quality assessment

Moulds. Use the stab inoculation procedure. Growth rate should be within 30% of the figure given.

Test strains	IMI No.	ATCC No.	Colony diameter (mm) after 5 days at 25°C
Rhizopus stolonifer	61269		12.0
Aspergillus flavus	91856ii		11.0
Eurotium amstelodami	17455	16018	3.3

Yeasts. Use the stab inoculation procedure. Growth rate should be within 30% of the figure given.

Test strains	NCYC No.	ATCC No.	Colony diameter (mm) after 5 days at 25°C
Saccharomyces cerevisiae	79	7754	0.6
Zygosaccharomyces rouxii	1522		0.5

Bacteria. Use streaking procedure (Appendix I, method C) with a loopful (10 ml) from a 24 h broth culture. No growth should be evident on sectors C, D or E of the streaked plate after incubation at 25°C for 5 days.

Test strain
Bacillus subtilis NCIMB 50018

References

Beckers, H.J., de Boer, E., Van Eikelenboom, C., Hartog, B.J., Kuik, D., Mol, N., Nooitgedagt, A.J., Northold, M.O., and Samson, R.A. (1982) Int. Stand. Org. Document ISO/TC34/SC9/N151.
Beuchat, L.R. and Hwang, C.-A. (1995) Evaluation of modified dichloran 18% glycerol (DG18) agar for enumerating fungi in wheat flour. Int. J. Food Microbiol. 29, 161–166.
Hocking, A.D. and Pitt, J.I. (1980) Dichloran-glycerol medium for enumeration of xerophilic fungi from low moisture foods. Appl. Env. Microbiol. 39, 488–492.
Olsen, M., Andersson, K. and Akerstrand, K. (1997) Quality control of two rose bengal and modified DRBC and DG18 media. Int. J. Food Microbiol. 35, 163–168.
Tapia de Daza, M.S. and Beuchat, L.R. (1992) Suitability of modified dichloran glycerol (DG18) agar for enumerating unstressed and stressed xerophilic molds. Food Microbiol. 9, 319–333.

Dichloran Rose Bengal Chloramphenicol (DRBC) agar

This monograph has been assessed by members of the IUMS-ICFMH Working Party on Culture Media and given 'Approved' status.

Description and history

This is a selective medium for the enumeration of moulds and yeasts in foods. King et al. (1979) showed that the combination of dichloran and rose bengal markedly restricts the size and height of mould colonies thus preventing overgrowth of luxuriant species and assisting accurate counting of colonies. The addition of chloramphenicol and the reduced pH of 5.6 serve to prevent the growth of most bacteria. This medium can be used to enumerate both toxigenic and non-toxigenic moulds. It is not diagnostic for detecting specific mycotoxin producers.

Composition (grams)

Mycological peptone	5.0
Glucose	10.0
Potassium di-hydrogen orthophosphate	1.0
Magnesium sulphate $\cdot 7H_2O$	0.5
Dichloran (2-6-dichloro-4-nitro-aniline)	0.002
Rose Bengal	0.025
Chloramphenicol	0.1
Agar	15.0
Distilled or deionized water	1000.0

Preparation

Suspend the ingredients except chloramphenicol in the water and bring to the boil to dissolve completely. Add 10 ml of a 1% ethanolic solution of chloramphenicol, mix and sterilize by autoclaving at 121°C for 15 min. Cool to below 50°C and dispense 15 ml amounts into sterile Petri dishes. Use immediately or store at 4 $\pm$ 2°C in the dark until required.

Physical properties

Appearance	Deep pink, without any significant precipitate.
pH	5.6 $\pm$ 0.2 at 25°C.

Shelf life

Ready to use medium	7 days at 4 $\pm$ 2°C in the dark.

Inoculation method for samples

Surface spreading over whole plate using 0.1 or 0.2 ml per 9 cm diameter plate.

Incubation method

At 25°C for 5 days in the dark. Where identification is required, prolong incubation until characteristic colonies are formed.

Reading of results and interpretation

Where separate counts of moulds and yeasts are required, identify by morphological appearance and, where necessary, microscopic examination of the two groups of microorganisms. Colonies of yeasts and bacteria can be confused and should be checked microscopically.

Quality assessment

Moulds. Use stab inoculation procedure. Growth rate should be within 30% of the figure given.

Test strains	IMI No.	ATCC No.	Growth rate (mm per day at 25°C)
Rhizopus stolonifer	61269		8.1
Aspergillus flavus	91856ii		4.5
Penicillium cyclopium	19759	16025	3.5

Yeasts. Use stab inoculation procedure. Growth rate should be within 30% of figure given.

Test strain	Growth rate (mm per day at 25°C)
Saccharomyces cerevisiae NCIMB 50105	1.4

Bacteria. Use streaking procedure (Appendix I, method C) with a loopful (10 :l) from a 24 h broth culture. No growth should be evident on sectors C, D or E of the streaked plate after incubation at 25°C for 5 days.

Test strain	*Bacillus subtilis* NCIMB 50018
Supplementary strain	*Escherichia coli* NCIMB 50034

Reference

King, D.A., Hocking, A.D. and Pitt, J.I. (1979) Dichloran-rose bengal medium for enumeration and isolation of molds from foods. Appl. Environ. Microbiol. 37, 959–964.

Differential Clostridial Agar (DCA)

This monograph has been reviewed by members of the IUMS-ICFMH Working Party on Culture Media and given 'Proposed' status.

Description and history

Differential clostridial agar was developed for the enumeration of spores of sulphite reducing clostridia in dried foods (Gibbs and Freame, 1965; Weenk et al., 1995). It consists of a nutritionally rich basal medium, including starch to promote spore germination. Resazurin is added as a redox indicator, turning red at high redox potentials, indicating aerobic conditions. Sulphite and an iron source are added as indicators. Sulphite-reducing clostridia produce sulphide from sulphite, which gives a black precipitate with the iron present in the medium. Sulphite reducing clostridia are enumerated as black colonies.

Composition (grams)

Starch	1.0
Casein peptone	5.0
Meat peptone	5.0
Meat extract powder	8.0
Yeast extract powder	1.0
Resazurin	0.002
Cysteine HCl	0.5
Glucose	1.0
Iron (III) ammonium citrate*	1.0
Sodium sulphite $\cdot 7H_2O$*	0.75
Agar	20.0
Distilled or deionized water	1000.0

Preparation

Dissolve all ingredients in the water, except those marked*. Autoclave at 121°C for 15 min, cool to about 48°C and aseptically add, just before use, appropriate volumes of solutions of iron (III) ammonium citrate (20% w/v, heat sterilized at 121°C for 15 min) and sodium sulphite (10% w/v, freshly prepared, filter sterilized).

Note: Where *Clostridium tyrobutyricum* is to be counted in the sample, the pH of the medium must be adjusted to 7.0 by the addition of 8.4 ml of a 60% solution of sodium lactate to 1000 ml of agar before sterilization (Senyk et al., 1989; Weenk et al., 1995).

Physical properties

Appearance	pale straw.
pH	7.6 ± 0.2

Shelf life

Basal medium	2 weeks at 4 ± 2°C.
Iron (III) ammonium citrate solution	1 month at 4 ± 2°C.
Complete medium	use on the day of preparation (do not remelt).

Inoculation method for samples

In order to facilitate spore germination and to kill competitive flora, a heat treatment of the spores/sample (80°C for 10 min) should be given before inoculation of the agar. For inoculation use the pour plate procedure, using 1 ml of sample per 9 cm diameter Petri dish. When set, plates are overlaid with sterile DCA tempered to 47 ± 2°C.

Incubation method

Anaerobically at 30°C for 3 days.

Reading of results and interpretation

Discrete black colonies of 1–5 mm diameter are considered to be presumptive sulphite reducing clostridia. When grey colonies are observed, the sodium sulphite $\cdot 7H_2O$ concentration may be increased to 1 g/l as not all sulphite reducing clostridia cause total blackening at the lower concentration. However increasing the sodium sulphite concentration may inhibit some sulphite sensitive strains (Mossel et al., 1959).

Bacillus licheniformis may, in some cases, also give rise to black colonies. When this is suspected, the identity of the colonies must be confirmed by checking their ability to grow aerobically and their sensitivity to metronidazole (Weenk et al., 1995).

Quality assessment

(i) *Productivity*

Test strains	*Clostridium perfringens* NCIMB 50027
	Clostridium bifermentans NCIMB 50026
	Clostridium sporogenes NCIMB 50099

Inoculation method	Modified Miles-Misra or streaking method using surface inoculation of pre-poured plates with subsequent overlay.
Criteria	Recovery on DCA should be within 0.3 $\log_{10}$ of the recovery on Reinforced Clostridial agar after 3 days at 30°C in an anaerobic jar.

(ii) *Characteristic appearance of colonies*
Discrete black colonies of 1–5 mm diameter.

References

Gibbs, B.M. and Freame, B. (1965) Methods for the recovery of clostridia from foods. J. Appl. Bacteriol. 28, 95–111.

Mossel, D.A.A., v. Goldstein Brouwers, G.W.M. and de Bruin, A.S. (1959) A simplified method for the isolation and study of obligate anaerobes. J. Pathol. Bacteriol. 78, 290–291.

Senyk, G.F., Scheib, J.A., Brown, J.M. and Ledford, R.A. (1989) Evaluation of methods for determination of spore-formers responsible for the late gas-blowing defect in cheese. J. Dairy Sci. 72, 360–366.

Weenk, G., Fitzmaurice, E. and Mossel, D.A.A. (1991) Selective enumeration of spores of *Clostridium* species in dried foods. J. Appl. Bacteriol. 70, 135–143.

Weenk, G.H., van den Brink, J.A., Struijk, C.B. and Mossel, D.A.A. (1995) Modified methods for the enumeration of spores of *Clostridium* species in dried foods. Int. J. Food Microbiol. 27, 185–200.

Enterobacteriaceae Enrichment (EE) broth

This monograph has been reviewed by members of the IUMS-ICFMH Working Party on Culture Media and given 'Proposed' status.

Description and history

The medium is, in essence, a modification of brilliant green bile lactose broth, which in turn was a modification of MacConkey's liquid medium (Mossel et al., 1963). EE broth therefore contains brilliant green and bile as the inhibitory agents for Gram positive organisms, glucose as the main energy source to give all Enterobacteriaceae an equal chance of rapid development and avoid a bias towards lactose positive types, and an increased amount of buffer to control inhibition of growth in the earlier stages of enrichment and autosterilization at the end.

Composition (grams)

Peptone	10.0
Glucose	5.0
Di-sodium hydrogen orthophosphate	6.45
Potassium di-hydrogen orthophosphate	2.0
Ox bile	20.0
Brilliant green	0.0135
Distilled or deionized water	1000.0

Preparation

Dissolve the ingredients in the water, distribute in the required volumes in flasks or tubes and heat at 100°C for 30 min only. Further heating must be avoided. Do not autoclave. Cool rapidly in cold running tap water.

Physical properties

Appearance	Green, clear.
pH	7.2 ± 0.2

Shelf life

Ready to use medium	1 month at 4 ± 2°C in screw capped containers.

Inoculation method for samples

Use food macerates or decimal dilutions and inoculate broth in the proportion 1:10. When large volumes of enrichment fluid are not appropriate, EE broth can be used at double strength; this makes strict adherence to the instructions for heat treatment (cf. above) imperative.

Incubation method

To detect mesophilic members of the Enterobacteriaceae incubate at 30°C for 24–48 h. Psychrotrophic Enterobacteriaceae can be detected by incubation at 4°C for 10 days.

Reading of results and interpretation

Turbidity and often change of colour of the medium towards yellowish-green provides presumptive evidence of the presence of Enterobacteriaceae. Isolation on a MacConkey type agar, e.g. violet red bile (lactose) agar and subsequent confirmatory tests must be carried out.

Quality assessment

(i) *Productivity*

Test strains	*Escherichia coli* NCIMB 50034 *Hafnia alvei* NCIMB 50037
Inoculation method	Dilution to extinction.
Criteria	Recovery in EE should be within one titre unit of the recovery in tryptone soya broth after 18–24 h at 30°C.

(ii) *Selectivity*

Test strains	*Enterococcus faecalis* NCIMB 50030 *Staphylococcus aureus* NCIMB 50080
Inoculation method	Dilution to extinction.
Criteria	Recovery in EE should be less than 5 titre units of the recovery in tryptone soya broth after 18–24 h at 30°C.

References

Mossel, D.A.A., Visser, M. and Cornelissen, A.M.R. (1963) The examination of foods for Enterobacteriaceae using a test of the type generally adopted for the detection of salmonellae. J. Appl. Bacteriol. 26, 444–452.

Mossel, D.A.A., Harrewijn, G.A. and Nesselrooy-van Zadelhoff, C.F.M. (1974) Standardisation of the selective inhibitory effect of surface active compounds used in media for the detection of Enterobacteriaceae in food and water. Health Labor. Sci. ll, 260–267.

Enterococcosel (ECS) agar/broth with or without vancomycin

This monograph has been reviewed by members of the IUMS-ICFMH Working Party on Culture media and given 'Proposed' status.

Description and history

This medium is based on the aesculin-bile salt-azide medium (ABA; Swan, 1954; Isenberg et al., 1970; Reuter, 1985) and has recently been used for the detection of vancomycin resistant or susceptible enterococci in foods (Klein et al., 1998) and clinical (faecal) samples (Ieven et al., 1999). The medium relies on the inhibitory properties of bile salts against most of the Gram positive microorganisms whereas enterococci are not inhibited. The growth of Gram negative species is reduced by sodium azide. As an indicator system, aesculin and ferric iron differentiate group D streptococci (enterococci) from other organisms. Enterococci are able to hydrolyse aesculin which, in the presence of iron ions, results in the formation of black iron phenolic compounds. Enterococci produce brown-black zones beneath and around brown-black colonies. In order to select glycopeptide-resistant enterococci (GRE) from a mixed population of enterococci vancomycin may be added in concentrations varying from 4 or 6 to 32 or even 64 mg/l. Lower concentrations are often used in liquid enrichment media (broth), higher concentrations in plates.

Composition (grams)

Pancreatic digest of casein	17.0
Peptic digest of meat	3.0
Yeast extract	5.0
Sodium chloride	5.0
Sodium citrate	1.0
Aesculin	1.0
Iron (III) ammonium citrate	0.5
Ox bile, dried	10.0
Sodium azide	0.25
Vancomycin	0.006 or 0.032*
Agar	13.5**
Distilled or deionized water	1000.0

Preparation

Suspend the ingredients except the vancomycin in the water. Heat carefully to boiling (at least 1 min) to dissolve completely. Dispense into bottles (or final containers) and sterilize at 121°C for 15 min. When cooled to approximately 50°C, add 1 ml of a filter-sterilized vancomycin solution to 1 litre of medium. To achieve maximum effect the sodium azide may be added after sterilization.

* Where high level vancomycin resistant enterococci (VRE) are expected a concentration of 32 mg/l is recommended (use 1 ml of a 0.32 % vancomycin solution). If a broader spectrum of VRE with lower levels of resistance is the focus of study, a concentration of 6 mg/l (1 ml of a 0.06 % solution) should be used. If it is required to isolate all enterococci, vancomycin should be omitted.

** For ECS broth the agar is omitted.

Physical properties

Appearance	Pale yellow-brown with a tinge of green, slightly opaque.
pH	7.1 ± 0.2

Shelf life

Ready to use medium	7 days at 4 ± 2°C.

Inoculation method for samples

Surface spreading over the whole plate using 0.1–0.25 ml suspension of the specimen per pre-dried 9 cm diameter plate. For the liquid medium add 0.8 ml sample to 5 ml broth in a test tube.

Incubation method

At 37 ± 0.5°C, in air, for 24 h (and 48 h if necessary).

Reading of results and interpretation

(i) *Plates*
Round black colonies about 2 mm diameter surrounded by brown-black haloes are considered to be enterococci (presumptive count) or VRE if vancomycin has been incorporated. Reading after 24 h is recommended as some lactobacilli and other Gram positives are able to grow slowly and also hydrolyse aesculin after 48 h (e.g. *Lactobacillus curvatus*). Confirmation should be carried out on 5–10 colonies by morphological and biochemical methods and MICs assessed for presumptive VRE.

(ii) *Broths*
If broth (with or without vancomycin) turns black after 24 h or at the latest after 48 h, subculturing to ECS and Columbia blood agar should be made.

Quality assessment

(i) *Productivity*

Test strains	*Enterococcus faecium* 70/90 (vancomycin MIC >256 mg/l) *Enterococcus faecalis* 1528 (vancomycin MIC >256 mg/l) *Enterococcus faecalis* ATCC 51299 (vancomycin MIC 16–32 mg/l)
Inoculation methods	
(plates)	Modified Miles-Misra, spiral plating or streaking/ecometry.
(broths)	Dilution to extinction.

(ii) *Selectivity*

Test strains	*Enterococcus faecalis* NCIMB 500112 (vancomycin MIC $\leq$ 4 mg/l) *Escherichia coli* NCIMB 50109 *Lactobacillus casei* ssp. *casei* ATCC 393 (vancomycin MIC >64 mg/l)
Supplementary strains	*Staphylococcus aureus* ATCC 29213 (vancomycin MIC $\leq$ 2 mg/l) *Pseudomonas fluorescens* NCIMB 50068
Inoculation methods	
(plates)	Modified Miles-Misra, spiral plating or streaking/ecometry.
(broths)	Dilution to extinction.

(iii) *Characteristic appearance of colonies*
Round brown or black colonies, about 2 mm diameter underlayed and surrounded by black haloes. Colonies without halo and/or brown or black colony colour are not considered to be enterococci. *Staphylococcus aureus* is able to grow without these features as white colonies.

References

Ieven, M., Vercauteren, E., Descheemaeker, P., van Laer, F. and Goosens, H. (1999) Comparison of direct plating and broth enrichment culture for the detection of intestinal colonization by glycopeptide-resistant enterococci among hospitalized patients. J. Clin. Microbiol. 37, 1436–1440.

Isenberg, H.D., Goldberg, D. and Sampson, J. (1970) Laboratory studies with a selective *Enterococcus* medium. Appl. Microbiol. 20, 433–436.

Klein, G., Pack, A. and Reuter, G. (1998) Antibiotic resistance patterns of enterococci and occurrence of vancomycin resistant enterococci (VRE) from raw minced beef and pork in Germany. Appl. Environ. Microbiol. 64, 1825–1830.

Reuter, G. (1985) Elective and selective media for lactic acid bacteria. Int. J. Food Microbiol. 2, 55–68.

Swan, A. (1954) The use of bile-esculin medium and of Maxted's technique of Lancefield grouping in the identification of enterococci (Group D streptococci). J. Clin. Pathol. 7, 160–163.

Swenson, J.M., Clark, N.C., Sahm, D.F., Ferraro, M.J., Doern, G., Hindler, J., Jorgensen, J.H., Pfaller, M.A., Reller, L.B., Weinstein, M.P., Zabransky, R.J. and Tenover, F.C. (1995) Molecular characterization and multilaboratory evaluation of *Enterococcus faecalis* ATCC 51299 for quality control of screening tests for vancomycin and high-level aminoglycoside resistance in enterococci. J. Clin. Microbiol. 33, 3019–3021.

FDA *Listeria* enrichment broth (1995)

This monograph has been reviewed by members of the ICFMH-IUMS Working Party on Culture Media and given 'Proposed' status.

Description and history

Originally developed by Lovett et al. (1987) from the formulae of Ralovich et al. (1971; 1972) this medium has been further modified (Hitchins, 1995) to increase its buffering capacity. Reduction of the acriflavine hydrochloride content has standardized the medium in respect of non-dairy and dairy foods. Addition of sodium pyruvate and the delayed addition of the selective agents assists in the recovery of injured cells.

Composition (grams)

Trypticase soy broth	30.0
Yeast extract	6.0
Potassium di-hydrogen orthophosphate	1.35
Di-sodium hydrogen orthophosphate	9.6
Acriflavine HCl	0.01
Nalidixic acid (sodium salt)	0.04
Cycloheximide	0.05
Sodium pyruvate	1.11
Distilled or deionized water	1000.0

Preparation

Add the ingredients, except those marked, to the water, distribute in 225 ml amounts and autoclave at 121°C for 15 min.

Before use aseptically add 2.5 ml of a 10% (w/v) filter sterilized solution of sodium pyruvate.

Prepare the acriflavine and nalidixic acid supplements as 0.5% filter sterilized solutions in distilled water and the cycloheximide as a 1.0% (w/v) solution in 40% (v/v) ethanol. These are added to the cultures four hours after incubation has commenced (see below).

Physical properties (basal broth)

Appearance	Clear, pale yellow.
pH	7.3 ± 0.1

Shelf life

Base medium	6 months at 4 ± 2°C.

Inoculation method for samples

Add 25 g of food to the basal broth with pyruvate. Blend or stomach as required. The mixture may be transferred to a 500 ml Ehrlenmeyer flask.

Incubation method

Aerobically at 30°C. After 4 h add aseptically the following stock solutions: acriflavine 0.455 ml, nalidixic acid 1.8 ml and cycloheximide 1.15 ml. Continue incubation for a further 44 h. At 24 and 48 h streak onto Oxford agar and either LPM, LPM aesculin iron or PALCAM agar.

Quality assessment

(i) *Productivity*

Test strains	*Listeria monocytogenes* serovar 1/2a (ATCC 35152/NCTC 7973) *Listeria monocytogenes* serovar 4b (ATCC 13932/NCTC 10527)
Inoculation method	Dilution to extinction.

(ii) *Selectivity*

Test strains	*Enterococcus faecalis* NCIMB 50030 *Proteus mirabilis* (ATCC 29906/CECT 4168/NCTC 11938)
Inoculation method	Dilution to extinction.

Comments

Selective properties of acriflavine may vary from lot to lot and manufacturer to manufacturer. Each new batch must be assayed in combination with other selective agents to be used in the medium to determine the optimum concentration for use, with regard to the efficiency of selectivity and absence of inhibition of *Listeria* spp. Storage of stock solutions of acriflavine for periods of more than one month is not recommended.

References

Hitchins, A.D. (1995) *Listeria monocytogenes*. FDA Bacteriological Analytical Manual, 8th edition, AOAC International, Gaithersburg MD, USA. 10.01–10.13.

Lovett, J., Francis, D.W. and Hunt, J.M. (1987) *Listeria monocytogenes* in raw milk: Detection, incidence and pathogenicity. J. Food Protect. 50, 188–192.

Ralovich, B., Forray, A., Mero, E., Malovics, I. and Szazados, I. (1971) New selective medium for isolation of *L. monocytogenes*. Zentralbl. Bakteriol. I. Abt.Orig. 216, 88–91.

Ralovich, B., Emody, L., Malovics, I., Mero, E. and Forray, A. (1972) Methods to isolate *L. monocytogenes* from different materials. Acta Microbiol. Acad. Sci. Hung. 19, 367–369.

Fraser broth and modified Half Fraser broth

This monograph has been reviewed by members of the IUMS-ICFMH Working Party on Culture Media and given 'Proposed' status.

Description and history

Fraser and Sperber (1988) developed the secondary enrichment broth (UVM Listeria Enrichment Broth II) used in the United States Department of Agriculture procedure for isolating *Listeria monocytogenes* from meats (McClain and Lee, 1988). The present formula is a subsequent modification with an increased amount of acriflavine (Cook, 1998). It gives a presumptive test for *Listeria* spp. as it contains an indicator for the hydrolysis of aesculin. This reaction is not exclusive to *Listeria* spp. so any microorganisms giving a positive reaction (blackening) must be isolated and further identified. The 'Half Fraser' broth used as a primary enrichment in the ISO (1997) method contains reduced amounts of the selective agents nalidixic acid and acriflavine to aid the recovery of injured cells.

Composition (grams)

Proteose peptone (peptic digest of animal tissue)	5.0
Tryptone (pancreatic digest of casein)	5.0
Beef extract	5.0
Yeast extract	5.0
Sodium chloride	20.0
Di-sodium hydrogen orthophosphate $\cdot 2H_2O$	12.0
Potassium di-hydrogen orthophosphate	1.35
Aesculin	1.0
Lithium chloride	3.0
Iron (III) ammonium citrate	0.5
Nalidixic acid	0.02[a] or 0.01[b]
Acriflavine	0.025[a] or 0.0125[b]
Distilled or deionized water	1000.0

[a] Fraser broth
[b] Half Fraser broth

Preparation

Basal broth

Suspend all of the ingredients, except the acriflavine and iron (III) ammonium citrate, in the water and heat to dissolve. Dispense 10 ml portions into 16 × 150 mm culture tubes with screw cap closures and sterilize at 121°C for 15 min.

Supplements

To each tube of the cooled basal broth, immediately before use, add 0.1 ml portions of filter sterilized aqueous solutions of 0.25% of acriflavine and 5.0% iron (III) ammonium citrate. For Half Fraser broth add 0.05 ml of 0.25% acriflavine and 0.1 ml of 5.0% iron (III) ammonium citrate.

Physical properties

Appearance	Light straw colour, clear.
pH	7.2 ± 0.2

Shelf life

Basal broth and supplement	14 days at 4 ± 2°C. Use complete medium immediately after addition of the supplements.

Inoculation method for samples

Food or environmental samples are cultured in a primary listeria enrichment broth (e.g. Half Fraser) and incubated at 30°C for 24 h. These cultures are gently mixed and 0.1 ml portions are added to 10 ml of complete Fraser broth.

Incubation

Half Fraser: at 30°C for 24 h.
Fraser: at 35°C for 24 and 48 h (Warburton et al., 1991), in air and in the dark.

Reading of results and interpretation

After incubation the tubes are compared to an uninoculated control against a white background. Blackened cultures are considered as presumptively positive for *Listeria* spp. However, confirmatory tests are necessary on all positive cultures. Cultures which retain the original straw colour are considered to be negative.

Quality assessment

(i) *Productivity*

Test strains	*Listeria monocytogenes* serovar 1/2a (ATCC 35152/ NCTC 7973) *Listeria monocytogenes* serovar 4b (ATCC 13932/ NCTC 10527) *Listeria ivanovii* serovar 5 NCIMB 50095
Inoculation method	Dilution to extinction.

(ii) *Selectivity*

Test strains	*Enterococcus faecalis* NCIMB 50030 *Proteus mirabilis* (ATCC 29906/CECT 4168/NCTC 11938)
Inoculation method	Dilution to extinction.

Comments

Selective properties of acriflavine may vary from lot to lot and manufacturer to manufacturer. Each new batch must be assayed in combination with other selective agents used in the medium to determine the optimum concentration for use, with regard to the efficiency of selectivity and absence of inhibition of *Listeria* spp. Storage of stock solutions of acriflavine for periods of more than one month is not recommended.

References

Cook, L.V. (1998) Chapter 8. Isolation and identification of *Listeria monocytogenes* from red meat, poultry, eggs and environmental samples (Revision 2; 11/08/99) *In:* USDA/FSIS Microbiology Laboratory Guidebook 3rd Edn. US Department of Agriculture, Washington, DC.

Fraser, J.A. and Sperber, W.H. (1988) Rapid detection of *Listeria* spp. in food and environmental samples by esculin hydrolysis. J. Food Protect. 51, 762–765.

ISO (1997) Microbiology of food and animal feeding stuffs – Horizontal method for the detection and enumeration of *Listeria monocytogenes.* Part 1. Detection method. ISO 11290–1:1997. International Organisation for Standardisation, Geneva.

McClain, D. and Lee, W.H. (1988) Development of USDA-FSIS method for isolation of *Listeria monocytogenes* from raw meat and poultry. J. Assoc. Off. Anal. Chem. 71, 660–664.

Warburton, D.W., Farber, J.M., Armstrong, A., Caldeira, R., Tiwari, N.P., Babiuk, T., Lacasse, P. and Read, S. (1991) A Canadian comparative study of modified versions of the 'FDA' and 'USDA' methods for the detection of *Listeria monocytogenes*. J. Food Protect. 54, 669–676.

Giolitti and Cantoni Broth with Tween 80 (GCBT)

This monograph has been assessed by members of the IUMS-ICFMH Working Party on Culture Media and given 'Approved' status.

Description and history

Giolitti and Cantoni broth was developed for the recovery of low numbers of coagulase-positive staphylococci in foods (Giolitti and Cantoni, 1966). It contains mannitol, glycine and sodium pyruvate to enhance repair of stressed staphylococci and to promote growth. Addition of Tween 80 is necessary for the successful recovery of *Staphylococcus aureus* (Chopin et al., 1985). Lithium chloride and potassium tellurite inhibit competitive microorganisms. Anaerobic growth conditions increase the selectivity of the medium.

Composition (grams)

Panceatic digest of casein	10.0
Meat extract	5.0
Yeast extract	5.0
Lithium chloride	5.0
Mannitol	20.0
Sodium chloride	5.0
Glycine	1.2
Sodium pyruvate	3.0
Sorbitan monooleate (Tween 80)	1.0
Potassium tellurite	0.1
Distilled or deionized water	1000.0

Preparation

The medium may be prepared at single strength (as above) or double strength (using double the quantities of solids).

Dissolve the ingredients, except the potassium tellurite, in water, heat and shake to obtain a complete solution. Cool to room temperature and adjust the pH to 6.9 ± 0.2. Dispense the medium in quantities of 10 ml into 16 × 160 mm test tubes in the case of single-strength or into 20 × 200 mm for double strength medium. Sterilize in an autoclave at 121°C for 15 min. Cool and add the filter sterilized (0.22 µm) 1 % (w/v) solution of potassium tellurite using 0.1 ml for single strength and 0.2 ml for double strength medium.

Note: When using commercially prepared media it may be necessary to add Tween 80 prior to autoclaving.

Physical properties

Appearance	Clear, light brown.
pH	6.9 ± 0.2

Shelf life

Basal medium	14 days at 4 ± 2°C. Before use, stored basal medium should be heated to 100°C for 15 min. to remove oxygen, cooled to approximately 47°C and the potassium tellurite added.
Complete medium	Use the same day.
Potassium tellurite solution	1 month at 4 ± 2°C. Discard if a white precipitate forms.

Inoculation method for samples

Use food macerates or decimal dilutions and inoculate 1 ml to single strength medium. To lower the detection limit, 10 ml of the test sample (liquid products) or of the first dilution (other products) may be inoculated in double strength medium. MPN procedures need at least three tubes for at least three dilution steps. If no anaerobic jar is available, overlay the broth with agar or molten paraffin (melting point 56°C).

Incubation method

At 37°C for 24–48 h, anaerobically.

Reading of results and interpretation

After 24 h subculture any tubes showing blackening or black precipitate by streaking onto Baird-Parker agar. Incubate the remainder of the tubes for a further 24 h and subculture all tubes showing growth (irrespective of blackening) to Baird-Parker agar. Examine growth on Baird-Parker agar as detailed in the monograph.

Quality assessment

(i) *Productivity*

Test strain	*Staphylococcus aureus* NCIMB 50081

Inoculation method	Dilution to extinction.
Criteria	Recovery in GCB should be within one titre unit of recovery in tryptose soya broth.

(ii) *Selectivity*

Test strains	*Escherichia coli* NCIMB 50034 *Proteus mirabilis* (ATCC 29906/CECT 4168/NCTC 11938)
Supplementary strain	*Bacillus licheniformis* NCIMB 50016
Inoculation method	Dilution to extinction.
Criteria	Recovery in GCB should be less than 5 titre units of the recovery in tryptone soya broth.

References

Chopin, A., Malcolm, S., Jarvis, G., Asperger, H., Beckers, H.J., Bertona, A.M., Cominazzini, C., Carini, S., Lodi, R., Hahn, G., Heeschen, W., Jans, J.A., Jervis, D.I., Lanier, J.M., O'Connor, F., Rea, M., Rossi, J., Seligmann, R., Tesone, S., Waes, G., Mocquot, G., and Pivnick, H. (1985) ICMSF Methods studies XV. Comparison of four media and methods for enumerating *Staphylococcus aureus* in powdered milk. J. Food Protect. 48, 21–27.

Giolitti, G. and Cantoni, C. (1966) A medium for the isolation of staphylococci from foodstuffs. J. Appl. Bacteriol. 29, 395–398.

IDF. (1997) Milk and milk-based products – Enumeration of coagulase-positive staphylococci – Most probable number technique. Provisional IDF Standard 60C: 1997. International Dairy Federation, Brussels.

 Handbook of Culture Media for Food Microbiology, J.E.L. Corry et al. (Eds.)

Haemorrhagic Colitis (HC) agar

This monograph has been reviewed by members of the IUMS-ICFMH Working Party on Culture Media and given 'Proposed' status.

Description and history

This medium was first described by Szabo et al. (1986) for the direct isolation of haemorrhagic colitis (HC) strains of *Escherichia coli* from food. It is especially useful for the differentiation of *Escherichia coli* O157 from other *Escherichia coli* strains. Most *Escherichia coli* O157 strains do not ferment sorbitol and are unable to produce glucuronidase which hydrolyses 4-methylumbelliferyl-β-D-glucuronide (MUG). Sodium chloride is used to protect *Escherichia coli* O157 from the elevated incubation temperature and bile salts are incorporated to inhibit non-enteric organisms.

Composition (grams)

Pancreatic digest of casein	20.0
Bile salts No. 3	1.12
Sodium chloride	5.0
Sorbitol	20.0
4-methylumbelliferyl-β-D-glucuronide	0.05
Bromocresol purple	0.015
Agar	15.0
Distilled or deionized water	1000.0

Preparation

Dissolve all the ingredients in the water and sterilize the medium at 121°C for 15 min. Cool to 45–50°C and pour 15–20 ml quantities into 9 cm diameter Petri dishes.

Physical properties

Appearance	blue/pale purple.
pH	7.2 ± 0.1

Shelf life

Prepared plates	14 days at 4 ± 2°C.

Inoculation method for samples

Surface streaking of pre-dried plates from selective enrichment broths.

Incubation method

At 41 ± 1°C for 16–18 h, in air.

Reading of results and interpretation

Typical colonies of *Escherichia coli* O157 do not ferment sorbitol, are negative for β-D-glucuronidase and positive in the indole reaction. These microorganisms are blue on this medium and do not fluoresce under UV light (366nm). Other *Escherichia coli* strains are yellow as a result of sorbitol fermentation and fluoresce.

Quality assessment

(i) *Productivity*

Test strains	*Escherichia coli* O157:H7 NCIMB 50139 – this strain is non-toxigenic
Supplementary strains	*Escherichia coli* O157:H7 (NCTC 12079) *Escherichia coli* O157:H- (NCTC 12080)
Inoculation method	Modified Miles-Misra, spiral plating or streaking/ecometry.

(ii) *Selectivity*

Test strains	*Escherichia coli* NCIMB 50034 *Proteus mirabilis* (ATCC 29906/NCTC 11938) *Hafnia alvei* NCIMB 50037 *Enterococcus faecalis* NCIMB 50030
Supplementary strain	*Escherichia coli* NCIMB 50109
Inoculation method	Modified Miles-Misra, spiral plating or streaking/ecometry.

(iii) *Characteristic appearance of colonies*
Colourless/blue, non-fluorescent colonies of more than 1 mm diameter.

References

Bülte, M. and Reuter, G. (1989) Glucuronidase-Nachweis und Indol-Kapillartest als zuverlässige Schnellidentifizierungsverfahren zur Erfassung von *E. coli* in Lebensmitteln-toxinogene Stämme eingeschlossen. Zbl. Hyg. 188, 284–293.

Szabo, R.A., Todd, E.C.D. and Jean, A. (1986) Method to isolate *Escherichia coli* O157:H7 from food. J. Food Protect. 49, 768–772.

Trumpf, T. (1990) Versuche zur Isolierung, Charakterisierung und Abgrenzung verotoxinogener *Escherichia coli* (VTEC) von anderen *E. coli*-Populationen aus der Darmflora von Rindern und die Erfassung von VTEC-Stämmen des Serovars O157: H7 aus Hackfleisch in Modellversuchen. Ph.D. Thesis, Veterinary Faculty of the Free University of Berlin.

Hektoen Enteric (HE) agar

This monograph has been assessed by members of the IUMS-ICFMH Working Party on Culture Media and given 'Approved' status.

Description and history

This medium for the isolation of *Shigella* and *Salmonella* was originally developed by King and Metzger (1968) of the Hektoen Institute. It relies on the use of bile salts for selective inhibition and two indicator systems: (i) bromothymol blue and acid fuchsin as indicators of carbohydrate dissimilation and (ii) ferric iron as an indicator of the formation of hydrogen sulfide from thiosulphate. HE agar allows good growth of *Shigella* spp. because the inhibition of these organisms by bile salts is reduced by the addition of relatively large amounts of peptone and carbohydrates. The medium provides good colonial differentiation and inhibits some coliforms and other non-lactose-fermenting bacteria, thereby facilitating the identification of *Salmonella* and *Shigella* from food products.

Composition (grams)

Proteose peptone	12.0
Yeast extract	3.0
Sodium chloride	5.0
Lactose	12.0
Sucrose	12.0
Salicin	2.0
Bromothymol blue	0.065
Acid fuchsin	0.08
Sodium thiosulphate	5.0
Iron (III) ammonium citrate	1.5
Bile salts	9.0
Agar	14.0
Distilled or deionized water	1000.0

Preparation

Suspend the components in the water. The medium is boiled for a few seconds until the ingredients are completely dissolved. Do not autoclave or overheat. Cool to 47°C and pour into Petri dishes. Final pH should be 7.5 ± 0.2.

For increased selectivity novobiocin can be added at a final concentration of 0.01 g/l. For this dissolve 100 mg of novobiocin in 10 ml of distilled water, sterilize by filtration and aseptically add 1 ml of this solution to the cooled medium, mix well and distribute into Petri dishes.

Physical properties

Appearance	Blue-green, transparent.
pH	7.5 ± 0.2

Shelf life

Ready to use medium	3 weeks at 4 ± 2°C.

Inoculation method for samples

1. From enrichment broths streak a loopful of the incubated selective broth onto the medium to attain isolated colonies.

2. For direct counting spread 0.5 ml of decimal dilutions of food homogenates over the whole plate.

Incubation method

At 35–37°C for 18–24 h, in air.

Reading of results and interpretation

On HE agar, *Salmonella* spp. produce transparent green or blue-green colonies, with or without black centres or may appear as almost completely black colonies (H_2S production). *Shigella* spp. produce green, transparent colonies. As other organisms can form colonies similar to *Salmonella* and *Shigella*, biochemical and serological confirmatory tests are necessary. Lactose, sucrose or salicin fermenting Gram negative bacteria may be inhibited or may produce salmon-coloured colonies.

Quality assessment

(i) *Productivity*

Test strains	*Salmonella virchow* NCIMB 50077 *Salmonella enteritidis* NCIMB 50073 *Shigella flexneri* NCIMB 50079
Inoculation method	Modified Miles-Misra, spiral plating or streaking/ecometry.

(ii) *Selectivity*

Test strain	*Enterococcus faecalis* NCIMB 50030
Inoculation method	Modified Miles-Misra, spiral plating or streaking/ ecometry.

(iii) *Characteristic appearance of colonies*
See above.

References

Andrews, W.H., Poelma, P.L. and Wilson, C.R. (1981) Comparative efficiency of brilliant green, bismuth sulfite, *Salmonella-Shigella*, Hektoen enteric, and xylose lysine deoxycholate agars for the recovery of *Salmonella* from foods: collaborative study. J. Assoc. Off. Anal. Chem. 64, 899–928.

King, S. and Metzger, W. (1968) A new plating medium for the isolation of enteric pathogens. I. Hektoen Enteric Agar. Appl. Microbiol. 16, 577–578.

Irgasan Ticarcillin Chlorate (ITC) broth

This monograph has been assessed by members of the IUMS-ICFMH Working Party on Culture Media and given 'Approved' status.

Description and history

This medium was formulated by Wauters et al. (1988) for the enrichment of *Yersinia enterocolitica* from meat products. It is derived from the enrichment broth for salmonellae described in 1956 by Rappaport et al. The basic selective agents are malachite green and magnesium chloride, however in a different ratio from that used in the original medium. Selectivity for *Yersinia enterocolitica* is enhanced by triclosan (irgasan), ticarcillin and chlorate, the latter being inhibitory for Enterobacteriaceae posessing type A nitratase. The medium performs well for the pathogenic *Yersinia enterocolitica* biotype 4 serotype 0:3, but is less appropriate for other serotypes.

Composition (grams)

Pancreatic digest of casein	10.0
Yeast extract	1.0
Magnesium chloride $\cdot 6H_2O$	60.0
Sodium chloride	5.0
Malachite green	0.01
Potassium chlorate	1.0
2,3,4'-Trichloro-2'-hydroxyldiphenyl ether	
(Irgasan, Ciba-Geigy)*	0.001
Ticarcillin*	0.001
Distilled or deionized water	1000.0

Preparation

Dissolve the ingredients except those marked * in water. Sterilize by autoclaving. Allow the medium to cool and add aseptically ticarcillin and irgasan. Aseptically distribute suitable volumes, usually 100 ml, in sterile containers. The surface should be small in comparison to the volume to minimise aeration. Erlenmeyer flasks form suitable containers.

Physical properties

Appearance	Blue-green, clear.
pH	6.8

Shelf life

Ready to use medium	At least 1 month at 4 ± 2°C in screw-capped containers.

Inoculation method for meat samples

A 20% (w/v) suspension of minced meat is prepared in peptone water and gently shaken without homogenizing by mixer or stomacher. The suspension is filtered through gauze to trap solid particles and a 5 ml volume of the filtrate is inoculated into 100 ml ITC broth. Inoculation of the meat itself, either minced or homogenized, results in a decreased recovery. Incubation should be carried out at 24 ± 1°C for 2 or 3 days.

Reading of results and interpretation

Since growth of yersiniae does not always result in obvious turbidity, the medium should always be subcultured by streaking one loopful either on SSDC agar (q.v.) (preferably) or on CIN agar.

Quality assessment

(i) *Productivity*

Test strain	*Yersinia enterocolitica* biotype 4, serotype 3 NCIMB 50087
Inoculation method	Dilution to extinction.
Criteria	Growth should be within 2 titre units of the growth in tryptone soya broth after 2 days, but even a slight turbidity in ITC should be recorded.

(ii) *Selectivity*

Test strains	*Escherichia coli* NCIMB 50034 *Pseudomonas aeruginosa* NCIMB 50067
Supplementary strains	*Proteus mirabilis* (ATCC 29906/CECT 4168/NCTC 11938) *Staphylococcus aureus* NCIMB 50080
Inoculation method	Dilution to extinction.

Criteria	Difference in growth should be equal to or less than 5 titre units of the growth in tryptone soya broth.

References

Rappaport, F., Konforti, N. and Navon, B. (1956) A new enrichment medium for certain salmonellae. J. Clin. Pathol. 9, 261–266.

Wauters, G., Goossens, V., Janssens, M. and Vandepitte, J. (1988) New enrichment method for isolation of pathogenic *Yersinia enterocolitica* serogroup O:3 from pork. Appl. Environ. Microbiol. 54, 851–854.

Iron sulphite agar

This monograph has been reviewed by members of the IUMS-ICFMH Working Party on Culture Media and given 'Proposed' status.

Description and history

The medium was first developed by Wilson and Blair (1924) to enumerate *Clostridium perfringens* (*B. welchii*) in water. The name is now given to a number of formulations used for different purposes. Iron sulphite agar used for the enumeration of mesophilic sulphite-reducing clostridia has the same composition as Tryptose Sulphite Cycloserine base (Harmon et al., 1971) but without the cycloserine (vide infra). For the enumeration of thermophilic sulphite-reducing organisms such as *Thermoanaerobacterium* (formerly *Clostridium*) *thermosaccharolyticum* and *Desulfotomaculum nigrificans*, a nutritionally poorer composition is used (Attenborough and Scarr, 1957; Donnelly and Graves, 1992). This is the medium described here. Both media utilize the ability of the genus *Clostridia* to reduce sulphite which reacts with the iron citrate to form ferrous sulphide, staining the colonies black.

Composition (grams)

Pancreatic digest of casein	10.0
Sodium sulphite (anhydrous)	0.5
Iron (III) citrate $\cdot 5H_2O$	0.5
Agar	15.0
Distilled or deionized water	1000.0

Preparation

Suspend ingredients other than sodium sulphite and iron (III) citrate in 980 ml water and boil to dissolve completely. Dispense in bottles or tubes and sterilize at 121°C for 15min. Cool to about 50°C and add 10 ml of a 5% sodium sulphite solution and 10 ml of 5% iron (III) citrate solution.

Physical properties

Appearance	Pale straw.
pH	7.0 ± 0.2

Shelf life

Ready to use medium	Day of preparation.

Inoculation method for samples

1. For enumeration of mesophilic sulphite-reducing organisms: pour plate procedure, using 1 ml per 9 cm diameter plate. When set plates are overlaid with sterile medium.

2. For enumeration of thermophilic sulphite-reducing organisms: inoculate 1–4 ml of appropriate dilutions into test tubes and pour about 12–15 ml of the finished medium into the tubes. Mix carefully and allow agar to solidify, then add a further 2–3 ml of molten medium, tempered to $47 \pm 2\,^{\circ}C$ as a cover to prevent uptake of oxygen.

Incubation method

For mesophilic organisms incubate anaerobically at 37°C for 48 h. For thermophilic organisms incubate the tubes at 55°C for 48 h.

Reading of results and interpretation

Read the cultures after 1 and 2 days incubation. Discrete black colonies of 1–5 mm diameter are considered to be presumptive clostridia.

Bacillus licheniformis may, in some cases, also give rise to black colonies. When this is suspected, the identity of the colonies must be confirmed by checking their ability to grow aerobically and their sensitivity to metronidazole (Weenk et al., 1995).

Quality assessment

1. For mesophilic sulphite-reducing clostridia:

(i) *Productivity*

Test strains	*Clostridium sporogenes* NCIMB 50099 *Clostridium perfringens* NCIMB 50027
Inoculation method	Surface plating, modified Miles-Misra or streaking technique using pre-poured plates with subsequent overlay.

(ii) *Selectivity*

Test strains	*Bacillus subtilis* NCIMB 50018 *Escherichia coli* NCIMB 50034
Inoculation method	As above.

2. For thermophilic sulphite-reducing organisms:

(i) *Productivity*

Test strains	*Thermoanaerobacterium thermosaccharolyticum* ATCC 7956/DSM 571/NCIMB 9385 *Desulfotomaculum nigrificans* ATCC 19858/DSM 574/NCIMB 8395/IFO 13698
Inoculation method	Mix inoculum with agar and allow to solidify. Add further 2–3 ml of agar.

(ii) *Selectivity*

Test strains	*Bacillus stearothermophilus* ATCC 10149
Inoculation method	As above.

(iii) *Characteristic appearance of colonies*

Discrete black colonies of 1–5 mm diameter.

References

Andrews, W. (1995) Microbiological methods. *In:* AOAC official methods of analysis. Vol.1, Chapter 17, edited by P. Cunnif. AOAC, Arlington, VA.

Attenborough, S.J. and Scarr, M.P. (1957) The use of membrane filter techniques for control of thermophilic spores in the sugar industry. J. Appl. Bacteriol. 20, 460–466.

Donnelly, L.S. and Graves, R.R. (1992) Sulphide spoilage sporeformers. *In:* Compendium of Methods for the Microbiological Examination of Foods, 3rd edition, pp 317–323, edited by C. Vanderzant and D.F. Splittstoesser. American Public Health Association, Washington, DC.

Harmon, S.M., Kautter, D.A. and Peeler, J.T. (1971) Improved medium for the enumeration of *Clostridium perfringens*. Appl. Microbiol. 22, 688–692.

Weenk, G.H., van den Brink, J.A., Struijk, C.B. and Mossel, D.A.A. (1995) Modified methods for the enumeration of spores of *Clostridium* species in dried foods. Int. J. Food Microbiol. 27, 185–200.

Wilson, W.J. and Blair, E.M.M'V. (1924) The application of a sulphite-glucose-iron agar medium to the quantitative estimation of *B. welchii* and other reducing bacteria in water supplies. J. Pathol. Bacteriol. 27, 119–121.

Kanamycin Aesculin Azide (KAA) agar

This monograph has been assessed by members of the IUMS-ICFMH Working Party on Culture Media and given 'Approved' status.

Description and history

The history of this medium is summarized by Mossel et al. (1978). It utilises (i) the selective inhibitory components azide and kanamycin and (ii) the indicator system aesculin and ferric iron for the isolation and differentiation of group D streptococci. Growth of the majority of unwanted organisms is suppressed, while sought organisms hydrolyse aesculin, producing black zones around the colonies due to the formation of black iron phenolic compounds derived from the aglucon. The selectivity of the medium can be increased by incubation at 42 ± 0.5°C. The selective agents inhibit aesculin hydrolysis of species belonging to the *Enterococcus avium* and *Enterococcus caecorum* groups (Devriese et al., 1993). Aesculin positive *Lactobacillus* strains should be excluded by confirmatory tests (Reuter, 1985).

Composition (grams)

Pancreatic digest of casein	20.0
Yeast extract	5.0
Sodium chloride	5.0
Sodium citrate	1.0
Aesculin	1.0
Iron (III) ammonium citrate	0.5
Sodium azide	0.15
Kanamycin sulphate	0.02
Agar No. 1 (Oxoid)	12.0
Distilled or deionized water	1000.0

Preparation

Suspend the ingredients in the water. Bring to the boil to dissolve completely. Dispense into bottles (or final containers) and sterilize by autoclaving at 121°C for 15 min.

Physical properties

Appearance	Light pale straw.

pH 7.2 ± 0.2

Shelf life

Ready to use medium 7 days at 4 ± 2°C.

Inoculation method for samples

Surface spreading over whole plate using 0.1 ml per pre-dried 9 cm diameter plate.

Incubation method

At 37°C or 42 ± 0.5°C, in air, for 18–24 h. The higher temperature results in greater selectivity.

Reading of results and interpretation

Round white or grey colonies, about 2 mm diameter surrounded by black haloes are considered to be group D streptococci ('presumptive' count).

Quality assessment

(i) *Productivity*

Test strains	*Enterococcus faecalis* NCIMB 50030 *Enterococcus faecium* NCIMB 50032 *Streptococcus bovis* (DSM 20480/NCDO 597/NCTC 8177)
Inoculation method	Modified Miles-Misra, spiral plating or streaking/ecometry.

(ii) *Selectivity*

Test strains	*Escherichia coli* NCIMB 50034 *Lactococcus lactis* ssp. *lactis* NCIMB 50058
Supplementary strains	*Aerococcus viridans* NCIMB 50012 *Bacillus cereus* NCIMB 50014 *Staphylococcus aureus* NCIMB 50080
Inoculation method	Modified Miles-Misra, spiral plating or streaking/ecometry.

(iii) *Characteristic appearance of colonies*

Round, white or grey, about 2 mm diameter surrounded by black haloes of at least 1 cm diameter.

References

Devriese, L.A., Pot, B. and Collins, M.D. (1993) Phenotypic identification of the genus *Enterococcus* and differentiation of phylogenetically distinct enterococcal species and species groups. J. Appl. Bacteriol. 75, 399–408.
Mossel, D.A.A., Bijken, P.H.G., Eelderink, I. and van Spreekens, K.A. (1978) *In:* Streptococci edited by F.A. Skinner and L.B. Quesnel, SAB Symposium Series No. 7. Academic Press, London.
Reuter, G. (1985) Selective media for group D streptococci. Int. J. Food Microbiol. 2, 103–114.

Lactobacillus Sorbic acid (LaS) agar
(syn. Sorbic acid agar base)

This monograph has been assessed by members of the IUMS-ICFMH Working Party on Culture Media and given 'Approved' status.

Description and history

LaS agar is prepared according to the formulation of Reuter (1968; 1970). The combination of anaerobic incubation and reading of the results after 2 days enables selective enumeration of 'lactobacilli' (genus *Lactobacillus*). Acetate, citrate, sorbic acid and low pH suppress other bacterial flora, including enterococci and *Streptococcus salivarius* ssp. *thermophilus*. Productivity for *Lactobacillus delbrueckii* ssp. *bulgaricus* may be slightly reduced. This medium is good for isolating lactobacilli from a mixed flora, but can be inhibitory for some yoghurt cultures and for sublethally damaged bacteria (Reuter, 1985).

Composition (grams)

Tryptic digest of casein	10.0
Beef extract	10.0
Yeast extract	5.0
Glucose	20.0
Sorbitan monooleate (Tween 80)	1.0
Sodium acetate $\cdot 3H_2O$	5.0
Tri-sodium citrate	3.0
Magnesium sulphate $\cdot 7H_2O$ *	0.2
Manganese (II) sulphate $\cdot 4H_2O$ $^+$	0.05
Agar	20.0
Sorbic acid	0.4
Distilled or deionized water	1000.0

* 20 ml of 1% solution
$^+$5 ml of 1% solution

Preparation

Suspend all the ingredients except the sorbic acid in the water and boil to dissolve completely. Heat at 100°C for 1 h. Add the sorbic acid dissolved in about 10 ml 1N

NaOH, adjust pH to 5.2 with 10% HCl, boil for 5 min. Final pH 5.0 at 30°C. *This medium must not be autoclaved.*

Physical properties

Appearance	Pale with slight opalescence.
pH (final)	5.0 at 30°C.

Shelf life

Ready to use medium	Up to 1 week at 4 ± 2°C. Up to 2 weeks in sealed plastic bags at 4 ± 2°C.

Inoculation method

Surface spreading over whole plate or modified Miles-Misra.

Note: pH of a representative plate should be checked before use to confirm selective potential.

Incubation method

Two days under anaerobic or micro-aerobic (6% O_2; 10% CO_2 in N_2) conditions at 37°C for thermophilic and mesophilic species or 30°C for mesophilic and psychrotrophic species.

Reading of results and interpretation

All well grown colonies after 48 h incubation at 30 or 37°C are considered as lactobacilli. Pin point colonies should be ignored. Confirmation by Gram stain may be necessary to exclude some strains of lactic streptococci, leuconostocs and pediococci from certain habitats. *Carnobacterium piscicola* does not grow.

Quality assessment

(i) *Productivity*

Test strains	*Lactobacillus gasseri* NCIMB 50040 *Lactobacillus plantarum* NCIMB 50054 *Lactobacillus sakei* ssp. *sakei* NCIMB 50056
Supplementary strain	*Lactobacillus brevis* NCIMB 50043
Inoculation method	Modified Miles-Misra, spiral plating or streaking/ ecometry.
Criteria	Recovery on LaS agar should be within 1.0 $\log_{10}$ of the

recovery on Briggs agar after 48 h at 30 or 37°C.

(ii) *Selectivity*

Test strains	*Staphylococcus aureus* NCIMB 50080 *Saccharyomyces cerevisiae* NCIMB 50105
Supplementary strains	*Bacillus cereus* NCIMB 50014 *Enterococcus hirae* NCIMB 50031 *Escherichia coli* NCIMB 50034
Inoculation method	Modified Miles-Misra, spiral plating or streaking/ ecometry.
Criteria	Recovery on LaS-Agar should be 5.0 $\log_{10}$ below the recovery on Briggs agar. *Enterococcus hirae* produces pin-point colonies.

(iii) *Characteristic appearance of colonies*

Small white or greyish colonies, 0.5–2.5 mm diameter, flat or raised, smooth or rough. Non-lactic acid bacteria may be distinguished by colony size of > 2.5 mm after prolonged incubation at room temperature. Enterococci may appear as medium-sized colonies of 0.5–1 mm diameter.

References

Reuter, G. (1968) Erfahrungen mit Nährböden für die selektive mikrobiologische Analyse von Fleischerzeugnissen. Arch. Lebensmittelhyg. 19, 53–57 and 84–89.

Reuter, G. (1970) Laktobazillen und eng verwandte Mikroorganismen in Fleisch und Fleischerzeugnissen, 2. Mitteilung: Die Charakterisierung der isolierten Laktobazillenstamme. Fleischwirtsch. 50, 954–962.

Reuter, G. (1985) Elective and selective media for lactic acid bacteria. Int. J. Food Microbiol. 2, 55–68.

Lauryl sulphate MUG X-gal (LMX) broth

This monograph has been reviewed by members of the ICFMH-IUMS Working Party on Culture Media and given 'Proposed' status.

Description and history

LMX broth is a selective enrichment medium for the simultaneous detection of total coliforms and *Escherichia coli* in foods and water which was first described by Manafi and Kneifel (1989). Its use in the examination of soft cheeses was reported by Hahn and Whittrock (1991). The combination of the chromogenic compound 5-bromo-4-chloro-3-indolyl-β-D-galactopyranoside (X-gal) and of the fluorogenic compound 4-methylumbelliferyl-β-D-glucuronide (MUG) incorporated into lauryl sulphate broth provides a double indicator system. A modification (Ossmer, 1993) of the original broth improved substrate utilization and is now commercially available. The high nutritional quality of the broth and the phosphate buffer incorporated promote the growth of coliforms while the lauryl sulphate and novobiocin inhibit most Gram positive organisms. When X-gal is split by β-D-galactosidase, the broth turns to blue-green due to the conversion of the liberated aglycone to indigo. 1-isopropyl-β-D-1-thiogalactopyranoside (IPTG) stimulates synthesis and increases the activity of β-D-galactosidase. In the presence of β-D-glucuronidase the fluorescent 4-methylumbelliferone is split off from MUG. Light blue fluorescence in the broth under UV light (365 nm) indicates the presence of *Escherichia coli*. Indole production in this broth can be detected using Kovacs' reagent.

Methods for *Escherichia coli* identification based on β-glucuronidase detection have been reviewed by Hartman (1989) and Frampton and Restaino (1993).

Composition (grams)

Tryptose	5.0
Sodium chloride	5.0
Sorbitol	1.0
Tryptophan	1.0
Di-potassium hydrogen orthophosphate	2.7
Potassium di-hydrogen orthophosphate	2.0
Sodium lauryl sulphate	0.1
5-Bromo-4-chloro-3-indolyl-β-D-galactopyranoside (X-gal)	0.08
4-Methylumbelliferyl-β-D-glucuronide (MUG)	0.05

1-isopropyl-β-D-thiogalactopyranoside	0.1
Novobiocin	0.03*
Distilled or deionized water	1000.0

Agar may be added if a solid medium is required.

*Supplementation with novobiocin at a maximum concentration of 0.03 g/l is specifically recommended for the examination of food samples. Alternatively, cefsulodin at 0.01 g/l may be substituted for the inhibition of Gram positive flora. Cefsulodin will also inhibit aeromonads and flavobacteria (Brenner et al., 1993; Alonso et al., 1996).

Preparation

Suspend the ingredients except the novobiocin in the water and heat to dissolve. Mix well and dispense into tubes in appropriate quantities and sterilize at 121°C for 15 min. Temper to approximately 48°C and add filter-sterilized novobiocin if required.

Physical properties

Appearance	Clear, colourless or slightly yellow.
pH	6.8 $\pm$ 0.1 at 25°C.

Shelf life

Ready to use broth	4 weeks at 4 $\pm$ 2°C.

Inoculation method for samples

The MPN technique may be used for enumeration of total coliforms and *Escherichia coli* in food samples.

Incubation method

At 35–37°C for 24 h. Incubation may be continued for a further 24 h if required.

Reading of results and interpretation

Total coliforms: the X-gal reaction results in blue-green broth.
Escherichia coli: fluorescence can be read using a UV lamp (365 nm). Light blue fluorescence in the broth indicates the presence of *Escherichia coli* (MUG reaction). To confirm the latter reaction, the culture should be covered with about 5 mm of Kovacs' indole reagent. A cherry-red colour appearing on the reagent layer after 1–2 min confirms the presence of *Escherichia coli*.

Quality assessment

(i) *Productivity*

Test strains	*Escherichia coli* NCIMB 50034 *Hafnia alvei* NCIMB 50037
Supplementary strains	*Escherichia coli* NCIMB 50109 *Klebsiella pneumoniae* NCIMB 50108 *Enterobacter cloacae* NCIMB 50122 *Citrobacter freundii* (ATCC 6750) *Shigella flexneri* NCIMB 50140 *Salmonella typhimurium* (ATCC 14208)
Inoculation method	Dilution to extinction.
Criteria	*Escherichia coli* should produce fluorescent growth whilst growth of other strains should be without fluorescence.

(ii) *Selectivity*

Test strain	*Enterococcus faecalis* NCIMB 50030
Inoculation method	Dilution to extinction.

References

Alonso, J. L., Amoros, I. and Alonso, M.A. (1996) Differential susceptibility of aeromonads and coliforms to cefsulodin. Appl. Env. Microbiol. 62, 1885–1888.

Brenner, K.P., Rankin, C.C., Roybal, Y.R., Stelma, G.N., Scarpino, P.V. and Dufour, A.P. (1993) New medium for the simultaneous detection of total coliforms and *Escherichia coli* in water. Appl. Env. Microbiol. 59, 3534–3544.

Frampton, E.W. and Restaino, L. (1993) Methods for *Escherichia coli* identification in food, water and clinical samples based on beta-glucuronidase detection. J. Appl. Bacteriol. 74, 223–233.

Geissler, K., Manafi, M., Amorós, I. and Alonso, J.L. (2000) Quantitative determination of total coliforms and *E. coli* in marine waters with chromogenic and fluorogenic media. J. Appl. Microbiol. 88, 280–285.

Hahn, G. and Whittrock, E. (1991) Comparison of chromogenic and fluorogenic substances for differentiation of coliforms and *Escherichia coli* in soft cheese. Acta Microbiol. Hung. 38, 265–271.

Hartman, P.A. (1989) The MUG (glucuronidase) test for *Escherichia coli* in food and water. *In:* Rapid Methods and Automation in Microbiology and Immunology, edited by A. Balows, R.C. Tilton and A. Turano, Brixia Academic Press, Brescia, Italy, pp. 290–308.

Manafi, M. and Kneifel, W. (1989) A combined chromogenic-fluorogenic medium for the simultaneous detection of total coliforms and *E.coli* in water. Zbl. Hyg. 189, 225–234.

Ossmer, R. (1993) Simultaneous detection of total coliforms and *E.coli* – Fluorocult^{R} LMX broth. Abstr. Food Micro '93, Bingen, Germany, p. 202.

Lauryl tryptose broth

This monograph has been assessed by members of the IUMS-ICFMH Working Party on Culture Media and given 'Approved' status.

Description and history

A selective medium for the detection of coli-aerogenes bacteria in water, dairy products and other foods (Mallman and Darby, 1941). The surface active agent lauryl sulphate acts as the selective agent in restricting the growth of bacteria other than the coli-aerogenes group. The broth is specifically designed to allow rapid multiplication and copious gas production from a small inoculum of target organisms. Unlike other media it is possible to test directly for indole production.

Composition (grams)

Tryptose	20.0
Lactose	5.0
Sodium chloride	5.0
Di-potassium hydrogen orthophosphate	2.75
Potassium di-hydrogen orthophosphate	2.75
Sodium lauryl sulphate	0.1
Distilled or deionized water	1000.0

Preparation

Add the ingredients to the water and mix until completely dissolved. Dispense in 10ml amounts into suitable plugged test tubes or screw cap bottles and add a Durham tube to each tube or bottle. Sterilize by autoclaving at 121°C for 15 min. To avoid bubbles in the Durham tubes, allow temperature to drop to 75°C before opening the autoclave.

Physical properties

Appearance	Light amber, clear to slightly opalescent.
pH	6.8 ± 0.2

Shelf life

Ready to use medium	1 month at 4 + 2°C in screw capped containers.

Inoculation method for samples

Use MPN tests for presumptive enumeration of coli-aerogenes bacteria and *Escherichia coli* as described in the American Public Health Association Standard Methods for water and waste water (1998), dairy products (1993) and foods (1992).

Incubation method

To detect members of the coli-aerogenes group incubate at 35 or 37°C for 24–48 h. For dairy purposes and where it is desired to detect the full range of coliform bacteria, incubate at 30°C for 24–48 h. For *Escherichia coli* the temperature of 44 $\pm$ 0.1°C for 18 h is specifically recommended. Psychrotrophic coliforms can be detected by incubation at 4°C for 10 days.

Reading of results and interpretation

Turbidity provides presumptive evidence of the presence of coli-aerogenes bacteria, particularly when accompanied by gas formation. Further confirmation of the presence of *Escherichia coli* may be obtained from the indole reaction and subculture to VRB agar (q.v.)

Quality assessment

(i) *Productivity*

Test strain	*Escherichia coli* NCIMB 50034 *Hafnia alvei* NCIMB 50037
Inoculation method	Dilution to extinction.

(ii) *Selectivity*

Test strain	*Enterococcus faecalis* NCIMB 50030
Inoculation method	Dilution to extinction.

References

American Public Health Association (1992) Compendium of Methods for the Microbiological Examination of Foods, 3rd edn., American Public Health Association, Washington, DC.

American Public Health Association (1993) Standard Methods for the Examination of Dairy Products, 16th edn., American Public Health Association, Washington, DC.

American Public Health Association (1998) Standard Methods for the Examination of Water and Waste Water, 20th edn., American Public Health Association, Washington, DC.

Mallman , W.L. and Darby, C.W. (1941) Use of a lauryl sulphate tryptose broth for the detection of coliform organisms. Am. J. Publ. Hlth. 31, 127–134.

Lithium chloride Phenylethanol Moxalactam (LPM) agar

This monograph has been reviewed by members of the IUMS-ICFMH Working Party on Culture Media and given 'Proposed' status.

Description and history

LPM agar was developed in 1986 to recover *Listeria* spp. from very mixed microflora in UVM enrichment broths (q.v.) and it is more selective for listeria than other media developed previously (Lee and McClain, 1986). LPM's selective nature is used successfully by the USDA (McClain and Lee, 1988) and the FDA (Datta et al., 1988) to enumerate >100/g of *Listeria* spp. or *Listeria monocytogenes* from many kinds of contaminated foods by direct plating without enrichment. *Listeria* spp. will form colonies on LPM agar in 24 h at 30°C while most, but not all, of the competing microflora are suppressed. Listeria colonies can be recognised and selected for identification using the 45° transillumination of Henry (1933). *Listeria* spp. which have been frozen grow well on LPM agar but heat injured cells do not.

Composition (grams)

Pancreatic digest of casein	5.0
Peptone (peptic digest of animal tissue)	5.0
Beef extract	3.0
Sodium chloride	5.0
Lithium chloride	5.0
Glycine anhydride* (Sigma G-7251)	10.0
2-phenylethanol (Sigma P-6134)	2.5
Sodium or ammonium moxalactam	0.02
Agar	15.0
Distilled or deionized water	1000.0

*Note: Do not substitute glycine anhydride with glycine, which makes the LPM very inhibitory to growth of *Listeria* spp.

Preparation

Base

Suspend the ingredients except the moxalactam in bottles or flasks containing a magnetic stirring bar and autoclave the medium for 12 min at 121°C. Mix the medium gently after autoclaving. LPM is heat sensitive and so it is important to cool it promptly to 46°C in a water bath.

Moxalactam solution

Sodium or ammonium moxalactam (Eli Lilly or Sigma M-1900)	1.0 g
Potassium buffer (0.1M) pH 6.0	100.0 ml

(Made by adding 0.1 M mono-potassium phosphate solution to 0.1 M di-potassium hydrogen phosphate solution to reach pH 6.0). Sterilize by filtration through 0.2 μm filter. Store both the powder and 3–5 ml aliquots of the 1% moxalactam in a freezer at –60°C.

Complete medium

Add 2 ml 1% moxalactam solution per litre of the autoclaved and cooled molten base while stirring with a magnetic mixer. Dispense 12 ml per 9 cm diameter Petri dish.

Physical properties

Appearance	Almost clear and colourless.
pH	7.3 ± 0.2

Shelf life

Dehydrated agar base	Store at 2–8°C to prevent loss of selectivity. Use before manufacturer's expiry date.
Ready to use medium	2–3 weeks at 4 ± 2°C. It is essential to store plates in sealed plastic bags as the phenylethanol is volatile.

Inoculation method for samples

Surface streaking or modified Miles-Misra. For rapid semiquantitative detection of >100/g of *Listeria* spp. in foods, swab the surface of solid samples with a moistened (dry for liquid samples) swab, and inoculate half of a LPM plate. Streak the other half of the plate with a sterile loop (McClain and Lee, 1988). Liquid samples can be concentrated by centrifugation before testing.

Incubation method

LPM plates are usually placed in thin polyethylene plastic bags and incubated at 30°C in air for 20–24 h for isolation, and 40 h for counting colonies without magnification.

Reading of results and interpretation

Under optimum 45° transillumination the more isolated and larger (24h old) listeria colonies appear as whitish piles of crushed glass often showing mosaic-like internal structures occasionally having blue-grey iridescent areas. Smaller, more crowded, colonies have a more pronounced blue-grey iridescence that tends to sparkle. When growth becomes near confluent an even blue-grey iridescent sheen can be observed. For quantitative determination, count all the colonies that grow on LPM agar in 40 h and then identify the percentage of these colonies that are *Listeria* spp. or *Listeria monocytogenes*. For semiquantitative estimation, simply compare the density of all the colonies in the swabbed area with standard density photos of 10^2, 10^3, 10^4, 10^5, 10^6 and 10^7 cfu/ml of *Listeria monocytogenes* swabbed and streaked on LPM agar (similar to Oxoid Dip Slide) and then identify from the streaked area the percentage of colonies as *Listeria* spp. or *Listeria monocytogenes*. Alternatively, scan the swab and streak plates at 24 h using 45° transillumination, and quantitatively enumerate the positive samples using the decimal dilution plating technique.

Quality assessment

(i) *Productivity*

Test strains	*Listeria monocytogenes* serovar 1/2a (ATCC 35152/ NCTC 7973) *Listeria monocytogenes* serovar 4b (ATCC 13932/ NCTC 10527) *Listeria ivanovii* serovar 5 NCIMB 50095
Inoculation method	Modified Miles-Misra or spiral plating.

(ii) *Selectivity*

Test strains	*Enterococcus faecalis* NCIMB 50030 *Proteus mirabilis* (ATCC 29906/CECT 4168/NCTC 11938)
Inoculation method	Modified Miles-Misra or spiral plating.

(iii) *Characteristic appearance of colonies*
See above.

References

Datta, A.R., Wentz, B.A., and Hill, W. E. (1988) Identification and enumeration of beta-haemolytic *Listeria monocytogenes* in naturally contaminated dairy products. J. Assoc. Off. Anal. Chem. 71, 673–675.

Domjan Kovacs, H. and Ralovich, B. (1991) Model examination of selective media for isolation of *Listeria* strains. Acta Microbiol. Hung. 38, 141–145.

Henry, B.S. (1933) Dissociation in the genus *Brucella*. J.Infect.Dis. 52, 374–402.

Lee, W.H. and McClain, D. (1986) Improved *Listeria monocytogenes* selective agar. Appl. Environ. Microbiol. 52, 1215–1217.

McClain, D. and Lee, W.H. (1988) Development of the USDA-FSIS method for the isolation of *Listeria monocytogenes*. J.Assoc.Off. Anal. Chem. 71, 660–664.

L-S Differential (LSD) agar

This monograph has been reviewed by members of the IUMS-ICFMH Working Party on Culture Media and given 'Proposed' status.

Description and history

LSD is an elective medium which supports good growth and differentiation of thermophilic lactobacilli and streptococci in yoghurt products (Reuter, 1985). The reduction of triphenyltetrazolium chloride in connection with the casein reaction allows differentiation between lactobacilli and streptococci (especially *Lactobacillus delbrueckii* ssp. *bulgaricus* and *Streptococcus thermophilus*) by means of colony morphology. LSD is prepared according to the formulation of Eloy and Lacrosse (1976).

Composition (grams)

Peptone from casein	10.0
Peptone from soya	5.0
Beef extract	5.0
Yeast extract	5.0
Dextrose	20.0
Sodium chloride	5.0
1-Cysteine-hydrochloride	0.3
2,3,5-Triphenyltetrazolium chloride (2%)* (ml)	10.0
Milk powder (10%)* (ml)	100.0
Agar	13.0
Distilled or deionized water	890.0

Preparation

Suspend all the ingredients except those marked * in the water and boil to dissolve completely. Sterilize for 20 min at 121°C. Add to the cooled (50°C) medium 100 ml of a sterilized (121°C for 5 min) 10% milk powder solution and 10 ml of filter sterilized 2,3,5-triphenyltetrazolium chloride solution. These solutions should be pre-warmed to 50°C.

Physical properties

Appearance	Pale amber, opaque.

pH 6.1 ± 0.2

Shelf life

Ready to use medium 10 days at 4 ± 2°C.

Inoculation method for samples

Surface spreading over whole plate or modified Miles-Misra.

Incubation method

At 43°C for 48 h, in air.

Reading of results and interpretation

Brilliant red colonies with clear halo can be considered as lactic streptococci (*Streptococcus thermophilus* and *Lactococcus lactis* ssp. *lactis*) whilst pink colonies with a red centre are likely to be *Lactobacillus delbrueckii* ssp. *bulgaricus*. *Lactobacillus acidophilus* and other lactobacilli from milk sources may appear as red colonies with an opaque halo.

Quality assessment

(i) *Productivity*

Test strains	*Streptococcus thermophilus* NCIMB 50083 *Lactococcus lactis* ssp. *lactis* NCIMB 50058 *Lactobacillus delbrueckii* ssp. *bulgaricus* NCIMB 50050 *Lactobacillus acidophilus* (NCIMB 50101)
Inoculation method	Modified Miles-Misra, spiral plating or streaking/ecometry.
Criteria	Recovery on LSD agar should be within 1.0 $\log_{10}$ of the recovery on MRS agar or Briggs agar for *Streptococcus thermophilus*.

(ii) *Characteristic appearance of colonies*

Brilliant red colonies with clear halo:	*Streptococcus thermophilus* *Lactococcus lactis* ssp. *lactis* and ssp. *diacetilactis*.
Pink colonies with red centre:	*Lactobacillus delbrueckii* ssp. *bulgaricus*.

Red colonies with opaque halo:	*Lactobacillus acidophilus* and other lactobacilli from milk sources.

Unwanted species may be distinguished by colony appearance after further incubation at room temperature.

References

Eloy, C. and Lacrosse, R. (1976) Composition d'un milieu de culture destine a effectuer le denombrement des micro-organismes thermophiles du yoghourt. Bull. Rech. Agron. Gemblou 11, 83–86.

Reuter, G. (1985) Elective and selective media for lactic acid bacteria. Int. J. Food Microbiol. 2, 55–68.

 Handbook of Culture Media for Food Microbiology, J.E.L. Corry et al. (Eds.)

M 17 agar

This monograph has been reviewed by members of the IUMS-ICFMH Working Party on Culture Media and given 'Proposed' status.

Description and history

M 17 agar is an elective medium for the detection and differentiation of lactic streptococci and is prepared according to Terzaghi and Sandine (1975). The special buffering capacity of di-sodium-β-glycerophosphate enables a better growth of the lactic streptococci (Reuter, 1985). The addition of lactose allows differentiation between lactose-positive streptococci and lactose-negative mutants. Bacteriophages can also be detected by plaque formation. The medium is used for cultivation procedures in plasmid demonstration and molecular based investigations.

Composition (grams)

Peptone from casein	5.0
Peptone from meat	5.0
Peptone from soya	5.0
Yeast extract	2.5
α-lactose	5.0
Di-sodium-β-glycerophosphate	19.0
Magnesium sulphate $\cdot 7H_2O$	0.25
Ascorbic acid	0.5
Agar	11.0
Distilled or deionized water	1000.0

Preparation

Suspend the ingredients in the water and boil to dissolve completely. Sterilize for 20 min at 121°C.

Physical properties

Appearance	Light amber, clear.
pH (final)	7.2 ± 0.2

Shelf life

Ready to use medium 7 days at 4 ± 2°C.

Inoculation method

Surface spreading over whole plate or modified Miles-Misra.

Incubation method

At 37°C for 48 h, in air. For mesophilic strains incubation at 30°C for 48 h in air is recommended.

Reading of results and interpretation

M 17 agar is an elective medium that gives good colony counts and a characteristic colony size for thermophilic streptococci, lactococci and enterococci. Other microorganisms must be excluded by specific colony appearance and by confirmation tests e.g. Gram stain and catalase test.

Quality assessment

(i) *Productivity*

Test strains	*Streptococcus thermophilus* NCIMB 50083 *Lactococcus lactis* ssp. *lactis* NCIMB 50058 *Enterococcus faecalis* NCIMB 50030
Inoculation method	Modified Miles-Misra or spiral plating.
Criteria	Recovery on M17 agar should be within 1.0 $\log_{10}$ of the recovery on Briggs agar or another non-selective medium.

(ii) *Characteristic appearance of lactic streptococci*
Usually well-grown white colonies.

Size:

Streptococcus thermophilus and *Lactococcus lactis* with subspecies except ssp. *cremoris*	>3.0 mm diameter
lactose-negative mutants	<1.0 mm diameter
enterococci	1.0–1.5 mm diameter
unwanted species	<1.0 mm diameter

References

Reuter, G. (1985) Elective and selective media for lactic acid bacteria. Int. J. Food Microbiol. 2, 55–68.

Terzaghi, B.E. and Sandine, W.E. (1975) Improved medium for lactic streptococci and their bacteriophages. Appl. Microbiol. 29, 807–813.

de Man, Rogosa and Sharpe (MRS) agar

This monograph has been assessed by members of the IUMS-ICFMH Working Party on Culture Media and given 'Approved' status.

Description and history

MRS agar was developed by de Man, Rogosa and Sharpe (1960) primarily for the cultivation of lactobacilli from various sources with the intention of producing a defined medium as a substitute for tomato juice agar. It may be used for the cultivation of the whole group of lactic acid bacteria. The medium shows good productivity for nearly all lactic acid bacteria but the original version is not selective. It may be made selective for lactic acid bacteria by lowering the pH to 5.7 and the addition of 0.14% sorbic acid (see MRS-S agar). Some strains from dairy sources may show reduced growth rates (see Briggs agar).

Composition (grams)

Tryptic digest of casein	10.0
Beef extract	8.0
Yeast extract	4.0
Glucose	20.0
Sorbitan monooleate (Tween 80)	1.0
Di-potassium hydrogen orthophosphate	2.0
Magnesium sulphate $\cdot 7H_2O$*	0.2
Manganese (II) sulphate $\cdot 4H_2O$ +	0.05
Ammonium citrate	2.0
Sodium acetate $\cdot 3H_2O$	5.0
Agar	15.0
Distilled or deionized water	1000.0

* 20 ml of 1% solution

+ 5 ml of 1% solution

Preparation

Suspend the ingredients in the water and boil to dissolve completely. Sterilize for 15 min at 121°C.

Physical properties

Appearance	Light amber, clear.
pH	6.2 Adjust pH after dissolving completely at about 50°C.

Shelf life

Ready to use medium	14 days at 4 ± 2°C.

Inoculation method for samples

Surface spreading over whole plate or modified Miles-Misra.

Incubation method

This depends on the particular habitat of the organisms to be cultivated. Dairy strains should be incubated at 30°C for 2 days followed by 1 day at 22°C, meat strains at 25°C for 2 days, intestinal or yoghurt strains for 2 days at either 37 or 42°C. All incubations should be performed under anaerobic or microaerobic (6% O_2: 10% CO_2 in N_2) conditions.

Reading of results and interpretation

MRS agar is an elective medium that gives good colony counts and a characteristic colony size and morphology for lactobacilli and for other lactic acid bacteria. Other microorganisms must be excluded by specific colonial appearance and by confirmation tests e.g. Gram-stain and catalase test. *Carnobacterium* spp. either show weak growth or fail to grow.

Quality assessment

(i) *Productivity*

Test strains	*Lactobacillus gasseri* NCIMB 50040 *Lactobacillus sakei* ssp. *sakei* NCIMB 50056
Inoculation method	Modified Miles-Misra, spiral plating or streaking/ ecometry.
Criteria	Recovery on MRS agar should be within 1.0 $\log_{10}$ of the recovery on Briggs agar.

(ii) *Characteristic appearance of colonies*

Usually small greyish white colonies, flat or raised, smooth, rough or intermediate.

Reference

de Man, J.C., Rogosa, M. and Sharpe, M.E. (1960) A medium for the cultivation of lactobacilli. J. Appl. Bacteriol. 23, 130–135.

de Man, Rogosa and Sharpe agar with sorbic acid (MRS-S agar)

This monograph has been assessed by members of the IUMS-ICFMH Working Party on Culture Media and given 'Approved' status.

Description and history

MRS agar was developed by deMan, Rogosa and Sharpe (1960) primarily for the cultivation of lactobacilli from various sources with the intention of producing a defined medium as a substitute for tomato juice agar. It may be used for the cultivation of the whole group of lactic acid bacteria. The medium results in good productivity for nearly all lactic acid bacteria but it is not selective in the original version. Some strains from milk sources may show reduced growth rates (see Briggs agar). It becomes selective for lactic acid bacteria if the pH is lowered to 5.7 and sorbic acid is added in a concentration of 0.14% (= 0.2% potassium sorbate) (ISO, 1984; Reuter, 1985).

Composition (grams)

Tryptic digest of casein	10.0
Beef extract	8.0
Yeast extract	4.0
Glucose	20.0
Sorbitan monooleate (Tween 80)	1.0
Di-potassium hydrogen orthophosphate $\cdot 3H_2O$	2.0
Magnesium sulphate $\cdot 7H_2O$ *	0.2
Manganese (II) sulphate $\cdot 4H_2O$ +	0.05
Ammonium citrate	2.0
Sodium acetate $\cdot 3H_2O$	5.0
Sorbic acid	1.4
Agar	15.0
Distilled or deionized water	1000.0

* 20 ml of 1% solution

\+ 5 ml of 1% solution

Preparation

Suspend the ingredients except the sorbic acid in the water and boil to dissolve completely. Heat at 100°C for 1 h. Add the sorbic acid in about 10 ml 1N NaOH, adjust pH to 5.8 with 10% HCl and boil for 5 min. Final pH should be 5.7 at 30°C. *This medium must not be autoclaved.*

Physical properties

Appearance	Light amber, clear.
pH	5.7 at 30°C.

Shelf life

Ready to use medium	7 days at 4 ± 2°C.

Inoculation method for samples

Surface spreading over whole plate or modified Miles-Misra.

Incubation method

This depends on the particular habitat of the organisms to be cultivated. Dairy strains should be incubated at 30°C for 2 days followed by 1 day at 22°C, meat strains at 25°C for 3 days, intestinal or yoghurt strains for 2 days at either 37 or 42°C. All incubations should be performed under anaerobic or microaerobic (6% 0_2: 10% CO_2 in N_2) conditions.

Reading of results and interpretation

Reading the results should be performed after a defined incubation time (see above). All well grown colonies are considered as lactic acid bacteria. Some enterococci may show reduced growth. Single yeast strains may occur with delayed growth. *Carnobacterium* spp. either show weak growth or fail to grow.

Quality assessment

(i) *Productivity*

Test strains	*Lactobacillus gasseri* NCIMB 50040
	Lactobacillus sakei ssp. *sakei* NCIMB 50056
	Pediococcus damnosus NCIMB 50065
Supplementary strains	*Lactococcus lactis* ssp. *lactis* NCIMB 50058
	Leuconostoc mesenteroides NCIMB 50060

Inoculation method	Modified Miles-Misra, spiral plating or streaking/ ecometry.
Criteria	Recovery on MRS-S agar should be within 1.0 $\log_{10}$ of recovery on MRS agar. *Lactococcus lactis* ssp. *lactis* produces pin-point colonies.

(ii) *Selectivity*

Test strains	*Staphylococcus aureus* NCIMB 50080 *Bacillus cereus* NCIMB 50014
Supplementary strain	*Escherichia coli* NCIMB 50034
Inoculation method	Modified Miles-Misra, spiral plating or streaking/ ecometry.
Criteria	Recovery on MRS-S should be 5.0 $\log_{10}$ below the recovery on MRS. *Staphylococcus aureus* produces pin-point colonies.

(iii) *Characteristic appearance of colonies*
Small greyish-white colonies, flat or raised, smooth, rough or intermediate.

References

deMan, J.C., Rogosa, M. and Sharpe, M.E. (1960) A medium for the cultivation of lactobacilli. J. Appl. Bacteriol. 23, 130–135.

ISO/TC 34/SC 6/WG 15, No. 3 and No. 5 (1984) Draft report: Enumeration of Lactobacteriaceae in meat and meat products.

Reuter, G. (1985) Elective and selective media for lactic acid bacteria. Int. J. Food Microbiol. 2, 55–68.

ISO (1997) Microbiology of Food and Animal Feeding Stuffs. Enumeration of Lactic Acid Bacteria. Colony-count technique at 30°C. ISO/DIS 15214 International Organisation for Standardization, Geneva.

Mannitol Egg Yolk Polymyxin (MEYP) agar

This monograph has been assessed by members of the IUMS-ICFMH Working Party on Culture Media and given 'Approved' status.

Description and history

This medium was devised by Mossel et al. (1967) for the isolation of *Bacillus cereus*. It relies on the selective inhibitory component polymyxin and two indicator systems, mannitol and phenol red and egg yolk. Thus growth of many unwanted organisms is suppressed. *Bacillus cereus* does not attack mannitol but dissimilates egg yolk and consequently gives rise to typical bacilliform colonies with purple-red zones and white haloes. *Bacillus anthracis*, *Bacillus mycoides*, *Bacillus pseudomycoides*, *Bacillus thuringiensis* and *Bacillus weihenstephanensis* may not be distinguishable from *Bacillus cereus*.

Since its introduction, MEYP agar has been adopted as a productive medium by Lancette et al. (1980) and Harmon et al. (1984) and is popular in both the United States and Europe (Kramer and Gilbert, 1989).

Composition (grams)

Peptone	10.0
Meat extract	1.0
Sodium chloride	10.0
Mannitol	10.0
Phenol red	0.025
Polymyxin B (i.u.)*	100 000.0
Egg yolk emulsion – 20%* (ml)	100.0
Agar	15.0
Distilled or deionized water	900.0

Preparation

Suspend the ingredients except those marked * in 900 ml of water. Bring to the boil to dissolve completely and sterilize by autoclaving at 121°C for 15 min. Cool to about 50°C and add the polymyxin B suspended in 2 ml of sterile distilled water and 100 ml of a sterile 20% egg yolk emulsion, using aseptic precautions throughout.

Physical properties

Appearance	Light pink, opaque.
pH	7.2 ± 0.2

Shelf life

Ready to use medium	4 days at 4 ± 2°C.

Inoculation method for samples

Surface spreading over whole plate using 0.1 ml per 9 cm diameter plate.

Incubation method

At 30°C for 24 – 30 h, in air.

Reading of results and interpretation

Crenated colonies, about 5 mm in diameter, on a distinct red background and surrounded by an intense egg yolk precipitate are virtually always isolates of *Bacillus cereus*. As *Bacillus cereus*, *Bacillus anthracis*, *Bacillus mycoides*, *Bacillus pseudomycoides*, *Bacillus thuringiensis* and *Bacillus weihenstephanensis* may not be distinguishable from each other on this medium further identification may be necessary. Do not mistake lipase activity for lecithinase activity; lipase produces a fainter precipitate.

Quality assessment

(i) *Productivity*

Test strains	*Bacillus cereus* NCIMB 50014
Inoculation method	Modified Miles-Misra, spiral plating or streaking/ ecometry.

(ii) *Selectivity*

Test strains	*Escherichia coli* NCIMB 50034 *Micrococcus luteus* NCIMB 50063
Inoculation method	Modified Miles-Misra, spiral plating or streaking/ ecometry.

(iii) *Characteristic appearance of colonies*

Crenated, about 5 mm in diameter on a distinctly red background and surrounded by a copious egg yolk precipitate. Do not mistake lipase activity for lecithinase activity.

References

Harmon, S.M., Kautter, D.A. and McClure, F.D. (1984) Comparison of selective plating media for enumeration of *Bacillus cereus* in foods. J. Food Protect. 47, 65–67.
Kramer, J.M. and Gilbert, R.J. (1989) *Bacillus cereus* and other *Bacillus species. In:* Foodborne Bacterial Pathogens edited by M.P. Doyle. Marcel Dekker, New York, pp21–70.
Lancett, G.A. and Harmon, S.M. (1980) Enumeration and confirmation of *Bacillus cereus* in foods. J. Assoc. Offic. Anal. Chem. 63, 581–586.
Mossel, D.A.A., Koopman M.J. and Jongerius, E. (1967) Enumeration of *Bacillus cereus* in foods. Appl. Microbiol. 15, 650–653.

Mannitol Lysine Crystal violet Brilliant green (MLCB) agar

This monograph has been reviewed by members of the IUMS-ICFMH Working Party on Culture Media and given 'Proposed' status.

Description and history

MLCB agar is a selective and diagnostic agar for the isolation of salmonellae, other than *Salmonella typhi* and *Salmonella paratyphi*, from faeces and foods. It may be particularly useful when lactose fermenting strains of salmonellae are sought. First described by Inoue et al. (1968), it was not until van Schothorst et al. (1987) reported favourably on the use of the medium that it became better known. The inhibitory properties of the medium are conferred by the inclusion of brilliant green and crystal violet. The recognition of *Salmonella* colonies depends on the fermention of mannitol and the decarboxylation of lysine together with an indicator system (sodium thiosulphate and iron (III) ammonium citrate) for hydrogen sulphide detection. The medium is not suitable for the isolation of *Salmonella typhi* or *Salmonella paratyphi*, because of the inhibitory concentration of brilliant green, nor for other brilliant green-sensitive strains (Curtis and Clarke, 1994; Arroyo and Arroyo, 1995). Van Schothorst et al. (1987) found the medium excellent and with enhanced selectivity for the isolation of hydrogen sulphide positive salmonellae after enrichment in Rappaport-Vassiliadis broth containing soya peptone (RVS) in place of tryptone. Salmonellae grow as large purple-black colonies due to H_2S production. Atypical salmonellae that produce little or no H_2S grow as mauve-grey colonies and may develop a central black 'bull's-eye'. Contaminating organisms may grow as small colourless colonies although some strains of *Citrobacter* species may mimic the appearance of *Salmonella* and some *Proteus* species may swarm (van Schothorst et al., 1987; Bridson, 1990).

Composition (grams)

Yeast extract	5.0
Peptone	10.0
'Lab-Lemco' powder	2.0
Sodium chloride	4.0
Mannitol	3.0
L-Lysine hydrochloride	5.0
Sodium thiosulphate	4.0

Iron (III) ammonium citrate	1.0
Brilliant green	0.005
Crystal violet	0.01
Agar	15.0
Distilled or deionized water	1000.0

* concentration varies with formulation and activity of dye – see note at the end of this monograph.

Preparation

Brilliant green solution

Prepare 0.5% w/v aqueous solution of brilliant green.

Crystal violet solution

Prepare 0.5% w/v aqueous solution of crystal violet.

Complete medium

Suspend all the ingredients except brilliant green and crystal violet in the water. Add 1 ml brilliant green solution and 10 ml crystal violet solution. Mix and heat very carefully to boiling with frequent agitation to dissolve the ingredients completely. Cool to 50°C and aseptically adjust the pH value to 6.8 ± 0.2. Pour 15–20 ml quantities into 9 cm diameter Petri dishes and allow to set. Dry the prepared plates before use.

Do not autoclave or overheat the medium.

Physical properties

Appearance	Violet, clear.
pH	6.8 ± 0.2

Shelf life

Ready to use medium	5 days at 4 ± 2°C.

Inoculation method for samples

Surface streaking of the sample to obtain well-isolated colonies, preferably after enrichment in a salmonella selective enrichment broth such as RVS broth. A heavy inoculum may be advantageous.

Incubation method

At 35 (or 37)°C for 18–24 h, in air.

Reading of results and interpretation

Typical strains of salmonellae produce large purple-black colonies. Atypical (non-H_2S-producing) strains grow as mauve-grey colonies and may develop a central black 'bull's eye'. Contaminating organisms may grow on the medium as small colourless colonies.

Quality assessment

(i) *Productivity*

Test strains	*Salmonella enteritidis* NCIMB 50073 *Salmonella virchow* NCIMB 50077
Supplementary strains	*Salmonella saintpaul* NCIMB 50075 *Salmonella typhimurium* (NCTC 12190)
Inoculation method	Modified Miles-Misra, spiral plating or streaking/ ecometry.

(ii) *Selectivity*

Test strains	*Escherichia coli* NCIMB 50034 *Proteus mirabilis* (ATCC 29906/CECT 4168/NCTC 11938) – qualitative test for control of swarming.
Inoculation method	Modified Miles-Misra, spiral plating or streaking/ ecometry.

(iii) *Characteristic appearance of colonies*
Large purple-black colonies, 2–3 mm in diameter.

Specification for brilliant green

Bacteriological performance

Brilliant green should supress the spreading of most *Proteus* species while not inhibiting the growth of salmonellae (other than *Salmonella typhi* and *Salmonella paratyphi* and other brilliant green-sensitive strains).

Method of test

Prepare MLCB agar containing various concentrations of brilliant green viz. 4.5 mg/l to 6 mg/l.

Procedure

Inoculate a set of plates with different brilliant green concentrations with a pure culture of a swarming *Proteus* and another set with a pure culture of a *Salmonella* and incubate these plates at 37°C for no longer than 24 h.

A satisfactory concentration of the dye should allow typical growth of salmonellae and limited growth of *Proteus*. Prepare the brilliant green solution to contain one thousand times the final concentration in the complete medium and add 1 ml/l of medium i.e. if the selected concentration is confirmed as 0.005 g/l, prepare 0.5% w/v solution of brilliant green.

References

Arroyo, G. and Arroyo, J.A. (1995) Selective action of inhibitors used in different media on the competitive microflora of salmonella. J. Appl. Bacteriol. 75, 281–289.

Bridson, E.Y. (1990) MLCB agar. *In:* The Oxoid Manual, 6th edition. Pp. 2:157–158.

Curtis, G.D.W. and Clarke, L.A. (1994) Comparison of the MSRV method with an in-house conventional method for the detection of *Salmonella* in various high and low moisture foods. (Corresp.) Lett. Appl. Microbiol. 18, 239–240.

Inoue, T., Takagi, S., Ohnishi, A., Tamura, K. and Suzuki, A. (1968) Food-borne disease Salmonella isolation medium (MLCB). Paper presented at the 66th Meeting of the Society Veterinary Science of Japan.

van Schothorst, M., Renaud, A. and van Beek, C. (1987) Salmonella isolation using RVS broth and MLCB agar. Food Microbiol. 4, 11–18.

M-Enterococcus (ME) agar

This monograph has been assessed by members of the IUMS-ICFMH Working Party on Culture Media and given 'Approved' status.

Description and history

In 1957 Slanetz and Bartley described this improved version of the original medium of Slanetz et al. (1955). The medium relies upon the selective inhibitory properties of sodium azide and the incorporation of tetrazolium which most organisms growing on this medium will reduce to some extent under the conditions provided. Although devised originally for use with membrane filters, the medium can also be used for direct plating. Selectivity is increased by incubation at 37°C for 4 h followed by 44 ± 1°C for 44 h.

Composition (grams)

Tryptose	20.0
Yeast extract	5.0
Glucose	2.0
Di-sodium hydrogen orthophosphate $\cdot 2H_2O$	4.0
Sodium azide	0.4
2,3,5-Triphenyltetrazolium chloride	0.1
Agar	10.0
Distilled or deionized water	1000.0

Preparation

The first five ingredients are dissolved in the water and the pH value adjusted to 7.2. The agar is then added and the solution heated sufficiently to dissolve the agar. When cooled to about 50°C, add 1 ml of 1% filter-sterilized triphenyltetrazolium chloride solution per 100 ml of molten medium. Do not autoclave or overheat this medium.

Physical properties

Appearance	Pale straw.
pH	7.2 ± 0.2

Shelf life

Ready to use medium 7 days at 4 ± 2°C.

Inoculation method for samples

Surface spreading over whole plate using 0.1 ml per pre-dried 9 cm plate.

Incubation method

At 37°C for 48 h or at 37°C for 4 h followed by 44 ± 1°C for 44 h, in air.

Reading of results and interpretation

Round, pink to dark maroon-coloured colonies, 0.5–3 mm diameter, are considered to be Lancefield Group D streptococci (including enterococci).

Quality assessment

(i) *Productivity*

Test strains	*Enterococcus faecalis* NCIMB 50030 *Enterococcus faecium* NCIMB 50032
Inoculation method	Modified Miles-Misra, spiral plating or streaking/ ecometry.

(ii) *Selectivity*

Test strains	*Lactococcus lactis* ssp. *lactis* NCIMB 50058 *Escherichia coli* NCIMB 50034
Supplementary strains	*Bacillus cereus* NCIMB 50014 *Staphylococcus aureus* NCIMB 50080
Inoculation method	Modified Miles-Misra, spiral plating or streaking/ ecometry.

(iii) *Characteristic appearance of colonies*
Round, pink to dark maroon-coloured colonies, 0.5–3 mm in diameter.

References

Burkwall, M.K. and Hartman, P.A. (1964) Comparison of direct plating media for the isolation and enumeration of enterococci in certain frozen foods. Appl. Microbiol. 12, 18–23.

Slanetz, L.W. and Bartley, C.H. (1957) Numbers of enterococci in water, sewage and faeces determined by the membrane filter technique with an improved medium. J. Bacteriol. 74,

591–595.
Slanetz, L.W., Bent, D.F. and Bartley, C.H. (1995) Use of membrane filter technique to enumerate enterococci in water. Publ. Hlth. Reports 70, 67–72.
Taylor, E.W. and Burman, N.P. (1964) The application of membrane filtration techniques to the bacteriological examination of water. J. Appl. Bacteriol. 27, 294–303.

Muller Kauffmann tetrathionate broth

This monograph has been assessed by members of the IUMS-ICFMH Working Party on Culture Media and given 'Approved' status.

Description and history

This medium is used for the selective isolation of salmonellae. Selectivity is conferred by the combination of tetrathionate (from the reaction of thiosulphate and iodine) and residual thiosulphate (Palumbo and Alford, 1970) and by the addition of ox bile and brilliant green which are modifications introduced by Kauffmann (1935). The medium has been the subject of European trials (Edel and Kampelmacher, 1968; 1969) and is recommended by several organisations (van Leusden et al., 1982; Andrews et al., 1995; Flowers et al., 1992). The medium is included in the Draft International Standard ISO/DIS 6579 (Anon., 2000).

Composition (grams)

Meat extract	4.5
Peptone	9.0
Sodium chloride	2.7
Calcium carbonate	40.5
Sodium thiosulphate $\cdot 5H_2O$	50.0
Iodine	4.0
Potassium iodide	5.0
Ox bile desiccated	5.0
Brilliant green	0.01*
Distilled or deionized water	1072.0

* concentration varies with activity of dye.

Preparation

Base

Add the meat extract, peptone, sodium chloride and calcium carbonate or the dehydrated complete base to 900 ml water and boil until completely dissolved. Sterilize at 121°C for 20 min.

Sodium thiosulphate solution

Dissolve the sodium thiosulphate in about 50 ml of water. Dilute to 100ml. Sterilize at 121°C for 20 min.

Iodine solution

Dissolve 25 g potassium iodide in a minimal volume of water and add 20 g iodine. Dilute to 100 ml. Store the solution in a tightly closed container. Add 20 ml to the cooled base.

Brilliant green solution

Add 0.5 g brilliant green to 100 ml water. Store the solution for at least one day in the dark to allow auto-sterilization to occur. Add 2 ml to the cooled base.

Ox bile solution

Dissolve 10 g ox bile in 100 ml water by boiling. Sterilize at 121°C for 20 min. Add 50 ml to the cooled base.

Complete medium

Using aseptic precautions add the other ingredients (in the order listed above) to the base. Mix well. Distribute required volumes aseptically into sterile containers of appropriate size, keeping calcium carbonate suspended during distribution.

Physical properties

Appearance	Pale green milky opaque suspension which on standing gives a pale liquid over a heavy precipitate.
pH	7.4 ± 0.2

Shelf life

Ready to use medium	7 days at ± 2°C.

Inoculation method for samples

1. Where damaged cells are sought transfer food-inoculated and incubated pre-enrichment medium (PEM) into MK tetrathionate broth in ratio 1 part PEM to 10 parts tetrathionate broth. Use 10 ml PEM to 100 ml tetrathionate broth (or 1 ml to 10 ml).

or

2. Add comminuted sample to medium in the ratio 1 part sample to 9 parts medium

e.g. 25 g food to 225 ml medium.

Incubation method

At 42 ± 0.5°C or 37° C for 24–48 h, in air.

Reading of results and interpretation

Inoculated incubated medium is subcultured after 24 and 48 h onto selective diagnostic agar in such a way as to obtain well-isolated colonies. Suspect colonies are subcultured and their identity confirmed by appropriate biochemical and serological tests.

Quality assesment

(i) *Productivity*

Test strains	*Salmonella enteritidis* NCIMB 50073 *Salmonella virchow* NCIMB 50077
Inoculation method	Dilution to extinction.
Criteria	Recovery in Muller-Kauffmann tetrathionate broth should be within 3 titre units of the recovery in tryptone soya broth after 48h at 42°C.

(ii) *Selectivity*

Test strains	*Enterobacter cloacae* (ATCC 23355) *Pseudomonas aeruginosa* NCIMB 50067
Inoculation method	Dilution to extinction.
Criteria	Recovery in Muller-Kauffmann tetrathionate broth should be less than 5 titre units of the recovery in tryptone soya broth after 48 h at 42°C.

Comments

This medium may be markedly influenced by the following:-

1. Source of ingredients or dehydrated complete medium
2. The presence of food or other organic material
3. The physiological state of the organisms sought
4. The competing microflora
5. Temperature of incubation – 43°C may be very inhibitory to salmonellae; 41.5°C may be preferred
6. Possibly by combinations of the above

Successful performance of the medium tested in one set of standard conditions may not be achieved if *any* of those conditions are altered.

References

Andrews, W.H., June, G.A., Sherrod, P.S., Hammack, T.S. and Amaguana, R.M. (1995) FDA Bacteriological Analytical Manual, 8th edn., Chapter 5. Salmonella.

Anon. (2000) Microbiology of food and animal feeding stuffs – horizontal method for the detection of *Salmonella* spp. International Organisation for Standardization, Geneva.

Edel, W. and Kampelmacher, E.H. (1968) Comparative studies on *Salmonella* isolations in eight European laboratories. Bull.Wld. Hlth. Org. 39, 487–491.

Edel, W. and Kampelmacher, E.H. (1969) Salmonella isolation in nine European laboratories using a standardized technique. Bull.Wld.Hlth.Org., 41, 297–306.

Flowers, R.S., D'Aoust, J-Y., Andrews, W.H. and Bailey, J.S. (1992) Salmonella. *In:* Compendium of methods for the Microbiogical Examination of Foods. 3rd edn., American Public Health Association, Washington, DC.

Kauffmann, F. (1935) Weitere Erfahrungen mit dem kombinierten Anreicherungsverfahren für Salmonellabacillen. Ztschr. Hyg., 117, 26–32.

Palumbo, S.A. and Alford, J.A. (1970) Inhibitory action of tetrathionate enrichment broth. Appl. Microbiol. 20, 970–976.

van Leusden, F.M., van Schothorst, M. and Beckers, H.J. (1982) The standard *Salmonella* isolation method. *In:* Isolation and identification methods for food poisoning organisms, edited by J.E.L. Corry, D. Roberts and F.A. Skinner. SAB Technical Series, No. 17. Academic Press, London, pp. 35–49.

Oleandomycin Polymyxin Sulphadiazine Perfringens agar (OPSPA)

This monograph has been assessed by members of the IUMS-ICFMH Working Party on Culture Media and given 'Approved' status.

Description and history

This medium was devised by Handford (1974). It depends upon (i) the selective inhibitory properties of oleandomycin, polymyxin and sulphadiazine and (ii) an indicator system involving sulphite and ferric iron. Most unwanted organisms are suppressed while *Clostridium perfringens* and related species will reduce the sulphite and form black colonies due to the production of ferrous sulphide (Mead et al., 1982).

Composition (grams)

Tryptose	15.0
Soya peptone	5.0
Yeast extract	5.0
Sodium metabisulphite	1.0
Iron (III) ammonium citrate	1.0
Sulphadiazine	0.1
Oleandomycin phosphate	0.0005
Polymyxin B sulphate (i.u.)	10 000.0
Agar	15.0
Distilled or deionized water	1000.0

Preparation

Suspend ingredients other than the oleandomycin and polymyxin in 1 litre of water. Bring to the boil to dissolve completely. Dispense 100 ml amounts into bottles (or final containers) and sterilize at 121°C for 15 min. Concentrated stock solutions of the antibiotics are made by dissolving 0.5 g of oleandomycin phosphate in 100 ml of sterile distilled water and by dissolving 500 000 i.u. of polymyxin B sulphate in 5 ml of sterile distilled water. These concentrated solutions are further diluted 1 in 100 and just before pouring 1 ml of each diluted solution is added to the molten basal medium.

Physical properties

Appearance	Pale straw.
pH	7.6 ± 0.2

Shelf life

Ready to use medium	Day of preparation.
Concentrated antibiotic solutions	2 months at 4 ± 2°C.
Diluted antibiotic solutions	1 week at 4 ± 2°C.

Inoculation method for samples

Pour-plate procedure, using 1 ml per 9 cm diameter plate. When set, plates are overlaid with sterile medium.

Incubation method

At 37°C for 18–24 h, anaerobically.

Reading of results and interpretation

Discrete black colonies of 1–5 mm diameter are considered to be presumptive *Clostridium perfringens.*

Quality assessment

(i) *Productivity*

Test strains	*Clostridium perfringens* (haemolytic) NCIMB 50027 *Clostridium perfringens* (non-haemolytic) NCIMB 50028
Inoculation method	Surface plating using modified Miles-Misra or ecometry, using pre-poured plates with subsequent overlay.
Criteria	Recovery on OPSPA of typical colonies (see below) should be within 0.7 $\log_{10}$ of the recovery on Reinforced Clostridial Medium after 18–24 h at 37°C anaerobically.

(ii) *Selectivity*

Test strains	*Clostridium bifermentans* NCIMB 50026 *Proteus mirabilis* (ATCC 29906/CECT 4168/NCTC 11938)
Inoculation method	As (i) above.

Criteria	Recovery on OPSPA should be less than 5 $\log_{10}$ of the recovery on Reinforced Clostridial Medium after 18–24 h at 37°C anaerobically.

(iii) *Characteristic appearance of colonies*
Discrete black colonies of 1–5 mm diameter.

References

Handford, P.M. (1974) A new medium for the detection and enumeration of *Clostridium perfringens* in food. J. Appl. Bacteriol. 37, 559–570.

Mead, G.C., Adams, B.W., Roberts, T.A. and Smart, J.L. (1982) Isolation and enumeration of *Clostridium perfringens*. *In:* Isolation and identification methods for food poisoning organisms, edited by J.E.L. Corry, D. Roberts and F.A. Skinner. SAB Technical Series No. 17. Academic Press, London, pp. 99–110.

 Handbook of Culture Media for Food Microbiology, J.E.L. Corry et al. (Eds.)

Oxford agar

This monograph has been assessed by members of the IUMS-ICFMH Working Party on Culture Media and given 'Approved' status.

Description and history

The medium was developed by Curtis et al. (1989a) for the isolation of *Listeria monocytogenes* from faeces and other clinical samples. It has also been recommended for the isolation of *Listeria* spp. from foods (Anon., 1995; Warburton et al., 1991a; 1991b). It utilises (i) the selective inhibitory components lithium chloride (Ludlam, 1949), acriflavine, colistin, fosfomycin, cefotetan and cycloheximide and (ii) the indicator system aesculin and ferric iron for the isolation and differentiation of listeria. *Listeria* spp. hydrolyse aesculin, producing black zones around the colonies due to the formation of black iron phenolic compounds derived from the aglucon. Gram negative bacteria are completely inhibited. Most unwanted Gram positive species are suppressed but some coagulase negative staphylococci may appear as aesculin negative colonies. Some strains of enterococci grow poorly and exhibit a weak aesculin reaction, usually only after 40 h incubation.

Typical *Listeria monocytogenes* colonies are usually visible after 24 h but incubation should be continued for a further 24 h to detect slow growing strains.

Composition (grams)

Columbia agar base	39.0
Aesculin	1.0
Iron (III) ammonium citrate	0.5
Lithium chloride	15.0
Cycloheximide	0.4
Colistin	0.02
Acriflavine (Sigma)	0.005
Cefotetan	0.002
Fosfomycin	0.01
Distilled or deionized water	1000.0

Preparation

Suspend the ingredients except the cefotetan and fosfomycin in the water. Bring to the boil to dissolve completely. Sterilize by autoclaving at 121°C for 15 min. Cool to 50°C then add the cefotetan and fosfomycin aseptically as filter-sterilized solutions.

Physical properties

Appearance	Clear, yellow.
pH	7.0 ± 0.2

Shelf life

Ready to use medium	14 days in the dark at 4 ± 2°C.

Inoculation method for samples

As the medium is highly selective a heavy inoculum should be spread over the whole plate.

Incubation method

At 30 or 37°C for 48 h, in air. Where *Listeria seeligeri* or *Listeria ivanovii* are sought incubation at 30°C has been shown to give better results (Curtis et al., 1989b).

Reading of results and interpretation

After 24 h colonies of *Listeria monocytogenes* are small (1 mm) and surrounded by black haloes. Incubation should be continued for a further 24 h to allow slow growing strains to develop. Confirmatory tests are necessary as other *Listeria* species may grow with similar colonial appearances.

Quality assessment

(i) *Productivity*

Test strains	*Listeria monocytogenes* serovar 1/2a (ATCC 35152/NCTC 7973) *Listeria monocytogenes* serovar 4b (ATCC 13932/NCTC 10527) *Listeria ivanovii* serovar 5 NCIMB 50095
Supplementary strain	*Listeria monocytogenes* sv 4a (ATCC 19114/NCTC 5214)
Inoculation method	Modified Miles-Misra or spiral plating.

(ii) *Selectivity*

Test strains	*Enterococcus faecalis* NCIMB 50030 *Proteus mirabilis* (ATCC 29906/CECT 4168/NCTC

	11938)
Supplementary strains	*Bacillus cereus* NCIMB 50014 *Micrococcus luteus* NCIMB 50063 *Pseudomonas aeruginosa* NCIMB 50067 *Staphylococcus warneri* NCIMB 50082
Inoculation method	Modified Miles-Misra or spiral plating.

(iii) Characteristic appearance of colonies
Listeria monocytogenes forms 1 mm diameter black colonies surrounded by black haloes after 24 h. At 48 h colonies are 2–3 mm in diameter, black with a black halo and sunken centre. Other *Listeria* species show a similar appearance. When examined before 24 h growth of *Listeria* spp. is sometimes apparent but without the characteristic blackening. Some strains of species other than *Listeria monocytogenes* are inhibited on this medium when incubated at 37°C. Incubation at 30°C will allow growth of these strains.

Note:
Where greater selectivity is required, e.g. in the examination of environmental samples, the cefotetan content may be increased (up to 0.01 g/l). However, this will result in slower growth of *Listeria* spp. and incubation for 48 h becomes essential.

Comments

Selective properties of acriflavine may vary from lot to lot and manufacturer to manufacturer. Each new batch must be assayed in combination with other selective agents to be used in the medium to determine the optimum concentration for use, with regard to the efficiency of selectivity and absence of inhibition of *Listeria* spp. Storage of stock solutions of acriflavine for periods of more than one month is not recommended.

References

Anon. (1995) FDA Bacteriological Analytical Manual, 8th edition. Chapter 10, Listeria monocytogenes. AOAC International, Gaithersburg MD, USA.

Curtis, G.D.W., Mitchell, R.G., King, A.F. and Griffin, E.J. (1989a) A selective differential medium for the isolation of *Listeria monocytogenes*. Lett. Appl. Microbiol. 8, 95–98.

Curtis, G.D.W., Nichols, W.W. and Falla, T.J. (1989b) Selective agents for listeria can inhibit their growth. Lett. Appl. Microbiol. 8, 169–172.

Domjan Kovacs, H. and Ralovich, B. (1991) Model examination of selective media for isolation of *Listeria* strains. Acta Microbiol. Hung. 38, 141–145.

Ludlam, G.B. (1949) A selective medium for the isolation of *Staph. aureus* from heavily contaminated material. Monthly Bull. Min. Hlth. Publ. Hlth. Lab. Ser. 8, 15–20.

Warburton, D.W., Farber, J.M., Armstrong, A., Caldeira, R., Hunt, T., Messier, S., Plante, R., Tiwari, N.P. and Vinet, J. (1991a) A comparative study of the 'FDA' and 'USDA' methods for the detection of *Listeria monocytogenes* in foods. Int. J. Food Microbiol. 13, 105–118.

Warburton, D.W., Farber, J.M., Armstrong, A., Caldeira, R., Tiwari, N.P., Babiuk, T., Lacasse, P. and Read, S. (1991b) A Canadian comparative study of modified versions of the 'FDA' and 'USDA' methods for the detection of *Listeria monocytogenes*. J. Food Protect. 54, 669–676.

Oxford agar – Modified (MC)

This monograph has been assessed by members of the IUMS-ICFMH Working Party on Culture Media and given 'Approved' status.

Description and history

Modified Oxford agar (MC) is similar to Oxford agar (q.v.) developed by Curtis et al. (1989a). In MC the lithium chloride content is reduced to 12 g/l to permit the growth of some of the more sensitive *Listeria monocytogenes* strains (e.g. ATCC 35152 and CDC F4561). Ceftazidime (20 mg/l) replaces the cycloheximide, acriflavine, cefotetan and fosfomycin used in Oxford agar. Limited experience has indicated that *Staphylococcus* spp. and fungal growth from food samples are not a problem with MC. However, the *Listeria monocytogenes* strain ATCC 35152 and some strains of *Listeria seeligeri* and *Listeria ivanovii* are inhibited on MC at 35°C (Curtis et al., 1989b). Thus incubation at 30°C is preferred for the optimal recovery of *Listeria* spp. on this medium. A suitable MC agar should support good growth of *Listeria monocytogenes* strain ATCC 35152 at 30°C and suppress the growth of the more resistant strains of *Enterococcus faecalis* (e.g. FSIS 16a).

Composition (grams)

Columbia agar base (Oxoid CM331)	39.0
Aesculin	1.0
Iron (III) ammonium citrate	0.5
Lithium chloride	12.0
Colistin, methane sulphonate (Sigma) or sulphate	0.01
Ceftazidime pentahydrate (Glaxo)	0.02
Distilled or deionized water	1000.0

Preparation

Suspend all the ingredients except ceftazidime in a bottle or flask containing a magnetic stirring bar and autoclave the medium for 12 min at 121°C. Mix the medium after autoclaving and cool to 46°C. Add 2 ml of filter sterilized 1% ceftazidime solution per litre of the autoclaved, cooled molten base while stirring with a magnetic mixer. Dispense 12 ml per 9 cm diameter Petri dish.

Physical properties

Appearance	Light yellow or tan. Dark colour indicates overheating and decomposition of aesculin.
pH	7.2 ± 0.2

Shelf life

Ready to use plates	14 days at 4 ± 2°C.

Inoculation methods for samples

1. From enrichment broths, soak a sterile cotton swab in the enrichment broth and swab it on half of an MC agar plate, then streak for isolation with a sterile loop in two 90° directions.
2. For direct plating and counting of >100/g of *Listeria* spp. in foods, make decimal dilutions of the foods and spread plate 0.1 ml of each dilution on to MC agar.
3. For the rapid detection of >100/g of *Listeria* spp. growing in many kinds of refrigerated foods, wipe a cotton swab moistened with sterile pH 7.2 PBS or similar phosphate buffer over the interface of vacuum packaged refrigerated foods or just under the surface of soft ripened cheeses. Swabs may also be dipped into food homogenates. MC agar plates should then be inoculated by the swab and streak procedure described above. Isolated listeria-like colonies on the swab and streak plate can be recognised and picked for further identification. The number of listeria colonies on the swab and streak MC plates can be estimated by comparing the density of colonies on the inoculated area with that of standard inocula in a similar manner to the Dip Slide procedure of Guttmann and Naylor (1967).

Incubation method

At 30°C for 26 h and 40 h, in air.

Reading of results and interpretation

Typical *Listeria* spp. form distinctive small (1 mm) white hemi-spherical colonies after 26 h incubation at 30°C which can be recognized easily with some practice. It is best to ignore the blackening or black zones on MC agar entirely when working with mixed cultures as this can be caused by other bacteria and become confusing. In limited observations, no additional *Listeria monocytogenes* colonies were recovered from MC agar after 26 h incubation. The 40 h incubation is used to detect other, slow growing, *Listeria* spp. Additional confirmatory tests are necessary to identify various *Listeria* species.

Quality assessment

(i) *Productivity*

Test strains	*Listeria ivanovii* serovar 5 NCIMB 50095 *Listeria monocytogenes* serovar 1/2a (ATCC 35152/ NCTC 7973) *Listeria monocytogenes* serovar 4b (ATCC 13932/ NCTC 10527)
Supplementary strain	*Listeria monocytogenes* sv 4a (ATCC 19114/NCTC 5124)
Inoculation method	Modified Miles-Misra or spiral plating.

(ii) *Selectivity*

Test strains	*Enterococcus faecalis* NCIMB 50030 *Proteus mirabilis* (ATCC 29906/CECT 4168/NCTC 11938)
Supplementary strain	*Staphylococcus aureus* NCIMB 50080
Inoculation method	Modified Miles-Misra or spiral plating.

(iii) *Characteristic appearance of colonies*
See above.

References

Curtis, G.D.W., Mitchell, R.G., King, A.F. and Griffin, E.J. (1989a) A selective differential medium for the isolation of *Listeria monocytogenes*. Lett. Appl. Microbiol. 8, 95–98.

Curtis, G.D.W., Nichols, W.W. and Falla, T.J. (1989b) Selective agents for listeria can inhibit their growth. Lett. Appl. Microbiol. 8, 169–172.

Guttmann, D. and Naylor, G.R.E. (1967) Dip Slide: an aid to quantitative urine culture in general practice. Br.Med. J. 3, 343–345.

Oxytetracycline Glucose Yeast extract (OGY) agar

This monograph has been reviewed by members of the IUMS-ICFMH Working Party on Culture Media and given 'Proposed' status.

Description and history

A selective medium for the enumeration of moulds and yeasts in foods. Mossel et al. (1962; 1970) showed that this medium gave better recovery of fungi from a variety of foodstuffs than media which rely on low pH to suppress bacterial growth. It is not suitable for use with certain high protein foods. In addition, it may not restrict adequately the growth of fast spreading moulds. It neither selects for specific groups of fungi, nor is diagnostic for detecting specific mycotoxin producers.

Composition (grams)

Yeast extract	5.0
Glucose	20.0
Biotin	0.0001
Oxytetracycline	0.1
Agar	12.0
Distilled or deionized water	1000.0

Preparation

Suspend the basal ingredients in the water and bring gently to the boil to dissolve completely. Sterilize by autoclaving at 115°C for 10 min. Cool to 50°C and aseptically add 10 ml of oxytetracycline solution (10 mg/ml). Mix well and dispense 15 ml amounts into Petri dishes. Use immediately or store at 4 ± 2°C until required.

Physical properties

Appearance	Medium amber, slightly opalescent without significant precipitate.
pH	7.0 ± 0.2 at 25°C.

Shelf life

Ready to use medium 7 days at 4 ± 2°C.

Inoculation method for samples

Surface spreading over whole plate using 0.1 or 0.2 ml per 9 cm diameter plate.

Incubation

At 25 ± 0.5°C for 5 days. Where identification is required, prolong incubation until characteristic colonies are formed.

Reading of results and interpretation

Where separate counts of moulds and yeasts are required, identify the two groups of microorganisms by morphological appearance and, where necessary, microscopic examination.

Quality assessment

Moulds. Use stab inoculation procedure. Growth rate should be within 30% of the figure given.

Test strains	IMI No.	ATCC No.	Growth rate (mm per day at 25°C)
Rhizopus stolonifer	61269		25.0
Aspergillus flavus	91856ii		8.6
Penicillium cyclopium	19759	16025	3.9

Yeasts. Use stab inoculation procedure. Growth rate should be within 30% of the figure given.

Test strain		Growth rate (mm per day at 25°C)
Saccharomyces cerevisiae	NCIMB 50105	2.5

Bacteria. Use streaking procedure (Appendix I, method C) with a loopful (10 µl) from a 24 h broth culture. No growth should be evident on sectors C, D or E of the streaked plate after incubation at 25°C for 5 days.

Test strain
Bacillus subtilis NCIMB 50018

References

Mossel, D.A.A., Visser M. and Mengerink, W.H.J. (1962) A comparison of media for the enumeration of moulds and yeasts in foods and beverages. Lab. Prac. 11, 109–112.

Mossel, D.A.A., Kleynen-Semmeling, A.M.C., Vincentie, H.M., Beerens, H. and Catsaras, M. (1970) Oxytetracycline-glucose-yeast extract agar for selective enumeration of moulds and yeasts in foods and clinical material. J. Appl. Bacteriol. 33, 454–457.

Phenol red brilliant green agar (modified brilliant green agar)

This monograph has been assessed by members of the IUMS-ICFMH Working Party on Culture Media and given 'Approved' status.

Description and history

This is a selective and diagnostic agar for the isolation, following pre-enrichment and selective enrichment, of salmonellae other than *Salmonella typhi* from foods and feeds. Emanating from the Netherlands (Edel and Kampelmacher, 1968; 1969), this formulation has been widely assessed in Europe. It is used in the standard European Community method and in the International Standards Organization Standards (Anon., 1993; 1995).

Composition (grams)

Meat extract	5.0
Peptone	10.0
Yeast extract	3.0
Di-sodium hydrogen orthophosphate	1.0
Sodium di-hydrogen orthophosphate	0.6
Lactose	10.0
Sucrose	10.0
Phenol red	0.09
Brilliant green	0.005*
Agar	15.0
Distilled or deionized water	1000.0

*concentration varies with formulation and activity of dye – see note at end of this monograph.

Preparation

Base

Dissolve the meat extract, peptone, yeast extract and buffer salts in 900 ml water by boiling. Adjust the pH value so that it is 7.0 ± 0.1 at 25°C after sterilization. Sterilize the base at 121°C for 15 min.

Sugar/phenol red solution

Dissolve the ingredients in 100 ml water. Heat in a waterbath at 70°C for 20 min. Cool to 55°C and use immediately.

Brilliant green solution

Dissolve 0.5 g brilliant green in 100 ml water. Store the solution for at least one day in the dark to allow auto-sterilization to occur.

Complete medium

Under aseptic conditions add 1 ml brilliant green solution to the sugar/phenol red solution cooled to about 55°C. Add to the base at 55°C and mix. Distribute quantities of about 40 ml of the freshly prepared medium cooled to 45°C to sterile 14 cm diameter Petri dishes. If large dishes are not available pour about 15 ml into 9 cm diameter Petri dishes.

Physical properties

Appearance	Red, clear.
pH	6.9 ± 0.2

Shelf life

Ready to use medium	4 days at 4 ± 2°C.

Inoculation method for samples

Streak inoculation from selective enrichment broth by method specified by ISO 6579 1981(E) (see also van Leusden et al., 1982). Other streak techniques which yield well isolated colonies may also be used.

Incubation method

At 37°C for 20–24 h, in air.

Reading of results and interpretation

Typical colonies of *Salmonella* spp. are mostly smooth, very low convex, moist pink/red. Subculture for further biochemical and serological tests.

Quality assessment

(i) *Productivity*

Test strains	*Salmonella enteritidis* NCIMB 50073 *Salmonella virchow* NCIMB 50077
Supplementary strains	*Salmonella dublin* NCIMB 50072 *Salmonella saintpaul* NCIMB 50075
Inoculation method	Modified Miles-Misra, spiral plating or streaking/ecometry.

(ii) *Selectivity*

Test strains	*Proteus mirabilis* (ATCC 29906/CECT 4168/NCTC 11938) – qualitative test for control of swarming. *Escherichia coli* NCIMB 50034
Supplementary strain	*Morganella morgani* NCIMB 50064
Inoculation method	Modified Miles-Misra, spiral plating or streaking/ecometry.

(iii) *Characteristic appearance of colonies*

Assess only on well isolated colonies. Colonies of *Salmonella* are red, surrounded by bright red medium.

Specification for brilliant green

Bacteriological performance

Brilliant green should suppress the spreading of *Proteus* spp. while not inhibiting the growth of salmonellae (other than *S. typhi* and *S. paratyphi* and other brilliant green-sensitive strains).

Method of test

Prepare phenol red brilliant green agar containing various concentrations of brilliant green viz. 4.5 mg/l to 6 mg/l.

Procedure

Inoculate a set of plates with different brilliant green concentrations with a pure culture of a swarming *Proteus* and another set with a pure culture of a *Salmonella* and incubate these plates at 37°C for no longer than 24 h.

A satisfactory concentration of the dye should allow growth of salmonellae with typical pink colonies, 1 to 2 mm in diameter, and limited growth of proteus i.e. no spreading.

Prepare the brilliant green solution to contain one thousand times the concentration in the complete medium and add 1 ml/l of medium i.e. if the selected concentration is confirmed as 0.005 g/l prepare a 0.5% w/v solution of brilliant green.

References

Anon. (1993) Microbiology – General guidance on methods for the detection of *Salmonella.* ISO 6579–1993 (E). International Organisation for Standardization, Geneva.

Anon. (1995) Milk and milk products – detection of *Salmonella.* ISO 6785–1995. International Organisation for Standardization, Geneva.

Edel, W. and Kampelmacher, E.H. (1968) Comparative studies on *Salmonella* isolation in eight European laboratories. Bull. Wld. Hlth. Org. 39, 487–491.

Edel, W. and Kampelmacher, E.H. (1969) *Salmonella* isolation in nine European laboratories using a standardized technique. Bull. Wld. Hlth. Org. 41, 297–306.

van Leusden, F.M., van Schothorst, M. and Beckers, H.J. (1982) The standard *Salmonella* isolation method. *In:* Isolation and identification methods for food poisoning organisms edited by J.E.L. Corry, D. Roberts and F.A. Skinner. SAB Technical Series, No. 17. Academic Press, London, pp. 35–49.

Polymyxin Acriflavine Lithium chloride Ceftazidime Aesculin Mannitol (PALCAM) agar

This monograph has been assessed by members of the IUMS-ICFMH Working Party on Culture Media and given 'Approved' status.

Description and history

PALCAM, developed by van Netten et al. (1989), is a selective differential medium for the isolation and direct counting of *Listeria* spp. from faecal and biological specimens as well as food and heavily contaminated environmental samples. It is an improved modification of RAPAMY (van Netten et al., 1988a), ALPAMY (van Netten et al., 1988b) and Oxford (Curtis et al., 1989) media. Selectivity is achieved by the use of microaerobic incubation and the inhibitory components lithium chloride, acriflavine, polymyxin and ceftazidime. Its diagnostic properties are based on the use of two indicator systems: (i) aesculin and ferric iron and (ii) D-mannitol combined with phenol red. The selective system aims to suppress virtually all unwanted organisms, including most *Enterococcus* spp. The double diagnostic system allows easy distinction of *Listeria* spp. from enterococci and staphylococci. *Listeria* spp. are aesculin positive, but mannitol negative, producing green colonies with black haloes against a pink-purple background. When direct counting is applied to foods processed for safety, solid medium repair has to precede the use of PALCAM.

Composition (grams)

Columbia agar	39.0
D-glucose	0.5
D-mannitol	10.0
Aesculin	0.8
Iron (III) ammonium citrate	0.5
Phenol red	0.08
Polymyxin B (i.u.)*	100 000.0
Acriflavine HCl*	0.005
Lithium chloride	15.0
Ceftazidime*	0.02
Distilled or deionized water	1000.0

Preparation

Suspend the above ingredients except those marked * in 960ml water and bring to the boil to dissolve completely. Adjust pH to 7.2 ± 0.1. Sterilize by autoclaving at 121°C for 15 min. Cool to 47°C and aseptically add the required amount of the following filter-sterilized solutions: (i) 10 ml of 0.1% polymyxin B sulphate (Pfizer), (ii) 10 ml of 0.05% ethanolic acriflavine (Sigma) and (iii) 20 ml of 0.116 % sodium ceftazidime pentahydrate (Glaxo). Mix gently and pour about 40 ml of freshly prepared medium into sterile 14 cm diameter Petri dishes or 15 ml into 9 cm diameter dishes.

Physical properties

Appearance	Pink-purple, opaque.
pH	7.2 ± 0.1

Shelf life

Ready to use medium	At least 4 weeks at 4 ± 2°C.

Inoculation method for samples

Dry plates for 20 min at 55°C to remove surface moisture prior to use.

1. Subculturing after enrichment
   A loopful of an incubated non-selective or selective enrichment broth is streaked onto PALCAM to attain isolated colonies.

2. Direct counting
   Spread plate technique using 0.1 ml of decimal dilutions of food homogenates on 9 cm diameter Petri dishes or 1.0 ml on 14 cm dishes. Spread inoculum over the whole plate until surface appears dry.

Incubation method

At 30°C for 24 to 48 h, microaerobically in a jar containing 5–12% CO_2, 5–15% O_2 and 75% N_2.

Reading of results and interpretation

After microaerobic incubation allow plates to regain their pink purple colour by exposure to air for 1 h. Well-isolated greenish, smooth colonies of approximately 1.5 to 2 mm with black haloes against pink-purple background are virtually always *Listeria* spp. Further biochemical and serological identification is performed according to recognised microbiological procedure.

Quality assessment

(i) *Productivity*

Test strains	*Listeria monocytogenes* serovar l/2a (ATCC 35152/ NCTC 7973) *Listeria monocytogenes* serovar 4b (ATCC 13932/ NCTC 10527) *Listeria ivanovii* serovar 5 NCIMB 50095
Inoculation method	Modified Miles-Misra, spiral plating or streaking/ ecometry.

(ii) *Selectivity*

Test strains	*Enterococcus faecalis* NCIMB 50030 *Proteus mirabilis* (ATCC 29906/CECT 4168/NCTC 11938)
Inoculation method	Modified Miles-Misra, spiral plating or streaking/ ecometry.

(iii) *Characteristic appearance of colonies*

Smooth green colonies, 1.5–2 mm in diameter sometimes with black centres, but always with black haloes are *Listeria* spp.

White or yellow colonies, 1.5–3 mm in diameter with yellow haloes are staphylococci. Round smooth white-grey colonies <1 mm in diameter with green haloes are *Enterococcus* spp.

Comments

Selective properties of acriflavine may vary from lot to lot and manufacturer to manufacturer. Each new batch must be assayed in combination with other selective agents to be used in the medium to determine the optimum concentration for use, with regard to the efficiency of selectivity and absence of inhibition of *Listeria* spp. Storage of stock solutions of acriflavine for periods of more than one month is not recommended.

References

Curtis, G.D.W., Mitchell, R.G., King, A.F. and Griffin, E.J. (1989) A selective differential medium for the isolation of *Listeria monocytogenes*. Lett. Appl. Microbiol. 8, 95–98.

van Netten, P., van der Ven, A., Perales, I. and Mossel, D.A.A. (1988a) A selective and diagnostic medium for use in the enumeration of *Listeria* spp. in foods. Int. J. Food Microbiol. 6, 187–198.

van Netten, P., Perales, I. and Mossel, D.A.A. (1988b) An improved selective and diagnostic medium for isolation and counting of *Listeria* spp. in heavily contaminated foods. Lett. Appl. Microbiol. 7, 17–21.

van Netten, P., Perales, I., Curtis, G.D.W. and Mossel, D.A.A. (1989) Liquid and solid selective differential media for the detection and enumeration of *L. monocytogenes* and other *Listeria* spp. Int. J. Food Microbiol. 8, 299–316.

 Handbook of Culture Media for Food Microbiology, J.E.L. Corry et al. (Eds.)

Polymyxin Acriflavine Lithium chloride Ceftazidime Aesculin Mannitol egg Yolk (L-PALCAMY) broth

This monograph has been assessed by members of the IUMS-ICFMH Working Party on Culture Media and given 'Approved' status.

Description and history

L-PALCAMY developed by van Netten et al. (1989), is a selective differential enrichment medium used in procedures for the detection and isolation of *Listeria* spp. from faecal and biological specimens as well as food and heavily contaminated environmental samples. The selectivity of L-PALCAMY is dependent on the inhibitory action of lithium chloride, acriflavine, polymyxin and ceftazidime. The diagnostic traits of L-PALCAMY are based on the use of (i) aesculin and ferric iron and (ii) D-mannitol combined with phenol red. The selective system aims to suppress virtually all unwanted organisms, including most *Enterococcus* spp. Presence of aesculin positive and mannitol negative *Listeria* spp., after enrichment, is often indicated by a brown-black colour.

Composition (grams)

Special peptone (Oxoid)	23.0
Yeast extract powder	5.0
Lab lemco (Oxoid)	5.0
Peptonized milk (Oxoid)	5.0
Sodium chloride	5.0
D-mannitol	5.0
Aesculin	0.8
Iron (III) ammonium citrate	0.5
Phenol red	0.08
Polymyxin B (i.u.)*	100 000.0
Acriflavine HCl*	0.005
Lithium chloride	10.0
Ceftazidime*	0.03
Egg yolk emulsion* (ml)	25.0
Distilled or deionized water	1 000.0

The egg yolk emulsion may be omitted.

Preparation

Suspend the above ingredients except those marked * in 925 ml water and bring to the boil to dissolve completely. Adjust pH to 7.2 ± 0.1. Sterilize by autoclaving at 121°C for 15 min. Cool to 47°C and aseptically add the egg yolk emulsion if required and the following filter-sterilized solutions: (i) 10 ml of 0.1% polymyxin B sulphate (Pfizer), (ii) 10 ml of 0.05% ethanolic acriflavine (Sigma) and (iii) 30 ml of 0.116 % sodium ceftazidime pentahydrate (Glaxo). Mix gently and distribute in 90 ml or 225 ml volumes in sterile flasks.

Physical properties

Appearance	Pink-purple, opaque.
pH	7.2 ± 0.1

Shelf life

Ready to use medium	2 days at 4 ± 2°C.

Inoculation method for samples

Use macerates of foods, faeces, biological or environmental samples and inoculate broth in proportion 1:10.

Incubation method

At 30°C for 24 to 48 h, in air.

Reading of results and interpretation

Often change of colour of the medium to brown-black provides presumptive evidence for the presence of *Listeria* spp. However if only small numbers of *Listeria* spp. are present this colour change may not be produced and it is therefore essential to subculture all broths to PALCAM agar (q.v.).

Quality assessment

(i) *Productivity*

Test strains	*Listeria monocytogenes* serovar 1/2a (ATCC 35152/ NCTC 7973) *Listeria monocytogenes* serovar 4b (ATCC 13932/ NCTC 10527) *Listeria ivanovii* serovar 5 NCIMB 50095

Inoculation method	Dilution to extinction.
Criteria	Recovery in L-PALCAMY should be within one titre unit of the recovery in Columbia broth after 48 h at 30°C.

(ii) *Selectivity*

Test strains	*Enterococcus faecalis* NCIMB 50030 *Proteus mirabilis* (ATCC 29906/CECT 4168/NCTC 11938)
Inoculation method	Dilution to extinction.
Criteria	Recovery in L-PALCAMY should be less than four titre units of the recovery in Columbia broth after 48 h at 30°C.

Comments

Selective properties of acriflavine may vary from lot to lot and manufacturer to manufacturer. Each new batch must be assayed in combination with other selective agents to be used in the medium to determine the optimum concentration for use, with regard to the efficiency of selectivity and absence of inhibition of *Listeria* spp. Storage of stock solutions of acriflavine for periods of more than one month is not recommended.

References

Anon. (1987) Testing methods for use in quality assurance of culture media. Int. J. Food Microbiol. 5, 291–296.

van Netten, P., Perales, I., Curtis, G.D.W. and Mossel, D.A.A. (1989) Liquid and solid selective differential media for the detection and enumeration of *L. monocytogenes* and other *Listeria* spp. Int. J. Food Microbiol. 8, 299–316.

Polymyxin pyruvate Egg yolk Mannitol Bromothymol blue Agar (PEMBA)

This monograph has been assessed by members of the IUMS-ICFMH Working Party on Culture Media and given 'Approved' status.

Description and history

A diagnostic selective medium formulated by Holbrook and Anderson (1980) for the enumeration of *Bacillus cereus* in foods. Selectivity is attained with polymyxin and a critical concentration of nutrients. Cycloheximide may be used to inhibit the growth of moulds. *Bacillus cereus* is identified by colony form, colour and egg yolk hydrolysis and rapidly confirmed by cell and spore morphology. *Bacillus anthracis, Bacillus mycoides, Bacillus pseudomycoides, Bacillus thuringiensis* and *Bacillus weihenstephanensis* may not be distinguishable from *Bacillus cereus*. Incubation at 30°C allows the growth of psychrotolerant strains of this group.

Composition (grams)

Peptone	1.0
Mannitol	10.0
Magnesium sulphate $\cdot 7H_2O$	0.1
Sodium chloride	2.0
Di-sodium hydrogen orthophosphate	2.5
Potassium di-hydrogen orthophosphate	0.25
Bromothymol blue (water soluble)	0.12
Egg yolk emulsion 20% (ml)	50.0
Polymyxin (i.u.)	100 000.0
Sodium pyruvate	10.0
Cycloheximide (Actidione)	0.04
Agar	15.0
Distilled or deionized water	1000.0

Commercially available egg yolk emulsion may be used or it may be prepared by Billing and Luckhurst's method (1957).

Preparation

Basal agar

Add components except egg yolk and antibiotics to the water, soak then steam to dissolve. Adjust pH, dispense in 90 ml amounts and autoclave at 121°C for 15 min.

Complete medium

Melt and cool base to 50°C. Aseptically add egg yolk emulsion 5 ml; 100,000 i.u./ml polymyxin, 1 ml; 20% w/v sodium pyruvate, 5 ml and 0.4% w/v cycloheximide, 1 ml. Mix and pour in approximately 12 ml amounts into 9 cm diameter Petri dishes.

Physical properties

Appearance	Apple green, turbid almost opaque.
pH	7.2 ± 0.2

Shelf life

Prepared basal medium	3 months at 20°C. Commercial media containing pyruvate may have shorter shelf life.
Complete medium	4 days at 4 ± 2°C or 1 day at 20°C.

Inoculation method for samples

Surface spreading over whole plate using 0.1 ml per 9 cm Petri dish.

Incubation method

At 30°C for 18–24 h, in air, then if necessary at room temperature until next day.

Reading of results and interpretation

Bacillus cereus grow as crenate, fimbriate or slightly rhizoid colonies up to 5 mm diameter (at 24 h), turquoise to peacock blue in colour with flat ground glass surface and surrounded by a precipitate from hydrolysed egg yolk. At 48 h colonies are peacock blue with raised greyish centre. Occasionally weak or negative egg yolk reacting strains may be isolated. Do not mistake lipase activity for lecithinase activity. Lipase results in a fainter precipitate. *Bacillus cereus* may not be distinguishable from *Bacillus anthracis, Bacillus mycoides, Bacillus pseudomycoides, Bacillus thuringiensis* or *Bacillus weihenstephanensis* on this medium. Further identification may be necessary. *Bacillus megaterium* and *Bacillus coagulans* are completely inhibited. *Bacillus subtilis* and *Bacillus licheniformis* are readily differentiated from *Bacillus cereus* by colony form and colour. Confirmation of identity should be carried out.

Quality assessment

(i) *Productivity*

Test strain	*Bacillus cereus* NCIMB 50014
Inoculation method	Modified Miles-Misra, spiral plating or ecometry.

(ii) *Selectivity*

Test strains	*Escherichia coli* NCIMB 50034 *Micrococcus luteus* NCIMB 50063
Inoculation	Modified Miles-Misra, spiral plating or ecometry.

(iii) *Characteristic appearance of colonies*
see above.

References

Holbrook, R. and Anderson, J.M. (1980) An improved selective and diagnostic medium for the isolation and enumeration of *Bacillus cereus* in foods. Can. J. Microbiol. 26, 753–759.

Billing, E. and Luckhurst, E.R. (1957) A simplified method for the preparation of egg yolk media. J. Appl. Bacteriol. 20, 90.

Preston campylobacter selective agar

This monograph has been assessed by members of the IUMS-ICFMH Working Party on Culture media and given 'Approved' status.

Description and history

This medium was formulated by Bolton and Robertson (1982) to facilitate the isolation of thermophilic *Campylobacter* spp. from all types of specimens (human, animal, and environmental (Bolton et al., 1983)). The selective inhibitory components are polymyxin, rifampicin, trimethoprim, and cycloheximide (actidione). The majority of unwanted contaminating organisms are suppressed whilst most thermophilic *Campylobacter* spp. produce typical moist, grey, flat spreading colonies.

Composition (grams)

Beef extract	10.0
Peptone	10.0
Sodium chloride	5.0
Polymyxin B (i.u.)	5000.0
Rifampicin	0.01
Trimethoprim lactate	0.01
Cycloheximide (Actidione)	0.1
Agar	15.0
Distilled or deionized water	1000.0
Lysed horse blood (ml)	50.0

Preparation

Suspend the beef extract, peptone, sodium chloride and agar in 1 litre of water and bring to the boil to dissolve completely. Sterilize by autoclaving at 121°C for 15 min. Cool to 50°C. Add aseptically 50 ml of lysed horse blood and the selective ingredients to give final concentrations stated.

Physical properties

Appearance	Red, translucent.
pH	7.4 ± 0.2

Shelf life

Ready to use medium 10 days at $4 \pm 2°C$.

Inoculation method for samples

Plates should not be overdried but preferably left overnight at room temperature. This medium can be inoculated using 0.5 ml per 9 cm diameter plate for whole surface spreading methods or by conventional plating to produce discrete colonies. As with other media containing rifampicin caution should be exercised when using this medium where stressed organisms may be encountered.

Incubation method

At 42°C for 48 h, microaerobically in a jar with an atmosphere containing approximately 5% O_2, 10% CO_2 and 85% N_2 or H_2.

Reading of results and interpretation

Thermophilic *Campylobacter* spp. tend to produce moist, grey, flat spreading growth, which tends to coalesce. A pink or green hue may be exhibited by some strains.

Quality assessment

(i) *Productivity*

Test strains	*Campylobacter jejuni* NCIMB 50091 *Campylobacter coli* NCIMB 50092
Inoculation method	Modified Miles-Misra, spiral plating or streaking/ ecometry.

(ii) *Selectivity*

Test strains	*Escherichia coli* NCIMB 50034 *Proteus mirabilis* (ATCC 29906/CECT 4168/NCTC 19938)
Inoculation method	Modified Miles-Misra, spiral plating or streaking/ ecometry.

(iii) *Characteristic appearance of colonies*

Occasionally some contaminating organisms may grow on this medium but they are usually restricted to the area of the primary inoculum. These include: *Pseudomonas* spp., more resistant coliforms, *Streptococcus* spp. and yeasts.

References

Bolton, F.J. and Robertson, L. (1982) A selective medium for isolating *Campylobacter jejuni/coli*. J. Clin. Pathol. 35, 462–467.

Bolton, F.J., Coates, D., Hinchliffe P.M. and Robertson, L. (1983) Comparison of selective media for isolation of *Campylobacter jejuni/coli*. J. Clin. Pathol, 36, 78–83.

Preston enrichment broth

This monograph has been reviewed by members of the IUMS-ICFMH Working Party on Culture Media and given 'Proposed' status.

Description and history

This medium was formulated to improve the recovery of thermophilic campylobacters when present in small numbers in specimens containing large numbers of contaminants (Bolton and Robertson, 1982). It has been used successfully for the isolation of these organisms from human, animal, avian, food, milk and water samples (Bolton et al., 1982, 1983; Korhonen and Martikainen, 1990).

Composition (grams)

Beef extract	10.0
Peptone	10.0
Sodium chloride	5.0
Iron (II) sulphate *	0.25
Sodium metabisulphite *	0.25
Sodium pyruvate *	0.25
Polymyxin B (i.u.)	5000.0
Rifampicin	0.01
Trimethoprim lactate	0.01
Cycloheximide (Actidione)	0.1
Distilled or deionized water	1000.0
Lysed horse blood (ml)	50.0

* FBP supplement (George et al., 1978)

Preparation

Dissolve the beef extract, peptone and sodium chloride in the water and sterilize by autoclaving at 121°C for 15 min. Cool to 50°C. Add aseptically 50 ml of lysed horse blood and the selective agents. The FBP supplement is added to the cooled medium to give the final concentrations stated from sterile stock solutions of the individual ingredients.

Physical properties

Appearance	Dark red, translucent.
pH	7.4 ± 0.2

Shelf life

Ready to use medium	7 days at 4 ± 2°C.

Inoculation method for samples

Food samples are usually added to the broth in a ratio of 1 in 4 i.e. 10 g of food plus 30 ml of broth, etc. As with other media containing rifampicin, caution should be exercised when using this medium where stressed organisms may be encountered.

Incubation method

Broth dispensed in screw capped containers with the minimum of air space can be incubated aerobically at 42°C for 24–48 h. Alternatively the broth can be incubated microaerobically in an atmosphere containing approximately 5% O_2, 10% CO_2 and 85% N_2 or H_2 which may improve the sensitivity of the medium with some types of samples. Broths are subcultured to plates of Preston agar (q.v.)(Bolton and Robertson, 1982) or modified CCD agar (Hutchinson and Bolton, 1984).

Quality assessment

(i) *Productivity*

Test strains	*Campylobacter jejuni* NCIMB 50091 *Campylobacter coli* NCIMB 50092
Inoculation method	Dilution to extinction.

(ii) *Selectivity*

Test strains	*Escherichia coli* NCIMB 50034 *Proteus mirabilis* (ATCC 29906/CECT 4168/NCTC 19938)
Inoculation method	Dilution to extinction.

References

Bolton, F.J., Coates, D., Hinchliffe, P.M. and Robertson, L. (1982) A most probable number method for estimating small numbers of campylobacter in water. J. Hyg. 89, 185–190.

Bolton, F.J., Coates, D., Hinchliffe P.M. and Robertson, L. (1983) Comparison of selective media for isolation of *Campylobacter jejuni/coli*. J. Clin. Pathol. 36, 78–83.

Bolton, F.J. and Robertson, L. (1982) A selective medium for isolating *Campylobacter jejuni/coli*. J. Clin. Pathol. 35, 462–467.

George, H.A., Hoffman, P.S., Krieg N.R. and Smimbert, R.M. (1978) Improved media for growth and aerotolerance of *Campylobacter fetus*. J. Clin. Microbiol. 8, 36–41.

Hutchinson, D.N. and Bolton, F.J. (1984) An improved blood-free selective medium for the isolation of *Campylobacter jejuni* from faeces. J. Clin. Pathol. 37, 956–957.

Korhonen, L.K. and Martikainen, P.J. (1990) Comparison of some enrichment broths and growth media for the isolation of thermophilic campylobacteri from surface water samples. J. Appl. Bacteriol. 68, 593–599.

Rabbit Plasma Fibrinogen (RPF) agar

This monograph has been reviewed by members of the IUMS-ICFMH Working Party on Culture Media and given 'Proposed' status.

Description and history

This is a selective medium for the enumeration of *Staphylococcus aureus* in foods likely to contain other coagulase positive staphylococci which may form atypical colonies on Baird-Parker agar or in cases when high amounts of competing micro-organisms are to be expected, as in cheeses made from raw milk and certain raw meat products. The egg yolk in Baird-Parker medium is replaced by rabbit plasma, fibrinogen and trypsin inhibitor (Hauschild et al., 1979; Beckers et al., 1984) so that the coagulase reaction can be observed directly on the plate. Compared to Baird-Parker agar the potassium tellurite concentration is lowered from 100 to 25 mg/l (Sawhney, 1986). Selectivity is attained with potassium tellurite and lithium chloride. Addition of sulphamezathine (Baird-Parker, 1969) is advised only if *Proteus* species are suspected in the test sample. Sodium pyruvate is a critical component essential to both the recovery of damaged *Staphylococcus aureus* cells and their subsequent growth (Baird-Parker and Davenport, 1965).

Composition (grams)

Note: From the point of quality assurance only commercially prepared media conforming to the formulation given below should be used. Each batch of rabbit plasma fibrinogen supplement should be tested before use and the manufacturer's instructions for preparation of supplement and complete medium carefully followed.

Baird-Parker base medium:

Pancreatic digest of casein	10.0
Yeast extract	1.0
Meat extract	5.0
Sodium pyruvate	10.0
L-Glycine	12.0
Lithium chloride	5.0
Agar	12 to 20.0
Distilled or deionized water	900.0

Complete medium:

Base medium	90.0 ml
Potassium tellurite solution	0.25 ml
Bovine fibrinogen solution	7.5 ml
Plasma-trypsin inhibitor solution	2.5 ml
Sulphamezathine solution (if necessary)	2.5 ml

Preparation

Dissolve the base medium components or dehydrated base in the water by boiling. If necessary, adjust the pH so that after sterilization it is 7.2 ± 0.2 at 25°C. Transfer the medium in quantities of 90 ml to flasks or bottles and autoclave for 15 min at 121°C.

Prepare membrane-filtered (0.22 µm) aqueous solutions of potassium tellurite 1 % w/v and, if necessary, sulphamezathine (sulphamezathine, sulphadimidine) 0.2 % w/v. Performance of this medium is greatly dependant on the quality of the rabbit plasma and potassium tellurite. It is important to use a satisfactory brand of potassium tellurite, e.g. Merck or Sigma and to buy in small quantities as, once opened, the salt begins to deteriorate. The salt must be completely dissolved prior to filtration. Solutions with white precipitate must be discarded. The sulphamezathine solution is prepared by dissolving 0.2 g of pure sulphamezathine in 10 ml of 0.1 N NaOH and diluting to 100 ml with water. Both solutions may be stored for up to one month at 4 ± 2°C. Bovine fibrinogen solution (5–7 g bovine fibrinogen in 100 ml sterile water) and plasma-trypsin inhibitor solution (30 mg trypsin inhibitor in 30 ml coagulase plasma EDTA) must be prepared immediately before use.

To prepare the complete medium melt and cool the base medium to approximately 47°C by means of a water bath. To each 90 ml amount add 0.25 ml of potassium tellurite solution, 7.5 ml of bovine fibrinogen solution, 2.5 ml plasma-trypsin inhibitor solution and, if necessary, 2.5 ml sulphamezathine solution. Each solution should be warmed to 45–47°C before addition and the medium well mixed after each addition. Commercially available media provide the Baird-Parker base medium and lyophilised RPF supplement containing rabbit plasma, bovine fibrinogen, trypsin inhibitor and potassium tellurite. In every case the medium must be used *immediately* after preparation.

Physical properties

Appearance	Complete medium clear, cream/pale fawn in colour.
pH	Base medium 7.2 ± 0.2 at 25°C.

Shelf life

Complete medium	Use *immediately* after preparation.
Solutions	Potassium tellurite and sulphamezathine 1 month at 4 ± 2°C. Discard tellurite if a white precipitate forms.

Inoculation method for samples

Spread plate technique using 0.1 ml inocula on 9 cm diameter dishes or 1 ml on 14 cm diameter dishes in order to raise the detection limit by a factor of 10, if necessary. It is also possible to inoculate 1 ml fractionated into 0.3, 0.3 and 0.4 ml to the surface of three 9 cm diameter plates. Carefully spread the inoculum as quickly as possible over the surface of the plates using a sterile spreader. Allow the plates to dry with their lids on for about 15 min at room temperature.

Incubation method

At 37°C for 24–48 h, in air.

Reading of results and interpretation

After incubation for 24 h count grey to black colonies with a halo. Opaque or cloudy zones indicate coagulase activity. Reincubate all plates for a further 24 h and count any newly developed colonies with haloes. Confirmation of coagulase activity is only necessary in the case of colonies of doubtful appearance.

Quality assessment

(i) *Productivity*

Test strains	*Staphylococcus aureus* NCIMB 50081 *Staphylococcus aureus* NCIMB 50080
Inoculation method	Modified Miles-Misra or spiral plating with Heart Infusion agar as reference medium.
Criteria	Counts on rabbit plasma fibrinogen agar should be within 0.5 $\log_{10}$ of the counts on the non-selective medium. NCIMB 50081 will show a reduction in count (>0.5 $\log_{10}$) if the rabbit plasma fibrinogen agar containing sulphamezathine is deficient in pyruvate.

(ii) *Selectivity*

Test strains	*Escherichia coli* NCIMB 50034 *Proteus mirabilis* (ATCC 29906/CECT 4168/NCTC 11938)
Inoculation method	As above.
Criteria	Recovery on rabbit plamsa fibrinogen agar should be 3.0 $\log_{10}$ below that on Heart Infusion agar.

(iii) *Characteristic appearance of* Staphylococcus aureus

Grey to black colonies of approximately 0.5–1 mm diameter surrounded by 1–2 mm precipitation haloes after 24 h incubation and of approximately 1–2 mm diameter with 2–4 mm haloes after 48 h incubation, respectively. On this medium all coagulase-positive staphylococci, including *Staphylococcus intermedius* and coagulase-positive strains of *Staphylococcus hyicus* show precipitation of fibrin around the colonies.

References

Baird-Parker, A.C. (1962) An improved diagnostic and selective medium for isolating coagulase positive staphylococci. J. Appl. Bacteriol. 25, 12–19.

Baird-Parker, A.C. (1969) *In:* Isolation methods for microbiologists, eds. D.A.Shapton and G.W.Gould. SAB Technical Series No. 3, Academic Press, London, pp. 1–8.

Baird-Parker, A.C. and Davenport, E. (1965) The effect of recovery medium on the isolation of *Staphylococcus aureus* after heat treatment and after storage of frozen or dried cells. J. Appl. Bacteriol. 20, 390–402.

Beckers, H.J., Van Leusden, F.M., Bindschedler, O. and Guerraz, D. (1984) Evaluation of a pour plate system with rabbit plasma-bovine fibrinogen agar for the enumeration of *Staphylococcus aureus* in food. Can. J. Microbiol. 30, 470–474.

Hauschild, A.H.W., Park, C.E. and Hilsheimer, R. (1979) A modified pork plasma for the enumeration of *Staphylococcus aureus* in foods. Can. J. Microbiol. 25, 1052–1057.

IDF (1997) Milk and milk-based products – Enumeration of coagulase-positive staphylococci – Colony count technique at 37°C. Provisional Standard 145A: 1997. International Dairy Federation, Brussels.

ISO (1999) Micobiology of foods and animal feeding stuffs – Horizontal method for the enumeration of coagulase-positive staphylococci (*Staphylococcus aureus* and other species) Part 2: Technique using rabbit plasma fibrinogen agar medium. ISO 6888–2: 1999. International Organisation for Standardization, Geneva.

Sawhney, D. (1986) The toxicity of potassium tellurite to *Staphylococcus aureus* in rabbit plasma fibrinogen agar. J. Appl. Bacteriol. 61, 149–155.

 Handbook of Culture Media for Food Microbiology, J.E.L. Corry et al. (Eds.)

Rambach (propylene glycol deoxycholate neutral red) agar

This monograph has been reviewed by members of the IUMS-ICFMH Working Party on Culture Media and given 'Proposed' status.

Description and history

This agar medium for the differentiation of *Salmonella* spp. from other members of the family Enterobacteriaceae was described by Rambach (1990). It exploits a novel phenotypic characteristic of *Salmonella* spp.: the formation of acid from propylene glycol. This characteristic is used in combination with a chromogenic indicator of β-galactosidase to differentiate *Salmonella* spp. from *Proteus* spp. and from other members of the Enterobacteriaceae. Deoxycholate is included in the medium as an inhibitor of Gram-positive organisms. Salmonellae other than *Salmonella typhi*, yield distinct, bright red colonies on the medium, allowing easy identification and unambiguous differentiation from *Proteus* spp.

Composition (grams)

Propylene glycol	10.0
Peptone	5.0
Yeast extract	2.0
Sodium deoxycholate	1.0
Neutral red	0.03
5-bromo-4-chloro-3-indolyl β-D-galactopyranoside	0.1
Agar	15.0
Distilled or deionized water	1000.0

Preparation

Suspend the ingredients in the water. Mix and heat very carefully to boiling, with frequent agitation, to dissolve soluble material. Cool to 45–50°C. Adjust pH to 7.4. Pour 15–20 ml quantities into 9 cm diameter Petri dishes.

Physical properties

Appearance	Pink, semi-opaque.

pH	7.4 ± 0.2

Shelf life

Prepared plates	14 days at 4 ± 2°C.

Inoculation method for samples

Streak pre-dried plates from selective enrichment broths to obtain well-isolated colonies.

Incubation method

At 37 ± 1°C for 18–24 h, in air.

Reading of results and interpretation

Typical colonies of salmonellae (other than *Salmonella typhi*) i.e. those that produce acid from propylene glycol, are bright red in colour. Colonies positive for β-galactosidase e.g. *Escherichia coli* are blue in colour; colonies producing both acid from propylene glycol and β-galactosidase e.g. *Citrobacter freundii*, are violet. Colonies producing neither acid from propylene glycol nor β-galactosidase e.g. *Salmonella typhi* and *Proteus mirabilis*, are colourless. *Pseudomonas* spp. may mimic the appearance of salmonellae from which they may be distinguished by an oxidase test.

Quality assessment

(i) *Productivity*

Test strains	*Salmonella enteritidis* NCIMB 50073 *Salmonella virchow* NCIMB 50077
Inoculation method	Modified Miles-Misra, spiral plating or streaking/ecometry.

(ii) *Selectivity*

Test strains	*Escherichia coli* NCIMB 50034 *Proteus mirabilis* (ATCC 29906/CECT 4168/NCTC 11938)
Inoculation method	Modified Miles-Misra, spiral plating or streaking/ecometry.
Criteria	*Escherichia coli* is not inhibited but should produce blue colonies clearly distinguishable from *Salmonella* spp.. *Proteus mirabilis* may be partially inhibited, pro-

duce colourless or orange to light brown colonies and should not swarm.

(iii) *Characteristic appearance of colonies*
See above.

Reference

Rambach, A. (1990) New plate medium for facilitated differentiation of *Salmonella* spp. from *Proteus* spp. and other enteric bacteria. Appl. Environ. Microbiol. 56, 301–303.

Rapid Perfringens Medium (RPM)

This monograph has been reviewed by members of the IUMS-ICFMH Working Party on Culture Media and given 'Proposed' status.

Description and history

This medium was developed by Erickson and Deibel (1978) for the detection and estimation of low numbers of *Clostridium perfringens* in foods and is especially useful for quality control at the manufacturing level. RPM is a liquid medium with a litmus milk base and is prepared in tubes. Selectivity is provided by the antibiotics polymyxin B sulphate and neomycin sulphate, coupled with an incubation temperature of 46°C. Detection of *Clostridium perfringens* is based on production of a stormy fermentation within 24 h. A 3-tube most probable number (MPN) procedure can be used to enumerate *Clostridium perfringens* in food samples and other products (Smith and Mood, 1983; Smith, 1985).

Composition (grams)

Solution A	
Litmus milk powder (Difco)	140.0
Neomycin sulphate	0.150
Polymyxin B sulphate	0.025
Distilled or deionized water	1000.0
Solution B	
Thioglycollate medium, fluid (Difco)	60.0
Gelatin	120.0
Peptone	10.0
Glucose	10.0
Di-potassium hydrogen orthophosphate	10.0
Yeast extract	6.0
Sodium chloride	3.0
Iron (II) sulphate	1.0
Distilled or deionized water	1000.0

Note: Crossley milk may be substituted for litmus milk in solution A.

Preparation

Prepare solution A by suspending the litmus milk powder in the water and bring to the boil. Adjust pH to 6.8. Sterilize by autoclaving at 121°C for 5 min. Cool to 50°C and aseptically add neomycin sulphate and polymyxin B sulphate. Prepare solution B by suspending the ingredients in the water and bring to the boil. Adjust pH to 7.1. Distribute 5 ml amounts in sterile screw-capped glass tubes. Sterilize by autoclaving at 121°C for 5 min. Cool to 50°C.

Prepare the final medium by aseptically adding 5 ml of solution A to each tube of solution B.

Cap the tubes tightly and invert several times to facilitate mixing. Store at 4 ± 2°C. Before use liquify the medium by placing the tubes in a waterbath at 45–50°C for 30 min. Alternatively the solutions may be stored separately at 4 ± 2°C and mixed and heated as above just before use.

Physical properties

Appearance	Light brown, opaque.
pH	7.0 ± 0.2

Shelf life

Ready to use medium	4 weeks at 4 ± 2°C.

Inoculation method for samples

Inoculate RPM tubes with 1 ml of macerates of food and mix well.

Incubation method

At 46°C for 18–20 h, in air.

Reading of results and interpretation

Detection of *Clostridium perfringens* is based on production of a stormy fermentation reaction. As stormy fermentation is a presumptive test, isolates should be confirmed by additional tests.

Quality assessment

(i) *Productivity*

Test strains	*Clostridium perfringens* (haemolytic) NCIMB 50027 *Clostridium perfringens* (non haemolytic) NCIMB 50028

Inoculation method	Inoculate medium with 100–1000 cells of the test strains.
Criteria	Stormy fermentation within 24 h at 46°C.

(ii) *Selectivity*

Test strains	*Clostridium bifermentans* NCIMB 50026 *Proteus mirabilis* (ATCC 29906/CECT 4168/NCTC 11938)
Inoculation method	Dilution to extinction.
Criteria	Difference in growth should be equal to or less than 5 titre units of the growth in tryptone soya broth. No stormy fermentation within 24 h at 46°C.

References

De Boer, E. and Boot, E. (1983) Comparison of methods for the isolation and confirmation of *Clostridium perfringens* from spices and herbs. J. Food Protect. 46, 533–536.

Erickson, J.E. and Deibel, R.H. (1978) New medium for rapid screening and enumeration of *Clostridium perfringens* in foods. Appl. Environ. Microbiol. 36, 567–571.

Smith, M. and Mood, T.J. (1983) Direct testing of gelatin hydrolysis in Rapid Perfringens Medium. J. Assoc. Off. Anal. Chem. 66, 1045–1046.

Smith, M. (1985) Comparison of Rapid Perfringens Medium and Lactose Sulfite Medium for detection of *Clostridium perfringens*. J. Assoc. Off. Anal. Chem. 68, 807–808.

Rappaport-Vassiliadis (RVS) broth

This monograph has been assessed by members of the IUMS-ICFMH Working Party on Culture media and given 'Approved' status.

Description and history

The efficiency of Rappaport-Vassiliadis enrichment medium for salmonellae is based on the following: (a) the ability of *Salmonella* spp. to multiply at relatively high osmotic pressures (concentration of hexahydrate magnesium chloride of 28.6 g/l in the final medium), at relatively low pH values, at a high temperature and with modest nutritional requirements, and (b) the suppression of the toxic effect of malachite green towards salmonellae by the presence of magnesium chloride.

The medium was first proposed by Rappaport et al. (1956) for the enrichment of selected *Salmonella* serotypes and was later modified by Vassiliadis et al. (1976; 1983), reducing the concentration of malachite green to one third. Owing to the unusual method of preparation there has been some confusion about the concentration of magnesium chloride - a final concentration of 35 g/l of medium is too high (Peterz et al., 1989; Busse, 1995). The ratio of inoculum to broth is normally 1:100 and incubation temperature 41–42°C because some strains, particularly *Salmonella dublin*, do not grow at 43°C (Peterz et al., 1989).

Composition (grams)

Soya peptone	4.5
Sodium chloride	7.2
Potassium di-hydrogen orthophosphate	1.44
Magnesium chloride ·$6H_2O$	28.6
Malachite green oxalate	0.036
Distilled or deionized water	1000.0

Preparation

Three solutions are required:

Solution A To 1000 ml of distilled water are added: soya peptone (Difco) 5 g; sodium chloride (AR) 8 g; potassium di-hydrogen orthophosphate (AR) 1.6 g. This solution is made on the day of the preparation of the RV medium. It is

	heated to about 80°C until the ingredients are completely dissolved.
Solution B	Contains 400 g of $MgCl_2 \cdot 6H_2O$ (AR) and 1000 ml of water (31.8% w/v). As this salt is very hygroscopic it is advisable to dissolve the entire contents of a newly opened container in distilled water. Solution B can be stored unsterilized in dark bottles at room temperature for at least 2 years.
Solution C	Contains 0.4 g of analytically pure malachite green oxalate (Merck No.1398) and 100 ml of water. Solution C can be kept for at least 8 months at room temperature in a dark bottle.

The final medium is prepared by adding to 1000 ml of solution A, 100 ml of solution B and 10 ml of solution C (final volume 1110 ml). It is then distributed in test tubes in 10 ml quantities or in three screw-capped bottles of 500 ml capacity each, and sterilized at 115°C for 15 min.

Physical properties

Appearance	Green-blue, transparent.
pH	5.2 ± 0.2

Shelf life

Ready to use medium	6 months in screw-capped bottles held at 4 ± 2°C (10ml is transferred to test tubes when needed).

Inoculation method for samples

A pre-enrichment of the sample should precede the enrichment in RV medium. For this purpose 25 g of the sample (food, water or other environmental sample) is added to 225 ml of a nutrient broth without sugars or, preferably, to buffered peptone water and incubated at 37°C for 16 to 24 h in air. From the pre-enrichment medium 0.1 ml is inoculated into 10 ml of Rappaport-Vassiliadis medium.

Incubation method

At 42°C for 24 and 48 h, in air.

Reading of results and interpretation

Appearance of turbidity: after 24 h and 48 h incubation of the RV medium, subcultures are made on selective plating media which are incubated for 24 h at 37°C.

Quality assessment

(i) *Productivity*

Test strains	*Salmonella virchow* NCIMB 50077 *Salmonella enteritidis* NCIMB 50073 *Salmonella typhimurium* (CECT 4156/NCTC 12190)
Inoculation method	Dilution to extinction.
Criteria	Recovery in RV broth should be within 2 titre units of recovery in tryptone soya broth after 48 h at 42°C.

(ii) *Selectivity*

Test strains	*Enterobacter cloacae* (ATCC 23355) *Pseudomonas aeruginosa* NCIMB 50067
Inoculation method	Dilution to extinction.
Criteria	Recovery in RV broth should be less than 5 titre units of the recovery in tryptone soya broth after 48 h at 42°C.

References

Busse, M. (1995) Media for salmonella. *In:* Culture Media for Food Microbiology, edited by J.E.L. Corry, G.D.W. Curtis and R.M. Baird. Elsevier Science BV, Amsterdam, pp. 187–201.

Peterz, M., Wiberg, C. and Norberg, P. (1989) The effect of incubation temperature and magnesium chloride concentration on growth of salmonella in home-made and in commercially available dehydrated Rappaport-Vassiliadis broths. J. Appl. Bacteriol. 66, 523–528.

Rappaport, F., Konforti, N. and Navon, B. (1956) A new enrichment medium for certain salmonellae. J. Clin. Pathol. 9, 261–266.

Vassiliadis, P. (1983) The Rappaport-Vassiliadis (RV) enrichment medium for the isolation of salmonellas: An overview. J. Appl. Bacteriol. 54, 69–76.

Vassiliadis, P., Paternaki, E., Papaiconomou, N., Papadakis, J.A. and Trichopoulos, D. (1976) Nouveau procède d'enrichessement de salmonella. Ann. Microbiol. Inst. Pasteur, 127B, 195–200.

Vassiliadis, P., Mavrommati, Ch., Efstratiou M. and Chronas, G. (1985) A note on the stability of Rappaport-Vassiliadis enrichment medium. J. Appl. Bacteriol., 59, 143–145.

Rappaport-Vassiliadis (MSRV) medium – Semisolid Modification

This monograph has been assessed by members of the IUMS-ICFMH Working Party on Culture Media and given 'Approved' status.

Description and history

The medium is a semisolid modification of Rappaport-Vassiliadis enrichment broth and is used as a motility enrichment in Petri dishes (De Smedt et al., 1986). It is a rapid and sensitive medium for the isolation of *Salmonella* spp. from food products and can be seeded directly after pre-enrichment or after 8 h enrichment in selective broth (De Smedt and Bolderdijk, 1987). Its efficiency is due to the ability of salmonellae to move through the highly selective medium.

Composition (grams)

Tryptose	4.59
Casein hydrolysate (acid)	4.59
Sodium chloride	7.34
Potassium di-hydrogen orthophosphate	1.47
Magnesium chloride $\cdot 6H_2O$	23.34
Malachite green oxalate	0.037
Novobiocin	0.02
Agar, technical (Oxoid L13)	2.7
Distilled or deionized water	1000.0

Preparation

Five solutions are required:

Solution A	To 600 ml of water are added: tryptose 5 g; casein hydrolysate (acid) 5 g; sodium chloride (AR) 8 g; potassium di-hydrogen orthophosphate (AR) 1.6 g. This solution is made on the day of the preparation of the MSRV medium. It is heated to about 80°C until the ingredients are completely dissolved.
Solution B	Contains 400 g of $MgCl_2 \cdot 6H_2O$ (AR) and 1000 ml of water

	(31.8% w/v). As this salt is very hygroscopic it is advisable to dissolve the entire contents of a newly opened container in water. Solution B can be stored unsterilized in dark bottles at room temperature for at least 2 years.
Solution C	Contains 0.4 g of analytically pure malachite green oxalate (Merck No. 1398) and 100 ml of water. Solution C can be kept for at least 8 months at room temperature in a dark bottle.
Solution D	Contains 3 g of agar in 400 ml of water. It is boiled until completely dissolved, then sterilized at 121°C for 15 min.
Solution E	Dissolve 100 mg of novobiocin in 5 ml of water. Sterilize by filtration.

The medium is prepared by adding to 600 ml of solution A, 80 ml of solution B and 10 ml of solution C. It is then sterilized at 115°C for 15 min. After cooling to 50°C, 400 ml of solution D at the same temperature and 1.1 ml of solution E are aseptically added. After mixing, the medium is poured into Petri dishes. As the gel of different lots of agar can vary, each new lot must be tested and if necessary the agar concentration adjusted. A gel stability between 5 and 7 g/cm^2 is optimal (De Smedt et al., 1987).

Physical properties

Appearance	Green-blue, transparent.
pH	5.2 ± 0.2

Shelf life

Ready to use medium	5 days at room temperature.

Inoculation method for samples

Three drops (approx. 0.1 ml) of incubated pre-enrichment broth are inoculated in separate spots on the surface of one plate of MSRV medium and air dried. Enrichment cultures are treated similarly.

Incubation method

At 42°C for 16–24 h, in air.

Reading of results and interpretation

After incubation the plates are examined for motile bacteria. A loopful of culture from the edge of migration is checked for purity and biochemical and serological identifications are made directly with culture from the edge of migration.

Quality assessment

(i) *Productivity*

Test strains	*Salmonella virchow* NCIMB 50077 *Salmonella enteritidis* NCIMB 50073
Supplementary strain	*Salmonella saintpaul* NCIMB 50075
Inoculation method	Dilution to extinction with inoculation of each dilution step.
Criteria	Recovery on MSRV should be within 4 titre units of the recovery in tryptone soya broth after 24 h at 42°C.

(ii) *Selectivity*

Test strains	*Proteus mirabilis* (ATCC 29906/CECT 4168/NCTC 11938) *Pseudomonas aeruginosa* NCIMB 50067
Supplementary strain	*Citrobacter freundii* NCIMB 50025
Inoculation method	Dilution to extinction with inoculation of each dilution step.
Criteria	Recovery on MSRV should be less than 6 titre units of the recovery on tryptone soya broth after 24 h at 42°C.

References

De Smedt, J.M., Bolderdijk, R.F., Rappold, H. and Lautenschlaeger, D. (1986) Rapid *Salmonella* detection in foods by motility enrichment on a modified semi-solid Rappaport-Vassiliadis medium. J.Food Protect. 49, 510–514.

De Smedt, J.M. and Bolderdijk, R.F. (1987) Dynamics of *Salmonella* isolation with modified semi-solid Rappaport-Vassiliadis medium. J. Food Protect. 50, 658–661.

Rogosa agar

This monograph has been assessed by members of the IUMS-ICFMH Working Party on Culture Media and given 'Approved' status.

Description and history

Rogosa agar (Rogosa et al., 1951) is a selective medium for the cultivation of lactobacilli (genus *Lactobacillus*) from various sources. High acetate concentration and low pH effectively suppress other bacteria while allowing lactobacilli to flourish.

Composition (grams)

Tryptic digest of casein	10.0
Yeast extract	5.0
Glucose	20.0
Sorbitan monooleate (Tween 80)	1.0
Potassium di-hydrogen orthophosphate	6.0
Ammonium citrate	2.0
Sodium acetate	15.0
Magnesium sulphate $\cdot 7H_2O$	0.575
Manganese (II) sulphate $\cdot 4H_2O$	0.14
Iron (II) sulphate $\cdot 7H_2O$	0.034
Agar	15.0
Acetic acid glacial (ml)	1.32
Distilled or deionized water	1000.0

Preparation

Suspend all the ingredients except the acetic acid in the water and heat to boiling to dissolve completely. Add the glacial acetic acid and mix thoroughly. Adjust pH to 5.5 at about 50°C. Heat to 95°C for 3 min. *Do not autoclave.* Cool to 50°C and distribute into sterile Petri dishes.

Physical properties

Appearance	Light amber, clear to slightly opalescent.
pH	5.5 ± 0.1 at 50°C.

Shelf life

Ready to use medium 5–7 days at 4 ± 2°C.

Inoculation method for samples

Surface spreading over whole plate or modified Miles-Misra.

Incubation method

This depends on the particular habitat of the organisms to be cultivated. Dairy strains should be incubated at 30°C for 2 days followed by 1 day at 22°C, meat strains at 25°C for 3 days, intestinal or yoghurt strains for 2 days at either 37 or 42°C. All incubations should be performed under anaerobic or microaerobic (6% O_2: 10% CO_2 in N_2) conditions.

Reading of results and interpretation

The results should be read after a defined incubation time (see above). Lactobacilli show good growth with the exception of some strains (*Lactobacillus sakei*, *Lactobacillus curvatus* = atypical streptobacteria), which occur in smaller colonies. *Carnobacterium* spp. do not grow.

Quality assessment

(i) *Productivity*

Test strains	*Lactobacillus gasseri* NCIMB 50040 *Lactobacillus sakei* ssp. *sakei* NCIMB 50056
Supplementary strains	*Lactobacillus brevis* NCIMB 50043 *Lactobacillus plantarum* NCIMB 50054
Inoculation method	Modified Miles-Misra, spiral plating or streaking/ecometry.
Criteria	Recovery on Rogosa agar should be within 1.0 $\log_{10}$ of recovery on MRS agar.

(ii) *Selectivity*

Test strains	*Lactococcus lactis* ssp. *lactis* NCIMB 50058 *Enterococcus hirae* NCIMB 50031
Supplementary strain	*Staphylococcus aureus* NCIMB 50080
Inoculation method	Modified Miles-Misra, spiral plating or streaking/

ecometry.

Criteria	Recovery on Rogosa agar should be 5.0 $\log_{10}$ below the recovery on MRS agar.

(iii) *Characteristic appearance of colonies*
Small greyish-white colonies, flat or raised, smooth or rough or intermediate.

Reference

Rogosa, M., Mitchell J.A. and Wiseman, R.F. (1951) A selective medium for the isolation of oral and faecal Lactobacilli. J. Bacteriol. 62, 132–133.

Rogosa agar modified (pH 6.2)

This monograph has been assessed by members of the IUMS-ICFMH Working Party on Culture Media and given 'Approved' status.

Description and history

Rogosa agar in the original version is primarily a selective medium for the cultivation of lactobacilli (genus *Lactobacillus*). High acetate concentration and low pH effectively suppress other bacteria but also many strains of other lactic acid bacteria. The modification of the pH to 6.2 instead of 5.5 alters the selectivity of the medium for the whole group of lactic acid bacteria (ISO, 1984; Reuter, 1985).

Composition (grams)

Tryptic digest of casein	10.0
Yeast extract	5.0
Glucose	20.0
Sorbitan monooleate (Tween 80)	1.0
Potassium di-hydrogen orthophosphate	6.0
Ammonium citrate	2.0
Sodium acetate	15.0
Magnesium sulphate $\cdot 7H_2O$	0.575
Manganese (II) sulphate $\cdot 4H_2O$	0.14
Iron (II) sulphate $\cdot 7H_2O$	0.034
Agar	15.0
Acetic acid glacial (ml)	1.32
Distilled or deionized water	1000.0

Preparation

Suspend all the ingredients except the acetic acid in cold water and heat to boiling to dissolve completely. Add the glacial acetic acid and mix thoroughly. Adjust pH to 6.2 at about 50°C. Heat to 95°C for 3 min. *Do not autoclave*. Cool to 45°C and distribute into Petri dishes.

Physical properties

Appearance	Light amber, clear to slightly opalescent.

pH	6.2 ± 0.1 at 50°C.

Shelf life

Ready to use medium	5–7 days at 4 ± 2°C.

Inoculation method for samples

Surface spreading over whole plate or modified Miles-Misra.

Incubation method

This depends on the particular habitat of the organisms to be cultivated. Dairy strains should be incubated at 30°C for 2 days followed by 1 day at 22°C, meat strains at 25°C for 3 days, intestinal or yoghurt strains for 2 days at either 37 or 42°C. All incubations should be performed under anaerobic or microaerobic (6% O_2: 10% CO_2 in N_2) conditions.

Reading of results and interpretation

Reading the results should be performed after a defined incubation time (see above). All well grown colonies may be considered as lactic acid bacteria. Enterococci and some pediococci show a reduced growth rate, some psychrotrophic leuconostocs from meat show slime production at 25°C. Some unwanted strains e.g. *Bacillus* spp, micrococci and yeasts may occur. *Carnobacterium piscicola* does not grow. Confirmation of identity must be done by Gram stain and catalase test.

Quality assessment

(i) *Productivity*

Test strains	*Lactobacillus gasseri* NCIMB 50040 *Lactobacillus sakei* ssp. *sakei* NCIMB 50056
Supplementary strains	*Leuconostoc mesenteroides* NCIMB 50060 *Lactobacillus plantarum* NCIMB 50054
Inoculation method	Modified Miles-Misra, spiral plating or streaking/ ecometry.
Criteria	Recovery on Rogosa agar (pH 6.2) should be within 1.0 $\log_{10}$ of recovery on MRS agar.

(ii) *Selectivity*

Test strains	*Staphylococcus aureus* NCIMB 50080 *Bacillus cereus* NCIMB 50014

Supplementary strain	*Escherichia coli* NCIMB 50034
Inoculation method	Modified Miles-Misra, spiral plating or streaking/ ecometry.
Criteria	Recovery on Rogosa agar (pH 6.2) should be 5.0 $\log_{10}$ below the recovery on MRS agar.

(iii) *Characteristic appearance of colonies*
Small greyish-white colonies, flat or raised, smooth, rough or intermediate. Unwanted strains usually attain a diameter >2.5 mm.

References

ISO/TC 34/SC 6/WG 15, No. 3 and No. 5 (1984) Draft reports: Enumeration of Lactobacteriaceae in meat and meat products.

Reuter, G. (1985) Elective and selective media for lactic acid bacteria. Int. J. Food Microbiol. 2, 55–68.

Rogosa, J., Mitchell J.A. and Wiseman, R.F. (1951) A selective medium for the isolation and enumeration of oral and fecal lactobacilli. J. Bacteriol. 62, 132–133.

Rose Bengal Chloramphenicol (RBC) agar

This monograph has been assessed by members of the IUMS-ICFMH Working Party on Culture Media and given 'Approved' status.

Description and history

A selective medium for the enumeration of moulds and yeasts in foods, developed by Jarvis (1973) and containing chlortetracycline. It is now more commonly used with chloramphenicol. The inclusion of rose bengal not only restricts the size and height of mould colonies but assists enumeration in that the colour is taken up by the fungi. Chloramphenicol is used as the selective agent to suppress most bacteria. RBC agar is suitable for use with proteinaceous foods and where higher than normal incubation temperatures (35°C) are required. It neither selects for specific groups of fungi, nor is diagnostic for detecting specific mycotoxin producers.

Composition (grams)

Mycological peptone	5.0
Glucose	10.0
Di-potassium hydrogen orthophosphate	1.0
Magnesium sulphate $\cdot 7H_2O$	0.5
Rose Bengal	0.05
Chloramphenicol	0.1
Agar	15.0
Distilled or deionized water	1000.0

Preparation

Suspend the basal ingredients in water and bring to the boil to dissolve completely. Add 10 ml of a 1% ethanolic solution of chloramphenicol, mix and sterilize by autoclaving at 121°C for 15 min. Cool to below 50°C and dispense approximately 15 ml amounts into 9 cm diameter Petri dishes. Use immediately or store at 4 ± 2°C in the dark until required.

Physical properties

Appearance	Deep pink, without any significant precipitate.
pH	7.2 ± 0.2 at 25°C.

Shelf life

Ready to use medium 7 days at 4 ± 2°C in the dark.

Inoculation method for samples

Surface spreading over whole plate using 0.1 or 0.2 ml per 9 cm diameter plate.

Incubation

At 25 ± 0.5°C in the dark for 5 days. Where identification is required, prolong incubation until characteristic colonies are formed.

Reading of results and interpretation

Where separate counts of moulds and yeasts are required, identify by morphological appearance and, where necessary, microscopic examination of the two groups of microorganisms. Colonies of yeasts and bacteria can be confused and should be checked microscopically.

Quality assessment

Moulds. Use stab inoculation procedure. Growth rate should be within 30% of the figure given.

Test strains	IMI No.	ATCC No.	Growth rate (mm per day at 25°C)
Rhizopus stolonifer	61269		13.2
Aspergillus flavus	91856ii		8.0
Penicillium cyclopium	19759	16025	4.6

Yeasts. Use stab inoculation procedure. Growth should be within 30% of figure given.

Test strain	Growth rate (mm per day at 25°C)
Saccharomyces cerevisiae NCIMB 50105	1.8

Bacteria. Use streaking procedure (Appendix I, Method C) with a loopful (10 μl) from a 24 h broth culture. No growth should be evident on sectors C, D and E of the streaked plate after incubation at 25°C for 5 days.

Test strain
Bacillus subtilis NCIMB 50018

Reference

Jarvis, B. (1973) Comparison of an improved rose-bengal-chlortetracycline agar with other media for the selective isolation and enumeration of moulds and yeasts in food. J. Appl. Bacteriol. 36, 723–727.

Salmonella Shigella Deoxycholate Calcium (SSDC) agar

This monograph has been assessed by members of the IUMS-ICFMH Working Party on Culture Media and given 'Approved' status.

Description and history

The medium, first described by Wauters (1973), is formulated for detection of pathogenic *Yersinia enterolitica*, especially biotype 4 serotype 0:3, and to some extent biotype 2 serotypes 0:9 and 0:5, 27. It may be less suitable or not suitable at all for other biotypes or other *Yersinia* species. The selectivity relies on the basic components of SS-agar supplemented by sodium deoxycholate, yeast extract and calcium chloride. As in SS-agar the indicative principle is the absence of lactose fermentation leading to colourless colonies, lactose fermenting colonies being pink. The medium is particularly suitable for subculturing from ITC enrichment (Wauters et al., 1988).

Composition (grams)

Beef extract	5.0
Peptone from meat or tryptose	5.0
Yeast extract	5.0
Lactose	10.0
Ox bile dried	8.5
Sodium deoxycholate	10.0
Sodium citrate	10.0
Sodium thiosulphate	8.5
Iron (III) citrate	1.0
Brilliant green	0.0003
Neutral red	0.025
Calcium chloride	1.0
Agar	15.0
Distilled or deionized water	1000.0

Preparation

Suspend the ingredients in the water. Bring to the boil to dissolve completely. Avoid prolonged heating. Pour into 9 cm diameter Petri dishes.

Physical properties

Appearance	Pale yellowish-brownish, clear.
pH	7.4 ± 0.1

Shelf life

Ready to use medium	7 days at room temperature in the dark. Do not refrigerate.

Inoculation method for samples

Surface spreading over whole plate using 0.1 ml on a 9 cm diameter pre-dried plate. Best results are achieved after ITC (q.v.) broth enrichment, by streaking one loopful on the surface of a whole plate.

Incubation method

At 30°–33°C for 24 h, in air.

Reading of results and interpretation

Yersinia enterocolitica appears as round colourless colonies, about 1 mm in diameter. By transillumination under a binocular (10×) lens they are finely granular. Confirmatory tests are required.

Quality assessment

(i) *Productivity*

Test strains	*Yersinia enterocolitica* biotype 4, serotype 0:3 NCIMB 50087
Inoculation method	Modified Miles-Misra, spiral plating or streaking/ ecometry.

(ii) *Selectivity*

Test strains	*Escherichia coli* NCIMB 50034 *Staphylococcus aureus* NCIMB 50080
Inoculation method	Modified Miles-Misra, spiral plating or streaking/ ecometry.

(iii) *Characteristic appearance of colonies*
Round, colourless about 1 mm in diameter.

References

Wauters, G. (1973) Improved methods for the isolation and the recognition of *Yersinia enterocolitica*. Contrib. Microbiol. Immunol. 2, 68–70.

Wauters, G., Goossens, V., Janssens, M. and Vandepitte, J. (1988) New enrichment method for isolation of pathogenic *Yersinia enterocolitica* serogroup 0:3 from pork. Appl. Environ. Microbiol. 54, 851–854.

Selenite cystine broth

This monograph has been assessed by members of the IUMS-ICFMH Working Party on Culture Media and given 'Approved' status.

Description and history

This is a selective enrichment medium used in procedures for the detection and isolation of salmonellae. Selectivity is conferred by the inclusion of sodium selenite in the medium, shown by Leifson (1936) to be much more inhibitory to other members of the Enterobacteriaceae than to many strains of *Salmonella*, in particular *Salmonella typhi* and *Salmonella paratyphi*. Selenite cystine broth is a modification (North and Bartram, 1953) of the original Leifson (1936) formula, differing only in the addition of L-cystine which is considered to enhance *Salmonella* growth by reduction of toxicity.

In view of developments in regulatory bodies such as ISO and AOAC this medium is expected to have a limited life.

Composition (grams)

Pancreatic digest of casein (tryptone)	5.0
Lactose	4.0
Di-sodium hydrogen orthophosphate .12H_2O	10.0
Sodium hydrogen selenite[1]	4.0
L-Cystine	0.01
Distilled or deionized water	1000.0

Preparation

Base

Dissolve the tryptone, lactose and disodium hydrogen orthophosphate in water by boiling for 5 min. After cooling add the sodium hydrogen selenite. Adjust the pH to 7.0 ± 0.1 at 20°C. Store at 4°C.

[1] Warning: Extreme care should be taken with the laboratory use of selenite solutions because of their potentially toxic effect.

L-Cystine solution

Add 0.1 g L-cystine to 15 ml N/l NaOH. Dilute to 100 ml with sterile water in a sterile flask. Do not heat.

Complete medium

Cool base and add L-cystine solution at the rate of 0.1 ml per 10 ml of base. Adjust the pH to 7.0 ± 0.1 at 20°C. Transfer the complete medium in appropriate quantities e.g. 10 ml, 100 ml, etc. to sterile containers, to give a depth of 50–60 mm.

Physical properties

Appearance	Pale straw coloured; clear or may have a slight precipitate. A brick red precipitate indicates overheating.
pH	7.0 ± 0.1 at 20°C.

Shelf life

Ready to use medium	Use on day of preparation.

Inoculation method for samples

1. Where damaged cells are sought transfer 10 ml (or 1 ml) of sample-inoculated and incubated pre-enrichment broth into 100 ml (or 10 ml) selenite cystine broth

or

2. Add comminuted (food) sample to selenite cystine broth in the ratio 1 part sample to 10 parts medium.

Incubation method

At 37°C for 24 h.

Reading of results and interpretation

Inoculated and incubated medium is subcultured after 24 h onto selective diagnostic agar media in such a way as to obtain well isolated colonies. Suspect colonies are subcultured and their identity confirmed by biochemical and serological tests.

Quality assessment

(i) *Productivity*

Test strains	*Salmonella enteritidis* NCIMB 50073 *Salmonella virchow* NCIMB 50077 *Salmonella typhimurium* (NCTC 12190)

Inoculation method	Dilution to extinction.
Criteria	Recovery in selenite cystine broth should be within one titre unit of the recovery in tryptone soya broth after 24 h at 37°C.

(ii) *Selectivity*

Test strains	*Enterobacter cloacae* (ATCC 23355) *Escherichia coli* NCIMB 50034
Inoculation method	Dilution to extinction.
Criteria	Recovery in selenite cystine broth should be less than 2 titre units of the recovery in tryptone soya broth after 24 h at 37°C.

Comments

This medium may be markedly influenced by the following:
1. Source of ingredients or dehydrated complete medium
2. The presence of food or other organic material
3. The physiological state of the organisms sought
4. The competing microflora
5. Possibly by combinations of the above

Successful performance of the medium tested in one set of standard conditions may not be achieved if any of these conditions are altered.

References

Leifson, E. (1936) New selenite enrichment media for the isolation of typhoid and paratyphoid (*Salmonella*) bacilli. Amer. J. Hyg. 24, 423–432.

North, W.R. and Bartram, M.T. (1953) The efficiency of selenite broth of different compositions in the isolation of *Salmonella*. Appl. Microbiol. 1, 130–124.

Skirrow campylobacter selective agar

This monograph has been assessed by members of the IUMS-ICFMH Working Party on Culture Media and given 'Approved' status.

Description and history

This medium was developed for the selective isolation of *Campylobacter jejuni* and *Campylobacter coli* from human faeces (Skirrow, 1977). It replaced the more cumbersome method of selective filtration through 0.65 μm pore size membranes. The medium relies on the extreme degree of resistance of campylobacters to trimethoprim and moderate resistance to vancomycin and polymyxin. It is necessary to include lysed horse blood in order to neutralise trimethoprim antagonists; failure of trimethoprim activity will result in growth of *Proteus* spp. which are among the few groups of Gram negative bacteria that are resistant to polymyxins.

Composition (grams)

Proteose peptone	15.0
Liver digest	2.5
Yeast extract	5.0
Sodium chloride	5.0
Vancomycin	0.01
Polymyxin B (i.u.)	2500.0
Trimethoprim	0.005
Agar	15.0
Distilled or deionized water	1000.0
Lysed horse blood (ml)	50.0

Preparation

Prepare base or Columbia agar without antibiotics and autoclave 121°C for 15 min. Cool to 50°C, add antibiotics and lysed horse blood.

Physical properties

Appearance	Dark red, clear.
pH	7.4 ± 0.2

Shelf life

Ready to use medium 7 days at 4 ± 2°C.

Inoculation method for samples

Suspend material in saline or broth or subculture from an enrichment broth. Inoculate section of agar and spread for discrete colonies.

Incubation method

At 42°C for 48 h microaerobically in a jar with an atmosphere containing approximately 5% O_2, 10% CO_2 and 85% N_2 or H_2.

Reading of results and interpretation

Typical colonies are flat, glossy and effuse, with a tendency to form a spreading film if the agar is moist. Mature colonies are low convex, often tan coloured.

Quality assessment

(i) *Productivity*

Test strains	*Campylobacter jejuni* NCIMB 50091 *Campylobacter coli* NCIMB50092
Inoculation method	Modified Miles-Misra, spiral plating or streaking/ecometry.

(ii) *Selectivity*

Test strains	*Escherichia coli* NCIMB 50034
Inoculation method	Modified Miles-Misra, spiral plating or streaking/ecometry.

(iii) *Characteristic appearance of colonies*

See above.

References

Skirrow, M.B. (1977) Campylobacter enteritis: a 'new' disease. Brit. Med. J. 2, 9–11.

Skirrow, M.B., Benjamin, J., Razi, M.H.H. and Waterman, S. (1982) Isolation, cultivation and identification of Campylobacter jejuni and C. coli. In: Isolation and identification methods for food poisoning organisms, edited by J.E.L. Corry, D. Roberts and F.A. Skinner, SAB Technical Series No. 17, Academic Press, London, pp. 313–328.

SM ID (Salmonella identification) medium

This monograph has been assessed by members of the IUMS-ICFMH Working Party on Culture media and given 'Approved' status.

Description and history

This medium for the isolation and differentiation of *Salmonella* spp. from other bacteria was developed by bioMerieux (Poupart et al., 1991). It relies on the use of bile salts and brilliant green for selective inhibition of Gram positive bacteria and contains two indicator systems. The acidification of gluconate combined with a coloured indicator (neutral red) produces a colour change to pink for *Salmonella* colonies. Two chromogenic substrates for β-galactosidase and β-glucosidase allow differentiation of *Salmonella* (negative) from other enterobacteria acidifying glucuronate (positive: blue to purple colonies). SM ID allows good growth of *Salmonella* spp. and reduces the growth of other bacteria. *Salmonella* spp. from both human and food samples, including *Salmonella typhi* and *Salmonella paratyphi*, produce deep pink colonies, allowing easy detection and unambiguous differentiation from other bacteria (Davies and Wray, 1994; Monnery et al., 1994).

Composition (grams)

Beef extract	3.0
Bio-Polytone	6.0
Yeast extract	2.0
Bile salts	4.0
Neutral red	0.025
Tris buffer	0.65
Brilliant green	0.0003
Sodium glucuronate	12.0
Sorbitol	8.0
5-Bromo-4-chloro-3-indolyl-β-D-galactoside (X-gal)	0.17
5-Bromo-4-chloro-3-indolyl-β-D-glucoside	0.025
Agar	13.5
Distilled or deionized water	1000.0

Preparation

Technical difficulties make this medium unsuitable for preparation in small amounts from basic ingredients.

Physical properties

Appearance	Orange to pink, transparent.
pH	7.6 ± 0.2

Shelf life

Ready to use medium	12 weeks at 4 ± 2°C.

Inoculation method for samples

Sreaking to obtain isolated colonies.

Incubation method

At 35–37°C for 16–24 h, in air.

Reading results and interpretation

Due to acidification of the glucuronate, colonies of *Salmonella* spp. are deep pink, round and sometimes show a colourless rim. Colonies positive for β-galactosidase and/or β-glucosidase, e.g. *Enterobacter agglomerans*, are blue; colonies producing both acid from glucuronate and β-galactosidase and/or β-glucosidase, e.g. *Escherichia coli, Klebsiella pneumoniae,* are violet. Colonies producing neither acid from glucuronate nor β-galactosidase nor β-glucosidase, e.g. *Proteus mirabilis* are colourless.

Some *Salmonella* serotypes, particularly *Salmonella arizona*, are not detected on this medium, growing as purple, blue or colourless colonies. It is possible to find pink coloured colonies which are not *Salmonella* spp; these are mainly strains of *Escherichia coli* without β-galactosidase and some strains of *Shigella, Yersinia* and *Morganella.*

Quality assessment

(i) *Productivity*

Test strain	*Salmonella enteritidis* NCIMB 50073
Supplementary strain	*Salmonella typhimurium* ATCC 25241
Inoculation method	Modified Miles-Misra, spiral plating or streaking/ecometry.

(ii) *Selectivity*

Test strain	*Escherichia coli* NCIMB 50034
Supplementary strain	*Escherichia coli* NCIMB 50109 (β-galactosidase negative)
Inoculation method	Modified Miles-Misra, spiral plating or streaking/ ecometry.
Criteria	*Escherichia coli* is not inhibited but should produce violet colonies clearly distinguishable from *Salmonella* spp.

(iii) *Characteristic appearance of colonies*
See above.

References

Davies, R.H. and Wray, C. (1994) Evaluation of SM ID agar for identification of *Salmonella* in naturally contaminated veterinary samples. Lett. Appl. Microbiol. 18, 15–17.

Monnery, I., Freydiere, A-M., Baron, C., Rousset, A.M., Tigaud, S., Boude-Chevalier, M., de Montclos, H. and Gille, Y. (1994) Evaluation of two new chromogenic media for detection of *Salmonella* in stools. Eur. J. Clin. Microbiol. Infect. Dis. 13, 257–261.

Poupart, M.C., Mounier, M., Denis, F., Sirot, J., Couturier, C. and Villeval, F. (1991) A new chromogenic ready-to-use medium for *Salmonella* detection. 5th European Congress of Clinical Microbiology and Infectious Diseases. Oslo, September 1991.

Starch Ampicillin Agar (SAA)

This monograph has been assessed by members of the IUMS-ICFMH Working Party on Culture Media and given 'Approved' status.

Description and history

This medium was developed for the isolation of *Aeromonas* spp. from food (Palumbo et al., 1985; Jeppesen, 1995). The selective principle of SAA is the antibiotic ampicillin. Differentiation is based on starch fermentation which is detected by the pH indicator (phenol red).

Composition (grams)

Beef extract*	1.0
Proteose peptone No. 3*	10.0
Sodium chloride*	5.0
Phenol red*	0.025
Agar*	15.0
Soluble starch, reagent grade	10.0
Ampicillin	0.01
Distilled or deionized water	1000.0

* these ingredients can be substituted by 31.0 g of Phenol red agar base (Difco).

Preparation

Suspend the ingredients except ampicillin in the water and heat to dissolve completely. Sterilize for 15 min at 121°C. To the cooled (50°C) medium add ampicillin dissolved in a very small quantity of distilled water. Mix well and pour plates.

Physical properties

Appearance	Red to orange, clear.
pH	7.4 ± 0.2

Shelf life

Basal medium	3 months at 4 ± 2°C.

Poured plates	7 days at 4 ± 2°C.

Inoculation method

Surface spreading over whole plate, using 0.1 ml inoculum per pre-dried 9 cm diameter Petri dish. Alternatively, for liquid samples membrane filtration may be used.

Incubation method

At 29 ± 1°C for 18–24 h, in air.

Reading of results and interpretation

Yellow to honey-coloured colonies, 2–3 mm in diameter, surrounded by a light halo (2–3 mm width) are considered to be *Aeromonas* spp. Optionally the presence of *Aeromonas* spp. can be verified by the addition of approximately 5 ml of Lugol's iodine solution to each plate. After this colonies will be surrounded by a clear zone of hydrolysed starch against a black background. Suspect colonies must be picked quickly at this step for verification since the iodine is rapidly lethal to the cells (Palumbo et al., 1992). Confirmatory tests are required.

Quality assessment

(i) *Productivity*

Test strains	*Aeromonas hydrophila* NCIMB 50013 *Aeromonas caviae* (ATCC 15468/NCIMB 13016)
Inoculation method	Modified Miles-Misra, spiral plating or streaking/ecometry.

(ii) *Selectivity*

Test strains	*Enterobacter cloacae* (ATCC 23355) *Escherichia coli* NCIMB 50034
Inoculation method	Modified Miles-Misra, spiral plating or streaking/ecometry.

(iii) *Characteristic appearance of colonies*
See above.

References

Jeppesen, C. (1995) Media for *Aeromonas* spp., *Plesiomonas shigelloides* and *Pseudomonas* spp. from food and environment. Int. J. Food. Microbiol. 26, 25–41.

Nordic Committee on Food Analysis. (1995) *Aeromonas* species, motile. Determination in foods.

NMKL Proposed Method No. 150.

Palumbo, S.A., Maxino, F., Williams, A.C., Buchanan, R.L. and Thayer, D.W. (1985) Starch-ampicillin agar for the quantitative detection of *Aeromonas hydrophila*. Appl. Environ. Microbiol. 50, 1027–1030.

Palumbo, S.A., Abeyta, C. and Stelma, G. Jr. (1992) *Aeromonas hydrophila* group. *In:* Compendium of Methods for the Microbiological Examination of Food, 3rd edition. American Public Health Association, Washington, DC. pp. 497–515.

Streptomycin Thallous Acetate Actidione (STAA) agar

This monograph has been assessed by members of the IUMS-ICFMH Working Party on Culture Media and given 'Approved' status.

Description and history

This medium, developed by Gardner (1966), is used for the quantitative enumeration of *Brochothrix thermosphacta* in meat and meat products. The combination of antibiotics aims to exclude all other organisms.

Composition (grams)

Peptone (Oxoid L37)	20.0
Yeast extract	2.0
Glycerol	15.0
Di-potassium hydrogen orthophosphate	1.0
Magnesium sulphate $\cdot 7H_20$	1.0
Streptomycin sulphate (as streptomycin)*	0.5
Cycloheximide (Actidione)*	0.05
Thallium (I) acetate*	0.05
Agar	13.0
Distilled or deionized water	1000.0

Preparation

All the ingredients except those marked * are dissolved in the water, pH adjusted to 7.0 and sterilized in measured amounts by autoclaving at 121°C for 15 min. The selective agents dissolved in distilled water are added to the cooled (45°C) melted base.

Physical properties

Appearance	Pale straw and slightly opaque.
pH	7.0 ± 0.2

Shelf life

Prepared plates	7 days at 4 ± 2°C.
Additive solutions	1 month at 4 ± 2°C.
Bottled basal medium	1 month at ambient.

These estimations are known to be useful but no definitive data exist on the maximum shelf life of the components.

Inoculation method for samples

Surface spreading over whole plate or modified Miles-Misra.

Incubation method

At 22°C for 48 h, in air.

Reading of results and interpretation

Occasional yeast and pseudomonad colonies may be found. The former are easily recognised and the latter also have a different appearance and may be confirmed by flooding the plate with Kovac's oxidase reagent: 1% (w/v) tetramethyl-*p*-phenylenediamine dihydrochloride in water; blue colonies should be subtracted from the total.

Quality assessment

(i) *Productivity*

Test strain	*Brochothrix thermosphacta* NCIMB 50019
Inoculation method	Modified Miles-Misra, spiral plating or streaking/ecometry.

(ii) *Selectivity*

Test strains	*Lactobacillus sakei* ssp. *sakei* NCIMB 50056 *Enterococcus faecium* NCIMB 50032
Inoculation method	Modified Miles-Misra, spiral plating or streaking/ecometry.

(iii) *Characteristic appearance of colonies*

White or semi-transparent convex, with or without an irregular margin and may exhibit a structure similar to masses of woven threads.

References

Gardner, G.A. (1966) A selective medium for the enumeration of *Microbacterium thermosphactum* in meat and meat products. J. Appl. Bacteriol. 29, 455–460.

Gardner, G.A. (1981) *Brochothrix thermosphacta (Microbacterium thermosphactum)* in the spoilage of meats: a review. *In:* Psychrotrophic microorganisms in spoilage and pathogenicity, edited by T.A. Roberts, G. Hobbs, J.H.B. Christian and N. Skovgaard. Academic Press, London, pp. 139–173.

Sulphite Cycloserine Azide (SCA) agar

This monograph has been reviewed by members of the IUMS-ICFMH Working Party on Culture Media and given'Proposed' status.

Description and history

SCA agar is a selective medium for the enumeration of sulphite-reducing clostridia in samples where the vegetative cells are predominant and which are therefore not subjected to heat shock (e.g. a pasteurisation step) before analysis. Selectivity is provided by D-cycloserine and sodium azide. The medium contains an indicator system involving sulphite and iron. Most unwanted species are suppressed, while *Clostridium perfringens* and related species will reduce the sulphite and form black colonies due to the production of ferrous sulphide (Eisgruber and Reuter, 1991; 1995). Originally the medium was prepared with higher amounts of D-cycloserine, sulphite and iron for the selective enumeration of *C. perfringens* (Hauschild and Hilsheimer, 1974). Modifications and the addition of sodium azide as a further selective component were first recommended by Eisgruber (1986). A proposal for a standardized medium was delivered by Eisgruber and Reuter (1995).

Composition (grams)

Tryptose	15.0
Peptone from soya meal	5.0
Meat extract	5.0
Yeast extract	5.0
Glucose	2.0
Di-sodium disulphite*	0.5
Iron (III) ammonium citrate*	0.5
D-cycloserine (97%)*	0.3
Sodium azide*	0.05
Agar	14.0
Distilled or deionized water	900.0

Preparation

Suspend ingredients except those marked * in the water and heat to boiling to dissolve completely. Sterilize by autoclaving at 121°C for 10 min then cool to 45–50°C. Add 100.0 ml of a filter-sterilized aqueous solution containing 0.5 g di-sodium disulphite,

0.5 g iron (III) ammonium citrate and 1.0 ml of a stock solution of D-cycloserine/ sodium azide.

The stock solution is prepared by suspending 1.5 g D-cycloserine and 0.25 g sodium azide in 50.0 ml distilled water. After sterilisation by filtration this solution can be stored at 4 ± 2°C for several months.

pH adjustment should be performed after addition of the solutions at 45°C.

Physical properties

Appearance	Pale amber.
pH	7.4 ± 0.2 at 45°C.

Shelf life

Ready to use medium	2 weeks in the dark at 4 ± 2°C.
Additive solution	several months at 4 ± 2°C.
Bottled basal medium	2 weeks in dark at 4 ± 2°C.

Inoculation method for samples

Pour plate method, using 1 ml suspension of sample-dilution per 9 cm diameter plate and mixing with about 15 ml SCA medium about 45–50°C.

Incubation method

At 37°C for 48 h, anaerobically.

Reading of results and interpretation

Discrete blackened colonies are considered to be presumptive sulphite reducing mesophilic clostridia. In doubtful cases confirm at least ten colonies from one plate by Gram staining (Eisgruber and Reuter, 1995).

Quality assessment

(i) *Productivity*

Test strains	*Clostridium perfringens* NCIMB 50027 *Clostridium sporogenes* NCIMB 50099
Inoculation method	As above.
Criteria	Recovery on SCA should be within 0.5 $\log_{10}$ of the recovery on a non- selective medium after 48 h at 37°C anaerobically.

(ii) *Selectivity*

Test strains	*Bacillus subtilis* NCIMB 50018 *Escherichia coli* NCIMB 50034 *Staphylococcus aureus* NCIMB 50080 *Lactobacillus delbrueckii* ssp. *bulgaricus* NCIMB 50050
Inoculation method	As above.
Criteria	The recovery of *Lactobacillus delbrueckii* ssp. *bulgaricus* on SCA should be less than 2 $\log_{10}$ of the recovery on a non-selective medium after 48 h at 37°C anaerobically without blackening of colonies. Other test strains should show no growth at all.

(iii) *Characteristic appearance of colonies*
Blackened colonies of 1–5 mm diameter.

References

Eisgruber, H. (1986) Prüfung von Verfahren zur Kultivierung und Schnellidentifizierung von Clostridien aus frischem Fleisch sowie aus anderen Lebensmitteln. Vet. Med. Diss. FU Berlin.

Eisgruber, H. and Reuter, G. (1991) SCA – ein Selektivnährmedium zum Nachweis mesophiler sulfitreduzierender lostridien in Lebensmitteln, speziell für Fleisch und Fleischerzeugnisse. Arch. Lebensmittelhyg. 42, 125–129.

Eisgruber, H. and Reuter, G. (1995) A selective medium for the detection and enumeration of mesophilic sulphite-reducing clostridia in food monitoring programs. Food Res. Int. 28, 219–226.

Hauschild, A.H.W. and Hilsheimer, R. (1974) Enumeration of food-borne *Clostridium perfringens* in egg-yolk free tryptose-sulphite cycloserine agar. Appl. Microbiol. 27, 521–527.

Thallous acetate Tetrazolium Glucose (TlTG) agar

This monograph has been assessed by members of the IUMS-ICFMH Working Party on Culture Media and given 'Approved' status.

Description and history

The medium incorporates thallous acetate as a selective inhibitory agent. Differentiation between *Enterococcus faecalis* and *Enterococcus faecium* depends upon the reduction of tetrazolium to the maroon formazan at pH 6.0. Selectivity of the medium can be increased by incubation at 44–45°C.

Composition (grams)

Proteose peptone	10.0
Beef extract powder	8.0
Glucose	10.0
Thallium (I) acetate	1.0
2,3,5-Triphenyltetrazolium chloride	0.1
Agar	14.0
Distilled or deionized water	1000.0

Preparation

The peptone and beef extract powder are dissolved by boiling in half the required volumc of water, making allowance for additions (a)–(c) below. The pH value is adjusted to 6.0–6.1. The agar is dissolved separately and the two solutions are mixed and distributed into bottles (or final containers) in 92 ml amounts for autoclaving at 121°C for 15 min. After cooling to 50–52°C, the following solutions are added to each 92 ml of molten basal medium:

a) 5ml of 20% (w/v) filter-sterilized glucose
b) 2ml of 5% (w/v) thallous acetate, autoclaved at 115°C for 15 min.
c) 1ml of 1% triphenyltetrazolium chloride (filter-sterilized).

Physical properties

Appearance	Pale straw.
pH	6.0–6.1

Shelf life

Ready to use basal medium	At least one month at 20 ± 2°C.
Poured plates	7 days at 4 ± 2°C.

Inoculation method for samples

Surface spreading over whole plate using 0.1 ml per 9 cm diameter pre-dried plate.

Incubation method

At 37°C for 24 h.

Reading of results and interpretation

Colonies are round and about 1 mm in diameter. Those which are characteristic of *Enterococcus faecalis* have a deep red centre with a narrow white periphery, whilst colonies of *Enterococcus faecium* are white or pink. Since certain streptococci can also grow on this medium at 37°C, usually forming very small red or white colonies, confirmatory tests may be necessary.

Quality assessment

(i) *Productivity*

Test strains	*Enterococcus faecalis* NCIMB 50030 *Enterococcus faecium* NCIMB 50032
Inoculation method	Modified Miles-Misra, spiral plating or streaking/ecometry.

(ii) *Selectivity*

Test strains	*Lactococcus lactis* ssp. *lactis* NCIMB 50058 *Escherichia coli* NCIMB 50034
Inoculation method	Modified Miles-Misra, spiral plating or streaking/ecometry.

(iii) *Characteristic appearance of colonies*

Round white/pale pink colonies (*Enterococcus faecium*) or colonies with a deep red centre and narrow white periphery (*Enterococcus faecalis*) each about 1 mm diameter.

Reference

Barnes, E.M. (1956) Methods for the isolation of faecal streptococci (Lancefield group D) from bacon factories. J. Appl. Bacteriol. 19, 193–203.

Handbook of Culture Media for Food Microbiology, J.E.L. Corry et al. (Eds.)

Thiosulphate Citrate Bile-salt Sucrose (TCBS) agar

This monograph has been reviewed by tmembers of the IUMS-ICFMH Working Party on Culture Media and given 'Proposed' status.

Description and history

This medium was originally developed by Kobayashi et al. (1963) for the isolation of *Vibrio parahaemolyticus* but is also suitable for the isolation of *Vibrio cholerae* and most other pathogenic *Vibrio* spp.from clinical specimens. The selective components of the medium are bile-salts, thiosulphate, citrate and a relatively high pH of 8.6. Differentiation is based on sucrose fermentation detected by the pH indicators and H_2S production from thiosulphate detected by the production of black iron sulphide from ferric citrate. *Vibrio* species do not produce H_2S but depending on the species may ferment sucrose. Variations in the performance of different commercial preparations of this formula were noted by Nicholls et al. (1976) and West et al. (1982) who also addressed problems relating to quality control of the medium.

Composition (grams)

Yeast extract	5.0
Peptone	10.0
Sodium thiosulphate	10.0
Sodium citrate	10.0
Ox bile	8.0
Sucrose	20.0
Sodium chloride	10.0
Iron (III) citrate	1.0
Bromothymol blue	0.04
Thymol blue	0.04
Agar	15.0
Distilled or deionized water	1000.0

Preparation

Prepare in a flask at least three times larger than the volume of medium being made. Suspend the ingredients in the water and slowly heat with constant stirring. On boiling remove from the heat immediately, cool to 50°C and pour into Petri dishes. *Do not autoclave.*

Physical properties

Appearance	Dark blue-green or green, clear.
pH	8.6 ± 0.2

Shelf life

Ready to use medium	14 days at 4 ± 2°C.

Inoculation method

Surface spreading over whole plate. If material is sufficiently dilute, 0.1 ml amounts of food suspensions may be spread on 9 cm diameter plates.

Incubation method

At 30–37°C for 18–24 h, in air. Longer incubation reduces the selectivity of the medium.

Reading of results and interpretation

Strains of different *Vibrio* spp. produce yellow or green colonies of various sizes usually with a minimum diameter of 0.5 mm. *Vibrio cholerae* usually has yellow (sucrosc fermenting) colonies of 1–3 mm diameter and *Vibrio parahaemolyticus* green colonies of 2–5 mm. Such colonies are considered to be presumptive *Vibrio* spp.

Quality assessment

(i) *Productivity*

Test strains	*Vibrio cholerae* (non 01)(NCTC 11348) *Vibrio parahaemolyticus* (NCTC 11344) *Vibrio fluvialis* NCIMB 50084
Inoculation method	Modified Miles-Misra, spiral plating or streaking/ecometry.

(ii) *Selectivity*

Test strains	*Escherichia coli* NCIMB 50034 *Proteus mirabilis* (ATCC 29906/CECT 4168/NCTC 11938) *Pseudomonas aeruginosa* NCIMB 50067

Inoculation method	Modified Miles-Misra, spiral plating or streaking/ ecometry.

(iii) *Characteristic appearance of colonies*

Round yellow (*Vibrio cholerae, Vibrio fluvialis*) or green (*Vibrio parahaemolyticus*) colonies without any evidence of blackening due to H_2S production.

References

Kobayashi, T., Enomoto, S., Sakazaki, R. and Kuwahara, S. (1963) A new selective medium for pathogenic Vibrios TCBS agar (modified Nakanishi's agar). Jap. J. Bacteriol. 18, 387–391.

Nicholls, K.M., Lee J.V. and Donovan, T.J. (1976) An evaluation of commercial thiosulphate citrate bile salt sucrose agar (TCBS). J. Appl. Bacteriol. 41, 265–269.

West, P.A., Russek, E., Brayton, P.R. and Colwell, R.R. (1982) Statistical evaluation of a quality control method for isolation of pathogenic *Vibrio* species on selected thiosulphate-citrate-bile salts-sucrose agars. J. Clin. Microbiol. 16, 1110–1116.

Tryptone Bile Agar (TBA)

This monograph has been assessed by members of the IUMS-ICFMH Working Party on Culture Media and given 'Approved' status.

Description and history

This medium forms part of a direct plating method for the rapid enumeration of *Escherichia coli* in food materials (Anderson and Baird-Parker, 1975; Holbrook and Anderson, 1982). Initially resuscitation of inocula, spread over cellulose acetate membranes overlaid on Minerals Modified Glutamate agar (MMGA), allows recovery of injured cells and also elimination of substances inhibitory to indole formation. The membrane is then transferred to TBA. Selectivity of TBA is achieved by a combination of bile salts and incubation at 44°C to suppress the majority of unwanted organisms. Specific identification of *Escherichia coli* relies on the demonstration of indole production by colonies formed on the membrane.

Composition (grams)

Panceatic digest of casein (tryptone)	20.0
Bile Salts No. 3 (Oxoid)	1.5
Agar	15.0
Distilled or deionized water	1000.0

Preparation

Add the ingredients to the water, soak and steam to dissolve agar. Adjust pH. Dispense in 100 ml amounts and sterilize by autoclaving at 121°C for 15 min.

Physical properties

Appearance	Clear straw.
pH	7.2 ± 0.2

Shelf life

Ready to use medium	4 days at 4°C or 1 day at 20°C.

Inoculation method for samples

Molten cooled media is poured in 12–15 ml amounts into 9 cm diameter Petri dishes. Surface moisture is removed by drying at 50°C immediately prior to use.

The inoculum (0.1–1.0 ml) adsorbed into the cellulose acetate membrane (pore size 450–1200 nm) overlayed on appropriately dried MMGA is incubated at 37°C for 4 h. The membrane is then transferred to the TBA; avoid trapping air bubbles, do *not* use a spreader.

Incubation method

At 44 ± 0.5°C for 18–24 h, in air, with membrane surface uppermost. Dishes in stacks of three or less.

Reading of results and interpretation

Membranes are removed from the agar surface into a Petri dish lid, containing 2 ml of Vracko and Sherris (1963) indole reagent, so that the whole of the lower surface is wetted. Remove excess reagent after 5 min and develop reaction in bright sunlight or under U.V. lamp. Count pink indole positive colonies.

Quality assessment

(i) *Productivity*

Test strain	*Escherichia coli* NCIMB 50035
Inoculation method	Dilution to extinction. Decimal dilutions in 0.1% peptone water from overnight culture in nutrient broth are prepared and 0.1 ml spread over membrane surface on MMGA, then to TBA as above.
Criteria	Following the resuscitation procedure this fastidious strain of *Escherichia coli* should give counts within 0.7 $\log_{10}$ of counts on Heart Infusion agar incubated at 37°C for 24 h. Colonies should be well coloured by the indole reaction.

(ii) *Selectivity*

Test strains	*Klebsiella oxytoca* NCIMB 50038 *Pseudomonas aeruginosa* NCIMB 50067 *Staphylococcus aureus* NCIMB 50080
Inoculation method	As above to give c. 200 colonies on membrane.
Criteria	Following resuscitation procedure, inhibition of growth

to give a negative indole reaction. The *Klebsiella oxytoca* strain may form small colonies at 44°C. It will not produce indole.

(iii) *Characteristic appearance of colonies*
Escherichia coli – from 1 to 3 mm diameter, depending on density, entire edge, pink stained on membrane by indole reaction.

Other Enterobacteriaceae – colourless to pale brown, 0.5 to 2 mm diameter, or completely inhibited.

References

Anderson, J.M. and Baird-Parker, A.C. (1975) A rapid direct plate method for enumerating *Escherichia coli* biotype I in foods. J. Appl. Bacteriol. 39, 111–117.

Holbrook, R. and Anderson, J.M. (1982) The rapid enumeration of *Escherichia coli* in foods by using a direct plating method. Society for Applied Bacteriology Technical Series No. 17, edited by J.E.L. Corry, D. Roberts and F.A. Skinner, Academic Press, London.

Vracko, R. and Sherris, J.C. (1963) Indole spot test in bacteriology. Amer. J. Clin. Pathol. 39, 429–432.

Tryptone Soya Broth with 10% NaCl & 1% Sodium Pyruvate (PTSBS)

This monograph has been assessed by members of the IUMS-ICFMH Working Party on Culture Media and given 'Approved' status.

Description and history

Tryptone soya broth with 10% sodium chloride and 1% sodium pyruvate (PTSBS) was developed by Lancette et al. (1986) for the recovery of low numbers of stressed *Staphylococcus aureus* in foods with a large population of competing bacteria. The addition of pyruvate to the medium allows for the growth and recovery of *Staphylococcus aureus* cells injured or stressed by sublethal heating, freezing or drying. PTSBS can be used for the qualitative, or quantitative (MPN) detection of low numbers of *Staphylococcus aureus* in foods (Lancette and Lanier, 1987). For more than 100 cfu/g of *Staphylococcus aureus* in foods, it is easier to count the bacteria by direct plating on Baird-Parker agar.

Composition (grams)

Tryptone soya broth	30.0
Sodium chloride	95.0
Sodium pyruvate	10.0
Distilled or deionized water	1000.0

Preparation

Dissolve all the ingredients, mix and adjust the pH to 7.3 if necessary. Dispense 10 ml to 16 × 150 mm tubes and sterilize the tubes for 15 min at 121°C.

Physical properties

Appearance	Light yellow broth.
pH	7.3 ± 0.2

Shelf life

Ready to use medium	30 days at 4 ± 2°C.

Inoculation method for samples

Use food macerates or decimal dilutions and inoculate 1 ml into the medium. MPN procedures need at least three tubes for at least three dilution steps.

Incubation method

At 35–37°C for 48 h, in air.

Reading of results and interpretation

Observe each PTSBS tube for signs of bacterial growth. To confirm the presence of *Staphylococcus aureus,* mix the contents of the positive tubes and streak a loopful of the growth on Baird-Parker agar. Only tubes from which confirmed *Staphylococcus aureus* are recovered are considered positive.

Quality assessment

(i) *Productivity*

Test strain	*Staphylococcus aureus* NCIMB 50080
Inoculation method	Dilution to extinction.

(ii) *Selectivity*

Test strains	*Enterococcus faecalis* NCIMB 50030 *Escherichia coli* NCIMB 50034
Supplementary strain	*Listeria monocytogenes* serovar 1/2a (ATCC 35152/ NCTC 7973)
Inoculation method	Dilution to extinction.

References

Lancette, G.A. (1986) Current resuscitation methods for recovery of stressed *Staphylococcus aureus* cells from foods. J. Food. Protect. 49, 477–481.

Lancette, G.A. and Lanier, J. (1987) Most probable number method for isolation and enumeration of *Staphylococcus aureus* in foods: Collaborative study. J. Assoc. Off. Anal. Chem. 70, 35–39.

Lancette, G.A., Peeler, J.T. and Lanier, J.M. (1986) Evaluation of an improved MPN medium for recovery of stressed and non-stressed *Staphylococcus aureus*. J. Assoc. Off. Anal. Chem. 69, 44–46.

Tryptose Sulphite Cycloserine (TSC) agar (without egg yolk)

This monograph has been assessed by members of the IUMS-ICFMH Working Party on Culture Media and given 'Approved' status.

Description and history

The medium utilizes the selective inhibitory properties of D-cycloserine and an indicator system involving sulphite and ferric iron. Most unwanted organisms are suppressed, while *Clostridium perfringens* and related species will reduce the sulphite and form black colonies due to the production of ferrous sulphide. Originally the medium was used with the addition of egg yolk (Harmon et al., 1971) but the egg yolk-free modification is more convenient (Hauschild and Hilsheimer, 1974a, b).

The medium without D-cycloserine is also known as Iron Sulphite agar (q.v.) and used for the enumeration of mesophilic sulphite-reducing clostridia.

Composition (grams)

Tryptose	15.0
Soytone	5.0
Yeast extract	5.0
Sodium metabisulphite	1.0
Iron (III) ammonium citrate	1.0
D-Cycloserine	0.4
Agar	15.0
Distilled or deionized water	1000.0

Preparation

Suspend ingredients other than D-cycloserine in 960 ml water. Bring to the boil to dissolve completely. Dispense 96 ml amounts into bottles (or final containers) and sterilize by autoclaving at 121°C for 10 min. After cooling to 50–52°C, add 4 ml of 1% filter-sterilized D-cycloserine.

Note: 4-methylumbelliferyl-phosphate di-sodium salt (0.1 g/l) may be added to the medium at the same time as the D-cycloserine.

Physical properties

Appearance	Pale straw.
pH	7.6 $\pm$ 0.2

Shelf life

Ready to use medium	Day of preparation.

Inoculation method for samples

Pour plate procedure, using 1 ml per 9 cm diameter plate. When set, plates are overlaid with sterile medium.

Incubation method

At 37°C for 18–24 h, anaerobically.

Reading of results and interpretation

Discrete black colonies of 1–5 mm diameter are considered to be presumptive *Clostridium perfringens*. Some other *Clostridia* spp. are able to grow on TSC and confirmation is necessary. However, acid phosphatase production is a highly specific indicator of *Clostridium perfringens* and is indicated by fluorescence under long wave U-V light if the fluorogenic substrate was added to the medium.

Quality assessment

(i) *Productivity*

Test strains	*Clostridium perfringens* (haemolytic) NCIMB 50027 *Clostridium perfringens* (non-haemolytic) NCIMB 50028
Inoculation method	Modified Miles-Misra, spiral plating or streaking method using surface inoculation of pre-poured plates with subsequent overlay.
Criteria	Recovery on TSC should be within 0.7 $\log_{10}$ of the recovery on Reinforced Clostridial Medium after 18–24 h at 37°C in an anaerobic jar.

(ii) *Selectivity*

Test strains	*Proteus mirabilis* (ATCC 29906/CECT 4168/NCTC 11938) *Enterococcus faecium* NCIMB 50032

Inoculation method	As (i) above
Criteria	Recovery on TSC (cf. above) should be less than 5 $\log_{10}$ of the recovery on Reinforced Clostridial Medium after 18–24 h at 37°C in an anaerobic jar.

(iii) *Characteristic appearance of colonies*
Discrete black colonies of 1–5 mm diameter surrounded by fluorescence if the medium was prepared with 4-methylumbelliferyl phosphate.

References

Harmon, S.M., Kautter, D.A. and Peeler, J.T. (1971) Improved medium for enumeration of *Clostridium perfringens*. Appl. Microbiol. 22, 688–692.

Hauschild, A.H.W. and Hilsheimer, R. (1974a) Evaluation and modification of media for enumeration of *Clostridium perfringens*. Appl. Microbiol. 27, 78–82.

Hauschild, A.H.W. and Hilsheimer, R. (1974b) Enumeration of food-borne *Clostridium perfringens* in egg yolk-free tryptose-sulphite-cycloserine agar. Appl. Microbiol. 27, 521–526.

Mead, G.C., Adams, B.W., Roberts T.A. and Smart, J.L. (1982) Isolation and enumeration of *Clostridium perfringens*. *In:* Isolation and identification methods for food poisoning organisms, edited by J.E.L. Corry, D. Roberts and F.A. Skinner. SAB Technical Series No. 17 Academic Press, London, pp. 99–110.

Mossel, D.A.A. and Pouw, H. (1973) Studies on the suitability of sulphite cycloserine agar for the enumeration of *Clostridium perfringens* in food and water. Zbl. Bakteriol. Hyg. I. Abt. Orig. A 223, 559–561.

University of Vermont (UVM) broths I & II

This monograph has been reviewed by members of the IUMS-ICFMH Working Party on Culture Media and given 'Proposed' status.

Description and history

UVM (University of Vermont) medium was originally reported by Donnelly and Baigent (1986) as a modification of the Listeria Enrichment Broth of Dominguez-Rodriguez et al. (1984). This medium was more recently modified to its present form (McClain and Lee, 1988) for use in the official United States Department of Agriculture Food Safety Inspection Service (FSIS) method for meats and poultry. The difference between UVM broths I and II is simply the inclusion of a higher concentration of acriflavine in the latter (McClain and Lee, 1988). Further work has shown the effectiveness of the substitution of UVM II with modified Fraser broth in the modified USDA-FSIS enrichment method (McClain and Lee, 1989; Warburton et al., 1991).

Composition (grams)

Proteose peptone (Difco)	5.0
Panceatic digest of casein (Difco)	5.0
Yeast Extract	5.0
Lab Lemco Powder (Oxoid)	5.0
Sodium chloride	20.0
Di-sodium hydrogen orthophosphate $\cdot 2H_2O$	12.0
Potassium di-hydrogen orthophosphate	1.35
Aesculin (Sigma)	1.0
Nalidixic Acid (Sigma)	0.02
Acriflavine HCl (Sigma)	0.012[a] or 0.025[b]
Distilled or deionized water	960.0

[a] UVM I
[b] UVM II

Preparation

Base

All ingredients except aesculin, nalidixic acid, and acriflavine HCl are combined with 960 ml of water and autoclaved at 121°C for 15 min.

Additional solutions

Prepare a 0.25% stock solution of acriflavine HCl (0.25 g acriflavine HCl in 100 ml distilled water) and a 5% stock solution of aesculin (5 g aesculin in 100 ml distilled water) and sterilize both solutions by autoclaving at 121°C for 15 min. Prepare a 0.4% nalidixic acid solution (0.4 g nalidixic acid in 100 ml of 0.1 N NaOH and sterilize by filtration through a 0.22 μm filter.

Complete medium

Add 4.8 ml (0.012 g) of the acriflavine solution plus 5.2 ml of sterile distilled water (UVM I) or 10 ml of the acriflavine solution (UVM II) to the cooled base to give final concentrations of 0.012 g/l (UVM I) or 0.025 g/l (UVM II), respectively. Add 20 ml (1.0 g) of the sterile aesculin solution to the sterile base to give a final concentration of 1 g/l and 10 ml of the filtered nalidixic acid solution to give a final concentration of 0.04 g/l.

Physical properties

Appearance	Yellowish, opaque.
pH	7.2 ± 0.2

Shelf life

Ready to use medium	5–7 days at 4 ± 2°C.

Inoculation method for samples

Primary enrichment is accomplished by mixing 25 g of food sample with 225 ml of UVM I and blending or mixing for 2 min. The mixture is then incubated at 30°C. After 4 h incubation, 0.2 ml of the UVM I enrichment broth is spread on LPM agar to determine the presence of presumptive *Listeria monocytogenes* colonies. After 24 h incubation UVM I is again spread on LPM agar. Isolated colonies are then recovered by streaking onto the uninoculated half of the LPM plate (McClain and Lee, 1988). Also after 24 h incubation, 0.1 ml of UVM I is added to 10 ml UVM II and incubated for 24 h before the sample is again swabbed onto LPM as with UVM I.

Incubation method

30°C for 24 h for each enrichment step.

Quality assessment

(i) *Productivity*

Test strains	*Listeria monocytogenes* serovar 1/2a (ATCC 35152/

	NCTC 7973) *Listeria monocytogenes* serovar 4b (ATCC 13932/ NCTC 10527) *Listeria ivanovii* serovar 5 NCIMB 50095
Inoculation method	Dilution to extinction.
(ii) *Selectivity*	
Test strains	*Enterococcus faecalis* NCIMB 50030 *Proteus mirabilis* (ATCC 29906/CECT 4168/NCTC 11938)
Inoculation method	Dilution to extinction.

Comments

Selective properties of acriflavine may vary from lot to lot and manufacturer to manufacturer. Each new batch must be assayed in combination with other selective agents to be used in the medium to determine the optimum concentration for use, with regard to the efficiency of selectivity and absence of inhibition of *Listeria* spp. Storage of stock solutions of acriflavine for periods of more than one month is not recommended.

References

Dominguez-Rodriguez, L., Suarez-Fernandez, G., Fernandez-Garayzabal, J.F. and Rodriguez-Ferri, E. (1984) New methodology for the isolation of *Listeria* microorganisms from heavily contaminated environments. Environ. Microbiol. 47, 1188–1190.

Donnelly, C.W. and Baigent, G.K. (1986) Method for flow cytometric detection of *Listeria monocytogenes* in milk. Appl. Environ. Microbiol. 52, 689–695.

McClain, D. and Lee, W.H. (1988) Development of USDA-FSIS method for isolation of *Listeria monocytogenes* from raw meat and poultry. J. Assoc. Off. Anal. Chem. 71, 660–664.

McClain, D. and Lee, W.H. (1989) FSIS method for the isolation and identification of *Listeria monocytogenes* from processed meat and poultry products. Laboratory Communication No. 57 (revised). USDA-FSIS. Beltsville MD.

Warburton, D.W., Farber, J.M., Armstrong, A., Caldeira, R., Tiwari, N.P., Babiuck, T., Lacasse, P. and Read, S. (1991) A Canadian comparative study of modified versions of the 'FDA' and 'USDA' methods for the detection of *Listeria monocytogenes*. J. Food Protect. 54, 669–676.

Handbook of Culture Media for Food Microbiology, J.E.L. Corry et al. (Eds.)

Violet Red Bile Glucose (VRBG) Agar

This monograph has been assessed by members of the IUMS-ICFMH Working Party on Culture Media and given 'Approved' status.

Description and history

This medium was designed for the enumeration of Enterobacteriaceae (Mossel et al., 1978). It relies on the use of the selective inhibitory components crystal violet and bile salts and the indicator system glucose and neutral red. Thus growth of many unwanted organisms is suppressed, while sought bacteria will dissimilate glucose and produce purple zones around the colonies. Some other related Gram negative bacteria may grow but may be suppressed by the overlay procedure. The selectivity of the medium can also be increased by incubation under anaerobic conditions and/or at elevated temperature, i.e. equal to or above 42°C (Mossel et al., 1986).

Composition (grams)

Peptone	7.0
Yeast extract	3.0
Sodium chloride	5.0
Glucose	10.0
Bile salts No. 3	1.5
Crystal violet	0.002
Neutral red	0.03
Agar	15.0
Distilled or deionized water	1000.0

Preparation

Suspend the ingredients in the water. Bring to the boil to dissolve completely. Further heating is neither necessary nor desirable. Mix well and dispense into bottles or tubes.

Physical properties

Appearance	Light purple-violet.
pH	7.4 ± 0.2

Shelf life

Ready to use plated medium 5 days at 4 ± 2°C.

Inoculation method for samples

Spread drop plate or poured plate procedure with or without overlay. Medium for the pour plate method should be freshly prepared, tempered to 47°C and used within 3 h. The medium can also be used as an overlayer for spread-drop plates of a rich primary plating medium, used for allowing resuscitation of sublethally stressed populations of Enterobacteriaceae. Stab inoculation procedures may also be carried out using this medium.

Incubation method

To detect mesophilic members of the Enterobacteriaceae, incubate at 30°C for 24 h. Psychrotrophic Enterobacteriaceae can be detected by incubation at 4°C for 10 days. Increased selectivity may be obtained by incubation at 42°C.

Reading of results and interpretation

Dark purple colonies, 1–2 mm diameter surrounded by purple haloes are included in the presumptive count. Confirmation may be obtained by the use of 'Gram-negative diagnostic tubes' (Mossel et al., 1977).

Quality assessment

(i) *Productivity*

Test strains	*Escherichia coli* NCIMB 50034 *Hafnia alvei* NCIMB 50037
Inoculation method	Modified Miles-Misra, spiral plating or streaking/ecometry.

(ii) *Selectivity*

Test strains	*Enterococcus faecalis* NCIMB 50030 *Staphylococcus aureus* NCIMB50080
Inoculation method	Modified Miles-Misra, spiral plating or steaking/ecometry.

(iii) *Characteristic appearance of colonies*

Round, purple, 1–2 mm diameter surrounded by purple haloes.

References

Mossel, D.A.A., Eelderink, I. and Sutherland, J.P. (1977) Development and use of single, "polytropic" diagnostic tubes for the approximate taxonomic grouping of bacteria, isolated from foods, water and medicinal preparations. Zbl. Bakt. Hyg. I, Orig. A, 278, 66–79.

Mossel, D.A.A., Eelderink, I., Koopmans, M. and van Rossem, F. (1978) Optimalisation of a MacConkey-type medium for the enumeration of Enterobacteriaceae. Lab. Pract. 27, 1049–1050.

Mossel, D.A.A., Eelderink, I., Koopmans M. and van Rossem, F. (1979) Influence of carbon source, bile salts and incubation temperature on the recovery of Enterobacteriaceae from foods using MacConkey type agars. J. Food Protect. 42, 470–475.

Mossel, D.A.A., van der Zee, H., Hardon A.P. and van Netten, P. (1986) The enumeration of thermotropic types amongst the Enterobacteriaceae colonizing perishable foods. J. Appl. Bacteriol. 60, 289–295.

Violet Red Bile (VRB) agar (syn. violet red bile lactose agar)

This monograph has been assessed by members of the IUMS-ICFMH Working Party on Culture Media and given 'Approved' status.

Description and history

This medium, designed for the enumeration of bacteria of the coli-aerogenes group is derived from MacConkey's original formula (1905). It relies on the use of the selective inhibitory components crystal violet and bile salts and the indicator system lactose and neutral red. Thus the growth of many unwanted organisms is suppressed, while tentative identification of sought bacteria can be made. Organisms which rapidly attack lactose produce purple colonies surrounded by purple haloes. Non-fermenters or late lactose fermenters produce pale colonies with greenish zones. Some other related Gram-negative bacteria may grow but may be suppressed by the overlay procedure. Selectivity can also be increased by incubation under anaerobic conditions and at elevated temperatures, i.e. > 42°C (Mossel and Vega, 1973).

Composition (grams)

Peptone	7.0
Yeast extract	3.0
Sodium chloride	5.0
Lactose	10.0
Bile salts No. 3	1.5
Crystal violet	0.002
Neutral red	0.03
Agar	15.0
Distilled or deionized water	1000.0

Preparation

Suspend the ingredients in the water. Bring to the boil to dissolve completely. Further heating is neither necessary nor desirable. Mix well and distribute into bottles or tubes.

Physical properties

Appearance	Light purple-violet, transparent.
pH	7.4 ± 0.2

Shelf life

Ready to use medium	5 days at 4 ± 2°C.

Inoculation method

Spread drop plate or poured plate procedure with or without overlay. Medium for the pour plate method should be freshly prepared, tempered to 47°C and used within 3 h. The medium can also be poured as an overlayer for spread drop plates of a rich primary plating medium used for allowing resuscitation of sublethally injured populations of Enterobacteriaceae, though discriminating between lactose +ve and lactose -ve types must then be done with care.

Incubation method

To detect members of the coli-aerogenes group count typical colonies after incubation at 35 or 37°C for 24 h. Where it is desired to count the full range of coliform bacteria, incubate at 30°C for 24 h and count all colonies present. Psychrotrophic coliforms can be detected by incubation at 4°C for 10 days. For *Escherichia coli* a temperature of 44 ± 0.1°C is specifically recommended.

Reading of results and interpretation

Dark purple colonies, 1–2 mm diameter surrounded by purple haloes and pale colonies with greenish zones are included in presumptive counts. Confirmation may be obtained by the use of 'Gram-negative diagnostic tubes' (Mossel et al., 1977).

Quality assessment

(i) *Productivity*

Test strain	*Escherichia coli* NCIMB 50034
Inoculation method	Modified Miles-Misra, spiral plating or streaking/ecometry.

(ii) *Selectivity*

Test strains	*Enterococcus faecalis* NCIMB 50030 *Staphylococcus aureus* NCIMB 50080
Inoculation method	Modified Miles-Misra, spiral plating or streaking/

ecometry.

(iii) *Characteristic appearance of colonies*
Round, purple l–2 mm diameter, surrounded by purple haloes (lactose +ve types), or pale with greenish zones (lactose -ve types).

References

MacConkey, A. (1905) Lactose fermenting bacteria in faeces. J. Hyg. 5, 333–379.

Mossel, D.A.A., and Vega, C.L. (1973) The direct enumeration of *Escherichia coli* in water using MacConkey's agar at 44°C in plastic pouches. Health Labor. Sci. 11, 303–307.

Mossel, D.A.A., Eelderink, I. and Sutherland, J.P. (1977) Development and use of single, "polytropic" diagnostic tubes for the approximate taxonomic grouping of bacteria, isolated from foods, water and medicinal preparations. Zbl. Bakt. Hyg. I., Orig., A, 278, 66–79.

Mossel, D.A.A., Eelderlink, I., Koopmans, M. and van Rossem, F. (1979) Influence of carbon source, bile salts and incubation temperature on the recovery of Enterobacteriaecae from foods using MacConkey type agars. J. Food Protect. 42, 470–475.

Mossel, D.A.A., van der Zee, H., Hardon, A.P. and van Netten, P. (1986) The enumeration of thermotropic types amongst the Enterobacteriaceae colonizing perishable foods. J. Appl. Bacteriol. 60, 289–295.

Xylose Lysine Deoxycholate (XLD) agar

This monograph has been assessed by members of the IUMS-ICFMH Working Party on Culture Media and given 'Approved' status.

Description and history

The medium was introduced by Taylor (1965) for the isolation of salmonellae and shigellae (Taylor and Schelhart, 1971). It relies on the use of the selective inhibitory component sodium deoxycholate and three indicator systems, i.e. xylose, lactose and sucrose combined with phenol red; lysine hydrochloride and again phenol red; sodium thiosulphate and iron. Thus the growth of many unwanted organisms is suppressed, while sought bacteria can be tentatively grouped by reading the net effect of carbohydrate dissimilation and lysine decarboxylation and the formation of hydrogen sulphide from thiosulphate, leading to black colonies.

Composition (grams)

Yeast extract	3.0
Sodium chloride	5.0
Xylose	3.75
Lactose	7.5
Sucrose	7.5
L-Lysine hydrochloride	5.0
Sodium thiosulphate	6.8
Iron (III) ammonium citrate	0.8
Phenol red	0.08
Sodium deoxycholate	1.0
Agar	15.0
Distilled or deionized water	1000.0

Preparation

Suspend the ingredients in the water. Heat, with frequent agitation, until the medium starts boiling. Avoid overheating. Transfer immediately to a water bath tempered at about 50°C, continue to agitate until the medium has cooled to approximately 50°C and pour into plates. Preparation of large volumes, requiring prolonged heating, is to be avoided.

Physical properties

Appearance	Light rose, transparent.
pH	7.4 ± 0.2

Shelf life

Ready to use medium	5 days at 4 ± 2°C in the dark.

Inoculation method for samples

Surface spreading over whole plate using 0.1 ml per 9 cm diameter plate and 0.5 ml per 14 cm diameter plate.

Incubation method

At 37 or 43°C for 18–24 h, in air, dependent on degree of selectivity to be attained.

Reading of results and interpretation

A tentative grouping of isolates can be made from colonial appearance on XLD. Roughly three groups of Enterobacteriaceae can be discriminated: (i) colonies with red zones and black centre: *Salmonella, Arizona*, and *Edwardsiella*; (ii) colonies with red zones and red centre: *Shigella, Providencia* and hydrogen sulphide negative salmonellae; (iii) colonies with yellow haloes and yellow centres: the genera *Escherichia, Enterobacter, Citrobacter, Kluyvera, Klebsiella, Hafnia, Serratia* and *Proteus*, and the species *Yersinia enterocolitica*. For confirmation of this presumptive evidence the usual biochemical tests must be carried out.

Quality assessment

(i) *Productivity*

Test strains	*Salmonella enteritidis* NCIMB 50073 *Salmonella virchow* NCIMB 50077 *Shigella flexneri* NCIMB 50079 *Shigella sonnei* (ATCC 29930)
Supplementary strain	*Salmonella typhimurium* NCIMB 50100
Inoculation method	Modified Miles-Misra, spiral plating or streaking/ ecometry.

(ii) *Selectivity*

Test strains	*Enterococcus faecalis* NCIMB 50030 *Proteus mirabilis* (ATCC 29906/CECT 4168/NCTC

	11938) – qualitative test for control of swarming
Supplementary strains	*Bacillus cereus* NCIMB 50014 *Hafnia alvei* NCIMB 50037 *Lactococcus lactis* ssp. *lactis* NCIMB 50058 *Micrococcus luteus* NCIMB 50063 *Staphylococcus aureus* NCIMB 50080
Inoculation method	Modified Miles-Misra, spiral plating or streaking/ ecometry.

(iii) *Characteristic appearance of colonies*
See above.

References

Mossel, D.A.A., Bonants-van Laarhoven, D.M.G., Ligtenberg-Merkus, A.M.Th. and Werdler, M.E.B. (1983) Quality assurance of selective culture media for bacteria, moulds and yeasts: an attempt at standardization at the international level. J. Appl. Bacteriol. 54, 313–327.

Taylor, W.I. (1965) Isolation of shigellae. I. Xylose lysine agars: new media for isolation of enteric pathogens. Amer. J. Clin. Pathol. 44, 471–475.

Taylor, W.I. and Schelhart, D. (1971) Isolation of shigellae. VIII. Comparison of xylose lysine deoxycholate agar, Hektoen enteric agar, Salmonella-Shigella-agar and eosin methylene blue agar with stool specimens. Appl. Microbiol. 21, 32–37.

Xylose Lysine Tergitol 4 (XLT4) agar

This monograph has been reviewed by members of the ICFMH-IUMS Working Party on Culture Media and given 'Proposed' status.

Description and history

This medium was developed by Miller et al. (1991) as a modification of Xylose Lysine Deoxycholate agar for the isolation of non-*typhi Salmonella* spp. The selectivity relies on the use of the detergent Niaproof 4 (7-ethyl-2-methyl-4-undecanol-hydrogen sulphate, sodium salt) formerly Tergitol 4. Its diagnostic properties are based on the use of three indicator systems, xylose, lactose and sucrose combined with phenol red; lysine HCl and again phenol red; sodium thiosulphate and iron. The low concentration of proteose peptone No. 3 incorporated in the medium produces blacker salmonella colonies in shorter incubation times (Miller et al., 1995). XLT4 inhibits many Gram negative bacteria and provides good colonial differentiation between *Salmonella* and *Citrobacter* spp.

Composition (grams)

Proteose peptone No. 3	1.2
Yeast extract*	3.0
Sodium chloride*	5.0
Xylose*	3.75
Lactose*	7.5
Sucrose*	7.5
L-Lysine hydrochloride*	5.0
Sodium thiosulphate	6.8
Iron (III) ammonium citrate	0.8
Phenol red*	0.08
Niaproof 4 (ml)	4.6
Agar* (15 g)	18.0
Distilled or deionized water	1000.0

Preparation

To one litre of water add 4.6 ml of Niaproof type 4 (Sigma N-1404). To this add 47 g of XL base (Difco 0555; ingredients marked *), 1.2 g proteose peptone No. 3 (Difco 0122), 0.8 g iron (III) ammonium citrate, 6.8 g sodium thiosulphate and 3 g agar. Mix for 5 min then heat with frequent agitation until the medium starts to boil then mix again for 5 min. Avoid overheating. Transfer immediately to a water bath at about 50°C, continue to agitate until the medium has cooled to approximately 50°C and pour into plates to a thickness of at least 4 mm. Preparing large volumes, requiring prolonged heating, is to be avoided.

Physical properties

Appearance	Light rose, transparent.
pH	7.3 ± 0.2

Shelf life

Ready to use medium	3 months at 4 ± 2°C.

Incubation method for samples

1. Direct counting
   Surface spreading over the whole plate using 0.1 ml of decimal dilutions of food homogenate per 9 cm diameter plate.

2. Subculturing after enrichment
   A loopful of an incubated selective broth is streaked onto the medium to attain isolated colonies.

Incubation method

At 35–37°C for 20–24 h, in air. Plates that appear salmonella negative are incubated for an additional 24 h.

Reading results and interpretation

On XLT4 agar, *Salmonella* spp. produce black or black-centered colonies (H_2S production), with a yellow or pink periphery. Other Gram negative bacteria are inhibited or may produce yellow or pink colonies without black coloration.

Quality assessment

(i) *Productivity*

Test strains	*Salmonella virchow* NCIMB 50077 *Salmonella enteritidis* NCIMB 50073
Inoculation method	Modified Miles-Misra, spiral plating or streaking/ecometry.

(ii) *Selectivity*

Test strains	*Proteus mirabilis* (ATCC 29906/CECT 4168/NCTC 11938) *Citrobacter freundii* NCIMB 50025
Inoculation method	Modified Miles-Misra, spiral plating or streaking/ecometry.

(iii) *Characteristic appearance of colonies*
See above.

References

Miller, R.G., Tate, C.R., Mallinson, E.T. and Scherrer, J.A. (1991) Xylose-Lysine-Tergitol 4: An improved selective agar medium for the isolation of *Salmonella*. Poult. Sci. 70, 2429–2432. *Erratum* (1992): Poult. Sci. 71, 398.

Miller, R.G., Tate, C.R. and Mallinson, E.T. (1995) Improved XLT4 agar: small addition of peptone to promote stronger production of hydrogen-sulfide by *Salmonellae*. J. Food Protect. 58, 115–119.

Appendix I

Testing methods for use in quality assurance of culture media

Note: A standard, based in part on some of these methods, is in the course of preparation by the Comité Européen de Normalisation (Anon., 1999).

Introduction

Seven testing methods are outlined in the following section. Familiarity with general microbiological techniques is assumed and therefore the methods are not given in exhaustive detail. Suitable test organisms are listed in the sections on individual media. The reference medium for bacteria is tryptone soya agar or broth unless another medium is specified in the monograph. The use of stationary phase cultures is suggested but the worker is at liberty to employ log phase cultures if a more exacting test is required and thought justified.

General criteria for assessment of bacteriological media are given in the Notes on the use of the monographs. Where these are inappropriate, specific criteria are given in individual monographs.

Definition of productivity ratios (PRs)

Where the method employed entails total counts these are expressed as N_s/N_o where N_s = count on test medium and N_o = count on reference medium.

Definition of selectivity factor (SF)

This expression of the degree of selectivity of unwanted strains on selective media tested quantitatively by either the modified Miles-Misra or dilution to extinction methods is calculated as the difference between the highest dilution showing growth on the reference medium (D_o) and the highest dilution showing comparable growth on the test medium (D_s). It is expressed as $\log_{10}$ (Anon., 1998).

Absolute and relative growth indices (AGI and RGI)

When using the ecometric streaking method, the AGI and RGI may be calculated as described by Mossel et al. (1983). Strains growing on all five streaks have an AGI of 5, those growing only on the first three streaks have an AGI of 3, and so on. The RGI for a given strain is the ratio of the AGI on the test medium to the AGI on the reference medium. The AGI and RGI are used when testing solid media by semi-quantitative techniques (see method B).

Methods

Plated media

A. *Modified Miles-Misra method*

This is a general method suitable for most bacteriological agars.

1. Using an aseptic technique prepare dilutions of a freshly raised and well mixed stationary phase culture of the test organism in saline peptone diluent.
2. Prepare plates according to the recommended formula with a minimum depth of agar of 4 mm and an adequately dried surface. Plates of the reference medium should be similarly prepared.
3. Using the same dropper for all the dilutions of one organism and commencing with the highest dilution to be tested, dispense four drops from each appropriate dilution onto the surface of four test and four reference plates. (Other dropping systems or a spiral plater may be used, see Corry, 1982.)
4. Spread each drop over 1/4 of the plate using a sterile spreader for each plate and starting at the highest dilution.
5. Incubate plates under conditions specified in individual monographs.
6. Assess results visually for numbers and appearance.

B. *Spiral plating method*

The spiral plater deposits liquid samples in an Archimedes spiral onto a solid agar surface achieving the equivalent of a 1000-fold dilution factor without the need for serial dilutions. Modern spiral platers offer the option to dispense variable sample volumes ranging from 10 to 400 μl. Depending on the chosen deposition volume it is thus possible to count colonies in the range 30 to 4×10^5 cfu/ml without the need to perform serial dilutions.

Agar plates should be prepared according to the manufacturer's recommendations. Repetitive and predictable deposition of a solution on the surface of a prepared plate is a primary requirement of spiral plating. Although the spiral plater, when properly adjusted, will deliver a precise volume from the stylus tip, the amount and distribution of the sample on the agar surface is influenced by the condition and character of the surface itself. This is especially apparent on areas of the plate where the volume dispensed is very low. Ideally the agar surface should be smooth, without wrinkles, pits or air bubbles. The agar should be of uniform depth over the entire plate and no free water droplets should be on the surface. There should be no areas of dehydration nor contamination on the plate. Plates of the reference medium should be similarly prepared.

1. Take a freshly raised well mixed stationary phase culture of the test organism(s).
2. Select the deposition or inoculum volume on the basis of the expected count of the culture and the recommendations of the manufacturer of the spiral plater as to restrictions on inoculum volume. For guidance, figures for the WASP plater (Don Whitley Scientific Ltd., Shipley, UK) are shown below.

Culture (cfu/ml)	Inoculum volume	Dilution of culture
10^6	10 µl	2-fold
10^6	50 µl	10-fold
10^5	50 µl	Not required
10^4	100 µl	Not required
10^3	200 µl	Not required

3. Incubate plates under conditions specified in the individual monographs.
4. Count the plates manually or using an appropriate image analysis system.

In order to explain the count limits it is necessary to understand the method of counting for spiral plates. All spiral plates are prepared in an identical manner whereby a fixed volume of sample is deposited on the surface of the agar at a rate decreasing from the centre to the edge, thus fixing the volume/area relationship for any portion of the plate. In a mode where 50 µl is dispensed, a dilution of 1000:1 is created across the plate which will produce a countable plate from any sample containing between 400 and 4×10^5 cfu/ml without the need for further dilution. After incubation colonies appear along the track made by the deposited sample, with spacing between colonies increasing from the centre to the edge. The viewing grid for counting spiral plates and a typical inoculated plate are shown below (Figs. 1 and 2). The major divisions on this grid are the concentric circles and the eight pie-shaped wedges or sectors which result in a number of annular segments.

Counting is done by placing the Petri dish over the grid and counting along the pie-shaped wedge, from the outside towards the centre, until at least 20 colonies are counted then completing the count in the segment in which the twentieth colony was found. The opposite segment is then counted. The sum of these two counts, divided by the known segment pair volume, gives the concentration.

The requirement for a minimum count of 20 is explained by reference to the co-efficient of variation (CV) applied to the counts. This is the standard deviation of replicate counts from separate plates divided by the mean counts and expressed as a percentage. The CV for spiral plate counts in three ranges (0–1, 6–20 and 21–200 cfu) are:

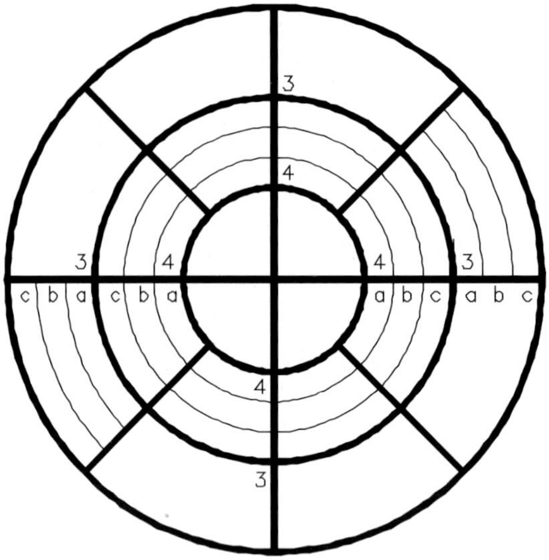

Fig.1.

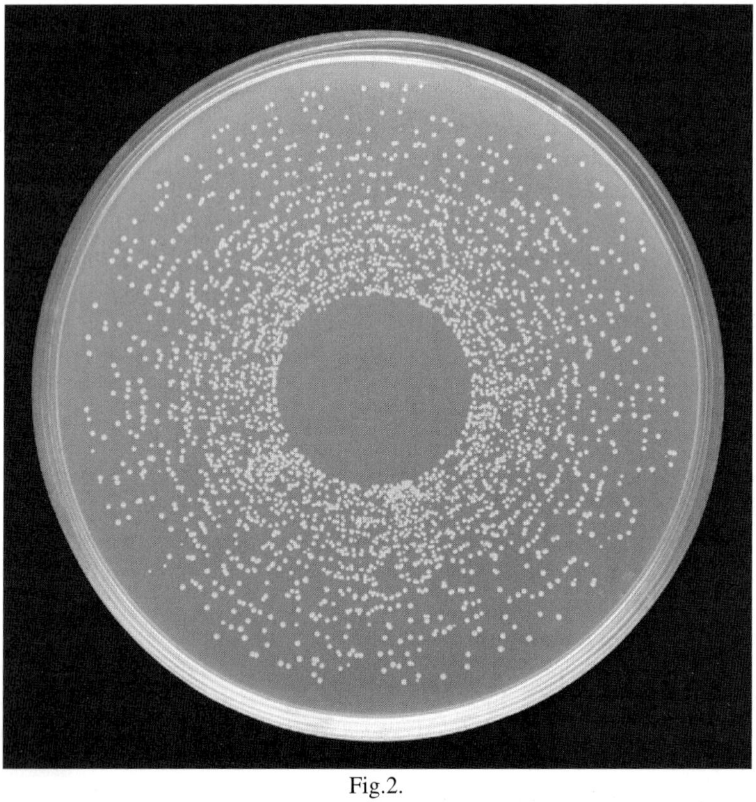

Fig.2.

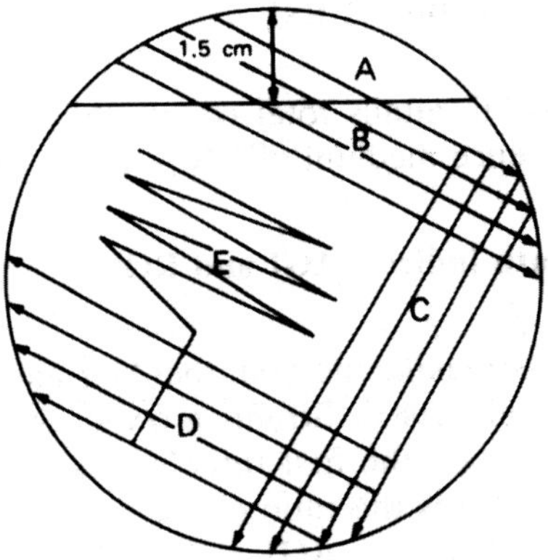

Fig.3.

Colonies counted per plate	CV (%)
0–5	78
6–20	25
21–200	15

Comparing these figures it can be seen that until the number of colonies counted reaches 20, the accuracy of the count is low. Above 20, accuracy compares favourably with the standard plate count.

An upper limit of 75 colonies in each sector is set as, above this number, the accuracy of the count will be low due to coincidence error associated with the crowding of colonies. However, to count 75 colonies in a sector assumes regular-sized 1–2 mm colonies and would not apply with large irregular colnies which will overlap at lower densities.

5. Counts and colony appearance should be recorded and compared to the reference medium.

C. *Streaking method (Ecometry)*

This may be a suitable alternative to method A when quantitative results are not required. Experience may show the need for initial dilution of the stationary phase culture for method (ii) below. This may be performed according to ISO 6887 (1983).

1. Take a freshly raised stationary phase culture in liquid medium.
2. Prepare plates as in A2.
3. *Either* (i) Using a standard loop streak plates of test and reference media according to the ecometric method of Mossel et al. (1983) *or*: (ii) Deposit a standard loopful (5 µl) of culture over sector A of the agar surface (Fig. 3.) Draw four parallel lines approximately 5 mm apart across the area of the spread culture in sector A and across sector B. Repeat for sectors C and D. Finally, draw a single line across sector D at right angles to the lines of inoculation and continue streak-

ing into sector E. It is essential to flame and cool the loop in between each sector in order to obtain well isolated colonies. Repeat on reference medium.

4. Incubate plates under conditions specified in individual medium protocols.
5. Assess results visually for numbers and appearance of colonies.
   (a) where the ecometric method has been used the absolute and relative growth indices may be calculated as described by Mossel et al. (1983).
   (b) a similar method of assessment can be employed for the streaking method or each sector may be scored separately with a minimum score of 1 for well spaced discrete colonies and a maximum of 4 for confluent growth along the lines of inoculation. The sum of the score for each sector may then be calculated and test and reference agars compared.

D. *Stab inoculation method*

This technique is suitable for assessing the performance of mycological media.

1. Prepare a heavy spore suspension from a 7–10 day old culture of the test organism by mixing a loopful of growth in 0.5 ml of 0.25% agar.
2. Prepare plates according to the recommended formula with a minimum depth of agar of 4 mm and an adequately dried surface.
3. Take three test plates and stab inoculate the medium in each plate from below to a depth of 1–2 mm with the wire loaded with spore suspension. The point of inoculation should be approximately midway between the rim and the centre of the plate. Sterilize the wire and reload with suspension for each inoculation.
4. Repeat operation 3 with three more test organisms, stab inoculating the same plates on the remaining three quadrants.
5. If required, use a further three plates and inoculate in a similar manner for other test organisms.
6. Incubate plates, inverted, under conditions specified in individual monographs and measure the diameter of each colony at daily intervals.
7. Plot the average colony diameter against time for each test organism and calculate the growth rate. Alternatively, determine the mean colony diameter on one occasion after a suitable incubation period.

Liquid media

The interactions leading to the successful isolation of organisms from liquid selective media are complex (Mackey, 1985; Beckers et al., 1987), hence devising quality control methods is less straightforward than with solid media.

Often, as in the isolation of salmonellae, selective enrichment in liquid medium is only one step in a complex isolation procedure. The performance of the entire system can be monitored by testing for the recovery of low numbers of a marked strain inoculated into appropriate sample material (Beckers et al., 1985). However it may also be desirable to test the performance of individual media. For this purpose the following methods using pure cultures of standard test strains are proposed.

E. *Assessment of productivity and selectivity of liquid selective media by dilution to extinction.* (Official French dilution method, Richard, 1982; Mossel et al., 1974).

1. Prepare stationary phase cultures of test strains.
2. Dilute 10^{-1} to 10^{-12} in tryptone soya broth. (If it is desired to count the inoculum, aliquots (0.1 ml) of appropriate dilutions should be spread over the surface of a suitable plated medium).
3. Add 1 ml volumes to 9 ml selective enrichment medium and a non-selective broth. (Other volumes may be found to be more suitable).
4. Incubate at the appropriate temperature for 18–24h.
5. Record highest dilutions showing turbidity. (Count the spread plates and calculate the smallest inoculum giving rise to growth in the test and reference broths).

F. *Assessment of the performance of liquid selective media with mixed cultures of wanted and unwanted organisms*

This method was devised for testing enrichment media for salmonellae. When used for this purpose the nalidixic acid resistant strain of *Salmonella typhimurium* (NCTC 12190) should be used as the wanted organism and plates of tryptone soya agar or other non-selective agar containing 20 μg/ml nalidixic acid prepared for use as described in sections 4 and 7 below.

1. Prepare stationary phase cultures of test strains (suitable unwanted strains for salmonella enrichment broths are *Citrobacter freundii* 50025, *Escherichia coli* 50034, *Proteus mirabilis* (ATCC 29906/CECT 4168/NCTC 11938), *Pseudomonas aeruginosa* 50067.
2. Mix together 1 ml volumes of the unwanted organisms and dilute to 10^{-1} in peptone saline.
3. Dilute the wanted organism 10^{-1} to 10^{-9} in peptone saline.
4. Estimate the viable count of the wanted organism by surface plating on tryptone soya agar (with and without nalidixic acid).
5. To determine the ability of the enrichment broth to recover small numbers of the wanted organism, inoculate 1 ml from each peptone saline dilution into 9 ml enrichment broth, incubate at the apropriate temperature for 18–24 h then streak onto plates of tryptone soya agar.
6. Examine tubes and plates from 4 and 5 above and, from the viable count (steps 3 and 4), calculate the minimum number of organisms required to initiate growth in the enrichment broth. Check that growth of the nalidixic acid resistant strain is not inhibited on the nalidixic acid containing medium.
7. To determine the ability of enrichment broth to recover small numbers of the wanted organism from mixed cultures, inoculate 1 ml from each peptone saline dilution (step 3) into a separate tube containing 10 ml enrichment broth and 0.02 ml of the 10^{-1} dilution of unwanted organisms (step 2). Incubate at the appropri-

ate temperature for 18–24 h then streak onto plates of tryptone soya agar with and without nalidixic acid and, if required, other selective agars (e.g. SS, XLD).
8. Examine plates from 7 above for typical colonies of the wanted strain. Record the highest dilution from which the wanted strain could be detected.
9. Interpretation of results.
i) *Productivity.* Liquid selective media should support the growth of a pure culture of the wanted organism from an inoculum of 20 or less. By plating on selective and non-selective agars any interaction between liquid and solid selective media may also be detected.
ii) *Selectivity.* Growth of the wanted strain in enrichment broth containing competing (unwanted) organisms should be detectable within one tenfold dilution of the highest dilution supporting growth of the pure culture.

G. *Method of assessing growth rate in broth*

This is intended as a general method for use with non-selective or selective (enrichment) liquid media. In the latter case it might also be necessary to add sample material comparable to that to be tested.

1. Prepare stationary phase culture, mix well and dilute in saline peptone water at room temperature to an approximate concentration of 10^5 cfu/ml.
2. Distribute 10 ml quantities of reference and test broths (see individual monographs) into 25–30 ml screw-capped containers and sterilize. Equilibrate broths to incubation temperature.
3. Inoculate a pair of broths for each test organism with 0.1 ml of the 10^5 cfu/ml suspension and mix.
4. Remove 0.1 ml of the broth to a non-inhibitory agar plate and spread over half of the surface. Repeat on the other half of the plate with 0.1 ml from the reference broth.
5. Incubate broths and inoculated plates under required conditions
6. At suitable time intervals remove samples from each broth with a standard loop or pipette. Dilute if necessary and spread over a quarter of a plate of the same medium as used in 4.
7. Results may be assessed visually and may be plotted as counts vs. time.

References

Anon. (1998) A new expression for selectivity of liquid and solid media – statement from the IUMS-ICFMH working party on culture media. Int. J. Food Microbiol. 45, 65.

Anon. (1999) Microbiology of food and animal feeding stuffs. Guideline on quality assurance and performance testing of culture media. ENV ISO/TR 11133, AFNOR, Paris.

Beckers, H.J., van Leusden, F.M., Meijssen, M.J.M. and Kampelmacher, E.H. (1985) Reference material for the evaluation of a standard method for the detection of salmonellas in food and feeding stuffs. J. Appl. Bacteriol. 59, 507–512.

Beckers, H.J., Heide, J.V.D., Fenigsen-Narucka, U. and Peters, R. (1987) Fate of salmonellas and competing flora in meat sample enrichments in buffered peptone water and in Muller-Kauffmann's tetrathionate medium. J. Appl. Bacteriol. 62, 97–104.

Corry, J.E.L. (1982) Quality assessment of culture media by the Miles-Misra method. *In:* Quality assurance and quality control of microbiological culture media, edited by J.E.L.Corry, G.I.T.-Verlag, Darmstadt, pp. 21–37.

ISO 6887 (1983) Microbiology – General guidance for the preparation of dilutions for microbiological examinations, International Standards Organisation, Geneva.

Mackey, B.M. (1985) Quality control monitoring of liquid selective enrichment media used for isolating salmonellae. Int. J. Food Microbiol. 2, 41–48.

Mossel, D.A.A., Harrewijn, G.A. and Nesselrooy-van Zadelhoff, C.F.M. (1974) Standardisation of the selective inhibitory effect of surface active compounds used in media for the detection of Enterobacteriaceae in foods and water. Health Lab. Sci.11, 260–267.

Mossel, D.A.A., Bonants-van Laarhoven, T.M.G., Ligtenberg-Merkus A.M.T. and Werdler, M.E.B. (1983) Quality assurance of selective culture media for bacteria, moulds and yeasts: an attempt at standardisation on the international level. J. Appl. Bacteriol. 54, 313–327.

Richard, N. (1982) Monitoring the quality of selective liquid media using the official French dilution technique for the bacteriological examination of food. *In:* Quality assurance and quality control of microbiological culture media, edited by J.E.L.Corry, G.I.T.-Verlag, Darmstadt, pp. 51–57.

Appendix II

Test strains

Where a strain is known to be in more than one collection, all known reference numbers are given. Absence of a reference number for a particular collection does not necessarily imply absence of the strain from that collection. Organisms appearing in these monographs only as supplementary strains are marked with an asterisk (*).

Abbreviation key:

A-NRRL	USDA Northern Utilization Research and Development Division, Peoria, Illinois, U.S.A. *e-mail: info@ncaur.usda.gov*
ATCC	American Type Culture Collection, 10801 University Boulevard, Manassess, VA 20110–2209, U.S.A. *e-mail: news@atcc.org*
BUCSAV	Biologiscky Ustav, Ceskoslovenska Akademie Ved (Institute of Biology, Czechoslovak Academy of Sciences), Prague, Czech Republic. *e-mail: wdcm.nig.ac.jp/msdn/ccfi*
CBS	Centraalbureau voor Schimmelcultures, Uppsalalaan 8, 3584 Utrecht, The Netherlands. *e-mail: info@cbs.knaw.nl*
CCM	Czechoslovak Collection of Microorganisms, Masaryk University, Tvrdeho 14, 602 00 Brno, Czechoslovakia. *e-mail: ccm@sci.muni.cz*
CCUG	CCUG Department of Clinical Bacteriology, University of Goteborg, Mikrobiologen, Guldhedsgaten 10, 6tr, S-413 46, Goteborg, Sweden. *e-mail: ccugef@ccug.gu.se*
CECT	Coleccion Espanola de Cultivos Tipo, Universidad de Valencia, Edificio de Investigacion, Campus de Burjassot, 46100 Burjassot, Valencia, Spain. *e-mail: cect@uv.es*
CIP	Institute Pasteur CIP, BP 52, 25 rue du Docteur Roux, 75724 Paris Cedex 15, France. *e-mail: www@pasteur.fr*
CSIRO	Commonwealth Scientific and Industrial Research Organisation, PO

	Box 52, North Ride, New South Wales, 2113, Australia. *e-mail: csiro.au/*
DSM	Deutsche Sammlung von Mikroorganismen und Zellkulturen GmbH, (German Collection of Microorganisms and Cell Cultures), Mascheroder Weg 1b, D-38124 Braunschweig, Germany. *e-mail: help@dsmz.de*
HNCMB	Hungarian National Collection of Medical Bacteria, National Institute of Hygiene, Gyali ut 2–6, H-l097 Budapest, Hungary. *e-mail: biroe@budapest.hu*
IAM	Institute of Applied Microbiology Culture Collection, Centre for Cellular and Molecular Research, Institute of Molecular and Cellular Biosciences, University of Tokyo, Yayoi 1-1-1, Bunkyo-ku, Tokyo 113-0032, Japan. *e-mail: iamcc@iam.u-tokyo.ac.jp*
IFO	Institute for Fermentation, 17-85 Juso-honmachi, 2-chome, Yadogawa-ku, Osaka 532-8686, Japan. *e-mail: ifo@mb.infoweb.ne.jp*
IMI	CABI Bioscience UK Centre (formerly Imperial Mycological Institute), Bakeham Lane, EGHAM, Surrey, TW20 9TY, UK. *e-mail: d.smith@cabi.org*
JCM	Japan Collection of Microorganisms, RIKEN (The Institute of Physical and Chemical Research), 2-1 Hirosawa, Wako, Saitama 351-0198, Japan. *e-mail: curator@jcm.riken.go.jp*
NCDO/ NCFB	National Collection of Food Bacteria (formerly National Collection of Dairy Organisms). This collection has now been merged with NCIMB.
NCIMB	National Collections of Industrial, Food and Marine Bacteria Ltd., 23, St. Machar Drive, Aberdeen, AB24 3RY, U.K. *e-mail: www.enquiries@ncimb.co.uk*
NCTC	National Collection of Type Cultures, Central Public Health Laboratory, 61 Colindale Avenue, London, NW9 5HT, U.K. *e-mail: www.bholmes@phls.nhs.uk*
NCYC	National Collection of Yeast Cultures, Institute of Food Research, Norwich Laboratory, Colney Lane, Norwich NR4 7UA, U.K. *e-mail: www.ian.roberts@bbsrc.ac.uk*

Bacteria	NCIMB QC Collection	ATCC	CCM	CCUG	CECT	CIP	DSM	NCIMB	NCTC	Other
Acinetobacter sp.	50 011							10 764		
**Aerococcus viridans*	50 012	11 563	1 914	4 311	978	54.145	20 340	11 775	8 251	HNCMB 86 001; IFO 12 219
Aeromonas caviae		15 468		25 939		76.16	30 190	13 016		
Aeromonas hydrophila	50 013	7 966		14 551		76.14	30 157	9 240	8 049	JCM 1 027
Aeromonas sobria		35 993		14 550					11 215	
Bacillus cereus	50 014	11 778	869	10 781	193	64.52	345	8 012	10 320	HNCMB 100 003
**Bacillus licheniformis*	50 016	14 580	2 145	7 422	20	52.71	13	9 375	10 341	IFO 12 200; JCM 2 505
Bacillus subtilis	50 018							13 061	5 398	
Bacillus stearothermophilus		10 149								
Brochothrix thermosphacta	50 019	11 509		35 132	847	103251	20 171	10 018	10 822	
Campylobacter coli	50 092			11 283		7 0.80	4 689		11 366	
Campylobacter jejuni	50 091			6 824					11 168	
Campylobacter upsaliensis		43 953							11 540	
Carnobacterium divergens	50 021	35 677	4 117	30 094	4 016	101029	20 623	11 952		JCM 5 816
Carnobacterium piscicola	50 020						20 624	13 093		
Citrobacter freundii	50 025							8 645	6 272	
**Citrobacter freundii*		6 750					30 040	8 173		
Clostridium bifermentans	50 026				550			506	506	
Clostridium perfringens (haem)	50 027	13 124		1 795	376	103409	756	6 125	8 237	JCM 1 290
Clostridium perfringens (non-haem)	50 028	12 916						13 079	8 238	
Clostridium sporogenes	50 099	19 404				79.03	1 664	532	532	
Desulfotomaculum nigrificans		19 858					574	8 395		IFO 13 698
Enterobacter cloacae		23 355								
**Enterobacter cloacae*	50 122	13 047		6 323		60.85	30 054	10 101	10 005	IFO 13 535; JCM 1 232
Enterococcus faecalis	50 030				4 176			8 260	8 213	
Enterococcus faecalis		51 299				104676				
Enterococcus faecalis	50 112	29 212				103214	2 570			JCM 2 875
Enterococcus faecium	50 032				4 102		2 918	12 672		
Enterococcus hirae	50 031	8 043	2 423	1 332	279	53.48	20 160	8 123	6 459	
Escherichia coli	50 034	11 775	5 172	24	515	54.8	30 083	11 943	9 001	

Bacteria	NCIMB QC Collection	ATCC	CCM	CCUG	CECT	CIP	DSM	NCIMB	NCTC	Other
Escherichia coli	50 035						13 059			
Escherichia coli	50 109	25 922				76.24	1 103	12 210	12 241	JCM 5 491
Escherichia coli O157:H7 (non-toxigenic)	50 139								12 900	
**Escherichia coli*									12 079	
**Escherichia coli*									12 080	
Hafnia alvei	50 037	13 337		15 720	158	57.31	30 163	11 999	8 105	JCM 1 666
Klebsiella oxytoca	50 038							12 819		
**Klebsiella pneumoniae*	50 108	13 883		225		82.91	30 104	13 281	9 633	JCM 1 662
Lactobacillus acidophilus	50 101	4 356		31 450		76.13	20 509	11 716	8 690	
Lactobacillus brevis	50 043	14 869		30 670		102806	20 054	11 973		JCM 1 059
Lactobacillus casei ssp. *casei*		393		21 451		103137	20 011	11 970		JCM 1 134
*ssp. *rhamnosus*	50 045	7 469	1 825			A 157	20 021	8 010		BUCSAV 227; IAM 1 118; IFO 3 425
Lactobacillus delbrueckii ssp. *bulgaricus*	50 050	11 842					20 081	11 778		
Lactobacillus gasseri	50 040	19 992		39 972		103699	20 077	13 081		
Lactobacillus plantarum	50 054	14 917		30 503	748	103151	20 174	11 974		IAM 12 477; JCM 1 149
Lactobacillus sakei ssp. *sakei*	50 056	15 521		30 501	906	103139	20 017	13 090		JCM 1 157
Lactobacillus viridescens	50 057	12 706		56		283	20 410	1 655		JCM 1 174
Lactococcus lactis ssp. *lactis*	50 058	19 435	1 877	7 980	185	70.56	20 481	6 681	6 681	BUCSAV 302; HNCMB 80 146
**Leuconostoc mesenteroides*	50 060	9 135	1 851				20 241	6 992		BUCSAV 312
Listeria ivanovii ssp. *ivanovii* sv 5	50 095	19 119	5 884	15 528	913	78.42	20 750		11 846	
Listeria monocytogenes sv l/2a		35 152	5 576	15 527		104794	12 464		7 973	
**Listeria monocytogenes* sv 4a		19 114			934	60.88			5 214	
Listeria monocytogenes sv 4b		13 932			935	59.53			10 527	
Micrococcus luteus	50 063	4 698	169	5 898	51	A 270	20 030	9 278	2 665	IAM 1 056; IFO 3 333; JCM 1 464
**Morganella morganii*	50 064	25 830		6 328	173	A 231	30 164	235	235	IFO 3 848
Pediococcus damnosus	50 065	29 358	3 453	32 251	793	102264	20 331	12 010		JCM 5 886
Photobacterium leiognathi	50 103	25 521		16 229		665		2 193		

Bacteria	NCIMB QC Collection	ATCC	CCM	CCUG	CECT	CIP	DSM	NCIMB	NCTC	Other
Proteus mirabilis		29 906		26 767	4 168	103181	4 479		11 938	
Pseudomonas aeruginosa	50 067	25 668			118			13 063	10 662	
Pseudomonas fluorescens	50 068	13 525	2 115	1 253	378	69.13	50 090	9 046	10 038	IAM 12 022; IFO 14 160
Pseudomonas fragi	50 069	4 973	1 974	556	446	55.4	3 456	8 542	10 689	IFO 3 458
**Pseudomonas putida*	50 070	12 633		12 690	324	52.191	291	9 494	10 936	IFO 14 164
Salmonella dublin	50 072				4 152				9 676	
Salmonella enteritidis	50 073				4 155				5 188	
Salmonella saintpaul	50 075				4 153				6 022	
**Salmonella typhimurium*	50 100	13 311		11 732	443		5 569		74	
Salmonella typhimurium									12 190	
**Salmonella typhimurium*		14 208							12 023	
**Salmonella typhimurium*		25 241								
Salmonella virchow	50 077				4 154				5 742	
Shigella flexneri	50 079	29 903				82.48	4 782			
**Shigella flexneri*	50 140	12 022				104222				
Shigella sonnei		29 930			4 887	82.49			12 984	
Staphylococcus aureus	50 080	6 538P			240	53.156	346	8 625	7 447	
Staphylococcus aureus	50 081							12 820		
**Staphylococcus aureus*		29 213				103429	2 569		12 973	JCM 2 874
**Staphylococcus warneri*	50 082							13 078	10 518	
Streptococcus bovis		33 317		17 828	213	102302	20 480		8 177	
Streptococcus thermophilus	50 083	19 258			986			8 510		
Thermoanaerobacterium thermosaccharolyticium		7 956				571	9 385			
Vibrio cholerae									11 348	
Vibrio fluvialis	50 084	33 809		13 622		103355		2 249	11 327	
Vibrio parahaemolyticus									11 344	
Vibrio vulnificus	50 104	27 562		13 448			10143	2 046		
Yersinia enterocolitica										
biotype 1 serotype O:8	50 085	9 610		11 291		80.27	4 780		12 982	
biotype 4 serotype O:3	50 087			4 586						

Fungi	NCIMB QC Collection	A-NRRL	ATCC	IMI	CSIRO	NCYC	Other
Aspergillus flavus				91 856 ii			
Aspergillus flavus		3 251			3 084		
Aspergillus parasiticus		2 999			2 744		
Aspergillus niger		3 361			2 522		
Eurotium amstelodami			16 018	17 455			CECT 2 586
Penicillium cyclopium			16 025	19 759			
Rhizopus stolonifer				61 269			
Saccharomyces cerevisiae	50 105		7 754			79	CBS 1 368
Zygosaccharomyces rouxii	50 106					1 522	

Subject Index

No attempt has been made to cite in the Index all the media mentioned in Part 1 (Review Section). The media cited are mostly those covered in the monographs and are to be found in Part 2 (397–638). Other media mentioned in the reviews are either of mainly historical interest or new, and as yet, untested and hence are not cited separately in the Subject Index.